APPLIED MECHANICS

PROCEEDINGS OF
THE TWELFTH INTERNATIONAL CONGRESS OF
APPLIED MECHANICS

STANFORD UNIVERSITY
AUGUST 26-31, 1968

EDITORS

M. HETÉNYI · W. G. VINCENTI

SPRINGER-VERLAG BERLIN HEIDELBERG NEW YORK 1969

Miklós Hetényi

Professor and Chairman of the Department
of Applied Mechanics
Stanford University, Stanford, CA 94305, USA

Walter G. Vincenti

Professor of Aeronautics and Astronautics
Stanford University, Stanford, CA 94305, USA

ISBN 978-3-642-85642-6   ISBN 978-3-642-85640-2 (eBook)
DOI 10.1007/978-3-642-85640-2

With 318 figures

Preface

This volume contains the Proceedings of the Twelfth International Congress of Applied Mechanics, held at Stanford University on August 26 to 31, 1968. The Congress was organized by the International Union of Theoretical and Applied Mechanics; members of the IUTAM Congress Committee and Bureau are listed under Congress Organization. The members of the Stanford Organizing Committee, which was responsible for the detailed organization of the Congress, are also given, as are the names of the sponsors and the industrial and educational organizations that contributed so generously to the financial support of the meeting. Those attending the Congress came from 32 countries and totaled 1337 persons, plus wives and children. A list of the registered participants is included in the volume.

The technical sessions of the Congress comprised four General Lectures and 281 contributed papers, the latter being presented in groups of five simultaneous sessions. The final choice of the contributed papers was made on the basis of abstracts by an International Papers Committee of IUTAM consisting of G. K. BATCHELOR, E. BECKER, N. J. HOFF, and W. T. KOITER. Initial review and recommendation of the contributions from France, West Germany, Poland, the U.S.S.R., and the United Kingdom were made by national committees from each of these countries. Contributions from the U.S.A. and all other countries were similarly handled by a committee of the U.S.A. In the case of papers from the U.S.A., the number of contributions was such that an average of only one out of five could be accepted. A considerable number of technically worthy papers therefore had to be declined. On the basis of the recommendations, a limited number of the contributed papers were selected for 30-minute presentation; the large majority were given 15-minute delivery. The texts of the General Lectures and the 30-minute papers are included in this volume and will not appear elsewhere*. The 15-minute papers are listed by author and title; most of them will presumably be published in the appropriate technical journals. Since division according to subject matter did not seem feasible, the papers within each category of presentation are arranged alphabetically according to first author. For the 30-minute papers, a complete list of authors is included in the volume; for the 15-minutes papers, the location of at least one of the authors (the one presenting the paper) can be found in the List of Participants.

In choosing the papers for the Congress no attempt was made to represent the various fields of applied mechanics proportionately. As between solid and fluid mechanics, the former appears to be more heavily represented than the relative level of activity throughout the world would indicate. Perhaps more meetings are available for the presentation of papers on fluids. Within each of the two major categories, however, the distribution of papers is probably representative of the current level of interest and activity in the various specialties.

The publication of these Proceedings has been handled very promptly and capably by Springer-Verlag. We are most grateful to them and to their editors.

Stanford University, California
May 1969.

Miklós Hetényi   **Walter G. Vincenti**

---

* One of the 30-minute papers (MORRISON, W. R. B., and R. E. KRONAUER, Structural Similarity for Fully Developed Turbulence in Smooth Tubes) has had to be omitted because of its length. It will be published elsewhere.

Front of Outer Quadrangle of Stanford University.

Opening Session conducted by NICHOLAS J. HOFF, President of the Congress.

Part of the Congress Banquet under the evergreen oak trees native to Northern California.

Informal gathering during sessions in Dinklespiel Auditorium.
(Photographs by R. ROSEWOOD.)

Contents

Résumé of the Congress

NICHOLAS JOHN HOFF, President of the Congress

General Lectures

Contributed Papers (30-minute presentation)

List of Other Contributed Papers (15-minute presentation)

Congress Organization

Congress Committee of the International Union of Theoretical and Applied Mechanics

J. ACKERET, Zürich
*G. K. BATCHELOR, Cambridge, UK
J. M. BURGERS, College Park, Md.
A. CAQUOT, Paris
G. COLONNETTI, Torino
G. A. CROCCO, Rome
S. GOLDSTEIN, Cambridge, Mass.
*H. GÖRTLER, Freiburg i. Br.
J. P. DEN HARTOG, Cambridge, Mass.
*W. T. KOITER, Delft
N. I. MUSKHELISHVILI, Tbilisi

F. K. G. ODQVIST, Stockholm
*M. ROY, Paris
W. RUBINOWICZ, Warsaw
L. I. SEDOV, Moscow
H. SOLBERG, Oslo
I. TANI, Tokyo
G. I. TAYLOR, Cambridge, UK
G. TEMPLE, Oxford
S. P. TIMOSHENKO, Stanford, Calif.
W. TOLLMIEN, Göttingen
W. WEIBULL, Brösarps Station

* Members of Executive Committee.

Bureau of the International Union of Theoretical and Applied Mechanics

M. ROY, President, Paris
G. TEMPLE, Vice-President, Oxford
H. GÖRTLER, Secretary, Freiburg i. Br.
W. T. KOITER, Treasurer, Delft

N. J. HOFF, Stanford, Calif.
W. OLSZAK, Warsaw
H. PARKUS, Vienna
L. I. SEDOV, Moscow

Stanford Organizing Committee

NICHOLAS J. HOFF, President
DANIEL BERSHADER, Vice President
CHARLES R. STEELE, General Secretary

Advisory Council

H. JULIAN ALLEN
W. FLÜGGE
I. FLÜGGE-LOTZ
J. N. GOODIER

W. C. GRIFFITH
S. J. KLINE
E. H. LEE
D. H. YOUNG

Congress Organization

Executive Committee

Publications
MIKLÓS HETÉNYI
WALTER G. VINCENTI

Entertainment
MILTON VAN DYKE

Technical Services
DANIEL DEBRA
KEVIN FORSBERG

Ladies' Program
JOYCE VINCENTI

General Services
WILLIAM REYNOLDS
SAMUEL McINTOSH

Finances
DONALD L. PUTT
RICHARD REMMEL

Executive Secretary
DONNA LOUBSKY

Director of Tresidder Memorial Union

FORREST TREGEA

Conference Coordinator, University Housing

LOIS M. BUTTS

Sponsors

Office of Naval Research of the U.S. Navy
    Representatives: H. LIEBOWITZ, N. PERRONE, J. M. CROWLEY, R. H. COOPER

Office of Scientific Research of the U.S. Air Force
    Representatives: M. ROGERS, J. POMERANTZ

National Science Foundation
    Representatives: R. M. ROBERTSON, J. M. IDE, M. P. GAUS

U.S.National Committee on Theoretical and Applied Mechanics

Representatives of Technical Societies

A. ACRIVOS	American Institute of Chemical Engineers
G. B. SCHUBAUER	American Physical Society
E. H. LEE	Society of Rheology
G. F. CARRIER	The American Society of Mechanical Engineers
W. PRAGER	American Mathematical Society
T. J. DOLAN	American Society for Testing and Materials
E. P. POPOV	American Society of Civil Engineers
D. C. DRUCKER	Society for Experimental Stress Analysis
N. J. HOFF	American Institute of Aeronautics and Astronautics

Ex-Officio Representatives of the Academy of Sciences

J. A. HUTCHESON	Engineering
R. H. BING	Mathematics
E. L. GOLDWASSER	Physical Sciences

Other Members

J. M. BURGERS	Y. H. KU
F. N. FRENKIEL	H. W. LIEPMANN
S. GOLDSTEIN	H. ROUSE
M. HETÉNYI	S. P. TIMOSHENKO
O. B. SCHIER II, Secretary	

Contributors

Ames Research Center of NASA
The Boeing Company
Brown University
California Institute of Technology
Case Western Reserve University
Food Machinery and Chemical Company
General Electric Missiles and Space Division
General Motors Corporation
Illinois Institute of Technology
International Business Machines Corporation
Kaiser Engineers
Litton Industries
Lockheed Aircraft Corporation
North American Rockwell Corporation

Northwestern University
Ohio State University
Rensselaer Polytechnic Institute
Shell Development Company
Standard Oil Company of California
TRW Systems, Inc.
United Aircraft Research Laboratories
United Air Lines
University of California at Berkeley
University of Iowa
University of Michigan
University of Wisconsin
Wayne State University
Westinghouse Electric Corporation

List of Authors of Papers in this Volume

Prof. ASHLEY, H., Department of Aeronautics and Astronautics, Stanford University, Stanford, CA 94305 (USA).

Prof. ATABEK, H. B., Department of Space Science and Applied Physics, Catholic University of America, Washington, DC 20017 (USA).

Dr. BAUMGARTE, J., Ingenieurakademie Wolfenbüttel, 3340 Wolfenbüttel, Lärchenweg 1 (Germany)

Dr. BEJDA, J., Institute of Basic Technical Problems Polish Academy of Sciences, Warsaw (Poland).

Prof. BERNDT, S. B., Kungliga Tekniska Högskolan, Stockholm 70 (Sweden).

Prof. CARMODY, J. J., Department of Space Science and Applied Physics, Catholic University of America, Washington, DC 20017 (USA).

Dr. CARRIERE, P., Office National d'Etudes et de Recherches Aérospatiales (ONERA), 29 Avenue de la Div. Leclerc, 76 Chatillon-sous-Bagneux/ Seine (France).

Prof. CHEN, YOUL-NAN, Department of Aeronautics and Applied Mechanics, Polytechnic Institute of Brooklyn, Brooklyn, NY 11201 (USA).

Prof. CLIFTON, R. J., Division of Engineering, Brown University, Providence, RI 02912 (USA).

Prof. DEBLER, W., Department of Engineering Mechanics, University of Michigan, Ann Arbor, MI 48104 (USA).

Dr. DREXLER, J., Aeronautical Research and Test Institute, Prague 9-Letňany (ČSSR).

Dr. DUMAS, R., Institut de Mécanique, Faculté des Sciences de Marseille, 12, Avenue Général Leclerc 13 Marseille IIIe (France).

Prof. FAVRE, A., Faculté des Sciences de Marseille, 12, Avenue Général Leclerc, 13 Marseille IIIe (France).

Prof. Dr.-Ing. FLÜGGE, W., Department of Applied Mechanics, Stanford University, Stanford, CA 94305 (USA).

Prof. FREUDENSTEIN, F., Department of Mechanical Engineering, Columbia University, New York, NY 10022 (USA).

FÜNER, E., M. Sc., Siemens AG, Referat Aus- und Weiterbildung, Nonnendammallee 101, 1000 Berlin 13 (Germany).

Prof. GAUTHIER, L., Université de Paris, Faculté des Sciences de Paris, 9, Quai St. Bernard, 75-Paris Ve (France).

Prof. GERDEEN, J. C., Michigan Technological University, Houghton, MI 49931 (USA).

Dr. HAHN, G. T., Battelle Memorial Institute, 505 King Avenue, Columbus, OH 43201 (USA).

Dr. HULBERT, L. E., Battelle Memorial Institute, 505 King Avenue, Columbus, OH 43201 (USA).

Dr. KANNINEN, M. F., Battelle Memorial Institute, 505 King Avenue, Columbus, OH 43201 (USA).

Prof. KAPLAN, R. E., School of Engineering, University of Southern California, University Park, Los Angeles, CA 90007 (USA).

Prof. KEMPNER, J., Department of Aeronautics and Applied Mechanics, Polytechnic Institute of Brooklyn, Brooklyn, NY 11201 (USA).

Dr. rer. nat. KIRCHGÄSSNER, K., Institut für angewandte Mathematik, und Mechanik der DVLR, Universität Freiburg, Hebelstraße 27, 7800 Freiburg/Brsg. (Germany).

Dr. KOLBE, O., Lehrstuhl A und Institut für Mechanik in der Fakultät für Maschinenwesen, Keplerstr. 17/9, Universität Stuttgart, 7000 Stuttgart (Germany).

Prof. Dr. KRÖNER, E., Institut für Theoretische Physik, Leibnizstraße 4, Technische Universität Clausthal, 3392 Clausthal-Zellerfeld (Germany).

Dr. KROPÁČ, O., Aeronautical Research and Test Institute, Prague (ČSSR).

Prof. KUNIO, T., Division of Mechanical Engineering, Keio University, Koganei-shi, Tokyo (Japan).

Prof. LAUFER, J., School of Engineering, University of Southern California, University Park, Los Angeles, CA 90007 (USA).

Acad. LAVRENTIEV, M. A., USSR Academy of Science, Novosibirsk 90 (USSR).

Prof. LING, S. C., Department of Space Science and Applied Physics, Catholic University of America, Washington, DC 20017 (USA).

Prof. LIPKIN, J., Division of Engineering, Brown University, Providence, RI 02912 (USA).

Dr. LYUBIMOV, A. N., Institute for Applied Mathematics, Miusskaia Pl. 4, Moscow A-47 (USSR).

Dr. MANSFIELD, E. H., Royal Aircraft Establishment, Structures Department, Farnborough, Hants (England).

Dr. McINTOSH, S. C., jr., Department of Aeronautics and Astronautics, Stanford University, Stanford, CA 94305 (USA).

Prof. MEI, C. C., Civil Engineering Department, Massachusetts Institute of Technology, Cambridge, MA 02139 (USA).

Prof. MILES, J. W., Institute of Geophysics and Planetary Physics, University of California, La Jolla, CA 92037 (USA).

Dr. MIYANO, Y., Division of Mechanical Engineering Keio University, Koganei-shi, Tokyo (Japan).

Prof. VAN DER NEUT, A., Aeronautical Engineering Department, Technological University of Delft, 1 Kluyverweg, Delft (Netherlands).

Prof. Dr. ODQVIST, F. K. G., Royal Institute of Technology, Torstensonsvagen 7 D, 18264 Djursholm (Sweden).

Prof. PEDERSEN, P. T., Department of Applied Mechanics, Technical University of Denmark, 13, Rigensgade, Copenhagen K (Denmark).

Acad. RABOTNOV, Y. N., Institute for Mathematics and Mechanics, Moscow State University, Moscow B 234 (USSR).

Prof. ROBSON, J. D., Department of Mechanical Engineering, University of Glasgow, Glasgow W 2 (UK).

Dr. ROSENFIELD, A. R., Battelle Memorial Institute, 505 King Avenue, Columbus, OH 43201 (USA).

Dr. RUSANOV, V. V., Institute for Applied Mathematics, Miusskaia Pl. 4, Moscow A-47 (USSR).

Prof. SCHAAF, B., Department of Engineering Mechanics, University of Michigan, Ann Arbor, MI 48104 (USA).

Dr.-Ing. SCHIEHLEN, W., Lehrstuhl und Institut B für Mechanik, Technische Universität München, Augustenstraße 77, 8000 München 2 (Germany).

Dr. SIRIEIX, M., Office National d'Etudes et de Recherches Aérospatiales (ONERA), 29 Avenue de la Div. Leclerc, 76 Chatillon-sous-Bagneux/Seine (France).

Dr. SOLIGNAC, J.-L., Office National d'Etudes et de Recherches Aérospatiales (ONERA), 29 Avenue de la Div. Leclerc, 76 Chatillon-sous-Bagneux/Seine (France).

Dr. SORGER, P., Institut für angewandte Mathematik, und Mechanik der DVLR Hebelstraße 27, Universität Freiburg, 7800 Freiburg/Brsg. (Germany).

Prof. TAYLOR, G. I., Farmfield, Huntingdon Road, Cambridge (UK).

Dr. VEROLLET, E., Faculté des Sciences de Marseille, 12, Avenue Général Leclerc, 13 Marseille IIIe (France).

Dr. WOO, L. S., IBM New York Scientific Center, 590 Madison Avenue, New York, NY 10022 (USA).

List of Participants

The following is a list of registered participants in the Congress grouped according to country. The asterisk indicates an author of a paper presented to the Congress.

Australia

*CRISP, J. D. C., Clayton, Victoria
DE VAHL DAVIS, G., Kensington, New S. Wales
LEAHY, T., Camp Hill, Brisbane, Queensland
*LUXTON, R. E., Sydney, New S. Wales
*MORRISON, W. R. B., St. Lucia, Brisbane, Queensland
MORTON, B. R., Clayton, Victoria
ROZVANY, G. I. N., Clayton, Victoria
SAG, T. W., Bedford Park, S. Australia
SHAW, F. S., Kensington, New S. Wales
THOMPSON, B. W., Parkville, Victoria
TRAHAIR, N. S., Sydney, New S. Wales
*WAECHTER, R. T., Bedford Park, S. Australia
WOOD, W. W., Bentleigh, Victoria

Austria

BARGMANN, H., Vienna
BEDNARCZYK, H., Vienna
*PARKUS, H., Vienna
TUNGL, E., Vienna
*ZIEGLER, F., Vienna

Belgium

*CROCHET, M. J., Heverlee
*DEHOUSSE, M. E., Bruxelles
*EBBENI, J., Bruxelles
HAUS, F. C. H., Bruxelles
JAUMOTTE, A. L., Bruxelles
KESTENS, J., Bruxelles
NIELSEN, H. L., Rhode St. Genese
REZETTE, Y., Bruxelles

Brazil

ALTMAN, W., Sao Paulo
*DIAZ, E. K., Rio de Janeiro

Canada

ANDERSON, D. L., Vancouver, British Columbia
*ARIARATNAM, S. T., Waterloo, Ontario
*BENTALL, R. H., Vancouver, British Columbia
BLACKWELL, J. H., London, Ontario
BOURASSA, P., Sherbrooke, Quebec
BREACH, D. R., Toronto, Ontario
CHENG, K. C., Edmonton, Alberta
COLBOURNE, J. R., Edmonton, Alberta
COWPER, G. R., Ottawa, Ontario
DENNIS, S. C. R., London, Ontario
DOIGE, A. G., Calgary, Alberta
*DUBEY, R. N., Waterloo, Ontario
DUBUC, J., Montreal, Quebec
FORD, G., Edmonton, Alberta
*GARG, S. K., Toronto, Ontario
GLOCKNER, P. G., Calgary, Alberta
JAMES, D. F., Toronto, Ontario
*KEFFER, J. F., Toronto, Ontario
LARDNER, R. W., Coquitlam, British Columbia
LENNOX, W., Waterloo, Ontario
*LEUTHEUSSER, H., Toronto, Ontario
LEVINSON, M., Hamilton, Ontario
LEWIS, E., London, Ontario
LIND, N. C., Waterloo, Ontario
MANUEL, P. W., London, Ontario
MIRZA, S., Calgary, Alberta
MURTHY, D. N. S., Waterloo, Ontario
NEALE, K. W. F., Waterloo, Ontario
NEIS, V. V., Saskatoon, Saskatchewan
OEN, S., Waterloo, Ontario
ORLIK-RUCKEMANN, K. J., Ottawa, Ontario
RAY, A. K., Ottawa, Ontario
RIMROTT, F. P. J., Toronto, Ontario
SHOEMAKER, E. M., Vancouver, British Columbia
SINGH, M. C., Calgary, Alberta
SMITH, A. C., Windsor, Ontario
*SMITH, S. H., Toronto, Ontario
*SRIKANTAIAH, T. K., Waterloo, Ontario
*TABARROK, B., Toronto, Ontario
TAYLOR, P. A., Toronto, Ontario
*VAUGHAN, H., Vancouver, British Columbia
WEAVER, D. S., Waterloo, Ontario
WENTZELL, R. A., Waterloo, Ontario
WU, J. H. T., Montreal, Quebec
*YONG, R. N., Montreal, Quebec

Czechoslovakia

*DREXLER, C. J., Prague
*FRYBA, L., Prague
*KRUPKA, V., Brno
*SEJVL, M., Plzeń

Denmark

*BJØRNØ, L., Copenhagen
*CHRISTIANSEN, S., Lyngby
*HANSEN, E., Lyngby
NIORDSON, F., Copenhagen
*OLHOFF, N., Copenhagen
*PEDERSEN, P. T., Copenhagen
REFSLUND, K., Copenhagen

France

ALZIARY DE ROQUEFORT, T., Poitiers
BASS, J., Paris
BEXIER, P., Tours
BINDER, G., Grenoble
BISMUTH, W., Marseille
 BRUN, L., Paris
*BRUN, L., Paris
*BRUN. R., Marseille
 CERNEAU, S., Paris
 CHARLES, A., Villers-les-Nancy
 CHEN, C. P., Orsay
 CHEZEAUX, M., Marseille
 CLARION, C., Marseille
*COMTE-BELLOT, G., Ecully
*CONTENSOU, P., Paris
*COUPRY, G., Chatillon-sous-Bagneux (Seine)
*DARROZES, J. S., Chatillon-sous-Bagneux (Seine)
 DEROUET, J., Poitiers
 DUMAS, R., Marseille
*DURAND, M., Suresnes
 EDOUARD, M., Antony
*FAVRE, A., Marseille
 FONTAINE, B. B., Marseille
 FORTIER, A., Clamart
*GAUTHIER, L., Paris
 GERMAIN, P., Paris
*GERMAIN, J. P., Grenoble
*GUFFROY, D., Marseille
*GUIBERGIA, J. P., Marseille
 GUILLAUME, L. E., Gif-sur-Yvette
*HUBERT, J., Paris
*KAMPE DE FERIET, J., Lille
 LAGARDE, A., Poitiers
*LE FUR, B. R., Colombes
 LE MANACH, J., Paris
*LEGENDRE, R., Chatillon-sous-Bagneux (Seine)
 MAISONNEUVE, O., Poitiers
*MARCILLAT, J., Marseille
*MARCOU, C., Grenoble
*MARECHAL, J. M., Grenoble
*MARMEY, R., Marseille
 MASSIGNON, D., Gif-sur-Yvette
 MATHURIN, C., Poitiers
*MICHEL, R., Chatillon-sous-Bagneux (Seine)
*MOREAU, J. J., Montpellier
 NAYROLES, B. J. P., Poitiers
 PARRAUD, P. P., Marseille
 PEUBE, J.-L., Poitiers
 RADENKOVIC, D., Paris
*RIGAUT, F., Chatillon-sous-Bagneux (Seine)
*ROBERT, A. J., Paris
*ROUX, B., Marseille
 ROY, M., Paris
*SOLIGNAC, J. L., Chatillon-sous-Bagneux (Seine)
 THIRY, Y., Paris
*TOURNEMINE, G., Rennes
 TSEN, L., Poitiers (Vienne)
 TURBAT, C., Paris
*VALID, R., Chatillon-sous-Bagneux (Seine)
 WEBER, J.-D., Strasbourg
*ZEYTOUNIAN, R. KH., Chatillon-sous-Bagneux (Seine)

Germany

ADAMS, E., Grünwettersbach, Karlsruhe
ADOMEIT, G., Aachen
*AZMEH, M., Braunschweig
*BAUMGARTE, J., Wolfenbüttel
*BEHR, D., Clausthal-Zellerfeld
 BLENDERMANN, W., Hoisbuettel, Hamburg
*BRAKHAGE, H., Karlsruhe
 BUFLER, H., Stuttgart
*BURGER, W., Darmstadt
*BUSSE, F. H., München
 EPPLER, R., Stuttgart
 ESSLINGER, M., Braunschweig-Flughafen
 EUJEN, E., Lamme bei Braunschweig
 FELSCH, K. O., Karlsruhe
 FERNHOLZ, H. H., Berlin
*FORSCHING, H., Göttingen
 GEROPP, D., Karlsruhe
*GERSTEN, K., Bochum-Querenburg
 GÖRTLER, H., Freiburg
*HAHN, H. G., Augsburg, München
*HINDELANG, F. J., München
*KOLBE, O., Stuttgart
 KRAEMER, E., Mannheim-Kaefertal
 KRESS, R., Darmstadt
*LEHMANN, TH., Hannover
*LIPPMANN, H., Braunschweig
 LOER, S. M., Zorneding
*MAHRENHOLTZ, O., Hannover
 METTLER, E., Karlsruhe-Duriach
*MOHRING, W. F., Göttingen
*MÜLLER, E. A., Göttingen
*MUSCHNER, W. H., Hamburg
*NEUBER, H., München
 RIEGELS, F. W., Göttingen
 ROTTA, J., Göttingen
*SCHAEFER, M., Clausthal-Zellerfeld
*SCHIEHLEN, W., München
*SCHMIDT, B., Karlsruhe
*SCHMIEDER, L., Oberpfaffenhofen
*SCHNEIDER, W. A., Aachen
 SCHONAUER, N., Karlsruhe
 SCHWARZENBERGER, R., Aachen
*SONNTAG, G., München
*SORGER, P., Freiburg
 STUKE, B., München
*VON BREDOW, H. J., Hannover
 VON SEEBACH, L., Göttingen
 WALZ, A. H., Berlin
*WEDEMEYER, E. H., Göttingen
*ZIEREP, J., Karlsruhe

Hungary

*KALISZKY, S., Budapest

India

*CHAKRABORTY, B. B., Bombay
 JAIN, P. C., Bombay
*RAO, A. K., Bangalore
 REDDY, R. V., Madras

Iran

MIRZAD, S. H., Tabriz

Israel

BETSER, A. A., Haifa
*BODNER, S. R., Haifa
GILLIS, J., Rehovoth
*SINGER, J., Haifa

Italy

AGOSTINELLI, C., Torino
*AUGUSTI, G., Naples
BIANCHI, G., Milano
BNZO, T., Milano
COLOMBO, G., Milano
*COMO, M., Naples
*CRAWFORD, D. R., Naples
GALLETTO, D., Padova
*ROMITI, A., Torino
*TONTI, E., Milano

Japan

*FURUYA, Y., Nagoya
HASIMOTO, H., Tokyo
*HAYASHI, K., Tokyo
*HAYASHI, T., Tokyo
*IMAI, I., Tokyo
*INOUE, J., Kyushu
IROBE, M., Tokyo
KAWAI, T., Tokyo
*KONDO, J., Tokyo
*KUNIO, T., Tokyo
*MIYANO, Y., Tokyo
NAKAMURA, T., Kyoto
OHASHI, Y., Nagoya
OHIRA, H., Kyushu
*OSHIMA, K., Tokyo
*SUNAKAWA, M., Tokyo
TAKAISI, Y., Ehime
TAKAMI, H., Tokyo
*TANI, I., Tokyo
*TATSUMI, T., Kyoto
UEMURA, M., Tokyo

Korea

WON, T. S., Seoul

Netherlands

*ALBLAS, J. B., Eindhoven
BARTELDS, G., Emmeloord
BENTHEM, J. P., Amsterdam
BESSELING, J. F., Delft
VAN BOMMEL, P., Utrecht
DE PATER, A. D., Delft
DE PATER, C., Enschede
ECKHAUS, W., Delft
FLOOR, W. K. G., Delft
GEERLINGS, J. J. P., Delft
GEERTSMA, J., Rijswijk (ZH)
HINZE, J. O., Delft
HOMMEL, G., Delft
VAN HORSSEN, W., Delft
VAN INGEN, J. L., Delft
*KALKER, J. J., The Hague

KAMINSKI, A., Delft
*VAN KEMPEN, H. P. M., Amsterdam
KOITER, W. T., Delft
KOSTER, W. E., Pijnacker
KUIKEN, G. D. C., Rotterdam
LEKKERKERKER, J., Delft
LUNGAGNANI, V., Petten
*MANDL, G., Rijswijk (ZH)
MEIJERS, P., Delft
MERK, H. J., Delft
MINKHORST, J. H. K., Geleen
*VAN DER NEUT, A., Delft
OTTO, P., Heerlen
*PEUTZ, M. G. F., Amsterdam
TENHOEVE, G., Enschede
TRAAS, C. R., Amsterdam
VANDERWERFF, K., Delft
VOOREN, J. V. D., Emmeloord
*WIJNGAARDEN, L. V., Enschede
ZANDBERGEN, P. J., Enschede

New Zealand

NIELD, D. A., Auckland

Nigeria

*SOBOYEJO, A. B. O., Lagos

Norway

*SVARDAL, A., Bergen
*TJOTTA, S., Bergen

Peru

DEL RIO, C. J., Lima

Poland

*BEJDA, J., Warsaw
*BURNAT, M., Warsaw
KALISKI, S., Warsaw
*KLEPACZKO, J., Warsaw
*LUKASIEWICZ, S., Warsaw
MOSSAKOWSKA, Z., Warsaw
NOWACKI, W., Warsaw
OLESIAK, Z., Warsaw
*OLSZAK, W., Warsaw
*PERZYNA, P., Warsaw
RYCHLEWSKI, J., Warsaw
*SAWCZUK, A., Warsaw
ZAHORSKI, S., Warsaw
ZORSKI, H., Warsaw
*ZYCZKOWSKI, M., Warsaw

Rumania

*CRISTESCU, N., Bucharest
*GHEORGHITZA, S. I., Bucharest
ION, S., Bucharest
MISICU, M. V., Bucharest
PELECUDI, C., Bucharest
PETRE, A. L., Bucharest
PREDELEANU, M., Bucharest
TEODORESCU, P. P., Bucharest

Senegal

*FAURE, R., Dakar

Sweden

AKESSON, B. A., Göteborg
*BERNDT, S. B., Stockholm
CARLSSON, J., Stockholm
DROUGGE, G., Bromma, Stockholm
FROSSLING, N., Göteborg
HOLMBERG, B. R., Göteborg
ISAKSSON, F. M., Stockholm
LANDAHL, M., Stockholm
LOYD, D., Göteborg
MAGI, M., Göteborg
MAGNUSSON, E., Stockholm
*ODQVIST, F., Djursholm
OHLSON, N. G., Stockholm
RAND, T., Stockholm
*RHOLIN, L. H., Uppsala
RYHMING, I. L., Stockholm
SAMANTA, S. K., Stockholm
SODERQUIST, B., Stockholm
STORAKERS, B., Stockholm
WALLER, I., Uppsala

Switzerland

KELLENBERGER, W., Wettingen
*ROTT, N., Zürich
*WEHRLI, C., Zürich

Turkey

*DOKMECI, M. C., Istanbul

Union of Soviet Socialist Republic

*ARUTIUNIAN, N. KH., Erevan
BAKHTURIN, S. G., Moscow
CHERNYI, G. G., Moscow
CHOGOSHVILI, G. S., Tbilisi
DURGARYAN, S. M., Erevan
GARIPOV, R. M., Novosibirsk
KALININ, V. S., Leningrad
*KERIMOV, K. A., Baku
*KHARLAMOV, P. V., Donetsk
KORYAVOV, P. O., Moscow
KOVALENKO, A. D., Kiev
KOVREVSKII, A. P., Kharkov
*KRASILSHIKOVA, E. A., Moscow
*KUNIN, I. A., Novosibirsk
KUSHELEV, N. G., Leningrad
KUZNETSOV, V.
*LAVRENTIEV, M. A., Novosibirsk
LEVCHENKO, V. A., Novosibirsk
*LOITSYANSKII, L. G., Leningrad
*LUNKIN, Y. P., Leningrad
*LYUBIMOV, A. N., Moscow
MALININ, N. N., Moscow
MANDZAVIDZE, G. F., Tbilisi
MARTYNENKO, O. G., Minsk
MIGIRENKO, G. S., Novosibirsk
MOSSAKOVSKII, V. I., Dnepropetrovsk

MUSKHELISHVILI, N. I., Tbilisi
*NEMIROVSKY, Y. V., Novosibirsk
*NIGUL, U. K., Tallinn
OBOLASHVILI, E. I., Tbilisi
*OVSYANNIKOV, L. V., Novosibirsk
PISARENKO, G. S., Kiev
PONOMAREV, S. D., Moscow
*PROSKURYAKOV, M. N., Moscow
*RAKHMATULIN, K. A., Moscow
*RUMYANTSEV, V. V., Moscow
*RUSANOV, V. V., Moscow
*SEDOV, L. I., Moscow
*SERENSEN, S. V., Moscow
*SHACHNEV, V. A., Moscow
*SHEMYAKIN, E. I., Novosibirsk
*SINITSYN, A. P., Moscow
*STRUMINSKII, V. V., Novosibirsk
TETERS, G. A., Riga
TRESHCHEVSKY, V. N., Leningrad
*VEKUA, I. N., Tbilisi
VEKUEV, N.
VERISHENKO, E.

United Kingdom

ADLER, J., London
*ALLISON, I. M., London
ATKINS, A. G., Oxford
BAINES, P. G., Cambridge
BARR, A. D. S., Edinburgh
*BARTA, T. A., London
BATCHELOR, G. K., Cambridge
BISHOP, R. E. D., London
*BOOCOCK, D., Derby
BRADSHAW, P., Teddington, Middlesex
*BROWN, E. H., S. Kensington, London
CHADWICK, P., Norwich, Norfolk
CHESTER, W., Bristol
*CHIGIER, N. A., Sheffield
CRAVEN, A. H., Falmer, Brighton, Sussex
*DAVIES, R. M., London
DIXON, C., Dundee, Scotland
*ELDER, J. W., Cambridge
ELLEN, C. H., London
ENGLAND, A. H., Nottingham
FAULKNER, D., British Navy Staff
   (Washington, D.C., USA)
FFOWCS-WILLIAMS, J. R., London
*GALLETLY, G. D., Liverpool
*GLADWELL, G. M. L., Southampton, Hants
GLAUERT, M. B., Norwich
*HALL, I. M., Manchester
*HALL, M. G., Farnborough, Hants
HOPKINS, H. G., Manchester
HSU, T.-C., Aston, Birmingham
JARVIS, R. J., Dundee, Scotland
JASWON, M. A., London
JOHANNESEN, N. H., Manchester
*JOHNS, K. C., London
KUCHEMANN, D., Farnborough, Hants
*LECKIE, F. A., Leicester
LIGHTHILL, M. J., London
LILLEY, D. G., Sheffield
LILLEY, G. M., Southampton, Hants
LOCKETT, F. J., Teddington, Middlesex

UK (Cont'd)

MacIsaac, J., Strathclyde, Glasgow, Scotland
Mangler, K. W., Farnborough, Hants
*Mansfield, E. H., Farnborough, Hants
Markland, E., Belfast, N. Ireland
*Maunder, L., Newcastle upon Tyne
Milne-Thomson, L. M., Sevenoaks, Kent
Moffatt, H. K., Cambridge
Morley, A. W., London
Morley, L. S. D., Farnborough, Hants
Morris, R. M., Cardiff, Wales
Musgrave, M. J. P., London
O'Connor, J. J., Oxford
Packham, B. A., London
Palmer, A. C., Cambridge
Price, H. L., Leeds
Riley, N., East Anglia, Norwich
Robinson, J. L., Bristol
*Robson, J. D., Glasgow, Scotland
*Roe, P. L., Farnborough, Hants
Rosenblat, S., London
Schmitz, P. D., Cambridge
Shail, R., London
*Shercliff, J. A., Coventry, Warwickshire
Sleeman, B. D., Dundee, Scotland
Smith, F. I. P., St. Andrews, Fife, Scotland
*Spence, D. A., Oxford
Stewartson, K., London
*Taylor, G. I., Cambridge
Thomas, D. P., Dundee, Scotland
*Tillman, S. C., London
Tooth, A. S., Glasgow, Scotland
*Warburton, G. B., Nottingham
*Wheatley, M. J., Sheffield
*Williams, J. G., London

United States of America

Ablow, C. M., Portola Valley, Calif.
Abramson, H. N., San Antonio, Texas
*Achenbach, J. D., Evanston, Ill.
Ackerberg, R. C., Farmingdale, N.Y.
Acrivos, A., Stanford, Calif.
Adams, D. F., Santa Monica, Calif.
Adler, W. F., Columbus, Ohio
Agaskar, V., Palo Alto, Calif.
Aggerwal, H. R., Menlo Park, Calif.
Agrawl, G. L., Boulder, Colo
Aguirre, R. G., Hunstville, Ala.
Ahmed, N., Princeton, N.J.
Albright, G., Wichita, Kan.
Allen, H. J., Moffett Field, Calif.
Almroth, B., Sunnyvale, Calif.
*Amazigo, J. C., Cambridge, Mass.
*Ames, W. F., Iowa City, Iowa
Anderson, A. E., Moffett Field, Calif.
*Anderson, D. G. M., Cambridge, Mass.
Anderson, W. M., New York, N.Y.
Anderson, W. E., Moffett Field, Calif.
Anliker, M., Stanford, Calif.
*Appleby, E. J., Monroe, Penn.
*Arbocz, J., Pasadena, Calif.
Ariman, T., Notre Dame, Ind.
Armand, J.-L., Woodside, Calif.

Arnold, K. W., Malibu, Calif.
Aroesty, J., Santa Monica, Calif.
Asch, V., Los Angeles, Calif.
*Ashley, H., Stanford, Calif.
Ashurst, W. T., Livermore, Calif.
Aspinwall, D. M., Sunnyvale, Calif.
Atkin, R. J., Baltimore, Md.
Au, N. N., Encino, Calif.
Au, S.-S., Menlo Park, Calif.
Axelson, J. A., Mountain View, Calif.
*Babcock, Jr., C. D., Pasadena, Calif.
Bader, F. E., Columbus, Ohio
Bagalman, P. M., Stanford, Calif.
*Baganoff, D., Stanford, Calif.
Bainum, P. M., Silver Springs, Md.
Baker, B., Menlo Park, Calif.
Bakke, A. P., Mountain View, Calif.
Baldwin, Jr., B. S., Mountain View, Calif.
Ballal, B. Y., Berkeley, Calif.
Banaugh, R. P., Kent, Wash.
Barba, P. M., Palo Alto, Calif.
*Barker, L. M., Albuquerque, N. Mex.
Baron, M. L., New York, N.Y.
*Bartel, D. L., Iowa City, Iowa
Barton, M. V., Redondo Beach, Calif.
Basdekas, N. L., Washington, D.C.
Bastianon, R. A., Sunnyvale, Calif.
Batdorf, S. B., San Bernardino, Calif.
Baum, D. W., San Leandro, Calif.
Beadle, C. W., Davis, Calif.
Beam, R., Moffett Field, Calif.
Becker, E. B., Austin, Texas
Beer, F. P., Bethlehem, Pa.
Beighley, C. M., Placerville, Calif.
Bencze, D. P., Moffett Field, Calif.
Bendick, P., Stanford, Calif.
Benedikt, E. T., Los Angeles, Calif.
*Benton, E. R., Boulder, Colo.
Berger, S. A., Berkeley, Calif.
Bernstein, T. N., New Carlisle, Ohio
*Bershader, D., Stanford, Calif.
Bertwell, W., Oakland, Calif.
Birnbaum, M. R., Alamo, Calif.
*Bluman, G. W., Pasadena, Calif.
Bogdonoff, S. M., Princeton, N.J.
*Boley, B. A., New York, N.Y.
Bolstad, J. H., Stanford, Calif.
Bongved, L., Bell Telephone Labs.
Botwin, M. R., Toledo, Ohio
Bowers, H. M., Sunnyvale, Calif.
Bozich, W. F., Stanford, Calif.
Bradley, J. E., Grandview, Mo.
Brandt, R., Canoga Park, Calif.
Breakwell, J. V., Palo Alto, Calif.
Brinson, H. F., Blacksburg, Va.
*Brull, M. A., Philadelphia, Pa.
Brush, D. O., Davis, Calif.
Bryson, A. E., Stanford, Calif.
*Budiansky, B., Cambridge, Mass.
Burgers, J. M., College Park, Md.
Burgess, J. C., Honolulu, Hawaii
Burggraf, O. R., Columbus, Ohio
Burnham, N. W., Rocky Flats, Colo.
Burrill, N. W., Palo Alto, Calif.

USA (Cont'd)

BUSEMANN, A., Boulder, Colo.
BUSHNELL, D., Palo Alto, Calif.
BUSSMAN, D. R., Kettering, Ohio
CAGLE, B., Pasadena, Calif.
CALVERT, D. L., Arlington, Va.
CALVIT, H. H., Austin, Texas
CAMARGO-MORA, G., Los Altos Hills, Calif.
CANNON, R. H., Stanford, Calif.
CAPIAUX, R., Palo Alto, Calif.
CAREY, J. J., Argonne, Ill.
CARMICHAEL, R. L., Moffett Field, Calif.
*CARRIER, G. F., Cambridge, Mass.
CASTELLANO, C. R., Moffett Field, Calif.
CATTANEO, A. G., Sunnyvale, Calif.
CATTON, I., Los Angeles, Calif.
CAUGHEY, D. A., Princeton, N.J.
*CAULK, R. H., Stanford, Calif.
CEBECI, T., Long Beach, Calif.
CHABRA, D. S., Stanford, Calif.
CHALUPUNIK, J. D., Seattle, Wash.
CHAN, S. P., Atlanta, Ga.
CHANG, C.-N., Stanford, Calif.
CHANG, I-D., Stanford, Calif.
CHANG, S. S., Sunnyvale, Calif.
CHANNAPRAGADA, R. S., Mountain View, Calif.
CHAO, C.-C., Stanford, Calif.
CHAPMAN, D. R., Moffett Field, Calif.
CHAPMAN, G. T., Sunnyvale, Calif.
CHASE, K. W., Berkeley, Calif.
CHEN, C. F., New Brunswick, N.J.
CHEN, W. T., Endicott, N.Y.
*CHEN, P. J., Albuquerque, N. Mex.
*CHEN, P., Davis, Calif.
*CHEN, Y. N., Brooklyn, N.Y.
CHENG, A. P., Houston, Texas
*CHENG, D. H., New York, N.Y.
*CHENG, H. K., Los Angeles, Calif.
CHENG, P., Palo Alto, Calif.
CHENG, S. L., Newark, N.J.
CHENG, S.-I., Princeton, N.J.
CHENG, Y. F., Seattle, Wash.
CHESTNUT, P. C., Palo Alto, Calif.
CHEVALIER, H. L., Sunnyvale, Calif.
CHEVRAY, R., Baltimore, Md.
CHIA, B. S., Palo Alto, Calif.
CHIARULLI, P., Oak Park, Ill.
CHILDRESS, J. M., Stanford, Calif.
CHILTON, E. G., Palo Alto, Calif.
*CHITALEY, A. D., Brecksville, Ohio
CHIU, S. S., Livermore, Calif.
CHOU, P. C., Philadelphia, Penn.
CHOU, T.-W., Stanford, Calif.
CHRISTENSEN, R. M., Emeryville, Calif.
CHU, C. K., New York, N.Y.
CHUANG, T.-Y., Berkeley, Calif.
CITERLEY, R. L., San Carlos, Calif.
CLAERBOUT, J. F., Stanford, Calif.
CLAPPER, R. B., Moffett Field, Calif.
*CLARK, R. A., Cleveland, Ohio
*CLIFTON, R. J., Providence, R.I.
CLINE, G. B., Palo Alto, Calif.
COALE, C. W., Palo Alto, Calif.
COE, C. F., Moffett, Field, Calif.

COHAN, H., Sunnyvale, Calif.
COHEN, D. S., Pasadena, Calif.
*COLE, J. D., Pasadena, Calif.
COLES, D., Pasadena, Calif.
COLLINS, R., Los Angeles, Calif.
COMPTON, D. L., Sunnyvale, Calif.
*CONTI, R. J., Palo Alto, Calif.
COON, M. D., Seattle, Wash.
CORCOS, G. M., Lafayette, Calif.
COTTRELL, M. M., Albuquerque, N. Mex.
COVINGTON, M. A., Moffett Field, Calif.
CRAIG, R. R., Austin, Texas
CRANCH, E. T., Ithaca, N.Y.
*CRANDALL, S. H., Cambridge, Mass.
CROWE, C. T., Sunnyvale, Calif.
CRUSE, T. A., Pittsburgh, Penn.
*CUMBERBATCH, E., Lafayette, Ind.
CUNNINGHAM, D. M., Berkeley, Calif.
CURRAN, R. L., Toledo, Ohio
DAIJ, J. L., Minneapolis, Minn.
DAILY, J. W., Ann Arbor, Mich.
*DALTON, C., Houston, Texas
DAMLE, S. K., Boulder, Colo.
DANNENBERG, R. E., Moffett Field, Calif.
DAVIDS, N., University Park, Penn.
DAVIS, L. R., San Jose, Calif.
*DAVIS, R. T., Blackburg, Va.
DAVIS, R. E., Del Mar, Calif.
DAVISON, L., Albuquerque, N. Mex.
*DEBLER, W. R., Ann Arbor, Mich.
DEIWERT, G., Moffett Field, Calif.
DELAURIER, J. D., Los Gatos, Calif.
DERESIEWICZ, H., New York, N.Y.
DERUNTZ, J. A., Sunnyvale, Calif.
*DESILVA, C. N., Detroit, Mich.
DEVAN, L., Lawrence, Kan.
DEVRIES, K. L., Salt Lake City, Utah
DHARAN, H., Berkeley, Calif.
DI PRIMA, R. C., Troy, N.Y.
DICKEY, R. W., Madison, Wis.
*DILLON, Jr., D. W., Lexington, Ky.
DISTEFANO, N., Kensington, Calif.
DODS, J. B., Los Altos, Calif.
DONNELL, L. H., Palo Alto, Calif.
DORÉ, R., Stanford, Calif.
DRAKE, J. L., Vicksburg, Miss.
*DRUCKER, D. C., Providence, R.I.
DUBLER, R. D., Iowa City, Iowa
DUDLEY, W. M., Sacramento, Calif.
*DUGAN, J. P., Baltimore, Md.
DUNHAM, R. S., Berkeley, Calif.
DUNLAP, R., Sunnyvale, Calif.
DUVALL, G. E., Pullman, Wash.
DUVVURI, T., Mountain View, Calif.
DWYER, T. J., San Jose, Calif.
EDELSTEIN, W. S., Chicago, Ill.
EDNEY, B., Buffalo, N.Y.
EISENBERG, M., Gainesville, Fla.
ELLIOT, Jr., J., San Diego, Calif.
ELLIS, A. T., La Jolla, Calif.
EMERY, R. R., San Jose, Calif.
EMERY, A. F., Livermore, Calif.
EMMONS, H. W., Cambridge, Mass.
ENG, G. H., Pittsburg, Pa.

USA (Cont'd)

ENGELBERT, D. F., Cupertino, Calif.
ENINGER, J., Stanford, Calif.
ERDOGAN, F., Bethlehem, Penn.
ERICKSON, A. L., Moffett Field, Calif.
ERICKSON, L. L., Moffett Field, Calif.
ERNSTEIN, N., Wright-Patterson AFB, Ohio
ESCUDIER, M. P., Cambridge, Mass.
ESSENBERG, F., Boulder, Colo.
*EVAN-IWANOWSKI, R. M., Syracuse, N.Y.
EVANS, R. J., Bainbridge Island, Wash.
EVENSEN, H. A., Moffett Field, Calif.
FAKHARI, M., Palo Alto, Calif.
*FARELL, C., Iowa City, Iowa
FARQUAR, R., Stanford, Calif.
FENZ, E., Berkeley, Calif.
FETTHALIOGLU, O. A., Stanford, Calif.
FIELD, F. A., San Bernardino, Calif.
FINE, A. D., East Hartford, Conn.
*FINLAYSON, B. A., Seattle, Wash.
FINNIE, I., Berkeley, Calif.
FISCHER, F. J., Houston, Texas
FISHER, G. M. C., Holmdel, N.J.
FLEISCHHACKER, J. E., Livermore, Calif.
*FLORENCE, A. L., Menlo Park, Calif.
FLÜGGE-LOTZ, I., Stanford, Calif.
*FLÜGGE, W., Stanford, Calif.
FOERSTER, R. E., Stanford, Calif.
FOGELSON, S., Tarzana, Calif.
FORSBERG, K. J., Sunnyvale, Calif.
FOSDICK, R. L., Chicago, Ill.
FOUSSE, E., Stanford, Calif.
FOX, J. L., Palo Alto, Calif.
*FRANCIS, P. H., San Antonio, Texas
FRAUENTHAL, J. C., Cambridge, Mass.
FRENKIEL, F. N., Washington, D.C.
*FREUDENSTEIN, F., New York, N.Y.
FREUND, L. B., Providence, R.I.
FRIEDMAN, M. D., San Jose, Calif.
FRIEHE, C. A., Palo Alto, Calif.
FROMME, J. A., Colorado Springs, Colo.
FRYMOYER, E. M., Newport Beach, Calif.
FUCHS, H. O., Stanford, Calif.
FUKUSHIMA, T., Stanford, Calif.
*FUNG, Y. C., La Jolla, Calif.
FYFE, I. M., Seattle, Wash.
GAKENHEIMER, D. C., Pasadena, Calif.
GARBELL, M. A., San Francisco, Calif.
GARRICK, I. E., Hampton, Va.
GASPERS, P. A., Moffett Field, Calif.
GAUS, M., Washington, D.C.
GAWAIN, T. H., Monterey, Calif.
GAWIENOWSKI, J. J., Moffett Field, Calif.
*GERDEEN, J. C., Columbus, Ohio
GERE, J. M., Stanford, Calif.
GHISTA, D. N., Moffett Field, Calif.
GILL, S. P., Atherton, Calif.
GILLICH, W. J., Aberdeen Proving Grounds, Md.
*GILLMAN, P. A., Boulder, Colo.
GLENN, H. D., Livermore, Calif.
GLUCKMAN, P. M., Stanford, Calif.
GOLDBURG, A., Seattle, Wash.
*GOLOBIC, R. A., Columbus, Ohio
GOMEZ, M. P., Sunnyvale, Calif.

GOODIER, J. N., Stanford, Calif.
GOODWIN, G., Moffett Field, Calif.
GOTO, J. T., Pasadena, Calif.
*GRADOWCZYK, M. H., Cambridge, Mass.
*GRAEBEL, W. P., Ann Arbor, Mich.
GRAY, M. H., Benton Harbor, Mich.
*GREEN, S., Warren, Mich.
GRIFFITH, W. C., Sunnyvale, Calif.
GROSS, J. F., Santa Monica, Calif.
GUDERLEY, K. G., Wright-Patterson AFB, Ohio
GUIST, L. R., Campbell, Calif.
HAKKINEN, R. J., Santa Monica, Calif.
HAN, L. S., Colombus, Ohio
*HANAGUD, S., Menlo Park, Calif.
HANDELMAN, G. H., Troy, N.Y.
HANLY, R. D., San Jose, Calif.
*HANSON, R., Stanford, Calif.
HARIRI, R., Ann Arbor, Mich.
HARRIS, D. D., Stanford, Calif.
HARTUNG, R. F., Palo Alto, Calif.
HARVEY, T. J., Sunnyvale, Calif.
HASSAN, S. D., College Station, Texas
HAYASI, N., Marietta, Ga.
HAYES, W. D., Princeton, N.J.
HEER, E., Pasadena, Calif.
HEGEMIER, G. A., La Jolla, Calif.
HEMP, G. W., Gainsville, Florida
*HERAKOVICH, C. T., Blacksburg, Va.
HERMSEN, R. W., Sunnyvale, Calif.
*HERRMANN, G., Evanston, Ill.
HETÉNYI, M., Stanford, Calif.
HICKS, R. M., Moffett Field, Calif.
HILL, H. T., San Francisco, Calif.
*HODGE, Jr., P. G., Chicago, Ill.
HOFF, N. J., Stanford, Calif.
HOFFMAN, G. H., Stanford, Calif.
HOFFMAN, O., Sunnyvale, Calif.
HOLDREN, J. P., Stanford, Calif.
HOLMES, A. M., Palo Alto, Calif.
HOLMES, R. E., Mountain View, Calif.
HOLT, A. C., Livermore, Calif.
HOLT, M., Berkeley, Calif.
HONIKMAN, T. C., Stanford, Calif.
HOPKINS, E. J., Atherton, Calif.
HORSTMAN, C., Moffett Field, Calif.
HOVIS, D. W., Stanford, Calif.
HOWE, J. T., Washington, D.C.
HSIA, H. T.-S., Palo Alto, Calif.
HSU, C.-C., Wright-Patterson AFB, Ohio
*HSU, C. S., Berkeley, Calif.
HSU, Y. C., Albuquerque, N. Mex.
*HUANG, N.-C., La Jolla, Calif.
HUANG, T. C., Madison, Wis.
*HUNG, J. Y., Appleton, Wis.
HUNTER, A. R., Palo Alto, Calif.
HUNTON, L. W., Moffett Field, Calif.
HUPPERT, D. W., Sunnyvale, Calif.
HURLEY, D. G., Baltimore, Md.
HUSSAIN, F., Stanford, Calif.
*HUSSAIN, M. A., Watervliet, N.Y.
HUTCHINSON, J. R., Davis, Calif.
*HUTCHINSON, J. W., Cambridge, Mass.
HWANG, C., Palos Verdes Peninsula, Calif.
HWU, G.-L., Stanford, Calif.

USA (Cont'd)

HYMAN, B., Washington, D.C.
INOUYE, M., Moffett Field, Calif.
IRICK, J. T., Houston, Texas
*ITO, M., Los Angeles, Calif.
IWANCIOW, B. L., Sunnyvale, Calif.
*JAFFRIN, M. Y., Cambridge, Mass.
*JAHSMAN, W. E., Boulder, Colo.
JAN, J. A., Harmarville, Penn.
*JANKOVICH, E., Seattle, Wash.
JEDLICKA, J. R., Moffett Field, Calif.
JIN, R.-S., Sunnyvale, Calif.
JOHANSEN, K. F., Palo Alto, Calif.
JOHNSON, A. R., Redwood City, Calif.
JOHNSON, D. B., Dallas, Texas
JOHNSON, R. E., Cincinnati, Ohio
*JOHNSON, J. N., Albuquerque, N. Mex.
JONES, J. L., Moffett Field, Calif.
JONES, J. P., Los Angeles, Calif.
JONES, E., San Jose, Calif.
JONES, J. B., Mountain View, Calif.
JORDAN, S., Stanford, Calif.
JORDAN, P. F., Baltimore, Md.
JORGENSEN, L., Moffett Field, Calif.
JOSEPH, D. D., Minneapolis, Minn.
JU, F. D., Albuquerque, N. Mex.
JUHASZ, S., San Antonio, Texas
KADLEC, R., Stanford, Calif.
KAISER, J. E., Blacksfurt, Va.
*KALNINS, A., Bethlehem, Penn.
*KANE, T. R., Stanford, Calif.
*KANNINEN, M. F., Columbus, Ohio
*KAPLAN, R. E., Los Angeles, Calif.
KARAMCHETI, K. K., Palo Alto, Calif.
KATZEN, E. D., Palo Alto, Calif.
KEARNEY. D. W., Stanford, Calif.
KELLY, R. E., Los Angeles, Calif.
KELSEY, S., Notre Dame, Ind.
*KEMPNER, J., Brooklyn, N.Y.
KERNEY, K. P., Washington, D.C.
KIANG, R. L., Stanford, Calif.
KING, L. S., Moffett Field, Calif.
KLEBANOFF, P. S., Washington, D.C.
KLEIN, E. J., Moffett Field, Calif.
KLEMP, J. B., Stanford, Calif.
KLINE, S. J., Stanford, Calif.
KLINGBEIL, W. W., Wayne, N.J.
KNOWLES, J. K., Pasadena, Calif.
*KO, D., Pasadena, Calif.
KOBASHI, Y. J., University Park, Penn.
KOELLER, R. C., Boulder, Colo.
KOGA, T., Stanford, Calif.
KOLSKY, H., Providence, R.I.
KOOPMAN, D. C. A., San Diego, Calif.
*KOVASZNAY, L. S. G., Baltimore, Md.
*KRATOCHVIL, J., Lexington, Ky.
*KRONAUER, R. E., Cambridge, Mass.
KU, Y. H., Philadelphia, Penn.
KUANG, J.-G., Tulsa, Okla.
*KUBOTA, T., Pasadena, Calif.
KUDIRKA, A. A., San Jose, Calif.
*KUIPER, R. A., Stanford, Calif.
KULGEIN, N. G., Palo Alto, Calif.
KWOK, M., Stanford, Calif.

LAGERSTROM, P. A., Pasadena, Calif.
LAMAR, I. D., Sunnyvale, Calif.
*LANDIS, F., Bronx, N.Y.
LANGLOIS, W. I., San Jose, Calif.
LARISCH, E., Iowa City, Iowa
LARSON, H. K., Moffett Field, Calif.
LARSON, J. C., Minneapolis, Minn.
*LAUFER, J., Los Angeles, Calif.
LAUPA, A., Pacific Palisades, Calif.
LAURMANN, J. A., Santa Barbara, Calif.
LAVAN, Z., Chicago, Ill.
LEAL, L. G., Atherton, Calif.
LEE, A. L.-M., Stanford, Calif.
LEE, C. W., Knoxville, Tenn.
LEE, C. M., Washington, D.C.
LEE, E. H., Stanford, Calif.
LEE, G., San Jose, Calif.
LEE, Jr., J. T., Sunnyvale, Calif.
LEE, J.-S., La Jolla, Calif.
*LEE, M. S., Wappingers Falls, N.Y.
LEE, L. H. N., Notre Dame, Ind.
*LEE, R. S., Santa Monica, Calif.
LEE, S. K., Syracuse, N.Y.
LEGNER, H. H., Palo Alto, Calif.
LEIBOLD, A., Dallas, Texas
LEIGH, D. C., Lexington, Ky.
LEKO, T., New York, N.Y.
LEMPRIERE, B., Palo Alto, Calif.
LENTZ, R. A., Oakland, Calif.
LENZEN, K. H., Lawrence, Kan.
LEONARD, R. W., Williamsburg, Va.
LEPPINGTON, F. G., Palo Alto, Calif.
LERNER, J. I., Sunnyvale, Calif.
LESSER, M. B., New York, N.Y.
LEVE, H. L., Long Beach, Calif.
LEVI, I. M., Palo Alto, Calif.
LEVIN, A. D., Moffett Field, Calif.
LEVY, A. M., Sunnyvale, Calif.
LEVY, Jr., L. L., Moffett Field, Calif.
*LEW, H. S., La Jolla, Calif.
LEWIS, R. E., Sunnyvale, Calif.
LIANG, S. F., Stanford, Calif.
*LIEBOWITZ, H., Washington, D.C.
LIEPMANN, H. W., Pasadena, Calif.
LIKINS, P. W., Los Angeles, Calif.
*LIN, T. H., Pacific Palisades, Calif.
LIND, R. C., Stanford, Calif.
LINDBERG, A. E., Menlo Park, Calif.
LINDHOLM, U. S., San Antonio, Texas
LINDSEY, K., Palo Alto, Calif.
LING, C.-H., Stanford, Calif.
LING, C. B., Blacksburg, Va.
*LING, S. C., Washington, D.C.
LINLOR, W. I., Moffett Field, Calif.
*LIPKIN, J., Providence, R.I.
LIU, J., Sunnyvale, Calif.
LIU, Y. K., Ann Arbor, Mich.
LIU, S.-C., Whippany, N.J.
LO, C. C., Columbus, Ohio
LOCK, M. H., El Segundo, Calif.
LOCKMAN, W. K., Moffett Field, Calif.
LOEBER, J. F., Scotia, N.Y.
LOGAN, T. R., Stanford, Calif.
LOMAX, H., Moffett Field, Calif.

USA (Cont'd)

LONDON, A. L., Stanford, Calif.
LOUBSKY, W. J., Menlo Park, Calif.
LOUTZENHEISER, C. B., Whippany, N.J.
LU, Z. A., San Jose, Calif.
LUBKIN, J. L., East Lansing, Mich.
LUBLINER, J., Berkeley, Calif.
LUCAS, R. D., Stanford, Calif.
LUCK, L. D., Pullman, Wash.
*LUDFORD, G. S. S., Ithaca, N.Y.
LUGT, H. J., Washington, D.C.
LUMLEY, J. L., University Park, Penn.
MAGERS, R. W., Moffett Field, Calif.
MAH, G. B. J., Sunnyvale, Calif.
*MAJERUS, J., Huntsville, Ala.
MALVERN, L. E., East Lansing, Mich.
*MANN-NACHBAR, P., La Jolla, Calif.
MANDEL, L., Portola Valley, Calif.
MARCHERTAS, A., Argonne, Ill.
MARGUILIES, G., Palo Alto, Calif.
MARK, R. M., Sunnyvale, Calif.
MARLOW, W. C., Sunnyvale, Calif.
MARLOWE, M. B., Santa Clara, Calif.
MARSH, M. C., Moffett Field, Calif.
MARTIN, C. J., East Lansing, Mich.
MARTIN, E. D., Moffett Field, Calif.
MARVIN, J. G., Moffett Field, Calif.
MASSARD, J. M., Palo Alto, Calif.
MASSON, B. S., Hawthorne, Calif.
MATTHEWS, D. R., Sunnyvale, Calif.
MATTOX, R. J., Pleasanton, Calif.
MAUTNER, S. E., Burbank, Calif.
*MAXWORTHY, T., Silver Bay, N.Y.
MAYERS, J., Stanford, Calif.
McNIVEN, H. D., Berkeley, Calif.
*McQUILLEN, E. J., U.S.N. Air Devel. Ctr, Penn.
McALISTER, K. W., Mountain View, Calif.
*McCLINTOCK, F. A., Cambridge, Mass.
McCORMICK, J. M., New York, N.Y.
McCOY, J. J., Philadelphia, Penn.
*McDEVITT, J., Moffett Field, Calif.
McFERON, D. F., Stanford, Calif.
*McINTOSH, S. C., Stanford, Calif.
*McIVOR, I. K., Ann Arbor, Mich.
McKEY, M. W., San Leandro, Calif.
McMILLAN, D. J., Stanford, Calif.
McMUN, J., Manhattan Beach, Calif.
*MEDICK, M. A., East Lansing, Mich.
*MELLENTHIN, J. A., Moffett Field, Calif.
MELLER, E., Stanford, Calif.
MELVIN, J. W., Ann Arbor, Mich.
MERCHANT, H. C., Seattle, Wash.
MESMER, G., Saint Louis, Mo.
*MIKLOWITZ, J., Pasadena, Calif.
*MILES, J. W., La Jolla, Calif.
MILES, J. B., Columbia, Miss.
MINDESS, S., Stanford, Calif.
*MINDLIN, R. D., Katonah, N.Y.
MINGORI, D. L., Manhattan Beach, Calif.
*MIURA, R. M., New York, N.Y.
*MOK, C.-H., Philadelphia, Penn.
MOLSTEAD, R. R., Sunnyvale, Calif.
MOOK, D. T., Blacksburg, Va.
MOORE, G. A., Sunnyvale, Calif.

MORAN, M. J., Columbus, Ohio
MORKOVIN, M. V., Baltimore, Md.
MORLAND, L. W., La Jolla, Calif.
MORRISON, J. A., Murray Hill, N.J.
MOSTAGHEL, N., Berkeley, Calif.
MOW, C. E., Santa Monica, Calif.
MUHLSTEIN, L., Moffett Field, Calif.
MUKI, R., Los Angeles, Calif.
MUNTER, A. R., Sunnyvale, Calif.
MURA, T., Evanston, Ill.
MURCH, S. A., San Jose, Calif.
MURPHY, J. D., Los Altos, Calif.
MURRELL, Jr., E. F., East Palo Alto, Calif.
MUSTER, D., Houston, Texas
MYERS, M. K., New York, N.Y.
*NACHBAR, W., La Jolla, Calif.
NACHTSHEIM, P. R., Moffett Field, Calif.
NADIR, S., Hawthorne, Calif.
NAGHDI, P. M., Berkeley, Calif.
NAKAJIMA, I., Stanford, Calif.
NAN, N., Stanford, Calif.
NARASIMHAN, K. Y., Argonne, Ill.
NARIBOLI, G. A., Ames, Iowa
NASH, W. A., Amherst, Mass.
NATHENSON, M., Stanford, Calif.
NEE, V. W., South Bend, Ind.
NELSON, C. W., Endicott, N.Y.
NELSON, F. C., Medford, Mass
NEMAT-NASSER, S., La Jolla, Calif.
*NEREM, R. M., Columbus, Ohio
NEVILL, Jr., G. E., Gainesville, Florida
NEWELL, A., Los Angeles, Calif.
NEWMAN, J. B., West Mifflin, Penn.
NG, L. S., Moffett Field, Calif.
NICHOLAS, T., Wright-Patterson AFB, Ohio
*NIILER, P. P., Fort Lauderdale, Fla.
NISHIOKA, K., Moffett Field, Calif.
NIU, H. P., Rapid City, S.D.
NORWOOD, F. R., Albuquerque, N. Mex.
NOTON, B. R., Stanford, Calif.
NOVAK, M., Port Huron, Mich.
OBERLANDER, K., Stanford, Calif.
ODEN, J. T., Huntsville, Ala.
O'DRISCOLL, P. F., Sunnyvale, Calif.
OGAWA, T., Stanford, Calif.
OLMSTEAD, W. E., Chicago, Ill.
OLSEN, E. M., Sunnyvale, Calif.
OPPENHEIM, A. K., Berkeley, Calif.
ORNE, D., San Jose, Calif.
ORSIGLIA, V. R., Moffett Field, Calif.
ORSZAG, S. A., Cambridge, Mass.
OSBORN, R. B., Seattle, Wash.
OSTRACH, S., Cleveland, Ohio
OWEN, D. R., Pittsburgh, Penn.
OWEN, Jr., G. N., Berkeley, Calif.
PAEK, U.-C., Berkeley, Calif.
PAI, S. I., College Park, Md.
PAN, Y. S., Los Angeles, Calif.
PANTON, R. L., Stillwater, Okla.
PAO, Y. C., Lincoln, Neb.
PAO, Y.-H., Seattle, Wash.
*PAO, Y.-H., Ithaca, N.Y.
PAPPAS, C. C., Menlo Park, Calif.
PARK, C., San Jose, Calif.

USA (Cont'd)

*PARKER, D. F., New Haven, Conn.
PARTHASARATHY, R., Stanford, Calif.
PAYTON, R. G., New York, N.Y.
PEARSON, W. E., Mountain View, Calif.
PECK, J. C., Santa Monica, Calif.
PENG, S.-D., Stanford, Calif.
PENNING, F. A., Denver, Colo.
PEREZ, M., W. Nyack, N.Y.
PERKINS, E. W., Moffett Field, Calif.
*PERRONE, N., Washington, D.C.
PETERSON, V. L., Los Altos, Calif.
PFENNINGER, W., Seattle, Wash.
*PHILLIPS, J. W., Providence, R.I.
PHILLIPS, A., New Haven, Conn.
PIAN, T. H. H., Cambridge, Mass.
PISTER, K. S., Berkeley, Calif.
PITTS, W. C., Moffett Field, Calif.
PLAUT, R. H., San Francisco, Calif.
PLUNKETT, R., Minneapolis, Minn.
POHLE, F. V., Garden City, N.Y.
POMERANTZ, J., Arlington, Va.
POPOV, E. P., Berkeley, Calif.
POTTER, M. C., East Lansing, Mich.
POWELL, W. R., Palo Alto, Calif.
PRADO, M. E., Livermore, Calif.
PRAGER, D. J., Whippany, N.J.
*PRAGER, W., La Jolla, Calif.
PRASAD, S. N., University, Miss.
PRESLEY, L. L., Moffett Field, Calif.
PRIAN, V. D., Palo Alto, Calif.
PRITCHETT, J. W., San Francisco, Calif.
*PU, S.-L., Watervliet, N.Y.
PUTLAND, L. W., Sunnyvale, Calif.
RADER, D., New Haven, Conn.
RAJAPAKSE, Y., Menlo Park, Calif.
RAKICH, J. V., Moffett Field, Calif.
RAMAN, K. R., Moffett Field, Calif.
RAND, R., Ithaca, N.Y.
RASHID, Y. R., San Diego, Calif.
RASMUSSEN, M. L., Norman, Okla.
RATTAYYA, J. V., Santa Clara, Calif.
RAUCH, H. E., Los Altos, Calif.
RAY, D., Chicago, Ill.
REDDY, N. M., San Jose, Calif.
REED, Jr., R. E., Moffett Field, Calif.
*REHFIELD, L. W., Atlanta, Ga.
REINHARDT, W. A., Los Altos, Calif.
*REISSNER, E., Cambridge, Mass.
REMENYIK, C. J., Knoxville, Tenn.
REYNOLDS, W. C., Stanford, Calif.
RICHMOND, O., Monroeville, Penn.
RIEDINGER, L. A., Sunnyvale, Calif.
RIKS, E., Palo Alto, Calif.
*RIM, K., Iowa City, Iowa
RIMON, Y., Washington, D.C.
RIPLOG, P. M., Tucson, Arizona
RIPPERGER, E. A., Austin, Texas
RIVLIN, R. S., Bethlehem, Penn.
ROBE, T. R., Lexington, Ky.
ROBERTSON, C. R., Palo Alto, Calif.
ROBINSON, R. C., Moffett Field, Calif.
ROBINSON, R. D., Moffett Field, Calif.
ROBINSON, W. W., San Ramon, Calif.

ROBSON, J. J., Portola Valley, Calif.
ROCKWELL, R. L., Stanford, Calif.
ROGALLO, R. S., Los Altos, Calif.
ROSE, W. C., Saratoga, Calif.
ROSEMAN, J., Madison, Wis.
ROSENBERG, R. M., Lafayette, Calif.
*ROTH, B., Stanford, Calif.
ROWLAND, S. C., Andrews, Mich.
ROWLEY, J. C., Los Alamos, N. Mex.
RUBESIN, M., Moffett Field, Calif.
*RUDINGER, G., Buffalo, N.Y.
RUES, D. T., Mountain View, Calif.
RUTOWSKI, R. W., Palo Alto, Calif.
SACHS, H. K., Detroit, Mich.
SACKMAN, J. L., Berkeley, Calif.
SAEGER, J., Stanford, Calif.
SAKURAI, A., Vicksburg, Miss.
SARIC, W. S., Albuquerque, N. Mex.
SCAVUZZO, R. J., Toledo, Ohio
*SCHAAF, B., Ann Arbor, Mich.
*SCHAPERY, R. A., Lafayette, Ind.
SCHER, M. P., Stanford, Calif.
SCHIERLOH, F. L., Warren, Mich.
SCHIPPER, J. F., Sacramento, Calif.
*SCHMIDT, R., Detroit, Mich.
SCHNEIDER, W. C., Houston, Texas
SCHULER, K. W., Albuquerque, N. Mex.
SCHUMAN, W. V., Towson, Md.
SCHWIND, R. G., La Honda, Calif.
*SCIAMMARELLA, C. A., Brooklyn, N.Y.
SCOTT, E. D., Sunnyvale, Calif.
SECHLER, E. E., San Marino, Calif.
SECOR, G., Salt Lake City, Utah
SEEGMILLER, H. L., Moffett Field, Calif.
SZEGO, P. A., San Jose, Calif.
SEIDE, P., Los Angeles, Calif.
SEIFERT, H. S., Stanford, Calif.
SENDECKY, G., Cleveland, Ohio
SENDELBECK, R. L., Stanford, Calif.
SEROVY, G. K., Ames, Iowa
SHAH, P. M., Houston, Texas
SHARMA, M. G., University Park, Penn.
SHARPE, Jr., W. N., Sunnyvale, Calif.
SHAW, R. P., Buffalo, N.Y.
SHAW, W., Auburn, Ala.
*SHEN, M.-C., Madison, Wis.
SHEN, Y. C., Sierra Madre, Calif.
SHENG, G., Stanford, Calif.
SHERMAN, M., Santa Monica, Calif.
*SHIEH, R. C., La Jolla, Calif.
SHIELD, R. T., Pasadena, Calif.
SIDHU, G. S., Menlo Park, Calif.
SIERAKOWSKI, R. L., Gainesville, Florida
SKINNER, L. A., West Lafayette, Ind.
SLITER, G., Menlo Park, Calif.
SMITH, C. E., Palo Alto, Calif.
SMITH, G. F., Bethlehem, Penn.
SMITH, H. W., Witchita, Kan.
SMITH, R. C., Moffett Field, Calif.
SOBEL, L. H., Los Altos, Calif.
SOLECKI, R., Storrs, Conn.
SOLER, A. I., Philadelphia, Penn.
SOONG, T. T., Placentia, Calif.
SPIER, E. E., San Diego, Calif.

USA (Cont'd)

SPREITER, J. R., Stanford, Calif.
SREEKANTH, A. K., Moffett Field, Calif.
*STALLYBRASS, M. P., Atlanta, Ga.
*STEARMANN, R., Austin, Texas
STEELE, C. R., Stanford, Calif.
STEGEN, G. R., La Jolla, Calif.
STEIN, R. P., Palo Alto, Calif.
STEINLE, F. W., Moffett Field, Calif.
STERN, P., Sunnyvale, Calif.
STERNBERG, E., Pasadena, Calif.
*STEVENS, K. K., Columbus, Ohio
STEWART, D. A., Moffett Field, Calif.
STRICKLAND, G. E., Palo Alto, Calif.
STRONGE, W. J., Menlo Park, Calif.
STROUD, R. C., Sunnyvale, Calif.
STUIVER, W., Honolulu, Hawaii
STYER, E. F., Bellevue, Wash.
SUCKLING, J. H., Aberdeen Proving Ground, Md.
SULLIVAN, R. F., Stanford, Calif.
SUPPLE, W. J., Pasadena, Calif.
SWAN, G. W. T., Madison, Wis.
SWEDLOW, J. L., Pittsburgh, Penn.
SWENSON, D. D., Middletown, Conn.
SYVERTSON, C. A., Moffett Field, Calif.
SZEWCZYK, A. A., Notre Dame, Ind.

TABAKMAN, H. D., Pomona, Calif.
TABLOFF, R. S., Oakland, Calif.
TABORDA, P. J., Redwood City, Calif.
TADJBAKHSH, I. G., Putnam Valley, N.Y.
TANG, S., Sunnyvale, Calif.
*TANG, S.-C., Dearborn, Mich.
TANNER, R., Livermore, Calif.
TASI, J., Port Jefferson, N.Y.
*TAYLOR, J. E., Los Angeles, Calif.
TAYLOR, S. M., Aberdeen, Md.
TAYLOR, T. D., Newport Beach, Calif.
TENERELLI, D. J., Stanford, Calif.
*TENNEKES, H., University Park, Penn.
THEODORIDES, P. J., Silver Spring, Md.
THOMAS, H., Stanford, Calif.
THOMAS, J., Palo Alto, Calif.
THOMAS, J. H., Rochester, N.Y.
*THOMAS, P. D., Palo Alto, Calif.
THOMPSON, G. A., San Jose, Calif.
THOMPSON, M. J., Austin, Texas
*TING, L., New York, N.Y.
*TING, T. C. T., Stanford, Calif.
TOBAK, M., Moffett Field, Calif.
TODD, R. H., Stanford, Calif.
TONG, K. J., Stanford, Calif.
TONG, P., Cambridge, Mass.
TOURYAN, K. J., Albuquerque, N. Mex.
TRAUGOTT, S. C., Baltimore, Md.
TREON, S. L., Moffett Field, Calif.
TRILLING, L., Cambridge, Mass.
TRIPODI, R., Mountain View, Calif.
TSAO, D. W., Palo Alto, Calif.
TUBA, I. S., Irwin, Penn.
TZENG, S. T., Pittsburg, Penn.
UNAL, G., Akron, Ohio
UNDERWOOD, R. L., Stanford, Calif.
URBAN, R. L., Las Cruces, N. Mex.
URTIEW, P. A., Livermore, Calif.

VAGLIENTE, V. N., Stanford, Calif.
*VAGLIO-LAURIN, R., New York, N.Y.
VAHEDI, M. I., Palo Alto, Calif.
VALANIS, K. C., Thousand Oaks, Calif.
VALENTIN, R. A., Argonne, Ill.
VAN BUREN, W., Pittsburgh, Penn.
VAN DYKE, M., Stanford, Calif.
VARGA, L. A., Des Plaines, Ill.
VARLEY, E., Bethlehem, Penn.
VAROGLU, E., Stanford, Calif.
VERMA, G. R., Kingston, R.I.
VINCENTI, W. G., Stanford, Calif.
VITTE, W. J., Atherton, Calif.
VOJVODICH, N. S., Moffett Field, Calif.
VOLTERRA, E., Austin, Texas
WADDOUPS, M. E., Fort Worth, Texas
WAGNER, S., Moffett Field, Calif.
WALSH, E. K., Pittsburg, Penn.
*WAN, F. Y. M., Cambridge, Mass.
WANG, C.-Y., Pasadena, Calif.
WANG, C.-C., Houston, Texas
WANG, K. C., Baltimore, Md.
WANG, L.-S., Stanford, Calif.
WANG, T. T., Murray Hill, N.J.
WARE, III, A. G., Stanford, Calif.
WARNER, R. W., Moffett Field, Calif.
*WARREN, W. E., Albuquerque, N. Mex.
WARWIND, R., Moffett Field, Calif.
WATABE, M., Oakland, Calif.
WATSON, E., Moffett Field, Calif.
WATSON, V. R., San Jose, Calif.
WEHAUSEN, J. V., Berkeley, Calif.
WEINGARTEN, V. I., Los Angeles, Calif.
WEIS, H. J., Ames, Iowa
WEISS, R. T., Berkeley, Calif.
WEISSHAAR, C. A., Cupertino, Calif.
WEISSHAAR, T., Cupertino, Calif.
WEITSMAN, Y., Providence, R.I.
WELLS, L. T., Columbus, Ohio
*WEMPNER, G., Huntsville, Ala.
WEN, S.-S., Goleta, Calif.
WESENBERG, D., Stanford, Calif.
WEST, J. C., Chicago, Ill.
WIGGIND, L. E., Sunnyvale, Calif.
WILEY, R. J., Stanford, Calif.
WILKINS, E. W. C., San Francisco, Calif.
WILKINSON, J. P. D., King of Prussia, Penn.
WILKOV, M. A., Austin, Texas
WILLEKE, K., Stanford, Calif.
WILLIAMSON, A. S., Stanford, Calif.
WILLIAMS, H., Claremont, Calif.
*WILLIS, D. R., Berkeley, Calif.
WILLOUGHBY, P. G., Sunnyvale, Calif.
WILSON, K. H., Sunnyvale, Calif.
WINEMAN, A. S., Ann Arbor, Mich.
WINOVICH, W., Moffett Field, Calif.
WITTMANN, A., Hermosa Beach, Calif.
*WNUK, M. P., Brookings, S. D.
WONG, P. K., Stanford, Calif.
*WOO, L. S., New York, N.Y.
WOOD, A. D., Sunnyvale, Calif.
WOODS, L. C., Austin, Texas
WOOLDRIGE, C. E., Menlo Park, Calif.
WOOLLEY, J. P., Stanford, Calif.

USA (Cont'd)

WORLEY, W. J., Urbana, Ill.
WRENN, B. G., Sunnyvale, Calif.
WRIGHT, Jr., O. C., Sunnyvale, Calif.
WRIGHT, T. W., Aberdeen Proving Ground, Md.
*WU, E. M., Saint Louis, Mo.
WU, J. N. C., Alliance, Ohio
WYGNANSKI, I., Bellevue, Wash.
YACHMAI, S. S., Berkeley, Calif.
YANG, A. T., Davis, Calif.
YANG, J.-S., Stanford, Calif.
YANG, K.-T., Notre Dame, Ind.
YANG, T.-L., Washington, D.C.
YANG, W. H., Ann Arbor, Mich.
YATES, W. G., Stanford, Calif.
YATTEAU, J. D., Los Altos Hills, Calif.
YEH, G. C. K., Redondo Beach, Calif.
*YEN, D. H. Y., East Lansing, Mich.
*YEUNG, K. W., Buffalo, N.Y.
YIN, F. C.-P., La Jolla, Calif.
YOSHIOKA, M., Moffett Field, Calif.

YOUNG, D., San Antonio, Texas
YOUNG, D. H., Stanford, Calif.
YOUNGDAHL, C. K., Argonne, Ill.
YU, Y.-Y., Philadelphia, Penn.
ZASLAWSKY, M., Livermore, Calif.
ZEREN, R. W., Stanford, Calif.
ZLOOF, M. M., Berkeley, Calif.

Venezuela

KULKA, C. A., Maracaibo
LAMAR, S., Caracas
ROCA, R. V., Caracas
ZAGUSTIN, A., Caracas

Yugoslavia

MURSIC, M. V., Ljubljana
*ZDRAVKOVICH, M. M., Belgrade

Résumé of the Congress

A Report by

N. J. Hoff

President of the Congress, Stanford University

Opening Session

The Opening Session of the Twelfth International Congress of Applied Mechanics was held in Frost Amphitheater on the Stanford University campus. The session was opened by NICHOLAS J. HOFF, President of the Congress, at 9:30 AM on Monday, August 26, 1968, with the following address:

Ladies and Gentlemen:

It gives me great pleasure to open the Twelfth International Congress of Applied Mechanics and to greet this outstanding gathering of scientists and engineers which will insure the success of the Congress.

I would like to tell you a little about the history of the congresses of applied mechanics. Those of you who belong to the younger generation may not know that the origins of this congress can be found in a meeting in the city of Innsbruck, Austria, which was held in 1922. It dealt with problems of fluid dynamics and was called by four men: OSEEN of Uppsala, Sweden; LEVI-CIVITÁ of Rome, Italy; VON KÁRMÁN of Aachen, Germany; and PRANDTL of Göttingen, Germany. The meeting was a great success and in consequence it was decided that a regular series of congresses should be inaugurated, dealing not only with fluid mechanics but also with other fields of mechanics, in particular solid mechanics. VON KÁRMÁN, who later was made honorary president of the International Union of Theoretical and Applied Mechanics, was particularly active in organizing the congresses. He was aided by Professor BURGERS of Delft, Holland, and the two of them persuaded Professor BIEZENO, also of Delft, Holland, to be president of the First Congress of Applied Mechanics which was to be held in Delft, in 1924.

This was the beginning of an extremely successful series of congresses which were held, in general, at intervals of four years except for interruptions by World War II. The second congress took place in Zürich, Switzerland, in 1926, and was followed by congresses in Stockholm, Sweden, in 1930; in Cambridge, England, in 1934; in Cambridge, Massachusetts, United States, in 1938; in Paris, France, in 1946; in London, England, in 1948; in Istanbul, Turkey, in 1952; in Brussels, Belgium, in 1956; in Stresa, Italy, in 1960; in Munich, Germany, in 1964; and finally now at Stanford, California, United States.

Last week I took the time to go into the Timoshenko Room of our School of Engineering which is named after the great Russian scientist who did so much for applied mechanics, particularly in the United States. He is still alive and is living in Germany with his older daughter. He wrote me that he was very sorry that he could not be here with us today but his health did not permit it. In the Timoshenko Room we have most of his books and all his medals; you may want to see them during your stay at Stanford. The library includes an almost complete set of the Proceedings of the congresses. I read through some of these volumes and I think that all of you would enjoy doing the same.

It was a great pleasure to see how the congresses started, how much interest there was in them, and in particular to note that in the early days of the congresses mechanics was not quite so much separated from other parts of physics as it is today. One can find among the contributors to the first few congresses the names of many great physicists and mathematicians. I cannot mention them all but at random I would like to name the physicist DEBYE. Among the mathematicieans present I have already mentioned LEVI-CIVITÁ, but I would like to add now COURANT, TREFFTZ and VON MISES.

The first few volumes contain a number of most valuable articles.

I would like to mention the paper by TREFFTZ entitled ,,Ein Gegenstück zum Ritzschen Verfahren" which establishes a method of calculating lower limits instead of the upper limits obtainable by the Rayleigh, Ritz and Timoshenko procedures. At the Zürich Congress Beggs of the United States discussed in great detail his experimental method of determining the stress distribution in practical engineering structures.

It is not true, of course, that the meeting in Innsbruck was the first one that dealt with applied mechanics. There had been meetings held before, particularly in Paris, but their subject matter was more applied, more what we would today call design. The modern series of applied mechanics congresses certainly has its roots in the Innsbruck meeting and in Delft where the first congress was held.

It is of interest to note that many persons who participated in the organization of the first few congresses are still alive but have found it too troublesome to come to Stanford because of their age. But there are a number of outstanding exceptions. In this regard I would like to mention first Professor BURGERS who was the co-organizer of the First Congress of Applied Mechanics and who is sitting here on the platform. I would like to ask him to rise so that you can all see him. Next, I would like to call on another man who is beloved by all people in applied mechanics and whom we have the great honor of having with us. This man has another distinction: He has read a paper in every one of the first eleven congresses, and his name can be found on the program of this congress also with a scientific contribution. I would like to ask Sir GEOFFREY TAYLOR to rise.

Since we must keep the official proceedings as short a possible, I cannot introduce everyone on the platform. I would like to tell you, however, that they include the members of the Bureau of the International Union of Theoretical and Applied Mechanics. In America the Bureau would probably be called the Executive Committee. In addition, we have the members of the Congress Committee, of the Executive Committee of the Congress Committee and of the International Papers Committee. And then we have a number of persons who have aided us in the organization of the Congress here locally. Among the people who have helped us, I would like to mention in particular Professor KOITER, who is the Secretary of the Congress Committee and who has put a tremendous amount of work into the preliminaries that have to precede a congress of this kind. I would like to ask Professor KOITER to rise.

I would also like to mention that the number of papers received was so large that the National Committees which screened them and the International Papers Committee which made the final decisions regarding acceptance were put to a very difficult task. In particular from among the papers originating in the United States, which naturally provided the largest number of abstracts because of the ease of travel, we had to reject four out of five. We know very well that many of these papers were excellent; we simply could not put them on the program for lack of time. As you know we have five simultaneous technical sessions most of the time. If we had accepted all the good papers, we would have had to organize twenty simultaneous sessions, and I know that no one would have liked that kind of arrangement. In any case, the International Committee imposed upon us a definite limit, and I would like to apologize in the name of the International Papers Committee and in the name of the U.S. Committee for not accepting a number of extremely valuable papers. And if it comes to apologies, I would like to make use of this opportunity to tell you that we are extremely sorry that many of you were put to inconveniences in connection with travel from the airport to Stanford and registration at Stanford. We are very sorry that it happened. If you would want us to hold the next International Congress four years from now again at Stanford, I think we could avoid most of these difficulties.

I would like to mention two more names only. One of them is that of another man who has been very active in these international congresses; although he is not here at this moment, he is going to be with us tomorrow. This man is ADOLF BUSEMANN who took part in all the congresses before World War II and presented a number of valuable papers. The other man whom we miss today is Professor RABOTNOV who was in the hospital a week ago when the Aeroflot plane left Moscow, but I received a telegram from him yesterday with the information that he would arrive here tonight. So I think that we are going to have many of our good friends here with us in spite of all these difficulties.

The papers to be presented originate in twenty-seven different countries and the flags you see around the platform, arranged in alphabetical order of the names of the nations they represent, are the flags of all those countries from which papers have been accepted for presentation. The flags do not indicate that we want to make this a meeting of international political groups. Everyone who is here is here in his own name only. We are all individuals interested in science and we can all live together in friendship without any difficulty. We hope that through individual contacts we are going to contribute to ever better relations among the various nations of the world.

I would also like to say that I am very grateful to a number of organizations which have helped us with the Congress in various ways. In particular I want to mention the Bureau of IUTAM which has labored very much through the last four years in order to make this meeting a success, as we hope it will be. I am not going to mention any of my collaborators here at Stanford on the local organizing committee because I think it will be much more appropriate if I mention them at the closing session; then you will know exactly who was responsible for the mistakes you will have noticed. I would like to thank, though, the Executive Committee of the Congress Committee, the International Papers Committee, and the various national committees as well as the Office of Naval Research of the United States Navy, the Office of Scientific Research of the U.S. Air Force, and the U.S. National Science Foundation and the National Aeronautical and Space Administration which contributed financially or otherwise in order to make it possible for us to hold such a congress here. We have also received substantial support from a number of industrial companies and universities whose names appear on the Program. We are very grateful to all of them.

This is the end of my introduction but I would like you to hear from a number of persons who are sitting on this platform. To begin with I would like to ask the President of Stanford University to welcome you here. We are extremely grateful to him because he is willing to be here with us this morning and because he and his wife will give a recpetion to all of you this afternoon at 5:30 P.M. in the garden of the home of the President. Normally one could expect this, but the circumstances are unusual. Dr. STERLING has been one of the most successful presidents of Stanford University in its entire history. He is retiring on the day of the closure of this Congress. He is already in the process of moving out of his house and his furniture is all piled up inside the rooms. In spite of this, he is willing to see you this afternoon. I am very happy indeed to present to you President J.E. WALLACE STERLING of Stanford University.

President STERLING addressed the Congress in the following manner:

Mr. Chairman, Ladies and Gentlemen:

It is true that I have held this office for almost two decades. I hope it will come as no surprise to you that during that span of time I have indeed had the opportunity, and one which I enjoyed, of welcoming the members of a variety of congresses and associations to this campus. But since this is indeed my last week in office I am especially gratified that before retiring I have this opportunity to welcome to Stanford this 12th International Congress of Applied Mechanics. My wife and I do look forward to seeing you in the gardens of our otherwise disrupted home this afternoon. I hope very much Mr. Chairman, Ladies and Gentlemen that your deliberations both formal and informal will be pleasant and productive and that you will

1*

enjoy your stay on this campus. I ask you therefore to accept my greetings, my welcome and my very best wishes.

After President Sterling's talk Professor HOFF introduced Dean PETTIT with the following words:

Next I would like to call on Dean JOSEPH M. PETTIT who is the head of the School of Engineering at Stanford University and who himself is a noted scholar, particularly in the field of solid state electronic devices. In the last few years he has been mostly engaged in problems of education. It gives me great pleasure to present to you Dean PETTIT.

Dean Pettit's talk follows:

Distinguished Delegates and Guests:

It is my privilege to add another word of welcome to those you have already received. This I do on behalf of Stanford's School of Engineering within which are centered all of the studies in applied mechanics. The various fields of applied mechanics permeate seven out of the ten departments in the School including not only the Department of Applied Mechanics itself but also the Departments of Aeronautics and Astronautics, Chemical Engineering, Civil Engineering, Electrical Engineering, Material Science and Mechanical Engineering. This is a cooperative enterprise among all these units. For instance, we have a single seminar in fluid mechanics with participation by professors and students from all of the interested departments. I mentioned that there are ten departments in the School of Engineering. We have 600 undergraduate engineering students and 1,500 postgraduate students. This year we awarded 134 Bachelor's degrees, 565 Master's degrees, and 146 Doctor's degrees in all the fields of engineering together. I would describe our role in American education as that of serving a highly selected undergraduate student group within a comprehensive, privately supported residential university, and at the postgraduate level, a large and diversified student population, nine out of ten having done their undergraduate work elsewhere, and 600 coming from foreign countries. Five hundred of our postgraduate students are taking their work on a part-time basis attending daytime courses while being employed in local industry, government laboratories, and research organizations. The professors number more than 145 and I am very proud of this splendid group of colleagues. You will be hearing from several of them this week. I hope that they and the many distinguished scholars that share the platform at this congress will provide you with a stimulating program which in combination with the many activities provided for you by the local committee will give you a totally enjoyable stay at Stanford. Thank you very much.

After this talk, Professor HOFF continued in the following manner:

Ladies and Gentlemen:

Up to now we have all been using only one language, namely English. Undoubtedly English is understood best by most of you here. Nevertheless, we have a very strong group of people present whose mother tongue is not English. In an international congress it would be most inappropriate not to make use of some of the other languages as well. For this reason I am going to ask in some other languages a number of people who represent themselves essentially but who were elected at a somewhat random manner from various countries all over the world to say a few words at this opening session.

In particular, I would like to start now with a man who is also to speak here in another capacity, namely in the capacity of being President of the International Union of Theoretical and Applied Mechanics which is ultimately responsible for the organization of all these congresses. This man is M. MAURICE ROY of whom I am sure you have all heard. He is a scientist himself, a professor at the Ecole Polytechnique in Paris which was organized just about a hundred years before the organization of Stanford University. He has also been president of a number of very important national and international organizations, among them the French Academy of Sciences and COSPAR, which is an inter-union committee for space research.

Monsieur le Président de l'IUTAM,

Mesdames et Messieurs de langue française,

J'ai grand plaisir de vous acceuillir, Mesdames et Messieurs de langue française, quand je prie Monsieur Roy de prendre la parole. Je voudrais vous dire comme je suis content de voir une délégation aussi forte du point de vue du nombre aussi bien que de la qualité. Je n'ignore pas de tout l'essor impressionant de notre science qui a eu lieu entre le milieu du 18ème et du 19ème siècle, surtout en France. Je sais bien que vous continuez cet effort et j'espère que vous contribuerez d'une façon importante au progrès de la mécanique pendant votre séjour à Stanford.

Mesdames et Messieurs, soyez les bienvenus!

Monsieur Roy, cher ami, vous avez la parole.

Professor Roy's answer to these words of greeting follows:

Monsieur le Président, Mesdames, Messieurs,

Comme mon ami le Professeur Nicholas Hoff l'a déjà mentionné, le Congrès qui s'ouvre en ce moment à Stanford est l'une des manifestations organisées périodiquement par IUTAM, c'est-à-dire par l'Union Internationale de Mécanique Théorique et Appliquée.

Ce Congrès a, en effet, le même objet que cette Union elle-même à savoir la promotion du progrès de la science de la Mécanique par la coopération internationale. Tous les quatre ans, les congrès de cette sorte complètent et élargissent le cycle des symposiums spécialisés que IUTAM organise, à raison de trois environ chaque année et en continuelle consultation de toutes les organisations nationales adhérentes.

Puisque j'ai aujourd'hui l'honneur de parler ici au nom de cette union alors que va d'ailleurs s'achever mon mandat de Président élu, je suis particulièrement heureux de souhaiter à tous les participants du XIIème Congrès une active et féconde coopération et aussi de féliciter et de remercier nos amis Américains pour l'excellente préparation du Congrès de Stanford, y compris la participation d'un brillant soleil californien.

Je voudrais ajouter qu'il m'est aussi très agréable de pouvoir associer ici spécialement à ces voeux, à ces félicitations et à ces remerciements les nombreuses organisations adhérentes qui sont représentées ici et qui appartiennent à des pays situés dans une partie assez large du continent Européen.

Je vous remercie de votre attention.

When Professor Roy sat down, Professor Hoff stood up and said:

Next, I shall call upon Professor Nikolai I. Muskhelishvili. Again those of you who are working in science know his name extremely well. By age and by contributions to the science of mechanics there is no one we can call superior to him. You all have heard of, seen, or read some of his books which have been published in English translation. They deal with integral equations and with the well-known Muskhelishvili method of using complex variables in the solution of problems of two-dimensional solid mechanics. He is also a man of many honors and I cannot even tell you all the offices he has held. They include vice-presidentships and presidentships of the Academies of Sciences of the Soviet Union and of his home country, Georgia.

Товарищи, друзья, и глубокоуважаемый Николай Иванович! Я очень рад приветствовать делегацию Советского Союза в Станфорде. Когда я был в Москве в январе этого года, вы меня приняли щедрой рукой. Тогда я вам сказал нет в мире гостеприимство подобного русскому и грузинскому. С другой стороны я вам обещал лучшую погоду в Стаифорде в августе чем в Москве в январе.

Вот прекрасная погода!

Я надеюсь, что вы будете довольны Станфордом.

Я очень рад дать слово дорогому другу Академику Мусхелишвили.

Academician Muskhelishvili answered in French:

Monsieur le Président, Mesdames, Messieurs!

Cela m'est très difficile de parler français après Monsieur ROY, mais je suis obligé de le faire.

Permettez-moi au nom des savants de l'Union Soviétique de saluer chaleureusement tous les participants de ce Congrès. Nous attachons une très grande importance aux Congrès internationaux qui sont organisés par IUTAM, car ils nous donnent le moyen d'établir les contacts personnels avec nos collègues étrangers pour le bien du progrès des sciences. Nous souhaitons un grand succès au présent Congrès.

Nous remercions cordialement nos collègues américains pour l'excellente organisation de ce Congrès et pour leur remarquable hospitalité. Tout particulièrement, nous remercions le Président de ce Congrès, Monsieur le Professeur NICHOLAS HOFF, qui a tout fait avec ses collègues pour rendre notre séjour en Amérique à la fois utile et agréable. Je ne savais pas jusqu'ici que ce soleil que nous voyons ici, est aussi organisé par Monsieur HOFF.

Merci pour votre attention!

Professor HOFF then said:

Now we are going to travel very far from Europe, all the way to Asia and in particular to Japan. We have a large Japanese delegation present at the Congress and it gives me very great pleasure to introduce to you Professor ITIRO TANI who, as you will see, is still an extremely young looking man but, according to the rules of his university, he has just retired as Professor of the University of Tokyo where he has been active for many years in teaching as well as in aeronautical research. He is also the President of the National Committee of Theoretical and Applied Mechanics of Japan.

Mina-san soshite Tani Ichiro Sensei,

Nihon-no yūshūna daihyō-ni o-meni kakarete taihen ureshii desu. Go-nen mae-ni watakushi-wa Tōkyō-ni ikimashita. Tōkyō Daigaku-no kyaku-in kyō-ju deshita. Hima-na toki-ni Nihon-go-no benkyō-wo shimashita. Keredomo hotondo wasuremashita kara Nihongo-de hanasu-no-wo yame-yō to omoi masu. Dewa-... Tani Sensei o-negai shimasu.

Next Professor TANI addressed the gathering in English:

Mr. Chairman, Ladies and Gentlemen:

It is indeed a great honor for me to be given the opportunity to address you on the occasion of the Twelfth International Congress of Applied Mechanics. First of all, I would like to say how joyful it is to see our friends coming from all parts of the world with one point in common, which is applied mechanics. Applied mechanics is a field such that everybody working in some area of the field has some understanding of what others are doing. Furthermore, with von Kármán's terminology, applied mechanics has been modest enough never to interfere with the divine world such as creating quantities of enormous magnitude, except perhaps by speaking of Reynolds numbers of 10^7 and numbers of repeated stress of 10^8, and so it has rarely been involved in political affairs. These features seem to me one of the reasons why the Congress survived through the turbulence of the war and continued to attract so many participants from all over the world. As one of the participants from Asia, I strongly feel this and entertain the hope that the Congress be held in the not too distant future somewhere in Asia. With my clumsy English, may I congratulate the success of the Congress and appreciate the effort of the organizing committee with Professor NICHOLAS HOFF as President. Thank you.

After Professor Tani's address, Professor HOFF said:

Now I am in a very awkward position. I am going to introduce a man whose language I cannot speak. He is Professor F. S. SHAW of Australia. I did spend eight weeks in Sydney at his invitation and taught a course on creep at the University of New South Wales where he is

Chairman of the Structural Department, but my wife did not permit me to learn the Australian way of speaking, so I have to do what I can in my poor American-Hungarian version of English. Professor SHAW is also President of a newly organized congress which includes Australia and New Zealand, which is known as the Australasian Congress of Structures and Mechanics. He is no stranger to the United States, as he spent about three years at Brown University, four years at the Polytechnic Institute of Brooklyn and I believe roughly a year or a little more in two installments at the University of Illinois. He is a very old friend of mine, and I am very happy indeed to introduce to you Professor SHAW.

Professor SHAW made the following statement:

Mr. President of the Stanford Organizing Committee:

It is obviously necessary for me to reply in exactly the same language and therefore I must apologize to many of you who may not understand Australian. As Professor HOFF has said, we had the extreme pleasure of having him with us two years ago for a very short time, much too short, of course, but as everyone knows Professor HOFF made very many friends there. Unfortunately, it cannot be said that applied mechanics as such is studied very widely in Australia. Only over the last few decades or so has the study of applied mechanics become centralized in enginneerig schools, engineering departments and the like. But as Professor HOFF has indicated, the interest is growing and last year we held our first Australasian congress and we are now planning for the second one next year. Well, Mr. President of this organizing committee, I have very much pleasure in bringing you greetings from Australia and wishing the Congress every success.

Professor HOFF continued:

Our next speaker will be again a man whose language I do not speak. Fortunately, he speaks very good English. As a matter of fact, he received his Ph. D. degree in the Applied Mechanics Department of Stanford University a few years ago. He is now Professor in Lagos, and he is one of the few representatives we have here from Africa. Professor SOBOYEJO, would you please say a few words.

Next, Professor SOBOYEJO, delivered the following talk:

Honorable President of the International Union of Applied Mechanics,
Distinguished Members and Friends of the Union, Ladies and Gentlemen:

First and foremost, I thank the organizing committee for asking me to say a few words of greeting on behalf of the continent of Africa to the honorable members of the Twelfth International Congress of Applied Mechanics which is taking place here at Stanford University. In addition to saying a few words of greeting, I wish to bring out briefly two major problems confronting Africa today. These problems can broadly be classified into two categories. One, political problems of development; and two, human problems of development. African countries can benefit immensely from excellent ideas and help from men of good will everywhere in the solution of urgent technical problems in the areas of education, industrialization, agriculture, aviation, transportation, communication, housing, and tropical medicine. A great deal of work has to be done, and in addition there are the wish and determination of the African people to help themselves. The continent of Africa will need all the sincere friendship, help and good will of many well-meaning people for many years to come. Africa had already benefited immensely from many other countries, and there still remains much more to be done in the years that lie ahead. For the future, an important world organization such as this one and their members all over the world should think very seriously of ways of contributing something to the solution of some of the technical problems of development in Africa. Some of the inevitable human problems of development have contributed largely to the state of affairs which we are now finding in many African countries. It is hoped that people will learn from the past and prepare to work harder for a much better future of the continent. We hope that one day Africa will be able to have you as guests for one of the future congresses. I greet you all and bring to you our very best wishes for a successful Twelfth International Congress of Applied Mechanics.

After Professor Soboyejo's address, Professor HOFF said:

Since we have not been able to find anyone from Antarctica to come here and bring greetings, we have only one man left on the program in this series. He is Professor CARLOS DEL RIO from Peru and he, I am glad to say, is also a man who knows this country very well. Professor DEL RIO received his Ph.D. degree from Stanford University under Professor HETÉNYI, but he has been back in Peru teaching. I feel that we would like to have many more scientists from Central America and South America here with us, and I hope that the connections between the northern and southern parts of this hemisphere will develop much more to our mutual benefit in the future.

Señôras y Señores de la America Latina!

Me alegro de saludar a los congresistas de la América Latina. Siento que el número de los delegados sea muy pequeño. Espero que ustedes se encuentren muy felices en California y vuelvan en un futuro cercano.

Señor DEL RIO, tenga la bondad de decir unas cuantas palabras.

Professor del Rio's talk follows:

Mr. Chairman, Ladies and Gentlemen:

It is a privilege to attend this Congress and it is a great pleasure to be on this platform to address you in the name of the few Latin American representatives. We Latin Americans are aware of what is the role of science in general and of applied mechanics in particular. We are a small group now working in this field, and we believe that we will help in building up a better tomorrow without any fear in Latin America. But for doing that, I would like to make a plea to all of you who work in developed countries to help us to build a better society. You can send us your papers and let us know what you are doing. Thus you would motivate us to work harder and perhaps someday also to contribute to the brotherhood of mankind. To Stanford University I am grateful for what I owe to it, especially to my former professor to whom I owe so much. And to all of you my best wishes and my hope that some day we can say "Welcome to Latin America", not for a fiesta but for an applied mechanics congress. Thank you.

After Professor del Rio's words, Professor HOFF said:

Ladies and Gentlemen:

I would like to read to you a wire I received from the President of the U.S. National Academy of Sciences, FREDERICK SEITZ. "On behalf of the National Academy of Sciences which exercises membership in IUTAM through its U.S. National Committee, it is a great pleasure to welcome participants, especially those colleagues traveling from abroad, to the Twelfth International Congress of Applied Mechanics and to the General Assembly of IUTAM. Please accept the very best wishes of the members and officers of the National Academy of Sciences for the full success of your Congress, the first to be held in the United States since 1938." Signed — FREDERICK SEITZ, President.

I am also very happy to read to you parts of a letter which was sent to me by the President of the First International Congress of Applied Mechanics, Professor BIEZENO. It says: "My wife and I do hope that the Congress will be a great success and a renewed contribution to international collaboration, a subject that in so many other parts than the scientific field of human activity is lacking in such a regrettable way. Unnecessary to say that we still feel thankful and flattered by your generous proposal to attend the Congress and that we regret wholeheartedly not to be present at a manifestation which is a worthy continuation of a work done in 1924 on the initiative of VON KÁRMÁN, BURGERS and myself. To you and to the Congress as a whole it must be a pleasure to having in your midst J.M.BURGERS to whom the Congress is so very highly indebted for his scientific insight. A same reason for joy is the presence of G.I.TAYLOR who was nominated a member of the First International Congress Committee on the 23rd of April 1924. A long time has lapsed since, but scientific and friendly relations still exist between the living members of the first Congress. The ideals from which our institution originated persist in a very long future." Signed — C. B. BIEZENO.

This concludes essentially the program of this opening session. All that remains to be said is that I hope that you will enjoy your stay here, will benefit from the sessions and from the association with other people, and if you find anything at fault you will not hesitate to let us know because we would like to correct it. Before we adjourn, I would like to ask the man who is more responsible for everything you see at this Congress than anyone else, our very hard-working Secretary, Professor CHARLES STEELE. He wants to say a few words in order to make sure that everything will run properly. I would like to repeat once more that you are all welcome to appear at the reception to be given by the President of Stanford University this afternoon at 5:30.

Following information given by Professor STEELE regarding the location of the sessions, printed programs, the names of chairmen and vice-chairmen of the technical sessions, the social program and bus schedules, Professor HOFF declared the Twelfth International Congress of Applied Mechanics opened.

Technical Sessions and Excursions

Between Monday, August 26, and Saturday, August 31, inclusively, 285 technical papers were presented in 73 sessions. Most of the time there were five sessions held simultaneously. The chairmen of the sessions were selected from among the foreign delegates, and the vice-chairmen from among the U.S. delegates. The chairmen and vice-chairmen were aided by graduate students of the School of Engineering of Stanford University of whom over seventy volunteered for this work and for other necessary functions. In the afternoons of Tuesday, Thursday, and Friday, excursions were organized to the Ames Research Center of the National Aeronautics and Space Administration and to the Stanford Linear Accelerator for those who felt that they could absent themselves from the technical sessions.

No technical sessions were held in the afternoon of Wednesday, but arrangements were made for a boat trip on San Francisco Bay followed by dinner in Chinatown. The banquet of the Congress took place the evening of Friday in the open air under the old live oak trees of the area between Tresidder Union, Bowman Alumni House and the Faculty Club on the Stanford campus. For those who arrived early for the Congress, excursions were arranged to San Francisco on Saturday, August 24, and to the Monterey Peninsula on Sunday, August 25. The ladies' program contained a number of trips in the surroundings of Stanford University and San Francisco.

Closing Session

The Closing Session of the Congress was held in Memorial Auditorium at 2 PM on August 31. It was opened by Professor HOFF with the following words:

Ladies and Gentlemen:

At this Closing Session of the 12th International Congress of Applied Mechanics I have very little to say. I do not want to repeat my thanks to the various committees of IUTAM for the excellent work they have done and for the wonderful support they have given us in the organization of this Congress. What I would like to mention is that naturally a great deal of work had to be done and that I hope that the persons who did the work and who contributed many interesting and original ideas will not be forgotten.

Most of you must have seen on the second floor of Tresidder Union the various exhibits, and in particular the book exhibit. These were conceived and organized by Mrs. HOLT ASHLEY. The arrangements for room and board were made by Mrs. LOIS BUTTS. The printing of all our publications was done by the *Accent on Offset* printing company whose owners, Cho and Larry Cline, labored day and night to get out everything on time. Incidentally, if you find any mistakes in the list of delegates please let us know because we would like to have this list corrected for final publication in the Proceedings.

Next, I would like to give all the credit due to BILL BOZICH, one of our most hard-working graduate students, who organized a student group of approximately seventy to act as coordinators of the technical sessions and as guides on the excursions. We are very grateful to Mr. BOZICH and to all the students who took care of these jobs and of many other chores. A former graduate student of ours, SAM MCINTOSH, who is now Assistant Professor, contributed many good ideas and, in particular, was in charge of everything connected with transportation. The projection and sound effects were in the hands of JIM CLOUD and I think we should say "thank you" to him because he has done a good job.

The man who was in charge of all the technical services is KEVIN FORSBERG who donated his time even though he is not even connected with Stanford University. He is an employee of the Lockheed Company. I must admit though that he used to have connections with Stanford as he obtained his Ph. D. degree from Stanford University. Those of you who thought the banquet was satisfactory should remember JOYCE VINCENTI who contributed to it the artistic touch. What is even more important, she was in charge of the entire ladies' program.

BILL REYNOLDS was in charge of registration and statistics and made good use of Stanford's digital computers. The general arrangements for the banquet as well as the whole social program were entirely in the hands of a great specialist, not exactly in this field but in aerodynamics, MILTON VAN DYKE. We are extremely grateful to him not only for the work he has done but for all the original ideas he has contributed. And finally, I come to the two persons whose shoulders carried the main load of the Congress, my close collaborators Donna LOUBSKY, our Executive Secretary, and the General Secretary, whom you all know, CHARLES STEELE.

Now I would like to give you some statistical information on the countries represented at the Congress. Their names will be followed by the number of delegates they sent to Stanford: United Arab Republic, 1; Australia, 13; Austria, 3; Belgium, 8; Brazil, 2; Canada, 45; Czechoslovakia, 4; Denmark, 7; France, 62; Germany, 51; Hungary, 1; India, 4; Iran, 1; Israel 4; Italy, 10; Japan, 22; Korea, 1; Netherlands, 35; New Zealand, 1; Nigeria, 1; Norway, 2; Peru, 2; Poland, 12; Romania, 8; Senegal, 1; Sweden, 20; Switzerland, 3; United Kingdom, 72; Union of Soviet Socialistic Republics, 49; Venezuela, 4; Yugoslavia, 2. This adds up, if after a week's hard work my arithmetic can still be trusted, to 451 from foreign countries. To this figure we have to add 886 from the United States which gives a grand total of 1337 members of this Congress not counting the wives and children.

After having given information on the Proceedings of the Congress, Professor HOFF continued:

Next it is my very great pleasure to ask a man to say a few words in the name of the Congress Committee and in the name of the International Union for Theoretical and Applied Mechanics. He is well known to all of you, Professor at the Technical University of Delft, Holland, Secretary of the Congress Committee and up to now Treasurer of IUTAM, after the closure of this session he is going to be the new President of IUTAM. I am very glad to present to you Professor WARNER KOITER.

Professor Koiter's talk follows:

Thank you Professor HOFF:

First of all I have to announce that Monsieur ROY, the present President of IUTAM regrets very much that he is unable to attend this closing ceremony because he had to return to France this morning. He has asked me to express on behalf of IUTAM and its Congress Committee our heartfelt thanks to the organizers of the 12th International Congress of Applied Mechanics. It would be impossible for me to mention the names of all the people here at Stanford and in the United States in general who were involved in the organization of this Congress, the largest by far in the entire series of International Congresses of Applied Mechanics. I hope you will permit me to single out two names, those of Professor and Mrs. HOFF, the President and Mrs. President of this Congress, and Professor and Mrs. STEELE, the General Secretary and a lady who took very good care of the foreign ladies coming here. In view of the size of the present

Congress, I think I can honestly state that never before has an International Congress of Applied Mechanics owed so much to so few. I should also like to express the gratitude of IUTAM and its Congress Committee to several United States agencies which have contributed so much to make this Congress a truly international one. But for their generosity, it would have been impossible for many foreign participants to come here, and I think, in particular, of the charter flight from Europe which could not have been organized without the backbone support from U.S. agencies. In this connection you will perhaps permit me to name two more persons who have taken on a large part of the work involved in the organization of the charter flight. All passengers no doubt know the names of Mr. VAN HORSSEN and Mrs. KAMINSKI.

Now I should like to make some announcements on changes in the Bureau of IUTAM and its Congress Committee. By our statutes we have the immensely good fortune that M. ROY, as retiring president, will continue as a member of the Bureau of IUTAM in the capacity of vice-president, and he will also continue to represent our Union in the International Council of Scientific Unions. Professor GÖRTLER, who has served our Union now for four years as its secretary general and has in this capacity continued the very great performance of his predecessor, is retiring as secretary general but will remain on the Bureau as its treasurer. Then we have elected to the Bureau of IUTAM Professor NIORDSON as the new secretary general and Professor LIGHTHILL as a member of the Bureau.

It is with deep regret that I have to add to these appointments that we shall miss our friends Professor TEMPLE and Professor PARKUS as members of the future Bureau. Professor TEMPLE has served the International Union for a very long time; eight years as treasurer, four years as president and the past four years as vice-president. But he will now retire from the Bureau. I am very happy, however, to announce that Professor TEMPLE has been elected a personal member of the General Assembly of IUTAM. We also regret as I have mentioned already that we shall lose Professor PARKUS as a member of the Bureau, but you will presently hear that we may hope to benefit from Professor Parkus' experience in another capacity.

I am now coming to the Congress Committee of IUTAM which proposed to the General Assembly of IUTAM the appointment of four new members: Professors BECKER of Darmstadt, BOLEY of New York, HOFF of Stanford and PARKUS of Vienna. The Executive Committee of the Congress Committee for the next four years to come will consist of the President of IUTAM; Professor BECKER, who has accepted an appointment as Secretary of the Congress Committee; Professor BATCHELOR; and Professor HOFF. A further announcement is due with reference to the next International Congress of Applied Mechanics which will be held in 1972. Two invitations have lain before the Congress Committee, one old invitation from the Netherlands and a much newer one from the USSR. The Congress Committee unanimously preferred to postpone a decision on the place of meeting for the next Congress to next year. The announcement of the decision will be made in the course of the next year, and the very careful statistics prepared for the present congress will enable us to transmit information on this decision to all members of the present congress immediately.

Professor KOITER ended his talk with information on the return trip of the charter flight. Subsequently Professor HOFF said:

Thank you very much Professor KOITER:

I see the raised hand of one of the outstanding members of IUTAM, a former president for four years and vice-president afterwards for four years, and also one of our invited general lecturers. Most of you must have heard his paper presented in the morning of the first day of this Congress. I am very glad indeed to present to you Professor ODQVIST.

Professor ODQVIST delivered the following talk:

Mr. President, Ladies and Gentlemen:

By self-appointment, I would like to say a few words on behalf of the rank and file of the Congress Members. As a matter of fact, our new President, Professor KOITER, has said the most that could be said in appreciation of the Congress from a formal point of view, but still I

2*

have a feeling that particularly the foreign participants, and they were 450 in number, would like to express their appreciation of the hospitality and the generosity to which they have been subject during the days of the Congress. Anyone who has had anything to do with the organization of an international meeting of similar character knows that there are always a lot of unforeseen difficulties that crop up in the course of the work. Due to the skill and understanding of the leading persons here, everything has come out all right, and I will include not only the practical arrangements, which are perhaps most obvious to most of us, but also the scientific level of this Congress which I think has not been surpassed and is a challenge to future organizers of congresses. This is the last Congress which carries the name of Applied Mechanics only in its title. In the future they will be named Congresses of Theoretical and Applied Mechanics. I do not think that this means much but for practical reasons the change has been introduced. With the best wishes for the future and for reasons I just mentioned, I wish to concentrate our thanks to the President of the present Congress when we now say goodbye to Stanford.

After Professor ODQVIST concluded his words of thanks, Professor HOFF said:

Thank you very much for the token of appreciation you have just given me and all the men who have been working with me on the organization of the Congress. Obviously, it has been a great deal of work, but it has also been a great deal of fun. From the standpoint where I am, of course, the situation is somewhat distorted. What I see all the time are the mistakes, the things we should have done better. I am very grateful to you indeed if you take a more charitable attitude toward the events of the last week. I hope very much to see you all, and many others, at the next Congress of Theoretical and Applied Mechanics four years from now. I declare the Twelfth International Congress of Applied Mechanics closed.

Sur le flambement et les questions d'instabilité

Par

L. Gauthier

Université de Paris

C'est certainement en lisant la très belle conférence mémoriale prononcée par le président de notre congrès, le professeur NICHOLAS J. HOFF[1] que j'ai pris conscience, il y a une douzaine d'années, de l'intérêt considérable que présente le problème du flambement.

C'est lui qui a attiré mon attention sur l'origine historique de l'emploi des colonnes en architecture, dans l'ancien monde essentiellement: le temple d'Amon à Karnak près de Louxor en haute Egypte, édifié sous Sethi Ier, aux environs de 1300 avant J.C. possède une grande salle hypostyle, remarquable par ses colonnes de 69 pieds.

Les colonnes sont formées de sections superposées et j'avais, quant à moi, plus d'admiration pour les obélisques, dressés vers le ciel, d'une seule pièce. Quand on songe à la primitivité des moyens mécaniques dont on disposait à l'époque de Thetmosis, on reste médusé en face d'un monolithe d'une trentaine de mètres de hauteur, pesant environ 300 tonnes.

J'ai appris, depuis lors, qu'en Mésopotamie on utilisait des colonnes formées de faisceaux de roseaux, environ 3000 ans avant notre ère. Par ailleurs, il semble bien que les mille colonnes de la place centrale de Chichen-Itzà ne soient en rien redevables à une influence de l'ancien continent: j'ignore quand, dans le nouveau monde, est apparue l'idée d'utiliser les colonnes. En tout cas, j'ai tendance à croire que l'idée pouvait germer spontanément, dans l'esprit d'un ami de la nature, voyant de beaux arbres, au fût bien droit, d'utiliser le cylindre vertical à la fois comme support et comme élément esthétique. Et c'est bien là que je retrouve, dans mes souvenirs d'adolescent, mes premières rencontres, confuses, avec le phénomène du flambement; j'étais surpris, en tenant par sa base une longue perche en bois, du tremblement de sa partie supérieure, à peu près indépendant des efforts que je déployais à la maintenir bien verticale. Et le phénomène se reproduisait identiquement quand je retournais la perche bout pour bout, constatant ainsi qu'aucune des extrémités ne pouvait valablement être considérée comme tellement plus flexible que l'autre. Ce n'est que de nombreuses années plus tard que j'ai été à même de comprendre ce qui se passait.

Le point de vue d'Euler

C'est dans l'appendice sur les courbes élastiques, dans sa "Méthode d'étude des courbes jouissant de propriétés extrémales", publiée à Lausanne en 1744, que L. EULER a donné la première explication du phénomène de flambement.

Imaginons une poutre droite réduite à sa ligne élastique, idéalement droite, encastrée à sa partie inférieure, et chargée en bout par une force P exactement centrée, qui reste constante et verticale quand varie son point d'application. Le raisonnement de BURIDAN laisse penser que "par raison de symétrie" la poutre restera verticale, droite, quelle que soit la charge P.

[1] The forty-first Wilbur Wright memorial lecture: Buckling and Stability. N.J. HOFF, J. Roy. Aeronaut. Soc. Jan. 1954.

Euler étudie la possibilité d'une forme fléchie : l'équation différentielle linéarisée

$$\frac{1}{R} \sim y'' = \frac{M}{EI} = \frac{P}{EI}(f - y)$$

($\frac{1}{R}$ courbure linéarisée suivant y'', EI rigidité à la flexion supposée constante. M moment fléchissant, calculé au moyen de la flèche terminale f), s'intègre élémentairement

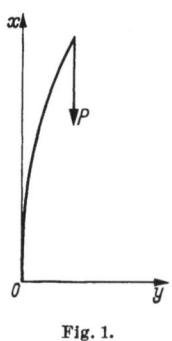

Fig. 1.

$$y = f + A \cos \omega x + B \sin \omega x$$

où

$$\frac{P}{EI} = \omega^2.$$

Les conditions aux limites ($y = y' = 0$ à l'encastrement) déterminent A et B :

$$y = f(1 - \cos \omega x)$$

et la définition de la flèche terminale f donne la condition (la longueur l de la poutre étant supposée invariable)

$$f \cos \omega l = 0.$$

On voit bien que si

$$\omega l < \frac{\pi}{2}$$

c'est-à-dire si

$$P < P_0 = \frac{\pi^2}{4} \frac{EI}{l^2}$$

la poutre reste droite, f étant nulle, et $y \equiv 0$.

Il y a un doute lorsque la charge atteint la valeur critique P_0, appelée charge d'Euler, mais le calcul laisse alors f indéterminée, ce qui n'est guère satisfaisant pour l'esprit.

En tout cas, par ce calcul de L. Euler, et d'autres analogues, la notion de *bifurcation de stabilité* a fait son apparition. Pour $P < P_0$ la forme d'équilibre $y = 0$ est la solution unique du problème : c'est donc celle qui est digne d'être appelée stable. Lorsque $P = P_0$ la solution perd, semble-t-il ce caractère d'unicité. Effectivement l'expérience montre que dès que P dépasse P_0, la poutre prend tout de suite une flexion importante, et la position droite est devenue "instable". Il est vrai que la déformation a lieu dans le plan de moindre résistance à la flexion, car la parfaite symétrie de révolution reste purement idéale.

Avant d'examiner avec plus d'attention le comportement de la poutre au voisinage de la charge critique, nous allons essayer de mieux comprendre la coopération d'une force axiale à l'effet de flexion en supposant la poutre console déviée de la position droite par une force transversale Q.

L'équation linéarisée de la ligne élastique

$$y'' = \frac{P}{EI}(f - y) + \frac{Q}{EI}(l - x)$$

s'intègre encore élémentairement, en posant

$$\frac{P}{EI} = \omega^2$$

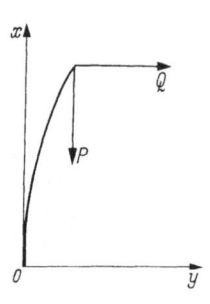

Fig. 2.

suivant

$$y = f + \frac{Q}{P}(l - x) + A \cos \omega x + B \sin \omega x$$

les conditions aux limites, à l'encastrement, déterminent A et B :

$$y = \left(f + \frac{Ql}{P}\right)(1 - \cos \omega x) + \frac{Q}{P\omega}(\sin \omega x - \omega x)$$

et la définition de la flèche terminale, donne en posant

$$\omega l = \varphi$$

la relation

$$f = \frac{Ql^3}{3EI} \frac{\operatorname{tg} \varphi - \varphi}{\dfrac{\varphi^3}{3}}$$

qui est assez instructive.

Lorsque P, c'est-à-dire φ, tend vers zéro, le second facteur tend vers 1 et nous trouvons la valeur de la flèche d'une poutre chargée transversalement.

L'influence de P sur cette flèche est exprimée par le facteur multiplicatif

$$\frac{\operatorname{tg}\varphi - \varphi}{\dfrac{\varphi^3}{3}} = 1 + \frac{2}{5}\,\varphi^2 + \frac{17}{105}\,\varphi^4 + \cdots$$

$$= 1 + \frac{2}{5}\frac{Pl^2}{EI} + \frac{17}{105}\left(\frac{Pl^2}{EI}\right)^2 + \cdots$$

qui tend vers l'infini lorsque φ tend vers $\frac{\pi}{2}$ c'est-à-dire lorsque P tend vers la valeur critique d'EULER.

On voit ainsi qu'à l'approche de la charge critique la perturbation apportée par la compression est très importante.

On voit surtout que, même pour les petites valeurs de la compression P, pour lesquelles l'emploi de l'équation linéarisée est valable, la flèche subit des modifications qui ne sont pas additives: le principe de superposition qui, avec le principe de proportionnalité, est à la base de la mécanique linéaire, n'étant pas vérifié, *le flambement*, qui consiste en l'apparition d'une flexion latérale sous l'action d'une poussée axiale suffisamment grande, *est un phénomène non linéaire*.

Revenons à la poutre uniquement chargée en bout, et supposons la charge P suffisante pour provoquer la flexion: puisque la déformation devient rapidement importante nous ne nous contenterons pas de l'équation linéarisée pour la ligne élastique: désignons par θ la rotation que subit la tangente au point courant pendant la déformation, et par s l'abscisse curviligne

$$\frac{1}{R} = \frac{d\theta}{ds} = \frac{P}{EI}\,(f - y)$$

puisque

$$dy = \sin\theta\, ds$$

nous formons l'équation en θ en dérivant:

$$\frac{d^2\theta}{ds^2} + \frac{P}{EI}\sin\theta = 0$$

(et nous continuerons à poser: $\dfrac{P}{EI} = \omega^2$).

Nous perdons ainsi la constante f, ce qui nous invite à noter la condition terminale

$$\frac{d\theta}{ds} = 0 \quad \text{pour} \quad s = l$$

et f sera au contraire défini par la condition initiale

$$\frac{P}{EI}\,f = \left(\frac{d\theta}{ds}\right)_{s=0}$$

les conditions d'encastrement étant $\theta = 0$, $y = 0$ pour $s = 0$.

L'équation pendulaire obtenue permet l'étude, au moyen des fonctions elliptiques de JACOBI, des déformées correspondant à toutes les valeurs de la charge P: on trouvera par exemple cette étude dans le livre classique de S. TIMOSHENKO (théorie de la stabilité élastique. Chapitre II. § 13).

Je me bornerai ici, pour étudier simplement le voisinage des conditions critiques, à utiliser l'intégration approchée, en tenant compte du "défaut d'isochronisme":

$$\theta = \alpha \sin(\Omega s + \varphi)$$

où les constantes d'intégration sont α et φ, et où

$$\Omega = \omega \left(1 - \frac{\alpha^2}{16}\right).$$

Les conditions aux limites donnent successivement

$$\varphi = 0; \qquad \Omega l = \frac{\pi}{2}$$

de sorte que l'angle α, déviation terminale de la tangente est donné par la relation

$$\frac{\pi^2}{4l^2} = \frac{P}{EI}\left(1 - \frac{\alpha^2}{8}\right).$$

Le facteur plus petit que l'unité montre que la déformation n'a lieu que lorsque la charge P dépasse la valeur critique d'EULER: Si $P < P_0$, la seule solution est $\theta \equiv 0$, $y \equiv 0$. Si $P = P_0(1 + \varepsilon)$ la déformation est donnée par

$$\alpha^2 = 8\varepsilon.$$

Le flambement d'une poutre est, comme le défaut d'isochronisme du pendule, un effet de non linéarité.

Le résultat obtenu pour la flèche terminale

$$\frac{f}{l} = \frac{2}{\pi} \sqrt{8\left(\frac{P}{P_0} - 1\right)}$$

est également caractéristique de la non linéarité du flambement: la flèche n'est pas du tout proportionnelle à la charge qui la produit.

La représentation graphique de cette formule, au moyen d'un arc de parabole au voisinage de son sommet montre bien que dès que la charge dépasse, même très peu, la charge critique,

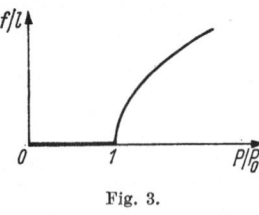

Fig. 3.

la déformation qui prend naissance, prend tout de suite une ampleur importante: lorsque la charge dépasse de 1% la charge critique d'EULER, la flèche atteint déjà 18% de la longueur totale de la poutre (si toutefois la rupture n'est pas survenue avant).

On peut expérimenter les grandes déformations correspondant à des charges qui dépassent largement la valeur critique, et les comparer aux valeurs calculées au moyen des fonctions elliptiques, en utilisant une lame élastique plate, très flexible dans le plan défini par son épaisseur. Lorsque la charge P devient très grande ($P \to \infty$) ou, ce qui revient au même, lorsque la rigidité de flexion EI devient très faible, l'équation pendulaire qui définit la déformée, montre que l'on a affaire à un problème de perturbation singulière: il y a au voisinage de l'appui encastré une "couche limite" c'est-à-dire un tout petit intervalle de longueur au cours duquel θ varie rapidement de o à π, puis sur tout le reste de la longueur θ demeure pratiquement égal à π. Il arrive, soit que la lame se casse au voisinage de l'encastrement, quand le rayon de courbure devient petit, soit que la lame subisse un pliage au voisinage de l'encastrement, et pende verticalement sous la traction de la forte charge P.

La Méthode des imperfections

L'étude de la poutre parfaite, réduite à l'approximation linéaire ramène la détermination des constantes d'intégration à partir des conditions aux limites, à un problème linéaire et homogène. La matrice est fonction de la charge: pour de petites valeurs de la charge la matrice est régulière et seule la position banale d'équilibre est solution, tandis que lorsque la charge prend la valeur critique, le déterminant de la matrice s'annule et le problème linéaire devient indéterminé.

Cette situation se généralise aux systèmes à un nombre fini de flèches, et aux systèmes comportant une infinité continue de positions admissibles: les équations aux dérivées partielles définissent une solution unique à partir des conditions aux limites, définie par un opérateur dont les charges critiques annulent la plus petite valeur propre.

La résolution d'un problème linéaire par l'inversion d'une matrice ne suppose aucunement le problème homogène, et la méthode des imperfections consiste précisément à remarquer que si les conditions de parfaite symétrie ne sont pas vérifiées (poutre non exactement droite en l'absence des charges; encastrement non horizontal; application excentrée de la charge), la charge critique est telle que la déformation évaluée sur les équations linéarisées, tend vers l'infini.

Reprenons par exemple la poutre encastrée, que nous supposerons chargée à une distance e de l'axe. L'équation linéarisée

$$y'' = \frac{P}{EI} \left(f + e - y \right)$$

ne contient que des données si on la dérive:

$$y''' + \omega^2 y' = 0$$

où

$$\frac{P}{EI} = \omega^2$$

avec trois conditions aux limites:

$$y(0) = 0$$

$$y'(0) = 0$$

$$y''(l) = \omega^2 e.$$

L'intégrale générale

$$y = A \cos \omega x + B \sin \omega x + C$$

comporte trois constantes d'intégration définies par le système

$$A + B + C = 0$$

$$B = 0$$

$$-A \cos \varphi - B \sin \varphi = e$$

en posant comme précédemment

$$\omega l = \varphi$$

le déterminant est $\cos \varphi$. Lorsqu'il est non nul la déformée est

$$y = \frac{e}{\cos \varphi} \left(1 - \cos \omega x \right)$$

et la flèche

$$f = e \left(\frac{1 - \cos \varphi}{\cos \varphi} \right)$$

qui tend vers l'infini lorsque $\varphi \to \frac{\pi}{2}$ c'est-à-dire lorsque P atteint la valeur critique.

Nous avons d'ailleurs déjà observé ceci lorsque nous avons montré le caractère non linéaire du flambement en combinant deux causes de flexion: nous aurions pu composer la charge horizontale Q avec P, la charge totale présentant alors une "imperfection" dans le choix de sa direction.

Considérée comme une méthode de détermination des charges critiques, l'introduction des imperfections n'apporte rien de substantiellement neuf aux idées d'EULER, mais son importance pratique est claire: la perfection n'existe pas et ce que l'on observe est la rapide croissance d'une déformation à peine perceptible initialement.

La Méthode de l'énergie

Il en va tout autrement de la méthode de l'énergie, dont le développement est surtout dû à S. P. TIMOSHENKO.

Cette méthode est bien connue à cause de sa puissance dans les techniques de calcul approché. Cependant, comme l'une des formules les plus couramment employées, auxquelles elle

18   L. GAUTHIER

aboutit, a fait l'objet d'une controverse récente, j'exposerai assez longuement la question, avec
l'intention de réhabiliter la formule critiquée.

Considérons une poutre quelconque (à section droite non nécessairement constante) chargée
axialement par une force $P > P_0$ supérieure à la charge critique, force qui reste constante et
verticale quand varie son point d'application. Cette poutre admet deux équilibres: l'équilibre
droit, instable, et l'équilibre fléchi, stable. Imaginons un saut faisant passer le premier équilibre au second sous l'effet d'une perturbation infinitésimale. L'énergie interne s'accroit de

$$W = \frac{1}{2} \int_0^l \frac{M^2}{EI} \, ds = \frac{1}{2} \int_0^l EI \left(\frac{d\theta}{ds}\right)^2 ds$$

et la force de compression fournit le travail

$$T = P \int_0^l (1 - \cos\theta) \, ds.$$

Le saut qui échange deux positions d'équilibre possibles sous la même charge se fait sans variation d'énergie et nous avons la relation

$$\int_0^l EI\theta'^2 \, ds = 4P \int_0^l \sin^2\frac{\theta}{2} \, ds.$$

Alors que l'équation différentielle de la ligne élastique définit la déformée connaissant la charge,
la relation que nous venons d'obtenir permet, connaissant la déformée $\theta(s)$, de calculer la
charge P correspondante. La charge critique P_0 est la borne inférieure des charges P correspondant à une déformation effective: on peut l'obtenir en se limitant, pour une petite déformation ,à l'approximation linéaire

$$ds \sim dx; \qquad \sin\theta \sim \theta \sim y'.$$

On obtient ainsi la formule de RAYLEIGH

$$P_0 = \frac{\int_0^l EIy''^2 \, dx}{\int_0^l y'^2 \, dx}. \tag{R}$$

TIMOSHENKO préfère (stabilité élastique § 15) calculer l'énergie interne en exprimant le moment M au moyen des charges.

$$M = P_0(f - y)$$

on obtient ainsi

$$P_0 = \frac{\int_0^l y'^2 \, dx}{\int_0^l \frac{(f - y)^2}{EI} \, dx} \tag{T}$$

que nous appellerons la formule de TIMOSHENKO.

Dans la stricte écriture que nous venons de leur donner, ces formules sont rarement utili-
sables puisqu'elles supposent connue la déformation y. Mais leur caractère intégral suggère
tout de suite l'idée de remplacer y par une expression approchée. Pour étudier cette question
nous appliquerons les intégrales qui figurent dans (R) et (T) à des fonctions $y(x)$ que nous
appellerons les déformations virtuelles admissibles: ce sont les fonctions admettant des dérivées secondes de carré sommable (EI est borné inférieurement et supérieurement) qui satisfont
à toutes les conditions aux limites. Dans le problème de la poutre encastrée étudié au début,
ces conditions sont:

$$y(0) = 0$$
$$y'(0) = 0$$
$$y(l) = f$$
$$y''(l) = 0$$

qui sont par exemple vérifiées par

$$y = f\left(1 - \cos\frac{\pi}{2}\frac{x}{l}\right)$$

et par

$$y = f\frac{x^2}{2l^3}(3l - x).$$

Bien entendu une déformation virtuelle admissible ne peut coïncider avec une déformation effective de la poutre que si on applique des charges telles que le moment fléchissant soit

$$M = EIy''$$

qui n'est habituellement pas égal à

$$P(f - y)$$

et ne correspond pas à une charge axiale.

Nous devons donc étudier le quotient de RAYLEIGH

$$\frac{\int\limits_0^l EIy''^2\, dx}{\int\limits_0^l y'^2\, dx}$$

qui est une fonctionnelle définie sur l'espace des déformations virtuelles admissibles. Les extrema du quotient sont obtenus par la méthode du multiplicateur de LAGRANGE, en écrivant l'extremum de

$$\int\limits_0^l (EIy''^2 - Py'^2)\, dx.$$

L'équation d'EULER du calcul des variations relative à cette intégrale

$$\frac{d}{dx}(EIy'') + Py' = 0$$

s'intègre suivant

$$EIy'' + Py = Pf$$

c'est l'équation de la ligne élastique, qui montre que *le quotient de* RAYLEIGH *est extremum lorsqu'il y a flambement, la valeur de ce quotient étant alors égale à la charge critique* P_0.

On peut également dire que la charge critique P_0 est la plus petite valeur du multiplicateur P, pour laquelle l'intégrale d'énergie

$$\int\limits_0^l (EIy''^2 - Py'^2)\, dx$$

cesse d'être définie positive.

Ce résultat explique pour quelle raison le quotient de RAYLEIGH, calculé pour une déformation virtuelle voisine de la déformation réelle fournit une bonne valeur approchée de la charge critique.

Nous avons remarqué que pour une déformation virtuelle admissible distincte de la déformation réelle

$$M \neq P(f - y)$$

P. F. DROZDOV[1] a montré comment ceci est lié à un paradoxe tiré de l'inégalité sur laquelle TIMOSHENKO fonde la méthode de l'énergie (loc. cit. § 14): si le travail virtuel fourni par la charge P est inférieur à l'énergie de déformation virtuelle, il manque de l'énergie pour réaliser la déformation et celle-ci ne peut pas avoir lieu; l'équilibre droit est stable. Avec l'expression

$$M = EIy''$$

du moment fléchissant, on trouve

$$\int\limits_0^l EIy''^2\, dx > P\int\limits_0^l y'^2\, dx$$

---

[1] DROZDOV, P. F.: Eng. Technic. Bull. 2 (1951).

et l'équilibre droit est stable pour tout P inférieur à P_0. Mais si l'on avait choisi

$$M = P(f - y)$$

on aurait trouvé

$$P \int_0^l \frac{(f-y)^2}{EI}\, dx > \int_0^l y'^2\, dx$$

et on aurait conclu, de façon absurde, que l'équilibre droit est stable pour tout P supérieur à P_0.

Le raisonnement que nous avons fait sur le déplacement d'équilibre à charge constante montre que lorsque y désigne la déformation effective, solution de l'équation différentielle de la ligne élastique, seule l'égalité du travail et de l'énergie de déformation a un sens. Lorsque, plus généralement, y désigne une déformation virtuelle compatible, l'inégalité elle-même a un sens, mais elle n'est justifiée que dans le cas de l'expression (R) de Rayleigh, car l'intégrale d'énergie qui figure dans l'expression (T) de Timoshenko, n'est plus l'énergie de déformation virtuelle.

Cependant, il ne faut pas abandonner inconsidérément la formule de Timoshenko: elle fournit pour la charge critique une valeur correcte lorsque y représente la déformée effective de la poutre, et il y a lieu d'examiner ses propriétés lorsque y parcourt l'espace des déformations virtuelles admissibles. Le quotient

$$\frac{\displaystyle\int_0^l y'^2\, dx}{\displaystyle\int_0^l \frac{(f-y)^2}{EI}\, dx}$$

est extremum lorsqu'il en est de même de l'intégrale

$$\int_0^l \left[y'^2 - \frac{P(f-y)^2}{EI} \right] dx$$

dans laquelle P est un multiplicateur homogène à une force. Cette intégrale a la signification d'une longueur: nout l'appellerons "l'affaissement" de Timoshenko. Elle est extrema lorsque y vérifie l'équation d'Euler:

$$\frac{d}{dx} y' + \frac{P}{EI} (y - f) = 0$$

qui est l'équation de la ligne élastique. La charge critique est la plus petite charge pour laquelle l'affaissement de Timoshenko cesse d'être défini positif.

Le principe variationnel que nous venons d'établir montre que la formule (T) de Timoshenko fournit comme la formule (R) de Rayleigh, un bon procédé d'approximation pour déterminer les charges critiques, et il y a lieu de les comparer. Je suivrai, pour le faire, une idée qui est due à Chi-teh-wang (Applied elasticity. Chap. 10).

Le premier membre de l'équation de la ligne élastique est un opérateur sur l'espace des déformations virtuelles admissibles, dont le noyau est non nul lorsque P est une valeur spectrale: la charge critique usuelle est la plus petite valeur spectrale. Soit $P_1, P_2, \ldots, P_n$, le spectre, et $y_1, y_2, \ldots, y_n, \ldots$ les fonctions propres. Ces fonctions constituent une base pour l'espace des déformations virtuelles admissibles: (ce théorème se réduit, pour EI constant, au théorème de décomposition en série de Fourier) qui a des propriétés d'orthogonalité faciles à démontrer à partir de l'équation différentielle:

$$\int_0^l \frac{1}{EI} (f_i - y_i)(f_k - y_k)\, dx = 0; \qquad \int_0^l y_i' y_k'\, dx = 0,$$

$$\int_0^l EI y_i'' y_k'' = 0$$

pour $i \neq k$ car, on a par exemple, avec une intégration par parties

$$\int_0^l EIy_i'' y_k'' \, dx = P_k \int_0^l (f_k - y_k) y_i'' \, dx = P_k \int_0^l y_i' y_k' \, dx$$

où il suffit, pour conclure, de permuter i et k supposés différents. Si au contraire on suppose $i = k$, on obtient les deux relations

$$\int_0^l EIy_k''^2 \, dx = P_k \int_0^l y_k'^2 \, dx = P_k^2 \int_0^l \frac{(f_k - y_k)^2}{EI} \, dx.$$

Considérons alors le polynôme du second degré

$$\Pi(\lambda) = \int_0^l EIy''^2 \, dx - 2\lambda \int_0^l y'^2 \, dx + \lambda^2 \int_0^l \frac{(f - y)^2}{EI} \, dx$$

où λ est une indéterminée et où y est une déformation virtuelle arbitraire que l'on exprime par la série

$$y = \sum_1^\infty a_k y_k$$

on obtient aisément

$$\Pi(\lambda) = \sum_1^\infty a_k^2 (P_k - \lambda)^2 \int_0^l \frac{(f_k - y_k)^2}{EI} \, dx$$

expression définie positive, qui ne s'annule que lorsque tous les a_k sont nuls sauf un. Le discriminant de $\Pi(\lambda)$ est donc négatif ou nul:

$$\frac{\int_0^l y'^2 \, dx}{\int_0^l \frac{(f - y)^2}{EI} \, dx} \leq \frac{\int_0^l EIy''^2 \, dx}{\int_0^l y'^2 \, dx}$$

et l'égalité n'a lieu, avec la valeur P_k, que pour

$$y = a_k y_k.$$

La même représentation d'une déformation virtuelle compatible permet de calculer le quotient de TIMOSHENKO (ou celui de RAYLEIGH):

$$\frac{\int_0^l y'^2 \, dx}{\int_0^l \frac{(f - y)^2}{EI} \, dx} = \frac{\sum_1^\infty a_k^2 P_k \int_0^l \frac{(f_k - y_k)^2}{EI} \, dx}{\sum_1^\infty a_k^2 \int_0^l \frac{(f_k - y_k)^2}{EI} \, dx}$$

formule qui montre que le rapport de TIMOSHENKO est le barycentre des valeurs spectrales $P_1, \ldots, P_k, \ldots$ affectées de coefficients positifs ou nuls

$$a_k^2 \int_0^l \frac{(f_k - y_k)^2}{EI} \, dx$$

il est donc compris dans l'enveloppe convexe du spectre: en particulier, il est supérieur ou égal à la charge critique P_0 qui est la plus petite valeur spectrale.

Le quotient T de TIMOSHENKO, et le quotient R de RAYLEIGH sont des valeurs approchées par excès de la charge critique d'EULER et T est une meilleure approximation que R.

L'affirmation heuristique par laquelle TIMOSHENKO justifiait sa méthode est donc mathématiquement justifiée.

Lorsque par exemple, on suppose EI constant, si l'on adopte

$$y = f\left(1 - \cos\frac{\pi}{2}\frac{x}{l}\right)$$

on obtient:

$$T = R = P_0 = \frac{\pi^2}{4}\frac{EI}{l^2}$$

et si l'on adopte

$$y = f\frac{x^2}{2l^3}(3l - x)$$

on obtient

$$T = \frac{42}{17}\frac{EI}{l^2}; \qquad R = \frac{5}{2}\frac{EI}{l^2}$$

et les coefficients sont bien dans l'ordre:

$$\frac{\pi^2}{4} = 2{,}467$$

$$\frac{42}{17} = 2{,}471$$

$$\frac{5}{2} = 2{,}500).$$

Un exemple intéressant de comparaison entre la méthode de RAYLEIGH et celle de TIMO-SHENKO est celui de l'étude des conditions critiques de la poutre encastrée à sa base, dans le cas où l'on tient compte de l'effet du poids propre de la poutre. Soit $Q = ql$ ce poids propre (nous supposerons EI et q constants).

En désignant par $\eta = \eta(\xi)$ l'équation de la ligne élastique entre les abscisses x et l, le moment fléchissant est

$$M = P(f - y) + \int_x^l q(\eta - y)\,d\xi$$

ce qui donne l'équation différentielle

$$EIy'' = P(f - y) + \int_x^l q(\eta - y)\,d\xi$$

que l'on peut dériver suivant

$$EIy''' + (P + Q - qx)\,y' = 0.$$

Suivant les idées de A. G. GREENHILL et de A. N. DINNIK, cette équation peut être intégrée au moyen des fonctions de BESSEL d'ordre $\pm 1/3$. Mais si nous remarquons que la déformation virtuelle

$$y = f\left(1 - \cos\frac{\pi}{2}\frac{x}{l}\right)$$

vérifie les quatre conditions aux limites, qui sont les mêmes qu'au début, il nous suffit de calculer

$$\int_0^l EIy''^2\,dx = \frac{\pi^4}{32}EI\frac{f^2}{l^3}$$

$$P\int_0^l y'^2\,dx + \int_0^l q\,dx\int_0^x \eta'^2\,d\xi = \frac{\pi^2}{8}\frac{f^2}{l}\left[P + \frac{Q}{2}\left(1 - \frac{4}{\pi^2}\right)\right]$$

et

$$\int_0^l \frac{M^2}{EI}\,dx = \frac{f^2 l}{2EI}\left[P^2 + PQ\left(1 - \frac{4}{\pi^2}\right) + Q^2\left(\frac{1}{3} + \frac{18}{\pi^2} - \frac{64}{\pi^3}\right)\right]$$

pour obtenir, dans le plan des variables P, Q les conditions critiques. Lorsqu'on emploie la méthode de RAYLEIGH, on obtient la droite

$$P + \frac{Q}{2}\left(1 - \frac{4}{\pi^2}\right) = P_0$$

lorsqu'on emploie la méthode de TIMOSHENKO, on obtient l'ellipse

$$P^2 + PQ\left(1 - \frac{4}{\pi^2}\right) + Q^2\left(\frac{1}{3} + \frac{18}{\pi^2} - \frac{64}{\pi^3}\right) = P_0\left[P + \frac{Q}{2}\left(1 - \frac{4}{\pi^2}\right)\right]$$

centrée au point $\left(\frac{P_0}{2}, 0\right)$ et très allongée.

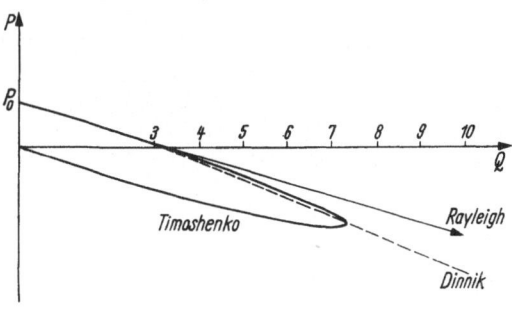

La droite de RAYLEIGH est la tangente à
l'ellipse au point $(P_0, 0)$: elle est au-dessus de
l'ellipse et pour chaque valeur de P, la méthode
de RAYLEIGH fournit une valeur approchée de
Q plus élevée que celle fournie par la méthode
de TIMOSHENKO.

Fig. 4.

Lorsqu'on choisit P nul (flambement de la
poutre sous le seul effet de son poids, c'est-à-
dire l'expérience même qui me taquinait dans
ma jeunesse, et qui soucia EULER pendant de
longues années), la formule de RAYLEIGH donne

$$Q = 3{,}37\,P_0$$

celle de TIMOSHENKO

$$Q = 3{,}20\,P_0$$

et le résultat calculé par DINNIK est

$$Q = 3{,}18\,P_0.$$

L'effet de courbure de l'ellipse, située au dessous de sa tangente, est encore la preuve du
caractère non linéaire du phénomène de flambement.

Les valeurs négatives de P sont les tractions qu'il faut faire subir en bout à une poutre qui
dépasse la longueur critique. Si par exemple le poids de la poutre vaut $Q = 5P_0$ la formule de
RAYLEIGH donne

$$P = -0{,}49\,P_0$$

celle de TIMOSHENKO

$$P = -0{,}62\,P_0$$

et le résultat calculé par DINNIK est

$$P = -0{,}63\,P_0.$$

Cependant, pour une poutre très longue et flexible, un ruban d'acier par exemple, la déformation virtuelle que nous avons utilisée est loin de la déformation réelle, et les formules approchées ne sont plus valables. C'est ainsi que pour $Q = 10P_0$, (ce qui correspond à une longueur
$1{,}46l_0$ (l_0 étant la longueur de la poutre critique), la formule de RAYLEIGH donne

$$P = -1{,}97\,P_0$$

la formule de TIMOSHENKO donne P imaginaire alors que le résultat calculé par DINNIK est

$$P = -2{,}78\,P_0.$$

Les vibrations transversales

Quittons pour quelques instants la statique et examinons quelques petits problèmes de
dynamique qui sont liés au phénomène du flambement. Désignons par $y(x, t)$ la ligne élastique,
en mouvement transversal, d'une poutre droite; pour l'étude des petits mouvements au voisinage de la position droite, l'équation linéarisée de la ligne élastique suffit:

$$\frac{\partial \theta}{\partial s} \sim \frac{\partial^2 y}{\partial x^2} = \frac{P}{EI}\,(f - y) - \frac{m}{EI}$$

où m désigne le moment fléchissant des effets d'inertie (principe de d'ALEMBERT). Le théorème
du moment cinétique appliqué à une tranche de matière, dont la rotation est θ, donne:

$$\frac{\partial m}{\partial x} = -\mu I\,\frac{\partial^2 \theta}{\partial t^2} + T$$

où μ est la masse par unité de volume, et T l'effort tranchant. Le théorème du centre d'inertie, dans les mêmes conditions, donne :

$$\frac{\partial T}{\partial x} = \mu S \frac{\partial^2 y}{\partial t^2}$$

où S est l'aire de la section. D'où l'équation aux dérivées partielles :

$$\frac{\partial^4 y}{\partial x^4} + \frac{P}{EI} \frac{\partial^2 y}{\partial x^2} + \frac{1}{EI} \left[-\mu I \frac{\partial^4 y}{\partial x^2 \partial t^2} + \mu S \frac{\partial^2 y}{\partial t^2} \right] = 0.$$

La méthode de séparation des variables montre immédiatement, en isolant l'opérateur $\frac{\partial}{\partial t^2}$, l'existence des vibrations transversales de la forme

$$y = f(x) \cos \Omega t$$

où $f(x)$ est solution de l'équation

$$EI y'''' + (P + \mu I \Omega^2) y'' - \mu S \Omega^2 y = 0$$

dont nous désignerons les racines caractéristiques par $\pm a$ et $\pm \omega i$:

$$y = A \sin \omega x + B \cos \omega x + C \operatorname{sh} ax + D \operatorname{ch} ax.$$

S'il s'agit par exemple d'une poutre reposant sur deux appuis simples, les conditions aux limites sont :

$$y = 0; \qquad y'' = 0 \quad \text{pour} \quad x = 0$$

qui entraine

$$B = D = 0$$

et

$$y = 0; \qquad y'' = 0 \quad \text{pour} \quad x = l$$

qui entraine

$$A \sin \omega l + C \operatorname{sh} al = 0$$

$$-A\omega^2 \sin \omega l + Ca^2 \operatorname{sh} al = 0$$

compatible pour une solution non nulle si

$$(a^2 + \omega^2) \sin \omega l \operatorname{sh} al = 0.$$

La flexion en cours de vibrations est donc définie par

$$\sin \omega l = 0$$

$$C = 0$$

$$y = A \sin \omega x.$$

Le mode fondamental correspond à

$$y = A \sin \pi \frac{x}{l}$$

avec l'équation aux pulsations :

$$EI \left(\frac{\pi}{l}\right)^4 - P \left(\frac{\pi}{l}\right)^2 - \mu I \Omega^2 \left(\frac{\pi}{l}\right)^2 - \mu S \Omega^2 = 0$$

en introduisant la charge critique

$$P_0 = \pi^2 \frac{EI}{l^2}$$

on obtient

$$\Omega^2 = \left(\frac{\pi}{l}\right)^2 \frac{P_0 - P}{\mu S + \mu I \left(\frac{\pi}{l}\right)^2}$$

(on néglige habituellement $I \frac{\pi^2}{l^2}$ devant S, le terme dû à l'inertie de rotation n'intervient effectivement que dans les modes d'ordre élevé qui ne sont pas de véritables harmoniques).

Si l'on désigne par Ω_0 la pulsation correspondant à l'absence de la charge axiale P, on obtient la relation simple

$$\frac{P}{P_0} + \left(\frac{\Omega}{\Omega_0}\right)^2 = 1.$$

La représentation graphique est une parabole dont le sommet est déterminé très élémentairement par la connaissance de deux points (la sous-normale au milieu d'une corde est égale au paramètre): ceci m'avait permis, autrefois d'élaborer une méthode non destructive d'évaluation des charges critiques en mesurant (au moyen d'une jauge) deux fréquences correspondant à deux charges faibles ou même négatives. Il est seulement nécessaire que les plateaux de la machine de traction-compression soient assez massifs pour ne pas participer aux vibrations de la poutre.

La charge critique correspond au sommet de la parabole des fréquences: la fréquence tend vers zéro quand la charge tend vers la charge critique.

Au-delà de la charge critique, la poutre fléchit et il y a deux types distincts de vibrations: les vibrations asymétriques, de petite amplitude, au cours desquelles la poutre ne revient pas jusqu'à la position droite, et les vibrations de grande amplitude, symétriques par rapport à la position droite. L'étude de l'équation non linéaire correspondante permet de faire dans le plan de phases un diagramme avec deux sommets symétriques par rapport à un col.

Le résultat de cette étude est très important, car il est la base d'une méthode d'analyse de la stabilité élastique, dite *méthode dynamique: la bifurcation de stabilité se produit lorsque la fréquence des vibrations autour de l'équilibre tend vers zéro*[1]. Cette méthode a amené à un profond élargissement des conceptions euleriennes.

Peut-être est-il bon de rappeler ici, au passage, le problème de E. METTLER (1940), qu'on pourrait aussi justement appeler le problème de N. M. BELIAEV (Léningrad 1924). Supposons que la charge P étant variable avec le temps, la déformée instantanée reste de la forme

$$y = f(t) \sin \pi \frac{x}{l}.$$

Un calcul analogue au précédent montre que la fonction $f(t)$ est solution de l'équation différentielle

$$f'' + \left(\frac{\pi}{l}\right)^2 \frac{P_0 - P(t)}{\mu S + \mu I \left(\frac{\pi}{l}\right)^2} f = 0.$$

Dans le mémoire que j'ai cité au début de cet exposé, N. J. HOFF étudie très complètement le cas où $P(t)$ est l'évolution de la charge imposée par une machine d'essais.

Le problème de METTLER-BELIAEV est celui dans lequel $P(t)$ est périodique, réduite à son premier harmonique

$$P = P_1 + P_2 \cos \omega t$$

et nous nous intéresserons à cette question dans le cas où P_2 est petit, que nous considérerons comme une fluctuation de la charge. L'équation qui définit les vibrations de la poutre est une équation de MATHIEU: on la ramène à sa forme canonique

$$\frac{d^2 f}{d\tau^2} + (a - 2q \cos 2\tau) f = 0$$

en posant

$$2\tau = \omega t$$

$$k = 4\left(\frac{\pi}{l}\right)^2 \frac{1}{\mu S + \mu I \left(\frac{\pi}{l}\right)^2}$$

$$a = k \frac{P_0 - P_1}{\omega^2}$$

$$2q = k \frac{P_2}{\omega^2}.$$

---

[1] Voir BOLOTIN, V. V.: Stabilité dynamique des systèmes élastiques. Moscou 1946.

Le diagramme classique de Strutt-Haines définit dans le plan (a, q) des zônes de stabilité (oscillations d'amplitude bornée) et des zônes d'instabilité (oscillations dont l'amplitude croit au-delà de toute limite). Bien entendu, lorsque $P_1 > P_0$ dépasse la charge critique, a est négatif, et la région $(a < 0, q$ faible) est une zone d'instabilité. Mais lorsque $P_1 < P_0$ et P_2 sont donnés, la droite

$$q = \frac{P_2}{2(P_0 - P_1)} a$$

coupe toute une suite de zônes d'instabilité. La plus voisine de l'origine

$$a - q < 1 < a + q$$

correspond à une bande de fréquences:

$$k\left(P_0 - P_1 - \frac{1}{2} P_2\right) <$$
$$< \omega^2 < k\left(P_0 - P_1 + \frac{1}{2} P_2\right)$$

entourant la fréquence

$$\omega_0 = \sqrt{k(P_0 - P_1)}$$

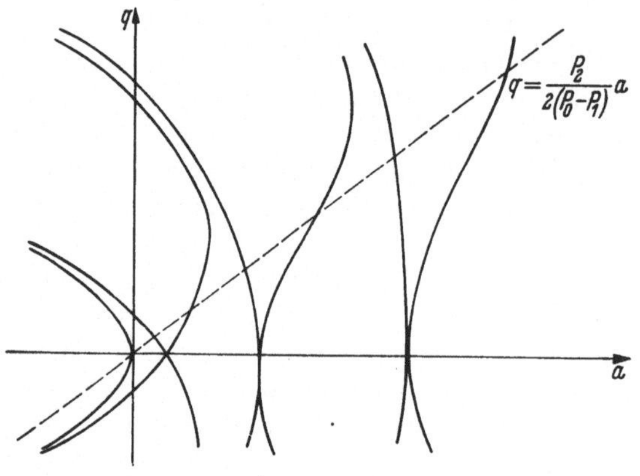

Fig. 5.

(on peut remarquer que c'est le double de la pulsation des vibrations transversales).

Ainsi, si petite que soit la fluctuation P_2 de la charge, il existe des bandes de fréquence donnant lieu à des vibrations divergentes, au voisinage des sous-harmoniques du double de la fréquence fondamentale des vibrations transversales.

Nous voyons ainsi que *le critère de stabilité dynamique est plus sévère que le critère de stabilité statique de* Euler.

Je n'entrerai pas au-delà[1] dans l'étude des oscillations à excitation paramétrique (dont l'exemple le plus suggestif est peut être celui de la balançoire, sur laquelle une personne meut périodiquement son centre de gravité), mais je voudrais rappeler, sur le sujet analogue de la statique du solide indéformable, en mécanique rationnelle, la belle expérience par laquelle P. L. Kapitsa[2] a réalisé la stabilisation d'un pendule dans sa position haute, en soumettant son articulation à des vibrations de très faible amplitude, mais de fréquence suffisamment grande.

Vitesse critique des arbres en rotation

Considérons un arbre cylindrique, éventuellement chargé en bout par une force P, soumis à une rotation uniforme de vitesse angulaire ω autour de son axe de révolution. Tant que l'axe reste droit, il n'est soumis à aucun effort latéral, mais si par suite d'une cause quelconque il prend une certaine flexion, dans un repère en rotation ω il est soumis tout le long à des forces centrifuges, qui si la vitesse est suffisante, entretiennent cette flexion.

En désignant par μ la masse par unité de longueur, par (ξ, η) un point de la ligne élastique, la contribution au moment fléchissant dûe aux forces centrifuges est

$$m = - \int_x^l \mu\omega^2\eta(\xi - x)\,d\xi.$$

Cette expression est la même que s'il s'agissait des forces d'inertie relatives à des vibrations transversales de pulsation ω. D'ailleurs la projection sur un plan méridien fixe de l'arbre en rotation représente l'arbre en vibration (puisque dans notre étude précédente les variables x et t se séparaient).

---

[1] Je voudrais cependant attirer l'attention sur les publications récentes de G. Schmidt (Berlin): Mathematische Nachrichten 1963 et 1964 (cette dernière consacrée aux systèmes rhéolinéaires); Archiwum mechaniki Stosowanej 1965 (cas de l'élasticité non linéaire).

[2] J. Phys. théor. exp. Vol. 21, No. 5 (1951).

Sans reprendre le calcul déjà fait nous trouvons donc

$$\frac{P}{P_0} + \left(\frac{\omega}{\omega_0}\right)^2 = 1$$

où P_0 est la charge d'EULER, et ω_0 la vitesse critique de l'arbre non chargé en bout: En supposant l'arbre reposant sur deux appuis et en négligeant l'inertie de rotation des sections droites, nous avons trouvé

$$\omega_0 = \Omega_0 = \frac{\pi^2}{l^2}\sqrt{\frac{EI}{\mu}}.$$

Si la notion de vitesse critique a été élucidée dès 1869 par RANKINE, et si la statique dans un repère tournant donne à cette vitesse critique la signification même de la bifurcation de stabilité au sens de L. EULER, il n'en va pas de même de l'aspect dynamique de la question, à cause des effets dûs à l'accélération de CORIOLIS. C'est pourquoi je voudrais m'arrêter quelques instants sur cette question.

Pour éviter quelques difficultés d'analyse et comprendre la véritable nature du phénomène avec le minimum d'appareil mathématique, je supposerai un arbre de machine flexible, reposant sur deux paliers simples, dont nous négligerons la masse propre par rapport à celle d'un disque plat placé par exemple en son milieu.

Nous supposerons l'axe et le disque entretenus en rotation *constante* de vitesse angulaire ω, au moyen d'un moteur muni d'un régulateur convenable.

Envisageons d'abord, dans le plan médiateur des paliers un repère fixe OXY: le rotor plat étant équilibré décrit dans ce plan fixe un mouvement plan sur plan, et nous désignerons par (XY) les coordonnées de G, par e l'excentration AG; l'arbre fléchi exerce en A une force de rappel

$$\vec{F} = -48\frac{EI}{l^3}\overrightarrow{OA} = -k\overrightarrow{OA}.$$

En désignant par M la masse du rotor et par J son moment central d'inertie, les théorèmes généraux de la dynamique en G donnent les deux équations:

$$J\frac{d^2\theta}{dt^2} = \overrightarrow{GA}\wedge\vec{F} + \vec{\Gamma} = 0$$

$$M\frac{d^2\overrightarrow{OG}}{dt^2} = \vec{F} = -k(\overrightarrow{OG} + \overrightarrow{GA})$$

dans lesquelles $\vec{\Gamma}$ est le couple du moteur qui assure l'uniformité de la rotation

$$\theta = \omega t + \varphi$$

le mouvement de G étant donné par

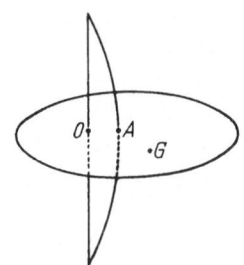

Fig. 6.

$$M\frac{d^2X}{dt^2} + kX = ke\cos(\omega t + \varphi)$$

$$M\frac{d^2Y}{dt^2} + kY = ke\sin(\omega t + \varphi).$$

Nous obtenons des équations d'oscillations forcées[1] pour la pulsation propre α

$$\alpha^2 = \frac{k}{M} = 48\frac{EI}{Ml^3}.$$

Il suffit d'introduire le point P défini par

$$\overrightarrow{GP} = \frac{\alpha^2}{\alpha^2 - \omega^2}\overrightarrow{GA}$$

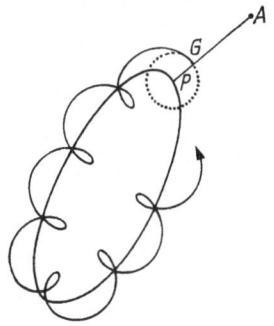

Fig. 7.

[1] On comparera utilement l'idée exposée ici à celle de LUCIEN BOREL. Thèse — Lausanne 1954. Voir aussi C. WEHRLI. Thèse — Zürich 1952 et l'article de H. ZIEGLER dans: Advances in applied mechanics Vol. 4 (1956).

4*

c'est-à-dire

$$\frac{PG}{PA} = \frac{\alpha^2}{\omega^2}.$$

Le point P décrit *l'ellipse de Lissajous* de centre O correspondant à la pulsation α, et le disque, entraîné en translation par le mouvement de P, tourne autour de ce point avec la rotation constante ω.

Il y a résonance lorsque $\omega = \alpha$ et α est la vitesse critique de l'arbre en rotation.

Lorsque $\omega < \alpha$ les points sont dans l'ordre PAG, et pour les faibles vitesses de rotation, P est voisin de A: le rotor tourne, pratiquement, autour du pied de l'axe. Lorsque $\omega > \alpha$ les points sont dans l'ordre AGP, et pour les très grandes vitesses de rotation, P est voisin de G: le rotor tourne, pratiquement, autour de son centre d'inertie G.

Nous voyons ainsi très clairement l'explication de la découverte faite en 1884 par LAVAL et PARSONS: dans le cas d'un rotor à axe flexible, tournant à vitesse angulaire croissante, si l'on franchit rapidement la vitesse critique on retrouve un régime stable. Et cette constatation différencie profondément l'instabilité élastique des arbres en rotation, de l'instabilité statique d'EULER.

D'ailleurs si nous changeons de repère, en prenant des axes Oxy qui tournent avec la vitesse angulaire constante ω, par rapport auxquels le disque aura seulement un mouvement de translation, nous obtenons:

$$Mx'' - 2M\omega y' + (k - M\omega^2)\, x = ke \cos \varphi$$

$$My'' + 2M\omega x' + (k - M\omega^2)\, y = ke \sin \varphi$$

qui définit un mouvement de LISSAJOUS excentré, correspondant aux pulsations

$$p = \pm\, \omega \pm \alpha.$$

L'instabilité dynamique correspond à l'annulation de p. Le couplage dû aux accélérations de CORIOLIS permet d'imputer la stabilité pour les grandes vitesses de rotation à *l'effet gyroscopique*.

Pour mieux comprendre comment la vitesse critique se trouve séparer deux zônes de stabilité, nous allons encore examiner un exemple, dans lequel l'imperfection sera d'une autre nature: nous supposons maintenant le rotor exactement centré ($e = 0$, G en A). Mais nous supposons que l'axe n'a plus la symétrie de révolution. La section droite présente alors deux moments d'inertie principaux $I' < I''$.

En posant

$$\alpha^2 = 48\, \frac{EI'}{Ml^3}$$

$$\beta^2 = 48\, \frac{EI''}{Ml^3}$$

et en prenant le repère tournant d'origine O, dont les axes sont parallèles aux directions principales de flexion de l'arbre, nous obtenons:

$$\begin{pmatrix} x'' \\ y'' \end{pmatrix} + \begin{pmatrix} 0 & -2\omega \\ 2\omega & 0 \end{pmatrix} \begin{pmatrix} x' \\ y' \end{pmatrix} + \begin{pmatrix} \alpha^2 - \omega^2 & 0 \\ 0 & \beta^2 - \omega^2 \end{pmatrix} \begin{pmatrix} x \\ y \end{pmatrix} = 0$$

équation d'un oscillateur classique à couplage gyroscopique, dont l'équation aux pulsations p est:

$$(\alpha^2 - \omega^2 - p^2)\,(\beta^2 - \omega^2 - p^2) - 4\omega^2 p^2 = 0$$

que l'on peut encore écrire:

$$p^4 - (\alpha^2 + \beta^2 + 2\omega^2)\, p^2 + (\alpha^2 - \omega^2)\,(\beta^2 - \omega^2) = 0$$

ou bien

$$(p^2 - \alpha^2 - \omega^2)\,(p^2 - \beta^2 - \omega^2) - 2\omega^2(\alpha^2 + \beta^2) = 0.$$

Ceci montre que pour $\omega < \alpha$ et $\omega > \beta$ la solution est purement trigonométrique, alors que, pour $\alpha < \omega < \beta$ il y a des exponentielles réelles.

La vitesse critique d'un arbre de révolution est donc la limite à laquelle se réduit un intervalle d'instabilité dynamique, compris entre deux domaines de stabilité.

Il ne faudrait pas croire que l'effet gyroscopique n'intervienne qu'en stabilité dynamique. Le problème de GREENHILL va nous fournir un exemple purement statique. Revenons à la poutre droite, idéale, chargée en bout par une force P, reposant à ses extrémités sur deux appuis simples et supposons qu'on la soumette en outre à un couple de torsion T, qui reste constant et vertical quand varie son point d'application. La déformation éventuelle peut être repérée sur deux axes rectangulaires y, z horizontaux. La contribution de T au moment fléchissant est

$$\vec{T} - (T \cdot t)\,\vec{t}$$

où $\vec{t}$ désigne le vecteur unitaire tangent à la déformée. L'équation linéarisée de la ligne élastique

$$EI \begin{pmatrix} y'' \\ z'' \end{pmatrix} + \begin{pmatrix} 0 & T \\ -T & 0 \end{pmatrix} \begin{pmatrix} y' \\ z' \end{pmatrix} + P \begin{pmatrix} y \\ z \end{pmatrix} = 0$$

où les dérivées sont prises par rapport à x, est exactement de même nature que celle d'un oscillateur à couplage gyroscopique. On peut aussi l'écrire

$$EIu'' - Tiu' + Pu = 0$$

au moyen de la variable complexe

$$u = y + iz.$$

Les pulsations sont les racines de l'équation

$$EI\omega^2 - T\omega - P = 0$$

et l'intégrale générale

$$u = A \exp i\omega_1 x + B \exp i\omega_2 x$$

comporte deux constantes complexes. La condition initiale

$$u = 0 \quad \text{pour} \quad x = 0$$

donne

$$u = A\,(\exp i\omega_1 x - \exp i\omega_2 x)$$

la condition terminale

$$u = 0 \quad \text{pour} \quad x = l$$

donne

$$i\omega_1 l - i\omega_2 l = 2i\pi$$

c'est-à-dire

$$(\omega_1 + \omega_2)^2 - 4\omega_1\omega_2 = 4\,\frac{\pi^2}{l^2}.$$

Les conditions de flambement sont donc

$$\left(\frac{T}{2EI}\right)^2 + \frac{P}{EI} = \frac{\pi^2}{l^2}$$

que l'on peut encore écrire

$$\frac{P}{P_0} + \left(\frac{T}{T_0}\right)^2 = 1$$

où P_0 est la charge critique d'EULER et

$$T_0 = 2\,\frac{\pi}{l}\,EI$$

est le couple de torsion critique de GREENHILL qui suffit, seul, à produire le flambement de la poutre[1].

Le cas d'une poutre encastrée à sa base et libre à l'autre bout mérite une étude attentive en fonction de la façon dont est réalisé le couple de torsion T: on trouvera un résumé de cette étude dans H. ZIEGLER (loc. cit. Advances in Applied Mechanics 1956).

---

[1] Le flambement sous l'effet conjugué d'une compression axiale, d'une torsion, et de la rotation d'un arbre a été étudié par Sir RICHARD SOUTHWELL en 1921.

Asservissements géométriques

La flexion d'un arbre qui tourne à sa vitesse critique est dûe, nous l'avons dit, à la répartition, tout le long de l'arbre, de forces centrifuges: ce système de forces, nul quand l'arbre est dans la position droite, est une fonctionnelle de la déformation et on peut dire qu'il est asservi à la variation géométrique. C'est un cas très simple car il y a proportionnalité de la force à la déformation, mais on rencontre, couramment des fonctionelles plus compliquées.

Imaginons que la poutre droite, sur deux appuis simples, qui faisait l'objet de nos études soit un tuyau dans lequel circule un fluide dont nous désignerons par μ la masse par unité de longueur, et par V la vitesse. Si, à la suite d'une perturbation, le tuyau prend une forme fléchie, en un point où le rayon de courbure est R, le fluide qui circule le long d'une paroi courbe exerce normalement une poussée

$$Y = \mu \frac{V^2}{R}.$$

En désignant comme précédemment par (ξ, η) un point de la ligne élastique, l'expression linéarisée du moment fléchissant

$$m = -\int_x^l \mu V^2 \eta''(\xi - x)\, d\xi$$

est une fonctionnelle portant sur la dérivée seconde de la déformation. L'équation de la ligne élastique

$$EIy'' + \int_x^l \mu V^2 \eta''(\xi - x)\, d\xi = 0$$

est ramenée par deux dérivations à l'équation différentielle

$$EIy'''' + \mu V^2 y'' = 0$$

avec les conditions aux limites:

$$y = 0; \qquad y'' = 0$$

pour

$$x = 0 \quad \text{et} \quad x = l.$$

En posant

$$\omega^2 = \mu \frac{V^2}{EI}$$

l'intégrale générale s'écrit

$$y = A \sin \omega x + B \cos \omega x + Cx + D$$

les conditions initiales entrainent

$$B = D = 0$$

et les conditions terminales

$$A \sin \omega l = 0; \qquad C = 0.$$

Si la vitesse V du fluide est suffisamment faible pour que

$$\omega l < \pi$$

le tuyau reste droit, en position stable. La vitesse critique, qui correspond à l'apparition de la flexion, est donc

$$V_0 = \frac{\pi}{l} \sqrt{\frac{EI}{\mu}}$$

et notre calcul est exactement analogue à celui du point de vue eulérien.

Un autre exemple d'asservissement géométrique va nous être fourni par la considération d'une plaque rectangulaire mince, encastrée à sa base, et soumise à un vent de bout, de vitesse V. En supposant la déformation cylindrique, et en gardant pour la section droite de la plaque les notations utilisées précédemment pour la ligne élastique, en admettant d'autre part, pour évaluer la pression exercée par le vent, la formule "piston", l'expression du moment fléchissant

$$m = \int_x^l kV\eta'(\xi - x)\, d\xi$$

est une fonctionnelle portant sur la dérivée première de la déformation. (La pression étant, comme on sait, proportionnelle à l'inclinaison du profil).

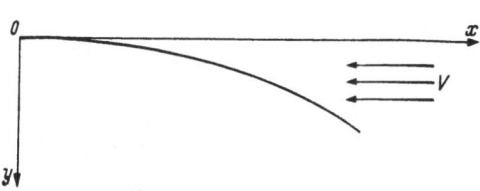

La section droite déformée est définie par l'équation, relative à une bande de largeur unitaire

Fig. 8.

$$\frac{Eh^3}{12(1 - \sigma^2)}\, y'' - \int_x^l kV\eta'(\xi - x)\, d\xi = 0$$

où h est l'épaisseur de la plaque mince et σ le coefficient de POISSON du matériau. On la ramène par deux dérivations à l'équation différentielle

$$\frac{Eh^3}{12(1 - \sigma^2)}\, y'''' - kVy' = 0$$

avec les conditions aux limites :

$$y = 0; \qquad y' = 0 \quad \text{pour} \quad x = 0$$
$$y'' = 0; \qquad y''' = 0 \quad \text{pour} \quad x = l.$$

En posant

$$\omega^3 = \frac{3}{2}\, kV \frac{1 - \sigma^2}{Eh^3}$$

l'intégrale générale s'écrit

$$y = A + Be^{2\omega x} + Cy_1 + Dy_2$$

où

$$y_1 = e^{-\omega x} \cos \omega x \sqrt{3}$$
$$y_2 = e^{-\omega x} \sin \omega x \sqrt{3}$$

et le procédé déjà maintes fois utilisé conduit, pour l'obtention d'une intégrale non identiquement nulle satisfaisant aux conditions aux limites, à la relation :

$$\cos \omega l \sqrt{3} = -\frac{1}{2} \exp - 3\omega l.$$

Etant donnée la décroissance rapide de l'exponentielle, on peut admettre, à 2% près la valeur approchée par défaut

$$\omega l \sqrt{3} = \frac{\pi}{2}.$$

Ce qui conduit à la vitesse critique approximative

$$V_0 = \left(\frac{\pi}{2}\right)^3 \frac{2}{9\sqrt{3}\, k} \frac{E}{1 - \sigma^2} \left(\frac{h}{l}\right)^3$$

définie, comme la précédente, d'un point de vue purement eulerien.

Mais je ne voudrais pas abandonner cet exemple tiré de l'aéroélasticité, sans revenir à la comparaison entre la stabilité d'Euler et la stabilité dynamique. L'exemple suivant, qui n'a, quant à l'étude des ailes d'avions, qu'une valeur purement schématique, nous suffira :

Considérons une longue plaque rectangulaire plane, suffisamment rigide pour être pratiquement indéformable. A l'équilibre horizontal dans un repère galiléen en translation horizontale uniforme de vitesse V, elle subit l'action de l'air ambiant. Pour lui donner deux degrés de liberté, nous la supposons rappelée vers son équilibre par des liens élastiques, exercés l'un sur le bord d'attaque, avec le coefficient k_1, l'autre sur le bord de fuite, avec le coefficient k_2.

En désignant par $2l$ la largeur entre ces deux bords, nous pouvons préciser les petits mouvements par le déplacement vertical y du milieu O, qui symbolisera la flexion de l'aile, et par la rotation φ autour de la ligne médiane de O, qui symbolisera la torsion de l'aile. L'action du vent est une portance verticale Y, appliquée à l'abscisse $x = a$ (souvent voisine de $\frac{l}{2}$), proportionnelle à l'angle d'attaque

$$Y = K\varphi$$

et au carré de la vitesse

$$K = \frac{\partial C}{\partial i} \varrho \frac{V^2}{2} S$$

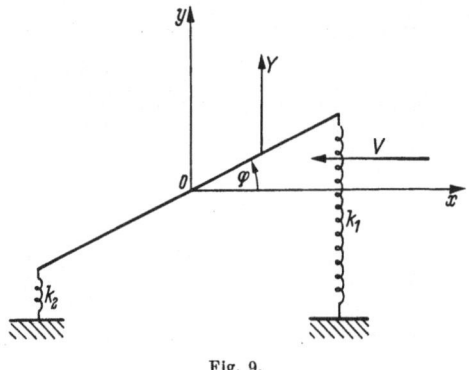

Fig. 9.

(où $\frac{\partial C}{\partial i}$ est un coefficient théoriquement voisin de 2π, ϱ la densité de l'air, S la surface de l'aile). En désignant par M la masse de la plaque et par I son moment d'inertie, les théorèmes généraux donnent les deux équations :

$$My'' = -k_1(y + l\varphi) - k_2(y - l\varphi) + K\varphi$$
$$I\varphi'' = -k_1 l(y + l\varphi) + k_2 l(y - l\varphi) + aK\varphi$$

en posant :

$$k_1 + k_2 = \alpha$$
$$(k_1 - k_2)\, l = \beta$$

nous obtenons

$$My'' + \alpha y + (\beta - K)\,\varphi = 0$$
$$I\varphi'' + \beta y + (\alpha l^2 - aK)\,\varphi = 0.$$

En l'absence de vent ($K = 0$) nous avons affaire à un oscillateur classique à couplage de rigidité :

$$My'' + \alpha y + \beta\varphi = 0$$
$$I\varphi'' + \beta y + \alpha l^2\varphi = 0$$

dont les pulsations, racines de

$$MI\omega^4 - \alpha(I + Ml^2)\,\omega^2 + 4k_1 k_2 l^2 = 0$$

sont ω_1 et ω_2, séparées, comme il est bien connu, par celles des degrés de liberté supposés découplés :

$$\omega_1^2 < \frac{\alpha}{M} < \frac{\alpha l^2}{I} < \omega_2^2.$$

En présence de vent, la matrice

$$\begin{pmatrix} \alpha & \beta - K \\ \beta & \alpha l^2 - aK \end{pmatrix}$$

montre que le couplage entre la flexion et la torsion devient dissymétrique. L'étude de la statique de la plaque dans le vent (linéarisée) est celle du système :

$$\begin{pmatrix} \alpha & \beta - K \\ \beta & \alpha l^2 - aK \end{pmatrix} \begin{pmatrix} y \\ \varphi \end{pmatrix} = 0.$$

Il y a bifurcation de stabilité lorsque la matrice devient irrégulière :

$$\alpha^2 l^2 - \beta^2 - (a\alpha - \beta)\,K = 0.$$

Comme

$$\alpha^2 l^2 - \beta^2 = 4k_1 k_2 l^2 > 0$$

ceci se produit lorsque les constantes de structure vérifient

$$a\alpha - \beta > 0.$$

Il y a alors une vitesse critique eulerienne, définie par

$$K = K_0 = \frac{4k_1k_2l^2}{a\alpha - \beta}$$

pour laquelle, en accord avec les notations des techniciens de l'aéronautique, nous dirons qu'il y a "divergence" de la plaque.

L'étude dynamique des petits mouvements de la plaque dans le vent conduit à l'équation aux pulsations:

$$\begin{vmatrix} \alpha - M\omega^2 & \beta - K \\ \beta & \alpha l^2 - aK - I\omega^2 \end{vmatrix} = 0$$

soit

$$MI(\omega^2 - \omega_1^2)(\omega^2 - \omega_2^2) - K[a\alpha - \beta - Ma\omega^2] = 0$$

ou

$$MI\omega^4 - [\alpha(I + Ml^2) - MaK]\omega^2 + 4k_1k_2l^2 - K(a\alpha - \beta) = 0$$

sur laquelle on voit que lorsque la vitesse atteint la valeur critique d'Euler, l'une des pulsations s'annule. Mais il ne suffit pas que la vitesse soit inférieure à cette valeur

$$K < K_0$$

pour que les racines de l'équation aux pulsations soient réelles, et lorsque ces racines sont complexes, la présence d'exponentielles croissantes montre qu'il y aura des oscillations dont l'amplitude croit au-delà de toute limite, c'est-à-dire, instabilité dynamique.

En posant

$$a\alpha - \beta = Ma\omega_0^2$$

le tracé de l'hyperbole

$$K = \frac{I}{a} \frac{(\omega^2 - \omega_1^2)(\omega^2 - \omega_2^2)}{(\omega_0^2 - \omega^2)}$$

représentative de K en fonction de ω^2, permet suivant les valeurs des constantes de structure, de discuter la question.

La plus petite des valeurs de K (si elles existent) pour lesquelles l'équation aux pulsations a des racines doubles (comprises entre ω_1 et ω_2) correspond à l'apparition de *l'instabilité dynamique, par confusion de fréquences*: en accord avec les notations des techniciens de l'aéronautique, nous dirons qu'il y a flottement (flutter) de la plaque.

Si nous supposons par exemple

$$a = \frac{l}{2}$$

et

$$k_2 < k_1 < 3k_2$$

on obtient

$$K_0 = \frac{8k_1k_2l}{3k_2 - k_1}$$

et

$$\omega_0^2 = \frac{3k_2 - k_1}{M}$$

avec

$$\omega_0 < \omega_1 < \omega_2$$

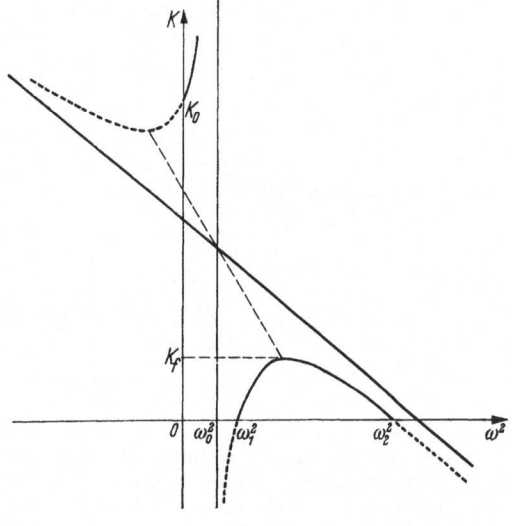

et pour $K = K_0$ l'équation aux pulsations a, outre la racine nulle, deux racines imaginaires. On voit sur l'hyperbole que le maximum K_f de K correspondant au flottement est très inférieur à K_0 correspondant à la divergence. La plaque n'est stable dans le vent que pour

$$K < K_f$$

c'est-à-dire pour des vitesses inférieures à la vitesse de flottement.

Fig. 10.

Nous avons ainsi un nouvel exemple dans lequel le critère dynamique de stabilité est beaucoup plus sévère que le critère d'Euler[1].

Asservissements dynamiques

Dans le premier exemple de flambement que nous avons étudié, la charge de compression P restait constante et verticale. Mais il y a de nombreux problèmes de stabilité élastique dans lesquels les charges ne forment pas un champ (c'est-à-dire ne sont pas des fonctions du seul point d'application): nous venons de rencontrer des fonctionnelles de la déformation, nulles avec celle-ci. Il peut arriver que cet aspect homogène disparaisse et que les charges soient asservies à la déformation sans s'annuler avec elle.

Le problème de Pflüger-Beck[2] va nous en fournir un exemple. Considérons de nouveau une poutre encastrée à sa base, et supportant, à son extrémité libre, un réacteur qui exerce une poussée constante P dans l'axe de la poutre: si la poutre fléchit, le réacteur reste solidaire de la tête de la poutre, et la force P qui reste constante en grandeur, est asservie à rester tangente à l'extrémité de la ligne élastique.

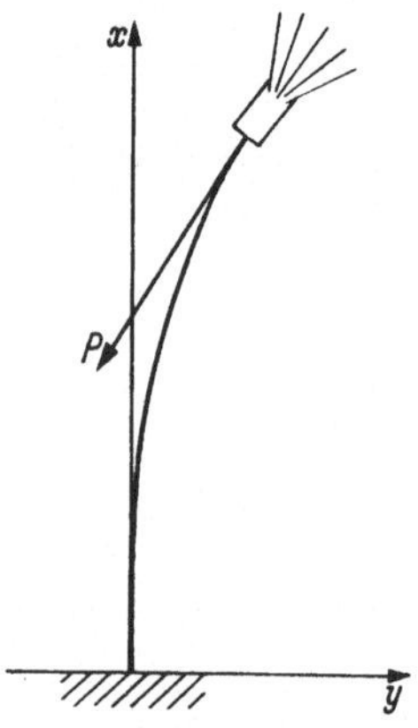

En désignant par M la masse du réacteur et par J son moment d'inertie, la poutre est, en cours de mouvement, soumise en bout, à trois charges: la force de compression, asservie P la force d'inertie transversale:

$$F = -M \frac{d^2 f}{dt^2}$$

le couple d'inertie

$$\gamma = -J \frac{d^2\theta}{dt^2}.$$

L'équation linéarisée de la ligne élastique

$$EIy'' = \gamma + (l - x)(F - P\theta) + P(f - y)$$

s'intègre élémentairement en posant

$$\frac{P}{EI} = \omega^2$$

Fig. 11.

suivant:

$$y = \frac{1}{P}\left[\gamma + Pf + (l - x)(F - P\theta)\right] + A\cos\omega x + B\sin\omega x$$

$$y' = -\frac{1}{P}(F - P\theta) - A\omega\sin\omega x + B\omega\cos\omega x.$$

Les constantes A et B sont déterminées par les conditions d'encastrement

$$A = -\frac{1}{P}(\gamma + lF) + l\theta -$$

$$B = \frac{F}{P\omega} - \frac{\theta}{\omega}$$

les conditions terminales, en introduisant l'angle

$$\omega l = \varphi$$

s'écrivent:

$$A\cos\varphi + B\sin\varphi + \frac{\gamma}{P} = 0$$

$$-A\sin\varphi + B\cos\varphi - \frac{F}{P\omega} = 0$$

_____

[1] Sans se reporter aux ouvrages spécialisés d'aéroélasticité, on peut trouver une étude de la stabilité des ailes d'avions beaucoup plus proche de la réalité dans Y. Rocard, L'instabilité en mécanique. Masson 1954.

[2] Pflüger, A.: Problèmes de stabilité. Berlin/Göttingen/Heidelberg: Springer 1950. — Beck, V. M.: cf. ZAMP 3, 225 (1952).

l'élimination de A et B permet le calcul de la déformation terminale (f, θ) en fonction des forces d'inertie (F, γ):

$$f + \frac{Fl}{P}\left(\cos\varphi - \frac{\sin\varphi}{\varphi}\right) + \frac{\gamma}{P}\left(1 - \cos\varphi - \varphi\sin\varphi\right) = 0$$

$$l\theta - \frac{Fl}{P}\left(1 - \cos\varphi\right) - \frac{\gamma}{P}\varphi\sin\varphi = 0.$$

En remplaçant $P = EI\left(\frac{\varphi}{l}\right)^2$ et les forces d'inertie par leurs valeurs nous obtenons l'équation du mouvement:

$$EI\begin{pmatrix}f\\\theta\end{pmatrix} = \begin{pmatrix} l^3\dfrac{\sin\varphi - \varphi\cos\varphi}{\varphi^3} & l^2\dfrac{\varphi\sin\varphi + \cos\varphi - 1}{\varphi^2} \\ l^2\dfrac{1 - \cos\varphi}{\varphi^2} & l\dfrac{\sin\varphi}{\varphi} \end{pmatrix}\begin{pmatrix}-Mf''\\-J\theta''\end{pmatrix}.$$

Supposons d'abord le réacteur inactif $(P = 0)$ nous obtenons

$$EI\begin{pmatrix}f\\\theta\end{pmatrix} = \begin{pmatrix}\dfrac{l^3}{3} & \dfrac{l^2}{2}\\[1mm]\dfrac{l^2}{2} & l\end{pmatrix}\begin{pmatrix}-Mf''\\-J\theta''\end{pmatrix}$$

ce sont, bien entendu, les équations du pendule élastique, oscillateur à deux degrés de liberté avec couplage par l'inertie.

Lorsque le réacteur est en fonctionnement $(P \neq 0)$, la matrice du second membre manifeste alors une dissymétrie du couplage entre la translation f et la rotation θ:

$$(1 - \cos\varphi) - (\varphi\sin\varphi + \cos\varphi - 1) > 0$$

pour

$$0 < \varphi < 2\pi.$$

L'équation aux pulsations Ω:

$$\begin{vmatrix} EI - M\Omega^2 l^3\dfrac{\sin\varphi - \varphi\cos\varphi}{\varphi^3} & -J\Omega^2 l^2\dfrac{\varphi\sin\varphi + \cos\varphi - 1}{\varphi^2} \\ -M\Omega^2 l^2\dfrac{1 - \cos\varphi}{\varphi^2} & EI - J\Omega^2 l\dfrac{\sin\varphi}{\varphi} \end{vmatrix} = 0$$

développée suivant:

$$MJ\Omega^4 l^4\frac{2 - 2\cos\varphi - \varphi\sin\varphi}{\varphi^4} - EI\Omega^2\left[Jl\frac{\sin\varphi}{\varphi} + Ml^3\frac{\sin\varphi - \varphi\cos\varphi}{\varphi^3}\right] + (EI)^2 = 0$$

montre qu'il n'y a pas de valeur de φ pour laquelle la pulsation Ω serait nulle: Le problème statique de la détermination d'une bifurcation de stabilité sous l'effet de la charge P asservie est donc en défaut. (Ce phénomène conduisant à une charge critique infinie, avait déjà été rencontré par E. L. NIKOLAI, dans l'étude de la torsion). C'est pourquoi nous appellerons ce phénomène le paradoxe de NIKOLAI.

En vérité, le produit des racines de l'équation aux pulsations change de signe pour

$$\varphi = 2\pi$$

$$P = 4\pi^2\frac{EI}{l^2}$$

qui correspond à une pulsation infinie, et on pourrait, en un certain sens, interpréter cette valeur comme une charge critique. Mais l'étude de la réalité des racines de l'équation aux pulsations va nous conduire à une valeur inférieure: bornons nous, ce qui est raisonnable, au cas où J est petit devant Ml^2 et posons

$$J = \varepsilon Ml^2.$$

La réalité des pulsations (qui sont réelles pour $\varphi = 0$) exige:

$$\varepsilon\varphi^2\sin\varphi + (\sin\varphi - \varphi\cos\varphi) > 0$$

$$\Delta > 0$$

(Δ étant le discriminant).

5*

On voit apparaître la racine

$$\varphi_0 = 4,493$$

de l'équation

$$\operatorname{tg} \varphi = \varphi$$

et nos deux conditions s'écrivent

$$\varphi < \varphi_0(1 - \varepsilon)$$

$$\varphi < \varphi_0(1 - 1,20 \sqrt{\varepsilon})$$

dont la seconde entraine la première: la stabilité dynamique disparait par confusion de fréquences lorsque la charge P atteint la valeur

$$P_0 = 4,493^2(1 - 2,40 \sqrt{\varepsilon}) \frac{EI}{l^2}.$$

Ce résultat est indépendant de la valeur de la masse terminale M, et il suffit de faire $\varepsilon = 0$ si on néglige complètement l'inertie terminale.

L'étude des vibrations transversales dans le cas d'une masse répartie tout le long de la poutre chargée en bout par la même force P asservie (éventuellement combinée à une force verticale) est exactement le problème de Pflüger-Beck.

L'expérience a été faite par N. Willems, mais dans des conditions qui ne sont pas tout à fait l'asservissement prévu: la force P passe par un point fixe de la verticale d'encastrement. Bien que cette propriété soit très près d'être vérifiée dans les déformations usuelles, et que la charge critique obtenue expérimentalement soit très peu inférieure (94%) à la charge critique théorique du problème de Beck, cette expérience n'a pas toujours été considérée comme probante: une force de grandeur constante dont le support passe par un point fixe appartient à un champ qui dérive d'un potentiel et la nature mécanique du problème s'en trouve considérablement simplifiée.

En vérité l'étude que nous venons de faire, comme la précédente, montre que la question est liée à l'apparition d'une dissymétrie dans les coefficients de couplage. La matrice des coefficients d'inertie (ou de rigidité, suivant le cas) est la somme d'une matrice symétrique et d'une matrice antisymétrique. La présence de cette matrice est liée à l'apport énergétique du vent dans le cas du flottement, du réacteur dans le cas du problème de Beck. C'est sans doute parce que la portance d'une aile, exprimée par un potentiel complexe non uniforme, est liée, comme l'a montré Joukowski, à la circulation, que les forces (effectives ou d'inertie) associées à la partie antisymétrique de la matrice, ont été appelées *circulatoires* (cf. H. Ziegler, op. cit.)[1,2,3].

Effets non linéaires — Sauts d'équilibres

Nous avons reconnu que le flambement est un phénomène essentiellement non linéaire, en ce sens qu'il ne vérifie pas le principe de superposition. Mais nous n'avons jusqu'ici étudié, du point de vue statique, que des phénomènes dans lesquels, lors du franchissement des charges critiques, la bifurcation de stabilité correspondait à un déplacement continu de l'équilibre. Un des aspects typiques de la mécanique non linéaire est le saut discontinu, dans l'évolution d'un phénomène, d'une branche à une autre d'une caractéristique multiforme: ce saut ne se produit d'ailleurs pas dans les mêmes conditions lorsque la caractéristique est décrite de façon ascendante ou descendante et il y a là une des explications de l'effet des cycles d'hystérésis.

Un des exemples les plus élementaires est peut être la ferme présentée par von Mises en 1925.

---

[1] La comparaison entre l'expérience de Willems et le problème de Beck a fait l'objet d'une excellente étude récente de Huang, Nachbar et Nemat-Nasser (J. Appl. Mech. Mars 1967).

[2] On lira avec intérêt l'article de G. Herrmann sur la stabilité d'un équilibre élastique soumis à des forces non conservatrices, dans Appl. Mech. Rev. Févr. 1967.

[3] Je voudrais recommander aussi la lecture de l'article de Lorenzo Contri, dans Méccanica 1966 No. 1, p. 61.

Considérons un assemblage formé de deux poutres identiques, articulées entre elles en A, et en leurs extrémités B, C avec un bâti fixe; le triangle ABC est isocèle. Pour réduire notre étude à l'essentiel, nous supposerons le bâti suffisamment rigide pour que les articulations B et C restent effectivement fixes, alors que l'assemblage est soumis à une charge P appliquée en A; ceci s'exprime par

$$l \cos \alpha = \text{cste} = l_0 \cos \alpha_0$$

où l_0 est la longueur de chaque barre non contrainte, et α_0 l'angle à la base du triangle ABC correspondant.

En charge sous l'action de la force P, chaque barre est soumise à l'effort normal

$$N = \frac{P}{2 \sin \alpha}$$

(compté positivement en compression).

En désignant par S la section d'une barre,

$$\frac{l_0 - l}{l_0} = \frac{\cos \alpha - \cos \alpha_0}{\cos \alpha} = \frac{N}{ES} = \frac{P}{2ES \sin \alpha} \; .$$

La caractéristique de l'assemblage, relation entre la charge P et la disposition géométrique définie par l'angle $\alpha \left(-\dfrac{\pi}{2} < \alpha < +\dfrac{\pi}{2} \right)$ est donc

$$\frac{P}{2ES} = \text{tg} \, \alpha (\cos \alpha - \cos \alpha_0)$$

le second membre est une fonction impaire, ce qui montre qu'en l'absence de charge, l'assemblage a trois positions d'équilibre:

$$\alpha = \alpha_0; \quad \alpha = 0; \quad \alpha = -\alpha_0.$$

Il est maximum lorsque l'angle α prend la valeur θ telle que

$$\cos^3 \theta = \cos \alpha_0$$

et sa valeur est alors

$$P_0 = 2ES \sin^3 \theta.$$

Le tracé de α en fonction de P permet de suivre l'évolution de l'assemblage sous charge lentement croissante: α décroit à partir de α_0, et le triangle ABC s'aplatit jusqu'à ce qu'on arrive à la valeur critique P_0 et l'angle de base θ. Une légère perturbation fait alors passer brusquement A au dessous de l'horizontale BC: les deux barres sont sous tension, avec l'angle γ défini par

$$\text{tg} \, \gamma (\cos \gamma - \cos^3 \theta) = \sin^3 \theta.$$

Lorsque P continue à croitre, l'angle de base augmente en valeur absolue, à partir de $|\gamma|$.

Si, poursuivant l'expérience nous faisons maintenant décroitre P, le point A reste au-dessous de BC et l'angle de base α prend la valeur $-\alpha_0$ lorsque P s'annule: l'assemblage a alors la disposition symétrique de la disposition initiale. Naturellement, si l'on donne à P des valeurs négatives, le triangle ABC s'aplatira jusqu'à ce que, pour $P = -P_0$, l'angle de base ait pour valeur $-\theta$. Si la charge dépasse $-P_0$, le point A repassera, par un saut brusque, au-dessus de l'horizontale BC, l'angle de base prenant la valeur $-\gamma$.

Nous voyons ainsi que la valeur critique de la charge $(\pm P_0)$ correspond à une bifurcation de stabilité qui se produit par un changement discontinu dans la configuration géométrique:

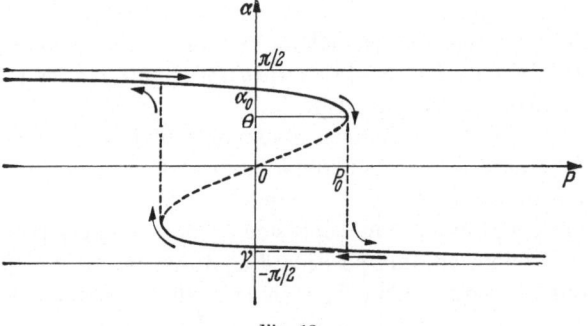

Fig. 13.

les dispositions du triangle ABC pour lesquelles

$$-\theta \leq \alpha \leq \theta$$

sont des équilibres instables.

L'étude dynamique, conduite en attribuant par exemple à l'articulation A une masse m utilise l'énergie de déformation des barres et l'énergie potentielle dont dérive la force constante

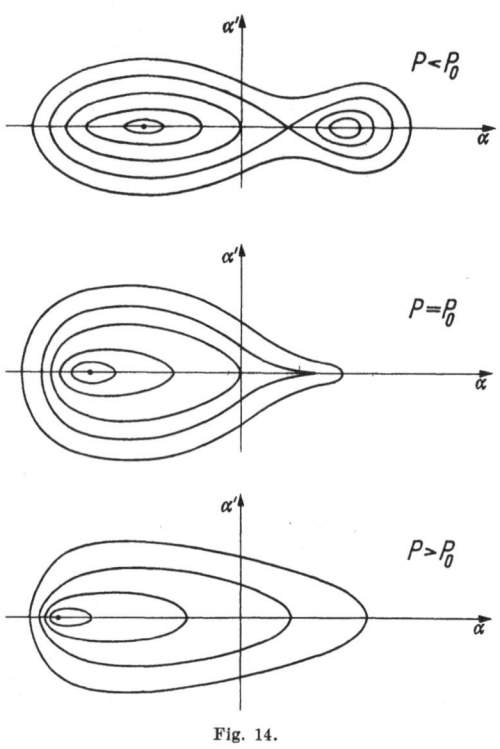

Fig. 14.

P. La configuration des trajectoires de phases revêt des aspects topologiques différents suivant les valeurs de P:

Pour $|P| > P_0$ il y a un seul point critique, correspondant à l'équilibre unique: c'est un sommet et l'équilibre est stable.

Pour $|P| < P_0$ il y a trois points critiques correspondant aux trois équilibres: deux sommets entre lesquels il y a un noeud, associé à l'équilibre instable. Ce noeud est le point double d'une séparatrice correspondant à *l'instabilité orbitale*. Il y a des petites oscillations entourant chacun des équilibres stables, et de grandes oscillations entourant l'ensemble des trois équilibres.

Le cas limite $|P| = P_0$ est celui où l'angle de base ayant la valeur θ, deux points critiques, un noeud et un sommet, confluent en un point de *bifurcation* (au sens de POINCARÉ).

Ainsi, *dans le cas d'un saut discontinu d'équilibre, la charge critique, qui exprime une bifurcation au sens de* POINCARÉ, *correspond du point de vue dynamique à l'apparition d'une séparatrice d'instabilité orbitale*. Le cas classique de EULER, c'est-à-dire la transition continue d'un équilibre à un autre est naturellement un cas limite du précédent, comme l'étude des vibrations transversales d'une poutre chargée en bout nous l'a bien montré (séparation, au-delà de la charge critique, entre les petites vibrations asymétriques et les grandes vibrations symétriques).

Les sauts discontinus d'équilibre se présentent fréquemment dans l'étude des milieux à deux ou trois dimensions. C'est ainsi que le flambement d'une enveloppe sphérique (balle de ping-pong par exemple) sous l'effet d'un choc est très facile à réaliser: un petit domaine de la calotte passant à l'intérieur par un brusque changement de sens de la concavité.

Quelques considérations sur le comportement en dehors du domaine élastique

Comme les charges susceptibles de produire le flambement d'une structure sont habituellement importantes, il arrive fréquemment qu'elles correspondent à un comportement du matériau qui soit en dehors du domaine élastique. C'est pourquoi il y a eu dans les trente dernières années une littérature abondante sur les problèmes de stabilité des matériaux élastoplastiques et viscoélastiques.

Je me bornerai, dans cet exposé qui veut rester élémentaire, au schéma de principe suivant: Reprenons la plaque plane qui nous a servi à étudier la divergence d'aile et le flottement, en la considérant maintenant comme la surface de base d'une poutre prismatique, dont nous désignerons (pour éviter une confusion de notations) la longueur par L.

Prenons comme position d'équilibre initial $y = 0$, $\varphi = 0$, la position de la poutre verticale sous la charge axiale P_0 pour laquelle le flambement, c'est-à-dire le basculement φ croissant, commence: le coefficient de rappel k_1 schématisera l'élasticité des fibres qui travaillent à la décompression (module d'YOUNG E) et le coefficient de rappel k_2 schématisera l'élasticité atténuée des fibres qui travaillent à la surcompression (module d'YOUNG $E' < E$).

Si, lorsque la charge est passée de P_0 à P, le basculement de la poutre est défini par $y < 0$ et φ, les équations d'équilibre, dans les notations déjà utilisées sont:

$$k_1(y + l\varphi) + k_2(y - l\varphi) + P - P_0 = 0$$

$$k_1 l(y + l\varphi) - k_2 l(y - l\varphi) - PL\varphi = 0$$

c'est-à-dire

$$\alpha y + \beta\varphi = -(P - P_0)$$

$$\beta y + (\alpha l^2 - PL)\,\varphi = 0$$

d'où l'on tire

$$\varphi = \frac{\beta(P - P_0)}{\alpha^2 l^2 - \beta^2 - \alpha PL}$$

valable lorsque la fibre relative à k_1 est décomprimée

$$y + \varphi l > 0$$

c'est-à-dire

$$\beta l - \alpha l^2 + PL > 0$$

et la fibre relative à k_2 surcomprimée

$$y - \varphi l < 0$$

c'est-à-dire

$$PL - \alpha l^2 - \beta l < 0.$$

Ainsi, nous sommes conduits aux trois expressions suivantes:

$$P_1 = \frac{\alpha l^2 - \beta l}{L} = 2k_2 \frac{l^2}{L}$$

qui est la charge critique calculée pour le module diminué k_2 (c'est-à-dire E')

$$P_2 = \frac{\alpha l^2 + \beta}{L} = 2k_1 \frac{l^2}{L}$$

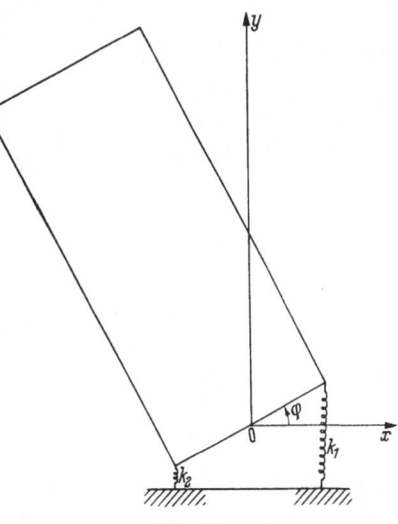

Fig. 15.

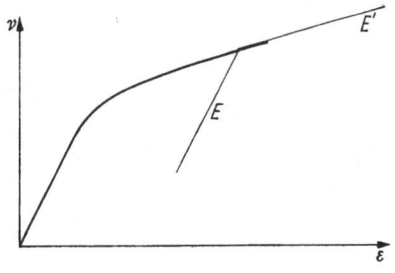

Fig. 16.

qui est la charge critique calculée pour le module de l'élasticité pure k_1 (c'est-à-dire E) c'est la charge d'EULER.

$$P_3 = \frac{\alpha^2 l^2 - \beta^2}{\alpha L} = \frac{4k_1 k_2 l^2}{(k_1 + k_2)\,L}$$

qui est la charge critique calculée pour le module mixte de T. VON KÁRMÁN (1909)[1]. Ces charges sont dans l'ordre

$$P_1 < P_3 < P_2$$

les conditions de validité de notre calcul

$$P_1 < P < P_2$$

et la formule homographique

$$\varphi = \frac{\beta}{\alpha L} \, \frac{P - P_0}{P_3 - P}$$

montrent que le basculement peut commencer à partir de toute charge P_0

$$P_1 \leq P_0 < P_3$$

et φ croit de o à $+\infty$ pendant que la charge croit de P_0 à P_3.

[1] Je voudrais rendre hommage ici à l'oeuvre considérable de T. VON KÁRMÁN, fondateur de l'IUTAM., décédé peu de temps avant le congrès de Munich.

Nous retrouvons ici la théorie présentée par F. R. Shenley (1946—1947) selon laquelle *en élastoplasticité la charge critique* d'Euler *fait place à une plage critique*

$$P_1 \leq P_0 < P_3$$

entièrement située au-dessous de la valeur eulérienne P_2, et bornée par les valeurs obtenues en substituant au module E, soit le module tangent E' (pour P_1) soit le module mixte (pour P_3)[1].

Lorsqu'un matériau présente du fluage il est intéressant d'étudier son comportement sur le même schéma élémentaire. En adoptant comme loi d'évolution

$$\varepsilon' = \frac{1}{E}\,\sigma' + k\sigma^m$$

où ε est la dilatation relative, σ la contrainte et où les accents désignent les dérivées par rapport au temps, (k et m sont, à température constante, deux constantes caractéristiques du matériau), on détermine l'évolution en fonction du temps de l'enfoncement y et du basculement φ:

$$\frac{1}{h}\,(y' + ls') = k\sigma_1^m + \frac{1}{E}\,\sigma_1'$$

$$\frac{1}{h}\,(y' - ls') = k\sigma_2^m + \frac{1}{E}\,\sigma_2'$$

où h désigne l'épaisseur, en l'absence de charges, de la zône au pied de la poutre, qui subit le fluage. On a posé

$$\sin \varphi = s.$$

Si P désigne la charge en bout, et S la surface de section, les contraintes σ_1, σ_2 sont données pour chaque valeur de φ par:

$$S(\sigma_1 + \sigma_2) = P$$

$$S(\sigma_1 - \sigma_2)\,l = PL \sin \varphi = PL\,s$$

c'est-à-dire

$$\sigma_1 = \frac{P}{2S}\left(1 + \frac{L}{l}\,s\right)$$

$$\sigma_2 = \frac{P}{2S}\left(1 - \frac{L}{l}\,s\right)$$

et l'équation différentielle qui donne l'évolution de φ:

$$2\left(\frac{l}{h} - \frac{P}{2ES}\,\frac{L}{l}\right)s' = k\left(\frac{P}{2S}\right)^m\left[\left(1 + \frac{L}{l}\,s\right)^m - \left(1 - \frac{L}{l}\,s\right)^m\right]$$

peut, puisque dans l'expérience courante $m > 1$, s'écrire sous la forme

$$\frac{ds}{sF(s)} = mk\left(\frac{P}{2S}\right)^m \frac{ES\,dt}{P_0 - P}$$

où $F(s)$ est une fonction continue sans racine réelle et

$$P_0 = 2ES\,\frac{l^2}{hL}$$

Si, à l'instant initial, $\varphi = 0$ la solution

$$s = \sin \varphi \equiv 0$$

montre que la poutre reste en équilibre droit. Mais si par suite d'une petite perturbation, φ est initialement différent de zéro:

$$s_0 = \sin \varphi_0 \neq 0$$

l'intégrale

$$\int_{s_0}^{1} \frac{ds}{sF(s)} = J$$

---

[1] Dans un autre ordre d'idée, je voudrais signaler un travail intéressant, en cours de publication, de Melle. Brigitte Messonnier (CNRS Paris) sur le flambement dans les milieux à la Cosserat.

indépendante de P, définit un temps d'écroulement de la colonne

$$T = \frac{J}{km}\left(\frac{2S}{P}\right)^m \frac{P_0 - P}{ES}.$$

Cet intervalle de temps, qui est très grand si P est petit, tend vers zéro lorsque la charge tend vers la valeur P_0.

Dans un matériau qui présente du fluage, la charge critique peut être définie comme la charge pour laquelle l'écroulement est immédiat.

Nous avons noté que l'étude précédente concerne un milieu qui évolue à température constante. Une discussion récente a opposé M. W. T. KOITER à MM. TRUESDELL et NOLL au sujet de la thermodynamique du flambement[1]. L'adjonction de la seule variable d'entropie suffit à faire l'étude énergétique et M. KOITER a heureusement remarqué que *l'énergie libre définit une fonction de* LIAPOUNOV *pour l'étude de la stabilité*. Lorsqu'on étudie un assemblage de charges critiques au sens d'EULER, il me parait certain que la thermodynamique qui est concernée est celle d'un *processus isotherme*: l'expérience courante des montages flexibles capables de grand flambement, dépassant largement la charge critique, montre en effet la possibilité d'appliquer les charges de façon croissante en suivant la déformation avec une lenteur arbitraire. Au contraire, il me semble que dans une étude dynamique de charge critique, définie par confusion de fréquences, il est essentiel de noter que des vibrations de très petite amplitude et de haute fréquence sont généralement du ressort de la thermodynamique des *processus adiabatiques*.

Il serait intéressant d'examiner expérimentalement de quelle nature thermodynamique sont les passages rapides d'un équilibre à un équilibre non voisin (saut discontinu), sous l'effet d'un choc, d'une percussion.

Quelques remarques en guise de conclusion

Je me suis borné à une présentation élémentaire et schématique du flambement dans les milieux élastiques à une dimension. Au risque de paraitre un peu "abstractionniste" dans un congrès de mécanique appliquée, je voudrais faire quelques remarques sur la philosophie de la question.

Un milieu continu à une, deux ou trois dimensions est décrit par un ensemble de fonctions numériques, généralement deux fois dérivables, de par la nature même des lois de la mécanique. Lorsqu'on effectue une perturbation dans l'état de ce milieu, il y a lieu d'évaluer l'ampleur de la perturbation produite (perturbation géométrique, dynamique, ou thermodynamique). Il y a donc lieu, dans l'espace fonctionnel concerné, de définir ce que sont des points voisins: le mathématicien dit qu'il a muni l'espace d'une topologie. Lorsqu'on parle de valeurs numériques, il est naturel de dire qu'un nombre z est voisin de a lorsqu'il appartient au disque $|z - a| < \varepsilon$. Lorsqu'on parle de fonctions, les exigences que l'on peut avoir pour définir des fonctions voisines sont très diverses: on peut exiger que le maximum de $|f - g|$ soit $< \varepsilon$; on peut exiger que la valeur moyenne de $|f - g|^2$ soit $< \varepsilon$. On peut aussi exiger que, en même temps que les fonctions, leurs dérivées jusqu'à un certain ordre vérifient un tel critère d'adaptation.

La notion de stabilité consiste à apprécier dans quelles conditions un milieu continu reste dans un état voisin d'un état d'équilibre (ou en cours d'évolution, si l'on veut généraliser à la mécanique des milieux continus la notion de stabilité d'un mouvement). Il est donc nécessaire, pour que cela ait un sens clair, que l'on précise le choix de la topologie. C'est ainsi que lorsqu'on

[1] TRUESDELL, C., et W. NOLL: The non-linear field theories of mechanics. Handbuch der Physik III/3, Berlin/Heidelberg/New York: Springer 1965.

KOITER, W. T.: Purpose and achievements of research in elastic stability. Delft.

KOITER, W. T.: On the thermodynamic background of elastic stability theory. Delft 1967.

Je voudrais, à cette occasion, remercier M. KOITER de m'avoir fait connaitre la traduction anglaise (NASA mars 1967) de sa thèse sur la stabilité élastique.

On trouvera un exposé élémentaire de l'introduction des potentiels thermodynamiques en élasticité dans R. SIESTRUNCK: Statique appliquée, chap. 4, Armand Colin 1967.

fait de la statique, et que l'on définit les voisinages par des distances géométriques, on aboutit aux définitions eulériennes de la stabilité. Lorsqu'on fait figurer le temps parmi les variables indépendantes, et que l'on considère un mouvement vibratoire de faible amplitude comme un état voisin du repos, on aboutit à la définition de la stabilité dite "dynamique". C'est une notion qui n'est pas très exigeante, car on ne fait pas intervenir la fréquence, de sorte qu'un mouvement voisin du repos peut admettre éventuellement de grandes accélérations.

On pourrait avoir d'autres notions de stabilité, intermédiaires (plus faibles que la stabilité dynamique) en ne considérant comme voisins du repos que des mouvements dont les accélérations restent petites, dont les variations de courbure restent petites. A leur tour les perturbations thermiques, les perturbations produites par de petits chocs, pourraient tout aussi bien être utilisées comme causes de troubles et fourniraient d'autres notions de stabilité.

Après avoir pénétré la nature topologique profonde d'une notion de stabilité, le mécanicien ne peut pas se contenter de raisonnements basés sur des flèches d'implication: il souhaite traduire au moyen d'inégalités explicites les conditions suffisantes d'instabilité. Pour atteindre ce but il est efficace de disposer d'une topologie définie par l'usage d'une norme. Monsieur Koiter utilise, dans ses travaux récents, la norme de l'espace L^2 qui est beaucoup utilisé en analyse. Il est bon alors de définir l'espace fonctionnel en engageant la continuité de certaines dérivées des fonctions de déformation, sinon, des discontinuités des dérivées sur un ensemble de mesure nulle seront négligées (pliage d'une plaque par exemple) sans qu'il s'agisse à proprement parler d'une petite déformation.

Une question de mécanique des milieux continus comporte des données aux limites et des données définies par des fonctionnelles de la déformation (c'était le cas dans les problèmes avec asservissement géométrique ou dynamique). Le passage de ces données à la détermination de l'équilibre, ou de l'état, du milieu continu est défini par un système d'équations fonctionnelles habituellement non linéaire, et la détermination des charges critiques est l'étude de la bifurcation de ce système[1].

Les progrès prodigieux qu'a fait récemment, sous l'impulsion de l'école soviétique[2], la théorie de la stabilité pour les systèmes mécaniques à un nombre fini de degrés de liberté[3] devraient raisonnablement, par le passage de la théorie des équations différentielles à celle des espaces fonctionnels, rejaillir prochainement sur le problème, dont la difficulté reste immense, de la stabilité des équilibres et des évolutions des milieux continus.

---

[1] Voir la conférence de G. Prodi sur les problèmes de bifurcation pour les équations fonctionnelles, au Congrès de l'Union Mathématique Italienne à Trieste en octobre 1967.

[2] Liapounov, Malkin, Chetaev (1961), Matrosov (1962), Pojaritskii (1961), Rumiantsev (1966), Arnold (en cours).

[3] Je recommande particulièrement la lecture de la conférence présentée sur la stabilité de l'équilibre en mécanique par Luigi Salvadori, professeur à Catane, au Congrès de l'Union Mathématique Italienne à Trieste en octobre 1967.

Les problèmes de l'hydrodynamique et les modèles mathématiques

Par

M. A. Lavrentiev

Académie des Sciences de l'URSS, Novosibirsk

Le progrès intense de la mécanique classique des milieux continus (liquides parfaits, théorie de l'élasticité) à la fin du siècle passé et au commencement du siècle présent a été très influencé par les nouveaux problèmes techniques et par la théorie des fonctions des variables complexes. Plus tard, les idées générales des mathématiques ont permis de considérer les problèmes mécaniques d'un point de vue plus général. Ceci a conduit aux progrès qui ont permis non seulement d'établir l'existence et l'unicité des solutions, mais encore d'élucider les propriétés générales des phénomènes. Parmi les plus remarquables propriétés obtenues, on peut citer celles concernant le degré de stabilité des solutions. Ces problèmes sont étroitement liés à la construction des algorithmes permettant d'effectuer des calculs approchés au moyen des calculatrices électroniques.

Actuellement, les recherches mathématiques sur l'étude des propriétés des milieux continus se développent essentiellement dans deux directions. La première est de caractère géométrique. Il s'agit de généraliser les méthodes géométriques de la théorie des fonctions d'une variable complexe au cas de deux équations aux dérivées partielles du premier ordre de caractère assez général. La seconde, de nature plutôt analytique, est fondée sur la théorie des équations intégrales singulières. Elle fait appel aux méthodes de l'analyse fonctionelle.

Mon propos n'est pas de comparer ces deux directions. Chacune a ses points forts et ses défauts. Je veux formuler une série de problèmes aux limites pour les équations aux derivées partielles qui contiennent les problèmes principaux de l'hydrodynamique pour des courants stationnaires. Ces problèmes seront formulés dans un langage géométrique.

I. La représentation quasi-conforme et les problèmes classiques de l'hydrodynamique

Rappelons les notions fondamentales. Etant donné un système d'équations aux dérivées partielles du premier ordre

$$F_1(x, y, u, v, u_x, u_y, v_x, v_y) = 0$$
$$F_2(x, y, u, v, u_x, u_y, v_x, v_y) = 0. \qquad (1)$$

Nous disons que la représentation homéomorphe du domaine D du plan (x, y) sur le domaine $\varDelta$ du plan (u, v), $u = u(x, y)$, $v = v(x, y)$ est une représentation quasi-conforme correspondante au système (1) si les fonctions $u = u(x, y)$, $v = v(x, y)$ qui réalisent cette représentation satisfont le système (1).

Si le système (1) est celui de CAUCHY-RIEMANN

$$\frac{\partial u}{\partial x} = \frac{\partial v}{\partial y}, \qquad \frac{\partial u}{\partial y} = -\frac{\partial v}{\partial x} \qquad (2)$$

la représentation quasi-conforme devient conforme. La construction de plusieures classes des courants stationaires d'un liquide

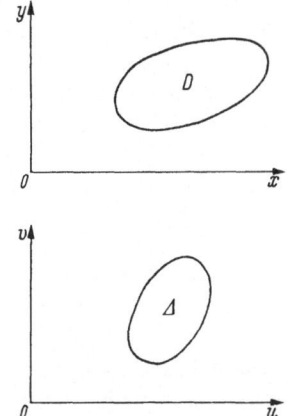

Fig. 1. Représentation du domaine D sur le domaine $\varDelta$.

idéal incompressible se réduit au problème de RIEMANN de la représentation conforme du domaine donné sur un domaine de caractère spécial; sur une bande rectiligne ou bien sur l'extérieur d'un segment, sur l'extérieur d'un cercle.

Dans le cas des gaz parfaits on a

$$\frac{\partial u}{\partial x} = \varrho(V)\,\frac{\partial v}{\partial y}, \qquad \frac{\partial u}{\partial y} = -\varrho(V)\,\frac{\partial v}{\partial x} \tag{3}$$

$$V^2 = u^2 + v^2$$

$\varrho(V)$ étant une fonction connue.

Considérons une représentation associée au système (1) et prenons la représentation linéaire tangente à cette représentation au point x_0, y_0. Soit Π_ν le parallélogramme du plan x, y qui se transforme en un carré unitaire dont un des côtés, l, forme un angle ν avec l'axe des u.

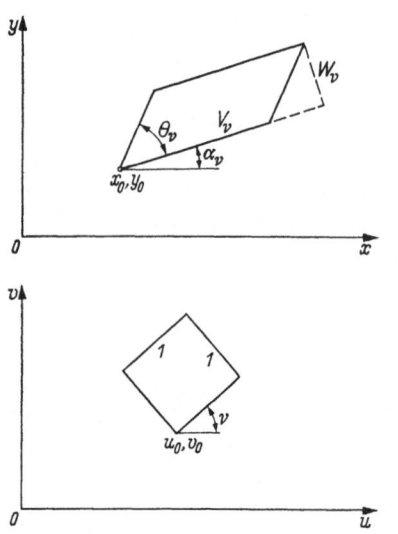

Désignons par V_ν le côté de Π_ν dont l'image est l et par α_ν l'angle que V_ν forme avec l'axe des X. Soit W_ν la hauteur de Π_ν orthogonale à V_ν et θ_ν l'angle du parallélogramme Π_ν au sommet x_0, y_0. Disons que les quantités introduites V, α, W, θ sont les caractéristiques de la représentation (1). Il est évident que toutes les caractéristiques peuvent être calculées à partir des coefficients de la transformation linéaire en question, c'est-à-dire à partir des fonctions u_x, u_y, v_x, v_y.

Le système (1) est équivalent aux deux équations suivantes

$$\begin{aligned} W_\nu &= \varphi_1(x, y, u, v, V_\nu, \alpha_\nu) \\ \theta_\nu &= \varphi_2(x, y, u, v, V_\nu, \alpha_\nu). \end{aligned} \tag{1'}$$

Fig. 2. Transformation d'un parallélogramme en un carré unitaire.

Si l'on désigne V_0, α_0, θ_0, W_0 par V, α, θ, W, les systèmes (2) ou (3) seront réduits à la forme

$$\left.\begin{aligned} W &= V \\ W &= F(V) \end{aligned}\right\} \theta = \pi/2.$$

Les systèmes (1) ou (3) sont dits *fortement elliptiques* s'il existe une constante $\varkappa$ ($\varkappa > 1$) telle que $\dfrac{1}{\varkappa} < \dfrac{\partial W_\nu}{\partial V_\nu} < \varkappa, \quad \dfrac{1}{\varkappa} < \theta_\nu < \pi - \dfrac{1}{\varkappa}$ pour toutes valeurs de ν et les valeurs des arguments.

Pour les équations fortement elliptiques, le théorème de RIEMANN sur l'existence et l'unicité d'une représentation quasi-conforme du domaine simplement connexe D sur un autre domaine simplement connexe Δ est applicable (sous les mêmes conditions pour le cas des représentations conformes).

Les équations de la dynamique des gaz sont fortement elliptiques lorsque V est inférieur à la vitesse de la propagation du son. On obtient ici un grand nombre de théorèmes d'existence et d'unicité des solutions pour des courants dont les vitesses sont inférieures à celle de la propagation du son.

Considérons, par example, un courant de gaz parfait dans une bande

$$y_0(x) < y < y_1(x)$$

les fonctions y_0 et y_1 étant différentiables deux fois et satisfaisant les conditions $0 \le y_0(x) \le \varkappa_0$, $0 < \varkappa_1 < y_1(x) - y_0(x) < \varkappa_2$, $\varkappa_0$, $\varkappa_1$, $\varkappa_2$ étant des constantes. Soit h le débit de courant. Il existe alors une constante h_0 telle que pour tous les $h \le h_0$ la solution existe et est unique. Dans le cas $h = h_0$, il existe nécessairement un point de la frontière où la vitesse devient égale à celle de la propagation du son.

Les démonstrations de théorèmes d'existence et d'unicité des solutions dépendent essentiellement du lemme suivant: le système étant fortement elliptique, les caractéristiques log V

et α satisfont à un système d'équations aux dérivées partielles elliptiques quasi-linéaires du premier ordre. Les recherches de caractère géométrique permettent de généraliser le principe de LINDELOF-MONTEL à une sous-classe très vaste de représentations quasi-conformes. Pour le cas de la dynamique des gaz, ce principe a le sens suivant: si la frontière y_1 s'élève et si la vitesse du courant est inférieure à celle du son, le débit h restant constant, toutes les lignes de courant s'élèvent et la vitesse du courant diminue sur toute la frontière inférieure. Ce principe est surtout important dans les études du mouvement d'un milieu gazeux à surface libre.

Mais il faut remarquer que les problèmes pesant à surface libre sont beaucoup plus difficiles non seulement dans le cas des gaz, mais encore dans le cas de liquides parfaits. Il s'agit de problèmes où la frontière n'est pas déterminée d'avance, mais que l'on doit déterminer à partir de quelques conditions spéciales. Voici un example; étant donnés une courbe Γ_0 et le système d'équations (1), on doit construire une autre courbe Γ qui détermine avec Γ_0 un domaine D tel que la représentation quasi-conforme du domaine D sur un domaine Δ, fixé d'avance, jouisse de la propriété suivante:

$$\varphi(x,\ y,\ u,\ v,\ u_x,\ u_y,\ v_x,\ v_y) = 0$$

partout sur la courbe Γ. Ce problème contient le problème classique de KIRCHHOFF sur le contournement hydrodynamique des corps avec détachement des jets, de même que la recherche du mouvement stationaire d'un liquide pesant possèdant une surface libre.

Ces problèmes sont encore bien loin d'être complètement résolus malgré les grands efforts qui leur ont été consacrés.

II. Quelques problèmes nouveaux

Je parle de problèmes classiques, d'une classe de problèmes nouveaux avec des nombreuses applications en physique, géologie, etc. Ce sont les problèmes nommés "incorrects". Il s'agit en particulier des représentations quasi-conformes qui correspondent à des systèmes d'équations aux dérivées partielles du type hyperbolique et du problème de CAUCHY pour des systèmes elliptiques. Les problèmes aux limites pour les équations de type mixte attirent un intérêt tout à fait exceptionnel. Ces problèmes contiennent comme cas particulier l'étude du mouvement du milieu gazeux dans des tuyères. Dans diverses parties de la tuyère les vitesses peuvent être inférieures ou bien supérieures à celle de la propagation du son.

On peut constater qu'à toutes les époques les succès scientifiques les plus considérables et les faits nouveaux les plus intéressants ont été obtenus dans les domaines dont on s'est peu occupé. Je crois que c'est justement le cas des problèmes spéciaux pour les systèmes d'équations aux dérivées partielles du premier ordre. Les résultats obtenus pendant les dernières années permettent d'espérer qu'on peut obtenir dans ce domaine, des théorèmes mathématiques suffisamment généraux et ayant des applications importantes en hydrodynamique. Les calculatrices sont appelées à jouer un grand rôle dans ces recherches.

Un des problèmes les plus importants de la théorie de la propagation des ondes dans des liquides est le problème suivant posé par STOKES. On produit une perturbation dans un bassin illimité. Il faut déterminer la surface à ciel ouvert de ce liquide à un moment arbitraire t. Ce problème a été résolu par Stokes lui-même, qui a su présenter la solution sous forme d'une intégrale. D'ailleurs c'est seulement de nos jours que l'on a pu obtenir des résultats concrets dans le cas où il y a des singularités au fond du bassin. Par example, supposons que pour une perturbation donnée, on ait au fond du bassin une chaîne de montagnes sous-marines de forme cylindrique. Alors, en faisant diverses hypothèses sur la chaîne de montagne, on peut obtenir les lois asymptotiques suivantes (dans la direction de la chaîne)

$$\zeta \sim \varkappa R^{-1/2}, \quad \zeta \sim \varkappa R^{-1/3}, \quad \zeta \sim \varkappa R^{-1/4},$$

ζ étant l'amplitude maximale des ondes, R leur distance de la perturbation initiale, $\varkappa = \text{const}$. Si le fond du bassin est plan, alors

$$\zeta \sim \varkappa R^{-1}.$$

Ce résultat est étroitement lié au problème de zounami. Dans cet ordre d'idées, il me paraît intéressant d'étudier l'influence des singularités au fond du bassin sur la propagation des ondes. On peut créer des fonds de basins où les ondes seront concentrées dans un endroit fixé d'avance, ou bien seront dispersées.

Je me permets de formuler encore trois problèmes essentiellement spatiaux.

1. On demande de démontrer l'existence et l'unicité des solutions du problème de la superposition de deux systèmes d'ondes se propageant dans deux directions différentes (sans faire appel à la linéarisation des équations).

2. Etant donné une surface Γ_0 différentiable un nombre suffisant de fois $z = z_0(x, y)$.

On demande de construire une autre surface $\Gamma: z = z(x, y)$ telle que, dans le mouvement de liquide dans le couche limitée par Γ_0 et Γ, en chaque point de la surface Γ la vitesse V du courant doit être constante, $V = 1$.

Le premier problème admet une solution pour des ondes suffisament petites à condition que les directions de leur propagation soient suffisamment proches l'une de l'autre.

Dans le deuxième problème, en empruntant des méthodes de la théorie de la représentation quasi-conforme, on peut démontrer le théorème d'existence et d'unicité lorsque Γ_0 est presque cylindrique dans la direction perpendiculaire à la vitesse du mouvement à l'infini.

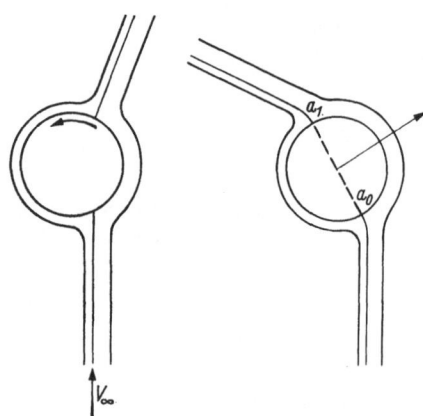

Fig. 3. Boule dans un jet aérien.

3. Avant d'exposer le troisième problème nous voudrions montrer une expérience. Voici une boule légère qui se trouve dans un jet aérien. La position de la boule set stable. Je pense que le phénomène est connu depuis des milliers d'années. D'ailleurs jusqu'à présent on ne savait pas l'expliquer de façon théorique.

Voici un modèle mathématique qui, en première approximation, peut être considéré comme la théorie de ce phénomène. Dans le cas d'un liquide parfait, la boule immobile et le jet aérien étant donné, il existe une infinité de solutions. Pour obtenir une solution unique, il suffit de fixer le point de formation du jet sur la surface de la boule opposé au point de la ramification du jet. D'ailleurs, si l'on considère la viscosité du liquide (quelle que petite qu'elle soit), on s'aperçoit sans peine que le point de ramification du courant et celui de la formation du jet doivent être des points opposés de la boule. La solution du problème devient unique. La déviation du courant produit une force qui ramène la boule vers l'axe du courant.

III. Nouveaux modèles mathématiques des problèmes classiques

On est obligé jusqu'a présent de construire des nouveaux modèles mathématiques des phénomènes hydrodynamiques pour d'une part pouvoir élucider les théories à moitié empiriques et obtenir de façon rigoureuse des résultats de caractère descriptif et d'autre part prévoir des phénomènes nouveaux. On peut se servir dans divers cas avec succès de la méthode du recollement des solutions pour des domaines partiels.

Cette méthode, quand elle est bien employée, donne de bonnes coïncidences aves les expériences. Voici un exemple. Considérons un courant à deux dimensions (mouvement plan). Supposons que le fond du bassin coïncide avec l'axe des x sauf le segment $(0, 1)$ où l'on a une cavité, dont la profondeur est de deux ou trois unités. On demande de déterminer le champ des vitesses. Il y a trois moyens classiques pour résoudre ce problème: 1. Les composantes des vitesses satisfont aux conditions de Cauchy-Riemann (le liquide étant supposé parfait). 2. Le courant est constant pour $y > 0$, tandis que dans la cavité le liquide ne bouge pas (le modèle de Kirchhoff). 3. On suppose avoir un contournement stationnaire satisfaisant aux équations de Navier-Stokes. Ces trois modèles conduisent quelquefois à des résultats éloignés de la réalité.

Il est plus souvent préférable d'adopter le modèle suivant. On décompose le domaine tout entier en quatre domaines partiels au moyen de courbes γ_1, γ_2, γ_3. La courbe γ_1 réunit les bords de la cavité, les autres courbes partagent la cavité en trois morceaux. Au-dessus de γ_1, nous avons un courant potentiel (laminaire). Dans tous les morceaux de la cavité on a un mouvement en circulation constante. Les courbes et les circulations doivent être choisies de telle façon que la vitesse du courant soit continue dans tout le domaine considéré.

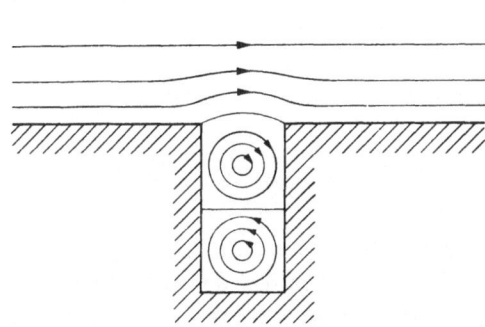
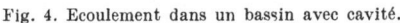
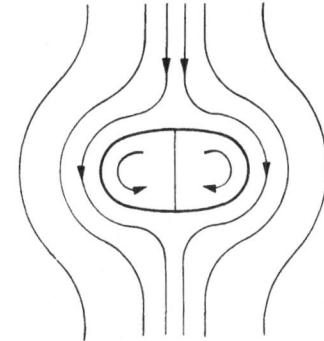

Fig. 4. Ecoulement dans un bassin avec cavité.   Fig. 5. Mouvement des anneaux de fumée.

De la même façon on peut obtenir une théorie approximative du mouvement des anneaux de fumée.

Parmi les problèmes de la mécanique pour lesquels on doit construire de nouveaux modèles mathématiques on doit citer en premier lieu le problème de la turbulence. Dans ce domaine il reste beaucoup de choses inattendues quoique l'on dispose d'une quantité immense d'informations empiriques que l'on ne parvient pas à systématiser dans le cadre d'une théorie suffisante.

IV. Les paradoxes des grandes vitesses et des grandes pressions

J'ai parlé de nouveaux problèmes de mathématiques et de leurs applications à l'hydrodynamique.

Maintenant, je veux parler de nouveaux problèmes de la mécanique des milieux continus liés aux phénomènes des explosions et qui paraissaient pradoxaux à première vue.

Un des problèmes les plus importants de l'artillerie est celui des obus perforants. Des milliers de laboratoires situés dans tous les pays ont effectué des dizaines de milliers d'expériences à ce sujet. A la fin du siècle passé, on a obtenu une formule empirique qui permet de calculer l'épaisseur maximale d'une couche perforée. Cette épaisseur doit être exprimée comme fonction du calibre, de la masse et de la vitesse de l'obus. La formule classique est

$$L = k\,\frac{m}{d^2}\,v^{1+\alpha},$$

m, d et v étant la masse, le calibre et la vitesse de l'obus. Les coefficients k et α dépendent de la forme et de la matière de l'obus et de la cuirasse. Vers la fin de la seconde guerre mondiale on a créé des obus à action cummulative. Cet objet ayant rencontré un obstacle, effectue une espèce de nouveau tir. Il envoie vers l'obstacle un nouvel obus ayant la forme d'un fil de fer de diamètre de 1—4 millimètres, d'une longueur de 200—800 millimètres. Ce fil possède une vitesse de trois à huit kilomètres par seconde. L'étude de la formulation et de l'action de ces obus a conduit à une série de paradoxes dont quelques-uns ne sont pas expliqués jusqu'à présent. Je veux signaler deux paradoxes parmi les plus frappants: 1. Au lieu de la formule (1) on a la formule (2)

$$L = l$$

l étant la longueur du fil de fer. L'erreur de cette formule ne dépasse pas quelques pour-cents.
2. On a mesuré la vitesse du fil de fer v_1 devant l'obstacle, la vitesse moyenne v_2 à l'intérieur de l'obstacle et sa vitesse derrière l'obstacle v_3 on a obtenu les relations suivantes:

$$v_1 = 2v_2 = v_3.$$

Un autre effet d'un grand intérêt technique est l'effet suivant.

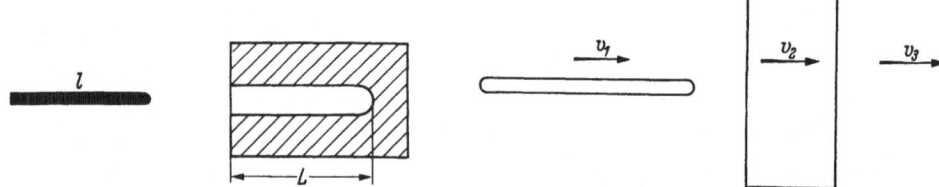

Fig. 6. Obus à action cumulative.   Fig. 7. Différentes vitesses du fil de fer.

Considérons deux couches métalliques situées vis-à-vis l'une de l'autre. Lançons la couche supérieure vers la couche inférieure au moyen d'une explosion. Les deux couches seront soudées. On peut souder de cette façon des couches de métaux différents qui ne peuvent être soudés par aucun autre moyen. La surface de la soudure a d'habitude un caractère ondulatoire. La longueur des ondes et leur amplitude peuvent être déterminées d'avance à partir des paramètres de l'expérience: épaisseur des couches, leur distance mutuelle, puissance de l'explosion, etc. Il est remarquable que l'on n'obtient pas de soudure si les couches métalliques se trouvent en contact avant l'explosion.

On obtient une explication théorique des phénomènes décrits en s'appuyant sur la théorie des liquides parfaits. Comme le mouvement du fil de fer à l'intérieur du corps métallique s'effectue avec la vitesse de quelques kilomètres par seconde, les pressions développées deviennent telles que la solidité et la viscosité du milieu sont négligeables en comparaison des forces d'inertie. De plus, la matière peut être considérée comme incompressible. Cela veut dire que la pénétration du jet d'un fil de fer dans un milieu métallique peut être considéré comme la pénétration d'un jet liquide dans un milieu liquide parfait. Il suffit donc de considérer le problème bien connu du choc de deux jets liquides en symétrie axiale. Les mêmes considérations peuvent expliquer l'effet de soudure des couches métalliques à l'aide d'explosions. L'effet de formation des ondes sur la surface de soudure est un problème plus compliqué. Pour le résoudre il faut employer des méthodes mathématiques modernes et les théorèmes de stabilité.

Parfois, il est important que le sol rejeté par l'explosion soit distribué d'une façon fixée d'avance. Pour cela il faut distribuer les matières explosives d'une façon tout à fait spéciale. Il n'y a pas longtemps, on a construit un barrage de 80 mètres pour défendre la ville d'Alma-Ata des torrents venant des montagnes à l'aide de deux explosions énormes. Les efforts unis des ingénieurs et des mathématiciens ont permis d'élaborer une théorie pour prédire d'une façon satisfaisante les actions de différentes explosions.

Depuis longtemps, on connaît la loi approchée de similitude des explosions:

Les dimensions linéaires des cavités obtenues sont proportionnelles au cube des charges. Quand les charges deviennent assez grandes (plus de 500 tonnes), cette loi cesse d'être rigoureuse. Les considérations citées plus haut montrent que le cube de la charge doit être remplacé par la puissance $3^1/_2$; cette nouvelle loi se trouve en accord avec l'expérience. Pour conclure, je veux encore considérer un groupe de problèmes d'une grande importance scientifique et technique. Il s'agit de l'étude de la stabilité des solutions des problèmes de la mécanique des milieux continus. Tout d'abord, il faut citer les problèmes de la destruction des corps solides, en particulier des roches. Ce problème est connu depuis plus de cent ans.

Imaginons qu'il se produise une explosion à une profondeur determinée dans un demi-espace formé de roches solides; on demande de calculer les dimensions et les distributions spatiales des débris des roches rejetées par l'explosion. Du point de vue pratique le problème inverse est surtout important: déterminer la distribution des charges de telle façon que la partie principale des débris ait des dimensions fixées d'avance. Le rapport de ces problèmes aux problèmes de stabilité peut-être élucidé par la considération de la destruction d'une barre sous l'action d'une pression appliquée à une extrémité.

Considérons le problème classique d'EULER : étant donnée une barre verticale qui est fixée d'un côté et qui est libre de l'autre côté. Une pression P verticale agit sur la barre de son côté libre.

P étant inférieure à la pression critique P_k la barre étant rectiligne reste stable. Dès que P devient supérieure à P_k l'équilibre de la barre devient instable. La barre devient sinusoïdale (1/2 de la période).

Lorsque P devient n fois plus grande que P_k l'état de la barre devient de nouveau instable. Elle prend une forme sinusoïdale mais seulement elle contient un nombre d'ondes de l'ordre de $\sqrt{n}$. On peut observer des effets analogues en sou-

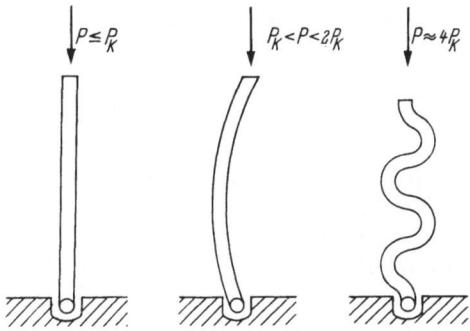

mettant un cylindre vide à une pression qui dé- passe n fois la pression critique. Ces effets peuvent être observés dans les expériences et être expliqués de façon théorique.

Supposons maintenant que la barre du raison- nement précédent se rompt lorsque sa courbure dépasse une grandeur fixée d'avance. Il est évident alors que sous l'action d'une pression qui dépasse n fois la pression critique, la barre doit se rompre, à peu près, en $\sqrt{n}$ morceaux.

Un problème encore plus difficile est celui de la translation des masses gazeuses ou liquides liés à la formation des cyclones.

Fig. 8. Courbes caractéristiques en fonction de P.

Les méthodes développées pendant les dernières années pour résoudre les problèmes de la dynamique de l'atmosphère à l'aide des calculatrices, présentent un grand succès pour le problème de la prédiction du temps mais il faut remarquer que cette méthode est valable seulement pour le cas du mouvement stable.

Dans le cas d'instabilité du mouvement, le problème réside d'abord en la détermination de la nature de l'instabilité et de son caractère. Ce problème a attiré plus de mille articles scien- tifiques. Malheureusement, beaucoup de déclarations affirmant que le problème est résolu n'ont pas été vérifiées. J'ai signalé plusieurs exemples pour lesquels la théorie des liquides parfaits permet de décrire non seulement les mouvements des milieux visqueux mais encore les mouve- ments des milieux solides, par exemple, les mouvements des milieux remplis de roches ou de métaux.

Je voudrais encore signaler un problème de l'hydrodynamique pure : l'étude des mouvements de quelques espèces de poissons où l'on peut considérer le fluide comme un corps solide. La force qui produit le mouvement dépend des efforts locaux produits par tous les éléments du corps du poisson.

Je vous remercie sincèrement pour votre attention.

Waves and wave drag in stratified flows

By

J. W. Miles

University of California, La Jolla Calif.

1. Introduction

Waves in a stratified flow are a prominent and dramatic feature in the lee of a mountain range that stands athwart the prevailing winds, such as the Sierra Nevada in California or the Southern Alps in New Zealand. These waves produce striking cloud patterns, have been responsible for world records and fatal crashes in sailplanes, and may be responsible for a significant fraction of the momentum exchange between the Earth and its atmosphere. They are basically gravity waves, in the sense that they owe their existence to the buoyant force of gravity in a stably stratified fluid, and are excited by flow over an obstacle in such a fluid. Examples of wave-induced clouds on the lee sides of the southern Andes, the Sierra Nevada, and Mount Fujiyama are shown in Figs. 1–3, in inverse order of scale[1]. The first two patterns are quasi-two-dimensional, whereas the third (Fujiyama) is essentially three-dimensional. The first is of continental scale, in consequence of which the CORIOLIS force associated with rotation of the Earth might be a significant factor.

The generation of waves by a fixed obstacle in a stratified flow or, equivalently, a moving body in a stratified fluid at rest, was first studied analytically by LORD RAYLEIGH (1883) and LORD KELVIN (1886, 1887), although preliminary studies had been made by JOHN SCOTT RUSSEL (1844), who remarked:

I am not aware that this species of standing wave in moving water has ever before been made the subject of philosophical examination. But I conceive that its study is highly important, especially in a theoretical view, as the means of conveying sound elementary conceptions of wave motion, as exhibiting the transition from the phaenomena of water currents to those of water waves ..., and as affording a basis from which we may commence, with some prospect of success, the application of the known principles and laws of motion to the investigation of the difficult theory of waves.

KELVIN considered both the stationary waves produced on the surface of a running stream (a stratified flow in which the density drops discontinuously to a negligible value across the air-water interface or free surface) by either an elevation or depression of the stream bed (1886, 1887) and, subsequently (1904/1905, 1905/1906), the wave pattern produced by a moving obstacle, such as a ship. Both RAYLEIGH and KELVIN dealt explicitly with the fact that the solution for an obstacle in a running stream is unique for speeds in excess of the *critical velocity*, $\sqrt{gd}$ (where g is the acceleration of gravity and d is the depth of the water), but that this solution contains an arbitrary train of gravity waves for speeds inferior to $\sqrt{gd}$. RAYLEIGH resolved this difficulty by introducing artificial frictional forces, KELVIN by considering the antecedent conditions from which the motion might develop; both concluded that gravity waves should not appear upstream of the obstacle, a conclusion subsequently supported by formal solutions

[1] See COLSON (1952), HOLMBOE and KLIEFORTH (1957), and QUENEY et al. (1960) for additional photographs.

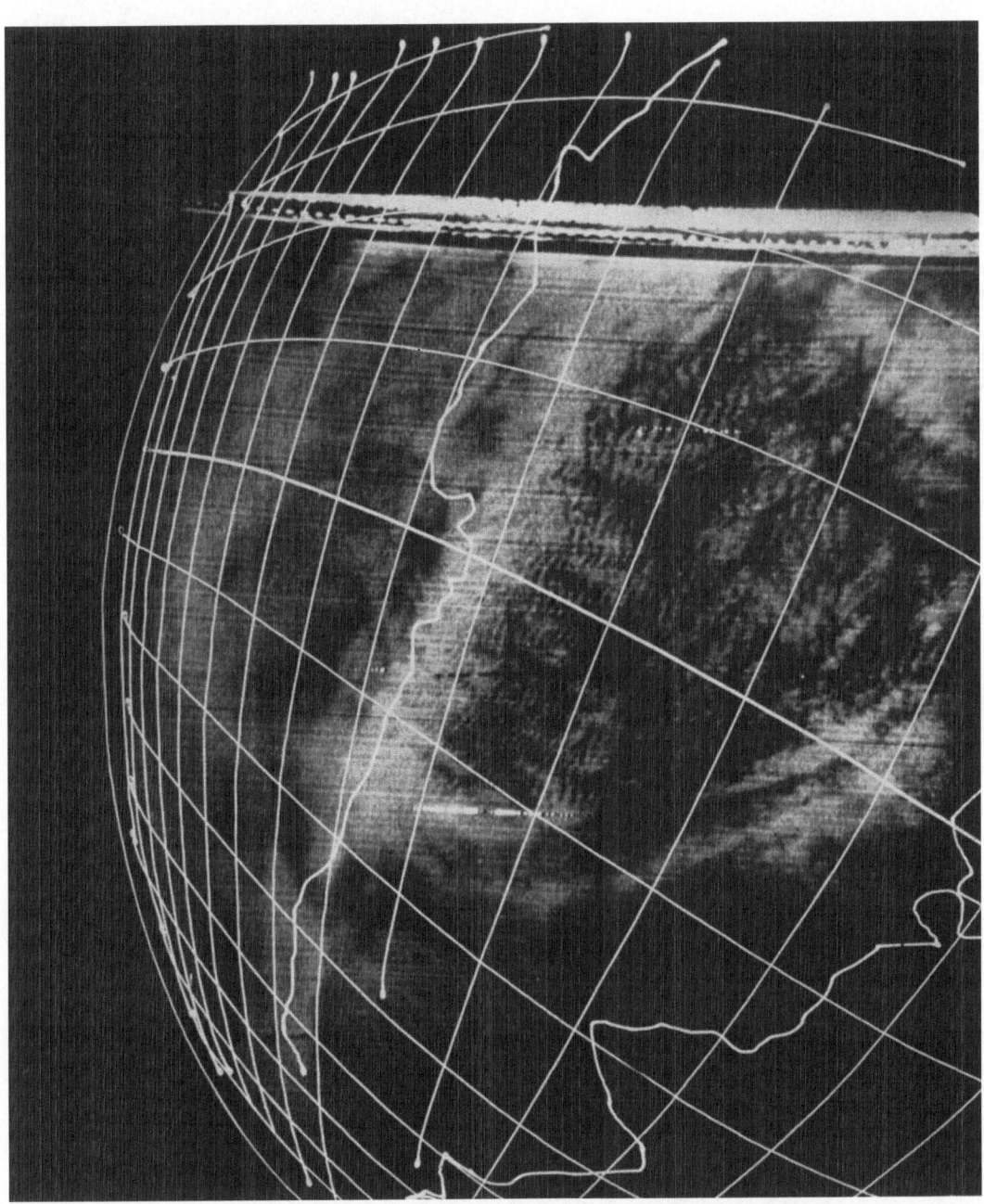

Fig. 1. Tiros-sattelite photograph of wave-induced clouds on the eastern side of the southern Andes. These waves are of continental
scale; nevertheless, they are basically gravity waves, although the effects of the Earth's rotation may not be negligible.

Fig. 2. Wave-induced clouds over Owens Valley in the lee of the Sierra Nevada (photographed by ROBERT SYMONS). The rising dust is in the updraft of a *rotor*.

Fig. 3. A crescent-shaped cloud induced by the essentially three-dimensional lee-wave field of Mount Fujiyama (ABE 1941).

of corresponding initial-value problems (HOILAND 1951, PALM 1953, STOKER 1953, WURTELE 1955)[1]. KELVIN also pointed out that the decisive role played by the critical velocity in the problem had been considered earlier by RUSSELL (1840) with special reference to the "Scottish system of 'fly-boat', carrying passengers on the Glasgow and Androssan Canal and between Edinburgh and Glasgow on the Forth and Clyde Canal, at speeds of from 8 to 12 or 13 miles an hour [2] by a horse, or a pair of horses, galloping along the bank." Russell's (1840) description of the original discovery of this basic phenomenon follows:

> As far as I am able to learn, the isolated fact was discovered accidentally on the Glasgow and Androssan Canal of small dimensions. A spirited horse in the boat of WILLIAM HOUSTON, Esq., one of the proprietors of the works, took fright and ran off, dragging the boat with it, and it was then observed, to Mr. Houston's astonishment, that the foaming stern surge which used to devastate the banks had ceased, and the vessel was carried on through water comparatively smooth, with a resistance very greatly diminished. Mr. HOUSTON had the tact to perceive the mercantile value of this fact to the Canal Company with which he was connected, and devoted himself to introducing on that canal vessels moving with this high velocity. The result of this improvement was so valuable in a mercantile point of view, as to bring, from the conveyance of passengers at a high velocity, a large increase of revenue to the Canal Proprietors. The passengers and luggage are conveyed in light boats, about sixty feet long and six feet wide, made of thin sheet iron and drawn by a pair of horses. The boat starts at a slow velocity behind the wave, and at a given signal it is by a sudden jerk of the horses drawn up on the top of the wave, where it moves with diminished resistance, at [a rate in excess of $\sqrt{gd}$].

Lee waves of geophysical interest, such as those produced by wind blowing over a mountain or tidal currents flowing over a submarine ridge, differ from lee waves on a running stream primarily in that the vertical stratification of the basic flow is gradual, rather than abrupt, and secondarily in consequence of shear (variation of flow speed with elevation). These differences imply that the critical speed may depend on both elevation and wavelength. A significant measure of the stratification associated with the density $\varrho(z)$, where z is the elevation, is the *intrinsic frequency* [see also (2.5) below]

$$N(z) = \left(- g \frac{d}{dz} \log \varrho \right)^{1/2}, \tag{1.1}$$

which may be identified as the angular frequency of small, vertical oscillations of a small parcel of fluid at a given elevation. The analog of the critical speed $\sqrt{gd}$ then is Nh, where h is an appropriate, vertical scale. The analog of $U/\sqrt{gd}$ is U/Nh, which is, in general, a function of z. The stratification is statically stable if and only if N is real.

The designation of physical parameters by the names of physical scientists is, I think, unfortunate. [For example, the intrinsic frequency N of (1.1) above appears to have been constructed independently by VÄISÄLÄ (1925), MILCH (1925), BRUNT (1927), and HESSELBERG (1929)[3]; it is typically designated as the VÄISÄLÄ frequency in Continental Europe, the BRUNT-VÄISÄLÄ frequency in the United Kingdom and the United States, and the VÄISÄLÄ-BRUNT frequency in the USSR.] Still, there would be considerable mnemonic virtue, and no great historical injustice, in designating the parameter U/Nh as the *Russell number*[4]. The principal

---

[1] The basic conclusion that waves should appear only upstream of an obstacle in a steady flow applies only to gravity waves or, more generally, waves for which the group velocity is less than the phase velocity; waves for which the reverse inequality holds appear upstream of the obstacle, as is evident from the capillary waves produced by a fish line (RAYLEIGH 1883).

[2] KELVIN, an early advocate for the adoption of the metric system in the United Kingdom, added the footnote that "One mile an hour is English and American reckoning of velocity, which, when not at sea, signifies 1.60933 kilometres per hour, or .44704 metre[s] per second."

[3] Dr. DRAZIN informs me that N appears in even earlier work of SCHWARZSCHILD, but I have been unable to document this.

[4] The designation *Froude number* for $U/\sqrt{g\,d}$ was introduced by M. WEBER (ROUSE 1957, p. 236), for whom the *Weber number* was subsequently named (WEBER also, but with greater historical justification, introduced the designation *Reynolds number* for Ul/ν). ROUSE (1957, p. 187) remarks that it is "One of the ironies of history ... that Froude's name has ... become inseparably associated with a law of similarity and a non-dimensional number, the first of which he did not originate, and the second of which he never even used. His very great contribution to boundary-layer research, on the contrary, is now seldom mentioned outside the field of naval architecture." BRESSE (1860; see ROUSE 1957, p. 169) appears to have been the first to use the parameter $U/\sqrt{g\,d}$ explicitly.

similarity parameters of geophysical fluid dynamics would then form a vowel-ordered set, namely:

$$Ra = -N^2 h^4 / \gamma \nu, \tag{1.2}$$

$$Re = Uh/\nu, \tag{1.3}$$

$$Ri = (N/U')^2 \qquad (U' \equiv dU/dz), \tag{1.4}$$

$$Ro = U/fh, \tag{1.5}$$

and

$$Ru = U/Nh, \tag{1.6}$$

where γ is the thermometric conductivity, ν the kinematic viscosity, and $\frac{1}{2}f$ the vertical component of the Earth's angular velocity. The Rayleigh number, Ra, is a ratio of buoyancy to diffusion effects (notice that $\varkappa$ and ν govern the diffusion of mass and momentum, respectively, and that $N^2 < 0$ is a necessary condition for free convection); the Reynolds number, Re, is a ratio of inertial to diffusion effects; the Richardson number, Ri, is a ratio of boundary to shear effects; the Rossby number, Ro, is a ratio of advective to Coriolis effects (both inertial); the Russell number, Ru, is a ratio of inertial to buoyancy effects. In fact, we find Nh/U analytically more convenient than U/Nh in the subsequent development. We also distinguish between horizontal and vertical length scales, so that h may take different values in (1.2)—(1.6).

The first, systematic observations of ascending currents on the lee sides of hills appear to have been those sponsored by the German Glider Research Institute, dating from 1928 (see QUENEY, CORBY, GERBIER, KOSCHMEIDER and ZIEREP 1960, p. 1) and those made by Count MASANAO ABE (1932, 1941) in his private observatory on Mount Fujiyama. KÜTTNER (1939a, b), who had made soaring flights as high as 8,000 m by 1937, and MANLEY (1945), who carried out an extensive field investigation in 1937—1939, appear to have been the first to identify these currents with lee waves, although the existence of such lee waves had been predicted theoretically by QUENEY (1936). KÜTTNER also called explicit attention to the similarity between these lee waves and those behind an obstacle in a running stream. The most extensive field study appears to have been the "Sierra Wave Project" (HOLMBOE and KLIEFORTH 1957), wherein lee-wave amplitudes of 2,000 m (peak-to-peak), vertical currents of 25 m/sec, and wave-induced clouds at altitudes of 30 km are reported. A survey of these and other observations, by both meteorologists and glider pilots, as well as of theoretical developments, has been given by QUENEY et al. (1960); see also the earlier survey by CORBY (1954). [The work of USSR scientists on lee waves is not covered in these surveys and is inadequately represented in the present paper; see KOZHEVHIKOV (1968), SOKHOV and GUTMAN (1968) and the Addendum below for partial bibliographies.]

The earliest theoretical work dealing explicitly with lee waves in stratified flows is that of QUENEY (1936, 1941, 1947, 1948) and LYRA (1940, 1943); it is discussed briefly in § 4 below. Both QUENEY and LYRA assumed that these waves should appear only downstream of the obstacle, as in the aforementioned surface-wave problem, in order to render their solutions unique. This *hypothesis of no upstream influence* appears to have been accepted by all writers on atmospheric lee waves with the exception of SCORER (1949, 1953, 1954, 1956, 1958), who appears to reject it. PALM (1953, 1958a), CORBY and SAWYER (1958), CRAPPER (1959), and FOLDVIK and WURTELE (1967) have provided additional, theoretical support for the hypothesis, including solutions for appropriate initial-value problems, and it is difficult to find any logical ground for Scorer's objections within the framework of a mathematical model that also rests on the hypothesis of small disturbances. It is certainly true (see below) that the flow upstream of an obstacle is influenced by sufficiently strong disturbances, but these effects are quite different than those predicted by SCORER on the hypothesis of small disturbances.

We limit our principal discussion to those aspects of the lee-wave problem that fall within the discipline of theoretical fluid mechanics and consider especially the *lee-wave régime*, in which the hypothesis of no upstream influence is well supported by theoretical considerations and both laboratory and field observations, and in which regular lee waves are the most prominent feature of the flow at points outside the immediate neighbourhood of the obstacle. The lower limit of this régime, in which $Nh/U \to 0$, corresponds to potential flow of a perfect fluid (see Figs. 5 and 12; viscous boundary layers are evident in Fig. 5, but there is little or no evidence of viscous separation; viscous separation might occur in the real-fluid counterpart of Fig. 12).

The upper limit of the lee-wave régime is marked by the appearance of local density inversions and flow reversals at some critical value, and closed streamlines at some higher value, of Nh/U. These regions are designated as *rotors* by meteorologists and are characterized by strong turbulence (see Figs. 7 and 8). They appear to be responsible for the large updraft of dust in Fig. 2, and LAWRENCE EDGAR, the pilot of a sailplane that was broken up during the exploration of the lee-wave field in this area on 25th April, 1955, reports that after drifting (in his parachute) a considerable distance to the East, at altitudes above 2,500 m, he was carried West at an estimated speed of 25 knots (13 m/sec) between the altitudes of 2,500 m and 1,200 m (ground level)[1].

We designate the régime in which closed streamlines appear as the *rotor régime*. Lee waves may still be a prominent feature in this régime (see Figs. 7 and 8), but turbulent energy dissipation in the rotors appears to decrease the lee-wave amplitudes (*vis-a-vis* the theoretical predictions based on the hypotheses of an ideal fluid and no upstream influence). Rotors also may induce viscous separation [SCORER (1955a) has suggested that eddies might be shed periodically from mountain crests; such a mechanism would yield a pattern similar to the Kármán vortex street].

The subsequent changes in the flow, as Nh/U increases, may lead to the formation of horizontal jets (see Fig. 9) or to phenomena resembling the hydraulic jump on a running stream (LONG 1954, 1956). Various hypotheses concerning such strong disturbances have been proposed [LONG (1954, 1962), QUENEY (1955), SCORER (1955a), SCORER and KLIEFORTH (1959), SCORER and WILSON (1963), KAO (1965)], and some theoretical progress has been made, but our understanding of his aspect of the problem is still quite imperfect.

The flow for very large values of Nh/U may be completely blocked upstream of the obstacle (see Fig. 10), and we designate the limit $Nh/U \to \infty$ as the *blocked-flow régime*. Some theoretical progress is possible in this limit, but the available solutions (see §§ 3 and 6 below) are singular, with implications that have not been quantitatively studied.

The lee waves generated by an obstacle transport momentum downstream; this momentum transport must be compensated by a downstream drag on the obstacle (Scorer's solutions, which admit upstream influence, may yield negative, or upstream, drag). The meteorological implications of this drag have been considered by MINTZ (1951), SAWYER (1959), and BLUMEN (1965a, b). SAWYER estimates that the wave drag associated with a distribution of 300 m hills should be comparable with the frictional drag over level terrain in a similar wind. This suggests that the wave drag on large mountain ranges, such as the Andes, the Sierra Nevada, and the Rocky Mountains in the Western hemisphere (each of which extends well into the troposphere and is roughly transverse to the prevailing westerlies from the Pacific) could dominate the momentum exchange between the solid Earth and its atmosphere (see MINTZ 1951 and MUNK and MACDONALD 1960).

We report here (§ 7) some calculations of wave drag on several cylindrical obstacles and one three-dimensional obstacle on the hypothesis of no upstream influence. This drag appears typically to be an increasing function of Nh/U, which is to say a decreasing function of speed. This paradoxical result reflects the fact that, as in the simpler example of the Scottish flyboat, an increase in speed may render the flow supercritical with respect to a certain portion of the lee-wave spectrum (which may be either continuous or discrete or both), but cannot render it subcritical with respect to any portion of that spectrum.

DRAZIN and MOORE (1967) obtained theoretical drag coefficients (drag divided by product of frontal area and dynamic pressure) of several thousand by retaining the hypothesis of no

---

[1] Edgar's accident is partially reported by HOLMBOE and KLIEFORTH (1957).

upstream influence in the rotor régime, but the maximum attainable wave-drag coefficient in the lee-wave régime (on the hypothesis of inviscid flow) appears to be roughly three (Huppert and Miles 1969). Rational calculations of the drag in the rotor and blocked-flow régimes are not available, but it is clear that the total drag in these régimes may be substantially influenced by such phenomena as viscous wakes and hydraulic jumps. These phenomena are essentially dissipative, but they also may act to increase the pressure on the lee side of the obstacle and may either increase or decrease the total drag.

The wave drag on a mountain in a stratified flow is similar in character to the wave drag on an obstacle moving through a stratified flow; however, the absence of the base plane in the latter problem implies the possibility of substantially lower drag by virtue of the ability of the fluid to move under, as well as over, the obstacle. The possibility of circulation in such problems presents a considerable difficulty, just as in the classical problem of homogeneous flow past a bluff body. The wave drag on a symmetrical body moving vertically through a stratified fluid has been considered by Warren (1960, 1961).

There is a close analogy between two-dimensional stratified flows and axisymmetric flows in a rotating fluid (Yih 1965, Chap. 6) and a mathematically exact analogy between two-dimensional flows in stratified and in rotating fluids (Trustrum 1964, Bretherton 1967). Long (1953b) has studied the problem of axisymmetric flow past an obstacle on the axis of a rotating fluid and provided experimental support for his hypothesis of no upstream influence for sufficiently small values of the inverse Rossby number[1]. Long also has conjectured that the upper limit of the lee-wave régime for such flows is determined by (an extension of) Rayleigh's (1916) criterion that the square of the circulation must be a monotonically increasing function of the cylindrical radius if the flow is to be stable. This criterion yields a critical inverse Rossby number, based on maximum body radius, of roughly two for both a sphere and a slender ellipsoid (Miles 1969).

2. Basic assumptions

We consider the steady flow of an ideal, incompressible, non-rotating fluid with preassigned, basic distributions of density and horizontal velocity, say $\varrho(z)$ and $U(z)$, and seek to determine the lee waves produced in this basic flow by, and the consequent drag on, an obstacle of prescribed shape. We define h as the height of the obstacle, l as a characteristic lateral scale ($l \equiv b$ for cylindrical obstacles of breadth $2b$ in §§ 4—6 below),

$$\varepsilon = h/l \tag{2.1}$$

as a characteristic slope or *slenderness ratio*, N_1 as a characteristic frequency, U_1 as a characteristic wind speed, and

$$\varkappa = N_1 h/U_1 \tag{2.2}$$

as the *reduced frequency* of the obstacle based on height.

The assumption of an ideal fluid requires

$$-Ra = N_1^2 h^4/\gamma \nu \gg 1 \tag{2.3}$$

and

$$Re = U_1 h/\nu \gg 1. \tag{2.4}$$

The restriction $Re \gg 1$ does not rule out the possibility of viscous separation, especially for a bluff body, but such separation may be less important in a stratified flow than in an unstratified flow by virtue of the tendency of a turbulent wake to collapse in a downstream distance of the order of U_1/N_1 (Schooley and Stewart 1963).

---

[1] W. Pritchard of Cambridge University has since observed flow past a sphere on the axis of a rotating fluid and concluded that upstream influence exists for all values of the Rossby number. It appears, however, that the observed effects could have been transient. (I am indebted to Dr. Brooke Benjamin for information on Pritchard's as yet unpublished work.)

Compressibility cannot be neglected in a real atmosphere; however, insofar as all velocities are small compared with the velocity of sound, we may incorporate compressibility in our calculations simply by replacing (1.1) by[1]

$$N = \left\{ \frac{g}{c_p} \frac{dS}{dz} \right\}^{1/2} = \left\{ \frac{g}{T} \left[\left(\frac{dT}{dz} \right) - \left(\frac{dT}{dz_a} \right) \right] \right\}^{1/2}, \tag{2.5}$$

where S denotes the entropy, c_p the specific heat of constant pressure, T the absolute temperature, and $-(dT/dz)_a$ the adiabatic lapse-rate. A typical value of N, based on a vertical lapse-rate of half the dry adiabatic, is 10^{-2} sec^{-1}.

The assumption of an ideal fluid also implies the neglect of condensation. In fact, condensation is not always negligible in a real atmosphere, as is evident from the cloud formations that often accompany lee waves. It may be especially significant in those regions in which local instability is incipient and may be partially incorporated in our model by using the wet adiabatic lapse-rate for $-(dT/dz)_a$ in (2.5).

The neglect of rotational effects requires
either

$$Ro = U_1/fl \gg 1 \tag{2.6a}$$

or

$$\varkappa Ro = \varepsilon N_1/f \gg 1, \tag{2.6b}$$

where

$$f = 2\Omega \sin \varphi \tag{2.7}$$

is the *Coriolis parameter* at latitude φ and Ω is the angular velocity of the Earth. The CORIOLIS force is dominated by inertial forces if (2.6a) is satisfied and by buoyancy forces if (2.6b) is satisfied. The Coriolis parameter is typically of the order of 10^{-4}, whereas N_1 is typically of the order of 10^{-2}, so that (2.6b) is typically satisfied. We remark that the proper lateral scale for the lee waves is U_1/N_1 and that $\varepsilon < 1$ for obstacles of geophysical interest; accordingly, (2.6b) implies (2.6a) in the primary context of the subsequent development, although it is possible that Coriolis forces could have a cumulative effect on a lee-wave train of continental scale, such as that of Fig. 1.

The dynamical similarity parameters for our mathematical model remain as ε, $\varkappa$, and an appropriately defined Richardson number. We proceed on the hypothesis that

$$Ri(z) \equiv \{N(z)/U'(z)\}^2 \gg 1 \tag{2.8}$$

throughout the flow. This hypothesis, which implies that the dynamical effects of shear are negligible, is generally a much stronger condition than is necessary for dynamic stability. A sufficient condition for stability of the basic flow with respect to *infinitesimal disturbances* is $Ri(z) > 1/4$ throughout the flow (HOWARD 1961, MILES 1961), but a rational extension of this condition for finite-amplitude disturbances is not available. We discuss this question further in the following section.

It remains to consider the assumption of a steady flow. The existence of a steady flow implies the existence of a well-posed, initialvalue problem, the asymptotic solution of which is both stable with respect to a non-trivial class of *small* disturbances and unique. It is possible that the asymptotic solution might not be unique in the sense that different initial conditions could imply different steady flows. The transition between two different, steady flows then could be initiated by a disturbance of finite amplitude. This question has been only partially explored, but it appears that a unique, steady flow typically exists for sufficiently small $\varkappa$ (see §§ 3 and 5 below) and that our basic assumptions are at least adequate for a qualitative description of such a flow, with viscous separation as the most serious departure of the real flow from its mathematical model. It is an open question as to whether our basic model is adequate for even a qualitative description of these flows in which closed streamlines appear ($\varkappa > \varkappa_c$ in § 5 below); it is almost certainly inadequate for $\varkappa \gg 1$.

---

[1] This equivalence is discussed in most textbooks on dynamic meteorology; see especially ECKART (1960, § 24).

3. Two-dimensional flows

We now consider (see Fig. 4) a two-dimensional flow over a cylindrical obstacle of pre-scribed cross section, say C, in a half-space, say $z \geq 0$, on the basis of the assumptions discussed in the preceding section. Let z_0 be the elevation of a streamline in the basic flow and $z_0 + \delta(x, z)$ its elevation in the disturbed flow, such that

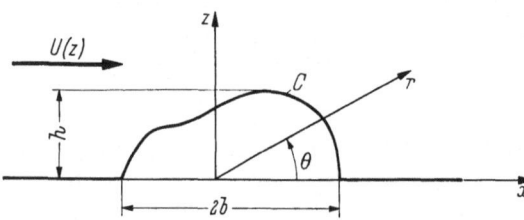

$$z - \delta(x, z) = z_0 \qquad (3.1)$$

along a streamline. The hypotheses of incompressibility and steady flow imply the existence of a stream function,

Fig. 4. The geometrical configuration of §§ 3 – 6.

$$\psi(z_0) = \int_0^{z_0} U(\xi)\, d\xi, \qquad (3.2)$$

from which the velocity may be derived according to

$$\vec{v} = \{\psi_z, -\psi_x, 0\} \qquad (3.3\text{a})$$

$$= U(z_0)\, \{1 - \delta_z, \delta_x, 0\}, \qquad (3.3\text{b})$$

where subscripts denote partial differentiation. These hypotheses also imply that density is conserved along streamlines:

$$\varrho = \varrho(z_0). \qquad (3.4)$$

We observe that ψ is a single-valued function of z_0, but that z_0 is a single-valued function of ψ only if $U(z_0)$ is positive definite. The equations of motion can be reduced to (DUBREIL-JACOTIN 1937, LONG 1953a)

$$(\vec{v} \cdot \triangledown) \left\{ p + \tfrac{1}{2}\, \varrho \vec{v}^2 + \varrho g z \right\} = 0 \qquad (3.5)$$

and

$$(\vec{v} \cdot \triangledown) \left\{ \triangledown^2 \psi + (d \log \varrho / d\psi) \left[\tfrac{1}{2}\, (\triangledown \psi)^2 + g z \right] \right\} = 0, \qquad (3.6)$$

where p is the pressure, and $\triangledown$ is the $\{x, z\}$-gradient operator. Substituting (3.1) and (3.2) into (3.6) and dividing the result through by $U\, |\vec{v}|$, we obtain

$$(\partial/\partial s) \left\{ \triangledown^2 \delta + (q'/q) \left[\delta_z - \tfrac{1}{2}\, (\triangledown \delta)^2 \right] + k^2 \delta \right\} = 0 \qquad [U(z_0) \neq 0], \qquad (3.7)$$

where s is a curvilinear coordinate measured along a streamline,

$$k = N(z_0)/U(z_0), \qquad q = \tfrac{1}{2}\, \varrho(z_0)\, U^2(z_0), \qquad (3.8\text{a, b})$$

and

$$q' \equiv dq/dz_0.$$

We now invoke the hypothesis that the basic flow remains undisturbed at a sufficient distance upstream of the obstacle and the supplementary hypothesis that every streamline originates in the basic, upstream flow. Integrating (3.5), we obtain the Bernoulli equation

$$p + \tfrac{1}{2}\, \varrho \vec{v}^2 + \varrho g z = p_h(z_0) + q(z_0) + \varrho g z_0, \qquad (3.9)$$

where

$$p_h(z) = p_h(0) - g \int_0^z \varrho(\xi)\, d\xi \qquad (3.10)$$

is the hydrostatic pressure. Integrating (3.7), we obtain

$$\triangledown^2 \delta + (q'/q) \left[\delta_z - \tfrac{1}{2}\, (\triangledown \delta)^2 \right] + k^2 \delta = 0 \qquad [U(z_0) \neq 0]. \qquad (3.11)$$

The requirements that the flow follow the lower boundary ($z_0 = 0$) and that no waves appear upstream of the obstacle yield the boundary conditions

$$\delta(x, z) = z \quad \text{on} \quad C \tag{3.12}$$

and

$$\delta(x, z) = O(r^{-1/2}) \qquad \left(r \to \infty, 0 \leq \theta < \frac{1}{2}\pi\right) \tag{3.13a}$$

$$= o(r^{-1/2}) \qquad \left(r \to \infty, \frac{1}{2}\pi < \theta \leq \pi\right), \tag{3.13b}$$

where r and θ are polar coordinates (a two-dimensional *wave* decays like $r^{-1/2}$ in a half-space). We infer from (3.13a) that $\delta_r = O(r^{-1/2})$ and $r^{-1}\delta_\theta = O(r^{-3/2})$ as $r \to \infty$ (although not uniformly with respect to $\left|\theta - \frac{1}{2}\pi\right|$), in consequence of which the lee-wave velocity field is *transverse* with respect to the radius $\vec{r}$.

The boundary-value problem posed by (3.11)—(3.13) is well set provided that $\varrho(z_0)$ and $U(z_0)$ are positive definite; accordingly, it has a unique solution within the limitations of the underlying hypotheses. There is, however, no a *priori* guarantee that this solution is stable. A rigorous discussion of the stability of finite-amplitude disturbances would be a formidable task, but it is at least plausible that a necessary condition for such stability is that the density (or, for a compressible fluid, the entropy) be a non-increasing function of elevation:

$$\frac{\partial \varrho}{\partial z} = \left(\frac{\partial \varrho}{\partial z_0}\right)(1 - \delta_z) \leq 0. \tag{3.14}$$

Invoking the requirement that the basic flow be statically stable, i.e. that $d\varrho/dz_0 < 0$ everywhere, we infer that

$$\delta_z \leq 1 \tag{3.15}$$

is a necessary condition for stability (LONG 1955, 1959). It is also likely to be a sufficient condition if (2.8) is satisfied, for shear instability then is precluded.

We remark that $\delta_z > 1$ implies not only that the density is locally increasing with elevation, but also (see (3.3b)) that there is a local reversal of the flow; closed streamlines arise at somewhat larger values of k, in consequence of which the assumption that all streamlines originate in the basic, upstream flow is violated. Moreover, the instability associated with the local density inversions is likely to render the flow at least locally turbulent (see discussion in § 1).

We designate the régime in which (3.15) is satisfied by a solution of (3.11)—(3.13) as the *lee-wave régime* and contrast it with the *blocked-flow* régime that obtains in the limit $\varkappa \to \infty$ and $R_i \to \infty$. The term $k^2\delta$ in (3.7) then dominates the preceding terms, and $\partial\delta/\partial s \to 0$ or, equivalently,

$$\delta \to \delta(z) \qquad (\varkappa \to \infty, Ri \to \infty). \tag{3.16}$$

The result (3.16) is the equivalent of the Taylor-Proudman theorem for rotating flows and reflects the tendency of stable stratification to inhibit vertical motion (YIH 1959). It yields the blocked flow

$$\vec{q} = \{u(z), 0, 0\} \qquad (z > h) \tag{3.17a}$$

$$= 0 \quad (z < h) \tag{3.17b}$$

for an obstacle of height h; this flow clearly violates the hypothesis that the basic flow remains undisturbed at a sufficient distance upstream of the obstacle. It also implies the existence of a singular layer near $z = h$.

BRETHERTON (1967) has studied the transient motion associated with the impulsively started translation of a circular cylinder in a stratified flow and has described the development of this singular layer (as well as the other singular layers that arise). The limiting form of the velocity is given by

$$U(z) \to U_0 z(z^2 - h^2)^{-1/2} \qquad (z > h, t \to \infty), \tag{3.18}$$

where U_0 is the velocity of the cylinder. It is clear that a significant boundary layer, in which both viscosity and heat conduction may be important, must exist in the neighbourhood of $z = h$. The character of this boundary layer does not appear to have been investigated analytically. There is clear evidence of a boundary layer adjacent to a region of blocked flow in Long's experiments; see Fig. 10, which is reproduced from his 1955 paper.

The lee-wave régime exists only for sufficiently small values of $\varkappa$, say $\varkappa < \varkappa_c$ (see § 5 below), the blocked-flow régime only for $\varkappa > \varkappa_b$, where $\varkappa_b \gg 1$. A completely satisfactory, mathematical model for $\varkappa_c < \varkappa < \varkappa_b$ has not been developed.

4. Small-disturbance models

The majority of the solutions to the boundary-value problem posed by (3.11)—(3.13) rest on the hypothesis of small disturbances, by virtue of which: (i) the term $\frac{1}{2}(\nabla \delta)^2$ may be neglected in (3.11); (ii) z_0 may be approximated by z in k and q; (iii) the left-hand side of (3.12) may be evaluated approximately at $z = 0$[1]. Invoking these approximations and introducing the change of variable

$$\delta(x, z) = \{q(z)/q(0)\}^{-1/2} \varphi(x, z) \tag{4.1}$$

in (3.11)—(3.13) and letting $z = \zeta(x)$ be the equation of the obstacle, we obtain

$$\nabla^2 \varphi + \{k^2(z) - \sigma(z)\} \varphi = 0, \tag{4.2}$$

$$\varphi(x, 0) = \zeta(x), \tag{4.3}$$

and

$$\varphi(x, z) = O(r^{-1/2}) \qquad \left(r \to \infty, 0 \leq \theta < \frac{1}{2}\pi\right) \tag{4.4a}$$

$$= o(r^{-1/2}) \qquad \left(r \to \infty, \frac{1}{2}\pi < \theta \leq \pi\right), \tag{4.4b}$$

where

$$\sigma(z) = q^{-1/2}(q^{1/2})'' \tag{4.5a}$$

$$\doteq U''/U, \tag{4.5b}$$

$q(z)$ and $k(z)$ are defined by (3.8), z_0 is approximated by z in each of q and k, and primes imply differentiation with respect to z. The approximation (4.5b) is essentially the *Boussinesq approximation*, in which the derivatives of $\varrho(z)$ are neglected except in the calculation of the buoyancy force. We emphasize, however, that the variation of the density still enters the problem through the transformation (4.1).

The two-dimensional, linearized boundary-value problem posed by (4.2)—(4.4) has been considered by LYRA (1940, 1943, 1952), QUENEY (1941, 1947, 1948), ZIEREP (1952, 1953, 1956), WURTELE (1953), CORBY (1954), CORBY and WALLINGTON (1956), HOLMBOE and KLIEFORTH (1957), WALLINGTON and PORTNALL (1958), PALM (1958), MERBT (1959), PALM and FOLDVIK (1960), QUENEY et al. (1960), SAWYER (1960), DÖÖS (1961), YIH (1965), GRAHAM (1966), and GRAHAM and GRAHAM (1966, 1967a, b). These studies differ principally in the assumed profiles for $\varrho(z)$ and $U(z)$. The problem also has been considered by SCORER (1949, 1953, 1954), but his solutions do not satisfy (4.4b); see the discussion in § 1 above.

An explicit, general solution of (4.2)—(4.4) may be constructed if the distributions of $\varrho(z)$ and $U(z)$ are such as to render $k^2 - \sigma$ constant, so that (4.2) reduces to the Helmholtz equation

$$\nabla^2 \varphi + k_1^2 \varphi = 0 \qquad (k^2 - \sigma \equiv k_1^2). \tag{4.6}$$

The simplest such model is an isothermal ($-\varrho'/\varrho$ constant) atmosphere with a uniform wind ($U' \equiv 0$), as studied originally in the present context by LYRA (1940, 1943); we designate it as *Lyra's model*. LYRA obtained the general solution to (4.6), (4.3) and (4.4) in a form equivalent to

$$\varphi(x, z) = \int_C \zeta(\xi) \, \varphi_1(x - \xi, z) \, d\xi, \tag{4.7}$$

---

[1] A detailed discussion of the approximations required for linearization is given by FOLDVIK (1968).

where $\varphi_1(x, z)$ is a dipole solution that exhibits the limiting behaviours

$$\varphi_1(x, z) \to (\pi r)^{-1} \sin \theta \qquad (k_1 r \to 0) \tag{4.8a}$$

$$\sim (2k_1/\pi r)^{1/2} \sin \theta \cos \left(k_1 r - \frac{1}{4} \pi\right) \qquad \left(k_1 r \to \infty, 0 < \theta < \frac{1}{2} \pi\right), \tag{4.8b}$$

$$\sim O(r^{-1}) \qquad \left(k_1 r \to \infty, \frac{1}{2} \pi < \theta < \pi\right). \tag{4.8c}$$

QUENEY (1948) obtained the solution to (4.6), (4.3) and (4.4) in a form equivalent to

$$\varphi(x, z) = \frac{1}{\pi} \mathcal{R} \int_0^\infty Z(\alpha) \exp \left\{i\alpha x + i(k_1^2 - \alpha^2)^{1/2} z\right\} d\alpha, \tag{4.9}$$

where

$$Z(\alpha) = \int_C \zeta(\xi) \, \varepsilon^{-i\alpha\xi} \, d\xi \tag{4.10}$$

is the Fourier transform of $\zeta(x)$, the operator $\mathcal{R}$ yields the real part of its operand, and the path of integration is indented under the branch point at $\alpha = k_1$. We remark that the lee-wave spectrum is confined to $\alpha = (0, k_1)$ — or, equivalently, to wavelengths greater than $2\pi/k_1$ — in the sense that the contributions from $\alpha > k_1$ to the integral of (4.9) are exponentially small as $r \to \infty$ with θ fixed. Approximating the integral over $\alpha = (0, k_1)$ by the method of stationary phase, we obtain

$$\varphi(x, z) \sim (2k_1/\pi r)^{1/2} \sin \theta \, \mathcal{R} \left\{Z(k_1 \cos \theta) \, e^{i(k_1 r - 1/4\pi)}\right\} \qquad \left(k_1 r \to \infty, 0 < \theta < \frac{1}{2} \pi\right). \tag{4.11}$$

There is no point of stationary phase if $\frac{1}{2} \pi < \theta < \pi$, in consequence of which there are no waves, and $\varphi = O(1/r)$, upstream of the obstacle.

LYRA (1943) evaluated (4.7) for a plateau ($\zeta = 0$ for $x < 0$, $\zeta = h$ for $x > 0$), corresponding to a point source at $x = 0$, and for a rectangle ($\zeta = h$ for $-b < x < b$), corresponding to point sources at $x = \pm b$. QUENEY (1948) evaluated (4.9) for the smooth profile (*Witch of Agnesi*) described by

$$\zeta(x) = h l^2 (x^2 + l^2)^{-1}. \tag{4.12}$$

MILES and HUPPERT (1969) obtained corresponding results for a semi-ellipse and an asymmetrical generalization there of.

LYRA (1952) also considered an isothermal atmosphere in which $U(z)$ is determined to render $k_1^2 \doteq k^2 - (U''/U)$ constant. The resulting profile for $U(z)$ is similar to that associated with the jet stream, but it is rather artificial in its dependence on $\varrho(z)$ and violates (2.8).

WURTELE (1953) considered an isothermal atmosphere in which U is linear in z. This model is of interest primarily because it yields a discrete spectrum of lee waves, in contrast to the continuous spectrum that is associated with the Helmholtz equation.

SCORER (1949, 1953, 1954), ZIEREP (1956), CORBY and SAWYER (1958), PALM (1958b), and PALM and FOLDVIK (1960) considered horizontally layered models of the atmosphere, in which $k^2 - \sigma$ takes different, constant values in different layers. Such models may possess both discrete and continuous spectra. SCORER (1949) states that some (continuous or layered) variation of $k^2 - \sigma$ is necessary for the existence of lee waves, presumably because he does not recognize the waves in the continuous spectra as lee waves; SCORER also admits waves upstream of the obstacle. PALM and FOLDVIK (1960) and DÖÖS (1961) considered models in which $U(z)$ and $T(z)$ are exponential.

QUENEY et al. (1960) gave an extended and comparative summary of the pre-1960 work.

A generalization of (4.9), which permits the introduction of more realistic profiles than those of Lyra's model, is provided by the WKB approximation

$$\varphi(x, z) = \frac{1}{\pi} \mathcal{R} \int_0^\infty Z(\alpha) \left\{\frac{k^2(0) - \sigma(0) - \alpha^2}{k^2(z) - \sigma(z) - \alpha^2}\right\}^{1/4} \exp \left\{i\alpha x + i \int_0^z [k^2(\xi) - \sigma(\xi) - \alpha^2]^{1/2} dz\right\} d\alpha. \tag{4.13}$$

This generalization does not appear to have been explored in the meteorological literature.

5. Finite-amplitude models

LONG (1953a) remarked that the nonlinear terms in (3.11) drop out if $q(z_0)$ and $\varrho'(z_0)$ are constant and that the resulting linear equation reduces to the Helmholtz equation

$$\nabla^2\delta + k^2\delta, \tag{5.1}$$

where

$$k = N/U \equiv (-g\varrho'/2q)^{1/2} \tag{5.2}$$

is constant. We designate the resulting model, including the boundary conditions (3.12) and (3.13), as *Long's model*. The essential mathematical feature of Long's model is that the linearity of (5.1) does not require the hypothesis of small disturbances. Other models for which finite-amplitude disturbances are governed by a linear differential equation have been suggested by LONG (1958), and YIH (1965), but they are less tractable. SCORER (1955b) and CLAUS (1964) have discussed similar models for compressible flows.

Long's model is of interest primarily because it permits an analytical examination of the effects of finite amplitude, with special reference to the question of stability. It also provides a good approximation to the flow of a stratified liquid over an obstacle in a channel (LONG 1955, 1959). On the other hand, the assumption that $q(z_0)$ is constant is not representative of real atmospheric flows.

Setting $z = \zeta$ in (3.12), we obtain the nonlinear boundary condition

$$\delta\{x, \zeta(x)\} = \zeta(x) \quad \text{on} \quad C, \tag{5.3}$$

which cannot be linearized without invoking the hypothesis of small disturbances. The boundary-value problem posed by (5.1), (5.3), and (3.13) cannot be solved explicitly with the same generality as the completely linearized problem, in which the left-hand side of (5.3) is replaced by $\delta(x, 0)$, but linear superposition of particular solutions to (5.1) remains possible.

LONG (1955) obtained a solution for an approximately sinusoidal obstacle in a channel of height H by first solving the linearized problem posed by (5.1) and the boundary conditions

$$\delta(x, 0) = \frac{1}{2} h\{1 + \cos(\pi x/b)\} \qquad (|x| < b) \tag{5.4a}$$

$$= 0 \qquad (|x| > b), \tag{5.4b}$$

$$\delta(x, H) = 0, \tag{5.4c}$$

and (3.13) and then substituting the resulting expression for $\delta(x, z)$ into (5.3) and determining $\zeta(x)$. This inverse method of solution is straightforward in principle, but the shape of the resulting obstacle depends on k. This disadvantage is relatively minor if ε is small, since the difference between $\delta(x, \zeta)$ and $\delta(x, 0)$ is $0(\varepsilon)$; on the other hand, it does prevent the calculation of the critical value of $\varkappa \equiv kh$ at which the stability criterion (3.15) is first violated for an obstacle of *prescribed* shape.

The counterpart of the asymptotic solution (4.11) for a channel is

$$\delta(x, z) \sim \mathcal{R} \sum_{n=1}^{n_k} A_n \exp\{i[k^2 - (n\pi/H)^2]^{1/2} x\}\sin(n\pi z/H), \tag{5.5}$$

where n_k, the number of independent lee-wave modes that can be excited in a channel of height H, is given by

$$n_k < kH/\pi < n_k + 1, \tag{5.6}$$

and A_n is the complex amplitude of the n'th mode. Long obtained inverse solutions for values of k corresponding to $n_k = 0, 1, 2, 3, 4$ and values of ε both smaller and larger than the critical value (Long's calculations yield the rough estimate $\varkappa_c \doteq 1/n_k$ provided that kH/π is not too close to an integer). LONG also carried out experiments, in which he obtained qualitative agreement (quantitative agreement for the wavelengths of the lee waves) with his theoretical calculations for $\varkappa < \varkappa_c$ (Figs. 5 and 6). The observed flows for values of $\varkappa$ somewhat larger than $\varkappa_c$ (Figs. 7 and 8) are similar to their theoretical counterparts, with local regions of turbulence

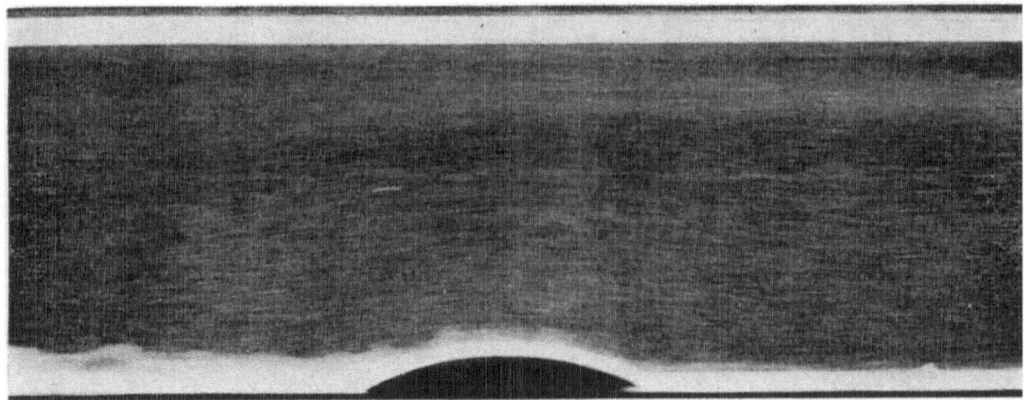

Fig. 5 a.

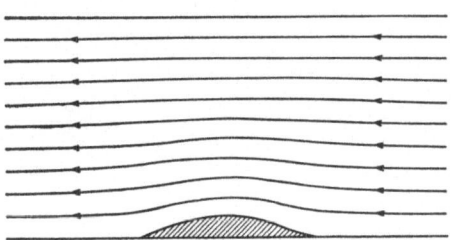

Fig. 5 b.

Fig. 5. Long's (1955) observed/calculated flow patterns for an approximately sinusoidal obstacle; $n_k = 0$, $\varkappa = 0.26/0.20$, and $\varepsilon = 0.22/0.25$. The flow is from right to left and is approximately irrotational. (Reproduced with permission of TELLUS.)

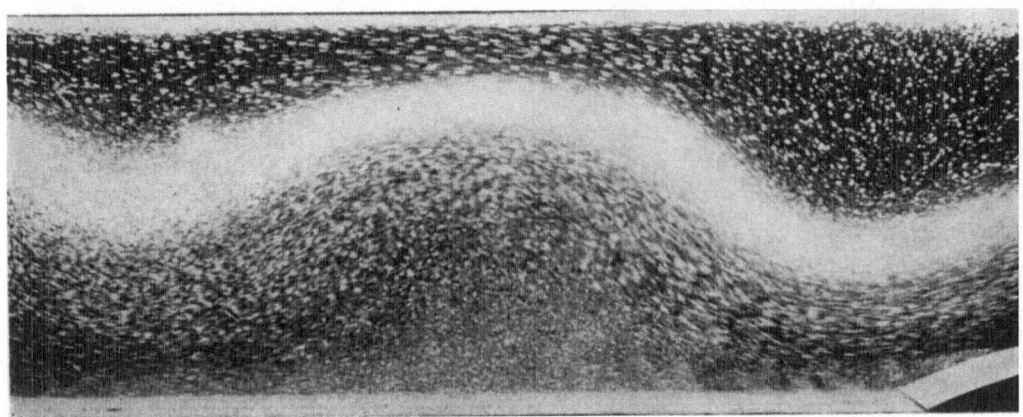

Fig. 6 a.

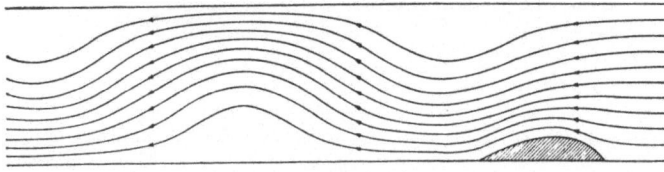

Fig. 6 b.

Fig. 6. Long's (1955) observed/calculated flow patterns for an approximately sinusoidal obstacle; $n_k = 1$, $\varkappa = 0.65/0.60$, and $\varepsilon = 0.27/0.38$. The flow is from right to left. (Reproduced with permission of TELLUS.)

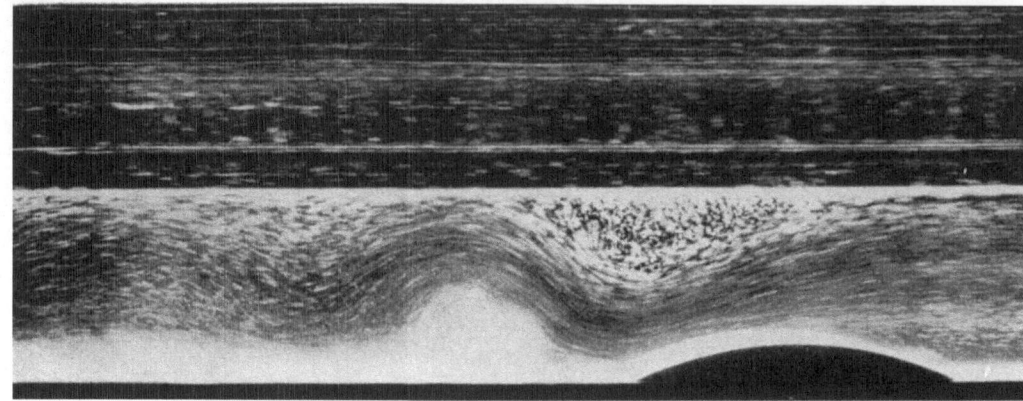

Fig. 7a.

Fig. 7b.

Fig. 7. Long's (1955) observed/calculated flow patterns for an approximately sinusoidal obstacle; $n_k = 1$, $\varkappa = 0.98/1.00$, and $\varepsilon = 0.23/0.20$. The flow is from right to left and contains both a well defined lee wave and a small rotor (adjacent to the upper boundary, just above the trailing edge in the observed flow). There is evidence of boundary-layer separation and turbulence below the first wave crest, and the subsequent wave crests are much weaker than their theoretical counterparts. (Reproduced with permission of TELLUS.)

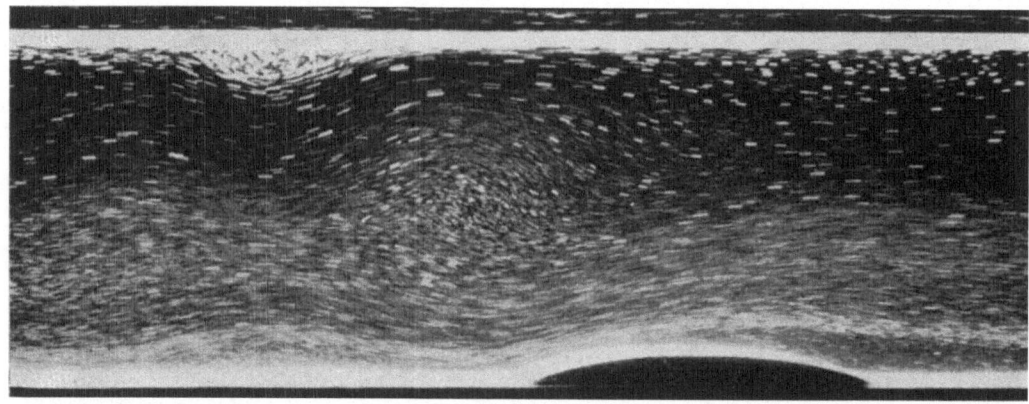

Fig. 8a.

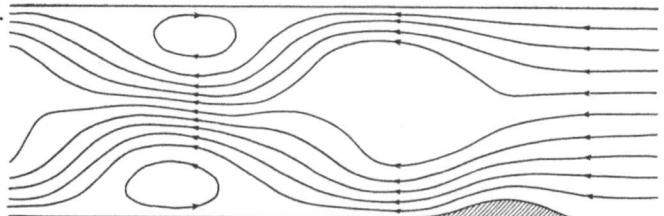

Fig. 8b.

Fig. 8. Long's (1955) observed/calculated flow patterns for an approximately sinusoidal obstacle; $n_k = 2$, $\varkappa = 0.65/0.70$, and $\varepsilon = 0.17/0.15$. The flow is from right to left and contains two rotors. The two lee-wave modes are much weaker than their theoretical counterparts. (Reproduced with permission of TELLUS).

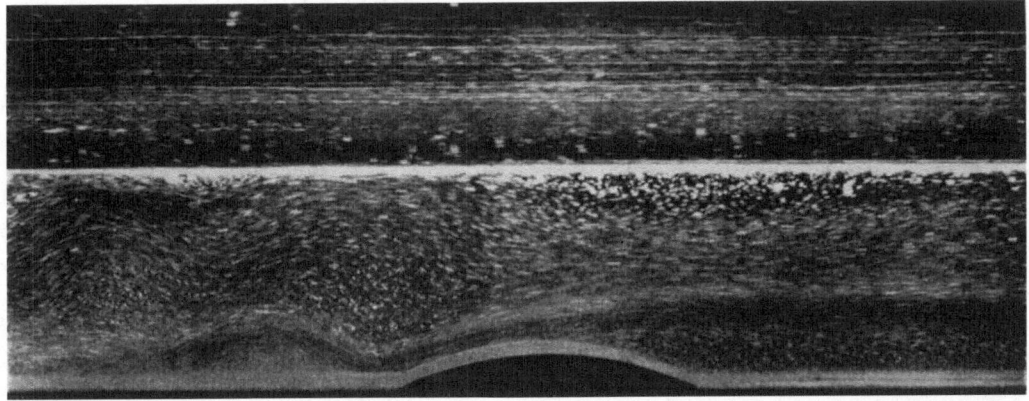

Fig. 9. Long's (1955) observed flow pattern for $n_k = 2$, $\varkappa = 1.17$ and $\varepsilon = 0.23$. The flow is from right to left and is partially blocked on the upstream side. There are turbulent eddies in the lee and a single, upstream jet. (Reproduced with permission of TELLUS.)

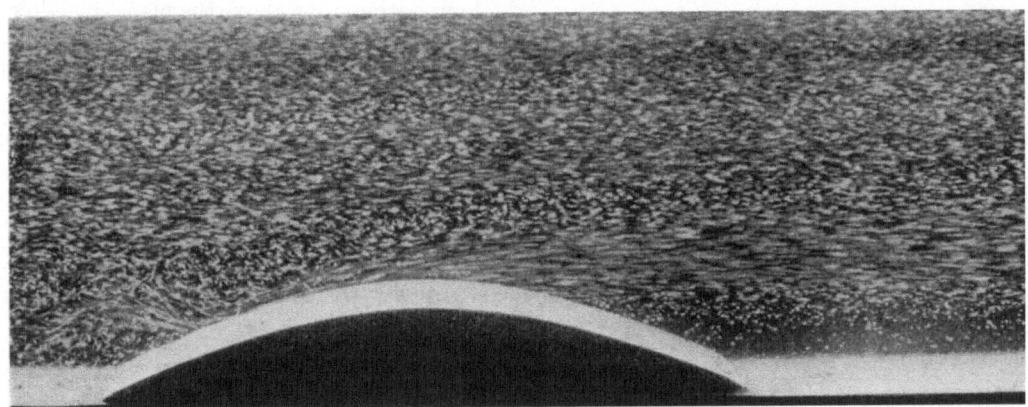

Fig. 10. Long's (1955) observed flow pattern for $n_k = 18$ $\varkappa = 8.82$ and $\varepsilon = 0.27$. The flow is from right to left and is completely blocked upstream of the obstacle. (Reproduced with permission of TELLUS.)

where the theory yields closed streamlines and lee waves of smaller amplitudes than those predicted by the theory. The observed flows for $\varkappa \gg \varkappa_c$ (Figs. 9 and 10) contain jets and regions of blocked flow upstream of the obstacle and are basically unsteady. This suggests that it may be impossible (with our present knowledge of turbulence) to fill the theoretical gap between the lee-wave régime ($\varkappa < \varkappa_c$) and the limiting régime ($\varkappa \to \infty$) of completely blocked flow.

DRAZIN and MOORE (1967) obtained a solution to Long's model for an infinitesimally thin, vertical plate of height h in a finite channel by matching Fourier expansions [in $\sin{(n\pi z/H)}$] of $\delta(x, z)$ on the two sides of the barrier. JONES (1967) obtained a formally exact solution for $h = \frac{1}{2} H$ (JONES also obtained an essentially numerical solution for a plateau). The results presented by DRAZIN and MOORE and by JONES typically contain closed streamlines; however, they did not discuss the question of stability, and the essentially numerical form of their solutions would render the determination of $\varkappa_c$ rather difficult. MILES (1968a) obtained a variational solution to Drazin and Moore's problem and concluded that $\varkappa_c$, based on barrier height, lies between 1 and 2 for $n_k = 1$.

MILES (1968a) also obtained a solution to Long's model for an infinitesimally thin, vertical plate in a half-space by separation of variables in elliptic-cylinder coordinates, which yields a complete, orthogonal set of lee-wave functions. The streamline pattern for the critical value of the reduced frequency, $\varkappa = \varkappa_c = 1.73$, is shown in Fig. 11. This pattern is qualitatively similar to that calculated by LYRA (1943) on the basis of (4.7) for a very narrow rectangle.

Solutions also may be obtained by separation of variables for semi-circular and semi-elliptical obstacles in a half-space, with both the former and the thin plate appearing as special cases of the latter. The individual lee-wave functions may be expressed as infinite series of elementary solutions to (5.1) and form a complete set, but they are orthogonal only in the limiting case of the infinitesimally thin plate. Miles (1968b) determined these functions for the semi-circular obstacle and obtained $\varkappa_c = 1.27$[1]. The streamline patterns for $\varkappa = 0.5$, 1, 1.3, 1.5, and 2 are shown in Figs. 12—16 (Huppert, Appendix to Miles 1968b), each of which is scaled for an obstacle of the same height as that of Fig. 11. The flow for $\varkappa = 0.5$ (Fig. 12) is

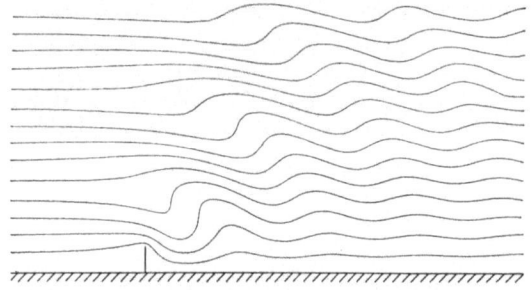

somewhat asymmetric, but otherwise departs only slightly from the corresponding potential flow. The flows for $\varkappa = 1.0$ and 1.3 (Figs. 13 and 14) are strongly asymmetric, and the amplitudes of the lee waves are comparable with the height of the obstacle. The critical point, at which the streamline achieves the vertical, is evident in Fig. 14 on the fourth streamline above the plane, just downstream from the obstacle. The flows for $\varkappa = 1.5$ and 2.0 (Figs. 15 and 16) contain regions of static instability and closed streamlines, although the contour interval is too

Fig. 11. The streamline pattern for flow over a thin, vertical barrier on the basis of Long's model for $\varkappa = \varkappa_c = 1.73$ (Huppert and Miles 1969).

coarse to reveal the closed streamlines in Fig. 15. Huppert and Miles (1969) obtained corresponding results for semi-elliptical obstacles; in particular, $\varkappa_c$ varies monotonically from 0.67 to 1.73 as ε varies from 0 to ∞. The streamline pattern for $\varepsilon = 0.3$ and $\varkappa = \varkappa_c = 0.93$ is shown in Fig. 17.

We obtain a general solution for Long's model simply by replacing k_1 by k, $\varphi(x, z)$ by $\delta(x, z)$, and $\zeta(\xi)$ by $f(\xi)$ in (4.7), where $f(x)$ is a dipole-distribution function:

$$\delta(x, z) = \int_C f(\xi)\, \varphi_1(x - \xi, z)\, d\xi. \tag{5.7}$$

Invoking the boundary condition (5.3), we obtain the integral equation

$$\zeta(x) = \int_C f(\xi)\, \varphi_1\{x - \xi, \varepsilon\zeta(x)\}\, d\xi \tag{5.8}$$

for the determination of $f(x)$. Substituting (4.8b) into (5.7) and carrying out the stationary-phase approximation to the result, we obtain the asymptotic lee-wave field in the form [cf. (4.11)]

$$\delta(x, z) \sim (2k/\pi r)^{1/2} \sin\theta \quad \mathcal{R}\{F(k\cos\theta)\, e^{i[kr - (\pi/4)]}\} \qquad (kr \to \infty), \tag{5.9}$$

where

$$F(\alpha) = \int_C f(x)\, e^{-i\alpha x}\, dx \tag{5.10}$$

is the Fourier transform of $f(x)$.

If ε is small, as it is in most cases of geophysical interest, we may seek a solution of (5.8) by expanding f and φ_1 about $\varepsilon = 0$. The first approximation yields

$$f(x) \to \zeta(x) \qquad (\varepsilon \to 0) \tag{5.11}$$

by virtue of (4.8a) and corresponds to Lyra's solution (4.7) in the context of Long's model. The second approximation has been obtained by Miles and Huppert (1969) for an obstacle that is not more blunt than a semi-ellipse, in which case $f(x) - \zeta(x)$ is $O(\varepsilon)$. The expansion of $f(x)$

---

[1] The semi-circular obstacle also has been studied by Kozhevnikov (1963, 1968) on the basis of Long's model. He used functions that satisfy the boundary condition on the obstacle identically and then determined the expansion coefficients to satisfy the condition of no upstream reflection approximately. This procedure appears to be less efficient than that described above, especially for the determination of $\varkappa_c$, and Kozhevnikov (1968) was able only to bound $\varkappa_c$ between 1 and 2.

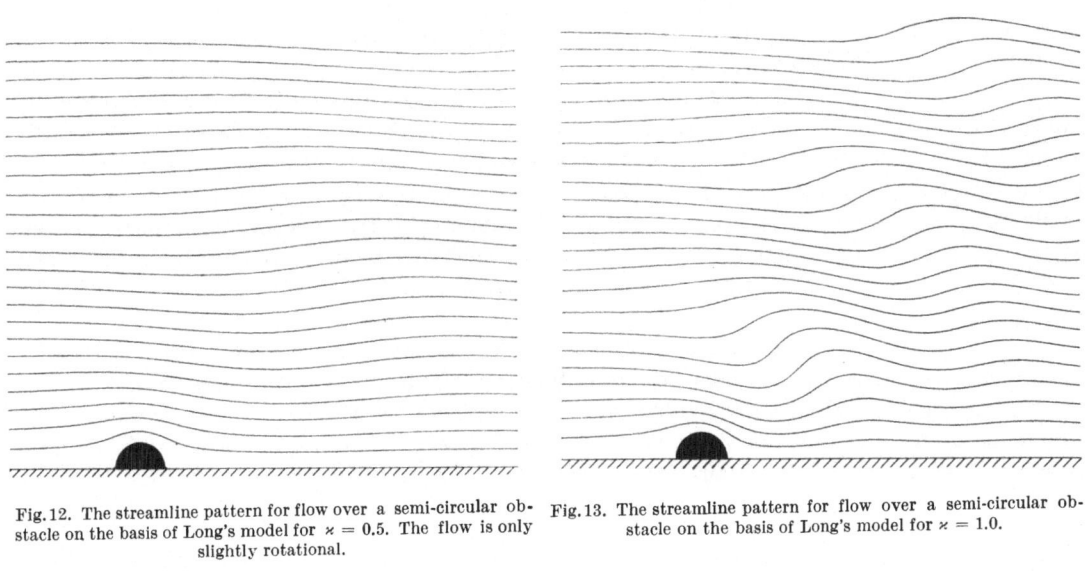

Fig. 12. The streamline pattern for flow over a semi-circular obstacle on the basis of Long's model for $\varkappa = 0.5$. The flow is only slightly rotational.

Fig. 13. The streamline pattern for flow over a semi-circular obstacle on the basis of Long's model for $\varkappa = 1.0$.

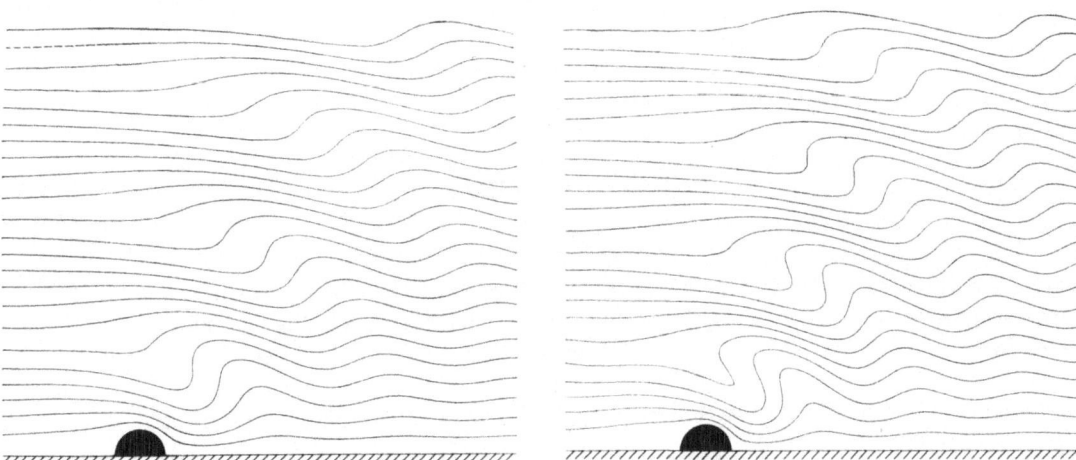

Fig. 14. The streamline pattern for flow over a semi-circular obstacle on the basis of Long's model for $\varkappa = \varkappa_c \doteq 1.3$.

Fig. 15. The streamline pattern for flow over a semi-circular obstacle on the basis of Long's model for $\varkappa = 1.5$ This flow contains density inversions in the regions of reversed flow and would be at least locally unstable.

Fig. 16. The streamline pattern for flow over a semi-circular obstacle on the basis of Long's model for $\varkappa = 2.0$. This solution is intended only to illustrate the unrealistic flows predicted on the hypothesis of no upstream influence with $\varkappa > \varkappa_c$; it would be quite unstable.

Fig. 17. The streamline pattern for flow over a semi-elliptical barrier for $\varepsilon = 0.15$ and $\varkappa = \varkappa_c \doteq 0.93$ (HUPPERT and MILES 1969).

9*

in ε is more difficult for an obstacle that is more blunt than an ellipse; for example, $f(x) - \zeta(x)$ is $0(\varepsilon \log \varepsilon)$ for a rectangular obstacle.

The expansion of $f(x)$ in ε, starting from the first approximation (5.11), is not uniformly valid in the neighbourhood of a stagnation point; in particular, $f(x)$ must vanish identically along a finite interval of the x-axis in such a neighbourhood. A systematic, uniformly valid expansion of the solution to (5.8) in powers of ε and $\log \varepsilon$ could be obtained for fixed kb by a modification of the method of HANDELSMAN and KELLER (1967); however, the approximations in the basic mathematical model scarcely warrant such an elaborate treatment, and we rest content with the remark that the flow in the neighbourhood of a stagnation point is locally potential if $\varkappa \ll 1$.

If $kl \ll 1$, where l is the maximum dimension of C, the flow is potential everywhere in the neighbourhood of the body. This potential flow constitutes an *inner approximation* to the solution of the original boundary-value problem as $k \to 0$ with r fixed. This inner approximation is not uniformly valid as $r \to \infty$, but we may continue it into $r \gg 1$ by invoking the approximation $|\xi| \ll |x|$ in (5.7) to obtain

$$\delta(x, z) \sim \mathcal{D}\delta_1(x, z) \qquad (k \ll 1), \qquad (5.12)$$

where

$$\mathcal{D} = \int_C f(x) \, dx = F(0) \qquad\qquad\qquad (5.13)$$

is the *dipole moment* of C with respect to a uniform potential flow in the half-space $z \geq 0$. The properties of $\mathcal{D}$ have been studied extensively (LAMB 1932, § 72a; PÓLYA 1947; MILES and HUPPERT 1969), and we note here only that

$$\mathcal{D} = A + (M/\varrho) \qquad\qquad\qquad (5.14\text{a})$$

$$\doteq A + \tfrac{1}{2}h^2, \qquad\qquad\qquad (5.14\text{b})$$

where A is the area under C, and M is the virtual mass of that area with respect to uniform, horizontal translation in an incompressible fluid of density ϱ. The approximation (5.14b) is exact for a semi-ellipse and generally within $10-20\%$ of the exact result for any finite obstacle.

Letting $k \to 0$ in (5.9) while holding kr fixed or, alternatively, substituting (4.8b) into (5.12), we obtain

$$\delta(x, z) \sim \mathcal{D}(2\pi k/r)^{1/2} \sin\theta \cos\left(kr - \frac{1}{4}\pi\right) \qquad (kr \to \infty, \, k \to 0), \qquad (5.15)$$

which we designate as the *Rayleigh-scattering approximation* (cf. RAYLEIGH 1897) to the lee-wave field.

We obtain an approximation to $f(x)$ that is uniformly valid as $k \to \infty$ with $\varkappa = kh$ fixed by determining the asymptotic approximation to $\varphi_1(x, z)$ as $k \to \infty$, which reduces (5.8) to the singular integral equation (MILES and HUPPERT 1969)

$$\zeta(x) = f(x) \cos\{k\zeta(x)\} + \frac{\sin\{k\zeta(x)\}}{\pi} \int_{-\infty}^{\infty} \frac{f(\xi) \, d\xi}{\xi - x}. \qquad (5.16)$$

The solution to (5.15) may be obtained by function-theoretic methods (MUSKHELISHVILI 1953, Chap. 2). The details are involved, and we report here only that the solution for $f(x)$ is uniformly valid with respect to k for $\varepsilon \ll 1$ (in particular, $f \to \zeta$ as $\varkappa \to 0$) and that the corresponding approximation to $\delta(x, z)$ implies

$$\max\{\delta_z\} > \varkappa \quad (\varepsilon \to 0). \qquad\qquad (5.17)$$

Invoking (3.15), we obtain

$$\varkappa_c < 1 \quad (\varepsilon \to 0); \qquad\qquad\qquad (5.18)$$

for example, $\varkappa_c = 0.67$ for a slender semi-ellipse.

6. Three-dimensional flows

A two-dimensional model is likely to be adequate for lee-wave generation by mountain ranges, such as the Sierra Nevada, but not for relatively isolated mountains, such as Fujiyama. The most important three-dimensional effect is the possibility of flow *around*, rather than *over*, an obstacle; this is especially significant at those large values of Nh/U for which a two-dimensional flow might be blocked.

Three-dimensional, stratified flow over an obstacle has been considered by SCORER (1956), SCORER and WILKINSON (1956), WURTELE (1957), PALM (1958b), CRAPPER (1959, 1962), and DRAZIN (1961). The first two papers are marred by incorrect boundary conditions (see comments in § 1). Only DRAZIN considered finite-amplitude effects.

WURTELE (1957) and CRAPPER (1959) considered the three-dimensional counterpart of Lyra's model, based on an isothermal atmosphere, a uniform wind, small disturbances, and the Boussinesq approximation. The resulting differential equation for the displacement, $\delta(x, y, z)$, of a stream surface relative to its position in the basic flow is

$$\frac{\partial^2}{\partial x^2}\left(\nabla^2\delta + k^2\delta\right) + k^2\frac{\partial^2\delta}{\partial y^2} = 0, \tag{6.1}$$

where ∇^2 is now the three-dimensional Laplacian operator, y is the transverse coordinate, and $k = N/U$ is constant.

WURTELE (1957) solved (6.1), subject to the condition of no upstream influence, for a three-dimensional, semi-infinite plateau of finite breadth (in the y-direction). Perhaps the most remarkable aspect of his results is the existence of crescent-shaped updraft areas that bear a qualitative resemblance to the crescent-shaped clouds frequently observed in the lee of isolated mountains (see Fig. 3). WURTELE also compared these wave patterns with the well known, wedge-shaped wake (KELVIN 1904/1905, 1905/1906) behind a ship. Similar comparisons were made by SCORER and WILKINSON (1956) and by PALM (1958b).

CRAPPER (1959) obtained the corresponding solution for a three-dimensional counterpart of the Witch of Agnesi with circular contours, namely

$$\zeta(x, y) = hl^3(x^2 + y^2 + l^2)^{-3/2}. \tag{6.2}$$

The corresponding cross-sectional area,

$$A(x) = \int_{-\infty}^{\infty} \zeta(x, y)\, dy = 2hl^3(x^2 + l^2)^{-1}, \tag{6.3}$$

is proportional to $\zeta(x)$ for the Witch of Agnesi, (4.12). CRAPPER concluded that the resulting lee waves are confined to a downstream strip of approximate width $2l$. He subsequently (1962) obtained three-dimensional solutions for more realistic distributions of $\varrho(z)$ and $U(z)$ and found that the lee waves then are typically confined to a wedge similar to that of the Kelvin wake. He also obtained solutions for obstacles with elliptical contours.

The general solution of (6.1), subject to the condition of no upstream influence, is given by

$$\delta(x, y, z) = \frac{1}{2\pi^2}\mathcal{R}\int_0^\infty dx \int_{-\infty}^\infty Z(\alpha, \beta)\exp\left\{i(\alpha x + \beta y) + i(k^2 - \alpha^2)^{1/2}(\alpha^2 + \beta^2)^{1/2}\alpha^{-1}z\right\}d\beta, \tag{6.4}$$

where

$$Z(\alpha, \beta) = \int_{-\infty}^\infty \int_{-\infty}^\infty \zeta(x, y)\, e^{-i(\alpha x + \beta y)}\, dx\, dy \tag{6.5}$$

is the Fourier transform of $\zeta(x, y)$. This generalization of Queney's solution (4.9) does not appear to have been explored systematically, although it is implicit in the work of CRAPPER (1959, 1962) and BLUMEN (1965a).

DRAZIN (1962) considered three-dimensional disturbances of finite amplitude and showed that the generalization of (3.11) is

$$(\nabla \times \vec{v}) \times \vec{v} + \left\{UU' + \frac{1}{2}(\varrho'/\varrho)(U^2 - \vec{v}^{\,2}) + N^2\delta\right\}\nabla z_0 = 0, \tag{6.6}$$

where z_0 is the elevation of a given stream surface far upstream of the obstacle, and ϱ and U are functions of z_0 alone. DRAZIN considered flows past a vertical cylinder and a hemisphere and attempted to obtain solutions as perturbation expansions in the reduced frequency $\varkappa$ (for $\varkappa \ll 1$) and $1/\varkappa$ (for $\varkappa \gg 1$). His results for $\varkappa \ll 1$ are of rather limited interest in that they are not uniformly valid at large distances from the obstacle and do not yield the lee-wave pattern (Drazin did not discuss this difficulty). His results for $\varkappa \gg 1$ yield a limiting flow around a hemisphere that is confined to horizontal planes and reduces to the basic flow in $z > h$ (where h is the height of the hemisphere). This limiting flow is the three-dimensional counterpart of the blocked flow described by (3.18), but differs from that flow in that the basic flow remains undisturbed in $z > h$. Neither approximation is uniformly valid near $z = h$, but infinite velocities appear in the three-dimensional solution only in the second approximation.

7. Wave drag

We calculate the wave drag, say D, on a cylindrical obstacle on the basis of Long's model. A momentum survey over a closed contour consisting of the obstacle boundary C, a semi-circle ($r = $ const.) that surrounds C, and connecting segments of the base plane ($z = 0$) yields

$$D = -\int_0^\pi (pr \cos \theta + \varrho u \psi_\theta)\, d\theta, \qquad (7.1)$$

where u is the horizontal component of $\vec{v}$, and $\varrho \psi_\theta\, d\theta$ is the mass flux across the semi-circle. Expressing ϱ, u, and ψ in terms of δ with the aid of (3.1)—(3.3), (3.9) and (3.10), invoking the assumptions that $q(z_0)$ and $\varrho'(z_0)$ are constant, letting $r \to \infty$, and invoking (3.13a, b), we tranform (7.1) to

$$D = q \lim_{r \to \infty} r \int_0^{1/2\pi} (\delta_r^2 + N^2 \delta^2) \cos \theta\, d\theta. \qquad (7.2)$$

Substituting (5.9) into (7.2) and introducing the change of variable $\alpha = k \cos \theta$, we obtain

$$D = (2q/\pi) \int_0^k |F(\alpha)|^2\, (k^2 - \alpha^2)^{1/2}\, \alpha\, d\alpha, \qquad (7.3)$$

which corresponds to results obtained by SAWYER (1959) and BLUMEN (1965a) in the context of Lyra's model. We obtain the corresponding result for the approximate solution of (4.13) by replacing q by $q(0)$, k by $\{k^2(0) - \sigma(0)\}^{1/2}$, and $F(\alpha)$ by $Z(\alpha)$ in (7.3).

The representation (7.3) exhibits the wave drag in terms of the power spectrum of the dipole-distribution, $|F(\alpha)|^2$, modified by a low-pass filter with the pass band $\alpha = (0, k)$. Letting $kl \to 0$ and invoking (5.12), we obtain the limiting form

$$D \to (2q/3\pi)\, \mathscr{D}^2 k^3 \qquad (kl \to 0), \qquad (7.4)$$

which implies that D is proportional to the square of the dipole moment of the obstacle and vanishes like $1/U$ as $U \to \infty$ with N fixed.

Substituting (5.10) into (7.3) and letting $k \to \infty$, we obtain

$$D \sim (2qk/\pi) \int_C \int_C f'(x)\, f'(\xi) \log |x - \xi|^{-1}\, d\xi\, dx \qquad (k \to \infty). \qquad (7.5)$$

[We infer from (7.5) that D vanishes like U as $U \to 0$ for fixed N and prescribed $f(x)$. Taken in conjunction with (7.4), this implies that the slender-body approximation ($f \doteq \zeta$) to D has a maximum with respect to U at some finite value of U. This inference is of only formal interest, since the joint limit $\varepsilon \to 0$, $kl \to \infty$ is singular. The results for $kl \to \infty$ with ε fixed imply that the drag typically increases with $1/U$ in the lee-wave régime (MILES and HUPPERT 1969).]

The aerodynamicist will not miss the analogy between (7.5) and both Prandtl's result for the vortex drag on a high-aspect-ratio wing and von Kármán's result for the wave drag on a slender body of revolution in a supersonic flow. We infer from this analogy that the variational problem of minimizing the wave drag on an obstacle of prescribed cross-sectional area and base in the ordered limit $\varepsilon \to 0$, $U \to 0$ yields a semi-ellipse, for which

$$D_{\min} \sim \pi q h \varkappa \qquad (\varepsilon \to 0, U \to 0). \qquad (7.6)$$

This result is of only formal interest, but it provides a convenient reference value for the drag on a slender obstacle.

A general result of more physical interest is that (7.3) is invariant under the transformation $f(x) \to f(-x)$. Observing that $\zeta(x) \to \zeta(-x)$ is equivalent to a reversal of the basic flow, we infer that D is invariant under such a reversal independently of the symmetry of the obstacle if $\varepsilon \ll 1$. We infer from (7.4) that this reverse-flow theorem holds also for $kl \to 0$ and non-small ε. It does not hold for $kl \to \infty$ with $\varkappa$ fixed unless the obstacle is actually symmetric.

The wave drag on a semi-ellipse of height h and base $2b$, relative to the reference value provided by (7.6),

$$D_* \equiv D/\pi q \varkappa h \equiv C_D/\pi \varkappa, \qquad (7.7)$$

is plotted as a function of kb in Fig. 18 with $\varepsilon = h/b$ as a parameter (HUPPERT and MILES 1969). It is clear from this plot that the linearized theory provides a reasonable approximation in $\varepsilon < 0.1$ and $kb < 3$, but is poor for all kb if $\varepsilon > 0.3$. The maximum value of the drag coefficient consistent with completely stable flow, namely C_D at $\varkappa = \varkappa_c$, is roughly 3 for all $\varepsilon < 1$.

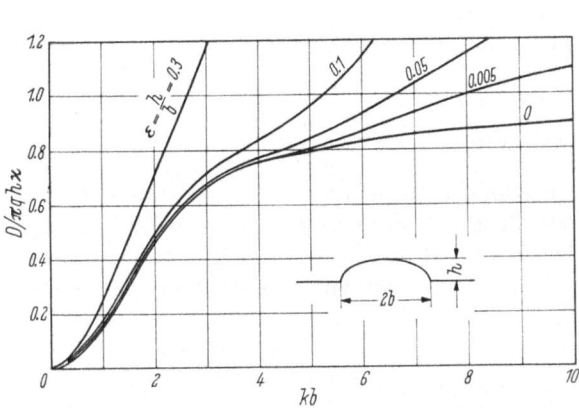

Fig. 18. The normalized wave drag on a semi-elliptical obstacle (HUPPERT and MILES 1969).

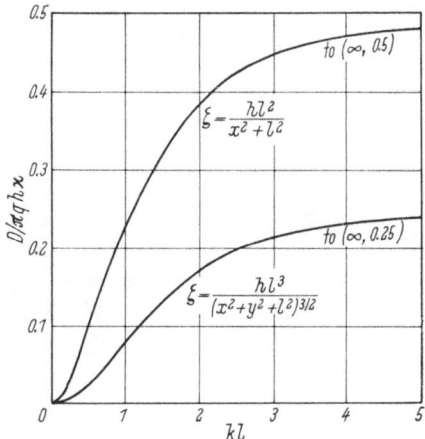

Fig. 20. The small-disturbance approximation to the normalized drag on a cylindrical obstacle with a Witch-of-Agnesi profile (4.12) and the corresponding, three-dimensional obstacle described by (6.2).

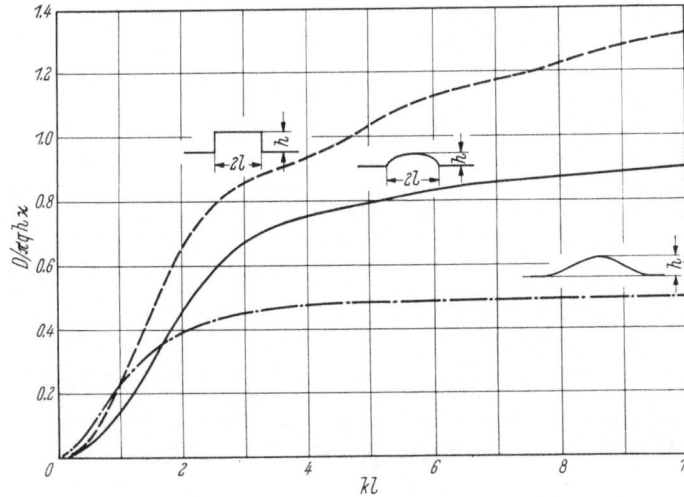

Fig. 19. The small-disturbance approximations to the normalized drag on a rectangular obstacle, a semi-elliptical obstacle, and the Witch of Agnesi (4.12).

The correspondingly normalized drags on a rectangular obstacle of height h and width $2l$ and on the Witch of Agnesi of (4.12) for $\varepsilon \ll 1$ are compared with that for a semi-ellipse in Fig. 19 (MILES and HUPPERT 1968). The drag on the rectangular obstacle is greater than that on the semi-ellipse for all kl in consequence of the infinitely steep shoulders. In fact, the disturbances induced by these shoulders are so strong that $\varkappa_c \to 0$ and $D = O(k \log k)$, rather than $0(k)$, as $kl \to \infty$. The drag on the Witch of Agnesi is larger than that on the semi-ellipse for sufficiently small kl, but is asymptotically smaller by a factor of two [this does not contradict (7.6), which refers only to obstacles of finite base] in consequence of the much gentler slope of the shoulders.

The drag on a three-dimensional obstacle for which (6.1) holds is given by (BLUMEN 1965a)

$$D = 4q \int_0^k \alpha^2 (k^2 - \alpha^2)^{1/2} \, d\alpha \int_{-\infty}^{\infty} (\alpha^2 + \beta^2)^{-1/2} \, |Z(\alpha, \beta)|^2 \, d\beta, \tag{7.8}$$

where $Z(\alpha, \beta)$ is the Fourier transform defined by (6.5). BLUMEN (1965a, b) evaluated (7.8) for the obstacle described by (6.2) and (6.3). We compare his result, in the normalized form [cf. (7.7)]

$$D_* = D/\pi q \varkappa A(0) \equiv C_D/\pi\varkappa, \tag{7.9}$$

with that for the Witch of Agnesi in Fig. 20. It appears from this comparison that the reduction in wave drag per unit frontal area in consequence of lateral flow around the obstacle for a given value of $\varkappa$ is always greater than 50% (the limiting reduction as $kl \to \infty$).

8. Conclusions

The hypothesis of no upstream influence provides a rational model for the calculation of the lee waves generated by, and the consequent drag on, an obstacle in the steady flow of an ideal, stratified fluid in a parametric régime that is limited by the appearance of closed stream-lines in the flow. Long's model, in which the dynamic pressure and density gradient are constant in the basic flow, permits the delineation of this régime, and of nonlinear effects within it, for two-dimensional flows. These results provide bounds for the parametric régime of models based on the additional hypothesis of small disturbances. Comparable results are not available for three-dimensional flows, but calculations such as those of CRAPPER (1962) may permit comparisons for families of three-dimensional obstacles that contain two-dimensional limits.

Rational models are not available for the calculation of stratified flows in that parametric régime in which the hypothesis of no upstream influence implies the existence of closed stream-lines. It may be that steady flows are impossible in this régime, at least in two-dimensions; moreover, two-dimensional models may be of limited significance for strongly stratified flows, in which flow *around* is likely to dominate flow *over* an obstacle.

Acknowledgment

This work was partially supported by the National Science Foundation, under Grant GA-849, and by the Office of Naval Research, under Contract Nonr-2216(29). I also am indebted to Professors R. R. LONG and M. G. WURTELE for several of the illustrations and to Dr. H. E. HUPPERT for some of the calculations.

References

ABE, M. 1932: The formation of cloud by the obstruction of Mount Fuji. Geophys. Mag., Tokyo 6, 1.
— 1941: Mountain clouds, their forms and connected air currents. Part II. Bull. Central Met. Observ. of Japan 7, 93.

BLUMEN, W. 1965a: A random model of momentum flux by mountain waves. Geofys. Publ. Norske Vid.-Acad. Oslo 26, No. 2.
— 1965b: Momentum flux by mountain waves in a stratified rotating atmosphere. J. Atmospheric Sci. 22, 529.

BRETHERTON, F. P. 1967: The time-dependent motion due to a cylinder moving in an unbounded rotating or stratified flow. J. Fluid Mech. 28, 545.

BRUNT, D. 1927: The period of simple vertical oscillations in the atmosphere. Quart. J. Roy. Meteor. Soc. 53, 30.

CLAUS, A. J. 1964: Large-amplitude motion of a compressible fluid in the atmosphere. J. Fluid Mech. 19, 267.

COLSON, D. 1952: Results of double-theodolite observations at Bishop. Bull. Amer. Meteor. Soc. 33, 107.

CORBY, G. 1954: The airflow over mountains. Quart. J. Roy. Meteor. Soc. 80, 491.

—, and J. S. SAWYER 1958: The air flow over a ridge—the effects of the upper boundary and high-level conditions. Quart. J. Roy. Meteor. Soc. 84, 25.

—, and C. E. WALLINGTON 1956: Airflow over mountains: the lee-wave amplitude. Quart. J. Roy. Meteor. Soc. 82, 266.

CRAPPER, G. D. 1959: A three-dimensional solution for waves in the lee of mountains. J. Fluid Mech. 6, 51.

— 1962: Waves in the lee of a mountain with elliptical contours. Phil. Trans., Roy. Soc. London (A) 254, 601.

DÖÖS, B. R. 1961: A mountain wave theory including the effect of the vertical variation of wind and stability. Tellus 13, 305.

DRAZIN, P. G. 1961: On the steady flow of a fluid of variable density past an obstacle. Tellus 13, 239.

—, and D. W. MOORE 1967: Steady two-dimensional flow of fluid of variable density over an obstacle. J. Fluid Mech. 28, 353.

DUBREIL-JACOTIN, M. L. 1937: Complément á une Note antérieure sur les ondes de type permanent dans les liquides hétérogénes. Atti. Accad. Lincei, Rend, Cl. Sci. Fis. Mat. Nat. 21, 344.

ECKART, C. 1960: Hydrodynamics of Oceans and Atmospheres. New York: Pergamon Press.

FOLDVIK, A. 1968: Linearization of the equations governing stratified shear flow with applications to finite amplitude flow over a barrier. Geophysical Institute, University of Bergen.

—, and M. G. WURTELE 1967: The computation of the transient gravity wave. J. R. astr. Soc. 13, 167.

GRAHAM, E. W. 1966: The two-dimensional flow of an inviscid density-stratified liquid past a slender body. Boeing Scientific Research Laboratories Document Dl-83-0550, Flight Sciences Laboratory Report No. 108.

—, and B. B. GRAHAM 1966: Further notes on the two-dimensional flow of an inviscid density-stratified liquid past a slender body. Boeing Scientific Research Laboratories Document Dl-82-0591, Flight Sciences Laboratory Report No. 112.

— — 1967a: The effect of Froude number on the two-dimensional flow of an inviscid density-stratified liquid past a slender body. Boeing Scientific Research Laboratories Document Dl-82-0614, Flight Sciences Laboratory Report No. 114.

GRAHAM, E. W. 1967b: The effect of a free surface on the two-dimensional flow of an inviscid density stratified liquid past a slender body. Boeing Scientific Research Laboratories Document Dl-82-0664, Flight Sciences Laboratory Report No. 121.

HANDELSMAN, R. A., and J. B. KELLER 1967: Axially symmetric potential flow around a slender body. J. Fluid Mech. 28, 131.

HESSELBERG, T. 1929: Die Stabilitätsbeschleunigung in Meere und in der Atmosphäre. Hydrgr. u. marit. Meteorol. 57, 273.

HOILAND, E. 1951: Fluid flow over a corrugated bed. (unpublished, but cited by WURTELE 1955; see also HOLMBOE and KLIEFORTH 1957).

HOLMBOE, J., and H. KLIEFORTH 1957: Investigation of mountain lee waves and the air flow over the Sierra Nevada. Department of Meteorology, University of California, Los Angeles.

HOWARD, L. N. 1961: Note on a paper of John W. Miles. J. Fluid Mech. 10, 509.

HUPPERT, H. E., and J. W. MILES 1969: Lee waves in a stratified flow. Part 3. Semi-elliptical obstacle. J. Fluid Mech. 35, 481.

JONES, O. K. 1967: Some problems in the steady two-dimensional flow of an incompressible, inviscid and stably stratified fluid. Ph. D. Thesis, University of Bristol.

KAO, T. W. 1965: The phenomenon of blocking in stratified flows. J. Geophys. Res. 70, 815.

KELVIN, LORD 1886, 1887: On stationary waves in flowing water. Phil. Mag. 22, 353, 445, 517; 23, 52; Mathematical and Physical Papers 4, 270.

— 1904/5: On deep water ship waves. Proc. Roy. Soc. Edin., June 20, 1904; Phil. Mag. 9, 733; Papers 4, 368.

— 1905/6: On deep sea ship waves. Proc. Roy. Soc. Edin., July 17, 1905; Phil. Mag. 11, 1; Papers 4, 394.

KOZHEVNIKOV, V. N. 1963: (cited by KOZHEVNIKOV 1968).

— 1968: Orographic perturbations in the two-dimensional stationary problem. Atmospheric and Oceanic Physics 4, 16—27.

KÜTTNER, J. 1939a: Zur Enstehung der Föhnwelle. Beitr. Phys. frei Atmos. 25, 251.

— 1939b: Moazagotl ud Föhnwelle. Beitr. Phys. frei Atmos. 25, 79.

LAMB, H. 1932: Hydrodynamics. Cambridge University Press.

LONG, R. R. 1953a: Some aspects of the flow of stratified fluids. I. A theoretical investigation. Tellus 5, 42.

— 1953b: Steady motion around a symmetrical obstacle moving along the axis of a rotating liquid. J. Meteor. 10, 197.

— 1953c: A laboratory model resembling the "Bishop Wave" phenomenon. Bull. Amer. Met. Soc. 34, 554.

— 1954: Some aspects of the flow of stratified fluids. II. Experiments with a two-fluid system. Tellus. 6, 97.

LONG, R. R. 1955: Some aspects of the flow of stratified fluids. III. Continuous density gradients. Tellus 7, 341.

— 1956: Models of small-scale atmospheric phenomena involing density stratification. Fluid Models in Geophysics. U. S. Govt. Printing Office, 135.

— 1958: Tractable models of steady-state stratified flow with shear. Quart. J. Roy. Meteor. Soc. 84, 159.

— 1959: The motion of fluids with density statification. J. Geophys. Res. 64, 2151.

— 1962: Velocity concentrations in stratified fluids. J. Hydraul. Div., Proc. ASCE 88, 9.

LYRA, G. 1940: Über den Einfluß von Bodenerhebungen auf die Strömung einer stabil geschichteten Atmosphäre. Beitr. Phys. frei Atmos. 26, 197.

— 1943: Theorie der stationären Leewellenströmung in freier Atmosphäre. Z. angew. Math. Mech. 23, 1.

— 1952: Bemerkungen über eine gewisse Klasse nicht-konstanter Anströmprofile beim Leewellenproblem. Beitr. Phys. Atmos. 31, 147.

MANLEY, G. 1945: The helm wind of Crossfell, 1937—1939. Quart. J. Roy. Meteor. Soc. 71, 197.

MERBT, H. 1959: Solution of the two-dimensional lee-wave equation for arbitrary mountain profiles, and some remarks on the horizontal wind component in mountain flow. Beitr. Phys. Atmos. 31, 152.

MILCH, W. 1935: Über Reibung und Austausch infolge von Turbulenz in der Atmosphäre. Ann. Hydr. u. marit. Meteorol. 53, 613.

MILES, J. W. 1961: On the stability of heterogeneous shear flow. J. Fluid Mech. 10, 496.

— 1968a: Lee waves in a stratified flow. Part 1. Thin barrier. J. Fluid Mech. 32, 549.

— 1968b: Lee waves in a stratified flow. Part 2. Semi-circular obstacle. J. Fluid Mech. 33, 803.

— 1969: The lee-wave régime in a rotating flow. J. Fluid Mech. (in press).

—, and H. E. HUPPERT 1969: Lee waves in a stratified flow. Part 4. Perturbation approximations. J. Fluid Mech. 35, 497.

MINTZ, Y. 1954: The geostrophic poleward flux of angular momentum. Tellus 3, 195.

MUNK, W. H., and G. J. F. MACDONALD 1960: The rotation of the Earth. Cambridge University Press, 135.

MUSKHELISHVILI, N. I. 1953: Singular Integral Equations. Groningen: P. Noordhoff.

PALM, E. 1953: On the formation of surface waves in a fluid flowing over a corrugated bed and on the development of mountain waves. Astrophys. Norveg. 5, 61.

— 1958a: Air flow over mountains: indeterminacy of solution. Quart. J. Roy. Meteor. Soc. 84, 464.

— 1958b: Two-dimensional and three-dimensional mountain waves. Geofys. Publ. 20, No. 3, 1.

—, and A. FOLDVIK 1960: Contribution to the theory of two-dimensional mountain waves. Geofys. Publ. 21, No. 6, 1.

PÖLYA, G. 1947: A minimum problem about the motion of a solid through a fluid. Proc. Nat'l. Acad. Sci. 33, 218.

QUENEY, P. 1936: Recherches Relatives a l'Influence du Relief sur les Eléments Météorologiques. La Météorologie, 334—53, 453—70.

— 1941: Ondes de gravité produites dans un courant aérien par une petite chaîne de montagnes. Comptes Rendus de l'Académie des Sciences de Paris 213, 588.

— 1947: Theory of Perturbations in Stratified Currents with Application to Airflow over Mountain Barriers. Department of Meteorology, University of Chicago, Misc. Report No. 23.

— 1948: The problem of air flow over mountains: a summary of theoretical studies. Bull. Am. Meteor. Soc., 29, 16.

— 1955: Rotor phenomena in the lee of mountains. Tellus 7, 367.

—, and G. CORBY, N. GERBIER, H. KOSCHMIEDER and J. ZIEREP 1960: The Airflow over Mountains. Geneva: World Meteorological Organization, Tech. Note 34.

RAYLEIGH, LORD 1883: The form of standing waves on the surfaces of running water. Proc. Lond. Math. Soc. 15, 69; Scientific Papers 2, 258.

— 1897: On the incidence of aerial and electric waves upon small obstacles. Phil. Mag. 44, 28; Scientific Papers 4, 305.

— 1916: On the dynamics of revolving fluids. Proc. Roy. Soc. London (A) 93, 148.

ROUSE, H., and S. INCE 1957: History of Hydraulics. State University of Iowa.

RUSSELL, J. S. 1840: Experimental researches into the laws of certain hydrodynamical phenomena that accompany the motion of floating bodies, and have not previously reduced into conformity with the known laws of the resistance of fluids. Trans. Roy. Soc. Edin. 19, 79.

— 1844: Report on waves. British Association Report, 311.

SAWYER, J. S. 1959: The introduction of the effects of topography into methods of numerical forecasting. Quart. J. Roy. Meteor. Soc. 85, 31.

— 1960: Numerical calculation of the displacements of a stratified airstream crossing a ridge of small height. Quart. J. Roy. Meteor. Soc. 86, 326.

SCHOOLEY, A. H., and R. W. STEWART 1963: Experiments with a self-propelled body submerged in a fluid with a vertical density gradient. J. Fluid Mech. 15, 83.

SCORER, R. S. 1949: Theory of waves in the lee of mountains. Quart. J. Roy. Meteor. Soc. 76, 41.

— 1953: Theory of airflow over mountains: II. The flow over a ridge. Quart. J. Roy. Meteor. Soc. 79, 70.

— 1954: Theory of airflow over mountains: III. Airstream characteristics. Quart. J. Roy. Meteor. Soc. 80, 417.

— 1955a: Theory of airflow over mountains: IV. Separation of flow from the surface. Quart. J. Roy. Meteor. Soc. 81, 340.

SCORER, R. S. 1955b: Theory of non-horizontal adiabatic flow in the atmosphere. Quart. J. Roy. Meteor. Soc. 81, 551.

— 1956: Airflow over an isolated hill. Quart. J. Roy. Meteor. Soc. 82, 75.

— 1958: Reply to Palm (1958a). Quart. J. Roy. Meteor. Soc. 84, 465.

—, and H. KLIEFORTH 1959: Theory of mountain waves of large amplitude. Quart. J. Roy. Meteor. Soc. 85, 131.

—, and M. WILKINSON 1956: Waves in the lee of an isolated hill. Quart. J. Roy. Meteor. Soc. 82, 419.

—, and S. D. R. WILSON 1963: Secondary instability in steady gravity waves. Quart. J. Roy. Meteor. Soc. 89, 532.

SOKHOV, T. Z., and L. N. GUTMAN 1968: The use of the long-wave method in the nonlinear problem of the motion of a cold air mass over a mountain ridge. Atmospheric and Oceanic Physics 4, 11—16.

STOKER, J. J. 1953: Unsteady waves on a running stream. Commun. Pure. Appl. Math. 6, 471.

TRUSTRUM, KATHLEEN 1964: Rotating and stratified fluid flow. J. Fluid Mech. 19, 415.

VÄISÄLÄ, V. 1925: Über die Wirkung der Windschwankungen auf die Pilotbeobachtungen. Soc. Scient. Fennica, Comm. Phys-Math II 19, 1.

WALLINGTON, C. E., and J. PORTNALL 1958: A numerical study of wavelength and amplitude of lee waves. Quart. J. Roy. Meteor. Soc. 84, 38.

WARREN, F. W. G. 1960: Wave resistance to vertical. motion in a stratified fluid. J. Fluid Mech. 7, 209

— 1961: The generation of wave energy at a fluid interface by the passage of a vertically moving slender body. Quart. J. Roy. Meteor. Soc. 87, 43.

WURTELE, M. G. 1953: Studies of lee waves in atomospheric models with continuously distributed static stability [unpublished; see QUENEY et al. (1960) for summary].

— 1955: The transient development of a lee wave. J. Marine Res. 14, 1.

— 1957: The three-dimensional lee wave. Beitr. Phys. Atmos. 29, 242.

YIH, C.-S. 1959: Effect of density variation on fluid flow. J. Geophys. Res. 64, 2219.

— 1965: Dynamics of Nonhomogeneous Fluids. New York: Macmillan.

ZIEREP, J. 1952: Leewellen bei geschichteter Anströmung. Berichte Deutsch. Wetterdien 35, 85.

— 1953: Bestimmung der Leewellenströmung bei beliebigem Anströmprofil. Z. Flugwiss. 1, 9.

— 1956: Das Verhalten der Leewellen in der Stratosphäre. Beitr. Phys. Atmos. 29, 1.

Addendum

I am indebted to Dr. R. Zeytounian of ONERA for the following references to recent works in the USSR on waves in stratified flows.

DORODNITSYN, A. A.: Perturbations d'un courant atmosphérique engendrées par les irrégularités de la surface du sol. Trudy du GGO, 23 (1938) (en russe). — Quelques problèmes d'écoulement de l'air au-dessus d'un relief terrestre. Trudy du GGO, 21 (1940) (en russe). — Influence du relief terrestre sur les courants atmosphériques. Trudy de l'Institut Central des prévisions, 21 (48) (1950) (en russe).

FAÏNBERG, E.: Ecoulement tridimensionnel d'une masse d'air froid au-dessus d'un obstacle infiniment petit. Doklady Acad. Sc. de l'URSS, t. 147, 2 (1962) (en russe).

FRANKL, F. I., et L. N. GUTMANN: Problème stationnaire sur le mouvement d'une masse d'air froid au-dessus d'un site montagneux. Doklady Acad. Sc. URSS, t. 141, 1 (1961) (en russe).

GODEV, N. G.: Etude des précipitations orographyques en approximation linéaire. Trudy du Centre Mondial Météorologique de Moscou, 6 (1965) (en russe).

KIBEL, I. A.: Sur l'emploi de la méthode des ondes longues en fluide compressible. Mécanique et Mathématique appliquée (PMM), t. VIII, 5 (1944) (en russe). — Problème tridimensionnel sur l'écoulement de l'air au-dessus d'un obstacle. Doklady Acad. Sc. de l'URSS, t. 100, 2 (1955) (en russe). — Short-range weather forecasting as a hydrodynamic problem. Applied mechanics

Proceedings of the eleventh international Congress of Applied Mechanics, Munich (Germany), 1964, H. GÖRTLER, Berlin/Heidelberg/New York: Springer 1966, p. 779—789.

KOSHEVNIKOV, V. N.: Au sujet d'un problème non-linéaire sur les perturbations orographiques d'un courant stratifié. Izvestia Acad. Sc. de l'URSS, série géophysique, 7, (1963) 1108—1116 (en russe).

KOTCHIN, N. E.: Problème tridimensionnel concernant les ondes sur une surface de séparation engendrées par un obstacle. Trudy de l'Obersvatoire Principal de Géophysique de Leningrad (GGO) 28 (1938) (en russe).

MUSAELIAN, CH. A.: Ondes de relief dans l'atmosphère. Edition hydrométéorologique de Leningrad, 1962 (une traduction anglaise éditée par Israel existe).

PEKELIS, E. M.: The numerical method for calculation of the finite-amplitude orographie disturbances (two-dimensional problem). Izvestia Acad. Sc. USSR; Atmospheric and oceanic physics, vol. 2, 11 (1966). — Sur la formulation des conditions aux limites pour la résolution numérique des problèmes d'écoulements stationnaires. Trudy du Centre Mondial Météorologique de Moscou, 14 (1966) (en russe). — Passage d'un obstacle isolé par un courant aérien (problème non-linéaire). Meteorologija i guidrologija 10 (1966) (en russe).

Trochu: Calcul d'un champ de vitesse verticale en mésométéorologie, applications. Etude de stage Ingénieur Elèves de l'Ecole de Météorologie, Météorologie Nationale, Paris, avril 1967.

Troubnikov, B.: Ecoulement tridimensionnel au-dessus d'un obstacle dans une atmosphère infinie. Doklady Acad. Sc. de l'URSS, t. 129, 4 (1959) (en russe).

Veltichev, I.: Sur l'interprétation de la méso-structure d'un champ de nébulosité. Trudy du Centre Mondial Météorologique de Moscou, 8 (1965) 45 (en russe).

Zeytounian, R. Kh.: Prise en compte des ondes courtes dans le problème non-linéaire de l'écoulement de l'air au-dessus d'un obstacle terrestre. Trudy du Centre de Calcul Météorologique de Moscou, Note No. 1, 1963 (en russe). — Problème non-linéaire sur la formation de la nébulosité en aval d'un obstacle. Izvestia Acad. Sc. de l'URSS, série géophysique 9, 1963 (en russe). — Hydrodynamic computation of an orographically induced cloud cover in a stable and unstable atmosphere. Doklady of the Acad. of Sc. of the USSR. Earth Sciences sections (English translation), vol. 148, 1—6. Washington, Am. geological Institute, nov. 1964. — Hydrodynamic calculation of lee wave. Bulletin (Izvestia) Academy of Sc. URSS, géophysics series (English edition) No. 9. Washington Am. Geophysical Union, septembre 1964, p. 865—867. — Sur l'influence de la variation du vent de base avec l'altitude dans les problèmes d'écoulements au-dessus d'un obstacle. Trudy du Centre Mondial Météorologique de Moscou, 6 (1965) (en russe). — Three-dimensional rotational motion in a compressible fluid. Izvestia Acad. of Sc. URSS; Atmospheric and oceanic physics, (English edition), vol. 2, 2. Washington Am. geophys. Union. fevr. 1966, p. 61—64. — Modèle hydrodynamique de prévisions de précipitations sur une région limitée. Communication à la Conférence Internationale l'EAU pour la PAIX. Washington, 23—31 mai 1967. — On a criterion for filtering parasite solutions in a numerical computation. Communication to the International Symposium on High-Speed computing in fluid dynamics; Monterey, California, 19—24 august 1968.

Zeytounian, R. Kh., and N. A. Fedotova: Répartitions des précipitations par "taches" dans une grande ville. Meteorologija i Gidrologija, No. 3 Moskva Gidrometeoizdat, 1965.

Non-linear solid mechanics, past, present and future

By

F. K. G. Odqvist

The Royal Institute of Technology, Stockholm

I. Introduction

1. Bounds set to our scope

First of all, the subject of "Applied Mechanics", or more particularly that of "Applied Solid Mechanics" must be somehow defined and its borders outlined. In what follows, when speaking about Solid Mechanics, I shall mean Solid Applied Mechanics. Such definition proves to be far from easy and has certainly resulted in different border lines from time to time.

In fact—and this will be a main point in this lecture—the demarcation line of our subject has to be revised repeatedly at the risk of stagnation and petrification. The line of demarcation, generally speaking, marks the border towards neighboring subjects. These are Mathematics, Physics and Chemistry and certainly also other branches of science. Just where the boundary has to be drawn at a given time may not be easily defined—must not be easily defined. Any strict definition is likely to become controversial. Very often inter-scientific fields offer the most profitable point of attack. By way of example, let us try to visualize where the indicated border line has been situated at some characteristic time periods.

In the eighteenth century, Euler's theory of columns was high mathematics devoid of all practical applications. Today it is a commonplace ingredient in most elementary text books on applied mechanics. The reason for this is not only that we know a little more mathematics than our ancestors five generations back, but first of all the fact that modern engineering structures are so slender as to make Euler's column theory more applicable in comparison with the heavy, wooden stanchions of his time. This fact certainly does not diminish his ingenuity and foresight.

In the 1920's, Mushkhelishvili's solution to two-dimensional problems of elasticity and Fredholm's integral equations were considered to be advanced mathematics only. Today both are indispensable parts of solid mechanics with useful applications, e.g., in fracture mechanics.

Very often encouragement for attack on new areas comes from practical considerations with consequences for the border line. The complicated trusses of early airplane structures had to give way to shell structures and shell theory—already in existence for many years—became a field of utmost actuality.

These examples could be multiplied many times but may be sufficient to prove my point. As a working principle and at the risk of being too restrictive, let us adopt a definition related to that advocated by the editor of Applied Mechanics Reviews: Remember that we are working for the benefit of the engineering profession. Perhaps we should extend this definition to include some other branches of human activity, such as meteorology, oceanography, seismics and astronautics. The main thing is that of a purpose, albeit a bit vague.

Perhaps this is the moment to remind you of a statement of RICHARD VON MISES, the founder of the journal ZAMM, Zeitschrift für angewandte Mathematik und Mechanik, which was to become leading journal for the development during the first decade of its existence. In his introductory article, written in 1921, v. Mises expressed opinions very much like those I have been propounding here. Some years later, when dealing with the scope of his journal in an editorial, he wrote (my free translation) "There should be no distinction between theoretical and experimental papers, ... All theoretical research depends in the end on observational facts. Experimental work is useless, unless it is undertaken in the view of some theory".

In the opinion of the German mathematician Felix Klein, "the goal of Natural Philosophy ('theoretische Naturwissenschaft') should be not merely a passive understanding but an active command of Nature", see letter from 1921.

2. What is to be required of a theory of applied mechanics?

There may be several, more or less elaborate answers to this question. For me it is enough to say that such theory shall—besides the minimum requirement of being free from inner contradictions—permit conclusions from one set of observations to another one. Thus it must be possible to predict the behavior of a material or structural element under prescribed conditions, knowing from previous experiments its behavior in some specified case. For example: knowing Young's modulus and Poisson's ratio from an ordinary tension test within the elastic range, it must be possible to predict the behavior of a rod of the same material under, say torsion. The answer, to some degree of approximation, is given by Saint-Venant's theory. But then, of course, it may be desirable to know the limits of load within which such theory is valid etc.

In principle, I do not think that applied mechanics can yield "explanations" much better than that. One may ask for reduction of the number of independent axioms, yes. But in my opinion, it is almost as important to ask for a reduction of the number of independent material constants. It may of course be necessary to take into account time effects, inertia effects, phase transformations, temperature changes etc.

3. Axiomatics of mechanics

What I have said so far may be a disappointment for all those who spend their time cultivating Mechanics *per se*. This is not meant to be a challenge to them. In fact, their efforts presently fill a very considerable part of our journals and periodicals. Some of their results have been useful indeed, as we shall see later. Mathematics and Mechanics have been stimulating each other mutually for at least 300 years. This has led to beautiful results in both sciences.

Certain theories, so necessary from a mathematical point of view, however, seem—at the first glance—to carry little substance to the cultivator of applied mechanics, such as here defined. The mathematicians, at least from C. G. J. Jacobi onwards, are used to follow the prescription "Man muß alles umkehren", i.e., it is necessary to try the reverse of every statement, proposition or theorem.

Consider e.g., an elementary example: Every isoceles triangle has equal angles at the end of the base. But the reverse is also true: All triangles with equal angles at the base are isoceles. These are well-known theorems from Euclidean geometry. Now we know today that Euclidean geometry is based upon certain axioms. If one of these axioms—the parallel axiom—is dispensed with, a more general theory ensues. In analyzing the axiomatics of plane geometry, the geometers of the 19th century paved the way for non-euclidean geometry and so created a theory which has been of importance for the further development also of Mechanics. More could be said about "geometrization" of Mechanics, old and modern, but would carry us beyond the scope of this lecture.

Nowadays, an enthusiastic group of young scientists has undertaken to find out which axioms in Mechanics, more particularly in Continuum Mechanics, that are indispensable. Thus they wish to enlarge the realm of validity of every theorem or, alternatively, to find out its utmost consequences. This work has to a great extent been initiated by Clifford Truesdell about 15 years ago. For simplicity, let us call this school the Truesdell group, even if most part of the work has been published by other people than Truesdell himself. His own witty and stimulating, albeit sometimes perhaps a little too vociferous, propagating of these ideas merits the group to be named after him.

To some of us, all this may seem to be dull matter. Sometimes, however, interesting results are encountered.

In summing up: the cultivators of Applied Mechanics ought to be grateful that this set of young brilliant mathematicians has undertaken the trouble to strengthen the foundations of our subject. I have mentioned the question of axiomatics already here, as it has proved to be of utmost importance when penetrating the complex subject of non-linear solid mechanics.

II. Non-linear solid mechanics during 50 years

1. The situation at the end of World War I

When students of my generation entered upon the study of Solid Mechanics—round about 1920—this subject was almost identical with the theory of elasticity such as presented e.g., by A. E. H. Love, Sedleian Professor of Natural Philosophy in Oxford University and so predecessor to our retiring vice-President of IUTAM, Professor G. Temple. In Love's treatise, linear theory predominates. From the phenomenological, physical standpoint, it had been brought to a certain degree of completion. Its fourth, i.e., its last authorized edition, appeared in 1927. The progress in linear elasticity, that has come since then might as well count as mathematics or applied mathematics.

As for non-linear teories—the subject of this lecture—I do not mean to say that progress was nonexistent at the time considered, i.e., about at the end of World War I. "There were heroes before Agamemnon"—as the classicists say. But I also think it is safe to say that most of the progress ever made in non-linear theories has been achieved since then, i.e., during the time of the congresses of applied mechanics.

Naturally, I shall have to content myself to refer to this development by way of examples. Certain aspects as the fertile field of non-linear vibrations and other problems were inertia forces are essentieal will be left aside entirely. I shall give preference to what may be termed physical non-linearities, cf. below. Geometric non-linearities such as appear in elasticity of finite deformations will be left out. Other modern applications of non-linear mechanics are in polymers, in fracture mechanics and rock mechanics where close collaboration with physical metallurgists, polymer chemists and geologists is essential. All these applications—interesting as such — have also to be ignored for lack of space.

I shall mainly concentrate upon questions of non-linear constitutive equations and related subjects. I am going to take up the behavior of solid material upon comparatively slow loading from the virgin state followed by at most one or a few unloadings and reloading. Shock and fatigue will be left out of consideration.

Even with these restrictions, there is no possibility to claim completeness.

2. Time-independent plasticity, 1921—1948

When speaking about non-linear solid mechanics it is natural, in the first place to think about the deviations from linearity as revealed in the ordinary uniaxial tensile test. Here, usually, force P or stress σ is plotted against displacement δ or strain ε between the ends of the gauge length l, see Fig. II.1. These deviations had been subject to extensive investigations during the 50 years period that preceded the one under consideration, see Th. v. Kármán (1914). In particular, it had been established that, for many structural metals, the so-called limit of proportionality very nearly coincides with the yield point $\sigma = \sigma_s$, i.e., the very load when residual strains appear upon unloading. The yield point in compression very nearly coincides numerically with that in tension so that the total elastic range amounts to about $2\sigma_s$, see Fig. II.2. For other solid substances such as polymers and organic substances, non- linear elastic behavior may occur. It had also been established that if the yield point σ_s had been once exceeded in tension and the test piece subsequently reloaded in compression, the yield point in compression will be absolutely lowered, i.e., will occur for a stress σ, for which $|\sigma| < \sigma_s$ (Bauschinger effect), see Fig. II.3. The elastic range remains approximately $2\sigma_s$, but for the effect of work-hardening.

In the ordinary tensile test with structural metals at room temperature, influence of time and strain rate are not considerable for tests carried out within a time period of minutes or at

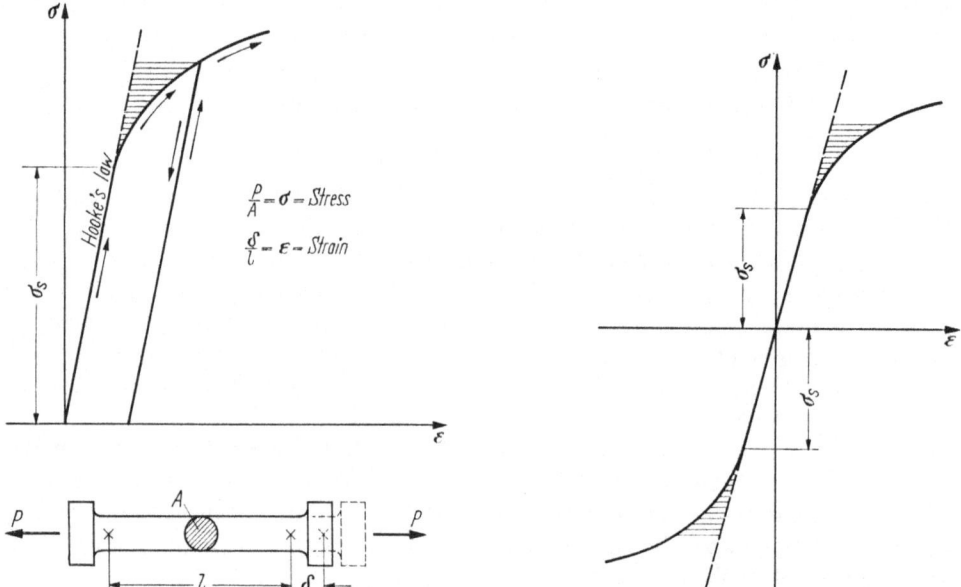

Fig. II.1. Uniaxial tension test: strain-hardening material.   Fig. II.2. Stress-strain diagram in tension and compression.

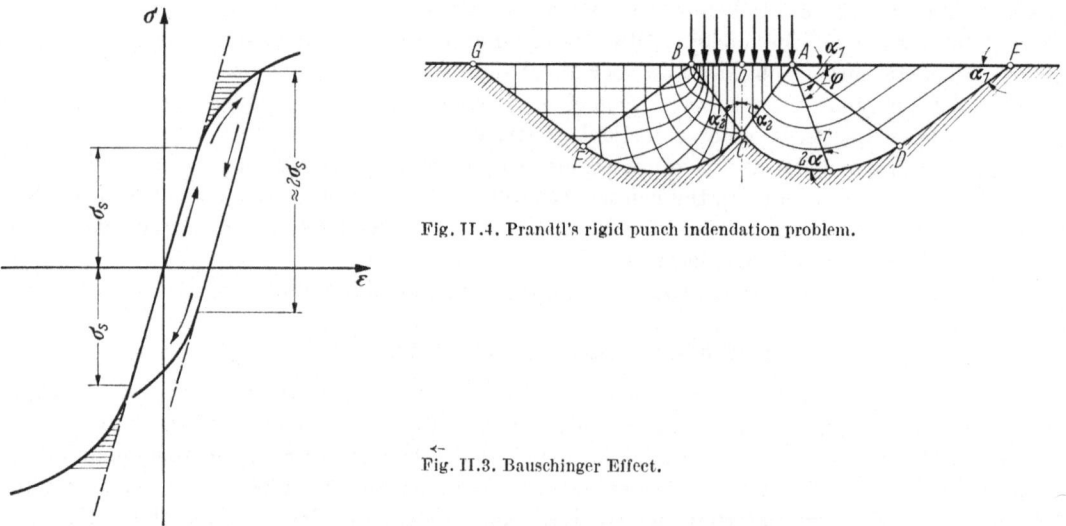

Fig. II.4. Prandtl's rigid punch indentation problem.

Fig. II.3. Bauschinger Effect.

most a few hours. It is reasonable and meaningful to introduce a conception of *time-independent plasticity*. Our task is to generalize the results from uniaxial tensile tests to three dimensions, i.e., to present the constitutive relations between strain tensor ε_{ij} and stress tensor σ_{ij}.

Prior to the time period under consideration, there existed already phenomenological theories of time-independent plasticity beginning with B. de Saint-Venant (1870, 1871), M. Levy (1871), R. v. Mises (1913) and others. But these theories were only in their infancy. Today's impressive theory of plasticity is a creation mainly by the last 50 years. There are many names to be mentioned.

Already in 1920 and later in the first volume of ZAMM, in 1921, LUDWIG PRANDTL gave his famous solution of the problem of indentation of a rigid punch into a semi-infinite body of plastic material under conditions of plan strain, see Fig. II.4. This paper—in itself not free from criticism, see R. HILL (1949), W. PRAGER and P. G. HODGE, Jr. (1951), M. SAYIR and H. ZIEGLER (1968)—and a few more by the same author, gave impetus to further work by H. HENCKY (1923, 1925) and E. REUSS (1930) and others who worked out the mathematical theory and treated special problems. Reference could be made to the book by R. HILL (1950) and to the article by A. FREUDENTHAL and H. GEIRINGER (1958) in the Encyclopedia of Physics.

In 1928, R. v. MISES presented his theory of crystal plasticity based upon the concept of a plastic potential φ, this being a homogeneous second degree function of the stress deviation tensor components $s_{ij} = \sigma_{ij} - \delta_{ij}\sigma_{kk}/3$, where δ_{ij} is Kronecker's symbol and summation convention for subscripts has been used. Thus v. MISES created the so-called flow rule, stating that the strain rate in every point of the yield surface, $\varphi =$ constant, is parallel to the normal of that surface in that very point. Important experimental investigations were carried out by A. NADAI (1923), W. LODE (1926), K. HOHENEMSER (1931) and by G. I. TAYLOR and collaborators (1931). More complete references to the early researches in time-independent plasticity may be found in the monograph by W. PRAGER (1937). It is remarkable that, albeit the general theory of perfectly plastic material had existed for more than half a century, the rich development of both theoretical and experimental research in the 1920's and 1930's obviously was initiated through the successful attack on special problems of practical engineering importance.

During the 1930's, important contributions to time-independent plasticity theory were given. F. ODQVIST treated isotropic strain hardening (1933), E. REUSS (1934) and W. PRAGER (1937) anisotropic work-hardening, generalizing v. Mises theory of the plastic potential. Reuss' theory meant a translation of the yield surface in stress space proportional to total strain, thus giving an explanation of the Bauschinger effect. The theory was further developed by E. MELAN (1938). Later PRAGER also gave numerous contributions to all branches of plasticity, e.g., to limit analysis and to the variational approach to plasticity problems. An excellent survey of the development was given by PRAGER himself in his general lecture before the VIIth International Congress of Applied Mechanics in London (1948). Many of the foremost today's workers in plasticity and adjacent fields have sprung from Prager's school at Brown University in the 1940's and 1950's.

3. Incremental versus finite theories of plasticity

There was at the time about 1950 a fight going on between advocates of the so-called incremental (or flow) theories and the finite (or deformation) theories of plasticity. At that time, it had become clear that time-independent plasticity theory, in order to be able to account for irreversibility, called for constitutive equations where stress is given in terms of strain rate rather than total strain, thus giving preference to the flow theories. As a matter of fact, this had been recognized from the outset by B. DE SAINT-VENANT and the other early theorists, but since then been disputed particularly by some experimentalists. For monotonic, proportional or even mainly unidirectional loading without sudden changes in the directions of principal stress, however, the errors committed are usually small, as pointed out, e.g., by B. BUDIANSKY (1959), who specified a large class of stress histories for which Drucker's postulate (see below) remains unmolested and the results of finite theory coincides with that of incremental theory. In order to present his theory in detail, BUDIANSKY makes use of a five-dimensional subspace in the six-dimensional vector space of the stress components already used by PRAGER (1948). PRAGER (1960) points out that, in order to interpret theoretically (more than very roughly) the above-mentioned experiments with thin-walled tubes in simultaneous torsion and tension by K. HOHENEMSER (1931), it is necessary to remember that the finite relationship between stress and strain need not be continuously differentiable.

4. Drucker's postulate. Stability in the small

About 1950 it had become clear that there was need for a more fundamental approach to the problems of time-independent plasticity. The reason for this was partly difficulties with uniqueness of solution at the edges and corners of the yield surfaec $\varphi =$ constant (in itself quite often an unnatural and unnecessary abstraction from reality), partly problems of stability. Here the work of R. HILL and D. C. DRUCKER was leading.

Drucker's stability postulate (D.P.) (1950, 1951), states that if any solid body (or structure) that is loaded on to the yield limit receives an additional load increment (finite or infinitely small), then the corresponding elastic and plastic deformations shall give rise to non-negative contributions to the accumulated and/or dissipated work.

Consider e.g., two homogeneous states of equilibrium A and B, with stresses σ_{ij}^A and σ_{ij}^B and the corresponding strain rates $\dot{\varepsilon}_{ij}^A$, $\dot{\varepsilon}_{ij}^B$. Then D.P. would take the form (summation convention for subscripts)

$$(\sigma_{ij}^A - \sigma_{ij}^B)\,(\dot{\varepsilon}_{ij}^A - \dot{\varepsilon}_{ij}^B) \geq 0. \tag{II.1}$$

What really matters here—let me repeat—is $d\varepsilon_{ij}$ in the differential quotient $\dot{\varepsilon}_{ij} = d\varepsilon_{ij}/dt$, as the time sequence of events is immaterial. If the state B is at a regular point of the yield surface in the six-dimensional vector space of σ_{ij}, there exists a tangent plane and a well-defined normal BN to that surface in B. If the arbitrary load increment vector BA shall point towards the interior of the plastic region, then the vector BA must form an acute angle with BN in order that the inequality (II.1) shall not be violated. Then $\dot{\varepsilon}_{ij}^B$ must point in the direction BN, as indicated symbolically in Fig. II.5. Thus D.P. generalizes v. Mises' flow rule, requiring *normality* of the flow increments at the yield surface, also for work-hardening materials.

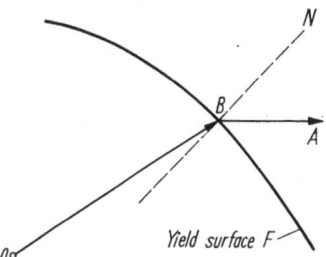

Fig. II.5. Drucker's postulate: normality.

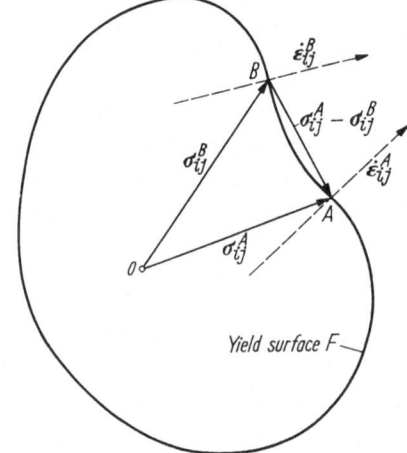

Fig. II.6. Drucker's postulate: convexity.

In Fig. II.6, F denotes the yield surface, symbolically. The points A and B are assumed to be at F, and the vectors $\sigma_{ij}^A - \sigma_{ij}^B$ and $\dot{\varepsilon}_{ij}^A - \dot{\varepsilon}_{ij}^B$ obviously will form an obtuse angle in the case when A and B are set at an outward concave part of the surface F. Hence (II.1) would be violated. In the case when the yield surface is convex from outside this is not so. D.P. requires *convexity* of the yield surface, i.e., F must be concave when looked upon from the origine in stress space and this origine must be inside F.

Instability, as normally understood, may arise when some but not all of the attributes of a system are considered. For example, when kinetic energy is ignored as in the treatment of problems as static, instability may indicate simply that potential energy is converted to kinetic. If small displacements only are treated, the appearance of instability may mean that deflections will be large. Stability is a matter of point of view. It depends upon which factors are supposed to be controllable or useful.

DRUCKER distinguishes between *stability in the large*, when the geometry of a whole structure may influence the stress distribution under given external loads, like the Euler column, and *stability in the small*, when an element of a structure or material at the yield point is being exposed to additional forces. Stability in the small, that is, stability for each possible infinitesimal change of the controllable variables, is directly related to the uniqueness of the solution and so appears to be a completely reasonable requirement for the actual configuration of a system. Stability in the large, on the other hand, deals with the response of a system to finite disturbances and has implications beyond that of the requirement of stability in the small for all loading or environmental paths. The reasonableness of requiring stability in the large is by no means certain despite its desirability from the experimental or design point of view. If the variables considered as controllable or useful are too few in relation to the physical behavior of the system, it may well be that the requirement of stability in the large will be too restrictive or completely unreasonable. Changes of geometry of elastic bodies under static compressive loading provide a good example. In the conventional static sense all such bodies are unstable in the large but may well be stable in the small.

Drucker's stability postulate has been generalized (1960) to the case of temperature changes and to structures like those in isothermal or adiabatic creep (see below). B. STORAKERS has recently published an application of D.P. to the instability of a thin-walled cylindrical tube under simultaneous loading with internal pressure, torsion and axial tension. A main result of this investigation is that instability may occur for combinations of load where anyone of the separate loads may be smaller than its absolute maximum value. It is also shown how superimposed viscous flow influences stability.

5. Variational approach. Uniqueness. Limit analysis

The constitutive equations of time-independent plasticity being in general non-linear, solutions of boundary value problems in special cases become extremely complicated. Variational theorems for direct attack within the frame of v. Mises' and Prandtl-Reuss' theories were given by PRAGER and HODGE (1948). About the same time, equivalent theorems for elastic-plastic materials were published in Russian by L. M. KACHANOV (1949). More general theorems were given by PRAGER and WANG (1953). Reference may be made also to the books by R. HILL (1950), W. PRAGER and PH. HODGE jr. (1951), F. ODQVIST and J. HULT (1962) and also F. ODQVIST (1966). It is possible to find upper and lower bounds for the potential of internal and external forces. In special cases, as e.g., torsion problems, bounds for quantities more directly connected with load or deformation may also be found.

In the case when a solution of a boundary value problem may be assumed to exist, it is also possible to ascertain its uniqueness, utilizing variational theorems, see HILL (1950). The situation is analogous to that of Dirichlet's problem in classical potential theory.

The literature on the theories of limit design or limit analysis goes back to G. KAZINCZY (1914). Later, there were independent fresh starts by M. GRUENING, F. NAKANISHI, H. BLEICH, K. W. JOHANSEN and possibly others too, all from different points of view. Most of these investigators were civil engineers and the subject is of considerable practical interest. The literature of limit design has lately grown immensely, see e.g., the books by B. G. NEAL (1956) and by A. SAWCZUK and TH. JAEGER (1963).

A thorough presentation of the results of time-independent plasticity with variational theorems, uniqueness, limit analysis and shakedown was given by W. T. KOITER (1960).

6. Prager's generalization of Drucker's postulate

PRAGER (1957) has generalized D.P. to non-isothermal deformation, introducing into the flow potential φ as "state variables", beyond stress σ_{ij} and strain ε_{ij}, the absolute temperature θ and a parameter α, allowing for irreversibility and dependence of the stress history

$$\alpha = \int\limits_0^t \sigma_{ij}\dot{\varepsilon}_{ij}\,dt. \tag{II.2}$$

11*

Similar parameters have been introduced for similar purposes by various writers: TAYLOR and QUINNEY (1931), ODQVIST (1933), NAMESTNIKOV and RABOTNOV (1961). For further references, see the books by R. HILL (1950) and L. I. SEDOV (1962/1965) and also Chapters II.(10) and III.(2) below.

PRAGER considers rigid-work-hardening materials and assumes $\varphi < 0$ to correspond to rigid behavior, $\varphi = 0$ for all plastic states, just as v. MISES. If in a plastic state, temperature and stress are given increments $\dot{\theta} \, dt$ and $\dot{\sigma}_{ij} \, dt$, where the dots as usual denote differentiations with respect to the time t, this will lead to a neighboring state which is rigid or plastic according to whether $\dot{\varphi}$ is negative or zero.

If $\dot{\varphi} < 0$, then changes of temperature and stress constitute unloading and we have $\dot{\varepsilon}_{ij} = 0$, $\alpha = 0$ (for the virgin state). Since

$$\dot{\varphi} = \frac{\partial \varphi}{\partial \theta} \, \dot{\theta} + \frac{\partial \varphi}{\partial \sigma_{ij}} \, \dot{\sigma}_{ij} + \left(\frac{\partial \varphi}{\partial \varepsilon_{ij}} + \frac{\partial \varphi}{\partial \alpha} \, \sigma_{ij} \right) \dot{\varepsilon}_{ij}$$

in this case we must have

$$\varphi = 0, \quad \frac{\partial \varphi}{\partial \theta} \, \dot{\theta} + \frac{\partial \varphi}{\partial \sigma_{ij}} \, \dot{\sigma}_{ij} < 0. \tag{II.3}$$

For neutral change of state, the inequality (II.3) changes to equality.

If on the other hand $\dot{\varphi} = 0$, and at the same time

$$\frac{\partial \varphi}{\partial \theta} \, \dot{\theta} + \frac{\partial \varphi}{\partial \sigma_{ij}} \, \dot{\sigma}_{ij} > 0 \tag{II.4}$$

we have plastic loading with $\dot{\varepsilon}_{ij} \not\equiv 0$.

It is now assumed that $\dot{\varepsilon}_{ij}$ be of the order one in the quantities $\dot{\theta}$ and $\dot{\sigma}_{ij}$. Thus Prager arrives at the constitutive equations

$$\dot{\varepsilon}_{ij} = \begin{cases} 0, & \text{if} \quad \varphi < 0 \\ A_{ij} \left(\frac{\partial \varphi}{\partial \theta} \, \dot{\theta} + \frac{\partial \varphi}{\partial \sigma_{kl}} \, \dot{\sigma}_{kl} \right), & \text{if} \quad \varphi = 0. \end{cases} \tag{II.5}$$

Conserving normality and convexity in the stress space, he puts

$$A_{ij} = \frac{1}{D} \frac{\partial \varphi}{\partial \sigma_{ij}}, \quad D = - \frac{\partial \varphi}{\partial \sigma_{ij}} \left(\frac{\partial \varphi}{\partial \varepsilon_{ij}} + \frac{\partial \varphi}{\partial \alpha} \, \sigma_{ij} \right). \tag{II.6}$$

In principle, integration of (II.5, 6) for given stress and temperature history, would determine the future deformation. Possibly, however, thermodynamical considerations might impose further restrictions on the constitutive equations or alternatively allow for more general expressions. The Drucker-Prager postulate may have yielded too general or too restrictive form of the constitutive equations. Also, visco-elasticity, which, as we shall see, is strongly non-linear in the stress tensor, may enter into play. As a matter of fact, DRUCKER was early aware of the shortcomings of D.P. for time-dependent plasticity, see DRUCKER (1960, 1962). This brings us on to the next chapter.

7. Time-dependent plasticity, creep etc. 1930—1952

Hereditary phenomena, after effects and similar time-dependent deformation of solids had been treated already in the 19th century. Linear theories of these phenomena existed and were due to J. C. MAXWELL, L. BOLTZMANN, V. VOLTERRA and others, see v. KÁRMÁN (1914). Here analytic methods have reached a high degree of perfection but are outside the scope of this lecture. These theories may well suffice to represent creep behavior of concrete and certain polymers but are certainly insufficient e.g., for the corresponding phenomena in structural metals.

Above a certain critical temperature of the order of half the melting point as measured on the absolute temperature scale, each metal, when loaded under constant uniaxial stress σ, shows plastic deformation—so-called creep—with strain ε, increasing progressively with time t as seen in Fig. II.7, reproducing, in principle, a "creep curve". The terminology "primary", "secondary" and "tertiary" stages of creep as shown in the figure, corresponds to decreasing (α), constant (β), and increasing (γ) strain rate $d\varepsilon/dt = \dot{\varepsilon}$, the so-called "creep rate". It stems from E. N. da Costa Andrade, who initiated systematic study of the phenomena of creep early in this century. The tertiary stage of creep ends up with creep rupture which may be more or less brittle. It is generally recognized that after effects of structural metals e.g., after unloading, when subject to creep conditions, are of the order of elastic deformations.

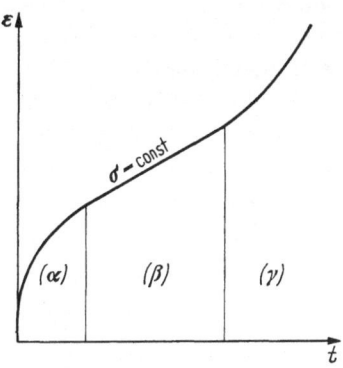

Fig. II.7. Creep curve for constant stress and temperature.

The real impetus to further development came from the fact that prime movers in the 1920's had reached a state, i.e., a working temperature, when creep asserts its influence on their vital parts. This would later become gradually still more accentuated in gas turbines, jet engines, nuclear and other chemical reactors and pressure vessels as well as in rockets. Thus technology has been a continuous challenge to engineers and scientists active in the field. One of the first to realize this was A. Stodola, grand old man of the theory and design of steam and gas turbines and professor of the Eidgenössische Technische Hochschule at Zürich.

Just before 1930, F. H. Norton of MIT had published results of his extensive uniaxial tests on structural steels at elevated temperatures, from which he had derived his exponential relation between creep rate $\dot{\varepsilon} = v$ and (constant) stress σ in secondary creep

$$v = k\sigma^n \qquad (\text{II.7})$$

where k and n are material constants depending on temperature, particularly so the constant k.

Stimulated by Stodola, the present speaker treated the problem of secondary creep under influence of combined states of stress, the first solution of which was presented before the IVth International Congress of Applied Mechanics, Cambridge 1934. In this theory the time t occurs as an "essential" variable. The creep rate tensor ε_{ij} in secondary creep was given in the form

$$\dot{\varepsilon}_{ij} = f(\sigma_e)\,s_{ij} \qquad (\text{II.8})$$

where s_{ij} is the stress deviation tensor defined in Chapter II(2) and $f(\sigma_e)$ is an exponential function of σ_e, the "effective stress", its second order invariant, defined through

$$\sigma_e^2 = \frac{3}{2}\,s_{ij}s_{ij}. \qquad (\text{II.9})$$

Here summation convention has been employed for the subscripts (except e) as before. Further the material has been considered as incompressible and elastic as well as primary creep deformation been neglected when using (II.8) as a constitutive equation. The strain rate tensor $\dot{\varepsilon}_{ij}$ in (II.8) has the form

$$\dot{\varepsilon}_{ij} = \frac{\partial\varphi}{\partial s_{ij}} \qquad (\text{II.10})$$

where φ is a function of σ_e only. This equation has the same form as v. Mises flow rule, but the time t now plays the part of an essential variable. The surfaces $\varphi = $ constant are convex, according to D.P. Thus the conditions of normality and convexity are preserved from time-independent plasticity.

A number of British and American investigators, working independently, were soon to follow. Sometimes (II.7) was replaced by other relationships between v and σ and the $f(\sigma_e)$ had to be correspondingly modified.

At about the same time as the speaker and independently, R. W. Bailey of the Metropolitan Vickers Co. gave a representation of the principal creep rates in secondary creep as functions of the second and third order invariants of the stress deviation tensor. According to the extensive experimental researches by his countryman A. E. Johnson (1951), the influence of the third order invariant is usually negligible, at least for small and moderate creep deformations. Later, in 1945, W. PRAGER and M. REINER, working independently, gave tensorial representations of Bailey's theory.

Reiner's interest was to a great deal focussed on rheological properties of concrete and a number of organic and inorganic substances which show a more or less dough-like behavior, something between solid and fluid. These substances do not belong to solid mechanics in the restricted sense adopted here.

In the 1950's the interest of primary and tertiary stages of creep became apparent to mechanical and aeronautical engineers. Primary creep had already in the 1930's drawn the attention of A. NADAI and his group at the Westinghouse Co. A controversy between "time-hardening" and "strain-hardening" theories of creep had been decided mainly in favor of the latter, see Fig. II.8. From an experimental creep curve, a relationship of the form

$$\varepsilon = A\sigma^n t^m \qquad (II.11)$$

may be found to hold in the primary stage of creep. Differentiating (II.11) we obtain

$$\frac{d\varepsilon}{dt} = \dot{\varepsilon} = mA\sigma^n t^{m-1} \qquad (II.12)$$

which represents the time-hardening type of relation. Eliminating t between (II.11) and (II.12) we obtain

$$\frac{d\varepsilon}{dt} = \dot{\varepsilon} = mA^{1/m}\sigma^{n/m}\varepsilon^{1-1/m} \qquad (II.13)$$

which represents strain-hardening and usually gives a more realistic representation of creep curves under changing stress, such as in Fig. II.8, see FINNIE (1959).

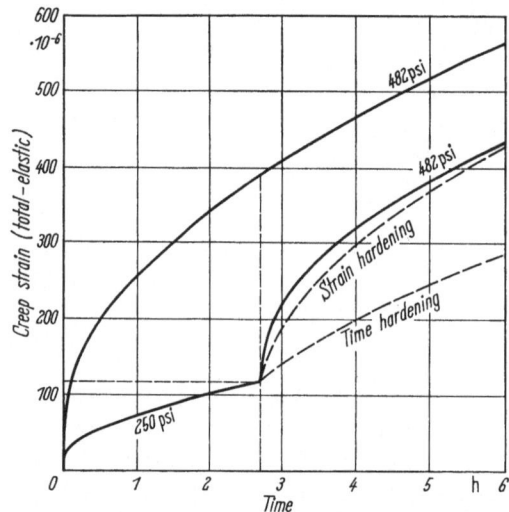

Fig. II.8. Time-hardening versus strain-hardening.

Full references of the material covered by this chapter may be found in the books by ODQVIST and HULT (1962) and ODQVIST (1966) and also in the survey article by I. FINNIE (1966).

8. Equation of state. Influence of primary creep. Elastic analogue

One way of looking at the phenomenon of creep is to investigate whether there exists a relation of the type of Eq. (II.13), viz.

$$\frac{d\varepsilon}{dt} = \dot{\varepsilon} = f(\sigma, \varepsilon) \qquad (II.14)$$

a so-called "equation of state" without the time included explicitly in the function f of the right side of (II.14). As shown by the present speaker, this equation is equivalent to the statement that the total amount of creep, resulting from a prescribed history of stress $\sigma_1, \sigma_2, \ldots, \sigma_m$ kept constant but different during time periods $t_1, t_2, \ldots, t_m$ is independent of the order in which they are applied. This "commutative law" holds fairly well for a number of structural metals. There are, however, discrepancies which may be accounted for by including a parameter similar to that introduced in (II.2), cf. NAMESTNIKOV and RABOTNOV (1961), see below, Chapter II(10).

At the VIIIth International Congress of Applied Mechanics in Istanbul 1952, the present speaker presented his theory of primary creep. Primary creep is here taken account of as an integral effect of the type of slip deformation to be added to the residual strain in the secondary stage.

Thus, in uniaxial creep, the creep rate $d\varepsilon/dt$ under a given stress $\sigma = \sigma(t)$ will be

$$\frac{d\varepsilon}{dt} = \dot{\varepsilon} = \frac{d}{dt}\,G(\sigma) + F(\sigma) \qquad (\text{II.15})$$

where the first term on the right side shall be retained only as long as $\sigma\,d\sigma > 0$. If unloading takes place at a certain time $t = t_a$ then the first term on the right side of (II.15) shall be left out until the time t_b, when σ again has reached the value σ_a, that stress from which unloading took place, see Fig. II.9. Integrating (II.15), we obtain

$$\varepsilon = G(\sigma) + \int_0^t F(\sigma)\,dt. \qquad (\text{II.16})$$

For a creep test under constant stress, (II.16) reduces to

$$\varepsilon = \varepsilon^{(0)} + vt \qquad (\text{II.17})$$

where $\varepsilon^{(0)} = G(\sigma)$ denotes the intercept at $t = 0$, and $v = F(\sigma)$ is the creep rate in secondary creep. The creep curve has degenerated into a straight line, which is unable to represent in detail the creep behavior in the primary stage but gives a fair representation in the secondary stage, see Fig. II.10.

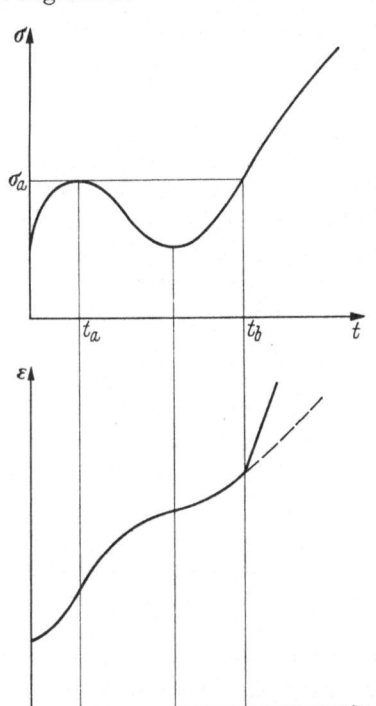

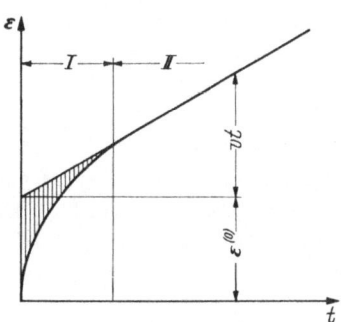

Fig. II.10. Odqvist's theory of primary creep.

Fig. II.9. Effect of unloading.

The functions $G(\sigma)$ and $F(\sigma)$ may often with advantage be taken in exponential form, e.g.,

$$G(\sigma) = \left(\frac{\sigma}{\sigma_0}\right)^{n_0}, \quad F(\sigma) = \left(\frac{\sigma}{\sigma_c}\right)^{n} \qquad (\text{II.18})$$

where σ_0, n_0, σ_c, n are material constants. In most cases it has been found that a relation of the form

$$n_0 \leq n \qquad (\text{II.19})$$

holds true.

The Eqs. (II.15) through (II.18) have been generalized to three dimensions in the same way as from (II.7) to (II.8). The applicability of these theories to real materials and structures has been discussed in some detail in the books by F. Odqvist and J. Hult (1962) and by F. Odqvist (1966).

In 1953, N. J. HOFF noticed the analogy between problems of non-linear elasticity and stationary flow in secondary creep. If a problem in non-linear elasticity is solved, then it is immediately possible to find the solution of the corresponding problem in secondary creep. For example: The stress distribution σ in the non-linear elastic bending of a beam of symmetric cross section was determined already in the 19th century as a function of bending moment M and sectional dimensions a, b, ..., if σ_0 and n_0 were known. For a point P of the cross section we may have

$$\sigma = \sigma(P; M, a, b, \ldots; \sigma_0 \, n_0).$$

Then, according to Hoff's elastic analogue, the corresponding stress distribution under conditions of secondary creep will be

$$\sigma = (P; M, a, b, \ldots; \sigma_c, n).$$

The analogue forms an excellent point of departure for developing variational theorems for the potential and complementary potential of internal and external forces, and for giving upper and lower bounds for these quantities.

Hoff's analogue is generally valid only for small creep rates and small total creep deformations. In this connection it is useful to distinguish between physical, geometrical and statical non-linearity. The physical non-linearity is contained in the constitutive equations of the type of (II.15) and its three-dimensional generalizations. This non-linearity is taken account of by the elastic analogue, as just stated. The geometric non-linearity stems from finite strain and the static non-linearity from taking account of changing form of the structure, such as in problems of stability in the large, e.g., column buckling. These two later non-linearities may not be taken into account when applying the elastic analogue, see ODQVIST (1960/1962, 1966). In the special case of finite creep of membranes, however, a generalization of Hoff's analogue holds true, see Z. BYCHAWSKI (1965). Relatively few problems of finite creep have been solved so far. A fairly complete catalogue of solved problems, see I. FINNIE (1966).

The first satisfactory solution of the problem of creep buckling of columns was given by HOFF (1954). Minor improvements to this theory have been added by ODQVIST (1954) and B. FRAEIJS DE VEUBEKE (1958). Problems of plate and shell buckling have been treated by T. H. LIN (1956) and E. SUNDSTRÖM (1957), respectively, and subsequently of a number of other writers. Of particular interest is the problem of buckling of circular cylindrical shells under axial compression, treated by A. SAMUELSON (1961, 1967). Experimenting with thin-walled cylinders of an aluminium alloy, corresponding to about 24-ST at 225°C, he found that influence of the boundary conditions was decisive. Constrained in radial direction at the boundary, the shell under axial compression develops the usual axisymmetric wave pattern, rapidly attenuated with distance from the boundaries. In the second half-wave compressive hoop stresses will be gradually developed, these giving rise to instantaneous non-symmetric buckling with a number of lobes, slightly less than for similar experiments carried out at room temperature. This might be expected due to the temperature dependence of Young's modulus.

Problems of tertiary creep are intimately related to material deterioration caused by creep, see Chapter II(10). These researches have brought us up to the present day. Their insufficiency in certain respects will be treated in more detail in Chapter II(10) and Section III.

9. Physical aspects. Dislocations

During the 19th century, the subject of mechanics—then believed to contain the essence of natural philosophy—turned more and more mathematical. It had developed into Rational Mechanics, such as presented in Paul Appell's Cours de Mécanique. Even today, there is a sharp trend of a similar character, albeit perhaps a little less pretentious, materialized above all in the efforts by the Truesdell group.

The ambitions of the Truesdell group to form a Rational Mechanics of Materials has been condensed —as far as constitutive equations are concerned—into the following three principles, see TRUESDELL (1966):

1) Principle of determinism. The stress in a body is determined by the history of the motion of the body. This may be evident close to triviality, but

2) Principle of local action, however, restricts the possibilities: The motion outside an arbitrarily small neighborhood of a particle may be disregarded in determining the stress of that particle. A material is called "simple" if the stress depends upon the deformation only through the deformation gradient.

3) Principle of material frame indifference (also called "principle of material objectivity"). If two observers consider the same motion in a given body they find the same state of stress. An "observer" in classical mechanics is a rigid frame which bears a clock.

To these principles come balance of linear and angular momentum, balance of energy and the more doubtful "principle of equipresence". All this resembles breathing very thin air. There is—as we shall see later in Section III—a long and strenuous path to go from these principles to results worth comparing with observations. It will be necessary to include also thermodynamical aspects.

In contrast to these tendencies, there has been efforts by other scientists who have emphasized the necessity of bringing in physical aspects—to keep Mechanics part of physical science. In fact, the present speaker must confess great sympathies with this trend. Among the founders of these congresses, TH. V. KÁRMÁN, R. V. MISES and G. I. TAYLOR have been strong supporters of this trend. Also L. PRANDTL and J. M. BURGERS. Even our grand old man S. TIMOSHENKO was greatly influenced by arguments of a physical nature. He thus broke away from the tendencies of an earlier period, as may be seen from his own historical writings.

From careful experiments with plastic deformation of single crystals of various metals, both his own and those of other workers, G. I. TAYLOR (1934) was led to the concept of dislocations in crystals, see Fig. II.11, which is taken from his paper before the IVth Congress in Cambridge, England. Thus he gave, for the first time, a physical explanation not only for the wide discrepancy between measured yield strength and that derived theoretically from the homogeneous crystal lattice, but also for the incipient plastic deformation.

Taylor's ideas were taken up by J. M. BURGERS (1939), and by N. F. MOTT (1951), F. C. FRANK and W. T. READ (1950), A. H. COTTRELL (1953) and many others in the 1940's and later, thus giving rise to a whole literature of undisputed importance for solid state physics. In these theories dislocations appear as discontinuities of the otherwise regular crystal lattice. It is, however, also possible to use the notion of continuously distributed crystal defects, see K. KONDO (1952) and later a number of other investigators, see E. KRÖNER (1958, 1960), and also B. A. BILBY (1960).

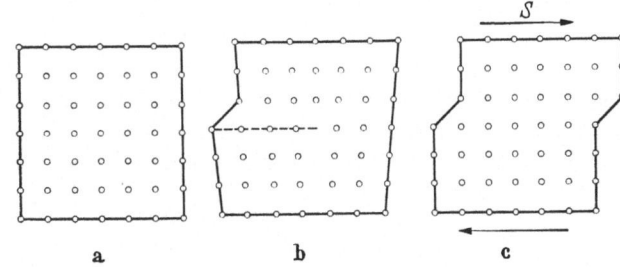

Fig. II.11 Taylor's concept of positive dislocation.

a   b   c

The concept of dislocations is necessary for a deeper understanding of the micro behavior of solid matter. Further, in the 1950's, dislocations have been susceptible to direct observations e.g., with electron microscopy. However, there still remains a gap between the dislocation theories and the constitutive equations of solid continuum mechanics. Bridging this gap unfortunately necessitates a number of more or less arbitrary assumptions. In fact, most of the development described in Chapters II(2)—II(8) has taken place irrespective of our present knowledge of micro behavior. Moreover, when it comes to metal creep, there is need for further complement, connected with the tertiary stage of creep, as defined in Fig. II.7.

10. Deterioration. Creep rupture

In 1958, L. M. KACHANOV presented an attempt to take account of the deterioration caused by creep and giving rise to brittle creep rupture. The insidious nature of creep rupture was known to practical engineers already in the 1930's. After at first having been ascribed to metallurgical effects, it was later recognized as a problem of non-linear solid mechanics. If initial stress σ_{10} is plotted against life time t_R in a double logarithmic plot for a uniaxial creep test, a "creep rupture curve" like that in Fig. II.12 comes out, showing in many cases two

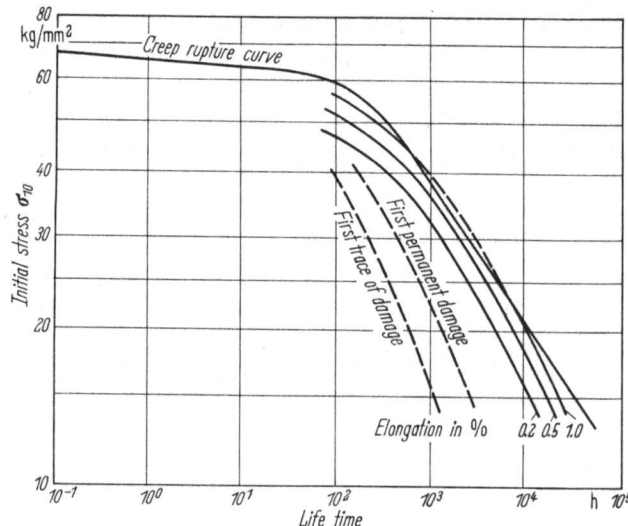

Fig. II.12. Creep rupture curve for a low-alloy steel at 500 °C (K. RICHARD).

distinct linear parts connected with a smooth curve. The left linear part, corresponding to higher initial stresses was recognized by HOFF (1953) as caused by instability, so-called ductile failure. The right linear part occurs for lower initial stress and longer life time and was accompanied by brittle fracture and notch sensitivity, hence its insidious nature. Kachanov's theory could account for the right linear part of the creep rupture curve and also, by combining with Hoff's theory, for the connection between the two linear parts. J. HULT and the present speaker could show (1961) that Kachanov's theory is equivalent with the theory of linear cumulative damage in creep, proposed for purely empirical reasons by E. L. ROBINSON (1952).

Material deterioration is described, according to KACHANOV (1958) by means of a function φ, called the "continuity" of the material. For this function he proposes, in the one-dimensional case

$$\frac{d\varphi}{dt} = - C \left(\frac{\sigma}{\varphi}\right)^{\nu} \tag{II.20}$$

where C and ν are material constants and σ the stress, which may depend on the time t. The continuity φ is $= 1$ at the virgin state, say at $t = 0$ and deterioration is complete $\varphi = 0$ at the time of rupture $t = t_R$. The quotient σ/φ may be considered as an "effective" stress. Integration of Eq. (II.20) yields

$$1 - \varphi^{\nu+1} = C(1 + \nu) \int_0^t \sigma^\nu \, d\tau \tag{II.21}$$

the variable of integration being denoted with τ. Equation (II.21) shows φ decreasing with t for a material the stress history of which is known $\sigma = \sigma(\tau) > 0$. For $t = t_R$, we have $\varphi = 0$ viz.

$$1 = C(1 + \nu) \int_0^{t_R} \sigma^\nu \, d\tau \tag{II.22}$$

which determines the life time t_R. For $\sigma = \sigma_k =$ constant, we have $t_R = t_k$ and (II.22) yields the relation

$$t_k \sigma_k^\nu = \frac{1}{C(1 + \nu)} = \text{constant} \tag{II.23}$$

which corresponds to the straight part to the right of the creep rupture curve in Fig. II.12. Evidently one oughts to have

$$v \leq n. \tag{II.24}$$

For combination with Hoff's theory of ductile creep rupture, KACHANOV makes the fundamental assumption that $\varepsilon = \varepsilon(t)$ be determined by creep action alone which is supposed to go on, irrespective of deterioration. KACHANOV neglects influence of primary creep and puts

$$\frac{d\varepsilon}{dt} = \dot{\varepsilon} = k\sigma^n. \tag{II.25}$$

For a test piece with homogeneous conditions of stress and strain along a gauge length l and with cross section A, incompressibility requires

$$Al = A_0 l_0 = \text{constant}$$

if A_0 and l_0 are the values of A and l for $t = 0$. Then we obtain

$$\dot{\varepsilon} = \frac{1}{l}\frac{dl}{dt} = -\frac{1}{A}\frac{dA}{dt}. \tag{II.26}$$

But $\sigma = P/A$ and P is kept constant, hence

$$\frac{d\sigma}{dt} = \dot{\sigma} = -\frac{P\dot{A}}{A^2} = \sigma\dot{\varepsilon} = k\sigma^{n+1}.$$

For $t = 0$, we have $\sigma = \sigma_{10}$, hence integration yields

$$\sigma = \sigma_{10}(1 - kn\sigma_{10}^n t)^{-1/n} \tag{II.27}$$

which shows that Hoff's time to ductile creep rupture will be t_H

$$t_H = (kn\sigma_{10}^n)^{-1}. \tag{II.28}$$

If, on the other hand, σ from (II.27) is substituted into (II.22), one obtains, setting $t_R = t_K^*$

$$t_K^* = t_H\left\{1 - \left(1 - \frac{n-v}{n}\frac{t_k}{t_H}\left(\frac{\sigma_k}{\sigma_{10}}\right)^v\right)^{n/(n-v)}\right\}. \tag{II.29}$$

KACHANOV now proposes to put $t_R = t_K^*$ in case $t_K^* < t_H$, otherwise $t_R = t_H$. It is seen from (II.29) that this limit corresponds to

$$\sigma_{10} = \bar{\sigma} = \sigma_k\left(\frac{n-v}{n}\frac{t_k}{t_H}\right)^{1/v}.$$

The result is seen in Fig. II.13, which shows at least qualitative agreement with Fig. II.12. As shown by ODQVIST (1962, 1966) also quantitative agreement may be obtained if the influence of primary creep is taken into account.

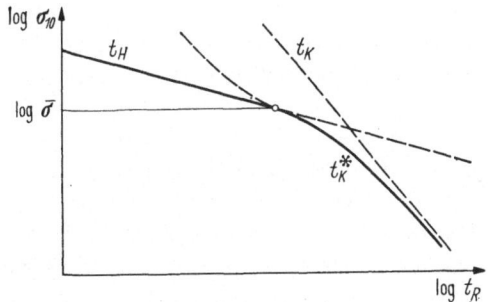

Fig. II.13. Kachanov's theory of combined ductile and brittle creep rupture.

Kachanov's theory as well as the improved theory may easily be applied to a number of special cases such as thin-walled cylinders under torsion and internal pressure where a homogeneous state of stress prevails. In (II.20) then σ has to be replaced by σ_{max}, the largest principal stress, supposed to be positive.

For inhomogeneous stress distributions, KACHANOV has extended his theory, distinguishing between the time periods of latent failure $t = 0 \cdots t_I$, and the time period of propagation of a failure front, $t = t_I \cdots t_{II}$, where t_{II} denotes the time of ultimate creep rupture. During the time period of latent failures, the continuity φ decreases from its initial value 1 and reaches the value zero in certain point or points of the structure under consideration. At $t = t_I$, a failure front, characterized by the condition $\varphi = 0$, starts to propagate through the structure and at the same time a redistribution of stress takes place if the external loads are supposed

12*

to remain constant. As an example of problems of this kind solved by KACHANOV may be mentioned the thick-walled tube, with internal pressure. Solving numerically the integro-differential equation of the propagation of the failure front, he found in special cases the ratio t_{II}/t_I to be at most 150 per cent. As shown by ODQVIST (1969) in this case, for certain combinations of tube dimensions and material constants, t_{II}/t_I may reach several hundred per cent.

Taking up anew the question of the existence of an equation of state, Y. N. RABOTNOV (1963) has suggested to include in the function f of Eq. (II.14) a series of parameters, acounting e.g., for stress history, compare (II.2). After having dismissed a number of such suggestions, he finally retains only the damage $1 - \varphi$, where φ is the continuity according to KACHANOV. We may here as well retain φ. For this quantity Rabotnov's equations would read

$$\frac{d\varepsilon}{dt} = a\sigma^n\varphi^{-q}$$
$$\frac{d\varphi}{dt} = -b\sigma^k\varphi^{-z}$$

(II.30)

where a, b, n, q, k and z are material constants. Thus he complies with Truesdell's "principle of equipresence", making both σ and φ appear on the right sides of (II.30). This system is in agreement with the fact that the slope of creep curves such as in Fig. II.7 may show a tertiary stage even without taking account of diminishing of the cross section, which is the reason for the increase of σ with time in Eq. (II.27). It must be remembered that this effect, which could be termed physical tertiary creep, by no means always exists. This observation would mean that $q = 0$, and would invalidate the principle of equipresence in this particular case.

So far Eqs. (II.30) have been applied only to the problem of rotating disks of uniform thickness, where the analysis is particularly simple. More research in this field is needed, taking account of anisotropic deterioration, caused by creep. KACHANOV himself has treated inherent anisotropic deterioration (1966), see also B. SÖDERQUIST (1968).

III. Present day non-linear solid mechanics

1. New trends. Necessary complements

Chapters II(2) through II(10) have brought us up to the present day. This is certainly not to be understood in such a way that these chapters are definitely finished. On the contrary, at present some details may be investigated very eagerly. One could mention e.g., A. Phillips' and his collaborators' penetration into the detailed behavior of the yield surface (1965), and a generalization of Drucker's postulate to the case when the origin of stress space is outside the yield surface (1968), or the improved theory of the Bauschinger effect by G. B. TALYPOV (1966). But still a certain degree of completion has been reached, at least in the case of time-independent plasticity, less so in time-dependent. Naturally, further research in plasticity will continue and may occasionally also become prosperous indeed.

When reaching present day research, it is much more difficult to claim an unbiased opinion of what is going on. Discoveries and new theories which later may prove to be of the utmost importance may easily be overlooked in today's market noise, where the outcry of new epoch making discoveries may be vociferous enough.

After these apologies, I will make my choice and am going to treat, in the sequel, mainly thermodynamical aspects. This does not mean that other, more or less interesting fields are lacking. One such is that of polar media, taking as point of departure the researches of E. and F. COSSERAT (1909). Problems connected with polar media have attracted a lot of attention during the last ten years, or so. So far very little has been seen about application of such media to explain observed facts. They may serve however, as an useful exercise.

2. Thermodynamical aspects. The work of B. D. Coleman

Elastic materials with one-to-one correspondence between stress and strain, have no dissipation. Thermodynamics does not put any restrictions upon the corresponding constitutive equations.

When plastic deformation occurs, work in general is dissipated and lost in being transformed into heat. This heat will be conducted or radiated or otherwise carried away, and temperature changes will take place. Such processes call for thermodynamical considerations. Attempts to introduce thermodynamics into solid mechanics are not new, cf. v. KÁRMÁN (1914). A decisive turn of events has been brought about by the Truesdell group. Leading journal has been the Archive for Rational Mechanics and Analysis, founded by TRUESDELL. Let us plunge directly into this modern development.

Just in order to try to give an idea of what is going on, we shall follow B. D. COLEMAN (1964) for a moment. He considers quite general motions of a simple material with memory, mapping an arbitrary point $x = x(X, t)$ of a body B at time t on X, a given reference frame, e.g., its position $B(0)$ at time $t = 0$. The gradient F of x with respect to X is assumed to be given $F = F(t - s)$ as a function of $t - s$, $0 \leq s < \infty$, expressing the memory of the material. The "principle of fading memory" states that the memory fades away in time, i.e., for a fixed t, $F(t - s)$ will decrease sufficiently rapid with s, this of course being expressed with all mathematical rigor for which there is no place here. The stress tensor T is assumed to be a general functional of F. In the body B, consisting of the points X, the quantities x, T, body force b, specific energy ε (not to be confused with strain in the other parts of this survey), specific entropy η, absolute temperature θ, heat flux q, and heat supply from outside r are considered as functions of X and t. Between these functions, the two laws of balance (B.L.) of linear momentum and of energy are supposed to hold. In order to specify a *thermodynamic process*, it is sufficient to prescribe the six quantities x, T, ε, q, η and θ. T is supposed to be symmetrical, satisfying the law of balance of angular momentum. The remaining quantities b and r are then uniquely determined from B.L..

A specific free energy ψ is introduced and defined by the relation $\psi = \varepsilon - \theta\eta$. The quantity ψ plays an important role in the theory. Further, as θ in general will vary throughout B, $g = \text{grad } \theta$ will cause heat exchange.

The quantities η, ε, T and q are assumed to depend on memory and to be given, non-linear functionals of F, θ, and g. *These are the constitutive relations, characterizing the material.* Object of Coleman's paper is to impose sufficient and necessary conditions upon these relations from thermodynamical point of view.

Author introduces the quantity γ, the specific rate of entropy production through

$$\gamma = \frac{d\eta}{dt} - \frac{r}{\theta} + \frac{1}{\varrho\theta} \text{div } q - \frac{1}{\varrho\theta^2} qg \qquad \text{(III.1)}$$

where ϱ is the density. Then COLEMAN postulates the Clausius-Duhem inequality in the form

$$\gamma \geq 0 \qquad \text{(C.D.)}$$

for all points in B and all values of t.

Through arguments which are — it must be admitted — a little beyond the speaker's comprehension, COLEMAN pretends to prove a series of theorems, the first and most important — as it seems to me — is the following one.

Under the conditions stated, the functionals ψ, η, ε and T do not contain g explicitly. The quantity $g = \text{grad } \theta$ will thus appear in the heat flux vector only. This statement is said to be a necessary and sufficient condition for C.D. to hold.

By specialization, COLEMAN later arrives at the constitutive equations of infinitesimal (linear) visco-elasticity e.g., as given by BOLTZMANN.

This may well be so, but alas where are the physical applications of the general theory? No mathematical theory will be better than its underlying fundamental assumptions. It is known that after effects in structural metals are of the order of elastic deformations, see

ODQVIST (1956). When COLEMAN speaks about dissipated work for inviscid materials with memory, this must be of the order of elastic work or less. If the dissipated work should have any influence on the constitutive equations for deforming metals, it must be considerably larger, and its influence is still likely to be small, see ODQVIST (1966).

There are however, also other approaches to the subject. The literature on this and related parts of applied mechanics is rapidly growing, its gradient is enormous and it has a fading memory for what used to be more or less established not long ago. I can touch only upon the matter—frankly speaking, I have barely had time to read and penetrate but a small fraction of the pertaining literature. Perhaps, in this connection, I should mention a method introduced by E. T. ONAT at the IUTAM Symposium on Thermoinelasticity at East Kilbide, 1968. Onat takes up Rabotnov's method of State variables (see Chapter II(10)), generalizing them from scalars to tensors and imposing conditions of partial frame indifference on the constitutive relations without reference to the C.D. inequality. Naturally, then his conditions may be sufficient but not necessary.

3. Alternative approaches. M. Biot. H. Ziegler

M. BIOT (1958) considers very general solid bodies, including multi-phase systems, characterized by a number m of generalized coordinates q_i ($i = 1, 2, \ldots, m$) having zero values at equilibrium and being allowed to assume small values only, so that the whole system appears linearized. To the coordinates q_i belong generalized velocities $\dot{q}_i = dq_i/dt$. In the following, as before, we use summation convention for subscripts.

Assuming Onsager's relations to hold ($b_{ij} = b_{ji}$), he proves the dissipation $D = \frac{1}{2} b_{ij}\dot{q}_i\dot{q}_j$ to be non-negative and proportional to the rate of entropy production for a system being isolated but for a connection with a large heat reservoir of constant temperature θ. *The generalized velocity vector $\dot{q}_i$ minimizes the entropy production* subject to the condition that *the power input $X_i\dot{q}_i = constant$.* Here X_i is given as the disequilibrium force, supposed to be fixed.

This result is illustrated symbolically in Fig. III.1, where $D = $ constant denotes a hyperellipsoid of m dimensions. If we consider $D = D_0$ as given, and $D_1 > D_0$, then D_1 is entirely outside D_0. If the plane $X_i\dot{q}_i = $ constant is given $= PP$, then evidently the smallest value of D, compatible with PP is the very hyperellipsoid D_0 that is tangent to PP or, otherwise stated, the vector $\dot{q}_i^{(0)}$ from the origin 0 to the point of tangency of D_0 with PP minimizes the entropy production for all vectors $\dot{q}_i$ outside D_0.

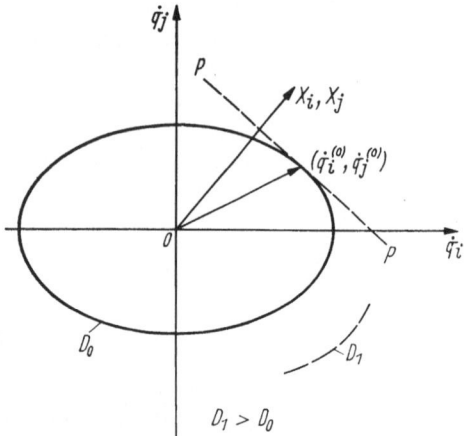

As pointed out by H. ZIEGLER (1957, 1966), this result may also be interpreted as a condition for maximum entropy production, X_i and $D = D_0$ given and letting $\dot{q}_i$ assume all possible values inside D_0. ZIEGLER has commented upon the relations of his extremum principles to that of BIOT as well as to those of I. PRIGOGINE (1946) and of S. R. DE GROOT (1959), cf. the books of PRIGOGINE (1961) and of DE GROOT and MAZUR (1962). More complete references to Ziegler's work may be found in his survey article, see ZIEGLER (1963).

Fig. III.1. Biot's theory of minimum entropy production.

Ziegler's point of departure is more general than that of BIOT, as well as that of PRIGOGINE and DE GROOT. Ziegler's principle is also, possibly, controversial. He does not restrict his variables $\dot{q}_i = w_i$ to take small values only but assumes dissipation, and so the entropy production, D to be given. We may as well assume the fictitious forces P_i to be given, and so write the rate of dissipation in the form

$$L = w_k P_k. \tag{III.2}$$

ZIEGLER now assumes the generalized velocity components w_k to include heat fluxes q_j, if only the corresponding generalized forces are defined as ($\theta =$ absolute temperature, as before)

$$P_j = -\frac{1}{\theta}\frac{\partial\theta}{\partial x_j} \qquad\qquad (III.3)$$

where $\partial\theta/\partial x_j$ is the temperature gradient in a cartesian coordinate system x_j. The expression $L =$ constant is a hyper plane in the space of m dimensions $P_1, P_2, \ldots, P_m$. In the case the constitutive equations

$$w_k = w_k(P_l) \qquad\qquad (III.4)$$

were known, even if these relations were non-linear, insertion in (III.2) would define a surface $L =$ constant, the "L-surface", *a hyper surface which Ziegler assumes to be closed and convex*, i.e., *L is supposed to be non-negative*.

This supposition is more restrictive than Clausius-Duhem's inequality C.D., as introduced by COLEMAN. From (III.1) it is seen that the last term in (III.1) corresponds to (III.3) and that ZIEGLER hence requires non-negativeness of the last term of (III.1) alone and also for the rest of the terms in (III.1), according to C.D. This is evidently a stronger requirement than C.D., and may eventually lead to non-necessary conditions for the constitutive equations.

For the constitutive equations, convexity and certain smoothness conditions of the L-surface, immediately leads to expressions of the form

$$w_i = L\frac{\partial L}{\partial P_i}\left(P_k\frac{\partial L}{\partial P_k}\right)^{-1}. \qquad\qquad (III.5)$$

In the special case of linear constitutive relations, (III.5) are equivalent to Onsager's relations. The Eqs. (III.5) could be put to crucial tests experimentally, as shown by ODQVIST (1967). For the case of uniaxial creep and simultaneous heat flux q in the x-direction, for an austenitic steel, the creep rate of which in secondary isothermal creep may be put

$$v = \left(\frac{\sigma}{\sigma_c}\right)^n \exp kT \qquad\qquad (III.6)$$

with σ_c, n and k material constants, and $T =$ excess temperature over θ, we thus obtain a coupled system for creep rate $d\varepsilon/dt$ and heat flux q

$$\frac{d\varepsilon}{dt} = \left(\frac{\sigma}{\sigma_c}\right)^n \exp kT - \frac{h_1}{\theta}\frac{dT}{dx}$$
$$q = \frac{2h_1\sigma}{3} - \frac{h}{\theta}\frac{dT}{dx}. \qquad\qquad (III.7)$$

Here h is the ordinary coefficient of heat conduction and h_1, a new material constant, expressing the couple effect. If from experiments, h_1 comes out $\neq 0$, it would speak in favor of Ziegler's theory. If, on the other hand, h_1 comes out zero, nothing would be proved, as Coleman's assumptions, as quoted above, are too restrictive with respect to character of the dissipation. It should be noted, however, that COLEMAN and NOLL (1963), from more general considerations, have claimed to prove "non-existence of a piezo-caloric effect", a result also quoted by them from a paper by A. C. PIPKIN and R. S. RIVLIN, which has been inaccessible to me.

There are a number of other approaches to the problems of this chapter, e.g., those of P. M. NAGHDI and A. E. GREEN (1965), of J. F. BESSELING (1967) and of L. I. SEDOV (1962/1965), etc. Lack of space does prevent me from entering upon them here. My general conclusion is that it must be safe to say that the influence of thermodynamics on the constitutive equations of structural metals — whatever it may be — is likely to be small.

4. Statistical aspects

Whenever in mechanics, our knowledge is incomplete in some respect, e.g., about initial or boundary conditions, it is customary ever since MAXWELL, GIBBS and BOLTZMANN to utilize statistical methods. So far, success of such methods has been foremost in fluid mechanics. But also in solid mechanics there have been attempts in the same direction. In most of these non-linear problems are involved.

One example is formed by W. Weibull's statistical theory of strength of brittle materials (1939). In 1967, H. PARKUS has treated influence of random temperature fluctuations on life time in ductile creep rupture. Recently, attempts have been made to correlate Kachanov's notion of deterioration with actually observed and statistically represented distribution of crack growth in austenitic steels at high temperatures, see U. LINDBORG (1968).

Thus, determinism has been forced to give way to statistical aspects. Is this not a sign of resignation? I do not think so. Even this way valuable results have been achieved in applied mechanics.

IV. Future aspects

1. New fields. Multi-phase systems

It is difficult to make predictions—particularly about the future—says Mark Twain. This is a true challenge to all prophets—major as well as minor. One must remember the golden rule for all critics: Moderation. And to judge all achievements of a spiritual character according to its art: "Nothing was created for its faults but for its merits", this being my free translation of THOMAS THORILD, a famous Swedish eighteenth century critic.

Did I try to make you believe in a crisis in present day Applied Mechanics? I hope not. I would rather have it the other way round. Without a perpetual crisis, there would in future be no further progress in our science. What crisis? There is a need for an ever existing influx of new material, be it of a mathematical, physical or possibly some other nature.

And how should the rank and file of our congress members react to such situation?

I think we are sometimes likely to follow a prescription by the British eighteenth century poet ALEXANDER POPE. I quote from memory: "we may not criticize but we may sleep".

What then could be expected in the near future? I will take the easy way and remain at what I happen to know, having listened to a lecture by CLIFFORD TRUESDELL, a few months ago.

TRUESDELL could then foresee, fairly soon, a generalization of Coleman's theory of homogeneous substances to multi-phase systems, including reactions between the phases and diffusion. In fact, such development does not seem unreasonable in view of the success of similar theories for fluid mixtures, see P. GLANSDORFF and I. PRIGOGINE (1954) and R. M. BOWEN (1967). See also the books by DE GROOT and MAZUR (1962), PRIGOGINE (1961), SEDOV (1962/1965) and SCHECHTER (1967).

Thus it might for instance become possible, in principle, to predict the behavior of an alloy steel at elevated temperature, including phase transformations, diffusion etc. Supposing this problem being solved in principle, then utilizing modern digital computers, one could think of computing the creep behavior of such materials as caused by complex stress and temperature histories. Still it may be necessary to introduce statistical aspects. The actual field of composite materials enhance the importance of this type of research.

One may add that further development may be expected in the border regions to neighboring subjects. Such are, as already mentioned, physical metallurgy, fracture mechanics and rock mechanics. These subjects may act as fertilizers, bringing in new physical aspects into the field of applied mechanics.

2. Conclusions

Mr. Chairman, I have tried to give you a balanced and unbiased opinion of the trends within our subject. Just as in world politics, this is a much more difficult task than to take an extreme point of view.

Modern theories in mechanics stand to engineering applications as Einstein to Newton. Both are indispensable.

We shall be critical—of course—but we must not forget to judge our colleagues and fellow workers with justice—to appreciate their merits and to forbear with their weaknesses.

References

BAILEY, R. W.: Inst. Mech. Engrs., London, Proc., Vol. 131, 1935, p. 260—396.

BESSELING, J. F.: Recent progress in applied mechanics, the Folke Odqvist Volume, Ed: B. Broberg, J. Hult and F. Niordson, Stockholm: Almquist and Wicksell, and New York: Wiley etc., 1967, p. 45—59.

BILBY, B. A.: Continuous distributions of dislocations, Progress in solid mechanics, Vol. I, Chapter IV, Ed: I. N. Sneddon and R. Hill, North-Holland 1960, p. 329—398.

BIOT, M.: IIIrd U.S. Nat. Congr. Appl.Mech. 1958, Proc. ASME 1958, p. 1—18.

BLEICH, H.: Bauingenieur 19, 20, 261—267 (1932) (in German).

BOWEN, R. M.: Arch. Rat. Mech. Anal. 24, 370—403 (1967).

BUDIANSKY, B.: J. Appl. Mech. 26, 259—264 (1959).

BURGERS, J. M.: Roy. Nederl. Acad. Sc., Amsterdam, Proc. Vol. 42, 1939, p. 293—325.

BYCHAWSKI, Z.: Arch. Mech. Stos. 4 (17), 593—600 (1965).

COLEMAN, B. D.: Arch. Rat. Mech. Anal. 17, 1—46, 230—254 (1964).

—, and W. NOLL: Arch. Rat. Mech. Anal. 13, 167—178 (1963).

COSSERAT, E. and F.: Théorie des corps déformables, Paris: Hermann 1909 (in French).

COTTRELL, A. H.: Dislocations and plastic flow in crystals, Oxford: Univ. Press 1953, pp. 223; new ed. 1961, pp. 223.

DRUCKER, D. C.: Quart. Appl. Math. 7, 411—418 (1950).

— Ist U.S. Nat. Congr. Appl. Mech. 1950, Proc. ASME 1951, p. 487—491.

— Plasticity, IInd Symposium on Naval Structural Mechanics, Brown University 1960, Proc. Ed: E. H. Lee and P. S. Symonds, Pergamon Press 1960, p. 170—184.

— IVth U.S. Nat. Congr. Appl. Mech. 1962, Proc. ASME 1962, Vol. I, p. 15—33.

FINNIE, I.: Applied mechanics surveys, Ed: H. N. Abramson, H. Liebowitz, J. M. Crowley and S. Juhasz, Washington, D.C. 1966, p. 373—387.

—, and W. R. HELLER: Creep of engineering materials, New York etc.: McGraw-Hill 1959, p. 116—118.

FRAEIJS DE VEUBEKE, B.: High temperature effects in aircraft structures, Chap. 13, Ed: N. J. Hoff, New York etc.: Pergamon Press 1958, p. 267—287.

FRANK, F. C., and W. T. READ: Phys. Rev. 79, 722 (1950).

FREUDENTHAL, A., and H. GEIRINGER: The mathematical theories of the inelastic continuum, Encyclopedia of Physics, Vol. VI, Ed: S. Flügge, Berlin/Göttingen/Heidelberg: Springer 1958, p. 228—433.

GLANSDORFF, P., and I. PRIGOGINE: Physica 20, 773 (1954).

GREEN, A. E., and P. M. NAGHDI: Arch. Rat. Mech. Anal. 18, 251 (1965).

DE GROOT, S. R., and P. MAZUR: Non-equilibrium thermodynamics, Amsterdam: North-Holland 1962; repr. 1963, pp. 512.

GRUENING, M.: Tragfähigkeit statisch unbestimmter Tragwerke ..., Berlin 1926 (in German).

HENCKY, H.: ZAMM 3, 241—251 (1923); ibidem 5, 115—124 (1925) (in German).

HILL, R.: The mathematical theory of plasticity, Oxford: Clarendon 1950, pp. 356.

HOFF, N. J.: J. Appl. Mech. 20, 105—108 (1953).

— Wilbur Wright Memorial Lecture, J. Roy. Aero. Soc., 58, 3—52 (1954).

— Quart. Appl. Math. 12, 49—55 (1954).

HOHENEMSER, K.: ZAMM 11, 15—19 (1931) (in German).

JOHANSEN, K. W.: Brudlinieteorier, Diss. Copenhagen 1943, pp. 191 (in Danish).

JOHNSON, A. E.: Inst. Mech. Engrs., London, Proc. Vol. 164, 1951, p. 432—447.

KACHANOV, L. M.: Some questions of creep theory, Leningrad-Moscow, 1949, pp. 164 (in Russian).

— Izv. Akad. Nauk, USSR, Nr. 8, 1958, p. 26 (in Russian); see also Problems of Continuum Mechanics, Contributions in Honor of N. I. Mushkelishvili's 70ieth Birthday, 16 Febr. 1961, Ed: J. R. M. Radok, Philadelphia 1961, p. 202—218.

— Recent progress in applied mechanics, the Folke Odqvist Volume, Ed: B. Broberg, J. Hult and F. Niordson, Stockholm: Almquist and Wicksell, and New York: etc. Wiley 1967, p. 329—336.

v. KÁRMÁN, TH.: Encykl. Math. Wiss., Vol. IV, 4 (5), Art. 31, Leipzig 1914, p. 695—770 (in German).

KAZINCZY, G.: Betonszemle, 2, Nr. 5, 6, 7, 1914 (in German).

KLEIN, F.: Letter Febr. 1921 to D. Meyer, published in an Obituary, ZAMM 5, 358—359 (1925) (in German).

KOITER, W. T.: General theorems for elastic-plastic solids, Progress in Solid Mechanics, Vol. I, Chap. IV, Ed: I. N. Sneddon and R. Hill, North-Holland 1960, p. 165—221.

KONDO, K.: IInd Jap. Nat. Congr. Appl. Mech., Proc. 1952, p. 41—47.

KRÖNER, E.: Kontinuumstheorie der Versetzungen und Eigenspannungen, Berlin/Göttingen/Heidelberg: Springer 1958, pp. 179 (in German); see also Applied Mechanics Surveys, Ed: H. N. Abramson, H. Liebowitz, J. M. Crowley and S. Juhasz, Washington, D.C. 1966, p. 189—199.

LEVY, M.: J. Math. Pur. Appl. Paris, 16, 369—372 (1871) (in French).

LIN, T. H.: J. Aero. Sc. 23, 883—887 (1956).

LINDBORG, U.: Creep cracks and the concept of damage, Manuscript 1968, to be published in the J. Mech. Phys. Sol.

LODE, W.: Z. Phys. 36, 913 (1926) (in German).

MELAN, E.; Ing.-Arch., 9, 116—126 (1938) (in German).

v. MISES, R.: Göttinger Nachr., math. phys. Klasse. 1913, p. 582—592 (in German).

— ZAMM 1, 1—15 (1921); ibidem 9, 433 (1929) (in German).

v. Mises, R.; ZAMM 8, 161—185 (1928) (in German).

Mott, N. F.: Phys. Soc. London, Proc. Ser. B, Vol. 64, 1951, p. 729; see also Phil. Mag., Vol. 43, 1952, p. 1151 and ibidem, 44, 1953, p. 742.

Nadai, A.: ZAMM 3, 442—454 (1923) (in German).

Nakanishi, F.: Rep. Aeronaut. Res. Inst., Tokyo 6, 83 (1931).

Namestnikov, V. S., and Y. N. Rabotnov: J. Appl. Mech. Techn. Phys. 3, 101—102 (1961) (in Russian).

Neal, B. G.: The plastic methods of structural analysis, New York: Wiley 1956, pp. 353.

Norton, F. H.: Creep of steel at high temperatures, New York: McGraw-Hill 1929, pp. 90, p. 67.

Odqvist, F. K. G.: ZAMM 13, 360—363 (1933) (in German).

— IVth Int. Congr. Appl. Mech. Cambridge 1934, Proc. Cambridge Univ. Press 1935, p. 228—229.

— VIIIth Int. Congr. Appl. Mech. Istanbul 1952, Proc. Vol. I, Istanbul 1953, p. 99; see also Roy. Inst. Tech. Stockholm, Trans. Nr. 66, 1953, pp. 18, and IUTAM Colloquium Creep in Structures, Stanford 1960, Proc., Ed: N. J. Hoff, Berlin/Göttingen/Heidelberg: Springer 1962, p. 137—160.

— Symposium on Plasticity, Varenna 1956, Memorie in Honor of Prof. Arturo Danusso, Zanichelli, Bologna 1957, p. 204—216.

— IUTAM Symposium Second-Order Effects in Elasticity, Plasticity and Fluid Dynamics, Haifa 1962, Proc. Ed: M. Reiner and D. Abir, Pergamon Press 1964, p. 295—313.

— Mathematical theory of creep and creep rupture, Oxford: Clarendon 1966, pp. 170.

— J. Appl. Math. Physics (ZAMP), Vol. 18, 1967, p. 762—764; see also L. I. Sedov 60ieth Anniversary Volume, to be published.

— A. E. Johnson Memorial Volume, to be published 1969.

—, and J. Hult: Roy. Acad. Sci. Stockholm, Arkiv för Fysik, Vol. 19, 1961, p. 379—382.

— — Creep strength of structural materials, Berlin/ Göttingen/Heidelberg: Springer 1962, pp. 303 (in German).

Onat, E. T.: Representation of inelastic thermomechanical behavior by means of static variables, IUTAM Symposium on Thermoinelasticity, East Kilbride 1968, Proc. Ed: B. Boley, to be published by Springer-Verlag, Vienna.

Parkus, H.: Recent Progress in Applied Mechanics, the Folke Odqvist Volume, Ed: B. Broberg, J. Hult and F. Niordson, Stockholm: Almquist and Wicksell, and New York: Wiley 1967, p. 391—397.

Phillips, A., and M. Eisenberg: Int. J. Non-Linear Mech., 1, 247—256 (1966); see also J. W. Justusson and A. Phillips, Acta Mech. II, 3, 251—267 (1966).

— Yield surfaces of pure aluminium at elevated temperatures, IUTAM Symposium on Thermoinelasticity, East Kilbride 1968, Proc. Ed: B. Boley, to be published by Springer-Verlag, Vienna.

Prager, W.: Mémorial des sciences mathématiques, 87, pp. 66 (1937) (in French).

— J. Appl. Phys. 16, 837—840 (1945).

— VIIth Int. Congr. Appl. Mech., London 1948, General Lecture; see also Journ. Appl. Phys. 20, 1949, p. 235—241.

— Office of Naval Research, Contract No. nr-562(20)/4, Nov. 1957, pp. 12.

— Plasticity, IInd Symposium on Naval Structural Mechanics, Brown University 1960, Proc. Ed: E. H. Lee and P. S. Symonds, Pergamon Press 1960, p. 608—610.

—, and Ph. Hodge, Jr.: J. Math. Phys. 27, 1—10 (1948).

— — Theory of perfectly plastic solids, New York: Wiley, 1951, pp. 264.

Prandtl, L.: Göttinger Nachr., Math.-phys. Klasse, 1920, p. 74—85 (in German).

— ZAMM 1, 15—20 (1921) (in German).

— Ist Int. Congr. Appl. Mech. Delft 1924, Proc. Delft 1925, p. 43—54.

Prigogine, I.: Acad. Roy. Belg., Bull. Classe sci., 32, 30 (1946).

— Thermodynamics of irreversible processes, Sec. Ed. New York: Wiley 1961, pp. 119.

Rabotnov, Y. N.: Joint Internat. Conf. Creep, London Sept./Oct. 1963, Paper 68, 2.117—122.

— Kinetics of creep and creep rupture, IUTAM Symposium on Irreversible Aspects of Continuum Mechanics, Proc., Ed: H. Parkus, Berlin/Heidelberg/New York: Springer 1968, p. 326—334.

Reiner, M.: Amer. J. Math. 67, 350—362 (1945).

Reuss, E.: ZAMM 10, 266—274 (1930) (in German).

— IVth Internat. Congr. Appl. Mech., Cambridge 1934, Proc. Cambridge: University Press 1935, p. 241.

Robinson, E. L.: Am. Soc. Mech. Engrs., Trans., Vol. 74, 1952, p. 777—780.

de Saint-Venant, B.: Acad. Sci., Paris, Comptes Rendus, Vol. 70, 1870, p. 473 (in French).

— J. Math. Pur. Appl., Paris, Vol. 16, 1871, p. 308—316 (in French).

Samuelson, A.: The Aeronautical Research Institute of Sweden (FFA), Rep HU902:1, 1961, Rep. Nr. 100, 1964, Rep. Nr. 108, 1967.

Sawczuk, A., and Th. Jaeger: Grenztragfähigkeitstheorie der Platten, Berlin/Göttingen/Heidelberg: Springer 1963, pp. 522 (in German).

Sayir, M., and H. Ziegler: Ing.-Arch. 36, 294—302 (1968) (in German).

Schechter, R. S.: The variational method in engineering, New York: McGraw-Hill 1967, pp. 287.

Sedov, L. I.: Introduction to the mechanics of a continuous medium, Moscow 1962 (in Russian); Engl. transl. Addison-Wesley, 1965, pp. 270.

Söderquist, B.: Acta Polytechnica Scandinavica, Phys. incl. neucl. ser., 1968, Nr. 51, pp. 61, Nr. 53, pp. 36, Nr. 58, pp. 16.

Stodola, A.: ZAMM 13, 143—146 (1933) (in German).

Storakers, B.: Int. J. Mech. Sci., 10, 1968, forthcoming, pp. 11.

Sundström, E.: Roy. Inst. Tech., Stockholm, Trans. Nr. 115, pp. 34.

Talypov, G. B.: Mechanics of Solids, Nr. 6, 1966, p. 81—88 (in Russian).

Taylor, G. I.: IVth Int. Congr. Appl. Mech., Cambridge 1934, Proc. Cambridge: University Press 1935, p. 113—125; see also Roy. Soc. London, Proc. Vol. A-145, 1934, p. 362—415.

Taylor, G. I., and H. Quinney: Roy. Soc. London, Phil. Trans. Vol. A-230, 1931, p. 323—362.

Timoshenko, S.: As I Remember (Transl. from the Russian original), New York: van Nostrand 1968, pp. 430.

Truesdell, C.: Six Lectures on Modern Natural Philosophy, Berlin/Heidelberg/New York: Springer 1966, pp. 117; see also Elements of Continuum Mechanics, ibidem, 1966, pp. 279; and Applied Mechanics Surveys, Ed: H. N. Abramson, H. Liebowitz, J. M. Crowley and S. Juhasz, Washington, D.C. 1966, p. 225—236. A fairly complete account of the development during 1945—1961 is given in Continuum Mechanics, Ed:

C. Truesdell, New York Ctr.: Gordon and Breach Sc. Publ. Vol. I, 1966, pp. 216, Vol. II, 1965, pp. 436, Vol. III, 1965, pp. 310.

Wang, A. J., and W. Prager: Brown University, Providence, Graduate Div. Appl. Math., A 11—102, Dec. 1953, pp. 8.

Weibull, W.: The Royal Swedish Institute for Engineering Research, Proc. Nr. 151, 1939, pp. 45 and Nr. 153, 1939, pp. 55; see also Applied Mechanics Surveys, Ed: H. N. Abramson, H. Liebowitz, J. M. Crowley and S. Juhasz, Washington, D.C., 1966, p. 397—400.

Ziegler, H.: Ing.-Arch. 25, 58—70 (1957); see also IUTAM Symposium on Irreversible Aspects on Continuum Mechanics, Vienna 1966, Proc., Ed: H. Parkus, Berlin/Heidelberg/New York: Springer 1968, p. 411—424.

— Some extremum principles in irreversible thermodynamics, Progress in Solid Mechanics. Vol. IV, Chap. II, Ed: I. N. Sneddon and R. Hill. North-Holland 1963, p. 91—193.

Application of aeroelastic constraints in structural optimization

By

H. Ashley and **S. C. McIntosh, Jr.**

Stanford University

I. Introduction

The optimum synthesis of structures, under preassigned conditions of static loading and various simplified criteria of elastic or plastic stability, has now been carried to a high degree of refinement. Total mass is normally chosen as the merit function for such designs, although occasionally one might seek a minimum of mass moment of inertia or some other quantity. To ensure practicability and arrive at a well-defined problem for mathematical optimization, the general configuration of members is chosen in advance and constraints are placed on shape or volume. As evidenced by examples like those shown in a recent paper by SCHMIT, MORROW and KICHER [1], the results are beginning to arouse keen interest wherever heavily-loaded structures must be built at the least penalty in weight. Not surprisingly, the aerospace industry is one of the strongest sources of stimulation for both research and applications of structural optimization.

In connection with the design of aerodynamically-sustained aircraft, however, recognition must be given to the fact that the member sizes of many lifting-surface structures are as critically affected by considerations of aeroelastic stability as they are by more conventional loading criteria. Freedom from flutter, divergence and buckling must be confidently guaranteed. Moreover, the redistribution of aerodynamic pressure, due to deformations, is now routinely accounted for during the airload analysis of high-speed aircraft. Signaled by the initial publication of TURNER [2], the formal introduction of such aeroelastic constraints into the optimization process has already formed the subject matter of several papers [3, 4, 5].

No purpose would be served here by summarizing the literature on conventional optimal design, so it will only be remarked that a stage of consolidation and ordering seems to have been reached. A survey published by GERARD [6] more than two years ago was already able to list 175 references.

One of several dimensions along which any piece of work can be categorized concerns whether the optimization technique involves an analytic search of function space, by means of the variational calculus, or the algebraic evaluation of a set of discrete real variables. The former approach, exemplified by PRAGER and his co-workers [e.g. Ref. 7, 8, 9], is limited to relatively simple applications but can produce generalizations of value for assessing the potentialities of optimization in a new area. The latter, which has been highly refined through the use of the digital computer by GELLATLY [10], MELOSH [11], SCHMIT [1] and many others, is almost certainly the route toward design procedures for realistic structures. The present authors have adopted the analytical approach as best suited to their objectives. But it must be noted that, in some of his examples [2, 5], TURNER has laid the foundations of aeroelastic optimization through the discrete-element idealization of more complex configurations.

As a framework for the determination of extremal solutions, it proves convenient to formulate each one-dimensional structure as a problem of BOLZA or LAGRANGE (see, for instance, MIELE [12]). The constraint equations on the physical dependent variables $q_i(x)$ can be solved explicitly for the first derivatives q_i' and arranged in the form

$$\vec{q}' = \vec{f}(\vec{q}, \, x), \tag{1}$$

$\vec{q}$ being a vector of n components q_i and x being the single independent variable. The analogy with optimal problems in modern control theory, whose formalism is especially well described in a recent book by BRYSON and HO [13], has proved invaluable.

II. Minimum weight for a given fundamental frequency of vibration

This is believed to be an old problem, although the earliest paper in recent literature is apparently that of NIORDSON [14]. Fixing the vibration frequency of a distributed elastic system is not exactly an aeroelastic constraint. TURNER has pointed out [2, 5], however, that the lowest torsional frequency of a cantilever wing has such a controlling effect on its critical flutter speed that fixing the former during optimization will often be closely equivalent to the more difficult constant-V_F solutions described in Sect. IV below. A summary of the solution [2] for longitudinal vibration of a nonuniform bar with tip mass (Fig. 1) also provides a simple illustration of the underlying mathematical apparatus.

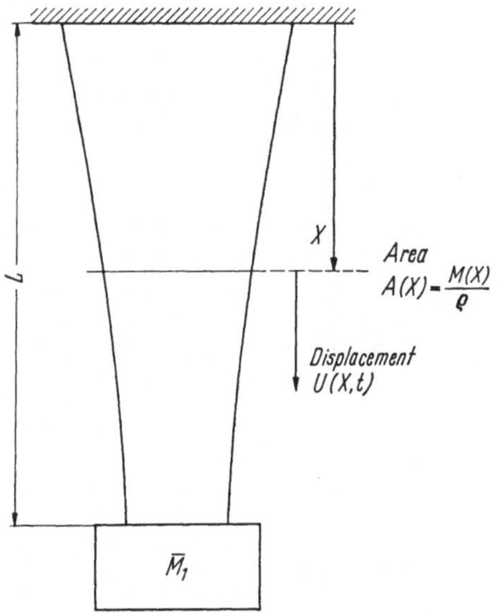

The dimensionless[1] differential equation governing the amplitude $u(x)$ of small, simple harmonic longitudinal displacements reads

$$(mu')' + \beta^2 mu = 0. \tag{2}$$

Here $m(x)$ is mass per unit length divided by ϱL^2, ϱ being density of the homogeneous linear elastic material. Also

$$\beta = \omega L \sqrt{\frac{\varrho}{E}}, \tag{3}$$

with E the Young's modulus and ω the circular frequency. Boundary conditions for the situation of Fig. 1, with unit tip amplitude, are

$$\left.\begin{array}{ll} u(0) = 0, & u(1) = 1 \\ (mu')|_{x=1} = \beta^2 \dfrac{\bar{M}_1}{\varrho L^3} & \end{array}\right\}. \tag{4}$$

In an effort to minimize

Fig. 1. Nonuniform elastic bar with tip mass.

$$\mathscr{I} = \int_0^1 m \, dx, \tag{5}$$

one has only the single control variable $m(x)$ to work with because the running mass and the longitudinal stiffness $EA(x)$ are directly proportional. ω is fixed at its value for the fundamental of a uniform bar carrying the same $\bar{M}_1$.

Necessary and sufficient conditions for an extremal of (5) are known to be completed by constructing the function

$$F = m(x) + \lambda_u(x)\left[\frac{y}{m} - u'\right] + \lambda_y(x)[-\beta^2 mu - y'] \tag{6}$$

[1] Capital and lower-case letters will be employed, respectively, to denote dimensional and corresponding dimensionless quantities. Length L of a one-dimensional elastic system is used throughout as a reference length, so that the bar, rod, beam, etc. extends from $x = 0$ to 1. Sectional properties, such as running mass, flexural rigidity, etc., of the optimum system are usually made dimensionless by division with the same property of a corresponding uniform system.

and applying the Euler-Lagrange differential equation

$$\frac{\partial F}{\partial q_i} - \frac{d}{dx}\left(\frac{\partial F}{\partial q_i'}\right) = 0 \qquad (7)$$

to $q_1 \equiv m$ and the state variables $q_2 \equiv u$ and $q_3 \equiv y = mu'$. λ_u and λ_y are Lagrange multipliers for the two first-order constraint equations. After some eliminations, Eqs.(7) can be reduced to

$$-u'\lambda_y' + \beta^2 u\lambda_y = 1 \qquad (8)$$

$$(mu')' + \beta^2 mu = 0 \qquad (9)$$

$$(m\lambda_y')' + \beta^2 m\lambda_y = 0. \qquad (10)$$

The so-called transversality condition [12, 15] yields an additional boundary condition

$$\lambda_y(0) = 0. \qquad (11)$$

Because u and λ_y satisfy identical differential equations, as well as the same boundary conditions (4), (11) at the built-in end, they must be proportional:

$$\lambda_y = -\frac{u}{C^2}. \qquad (12)$$

The control equation (8) therefore becomes

$$(u')^2 - \beta^2 u^2 = C^2 \qquad (13)$$

(12) is obviously solved, with $u(1) = 1$, by

$$u(x) = \frac{\sinh \beta x}{\sinh \beta}, \qquad (14)$$

and (9) can then be integrated to yield the optimal mass distribution, which should meet the third of conditions (4),

$$m(x) = \frac{\beta \bar{M}_1 \sinh 2\beta}{2\varrho L^3 \cosh^2 \beta x}. \qquad (15)$$

While finding the solution (15), one must assume $u'(x) \neq 0$. Thus the results apply only to the fundamental, and higher harmonic frequencies do not remain fixed as m is varied.

TAYLOR has recently proved [16] both the uniqueness of Turner's solution and that it applies equally well to the determination of the bar of given mass which has the highest fundamental ω. It is easy to show that the ratio of structural mass of the optimum bar to that of a uniform bar with equal β is $\sinh^2 \beta/\beta \tan \beta$. The practice of PRAGER will be followed here of treating such ratios as measures of what can be achieved by optimal design (that is, a uniform structure with the same stiffness or aeroelastic performance is taken as a standard of reference). Fig. 2 shows how this mass ratio varies with β.

In the foregoing example, the tip mass is introduced mainly as an artifice to prevent the optimal bar from shrinking to vanishing proportions. This same aim can be accomplished by altering the problem in either of two practically interesting ways, both of which have recently been studied by TAYLOR [17]: distributed nonstructural mass may be added as a burden along the bar's length or it may be required that the area exceed a specified minimum,

$$A(x) \geq A_0. \qquad (16)$$

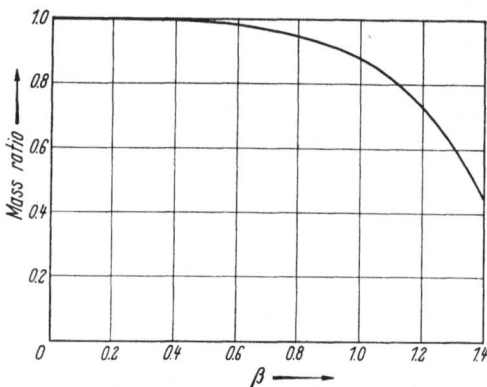

Fig. 2. Ratio of structural mass of optimum bar to that of a uniform bar with the same value of $\beta = \omega L \sqrt{\frac{\varrho}{E}}$ for the fundamental frequency of longitudinal vibration.

It is of more significance in aeroelasticity to present these results in terms of torsional vibration of a cantilever rod or wing, however, since the most important form of the latter problem is mathematically analogous to the bar.

As pointed out by PRAGER and TAYLOR [7], the greatest rigidity is attained within a given cross section or volume by locating structural material as close as possible to the bounding surface. (One-dimensional instances would be the sandwich beam or column with thin face sheets and the thin-walled tube in torsion). In the spirit of this observation, consider a wing of rectangular planform and span L, whose torsionally-effective material is concentrated in a single box of fixed cross-sectional shape and size. If the box skin thickness $T(x)$ is small compared with its depth, Bredt's formula shows that the torsional rigidity $GJ \sim T$.

Let the uniform reference wing have constant rigidity GJ_0, thickness T_0, and mass moment of inertia I_0 (per unit span about the elastic axis). The dimensionless differential equation for the torsional vibration amplitude $\theta(x)$ is

$$\theta'' + \Omega^2 \theta = 0 \tag{17}$$

where

$$\Omega = \omega L \sqrt{\frac{I_0}{GJ_0}} . \tag{18}$$

With cantilever boundary conditions

$$\theta(0) = 0, \quad \theta'(1) = 0 \tag{19}$$

the familiar quarter-sine-wave fundamental mode corresponds to $\Omega = \pi/2$.

For the optimization problem, note that

$$\frac{GJ(x)}{GJ_0} = \frac{T(x)}{T_0} \equiv t(x). \tag{20}$$

Let a fraction δ_1 of the running moment of inertia $I_\theta(x)$ be contained in the skin, and let the remaining inertia, *which is assumed to have the same radius of gyration as the skin box*, be equal to that of the reference wing. It follows that

$$\frac{I_\theta(x)}{I_0} = \delta_1 t(x) + \delta_2, \tag{21}$$

where $\delta_1 + \delta_2 = 1$. Moreover, the dimensionless differential equation and boundary conditions read

$$(t\theta')' + [\delta_1 t + \delta_2] \Omega^2 \theta = 0 \tag{22}$$

$$\theta(0) = 0, \quad (t\theta')|_{x=1} = 0 \tag{23}$$

with $\Omega = \pi/2$ held fixed during optimization.

Solution for a minimum value of

$$\mathscr{I} = \int_0^1 t \, dx \tag{24}$$

proceeds in a manner similar to the bar. For later purposes, it is mentioned that the control equation[1] takes the form

$$\frac{\lambda_\theta s}{t^2} + \left(\frac{\pi}{2}\right)^2 \delta_1 \lambda_s \theta = 1, \tag{25}$$

where $s = t\theta'$, and that (25) can be manipulated into

$$(\theta')^2 - \left(\frac{\pi}{2}\right)^2 \delta_1 \theta^2 = C^2. \tag{26}$$

The optimum vibration mode turns out to be (cf. (14))

$$\theta = \theta(1) \sinh\left(\frac{\pi}{2}\sqrt{\delta_1}\, x\right), \tag{27}$$

---

[1] The term "control equation" designates the algebraic relation obtained when (7) is applied to the control variable (t in the present problem).

with a corresponding thickness distribution

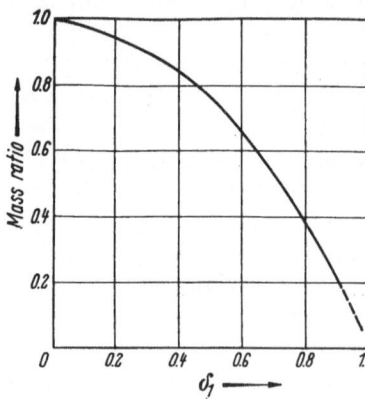

$$t(x) = \frac{\delta_2}{2\delta_1} \left\{ \left(\frac{\cosh\left(\frac{\pi}{2}\sqrt{\delta_1}\right)}{\cosh\left(\frac{\pi}{2}\sqrt{\delta_1}\,x\right)} \right)^2 - 1 \right\}. \qquad (28)$$

Recalling that masses and moments of inertia have been arranged to be in proportion, one finds for the overall mass ratio

$$\frac{\delta_1 \int_0^1 t(x) + \delta_2}{\delta_1 + \delta_2} = \frac{(1-\delta_1)}{2}\left[1 + \frac{1}{\pi\sqrt{\delta_1}}\sinh\left(\pi\sqrt{\delta_1}\right)\right]. \qquad (29)$$

Fig. 3. Ratio of mass of optimum cantilever rectangular wing that to of a uniform wing with the same fundamental torsional frequency, plotted vs. the fraction δ_1 of total mass devoted to skin material effective in torsion.

Equation (29) is plotted as a function of δ_1 in Fig. 3. The uniform-wing limit of unity when $\delta_1 \to 0$ is self-evident, whereas the case $\delta_1 \to 1$ is unrealistic for the same reason as the bar with $\overline{M}_1 = 0$.

III. Static aeroelastic examples

The static aeroelastic instabilities known as torsional divergence and supersonic chordwise divergence normally would occur at such high flight speeds as not to have a direct influence on structural design. Nevertheless, the speed V_D (or dynamic pressure $\frac{\varrho}{2}V^2$ for neutral stability) are good enough measures of the general stiffness level of a lifting surface that a mass minimization based on holding them constant could have practical interest. Several divergence solutions related to a wing like that in Fig. 4, with or without the body attached at the tip, have already been published [3, 4]. For example, McIntosh and Eastep [4] analyzed a straight-tapered planform whose variation of chordlength with spanwise distance is given by

$$C(x) = C_R[1 - \alpha x]. \qquad (30)$$

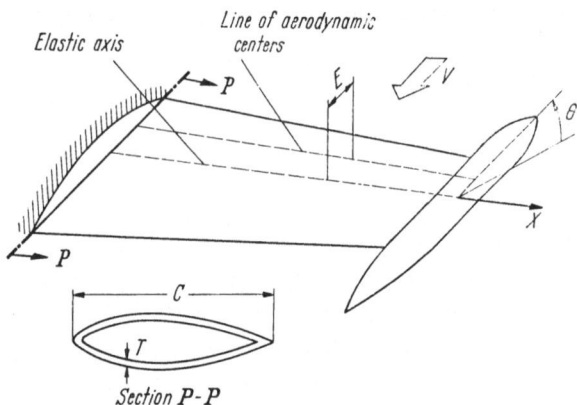

Fig. 4. Unswept cantilever wing used for torsional divergence calculations.

Only the skin mass was considered, and torsional rigidity was again assumed proportional to thickness $T(x)$, this time multiplied by C^2 to account for the sectional area variation. Aerodynamic induction was neglected.

The special case of a constant-chord, rectangular wing of fixed V_D was used to validate numerical procedures for integrating the optimizing equations. The constraint differential equations and boundary conditions for this problem are

$$\frac{s}{t} - \theta' = 0 \qquad (31)$$

$$\Omega^2\theta + s' = 0 \qquad (32)$$

$$\theta(0) = 0, \quad s(1) = 0 \qquad (33)$$

where

$$\Omega^2 = \frac{\varrho V_D^2 C^2 e L^2}{2 G J_0} \, C_{L_\alpha} = \left(\frac{\pi}{2}\right)^2. \qquad (34)$$

C_{L_α} is the lift-curve slope of the airfoil section, ϱ is the air density and e is the chord-fraction by which the elastic axis lies behind the sectional aerodynamic center. The Euler-Lagrange equations for the function

$$F = t + \lambda_\theta \left[\frac{s}{t} - \theta' \right] + \lambda_s [-\Omega^2 \theta - s']$$ (35)

are found to read

$$\lambda_\theta' - \Omega^2 \lambda_s = 0$$ (36)

$$\lambda_s' + \frac{\lambda_\theta}{t} = 0$$ (37)

$$\lambda_\theta \frac{s}{t^2} - 1 = 0.$$ (38)

Transversality yields two further boundary conditions,

$$\lambda_s(0) = 0, \quad \lambda_\theta(1) = 0.$$ (39)

Of particular interest is that considerations like those underlying (13) transform the control equation (38) into

$$(\theta')^2 = A^2.$$ (40)

It is a simple matter to construct an exact solution

$$\theta = Ax$$ (41)

$$t(x) = \frac{\Omega^2}{2} [1 - x^2]$$ (42)

corresponding to

$$\int_0^1 t \, dx = \frac{\pi^2}{12}$$ (43)

and a weight saving of 18%. For comparison purposes, the system[1] (31)—(34), (36)—(39) was integrated numerically on an IBM 360/67 computer by means of a Kutta-Merson predictor-corrector procedure. After examining several alternatives, the straightforward transition-matrix method described in Sect. 7.3.2. of BRYSON and HO [13] was found to be most effective. Convergence was found to be quite rapid, even from a starting solution corresponding to constant $T = T_0$. The success of the procedure can be inferred from Fig. 5, where numerically determined values of $t(x)$ are plotted alongside the exact result (42).

Figures 6 and 7 present the optimal thickness distributions and weight savings for divergence of rectangular cantilevers whose skin thickness is constrained to be greater than or equal to some fraction t_0 of the reference thickness T_0. Details of the solution will be given in Ref. [19]. The constraint is applied in the customary manner, which defines a new control variable whose square equals $(t - t_0)$. The Euler-Lagrange equations are then found to apply only to a variable-thickness segment inboard of $x = x_0$, whereas for $x_0 \leq x \leq 1$ the thickness is constant at t_0 and the uniform-wing solution is employed. The Weierstrass-Erdmann corner conditions allow no discontinuity in t at station x_0, whose location is determined as a function of t_0 by solving a simple transcendental equation.

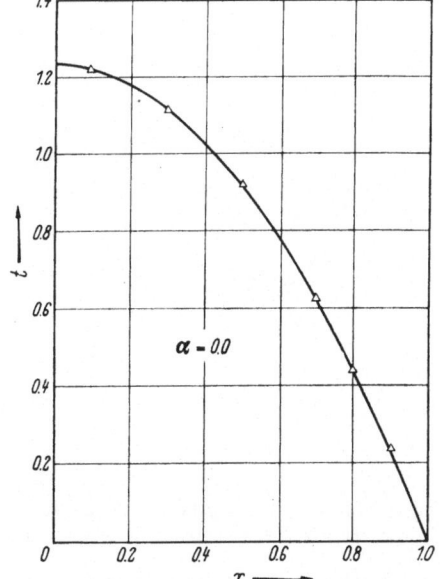

Fig. 5. Optimum dimensionless skin thickness distribution for a rectangular wing of fixed divergence speed. Exact solution is compared with numerical computations by the transition-matrix procedure.

---

[1] Clearly these differential equations have the standard form (1).

Aeroelastic stability of the sort of system illustrated in Fig. 8 was thoroughly investigated during the 1950's by Biot [20]. "Curling up" of such a leading edge can occur in supersonic flight because increasing the chordwise slope

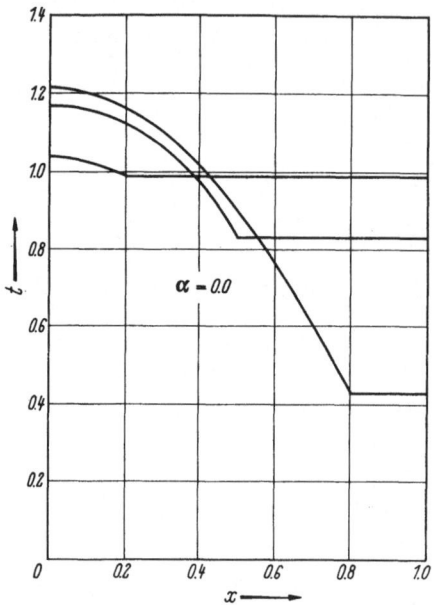

$$\alpha_e = \frac{dW}{dX} \qquad (44)$$

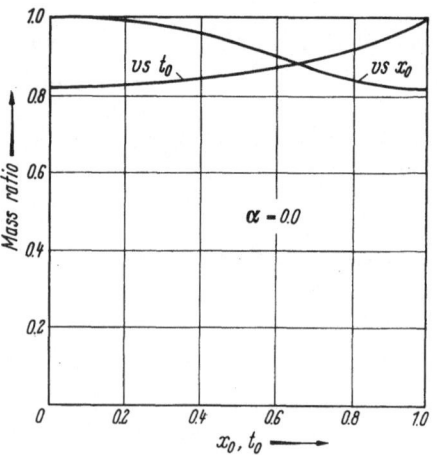

Fig. 6. Optimum dimensionless skin thickness distributions for torsional divergence of a rectangular cantilever wing with $\delta_1 = 1$ and various minimum values of skin thickness.

Fig. 7. Mass ratio vs. minimum skin gauge t_0 for the wings of Fig. 8. Same ratio is also shown vs. x_0, point where minimum tikchness begins.

locally produces an increase of upward loading. Without giving full details, it can be stated that α_e is governed by the differential equation

$$\frac{d^2}{dx^2}\left[\frac{EI}{\frac{\varrho}{2}V^2B_1^3(1-\nu^2)}\frac{d\alpha_e}{dx}\right]$$
$$= \frac{4}{\sqrt{M^2-1}}\left[1+\frac{\gamma+1}{2}M\frac{dZ_t}{dX}\right][-\alpha_e+\alpha_0]. \qquad (45)$$

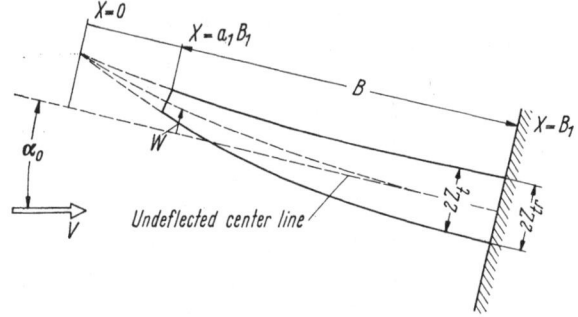

Fig. 8. Cross section of forward half of an airfoil undergoing chordwise bending in a two-dimensional supersonic airstream.

Here M is flight Mach number, I is area moment of inertia per unit span, ν is Young's modulus for the uniform plate material, and other symbols are defined in the figure. The homogeneous divergence problem is obtained by setting the initial incidence $\alpha_0 = 0$ and enforcing the boundary conditions

$$\alpha_e(1) = 0 \qquad (46)$$

plus zero shear and bending moment at the leading edge.

Of the various optimal problems that can be based on maintaining constant chordwise-divergence speed, one that has proved analytically tractable concerns the case of a honeycomb structure with constant depth Z_t and thin face sheets. (The question of airloads on the blunt leading edge is not addressed here.) This structure has a moment of inertia

$$I = 2Z_t^2 T. \qquad (47)$$

The dimensionless homogeneous differential equation and boundary conditions can therefore be written

$$[t\alpha_e']'' + k_D\alpha_e = 0 \tag{48}$$

$$t\alpha_e'|_{x=0} = 0, \quad (\alpha_e')'|_{x=0} = 0, \quad \alpha_e(1) = 0, \tag{49}$$

with $t = T(x)/T_0$ and

$$k_D = \frac{\varrho V_D^2(1 - \nu^2)}{E\sqrt{M^2 - 1}}\left(\frac{B_1^3}{T_0 Z_i^2}\right). \tag{50}$$

Biot [20] gives the fundamental eigenvalue $k_D = 6.33$ for the case of uniform thickness, $t = 1$.

The details of optimization, which will be found in Ref. [19], have certain interest because of the non-self-adjointness of the related pairs of constraint and Euler-Lagrange equations. It is discovered that the control equation can be put in the alternative forms

$$1 - \frac{\lambda_s \alpha_e}{t^2} = 1 + \lambda_s'\alpha_e' = 0, \tag{51}$$

where λ_s is the multiplier for the constraint on $s \equiv t\alpha_e'$. The assumption that both λ_s' and α_e' are constants finally leads to a simple, self-consistent solution:

$$\alpha_e = A[1 - x] \tag{52}$$

$$t = k_D\left[\frac{x^2}{2} - \frac{x^3}{6}\right]. \tag{53}$$

The mass ratio calculated from (53) is 6.33/8, so that there is a 21% weight saving relative to a uniform sandwich plate of the same aeroelastic stiffness.

IV. Dynamic aeroelastic examples

Flutter of a thin two-dimensional plate and of a cantilever lifting surface, idealized structurally as a beam in bending and rod in torsion, has formed the subject of several optimizations with fixed dynamic aeroelastic eigenvalues. The former problem will be discussed first, and for simplicity Hedgepeth's approximation [21] will be adopted of zero in-plane tension, quasi-steady supersonic airloading, and negligible aerodynamic damping. Under these circumstances, the dimensionless equation of motion for a homogeneous plate with thickness $T(x)$, like that shown in Fig. 9, becomes

$$[t^3 w'']'' + \lambda_0 w' - k_0\left(\frac{\omega}{\omega_0}\right)^2[\delta_1 t + \delta_2]w = 0. \tag{54}$$

Here λ_0 and k_0 are constant groups familiar in the literature [e.g. 21], the former being an aerodynamic parameter proportional to $\varrho V^2/\sqrt{M^2 - 1}$ and the latter proportional to panel inertia and the square of frequency. The factor $[\delta_1 t + \delta_2]$ is applied, as in free-vibration problems, to allow for the possibility of a uniform mass loading that is not affected by changes in t. The factor $(\omega/\omega_0)^2$ allows the flutter frequency to be varied during optimization, since only the speed need be held constant.

The nonlinear factor t^3 in the lead term of (54), which comes from the expression for bending rigidity of a homogeneous plate, results in systems of optimizing equations that can only be handled numerically. A more tractable case is the sandwich panel with thin face sheets. Under simple-support

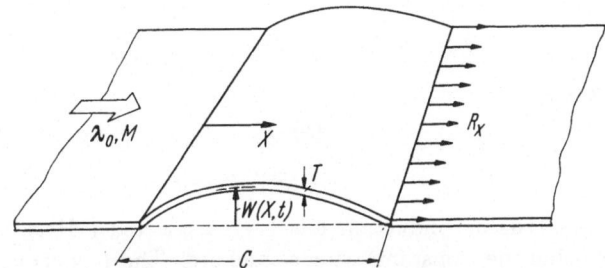

Fig. 9. Skin panel with one side exposed to a supersonic airstream. Panel may be a homogeneous plate of variable depth or a sandwich with equal thin face sheets of thickness $T(x)$.

14*

boundary conditions, the problem statement for the sandwich reads

$$[tw'']'' + \lambda_0 w' - k_0 \left(\frac{\omega}{\omega_0}\right)^2 [\delta_1 t + \delta_2] w = 0 \tag{55}$$

$$w(0) = 0, \quad w(1) = 0, \quad tw''|_{x=0} = 0, \quad tw''|_{x=1} = 0. \tag{56}$$

The reference thickness for t is now that of face sheet for a uniform sandwich. Also $\lambda_0 = 343$, $k_0 = (1.9\pi)^4$ — well-known eigenvalues given in Ref. [12] and elsewhere. For minimizing the integral (24), one applies (7) to the function

$$\begin{aligned} F = t + \lambda_w[p - w'] + \lambda_p \left[\frac{q}{t} - p'\right] + \lambda_q[r - q'] \\ + \lambda_r \left[k_0 \left(\frac{\omega}{\omega_0}\right)^2 (\delta_1 t + \delta_2)\, w - \lambda_0 p - r'\right]. \end{aligned} \tag{57}$$

In due course one finds that λ_r must satisfy

$$[t\lambda_r''] - \lambda_0 \lambda_r' - k_0 \left(\frac{\omega}{\omega_0}\right)^2 [\delta_1 t + \delta_2]\, \lambda_r = 0. \tag{58}$$

Since the boundary conditions yielded by transversality for λ_r are identical with (56), a relation of the form

$$\lambda_r(x) = \frac{w(1-x)}{A^2} \tag{59}$$

can be inferred. By means of (59) and other substitutions from the constraint and Euler-Lagrange equations, one is then able to convert the original control equation

$$1 - \frac{\lambda_p q}{t^2} + k_0 \left(\frac{\omega}{\omega_0}\right)^2 \delta_1 \lambda_r w = 0 \tag{60}$$

into a curious nonlinear differential-difference equation:

$$A^2 - w''(x)\, w''(1-x) + k_0 \left(\frac{\omega}{\omega_0}\right)^2 \delta_1 w(x)\, w(1-x) = 0. \tag{61}$$

In view of (56), it is clear that (61) can be reduced to a repeating system of nonlinear algebraic relations by assuming a Fourier series

$$w(x) = \sum_{n=1} C_n \sin n\pi x. \tag{62}$$

Although no example has been worked out, convergence of (62) has been demonstrated. $t(x)$ can be found from a similar substitution into (55), and the absolute minimum weight discovered by repeating the process for different values of $(\omega/\omega_0)^2$. From (60) it is evident that t must vanish at both ends of the panel, and more practically useful solutions are therefore anticipated from the additional constraint $t \geqq t_0$.

The final series of examples studied relate to flexure-torsion and pure-torsional flutter of the wing of Fig. 10. A neutrally stable condition of simple harmonic motion at the stability boundary is assumed, which is the usual approach to flutter analysis and takes special advantage of the available unsteady aerodynamic theory. As described, for instance, in Sect. 7−7(a) of Ref. [22], suitable equations for the dimensionless complex amplitudes of

$$y(x, \tau) = \bar{y}(x)\, e^{ik\tau}, \tag{63}$$

$$\theta(x, \tau) = \bar{\theta}(x)\, e^{ik\tau}, \tag{64}$$

would be

$$[t\bar{y}'']'' - [\alpha_1 t + \bar{\alpha}_2]\, \bar{y} - [\beta_1 t + \bar{\beta}_2]\, \bar{\theta} = 0, \tag{65}$$

$$[t\bar{\theta}']' + [\gamma_1 t + \bar{\gamma}_2]\, \bar{y} + [\delta_1 t + \delta_2]\, \bar{\theta} = 0. \tag{66}$$

(Aerodynamic induction has been neglected.) Here τ is dimensionless time and $k = \omega B/V$, B being the (constant) wing semichord. The forms of the eight inertia and aerodynamic coefficients α_1 through $\bar{\delta}_2$ are not essential for present purposes; they are listed *in extenso* by Ref. [19]. All are fixed for a given wing fluttering at given values of k and M. The assumed linear relationship between t and the stiffness and inertia properties ensures they will remain constants during optimization when ω is constrained. Variations of frequency would have to be studied as in the panel-flutter example.

When the Euler-Lagrange equations are constructed, there occurs a mixture of real and complex variables, the latter being identified in (65)—(66) by overlines. As first pointed out by TURNER [5], special care must be exercised during numerical computation to force $t(x)$ to remain real. This can be done by separating reals and imaginaries in the entire system or (more efficiently) by adding the conjugates of all the constraint equations into function F. A rigorous proof has been devised [19] of Turner's supposition that, when this is done, corresponding Lagrange multipliers must also be conjugates.

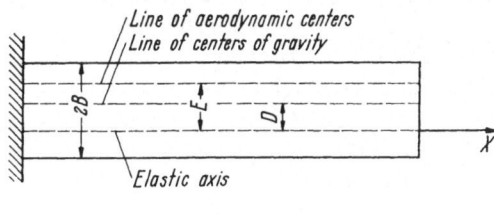

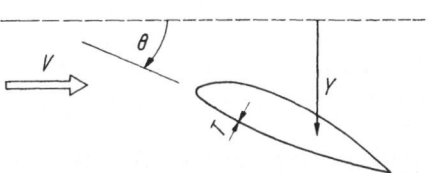

Fig. 10. Unswept cantilever wing, showing definitions of symbols used in analyzing flexure-torsion and pure torsional flutter.

Optimization based on (65)—(66) thus leads to a system of 25 real, first-order differential equations. A numerical demonstration has been undertaken for a uniform wing tested by RUNYAN and SEWALL [23] and flutter-analyzed by RUNYAN and WATKINS [24]. Although no results are yet available, one can be confident of obtaining something meaningful in the light of the example worked by TURNER [5], using three finite elements equally spaced along the wingspan. TURNER adopted, as the starting point for his algebraic minimization, a wing whose skin thickness distribution had already been optimized for free vibration at a fixed torsional frequency. Although he found very little structural weight reduction relative to this highly efficient initial step, one can estimate that this final design weighs about 82% of a uniform wing with the same speed of flexure-torsion flutter. This includes the contribution of the unvaried internal mass.

There is a special case of (65)—(66) whose simple exact solution emphasizes both some of the pitfalls and the extreme potentialities of structural optimization. This involves the (relatively rare) occurrence of pure-torsional flutter, for which the equation is (66) with the bending displacement $\bar{y} = 0$:

$$[t\bar{\theta}']' + [\delta_1 t + \bar{\delta}_2]\,\bar{\theta} = 0. \tag{67}$$

The cantilever boundary conditions read

$$\bar{\theta}(0) = 0, \quad t\bar{\theta}'|_{x=1} = 0. \tag{68}$$

The reference solution for $t = 1$ is easily seen to have mode shape

$$\bar{\theta}(x) \sim \sin\left(\sqrt{\delta_1 + \bar{\delta}_2}\right)x. \tag{69}$$

Moreover, the second of conditions (68) requires that the imaginary part of $\bar{\delta}_2$ be equal to zero and furnishes the fundamental eigenvalue

$$\delta_1 + \delta_2 = \left(\frac{\pi}{2}\right)^2. \tag{70}$$

(The overline has been removed to indicate real δ_2.)

When the flight speed is sufficiently low ($M \cong 0$), Smilg's solution [25] for the two-dimensional profile in incompressible flow furnishes information on elastic-axis locations and other wing properties that can satisfy (70). In particular, $\mathscr{I}m\{\delta_2\}$, which is just the component of aerodynamic moment out of phase with the torsional displacement, may vanish only when the elastic axis lies ahead of the quarter-chordline.

By immediate analogy with (27)—(28), the real-$\bar{\delta}_2$ optimal solution for (67)—(68) leads to

$$t(x) = \frac{\delta_2}{2\delta_1}\left[\left(\frac{\cosh\left(\sqrt{\delta_1}\right)}{\cosh\left(\sqrt{\delta_1}\,x\right)}\right)^2 - 1\right]. \tag{71}$$

Now, if all the weight of the wing is taken to lie in the variable-thickness skin, δ_2 is found to be proportional to the aerodynamic moment in phase with θ; Smilg's calculations show this always to be negative. Thus one arrives at the meaningless result that $t(x) < 0$ everywhere!

Although a physical explanation for this strange extremal can be offered, it seems more profitable to modify the problem in ways that might produce positive $t(x)$. One such approach is to allocate enough of the total wing mass to (fixed) internal structure that its contribution changes the sign of δ_2. For instance, a case was studied from Smilg [25] in which the rotational axis was at the leading edge and the flutter $k \cong 0.04$. Then, with 50% of the structural mass allocated to the skin of the reference wing, and with radii of gyration assumed equal, it turns out that $\delta_1 = 2.04$ and $\delta_2 = 0.43$. A simple computation then demonstrates that a 39% saving of total weight and a 78% saving of skin weight are achieved by going from the uniform to the optimum wing of equal flutter speed. Although this configuration has somewhat unrealistic physical properties, it constitutes the most dramatic theoretical success of all examples studied to date.

For appropriate ranges of t_0, the foregoing solution can also be rationalized by applying the constraint $t \geq t_0$. Thus it is found that $t(x)$ will assume reasonable positive values for all positive δ_2 and for negative δ_2 when $t_0 \geq |\delta_2|/\delta_1$. The reader is referred to Ref. [19] for details; Figs. 11−12 plot some numerical results.

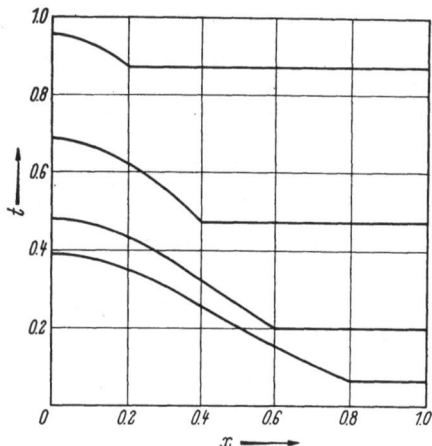

Fig. 11. Optimal skin thickness distributions for fixed pure-torsional flutter speed, with various constraints placed on minimum thickness. 50% of total mass effective in torsion, rotational axis at leading edge, $\delta_1 = 2.04$, $\delta_2 = 0.43$.

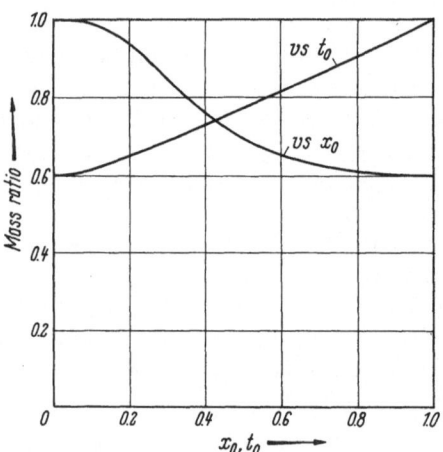

Fig. 12. Mass ratio vs. minimum skin gauge t_0 for the wings of Fig. 11. Same ratio is also shown vs. x_0, point where minimum thickness begins.

V. Discussion and conclusions

Prager and Taylor [7] have studied certain extremal problems, such as maximum buckling load and maximum vibration frequency for given mass, wherein the control variable enters linearly both the integrand of the merit function and the differential equation of equilibrium. They present general theorems, based on the variational principle underlying the equilibrium equation, by which simple nonlinear control equations can be written entirely in terms of the displacement function. Some, but not all, of the examples of this paper are encompassed by their results, which might be called "constant specific Lagrangian density" theorems. It is therefore instructive to summarize in a table the relevant forms of some of the previously-derived control equations.

In the first four problems listed, it is not difficult to return to dimensional variables and see that the quantity required to be constant is the difference between the internal potential energy amplitude and the kinetic energy amplitude, *each divided by the local structural stiffness-mass parameter* (i.e., $m(x)$ or $t(x)$). Thus, for the longitudinally vibrating bar, the potential and kinetic

Table 1

Physical Problem	Quantity which is constant in the optimal solution
Longitudinal vibration of bar	$(u')^2 - \beta^2 u^2$
Torsional vibration of rod or wing	$(\theta')^2 - \left(\dfrac{\pi}{2}\right)^2 \delta_1 \theta^2$
Torsional divergence of wing	$(\theta')^2$
Chordwise divergence	$(\alpha'_e)^2 \equiv (w'')^2$
Pure-torsional flutter of wing	$(\theta')^2 - \delta_1 \theta^2$
Sandwich panel flutter	$w''(x)\, w''(1-x)$ $-k_0 \dfrac{\omega^2}{\omega_j^2} \delta_1 w(x)\, w(1-x)$

energies per unit volume are involved—a fact which suggested to the authors the term "specific Lagrangian density".

For static problems, of course, only potential energy appears. From the physical point of view, it is certainly no surprise to find the specific potential energy uniform throughout the optimal configuration. In fact, one sees a parallel with the uniform strain energy solutions that have been proposed for strength design of structures under static load.

The theorems of Ref. [7] will not be reproduced here, but rather a related new result will be given which extends them to static cases where the external forces are not necessarily conservative. Any example of divergence or steady deformation under airload is thereby encompassed. For generality, consider a three-dimensional elastic solid occupying a volume V. (All integrals are referred to the unstrained positions of mass elements.) Let the density of structurally effective material be ϱ, the displacement vector $\vec{q}$ and the externally applied force per unit volume $\gamma\vec{R}$. γ is some parameter, such as dynamic pressure, which is held constant during optimization, whereas $\vec{R}$ may contain both terms dependent on and independent of the deformation state. Surface forces like aerodynamic pressure are included in $\vec{R}$ through a familiar application of the Dirac function.

For this system, Hamilton's principle may be written

$$\delta \int_V \varrho e(\vec{q})\, dV = \int_V \gamma\vec{R} \cdot \delta\vec{q}\, dV \qquad (72)$$

for any arbitrary smooth variation $\delta\vec{q}$. $e(\vec{q})$ is independent of ϱ and is the quantity called "specific elastic strain energy" in Ref. [7]; evidently, the investigation is limited to structures where stiffness is directly proportional to structurally effective mass.

Let subscript zero identify an optimal solution in the sense that, for all neighboring density distributions corresponding to the same γ,

$$\int_V [\varrho - \varrho_0]\, dV \geq 0 \qquad (73)$$

Hamilton's principle for ϱ_0 under the load system $\vec{R}_0$ reads

$$\delta \int_V \varrho_0 e(\vec{q}_0)\, dV = \int_V \gamma\vec{R}_0 \cdot \delta\vec{q}_0\, dV. \qquad (74)$$

It is also a consequence of this principle that, if the structure ϱ is strained into the kinematically admissible deformation shape $\vec{q}_0$, the energy variation will have the right-hand side of (74) as a lower bound:

$$\delta \int_V \varrho e(\vec{q}_0)\, dV \geq \int_V \gamma\vec{R}_0 \cdot \delta\vec{q}_0\, dV. \qquad (75)$$

Subtracting (74) from (75), one gets

$$\delta \int_V [\varrho - \varrho_0]\, e(\vec{q}_0)\, dV = \int_V [\varrho - \varrho_0] \frac{\partial e}{\partial \vec{q}_0} \cdot \delta \vec{q}_0\, dV \geq 0, \tag{76}$$

where the meaning of vector $\partial e / \partial \vec{q}_0$ will become evident from what follows.

For general forms of $e(\vec{q})$, no simple deduction has yet been found from (76). But *if e is a symmetrical homogeneous quadratic form*, as it is for all structures considered here, one can choose the particular variation $\delta \vec{q}_0 \sim \vec{q}_0$ and make use of the well-known fact that

$$\frac{\partial e}{\partial \vec{q}_0} \cdot \vec{q}_0 = 2e(\vec{q}_0). \tag{77}$$

Consequently, under this restriction, (76) implies

$$\int_V [\varrho - \varrho_0]\, e(\vec{q}_0)\, dV \geq 0. \tag{78}$$

The only way that (78) and (73) can be consistent for *all* ϱ neighboring the optimum is to have

$$e(\vec{q}_0) = \text{const.} \tag{79}$$

This result encompasses the third and fourth entries in Table 1 and a variety of other static aeroelastic cases.

From the form of the control equations derivable for sandwich panel flutter and certain other problems of simple harmonic motion under non-conservative airloads, one can speculate on the existence of a generalization of the foregoing theorem which might greatly simplify the construction of these equations for other such situations where the control variable appears linearly in the stiffness.

By way of conclusion, it must first be freely admitted that the optimal aeroelastic examples analyzed to date are oversimplified relative to anything that might be employed in the actual design of a flight vehicle. No such single stability or loading condition is adequate for sizing an aerospace structure, in which connection one should mention the recent proposal of Prager and Shield [26] for combining certain design conditions by requiring a sum of potential energy expressions to be constant. In view of the significant weight savings demonstrated by some elementary cases, however, the subject would certainly seem worthy of further investigation.

Acknowledgment

This work was supported by Air Force Office of Scientific Research, Contract No. F 44620-68-C-0036, and by National Aeronautics and Space Administration, Grant NGR-05-020-102. The authors wish to acknowledge valuable assistance by Mr. William J. Vitte.

References

1. Schmit, L. A., W. M. Morrow and T. P. Kicher: A Structural Synthesis Capability for Integrally Stiffened Cylindrical Shells, Paper No. 68—327, presented at AIAA/ASME 9th Structures, Structural Dynamics and Materials Conference, Palm Springs, Calif., April 1968.
2. Turner, M. J.: Design of Minimum Mass Structures with Specified Natural frequencies, AIAA J. 5, 406—412 (1967).
3. Ashley, H., and S. C. McIntosh: Expanding the Consciousness of the Aeroelastician, Proc. of ASME Symposium, Fluid-Solid Interaction, Pittsburgh, Nov. 1967, pp. 120—143.
4. McIntosh, S. C., and F. E. Eastep: Design of Minimum-Mass Structures with Specified Stiffness Properties, AIAA J. 6, 962—964 (1968).
5. Turner, M. J.: Optimization of Structures to Satisfy Flutter Requirements, unnumbered report of Commercial Airplane Division, The Boeing Co., 1968 (to appear in AIAA J.).
6. Gerard, G.: Optimum Strutural Design Concepts for Aerospace Vehicles, J. Spacecraft Rockets 3, 5—18 (1966).
7. Prager, W., and J. E. Taylor: Problems of Optimal Structural Design, ASME Paper No. 67-WA/APM-29 (to appear in Journal of Applied Mechanics).
8. Shield, R. T.: Optimum Design Methods for Multiple Loading, ZAMP 14, 38—45 (1963).
9. Prager, W.: Optimization in Structural Design, Mathematical Optimization Techniques, Ed.:

R. Bellman University of California Press, 1963, pp. 279—289.

10. GELLATLY, R. A.: Development of Procedures for Large Scale Automated Minimum Weight Design, U.S. Air Force AFFDL-TR-66-180, Wright-Patterson AFB, Ohio, Dec. 1966.

11. MELOSH, R. J., and R. LUIK: Approximate Multiple Configuration Analysis and Allocation for Least Weight Structural Design, U.S. Air Force AFFDL-TR-67-59, Wright-Patterson AFB, Ohio, April 1967.

12. MIELE, A.: Extremization of Linear Integrals by Green's Theorem, Chapter 3 of Optimization Techniques with Applications to Aerospace Systems, Ed.: G. Leitman. New York: Academic Press 1962.

13. BRYSON, A. E., and Y. C. HO: Applied Optimal Control, to be published by Blaisdell Publishing Co., Waltham, Mass., 1968.

14. NIORDSON, F. I.: On the Optimum Design of a Vibrating Beam, Quart. Appl. Math. 23, 47—53 (1965).

15. HALFMAN, R. L.: Dynamics, Vol. II. Systems, Variational Methods and Relativity, Reading, Mass.: Addison-Wesley (see Chaps. 10 and 11).

16. TAYLOR, J. E.: Minimum Mass Bar for Axial Vibration at Specified Natural Frequency, AIAA J. 5, 1911—1913 (1967).

17. — Optimum Design of a Vibrating Bar with Specified Minimum Cross Section, AIAA J. 6, 1379—1381 (1968)

18. MIELE, A,: The Calculus of Variations in Applied Aerodynamics and Flight Mechanics, Chapter 4

of Optimization Techniques with Applications to Aerospace Systems, Ed.: G. Leitman, New York: Academic Press 1962, pp. 104—105.

19. VITTE, W. J., S. C. McINTOSH and H. ASHLEY: Applications of Aeroelastic Constraints in Structural Optimization, Stanford University SUDAAR Report, 1968 (to be issued).

20. BIOT, M. A.: Divergence of Supersonic Wings Including Chordwise Bending, J. Aeronaut. Sci. 23, 237—251, 271 (1956).

21. HEDGEPETH, J. M.: Flutter of Rectangular Simply Supported Panels at High Supersonic Speeds, J. Aeronaut. Sci. 24, 563—573 (1957).

22. BISPLINGHOFF, R. L., and H. ASHLEY: Principles of Aeroelasticity, New York: Wiley 1962.

23. RUNYAN, H. L., and J. L. SEWALL: Experimental Investigation of the Effects of Concentrated Weights on Flutter Characteristics of a Straight Cantilever Wing, NACA Technical Note 1594, 1948.

24. RUNYAN, H. L., and C. E. WATKINS: Flutter of a Uniform Wing with an Arbitrarily Placed Mass According to a Differential-Equation Analysis and a Comparison with Experiment, NACA Report 966, 1950.

25. SMILG, B.: The Instability of Pitching Oscillations of an Airfoil in Subsonic Incompressible Potential Flow, J. Aeronaut. Sci. 16, 691—696 (1949).

26. PRAGER, W., and R. T. SHIELD: Optimal Design of Multi-Purpose Structures, unpublished report of University of California, San Diego, under Contract 31-124-ARO-D-257, August 1967.

$3n$-dimensional mechanics of generalized continua

By

J. Baumgarte[1]   and   **E. Kröner**

Ingenieur-Akademie,   Technische Universität,
Wolfenbüttel   Clausthal

1. Introduction

It has become clear from the work of MINDLIN, KRUMHANSL, KUNIN, TOUPIN, KRÖNER and others that several of the recent theories of generalized continua can be considered as the continuum limits of certain formulations of the atomic lattice theory.

Whereas the normal elasticity theory arises from the continuization of the theory of primitive lattices (Bravais lattices) with local response characteristics (next neighbour interaction) one obtains the theory of materials of grade 2, 3, 4, etc. when the interaction over farther distances is included. An important aspect of this extented theory is that new functional degrees of freedom are not needed in order to describe the state of the body. For instance, one can still use an ordinary displacement field for describing the geometric configuration of the particles.

On the other hand, if one continuises the theory of non-primitive lattices of n particles per cell, say, then additional functions must be introduced for a complete specification of the state. One way to describe the microstructure implied by the lattice cells has been presented by MINDLIN. Other possibilities were discussed by ERINGEN, by GREEN and RIVLIN and by others.

A different approach is presented in this work[2]. A model well-known from the atomic lattice theory is used. Non-primitive lattices are often described as a composition of n primitive lattices (later-on called sub-lattices) put into each other and coupled together. The continuisation leads to a generalized continuum built up from n sub-continua. To each sub-continuum can be applied the equations of normal elasticity (or, for instance, of the theory of grade 2, 3, 4 etc. if range effects are included), however with the following addition: the "external" forces acting on the sub-continuum must include the forces resulting from the interaction with the other sub-continua.

A displacement field will often be sufficient to specify the spatial configuration of the particles in the sub-continuum[3]. In order to specify the configuration of the whole continuum n displacement fields are needed. If the dimensionality of the sub-continuum is 3 the whole continuum can be said to be $3n$-dimensional. In fact, the n 3-dimensional displacement fields are equivalent to one $3n$-dimensional field. This observation explains the title of this work.

2. The Lagrangean

In order to derive the theory we shall apply the Lagrange formalism. In this section we shall set up the potential and kinetic energy. To keep the presentation clear we shall apply three simplifications:

---

[1] Now at Technische Universität Braunschweig. A more detailed treatment of the present topic will appear in the Abhandlungen der Braunschweigischen Wissenschaftlichen Gesellschaft 1970.

[2] In some way this approach (due to BAUMGARTE) can be considered as an extension of a very recent work of MINDLIN, see Ref. [1].

[3] This is not the case in situations where crystal defects play a role.

(a) the theory is prseented in the linearised form,

(b) the lattices are assumed to be infinitely extended and

(c) only cubic sub-lattices are considered.

The latter case occurs, indeed, in many materials.

The n sub-lattices are labelled by α ($= 1, 2, 3, \ldots, n$). Let U be the total potential energy (including the interaction) of the n sub-lattices. We normalise U in such a way that $U = 0$ in the perfect crystal configuration which by definition is the equilibrium configuration in the absence of external and inertia forces. U depends on the coordinates, i.e. on the displacements of all particles (lattice points). We denote the ith displacement vector component of the particle labelled by $\boldsymbol{m} = (m_1, m_2, m_3)$ in the lattice α by $\overset{m}{u_i^\alpha}$.

We now apply a procedure which is standard in lattice theory: We develop U around the equilibrium position. The linear terms then vanish. All terms of degree higher than 2 are neglected to make the theory linear. So we obtain

$$U = -\frac{1}{2} \sum_{\substack{m,n \\ \alpha,\beta}} \overset{mn}{\Phi_{ik}^{\alpha\beta}} \overset{m}{u_i^\alpha} \overset{n}{u_k^\beta}, \quad \overset{mn}{\Phi_{ik}^{\alpha\beta}} = -\frac{\partial^2 U}{\partial \overset{m}{u_i^\alpha} \, \partial \overset{n}{u_k^\beta}}. \tag{1}$$

Here the summation convention is used for the vector and tensor indices i, k. The quantities $\overset{mn}{\Phi_{ik}^{\alpha\beta}}$ are called the coupling parameters. It is convenient to make the transition to continuum now. To this end we introduce continuous functions $\Phi_{ik}^{\alpha\beta}(\boldsymbol{r}, \boldsymbol{r}')$ and $u_i^\alpha(\boldsymbol{r})$ where $\boldsymbol{r}$ denotes the position vector. These continuous functions coincide with the values of $\overset{mn}{\Phi_{ik}^{\alpha\beta}}$ and $\overset{m}{u_i^\alpha}$ at the positions $\boldsymbol{m}$ of the particles in the equilibrium configuration. Hereafter we apply the Euler-Maclaurin formula for converting sums into integrals.

The transition from lattice to continuum is crucial in all these kinds of reflections. Hence we add a few further remarks.

. The introduced continuous functions are not defined uniquely by the above prescription. The convergence of the Euler-Maclaurin formula will depend on the choice of these functions. We now agree to choose the continuous functions in such a way that those terms which contain derivatives of these functions in the Euler-Maclaurin formula contribute a minimum to the sums to be converted into integrals (for instance the sums in Eq.(1)).

Clearly, this agreement implies that the continuous functions go smoothly along the lattice points. When the lattice is finite rather than infinite, then the Euler-Maclaurin formula converts lattice sums into volume plus surface integrals as was shown in Ref. [2]. The above prescription means that surface integrals which contain derivatives of the functions contribute a minimum. It appears that it is impossible to make these integrals vanish exactly. The reason is that then we would obtain a rigourous continuum theory to describe a discrete manifold. This, however, cannot be.

The present continuisation procedure seems related to the one recently proposed by RIVLIN in Ref. [3].

The application of the Euler-Maclaurin formula now transforms Eq.(1) into

$$U = \frac{1}{2} \sum_{\alpha,\beta} \iint dV \, dV' \, \Phi_{ik}^{\alpha\beta}(\boldsymbol{r}, \boldsymbol{r}') \, u_i^\alpha(\boldsymbol{r}) \, u_k^\beta(\boldsymbol{r}'). \tag{2}$$

This is the potential energy to be used in the Lagrange formalism. Clearly, the kinetic energy has the form

$$T = \frac{1}{2} \sum_\alpha \int dV \, \varrho^\alpha \dot{u}_i^\alpha \dot{u}_i^\alpha \tag{3}$$

where ϱ^α is the mass density of the sub-continuum α. The mass density as well as the later introduced force per unit mass in the sub-continuum α is formed from the corresponding discrete quantities by further applications of the Euler-Maclaurin formula.

Subtracting (2) from (3) we obtain the Lagrangian

$$L = T - U. \tag{4}$$

3. The equations of motion

We denote by $F_i^\alpha(\mathbf{r})$ the external force per unit mass in the sub-continuum α. This force does not contain the interaction with the other sub-continua since this is included into the potential energy U. If we substitute the last three equations into the equations of motion

$$\frac{d}{dt}\frac{\partial L}{\partial \dot{u}_i^\alpha} - \frac{\partial L}{\partial u_i^\alpha} = \varrho^\alpha F_i^\alpha, \tag{5}$$

we obtain a system of $3n$ simultaneous integro-differential equations:

$$\varrho^\alpha \ddot{u}_i^\alpha(\mathbf{r}) - \sum_\beta \int dV' \, \Phi_{ik}^{\alpha\beta}(\mathbf{r}, \mathbf{r}') \, u_k^\beta(\mathbf{r}') = \varrho^\alpha F_i^\alpha(\mathbf{r}). \tag{6}$$

A special case of these equations has been written down before, namely that for $n = 1$ (see Ref. [2]). In this case the kernel $\Phi_{ik}(\mathbf{r}, \mathbf{r}')$ is not of a simple form in actual materials. We conclude that this also holds with the present kernels $\Phi_{ik}^{\alpha\beta}(\mathbf{r}, \mathbf{r}')$. To overcome this difficulty we develop the $\Phi_{ik}^{\alpha\beta}$ in terms of delta functions:

$$\Phi_{ik}^{\alpha\beta}(\mathbf{r}, \mathbf{r}') = [T_{ik}^{\alpha\beta} + T_{ikj}^{\alpha\beta} \, \partial_j + T_{ikjl}^{\alpha\beta} \, \partial_j \, \partial_l + \cdots] \, \delta(\mathbf{r}, \mathbf{r}'). \tag{7}$$

This series can be derived from the Taylor series of the Fourier transform of $\Phi_{ik}^{\alpha\beta}$, if one assumes that $\Phi_{ik}^{\alpha\beta}$ depends on $\mathbf{r} - \mathbf{r}'$ alone which is always true in the case of the infinite perfect crystal. That this series does, in fact, exist, is our assumption about the kernels $\Phi_{ik}^{\alpha\beta}$. It seems general enough to cover all cases of interest.

For our purposes it is advantageous to rewrite Eq. (7) in the form

$$\Phi_{ik}^{\alpha\beta}(\mathbf{r}, \mathbf{r}') = [T_{ik}^{\alpha\beta} + T_{ikj}^{\alpha\beta} \, \partial_j + T_{ikjl}^{\alpha\beta} \, \partial_j \, \partial_l] \, \delta(\mathbf{r}, \mathbf{r}') + \partial_j \, \partial_l t_{ikjl}^{\alpha\beta}(\mathbf{r}, \mathbf{r}') \tag{8}$$

where the $T_{ik}^{\alpha\beta}$, $T_{ikj}^{\alpha\beta}$ and $T_{ikjl}^{\alpha\beta}$ are constants (homogeneous case), and

$$t_{ikjl}^{\alpha\beta}(\mathbf{r}, \mathbf{r}') = [T_{ikjlm}^{\alpha\beta} \, \partial_m + T_{ikjlmn}^{\alpha\beta} \, \partial_m \, \partial_n + \cdots] \, \delta(\mathbf{r}, \mathbf{r}') \tag{9}$$

is the function corresponding to the series on the right of (8). It is clear from former investigations (see Ref. [2]) that the regular part of the kernels $t_{ikjl}^{\alpha\beta}(\mathbf{r}, \mathbf{r}')$ describes the long range cohesive forces of the material. No examples are known where singular parts of these kernels play a role.

If Eq. (8) is substituted in Eq. (6) one obtains after partial integration the equations of motion in the form

$$\varrho^\alpha \ddot{u}_i^\alpha - \sum_\beta [T_{ik}^{\alpha\beta} u_k^\beta + T_{ikj}^{\alpha\beta} u_{k,j}^\beta + T_{ikjl}^{\alpha\beta} u_{k,jl}^\beta + \partial_j \int dV' \, t_{ikjl}^{\alpha\beta}(\mathbf{r}, \mathbf{r}') \, u_{k,l}^\beta(\mathbf{r}')] = \varrho^\alpha F_i^\alpha \tag{10}$$

where $\alpha, \beta = 1, 2, 3, \ldots, n$.

If $n = 1$, these equations reduce to the form discussed in Ref. [2]. Note that in this case the terms with T_{ik} drop out because no response is created when all particles suffer the same displacement. Since a crystal with $n = 1$ is always centro-symmetric also $T_{ikj} = 0$ in the present case. In general $n \neq 1$ and the tensors $T_{ik}^{\alpha\beta}$ obey the condition

$$\sum_\beta T_{ik}^{\alpha\beta} = 0 \tag{11}$$

(no response under rigid translation).

Note that the Eqs. (10) have solutions which describe mutual rigid translational vibrations of the sub-lattices.

The expression $\sigma_{ji}^{\alpha\beta} = T_{ikj}^{\alpha\beta} u_k^\beta + T_{ikjl}^{\alpha\beta} u_{k,l}^\beta + \int dV' \, t_{ikjl}^{\alpha\beta}(\mathbf{r}, \mathbf{r}') \, u_{k,l}^\beta(\mathbf{r}') \tag{12}$

with $\alpha \neq \beta$ can be called the mutual stress between lattices β and α whereas it is called the self stress of sub-lattice if $\alpha = \beta$. The total stress in α will be defined as

$$\sigma_{ji}^\alpha = \sum_\beta \sigma_{ji}^{\alpha\beta}. \tag{13}$$

Using this abbreviation, one can rewrite the equations of motion in the more concise form

$$\varrho^\alpha \ddot{u}_i^\alpha - \sum_\beta T_{ik}^{\alpha\beta} u_k^\beta - \partial_j \sigma_{ji}^\alpha = \varrho^\alpha F_i^\alpha. \tag{14}$$

Of particular importance is the special case of local response. Then $t^{\alpha\beta}_{ikjl} = 0$ and the equations of motion assume the form

$$\varrho^\alpha \ddot{u}^\alpha_i - \sum_\beta [T^{\alpha\beta}_{ik} u^\beta_k + T^{\alpha\beta}_{ikj} u^\beta_{k,j} + T^{\alpha\beta}_{ikjl} u^\beta_{k,jl}] = \varrho^\alpha F^\alpha_i. \tag{15}$$

4. The special case of the Cosserat continuum with local response

The material described by the terms $T^{\alpha\beta}_{ik}$, $T^{\alpha\beta}_{ikj}$ and $T^{\alpha\beta}_{ikjl}$ will degenerate to a Cosserat continuum if extra conditions are introduced in such a way that the distances between the particles in a cell are conserved during the motion. Let us take $n = 2$ for simplicity. The Cosserat momentum equations are obtained by adding the two sets of Eqs. (15) corresponding to $\alpha = 1, 2$. If the difference of these equations is multiplied by $e_{imn}a_n$ where e_{imn} denotes the permutation tensor and a_n the vector which describes the mutual positions of the particles in the cell, then one obtains the Cosserat equations for the moment of momentum. Since these results are rather obvious we shall not derive them in detail.

It is clear that the Cosserat continuum is a reasonable approximation to a crystal with $n > 1$ if the binding between the particles in a cell is much stronger than the binding between the neighbouring cells.

5. Propagation of plane waves

In this section we consider the homogeneous equations of motion

$$\varrho^\alpha \ddot{u}^\alpha_i - \sum_\beta [T^{\alpha\beta}_{ik} u^\beta_k + T^{\alpha\beta}_{ikjl} u^\beta_{k,jl}] = 0 \tag{16}$$

and derive the dispersion relations of propagating plane waves.

The omitting of the term containing $T^{\alpha\beta}_{ikj}$ means that we assume centro-symmetry. The including of $T^{\alpha\beta}_{ikj}$ would admit to describe materials in which waves of different polarization have different sound velocities.

For simplicity we assume that the medium is isotropic. Then we can set

$$T^{\alpha\beta}_{ik} = \iota^{\alpha\beta} \delta^{ik}, \quad T^{\alpha\beta}_{ikjl} = \lambda^{\alpha\beta} \delta_{il}\delta_{jk} + \mu^{\alpha\beta}(\delta_{ik}\delta_{jl} + \delta_{ij}\delta_{kl}) \tag{17}$$

where $\iota^{\alpha\beta}$, $\lambda^{\alpha\beta}$, $\mu^{\alpha\beta}$ are elastic constants. The equations of motion now read, in an obvious notation,

$$\varrho^\alpha \ddot{u}^\alpha_i - \sum_\beta [\iota^{\alpha\beta} u^\beta_i + (\lambda^{\alpha\beta} + \mu^{\alpha\beta}) u^\beta_{k,ki} + \mu^{\alpha\beta} u^\beta_{i,kk}] = 0. \tag{18}$$

A plane wave propagating in x-direction is described by

$$u^\alpha_i = \overset{\circ}{u}{}^\alpha_i e^{i(kx-\omega t)}. \tag{19}$$

If this is substituted in Eq. (18) one obtains a system of linear simultaneous differential equations for the quantities $\overset{\circ}{u}{}^\alpha_i$. Nontrivial solutions exist only if the corresponding determinants vanish. Accordingly

$$\det \{\varrho^\alpha \omega^2 \delta_{\alpha\beta} + \iota_{\alpha\beta} - (\lambda_{\alpha\beta} + 2\mu_{\alpha\beta}) k^2\} = 0 \tag{20}$$

for the longitudinal waves and

$$\det \{\varrho^\alpha \omega^2 \delta_{\alpha\beta} + \iota_{\alpha\beta} - \mu_{\alpha\beta} k^2\} = 0 \tag{21}$$

for the transversal waves. Just as in the simple elasticity theory the number of transversal waves here is twice the number of longitudinal waves.

Each determinant leads to an equation of degree n in ω^2. Hence, given the wave number k, the number of longitudinal waves in n and that of the transversal waves is $2n$.

The behaviour of ω^2 for $k = 0$ is of particular importance. In view of Eq. (11) the sum over β of $\iota^{\alpha\beta}$ vanishes which implies that the determinant of $\iota^{\alpha\beta}$ vanishes, too. For this reason, $\omega^2 = 0$ is a solution of the determinantal equation with $k = 0$, both for longitudinal and for transversal waves. The branches $\omega(k)$ with $\omega(0) = 0$ are usually called acoustical branches.

They correspond to waves in which the particles swing in phase. There are exactly one longitudinal and two coinciding transversal acoustical branches.

Beside the 3 acoustical branches there exist $3(n-1)$ branches with $\omega(0) \neq 0$. They are known as optical branches and correspond to an antiphase motion of the particles in a cell. The case $k = 0$ $(\omega \neq 0)$ obviously describes the mentioned mutual rigid translational vibrations of the sub-lattices.

Note that the appearance of $\iota_{\alpha\beta}$, i.e. of $T^{\alpha\beta}_{ik}$, in the equations of motion is of fundamental importance for the existence of optical branches. In fact, there would not be any of these branches without $T^{\alpha\beta}_{ik}$.

The qualitative behaviour of the wave spectrum as discussed above has been discovered before in the lattice theory. It seems satisfactory that it comes out correctly from a continuum theory, too.

6. The problem of specific heat[1]

This problem was solved by P. Debye in 1912 using certain idealizations. He calculated the energy of the elastic vibrations and applied statistical mechanics, but he did not consider optical branches in this application of elasticity theory. We have reconsidered the problem by including the optical vibrations.

If N is the number of cells, V_c the volume of the (undeformed) cell, then the number dz_s of eigenvibrations in the intervall dk of a branch, labelled by s, of the spectrum is

$$dz_s = \frac{4\pi N V_c}{(2\pi)^3} k^2 \, dk, \quad s = 1, 2, 3, \ldots, 3n. \tag{22}$$

The total number of vibrations in branch s from $k = 0$ up to the Debye limit k_D is

$$N = \int dz_s = \frac{4\pi N V_c}{(2\pi)^3} \int_0^{k_D} k^2 \, dk = \frac{4\pi V_c}{3(2\pi)^3} N k_D^3, \tag{23}$$

hence

$$k_D = 2\pi \left(\frac{3}{4\pi V_c}\right)^{1/3} \tag{24}$$

and

$$Z_s(k) = \frac{dz_s}{dk} = \frac{3N}{k_D^3} k^2. \tag{25}$$

The energy $E_s(k)$ of the vibration of frequency $\omega_s(k)$ in branch s is

$$E_s(k) = \frac{\hbar \omega_s(k)}{\exp\left\{\frac{\hbar \omega_s(k)}{\varkappa T}\right\} - 1}, \quad \hbar = h/2\pi \tag{26}$$

where $\varkappa$ is the Boltzmann's constant, h is the Planck's constant and T the absolute temperature. Now $Z_s(k)$ and $E_s(k)$ are substituted into the energy expression

$$U_s = \int_0^{k_D} dk \, Z_s(k) \, E_s(k), \tag{27}$$

where U_s is the contribution to the internal energy U of the branch s. We thus obtain

$$U = \sum_{s=1}^{3n} U_s = \frac{3N \varkappa T}{k_D^3} \sum_{s=1}^{3n} \int_0^{k_D} dk \, k^2 \, \frac{\hbar \omega_s(k)/\varkappa T}{\exp\left\{\frac{\hbar \omega_s(k)}{\varkappa T}\right\} - 1}. \tag{28}$$

It is convenient to set

$$y_s = \frac{\hbar \omega_s}{\varkappa T}, \quad x = \frac{\hbar k}{\varkappa T}, \quad x_D = \frac{\hbar k_D}{\varkappa T}. \tag{29}$$

Then

$$U = \frac{3N}{\hbar^3 k_D^3} \cdot (\varkappa T)^4 \sum_{s=1}^{3n} \int_{x=0}^{x_D} dx \, x^2 \, \frac{y_s(x)}{e^{y_s(x)} - 1}. \tag{30}$$

[1] In this section we use U for the thermodynamical internal energy and T for the absolute temperature.

For a given dispersion function $\omega_s(k)$ one can easily find $y_s(x)$ and the integrals can be calculated with more or less trouble. Here we only treat the limiting cases of $T \to 0$ and of $T \to \infty$.

Case $T \to 0$

For the optical branches ω_s is never zero, hence $y_s(k) \to \infty$ for $T \to 0$. This implies that the contribution to U of the optical branches dies out exponentially near $T = 0$. Only the long waves of the acoustical branches are essential, the contribution of which decays as T^4 (observe that near $T = 0$ the acoustical branches have the same form in our theory as they have in elasticity theory). Hence, the Debye T^3-law for the specific heat is restored in our theory.

Case $T \to \infty$

Here due to the limiting process the integral in (30) reduced to and hence

$$U = 3nN\varkappa T. \tag{31}$$

This is the rule of DULONG and PETIT. Since $3nN$ is the number of degrees of freedom of the particles of the body, this is result coincides with the result of a Debye calculation on a body with nN particles where optical branches are not considered. This can well be understood in the following way: If the temperature is extremely high, then details of the vibration spectrum no longer play a role.

Essential deviations from the simple Debye behaviour are to be expected at medium temperatures where $\varkappa T$ is within the range of the $\hbar\omega_s$. The integrations may become rather involved in this case. If the various branches are widely separated then some of the $\hbar\omega_s/\varkappa T$ may be "nearly 0", others being "nearly ∞". In this case, the energy vs. temperature curve contains sections which resemble the Debye curve of a primitive lattice except that the T^3-behaviour occurs only near $T = 0$. The form of the curve is then conveniently described by means of a set of Debye temperatures defined by

$$\theta_s = \frac{\hbar}{\varkappa}\,\omega_s(0). \tag{32}$$

7. Conclusion

The results on the frequency spectra and specific heat and the comparison with former results in the atomic lattice theory show that the here developed $3n$-dimensional elasticity theory correctly describes phenomena which are far beyond the range of the normal (3-dimensional) elasticity theory. Since the existence of optical vibrations is proved experimentally in many cases the extra results of our theory are, in fact, physically relevant.

The frequency spectra reflect completely the elastic behaviour of the material. Hence measurements of the $\omega_s(k)$-curves would permit us to find all the elastic constants introduced in Sect. 5. Unfortunately, the specific heat, though being easily accessible to experiments, is not very sensitive to details of the frequency spectrum. On the other hand, direct measurements of the frequency spectra are rather involved. Notwithstanding this fact, progress has been made in this field in recent time, for instance by neutron diffraction investigations. So it can be hoped to determine the elastic constants experimentally with some degree of truth. The close relationship to lattice theory also allows us to calculate these constants from the often known basic atomic potentials. This fact makes the comparison of theory and experiment possible.

Any reasonable solution of the theory can be represented in terms of a Fourier sum over all the waves which are possible in the body. The optical waves, too, have to be regarded when the energy of the strain state is high enough. So the criterion whether the body can or cannot be treated as normally elastic (3-dimensionally elastic) is *not* so much one of a characteristic length but rather one of a characteristic energy or a characteristic frequency [1]. In fact, the optical waves have a high energy even near $k = 0$, i.e. at very long wave length.

A glance at the determinants (20, 21) shows that the relevant characteristic material parameters are the ratios of $\iota^{\alpha\beta}$, i.e. $T_{ik}^{\alpha\beta}$, to ϱ^α which have the dimension of the square of a fre-

---

[1] A restriction to small frequencies implies a restriction to small wave numbers whereas the opposite is not necessarily true.

quency. In contrast to these parameters, the ratios of $T^{\alpha\beta}_{ikjl}$ to $T^{\alpha\beta}_{ik}$ define the squares of some lengths which have the character of decay lengths.

Beside these lengths and beside the atomic distances which confine any continuum theory, further characteristic material lengths enter the theory if non-local response is considered. We have neglected such range effects in the calculation of the frequency spectra, but not in the general formulation of the theory. It is not difficult to include range effects in the frequency spectra. In this way a specific dispersion enters the theory. This dispersion is distinctive when the wave length becomes comparable to a certain length which is to be interpreted as an effective range of the cohesive forces. Again, the non-locality of the response is reflected in the $\omega_s(k)$-curves and hence is accessible to experimental investigation.

An important field of application of the new theory may be the elasticity theory of crystal defects. Near the defects the strain energy is very high. So we expect that optical waves play a role in the Fourier representation of the stress field of such a defect.

References

1. MINDLIN, R. D.: Mechanics of Generalized Continua, Proc. IUTAM — Symposium Freudenstadt-Stuttgart 1967. Ed.: E. Kröner, Berlin/Heidelberg/New York: Springer 1968.
2. KRÖNER, E.: Int. J. Solids Structures 3, 731 (1967).
3. RIVLIN, R. S.: Mechanics of Generalized Continua, Proc. IUTAM—Symposium Freudenstadt-Stuttgart 1967. Ed.: E. Kröner, Berlin/Heidelberg/New York: Springer 1968.

Propagation of two-dimensional stress waves in an elastic/viscoplastic material

By

J. Bejda

Polish Academy of Sciences, Warsaw

1. Introduction

In recent years much attention has been devoted to problems of wave propagation for loading conditions which produce plastic deformation. Many dynamic boundary value problems have been solved assuming plane, cylindrical or spherical symmetry of the body and of the pressure applied to its boundary. All these problems may be treated using a one-dimensional theory. Mathematically, the governing equations form a system of hyperbolic partial differential equations of the first order in two independent variables; a good review of these equations and their possible methods of solution is given in [1] and [2].

There are, however, many very important practical instances of boundary-value problems in which the assumption of uniform symmetrical pressure leads to unsatisfactory results. A more exact mathematical description of the behavior of real materials is also needed; such factors as plastic deformation, strain rate effects and work-hardening must be taken into consideration.

In this paper we shall discuss the propagation of two-dimensional stress waves in strain-rate-sensitive, elastic/plastic materials. The problem considered is that of a half-space $y \geq 0$ which is loaded on the boundary $y = 0$ by an arbitrary normal pressure $P(x, t)$ as shown in Fig. 1. This problem is interesting from the practical as well as from the mathematical point of view. It may have immediate application to the analysis of seismic waves; to the explosive forming of long tunnels, corridors and ditches under and on the surface of the earth in mining; and also in considering the dynamic response of structural elements under conditions of plane strain.

Because the governing equations are nonlinear, exact solutions are not to be expected and it is necessary to employ a numerical method for obtaining approximate solutions. In this paper numerical solutions of hyperbolic partial differential equations of argument t, x, y are obtained using finite difference equations obtained by integration along bicharacteristics. Only a few solutions of boundary-value problems have been obtained with this procedure; it was applied first by BUTLER [3] and then by BURNAT et al. [4] to gasdynamic problems, by RICHARDSON [5] to problems in fluid dynamics and by CLIFTON [6, 7], ESTRIN [8] and BALTOV [9] to elastic and plastic media.

The basic difficulty with this method is the stability and convergence of the difference solution. Papers concerning this problem have recently begun to appear in mathematical literature [10—14]; however, for mixed initial and boundary value problems for systems of partial differential equations of the type obtained here, usable stability and convergence criteria are not available. A practical and useful estimation of the convergence of the difference solution for the elastic case was presented by CLIFTON [7], based on energy considerations. The problem of the numerical treatment of unloading waves is also not completely solved; the surface which divides the plastic and elastic regions can be obtained only approximately.

Some reasonable additional assumption must be made in calculating unknown functions at points lying on the unloading surface or in its immediate neighborhood.

The transition from the two-dimensional to the one-dimensional theory is obtained automatically. The method used in this paper may easily be generalized for more than two independent variables. Its generalization to three-dimensional theory and its application to elastic solids and elasto-plastic soils have been given by RECKER [15] and SAUERWEIN [16], respectively.

There are other methods of solution for boundary-value problems which use difference schemes [17—19], but the method based on the characteristic surface and the integration of differential equations along bicharacteristics seems to be the most suitable for problems of stress wave propagation. Its advantage lies in the fact that the characteristic surfaces are at the same time the wave surfaces, and therefore a good illustration of the propagation, reflection and interaction of the shock waves produced by the sudden application of a finite load. Besides, this approach leads in a straight-forward way to the appropriate difference equations at boundary points.

An interesting approximation of the theory may be obtained using Southwell and Allen's assumption that both elastic and plastic strains in the z direction are independently equal to zero. The two theories are compared numerically for the case of mild steel (Fig. 13).

2. Derivation of the general equations

Consider a half-space $y \geq 0$ in the Cartesian system of coordinates x, y, z. At time $t = 0$ an arbitrary pressure $P(x, t)$ is applied on the plane $y = 0$ (Fig. 1). The resulting deformation is one of plane strain, with the deformation perpendicular to the x-y plane being equal to zero. Let the x, y, and z components of the displacement vector be denoted respectively by

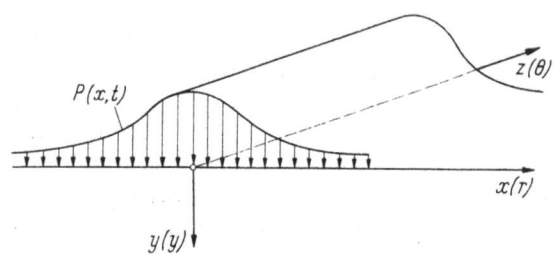

Fig. 1. Distribution of the pressure $P(x, t)$ on the boundary $y = 0$.

$$u_x(x, y, t), u_y(x, y, t), u_z = 0. \qquad (1)$$

Then, assuming small strains and rotations the components of the strain tensor are given by:

$$\varepsilon_x(x, y, t), \varepsilon_y(x, y, t), \gamma_{xy}(x, y, t). \qquad (2)$$

The state of stress is represented by four non-vanishing components:

$$\sigma_{xx}(x, y, t), \sigma_{yy}(x, y, t), \sigma_{zz}(x, y, t), \tau_{xy}(x, y, t). \qquad (3)$$

Constitutive equations for the dynamic behavior of the material are assumed to have the following form [20]

$$\dot{e}_{ij} = \frac{1}{2\mu} \dot{s}_{ij} + \gamma \Phi(F) \frac{s_{ij}}{\sqrt{J_2}} \; ; \quad \text{for} \quad F = \frac{\sqrt{J_2}}{k_0} - 1 \geq 0; \qquad (4)$$

$$\dot{e}_{ij} = \frac{1}{2\mu} \dot{s}_{ij}, \qquad \text{for} \quad F \leq 0, \qquad (5)$$

$$\dot{\varepsilon}_{ii} = \frac{1}{3K} \dot{\sigma}_{ii}; \qquad (6)$$

where ε_{ij} and σ_{ij} are the components of the strain and stress tensors, e_{ij} and s_{ij} are the components of the deviatoric strain and deviatoric stress tensors, μ, K, and k_0 are the Lamé constant, bulk modulus and a static value of the yield shear stress, respectively, and γ is the viscosity coefficient. The function F here denotes the static Huber-Mises condition of plasticity, while J_2 is the second invariant of the deviatoric stress tensor. Equation (4) holds in the elastic/viscoplastic region, Eq. (5) in the elastic region, and the compressibility condition, Eq. (6), holds in both regions.

The hardening of the material may also be taken into consideration; then the function F will take the form

$$F = F(\sigma_{ij}, \varepsilon_{ij}^p) = \frac{f(\sigma_{jj}, \varepsilon_{ii}^p)}{\varkappa} - 1$$

where ε_{ij}^p denotes the tensor of plastic strains and the work-hardening parameter $\varkappa$ may be expressed in a general form as a function of the plastic work, W_p:

$$\varkappa = \varkappa(Wp) = \varkappa\left(\int_0^{\varepsilon_{ij}^p} \sigma_{ij} \, d\varepsilon_{ij}^p\right).$$

We can write the constitutive equations for elastic, visco-perfectly plastic soils in the form [21]:

$$\dot{\varepsilon}_{ij} = \frac{1}{2\mu} \dot{s}_{ij} + \frac{1-2\nu}{E} \dot{s} \, \delta_{ij} + \gamma < \Phi(F) > \left(\alpha s_{ij} + \frac{s_{ij}}{2J_2^{1/2}}\right); \tag{7}$$

where the static yield function

$$F = \frac{\alpha J_1' + J_2^{1/2}}{k_0} ; \tag{8}$$

depends on J_2, the second invariant of the stress deviator and also on J_1' the first invariant of the stress tensor. E and ν here denote Young's modulus and Poisson's ratio, respectively, α is a constant characterizing the rate of dilatation of the soil and $s = \frac{1}{3} \sigma_{ii}$.

The constitutive Eqs. (4), (5), and (7) have the property that, when the method of bicharacteristics is applied to the solution of boundary-value problems, the geometry of the characteristic surfaces, the velocity of wave propagation and the bicharacteristics are the same as in a medium governed by Hooke's law (5). Moreover, the velocity of the waves is constant and the bicharacteristics are straight lines. These facts greatly simplify the method mentioned above. This remark also holds for the more general class of materials in which, for example, the static yield function [Eq. (8)] is more complex and depends on all three invariants of the stress state; i.e., when $F = F(J_1, J_2, J_3)$.

From Eqs. (1)—(3), the constitutive Eqs. (4)—(6) now become:

$$\dot{\varepsilon}_x = \frac{1}{2\mu} \dot{S}_x + \frac{1}{3K} \dot{\sigma}_m + DS_x, \tag{9a}$$

$$\dot{\varepsilon}_y = \frac{1}{2\mu} \dot{S}_y + \frac{1}{3K} \dot{\sigma}_m + DS_y, \tag{9b}$$

$$0 = \frac{1}{2\mu} \dot{S}_z + \frac{1}{3K} \dot{\sigma}_m + DS_z, \tag{9c}$$

$$\dot{\gamma}_{xy} = \frac{1}{\mu} \dot{\tau}_{xy} + 2D\tau_{xy}, \tag{9d}$$

which, together with the equations of motion

$$\frac{\partial \sigma_x}{\partial x} + \frac{\partial \tau_{xy}}{\partial y} = \varrho \frac{\partial v_x}{\partial t} , \tag{10a}$$

$$\frac{\partial \sigma_y}{\partial y} + \frac{\partial \tau_{xy}}{\partial x} = \varrho \frac{\partial v_y}{\partial t} , \tag{10b}$$

where ϱ is the density of the material, form the basic system of equations. We have introduced here the following notation:

$$S_x = \frac{1}{3}(2\sigma_x - \sigma_y - \sigma_z), \qquad S_y = \frac{1}{3}(2\sigma_y - \sigma_x - \sigma_z),$$

$$S_z = \frac{1}{3}(2\sigma_z - \sigma_x - \sigma_y), \qquad \sigma_m = \frac{1}{3}(\sigma_x + \sigma_y + \sigma_z).$$

$$D = \begin{cases} \gamma\Phi(J^{1/2}/k_0 - 1)/J_2^{1/2} & \text{for } J_2 \geq k_0, \\ 0 & \text{for } J_2 < k_0, \end{cases}$$

$$J_2^{1/2} = \frac{1}{3^{1/2}}(\sigma_x^2 + \sigma_y^2 + \sigma_z^2 - \sigma_x\sigma_y - \sigma_x\sigma_z - \sigma_y\sigma_z + 3\tau_{xy}^2)^{1/2},$$

$$v_x = \frac{\partial}{\partial t} u_x, \quad v_y = \frac{\partial}{\partial t} u_y.$$

We now introduce the dimensionless form of the basic system of Eqs. (9) and (10). Using Clifton's notation [6] we define the following dimensionless quantities: velocities u and v, time t, Cartesian coordinates x and y, plastic limit in simple shear k_0 and viscosity coefficient γ:

$$u = \frac{\hat{v}_x}{c_1}, \quad v = \frac{\hat{v}_y}{c_1}, \quad t = \frac{\hat{t}c_1}{b}, \quad x = \frac{\hat{x}}{b},$$

$$y = \frac{\hat{y}}{b}, \quad k_0 = \frac{\hat{k}_0}{\varrho c_1^2}, \quad = \gamma \frac{b}{c_1}\hat{\gamma}, \quad \Gamma = \frac{c_1}{c_2}.$$

Here a hat '^' denotes the corresponding dimensional quantity, c_1 is the velocity of propagation of elastic, dilatational waves, c_2 the velocity of elastic, shear waves, Γ their ratio and b an arbitrary characteristic length. The speeds c_1 and c_2 ar given by

$$c_1 = \left[\frac{3K + 4\mu}{3\varrho}\right]^{1/2}, \quad c_2 = \left[\frac{\mu}{\varrho}\right]^{1/2}.$$

We introduce dimensionless stresses p, q, τ, and r as follows:

$$p = \frac{1}{2}\frac{\sigma_x + \sigma_y}{\varrho c_1^2}, \quad q = \frac{1}{2}\frac{\sigma_x - \sigma_y}{\varrho c_1^2}, \quad \tau = \frac{\tau_{xy}}{\varrho c_1^2}, \quad r = \frac{\sigma_z}{\varrho c_1^2}.$$

Replacing the first two equations of Eqs. (9) by their sum and difference, respectively, we obtain six governing equations in dimensionless form:

$$u_t - p_x - q_x - \tau_y = 0, \tag{11a}$$

$$v_t - p_y + q_y - \tau_x = 0, \tag{11b}$$

$$-u_x - v_y + \frac{\Gamma^4}{3\Gamma^2 - 4}p_t + \frac{\Gamma^2(2 - \Gamma^2)}{3\Gamma^2 - 4}r_t = -\frac{2}{3}\langle D\rangle(p - r), \tag{11c}$$

$$-u_x + v_y + \Gamma^2 q_t = -2q\langle D\rangle, \tag{11d}$$

$$\Gamma^2\frac{(2 - \Gamma^2)}{3\Gamma^2 - 4}p_t + \frac{\Gamma^2 - 1}{3\Gamma^2 - 4}r_t = -\frac{2}{3}\langle D\rangle(r - p), \tag{11e}$$

$$-u_y - v_x + \Gamma^2\tau_t = -2\tau\langle D\rangle, \tag{11f}$$

where

$$\langle D\rangle = \gamma\Phi\left(\frac{\sqrt{J_2}}{k_0} - 1\right)\frac{1}{\sqrt{J_2}}, \quad \sqrt{J_2} = \frac{1}{\sqrt{3}}\sqrt{(p - r)^2 + 3(q^2 + \tau^2)}.$$

The first two equations of Eqs. (11) are equations of motion, the next two are the sum and difference mentioned above, and the last two are Eqs. (9c) and (9d) written in dimensionless form. The subscripts denote partial differentiation with respect to the corresponding variables.

Equations (11) may be written in a more compact matrix form:

$$L[\mathfrak{w}] = \mathfrak{A}^t\mathfrak{w}_t + \mathfrak{A}^x\mathfrak{w}_x + \mathfrak{A}^y\mathfrak{w}_y - \mathfrak{B} = 0, \tag{12}$$

where the vectors $\mathfrak{w}$, $\mathfrak{B}$ and tensors $\mathfrak{A}^t$, $\mathfrak{A}^x$, $\mathfrak{A}^y$ are:

$$\mathfrak{w} = \begin{bmatrix} u \\ v \\ p \\ q \\ r \\ \tau \end{bmatrix} \quad \mathfrak{B} = \begin{bmatrix} 0 \\ 0 \\ -\frac{2}{3}\langle D\rangle(p - r) \\ -2\langle D\rangle q \\ -\frac{2}{3}\langle D\rangle(r - p) \\ -2\langle D\rangle\tau \end{bmatrix} \quad \mathfrak{A}^t = \begin{bmatrix} 1 & 0 & 0 & 0 & 0 & 0 \\ 0 & 1 & 0 & 0 & 0 & 0 \\ 0 & 0 & M & 0 & Q & 0 \\ 0 & 0 & 0 & \Gamma^2 & 0 & 0 \\ 0 & 0 & Q & 0 & N & 0 \\ 0 & 0 & 0 & 0 & 0 & \Gamma^2 \end{bmatrix}$$

$$\mathfrak{A}^x = \begin{bmatrix} 0 & 0 & -1 & -1 & 0 & 0 \\ 0 & 0 & 0 & 0 & 0 & -1 \\ -1 & 0 & 0 & 0 & 0 & 0 \\ -1 & 0 & 0 & 0 & 0 & 0 \\ 0 & 0 & 0 & 0 & 0 & 0 \\ 0 & -1 & 0 & 0 & 0 & 0 \end{bmatrix} \quad \mathfrak{A}^y = \begin{bmatrix} 0 & 0 & 0 & 0 & 0 & -1 \\ 0 & 0 & -1 & 1 & 0 & 0 \\ 0 & -1 & 0 & 0 & 0 & 0 \\ 0 & 1 & 0 & 0 & 0 & 0 \\ 0 & 0 & 0 & 0 & 0 & 0 \\ -1 & 0 & 0 & 0 & 0 & 0 \end{bmatrix} \tag{13}$$

in which $M = \Gamma^4/(3\Gamma^2 - 4)$, $N = (\Gamma^2(\Gamma^2 - 1)/(3\Gamma^2 - 4)$, $Q = \Gamma^2(2 - \Gamma^2)/(3\Gamma^2 - 4)$. The matrices $\mathfrak{A}^x$ and $\mathfrak{A}^y$ are symmetric; the matrix $\mathfrak{A}^t$ is symmetric positive definite. Thus, Eqs. (12) constitute a symmetric hyperbolic system of partial differential equations.

3. Characteristic properties of the governing equations

The system of Eqs. (12) is a system of six first-order semilinear hyperbolic partial differential equations in three independent variables with constant coefficients for the terms involving derivatives. A theory of such equations is given, for instance, in [20]. The method of finite differences along bicharacteristics will be used for the solution of Eqs. (12). Thus the geometry of the characteristic surfaces associated with Eqs. (12) will now be investigated.

The condition that a surface $\varphi(t, x, y) = \text{const}$, be a characteristic surface of Eqs. (12) is equivalent to the condition that the determinant of the characteristic matrix $\mathfrak{A}$ be zero:

$$\text{Det } \mathfrak{A} = 0 \tag{14}$$

where

$$\mathfrak{A} = \mathfrak{A}^t \varphi_t + \mathfrak{A}^x \varphi_x + \mathfrak{A}^y \varphi_y. \tag{15}$$

Equation (14) is equivalent to

$$[\varphi_t^2 - (\varphi_x^2 + \varphi_y^2)]\left[\varphi_t^2 - \frac{1}{\Gamma^2}(\varphi_x^2 + \varphi_y^2)\right]\varphi_t^2 = 0. \tag{16}$$

The two terms in square brackets in Eq. (16) describe the propagation of dilatational and shear waves with dimensionless wave velocities c equal to ± 1 and $\pm 1/\Gamma$, respectively; these correspond to the dimensional velocities c_1 and c_2.

It must be emphasized here that discontinuities are propagated in an elastic/viscoplastic medium with elastic wave velocities. This is explained mathematically by the fact that in Eqs. (12) the nonlinear term $\mathfrak{B}$, describing the plastic and viscous effects, does not enter into the analysis of the characteristic surfaces.

The bicharacteristics of Eqs. (12) are the generators of the following characteristic cones passing through the point (t_0, x_0, y_0):

$$c^2(t - t_0)^2 = (x - x_0)^2 + (y - y_0)^2, \quad c = 1, \frac{1}{\Gamma}. \tag{17}$$

It is convenient to introduce the following parametrization of the characteristic cones in terms of the two parameters α and $\tilde{t}$:

$$x - x_0 = c\tilde{t} \cos \alpha, \qquad c = 1, \frac{1}{\Gamma}, \tag{18a}$$

$$y - y_0 = c\tilde{t} \sin \alpha, \tag{18b}$$

$$t - t_0 = \tilde{t}. \tag{18c}$$

Relations (18) give the desired equations of bicharactristics as the generators of the characteristic cones. The bicharacteristic strips associated with the bicharacteristic lines (18) are:

$$\varphi_t = c, \qquad c = 1, 1/\Gamma \tag{19a}$$

$$\varphi_x = -\cos \alpha, \tag{19b}$$

$$\varphi_y = -\sin \alpha. \tag{19c}$$

In order to determine the equations along bicharacteristics, the null vectors associated with the system (12) must first be determined. The null vectors $\mathfrak{l} = [l_1, l_2, l_3, l_4, l_5, l_6]$ are the solutions of the following homogeneous system of equations:

$$\mathfrak{l} \, \mathfrak{A} = 0. \tag{20}$$

Substituting for $\mathfrak{A}$ from Eqs. (15) and (19), solutions of Eqs. (20) corresponding to the dimensionless velocities $c = 1$ and $c = \frac{1}{\Gamma}$ are:

$$\mathfrak{l} = [-\Gamma^2 \cos \alpha, \quad -\Gamma^2 \sin \alpha, \quad \Gamma^2 - 1, \quad \cos 2\alpha, \quad \Gamma^2 - 2, \quad \sin 2\alpha] \tag{21a}$$

for $c = 1$, and

$$\mathfrak{l} = [\Gamma \sin \alpha, \quad -\Gamma \cos \alpha, \quad 0, \quad -\sin 2\alpha, \quad 0, \quad \cos 2\alpha] \quad \text{for} \quad c - 1/\Gamma. \tag{21b}$$

The desired differential equations along bicharacteristics are obtained from the equation

$$\mathfrak{l} \cdot L[\mathfrak{w}] = 0 \tag{22}$$

where the dot denotes the inner product. In Eqs. (22) the partial derivatives with respect to t, can be eliminated by use of

$$\mathfrak{w}_t = \frac{d\mathfrak{w}}{dt} = \mathfrak{w}_x \frac{dx}{dt} - \mathfrak{w}_y \frac{dy}{dt} \tag{23}$$

where $d\mathfrak{w}/dt$ is the total time derivative of $\mathfrak{w}$ taken along a bicharacteristic and dx/dt, dy/dt are obtained by differentiation of Eqs. (18). In order to have a backward drawn bicharacteristic in the positive x-direction correspond to $\alpha = 0$ as indicated in Fig. 2 we replace α by $\alpha + \pi$. Then, substituting Eqs. (21) and (23) in Eq. (22) gives

$$\cos \alpha \, du + \sin \alpha \, dv + dp + \cos 2\alpha \, dq + \sin 2\alpha \, d\tau = -S_1(\alpha) \, dt \tag{24a}$$

for $c = 1$, and

$$-\Gamma \sin \alpha \, du + \Gamma \cos \alpha \, dv - \Gamma^2 \sin 2\alpha \, dq + \Gamma^2 \cos 2\alpha \, d\tau = -S_2(\alpha) \, dt \tag{24b}$$

for $c = 1/\Gamma$, where $S_1(\alpha)$ and $S_2(\alpha)$ are

$$
\begin{aligned}
S_1(\alpha) = &\left[-\sin^2 \alpha + \frac{1}{\Gamma^2}(1 - \cos 2\alpha)\right] u_x + \left(\frac{1}{2} \sin 2\alpha - \frac{1}{\Gamma^2} \sin 2\alpha\right) u_y \\
&+ (-1 + \cos 2\alpha) q_x \cos \alpha + (1 + \cos 2\alpha) q_y \sin \alpha \\
&+ \left(\frac{1}{2} - \frac{1}{\Gamma^2}\right) v_x \sin 2\alpha + \left[-\cos^2 \alpha + \frac{1}{\Gamma^2}(1 + \cos 2\alpha)\right] v_y \\
&+ (\sin 2\alpha \cos \alpha - \sin \alpha) \tau_x + (\sin 2\alpha \sin \alpha - \cos \alpha) \tau_y \\
&+ \frac{2D}{\Gamma^2}\left(\frac{1}{3}(p - r) + \tau \sin 2\alpha + q \cos 2\alpha\right)
\end{aligned}
\tag{25a}
$$

and

$$
\begin{aligned}
S_2(\alpha) = &\frac{1}{2} \sin 2\alpha \, (u_x - v_y) - \cos^2 \alpha u_y + \Gamma \sin \alpha p_x \\
&+ \Gamma(\sin \alpha - \sin 2\alpha \cos \alpha) q_x + \Gamma \sin \alpha \, (1 + \cos 2\alpha) \tau_y \\
&+ \sin^2 \alpha v_x - \frac{1}{2} \sin 2\alpha \, v_y - \Gamma \cos \alpha p_y + (1 - 2 \sin 2\alpha) q_y \\
&- \Gamma \cos \alpha (1 - \cos 2\alpha) \tau_x - 2D(\sin 2\alpha q - \cos 2\alpha \tau).
\end{aligned}
$$

The spatial derivatives in Eqs. (25) correspond to derivatives in the direction tangential to the characteristic cone (cf. Fig. 2). In the next section a system of difference equations are obtained by integration of Eqs. (24) along bicharacteristics.

4. Solution by finite differences

We regard the half-space $y \geq 0$ as covered by a square mesh with mesh size h. Difference equations will be derived for computing the solution at a mesh point (t_0, x_0, y_0) (hereafter called simply 0) from known data at neighboring mesh points on the plane $t = t_0 - k$ (see Fig. 2). These equations are obtained by forming linear combinations of equations resulting from the integration of Eqs. (24) along bicharacteristics and the integration of Eqs. (12) along the line $x = x_0, y = y_0$. We follow the procedure introduced by BUTLER [3], and used by CLIFTON [7] for the elastic case.

Integrating Eqs. (24) along the bicharacteristic for which $\alpha = \alpha_i$, from the point 0 to the point $(t_0 - k, x_i, y_i)$ where this bicharacteristic intersects the plane $t = t_0 - k$, gives an equation involving the difference $\Delta w_i = w(t_0, x_0, y_0) - w(t_0 - k, x_i, y_i)$. It is convenient to eliminate Δw_i by use of the identity

$$\Delta w_i = \delta w + w(t_0 - k, x_0, y_0) - w(t_0 - k, x_i, y_i) \tag{26}$$

where

$$\delta w = w(t_0, x_0, y_0) - w(t_0 - k, x_0, y_0). \tag{27}$$

When Eqs. (26) and (27) are used, the difference equations resulting from integrating Eqs. (24) along the bicharacteristics for which $\alpha = \alpha_i$ are

$$\cos \alpha_i \, \delta u + \sin \alpha_i \, \delta v + \delta p + \cos 2\alpha_i \, \delta q + \sin 2\alpha_i \, \delta \tau = - \frac{k}{2} \, [S_1(\alpha_i)^0 + S_1(\alpha_i)_i] - W_1(\alpha_i) \tag{28a}$$

and

$$-\varGamma \sin \alpha_i \, \delta u + \varGamma \cos \alpha_i \, \delta v - \varGamma^2 \sin 2\alpha_i \, \delta q + \varGamma^2 \cos 2\alpha_i \, \delta \tau$$
$$= - \frac{1}{2} \, k[S_2(\alpha_i)^0 + S_2(\alpha_i)_i] - W_2(\alpha_i) \tag{28b}$$

for the exterior and interior cones respectively, where

$$W_1(\alpha_i) = \cos \alpha_i \, (u_0 - u_i) + \sin \alpha_i (v_0 - v_i) + \cos 2\alpha_i (q_0 - q_i) + \sin 2\alpha_i (\tau_0 - \tau_i) + p_0 - p_i,$$

$$W_2(\alpha_i) = \varGamma \sin \alpha_i (u_0 - u_i) + \varGamma \cos \alpha_i (v_0 - v_i) - \varGamma^2 \sin (q_0 - q_i) + \varGamma^2 \cos 2\alpha_i (\tau_0 - \tau_i).$$

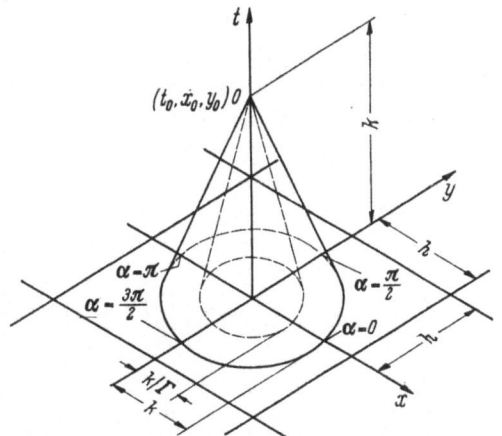

Fig. 2. Characteristic cones for the dynamic elastic, viscoplastic equations.

In Eqs. (28) the superscript $_0$ denotes evaluation of the function at the point 0; the subscript $_0$ denotes evaluation of the function at the point $(t_0 - k, x_0, y_0)$; the subscript i denotes evaluation of the function at the point where the bicharateristic α_i on the appropriate characteristic cone intersects the plane $t = t_0 - k$.

Six additional difference equations, obtained by integrating Eqs. (12) along the line $x = x_0$, $y = y_0$, are as follows

$$\delta u = \frac{k}{2} \, [(p_x + q_x + \tau_y)^0 + (p_x + q_x + \tau_y)_0]. \tag{29a}$$

$$\delta \tau = \frac{k}{2} \, [p_y - q_y + \tau_x)^0 + (p_y - q_y + \tau_x)_0], \tag{29b}$$

$$\delta p = \frac{k}{2} \, \left[\left(\frac{\varGamma^2 - 1}{\varGamma^2} (u_x + v_y) - \frac{2D}{3\varGamma^2} (p - r) \right)^0 + \left(\frac{\varGamma^2 - 1}{\varGamma^2} (u_x + v_y) - \frac{2D}{3\varGamma^2} (p - r) \right)_0 \right], \tag{29c}$$

$$\varGamma^2 \, \delta q = \frac{k}{2} \, [(u_x - v_y - 2qD)^0 + (u_x - v_y - 2qD)_0], \tag{29d}$$

$$\delta r = \frac{k}{2} \, \left[\left(\frac{\varGamma^2 - 2}{\varGamma^2} (u_x + v_y) - \frac{4D}{3\varGamma^2} (r - p) \right)^0 + \left(\frac{\varGamma^2 - 2}{\varGamma^2} (u_x + v_y) - \frac{4D}{3\varGamma^2} (r - p) \right)_0 \right], \tag{29e}$$

$$\varGamma^2 \, \delta \tau = \frac{k}{2} \, [(u_y + v_x - 2\tau D)^0 + (u_y + v_x - 2\tau D)_0] \tag{29f}$$

where the $_0$ superscripts and subscripts have the same meaning as in Eqs. (28).

All the terms on the right hand side of Eqs. (28) and (29) can be evaluated from data on the plane $t = t_0 - k$ except those terms having a superscript $_0$. Just as in the elastic case [7], all the terms involving partial derivatives at 0 can be eliminated by forming linear combinations of Eqs. (29), and the eight equations obtained by writing Eqs. (28) for $\alpha_i = (i - 1) \, \pi/2$, with

$i = 1, 2, 3, 4$. In this way we obtain a system of six equations which determine the six unknown increments δu, δv, δp, δq, δr, and $\delta \tau$. These equations involve terms, to be evaluated on the plane $t = t_0 - k$, which have the form

$$w(x_0 + ck, y_0) - w(x_0 - ck, y_0) \tag{30a}$$

$$w(x_0 + ck, y_0) + w(x_0 - ck, y_0) - 2w(x_0, y_0) \tag{30b}$$

$$k[w_x(x_0, y_0 + ck) - w_x(x_0 y_0 - ck)] \tag{30c}$$

$$k[w_x(x_0, y_0 + ck) + w_x(x_0, y_0 - ck) - 2w_x(x_0, y_0)]. \tag{30d}$$

The expressions (30) can be replaced, respectively, by

$$2ckw_x(x_0, y_0) \tag{31a}$$

$$(ck)^2 w_{xx}(x_0, y_0) \tag{31b}$$

$$2ck^2 w_{xy}(x_0, y_0) \tag{31c}$$

$$0 \tag{31d}$$

with an accuracy of order k^3. In Eqs. (30) and (31) c is equal to 1 for the exterior cones and to $1/\Gamma$ for the interior cones. The same relations hold if we change the roles of x and y in Eqs. (30) and (31). Making use of the substitution of expressions (31) for those appearing in (30), the six equations for the six unknown increments become

$$2\delta u = \frac{k^2}{\Gamma^2}\left[(\Gamma^2 - 1)\, v_{yx} + \Gamma^2 u_{xx} + u_{yy} - 2\left(D\left(\frac{p-r}{3} + q\right)\right)_x - (2D\tau)_y\right]_0 + 2k(q_x + p_y + \tau)_0, \tag{32a}$$

$$2\delta v = 2k(p_y - q_y + \tau_x)_0 + \frac{k^2}{\Gamma^2}\left[(\Gamma^2 - 1)\, u_{xy} + \Gamma^2 v_{yy} + v_{xx} - 2\left(D\left(\frac{p-r}{3} - q\right)\right)_y - (2D\tau)_x\right]_0, \tag{32b}$$

$$2\Gamma^2\, \delta\tau = 2k(v_x + u_y)^0 + k^2((2p_{xy})_0 + (\tau_{xx} + \tau_{yy})_0) - \frac{k^3}{\Gamma^2}((D\tau)_{xx} + (D\tau_{yy}))_0, \tag{32c}$$

$$2\Gamma^2\, \delta q = -\frac{k}{2}\,[(4Dq)^0 + (4Dq)_0] + 2k(u_x - v_y)_0 + k^2(q_{xx} + q_{yy} + p_{xx} - p_{yy})_0$$
$$-\frac{k^3}{\Gamma^2}\left[\left(D\left(\frac{p-r}{3} + q\right)\right)_{xx} - \left(D\left(\frac{p-r}{3} - q\right)\right)_{yy}\right]_0 \tag{32d}$$

$$2\frac{\Gamma^2}{\Gamma^2-1}\,\delta p = 2k(u_x + v_y)_0 - \frac{k}{2}\left[\left(\frac{4D(p-r)}{3(\Gamma^2-1)}\right)^0 + \left(\frac{4D(p-r)}{3(\Gamma^2-1)}\right)_0\right],$$

$$+ k^2[q_{xx} - q_{yy} + p_{xx} + p_{yy} + 2\tau_{xy}]_0 - \frac{k^3}{\Gamma^2}\left[\left(D\left(\frac{p-r}{3} + q\right)\right)_{xx} + \left(D\left(\frac{p-r}{3} - q\right)\right)_{yy}\right]_0, \tag{32e}$$

$$\delta r = \frac{\Gamma^2 - 2}{\Gamma^2 - 1}\,\delta p + \frac{k}{3\Gamma^2}\,\frac{3\Gamma^2 - 4}{\Gamma^2 - 1}\,\{[D(p - r)]^0 + [D(p - r)_0]\}. \tag{32f}$$

In order to convert Eqs. (32) to difference equations, the partial derivatives with respect to x and y are replaced by the corresponding centered difference equations. Thus, for example, the second partial derivative with respect to x is replaced by

$$W_{xx}(x_0, y_0) = \frac{1}{h^2}\left(w(x_0 + h, y_0) + w(x_0 - h, y_0) - 2w(x_0, y_0)\right). \tag{33}$$

The final difference equations correspond to a second order accurate differene method. That is, the error in the computed increments δu, δv, ..., δr for a single step is $0(k^3)$.

Because stresses at 0 are unknowns, the terms involving superscripts $_0$ in Eqs. (32) are not known initially. As a result, an iterative procedure must be used for solving these equations. For the numerical example presented in this paper, initial values for the stresses at 0 were taken to be the same as the stresses at the point $(t_0 - k, k_0, y_0)$. The solution was computed using a single iteration at each mesh point.

The same basic formulae, Eqs. (32), are used in both elastic and viscoplastic regions. In elastic regions the coefficient D is equal to zero. The unknown function values on the plane $t = t_0$ are determined from known values at neighboring points on the plane $t = t_0 - k$ and, where applicable, from values given on the boundary. These points may be all in the plastic region, all in the elastic region, or some in each region. The value of the plastic yielding function F is calculated at every mesh point. Depending on whether $F < 0$ or $F \geq 0$, the viscosity coefficient γ is put equal to zero or to γ. In this way, the computations can be carried out without explicitly locating boundaries between elastic and plastic regions.

Although the elastic-plastic boundary is not located explicitly by this approach it is possible to draw on each of the planes $t = \text{const.}$ an approximate curve Γ_i which divides elastic and elastic/viscoplastic regions by interpolating between neighboring mesh points, one of which corresponds to $(J_2)^{1/2} < k_0$ and the other corresponds to $(J_2)^{1/2} \geq k_0$. Superimposing these curves for successive times we obtain a picture of the motion of the elastic-plastic boundary as shown schematically in Fig. 3.

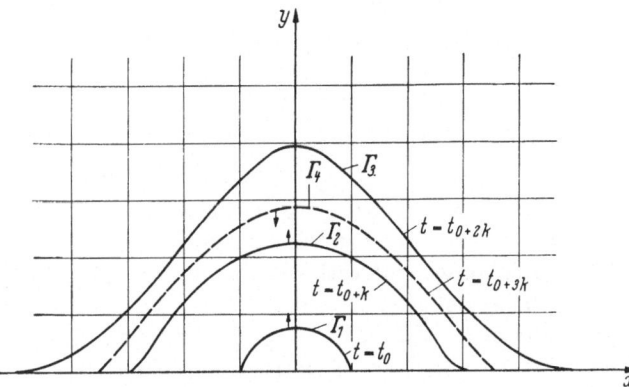

Fig. 3. Plastic zones on the x-y plane at fixed time.

So far, we have only considered difference equations for interior points of the half-space $y \geq 0$. We now derive appropriate equations for mesh points on the boundary $y = 0$. These equations are obtained by eliminating equations along bicharacteristics for which $\alpha = \dfrac{3\pi}{2}$ on both exterior and interior cones since these bicharacteristics intersect the plane $t = t_0 - k$ at points outside the region of interest (cf. [3, 7]). Combining Eqs. (28) and (29) as for interior points and then eliminating relations along the bicharacteristics corresponding to $\alpha = \dfrac{3\pi}{2}$ leads to the following equations for use at mesh points on the boundary $y = 0$.

$$2\delta u = (32\text{a}) + \frac{1}{\Gamma}(32\text{c}), \tag{34a}$$

$$2\delta v + \frac{2\Gamma^2}{\Gamma^2 - 1}\,\delta p = (32\text{b}) + (32\text{c}), \tag{34b}$$

$$-2\delta v + 2\Gamma^2\,\delta q = -(32\text{b}) + (32\text{d}), \tag{34c}$$

$$\delta r = (32\text{f}). \tag{34d}$$

The terms on the right-hand side indicate the right-hand side of the corresponding equation of Eqs. (32). Equations (34) constitute four equations in six unknowns. The remaining two equations come from the boundary conditions on the boundary $y = 0$. For the case being considered namely that of an applied normal pressure $P(x, t)$, the two additional equations are

$$\tau(x, t) = 0, \tag{35a}$$

$$p(x, t) - q(x, t) = -P(x, t). \tag{35b}$$

The derivatives with respect to y that appear in Eqs. (34) are replaced by forward differences of sufficient accuracy so that the $0(k^3)$ accuracy is preserved. Thus, the following difference approximations are used.

$$2ckw_y(x_0, 0) = 2c\frac{k}{h}\left[2(x_0w, h) - \frac{1}{2}w(x_0, 2h) - \frac{3}{2}w(x_0, 0)\right], \tag{36a}$$

$$(ck)^2 w_{yy}(x_0, 0) = \left(\frac{ck}{h}\right)^2 [w(x_0, 0) - 2w(x_0, h) + w(x_0, 2h)], \tag{36b}$$

$$2ck^2 w_{yx}(x_0, 0) = \left(\frac{ck}{h}\right)^2 \left[2w(x_0 + h, h) - \frac{1}{2}w(x_0 + h, 2h) - \frac{3}{2}w(x_0 + h, 0)\right.$$
$$\left. - 2w(x_0 - h, h) + \frac{1}{2}w(x_0 - h, 2h) + \frac{3}{2}w(x_0 - h, 0)\right]. \tag{36c}$$

When the applied load $P(x, t)$ is symmetric with respect to the y-axis only the region $x \geq 0$ needs to be considered. The appropriate difference equations for mesh points on the plane of symmetry $x = 0$ and the corner point $x = 0$, $y = 0$ can be obtained from Eqs.(32) and Eqs.(34), respectively, by imposing the following symmetry conditions:

$$u(0, y) = 0, \quad \tau(0, y) = 0, \quad u(-h, y) = -u(h, y),$$
$$\tau(-h, y) = -\tau(h, y), \quad v(-h, y) = v(h, y), \quad p(-h, y) = p(, h, y), \tag{37}$$
$$q(-h, y) = q(h, y), \quad r(-h, y) = r(h, y).$$

5. A numerical example

For numerical calculations we use Eqs.(32) and (34). The material constant γ and the function Φ are chosen so that the dynamic material behavior represented by relations (4)—(6) is in reasonable agreement with experimental results. In this paper the relaxation function Φ is assumed, for simplicity, to be linear. The material constant γ is determined from the experimental results of HARDING, WOOD and CAMPBELL [23].

The following data have been used for mild steel:

$$\mu = 0.82 \times 10^6 \text{ kg cm}^{-2}, \qquad K = 1.66 \times 10^6 \text{ kg cm}^{-2},$$
$$k_0 = 2.44 \times 10^3 \text{ kg cm}^{-2}, \qquad \varrho = 7.8 \times 10^{-6} \text{ kg cm}^{-3} \text{ sec}^2, \tag{38}$$
$$\nu = 0.29, \qquad \gamma = 750 \text{ sec}^{-1}.$$

For these data we have

$$c_1 = 0.5941 \times 10^6 \text{ cm sec}^{-1},$$
$$c_2 = 0.324 \times 10^6 \text{ cm sec}^{-1}, \tag{39}$$
$$\Gamma = c_1/c_2 = 1.83.$$

For this example, we consider a symmetrical, continuous distribution of the pressure given by the relation

$$\hat{P}(\hat{x}, \hat{t}) = \hat{P}_0 \hat{t}(\hat{t}_f - \hat{t}) e^{-(\hat{x}/\alpha)^2} \tag{40}$$

where $\hat{t}_f$ denotes the duration of the load and $\hat{P}_0$ and α are two parameters describing the magnitude and the shape of the load in the x direction, respectively. We assume also that the half-space was undisturbed before application of the load; thus all the stresses and velocities are equal to zero at $t = 0$.

Actual durations of explosive, blast or impact loading are measured in small fractions of seconds, i.e. on the order of $10^{-3} - 10^{-6}$ sec. For these calculations we assume dimensionless duration of loading $t_f = 0.9$ corresponding to a dimensional time $\hat{t}_f = 0.9 \times 10^{-5}$ sec. An arbitrary constant b is set equal to 5.941 cm. We take the time interval $k = 0.1$ and mesh dimension $h = 0.2$. Thus $\lambda = k/h = 0.5$, which is in agreement with present ideas concerning the stability of difference equations [7]. The dimensionless viscosity constant, γ, which depends on the dimension b, we take equal to 0.0075. The mesh function approximation of the general function $P(x, t)$ is shown in Fig. 4.

The calculation procedure is such that for every fixed time $t =$ const. the values of the unknown functions are computed at points along grid lines parallel to the x axis, starting from the boundaries $y = 0$ and $x = 0$ respectively. After the first time increment we have calculations along only one grid line; after the second, two grid lines, and so on. The solution on the

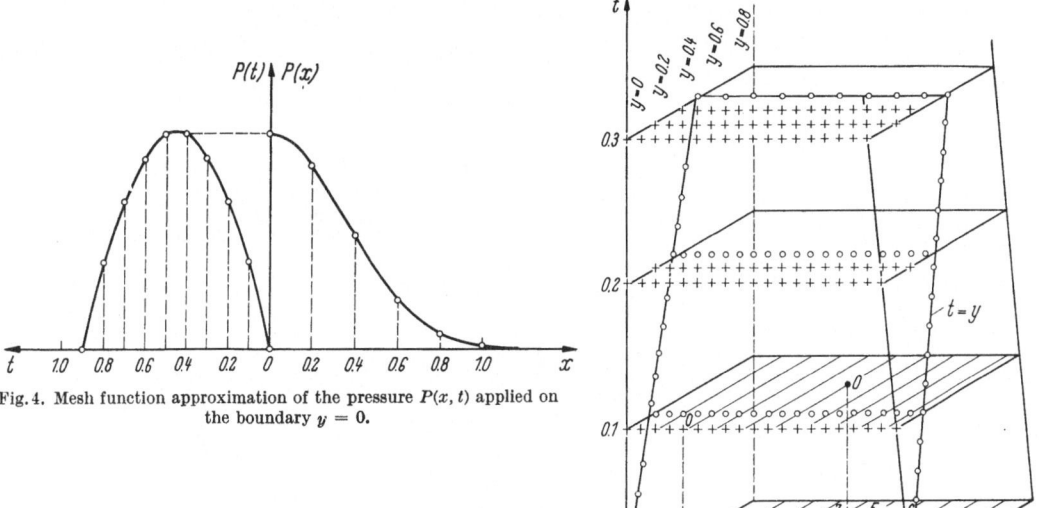

Fig. 4. Mesh function approximation of the pressure $P(x, t)$ applied on the boundary $y = 0$.

Fig. 5. Mesh points (∗) at which the solution is computed.

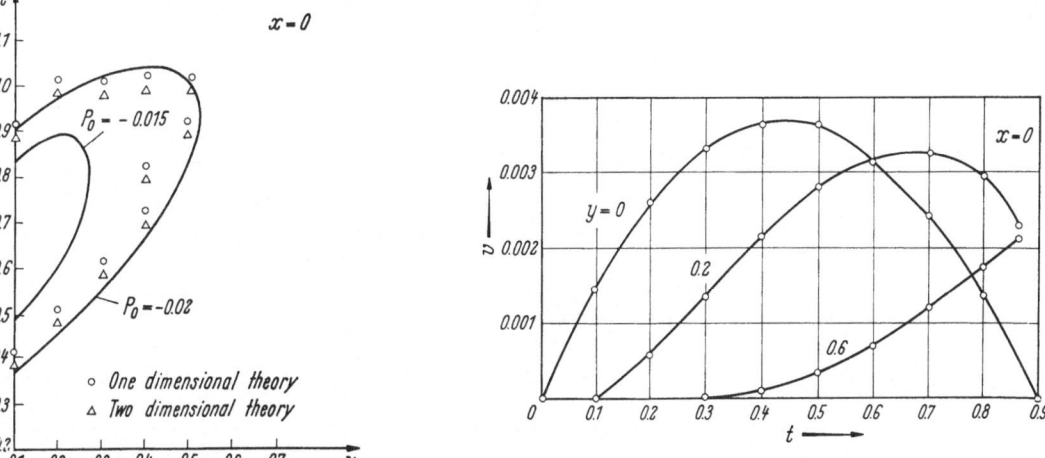

Fig. 6. Plastic zones on the t-y plane for $x = 0$ according to the one dimensional (○) and two dimensional theory (△).

Fig 7. Distribution of the velocity v in time for different fixed values of y and $x = 0$.

characteristic surface $y = t$ is prescribed to be zero. The mesh points at which the solutions is computed are indicated by asterisks in Fig. 5.

To check the theory, the one-dimensional case was first calculated numerically and then compared with the two-dimensional case, assuming uniform pressure distribution ($\alpha = \infty$) at the boundary $y = 0$ and that $p - r = 0$, $p = -q$, $u = t = 0$. The curves for the velocity v and dimensionless stress σ_{yy} versus t (at fixed y) or versus y (at fixed t) calculated according to both theories differ very little. The curves dividing the elastic and plastic regions are shown in Fig. 6. Circles and triangles indicate, respectively, points calculated from the one-dimensional theory and from the two-dimensional theory (with the assumptions mentioned above).

17*

For illustration, typical velocity and stress profiles, obtained by holding two of the three arguments t, x, y, fixed and varying the third are shown in Figs. 7—13. The most interesting and critical sections are in the vicinity of the point $y = 0$, $x = 0$ where the maximum load is

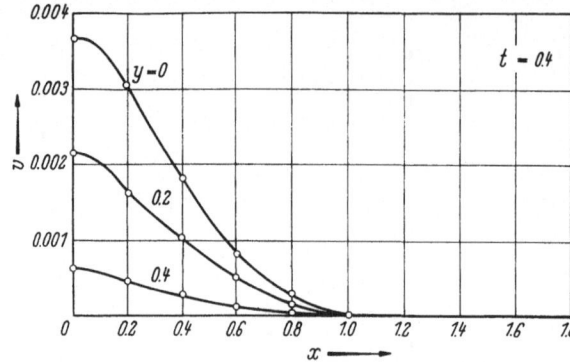

Fig. 8. Distribution of the velocity v in x direction for different fixed values of y and $t = 0.4$.

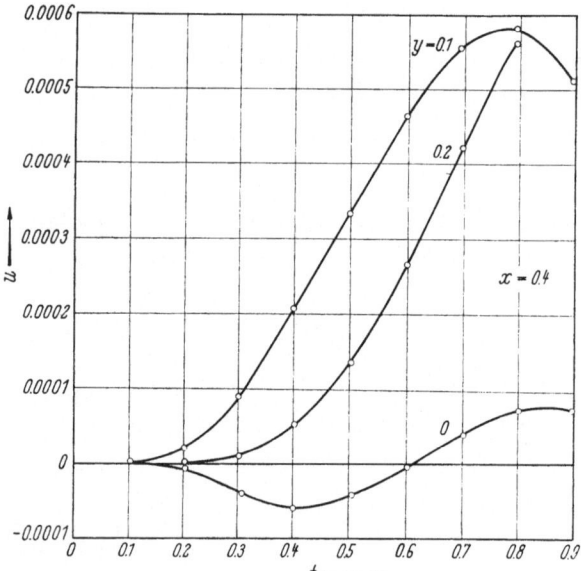

Fig. 9. Distribution of the velocity u in time for different fixed values of y and $x = 0.4$.

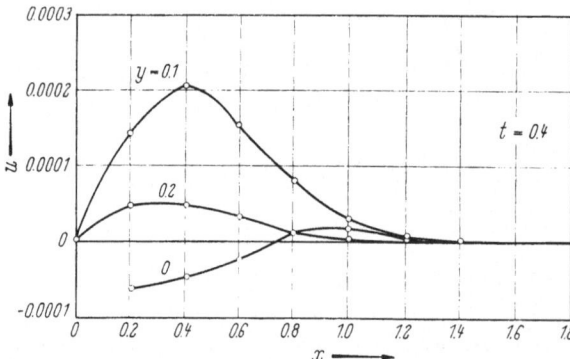

Fig. 10. Distribution of the velocity u in x direction for different fixed values of y and $t = 0.4$.

applied. The loading parameters $P_0 = -0.015$, $t_f = 0.9$, and $\alpha = b/2$ are assumed. The distribution of the velocity v with time at $x = 0$ for different fixed values of y is shown in Fig. 7, and in Fig. 8 the distribution of the same function in the x direction at fixed time $t = 0.4$ is shown. Curves of the velocity u versus t and x are given in Fig. 9 and 10 respectively. Figs. 11 to 13 show the distribution of the dimensionless stresses σ_{yy}, σ_{xx}, τ and σ_{zz} in the t, x and y

directions. There is a similarity in the distribution of the functions v, σ_{yy}, σ_{zz} and the functions u and τ in time and in the x and y directions.

From the results of these numerical calculations it may be seen that the method described in Sections 2—4 may be used successfully for very short durations of loading. In many cases of blast, explosive and impact loading this requirement is fulfilled.

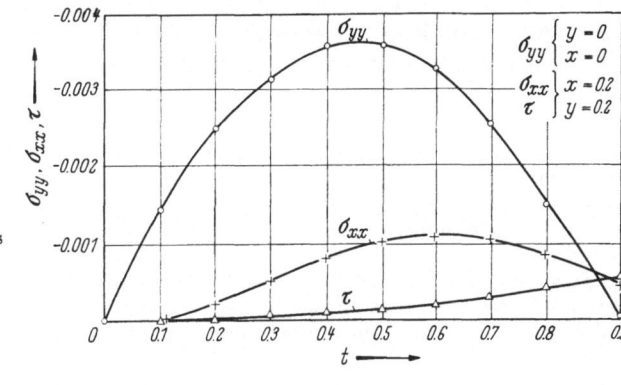

Fig. 11. Distribution of the dimensionless stresses σ_{yy}, σ_{xx}, τ in time for fixed values of x and y.

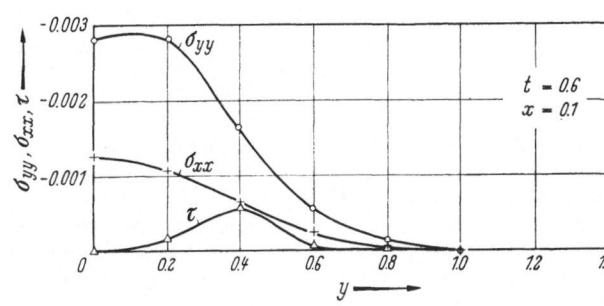

Fig. 12. Distribution of the dimensionless stresses σ_{yy}, σ_{xx}, τ in y for fixed $t = 0.6$, $x = 0.1$

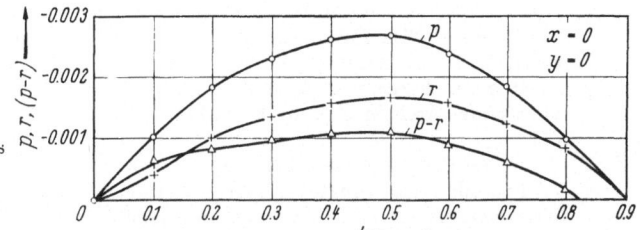

Fig. 13. Distribution of the dimensionless stresses p, r, $p - r$ in t for $x = 0$, $y = 0$.

6. Concluding remarks

The results presented here were obtained assuming smooth loading curves which increase from zero to a certain maximum value and then decrease to zero. When a sudden load of finite value is applied at $t = 0$ to the boundary $y = 0$, shock or discontinuity waves will be generated. The method of solution presented here does not include these effects. The analysis of the propagation and reflection of strong discontinuity waves, the method of integration of equations for the discontinuity surface $t = y$ and the problem of the interaction of different types of waves will be the subject of further work.

Using a cylindrical system of coordinates, we have also considered the propagation of axially symmetric cylindrical waves in an elastic/viscoplastic half-space $y \geq 0$. We assumed a non-uniform, radial, time-dependent pressure $p(r, t)$ applied to the surface $y = 0$. Derivation of six governing equations in six unknown functions v_r, v_y, σ_{rr}, $\sigma_{\varphi\varphi}$, σ_{yy} and τ_{yr}, analysis of characteristic surfaces, and the method of solution are similar to those in this paper and will not be presented.

From a practical point of view, it is of interest to consider cases in which a load is applied on a fixed strip of the plane $y = 0$ or is spreading out in the x-direction with a constant velocity.

One may also treat the situation in which on the part $|x| \leq x_0$ of the surface $y = 0$ the load is of the form $p(x, t)$, while on the other part $|x| > x_0$ it is of the form $p(t)$. The solution of the latter problem is obtained by the combined use of the difference methods for both two-dimensional and one-dimensional problems.

An analytical investigation of the problem of stability and convergence of the numerical solution is not considered here; we have assumed that all conclusions concerning this subject given by Clifton in [6, 7] are valid. In particular, we have assumed the same stability condition: that the ratio of time step to mesh spacing be not greater than 0.837.

Acknowledgment

It is a pleasure to acknowledge the advice and encouragement of Professor P.S.Symonds and several conversations with Professor R.J.Clifton, who read this manuscript and has given me many helpful suggestions.

Thanks are also due to Professor W.M.Jaworski for his assistance in programming and computing the final results.

This work was supported in part by Ballistic Research Laboratories, Maryland, through Contract DAADO 5-68-C-0083 with Brown University, while the author was a Ford Foundation Scholar at Brown University, Providence, Rhode Island during the academic year 1967—1968.

References

1. Hopkins, H. G.: The method of characteristics and its application to the theory of stress waves in solids, Engineering Plasticity, Papers for a conference held in Cambridge, March, 1968, Ed.: J. Heyman and F. A. Leckie, Cambridge University Press 1968.

2. Perzyna, P.: Fundamental problems in viscoplasticity, Advances in Applied Mechanics, Vol. 9, 1966.

3. Butler, D. S.: The numerical solution of hyperbolic systems of partial differential equations in three independent variables, Proc. Roy. Soc., London, 1962, A 255, 232—252.

4. Burnat, M., A. Kiełbasinski and A. Wakulicz: The method of characteristics for a multidimensional gas flow, Arch. Mech. Stos. 3, 16, 179—187 (1964).

5. Richardson, D. J.: The solution of two-dimensional hydrodynamic equations by the method of characteristics, Methods in Computational Physics, Vol. 3, Academic Press, 1964.

6. Clifton, R. J.: Analysis of dynamic deformation of elastic/plastic solids under conditions of plane strain, Ph. D. Thesis, Carnegie Institute of Technology, Pittsburgh, Pennsylvania, December, 1963.

7. — A difference method for plane problems in dynamic elasticity, Quart. Appl. Math., XXV, No. 1, April, 1967, Raznostnyi Metod v Ploskikoh Zadatchiakh Dinamitcheskoi Uprugosti. Mekchanika 1, 107, 103—122 (1968).

8. Estrin, M. I.: The equations of the dynamics of a compressible plastic medium, Soviet Phys., Dokladi, 5, 6, 1349—1352, 1961. (Translation of Dokladi Akad. Nauk SSR (N.S.) 135, 1, 36—39, Nov. 1960, by Amer. Inst. Phys. Inc., New York, N.Y.)

9. Baltov, A.: The plane problem for elastic/viscoplastic bodies, Archiwum Mechaniki Stosowanej, 2, 18 (1966).

10. Zhukov, A. I.: The application of the method of characteristics to the numerical solution of one dimensional problems of gas dynamics, Trudy Mathematitcheskogo Instituta im W. A. Cteklova, 58, 1960 (in Russian).

11. Lax, P. D., and B. Wendroff: Difference schemes with high order of accuracy for solving hyperbolic equations, Comm. Pure Appl. Math. 17, 381—398 (1964).

12. Prouse, G.: Sulla risoluzione del problema misto per le equazioni iperboliche non lineari mediante le differenze finite, Ann. di Mat. 46, 313—341 (1958).

13. Thomée, V.: A difference method for a mixed boundary problem for symmetric hyperbolic systems, Arch. Rat. Mech. Anal. 13, 122—136 (1963).

14. Keller, H. B., and V. Thomée: Unconditionally stable difference methods for mixed problems for quasi linear hyperbolic systems in two dimensions, Comm. Pure Appl. Math. 15, 63—73 (1962).

15. Recker, W. W.: A numerical solution of three-dimensional problems in dynamic elasticity, unpublished manuscript based on his Ph. D. thesis, Dpt. of Civil Engineering, Carnegie Mellon University, 1967.

16. Sauerwein, H.: Anisotropic waves in elastoplastic soils, Intern. J. of Engrg. Sci., Vol. 5, 1967, pp. 455—475.

17. Wilkins, M. L.: Calculation of elastic-plastic flow, Methods in computational physics, Vol. 3, Academic Press, 1964, pp. 211—263.

18. Maenchen, G., and S. Sack: The tensor code, Methods in computational physics, Vol. 3, Academic Press, 1964.

19. Zienkiewicz, O. C.: The finite element method in structural and continuum mechanics, McGraw-Hill, 1967.

20. Courant, R., and D. Hilbert: Methods of mathematical physics, Vol. II, Partial Differential Equations, New York: Interscience 1962.

An approach to the problem of axisymmetric sonic flow around a slender body

By

S. B. Berndt

Royal Institute of Technology, Stockholm

1. Introduction

To calculate the flow field set up by a slender body flying at sonic speed is a classical problem — a problem which, over the years, has been tried many times without receiving a definite solution. The dominating feature of the problem is of course its inherently nonlinear character: the linear acoustic approximation is unable to account for the sound-wave interaction which keeps finite the wave travelling with the body. There arises, in the first approximation for slender bodies, the well-known transonic differential equation

$$\Delta\varphi = \varphi_x\varphi_{xx} \tag{1}$$

for the perturbation velocity potential (the x-axis of the body-fixed reference frame pointing in the flow direction and the Laplacian Δ operating in planes $x =$ constant). In the special case of plane flow, this equation can be solved after transforming it to a linear one by the Legendre transformation, or can be attacked directly by efficient methods of approximation. In other cases, however, in particular the axisymmetric one to be treated here, the situation is less satisfactory.

Among the methods of approximation are those derived from the so-called parabolic method. In its basic form, originally explored by OSWATITSCH and KEUNE [1], this method consists in approximating φ_{xx} in (1) by a positive constant, thought of as a representative mean value of the flow acceleration *outside* the frontal part of the body; the resulting heat-conduction equation is easily solved. Later developments, however, are more concerned with the flow close to the body. They include the local-linearization method of SPREITER and ALKSNE [2, 3], where an approximation for φ_{xx} is permitted to vary slowly with x, the method of COLE and ROYCE [4, 5], where φ_x rather than φ_{xx} is approximated at the body, and, in a sense, the method of parametric differentiation, explored by RUBBERT and LANDAHL [6, 7], where similar approximations for φ_x as well as φ_{xx} are taken to vary freely with x while taken as constant across the flow.

In plane flow, φ_x and φ_{xx} at the body *are* constant across the flow to the appropriate order [6, 7]. In axisymmetric flow they are *not*; there is the well-known logarithmic singularity at the axis. This is an obvious objection, of course, to applying such local methods to bodies of revolution. In principle, it is an easily overcome objection, though, since the proper variation can be estimated from the slender-body approximation. The analysis would become much more complex, of course.

There is a more serious objection, however. As was observed already by OSWATITSCH and KEUNE [1], for smooth slender bodies the nonlinear term of the differential equation is negligible in the neighbourhood of the body and attains prime importance only at radial distances such that the radial velocity component has decreased sufficiently. In fact, the slender-body approximation, which is obtained precisely by neglecting the nonlinear term, is known from experiments to be valid out to considerable distance from the axis [8, 9]. This being the case, it does not seem to make sense to try to account for the nonlinear term by some local analysis close to the body. This makes it all the more remarkable that such methods do give good agreement with experimentally obtained pressure distributions on slender bodies of revolution [3]. Perhaps the agreement is responsible for the fact that no more satisfactory method of solution has been produced in spite of early criticism [1, 10].

The present paper is an attempt to survey the field for some rational method of computation based on the fact that the slender-body approximation is valid close to the body while farther away the nonlinear term must be retained. Since at large distance the Guderley expansion [11] for the far field is expected to be valid, it seems natural to try to establish an outer boundary condition at some finite distance by employing this expansion. This would seem to be a necessary step if a numerical method of integration is to be employed. Our basic goal thus is to determine the extent of the regions in which the slender-body approximation and the Guderley expansion are useful.

Methods of the type envisioned have been attempted earlier, for example by J. W. MILES [12], and no doubt by many others. YOSHIHARA [13], in treating the flow field of a cone-cylinder body by the relaxation method, used an outer boundary condition derived from the leading term of the Guderley expansion, and found that his solution was well represented by the slender-body approximation in the neighbourhood of the body. Recently U. MÜLLER [14] found it useful, in a case of plane flow, to use the corresponding outer solution to extend to larger distances the flow at the sonic line as determined by the parabolic method.

It should perhaps be stressed that what we have in mind is not a procedure for the determination of the body defined by the analytic continuation of a given Guderley expansion. From preliminary surveys of this approach by GUDERLEY and BREITER [15] and by RANDALL [26], it seems that great difficulties arise. It is indicated by the former authors that the analytic continuation might more easily be performed in the hodograph plane, as was in fact already tried approximately by GUDERLEY and YOSHIHARA [11] in the paper introducing the Guderley expansion.

2. Résumé of the first-order problem

With the sonic free-stream velocity as the unit of velocity and a body length, to be specified below, as the unit of length, the problem for the velocity potential $\Phi(x, r)$ of inviscid flow around a body of revolution of radius $\tau R(x)$ becomes

$$\left.\begin{aligned}
(a^2 - \Phi_x^2)\, \Phi_{xx} + (a^2 - \Phi_r^2)\, \Phi_{rr} + \frac{a^2}{r}\, \Phi_r - 2\Phi_x\Phi_r\Phi_{xr} = 0 \quad &\text{for} \quad r > \tau R(x), \\
\Phi_r = \tau R'(x)\, \Phi_x \quad &\text{for} \quad r = \tau R(x), \\
\Phi \to x \quad &\text{for} \quad r^2 + x^2 \to \infty.
\end{aligned}\right\} \tag{2a}$$

The thickness parameter τ is defined such that the maximum value of R is of the order of unity. For a perfect gas with constant specific heats, the square of the speed of sound is given by

$$a^2 = 1 - \frac{\gamma - 1}{2}\, (\Phi_x^2 + \Phi_r^2 - 1). \tag{2b}$$

The supersonic flow downstream of the limiting Mach wave is of no immediate concern; it can be treated subsequently by the method of characteristics.

In seeking a uniformly valid asymptotic expansion with respect to the parameter τ, in the limit of $\tau \to 0$, one would conclude from (2) that three different scales of r are involved, with r/τ, r, and τr, respectively, of order unity. It turns out, however, that to the first order the outer expansion, corresponding to $\tau r = 0(1)$, includes the inner as well as all intermediate expansions (while to the second order the inner expansion is significantly different) [16, 17, 18, 19]. Hence, introducing a perturbation potential $\varphi(x, \eta)$ by

$$\Phi(x, r) = x + \tau^2\varphi(x, \eta), \quad \eta = \sqrt{\gamma + 1}\,\tau r, \tag{3}$$

we may expect the solution of the resulting (outer) problem to be uniformly valid outside and at the body to the first order:

$$\left.\begin{array}{c} \eta^2\varphi_{\eta\eta} + \eta\varphi_\eta = \eta^2\varphi_x\varphi_{xx} \quad \text{for} \quad \eta > 0, \\[2mm] \displaystyle\lim_{\eta\to 0} \eta\varphi_\eta = s'(x), \\[2mm] \varphi_\eta, \varphi_x \to 0 \quad \text{for} \quad \eta^2 + x^2 \to \infty. \end{array}\right\} \tag{4}$$

The function $s(x)$, proportional to the cross-sectional area of the body, is given by

$$s(x) = R^2/2. \tag{5}$$

The nonlinear term $\varphi_x\varphi_{xx}$ is not of the first order in the inner expansion, nor in any intermediate expansion, so φ is expected to be of the form

$$\varphi = s'(x)\ln\eta + g(x) \tag{6}$$

in some neighbourhood of the body. This leads to the boundary condition in (4) for $\eta \to 0$, as well as to the following simple expression for the pressure coefficient at the body surface (in the inner limit):

$$C_p = -\tau^2[2s''\ln(\tau^2\sqrt{\gamma+1}R) + 2g' + R'^2]. \tag{7}$$

In essence, our problem is to find the function $g'(x)$. Since (6) is not valid for $\eta \to \infty$, it cannot be obtained by enforcing the outer boundary condition of (4). Note that difficulties show up at points where the body contour is not sufficiently smooth.

In view of the expression (6), the curve $\varphi_x = 0$ (not quite the sonic line, in the inner limit!) is expected to start from the axis at the point where $s'' = 0$. Therefore it seems natural to normalize the function s by locating the origin of x at the point where $s'' = 0$, and by choosing the unit of length and the thickness parameter[1] such that

$$s(0) = 1, \quad s'(0) = 1, \quad s''(0) = 0. \tag{8}$$

The function $s(x)$, assumed at least three times continuously differentiable up to and including a neighbourhood of the limiting characteristic, increases together with s' monotonically from zero upstream of the body to unity at the origin, where s'' is zero and s''' is negative. In practice some sort of discontinuity often will be present at the tip of the body, calling for special consideration.

From (6) and (7), the sonic point at the body surface is found to be located approximately at

$$x = -\frac{g'(0) + 1/4}{s'''(0)\ln[\tau^2\sqrt{2(\gamma+1)}]}, \tag{9}$$

while the point where $\varphi_x = 0$ is obtained by deleting the term $1/4$. It seems that $g'(0)$ is approximately -0.4, so both points will be located slightly downstream of the origin as indicated in Fig. 1. The limiting characteristic,

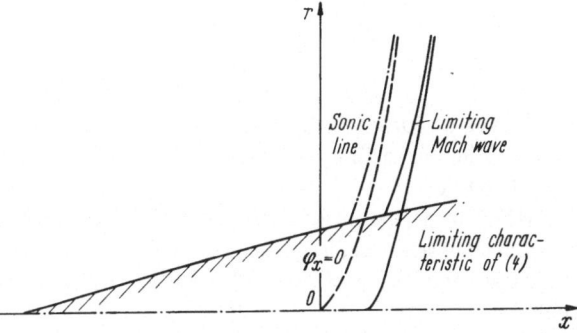

Fig. 1. The sonic line and the limiting Mach wave in the neighbourhood of the body. ($s'' = 0$ at $x = 0$).

again not quite the limiting Mach wave in the inner limit, is farther downstream, of course.

Thus we have to solve the nonlinear problem (4), which cannot be made linear by a Legendre transformation (as in the plane case). There is no apparent parameter present to permit us, by some expansion procedure, to circumvent the difficulty of solving a nonlinear partial differential equation. Nor is there any obvious way of determining the unknown function $g(x)$ of the inner approximation (6) (as would be sufficient for most purposes), except by solving the non-

---

[1] Hence a body of cross-sectional area $S(x)$ (with $S''(0) = 0$) has the unit of length $c = S(0)/S'(0)$, while the thickness parameter is $\tau = \sqrt{S(0)/(2\pi)}/c$.

linear problem with an outer boundary condition requiring that $\varphi_x \to 0$ for $\eta \to \infty$. Faced with the formidable task of integrating numerically over the infinite domain upstream of the limiting characteristic, it seems natural to try using the Guderley expansion for the flow far away from the body, as indicated in the Introduction.

3. Résumé of the Guderley expansion

The Guderley and Yoshihara [11] basic selfsimilar solution of the transonic Eq. (4), representing the flow far from a body, will for the present purpose be written in the form

$$\varphi(x, \eta) = C^3 \eta^{-2/7} f(\zeta); \qquad \zeta = C^{-1}(x - x_0) \eta^{-4/7}. \tag{10}$$

Here f is a known function of ζ, while C and x_0 are parameters related in an unknown way to the shape of the body, i.e., to the normalized function $s(x)$. Recently [20, 21, 22] the function $f(\zeta)$ was obtained in the following simple parametric form (here normalized as proposed in [15]):

$$\left.\begin{aligned} f &= -\frac{8}{9}\sigma^{1/7}(6 + 3\sigma - 2\sigma^2) \\ \zeta &= -\sigma^{-2/7}(1 - 2\sigma) \end{aligned}\right\} \sigma > 0. \tag{11}$$

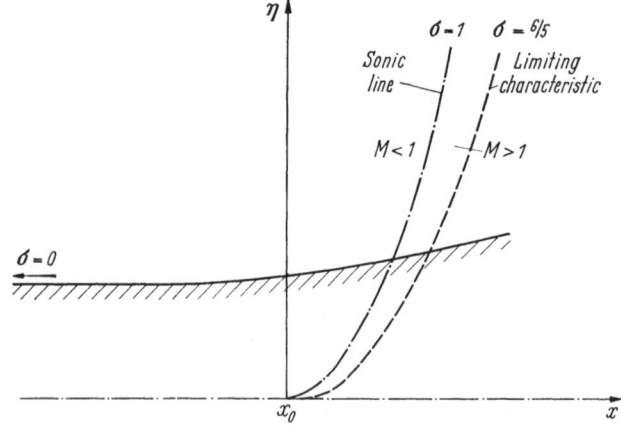

Fig. 2. The flow outside a quasi-cylindrical body as described by the basic axisymmetric solution of GUDERLEY and YOSHIHARA.

Curves of constant ζ, or constant σ, are parabolas of degree 7/4 with the vertex at $x = x_0$, $\eta = 0$. The sonic line is obtained for $\sigma = 1$ and the limiting characteristic for $\sigma = 6/5$ (see Fig. 2), while $\sigma = 0$ corresponds to upstream infinity and the x-axis upstream of x_0. The flow described by the solution (10) can always be interpreted as the flow outside a "quasi-cylinder" of a particular shape, as indicated in Fig. 2. At the vertex the solution is highly singular, and a neighbourhood thereof must always be excluded to comply with the assumption of small perturbations.

The Guderley expansion, then, consists of the basic solution (10) followed by an infinite sum of terms, each being a solution of the linear perturbation problem relative to the basic solution, that is

$$\varphi(x, \eta) = C^3 \left[\eta^{-2/7} f(\zeta) + \sum_{n=0}^{\infty} \alpha_n \eta^{-k/7} f_n(\zeta) \right]. \tag{12}$$

The functions $f_n(\zeta)$ are regular upstream of and across the limiting characteristic of the basic solution. They recently [22, 23] became available as simple expressions in terms of the parameter σ, with the exponents k given by

$$k = \sqrt{24n(n + 1) + 1} - 2n + 1 \qquad (n = 0, 1, 2, \ldots; \quad k = 2, 6, 9.04, 12, \ldots). \tag{13}$$

More specifically, each f_n is the product of a factor $\sigma^{k/14}$ and a polynomial of degree n in σ.

The first two terms of (12) describe perturbations equivalent to small variations of C and x_0, respectively, and so their coefficients, α_0 and α_1, can be put equal to zero (or used for correcting an erroneous initial choice of values for C and x_0). The succeeding coefficients α_n are related in an unknown way to the shape of the body.

EUVRARD [23] has shown how the expansion (12) can be extended to account for higher-order terms.

4. Fitting in the basic solution

Thus it may be assumed that upstream of the limiting characteristic the Guderley expansion, truncated after a small number of terms, is a valid approximation to the first-order (outer) solution for sufficiently *large* values of η, and similarly that the slender-body approximation (6) is a valid approximation for sufficiently *small* values of η. Presumably there is a gap between them. The crucial question now is: where is that gap located, and how wide is it? The precise answer will of course depend on how accurately we want to represent the solution. As a first step we shall consider only the most crude representation in trying to connect the basic solution directly to (6).

We propose to do this along the η-axis, which is located as close to the sonic line as we can take the slender-body approximation without having to consider the risk of crossing the limiting characteristic. This tends to minimize the errors of the slender-body approximation in estimating the streamline slope and the streamline-displacement area by neglecting the integrals to the right in the expressions

$$\eta \varphi_\eta = s'(x) + \int_0^\eta \varphi_x \varphi_{xx} \eta \, d\eta \,, \tag{14}$$

and

$$\int_{-\infty}^x \eta \varphi_\eta \, dx = s(x) + \int_0^\eta \frac{\varphi_x^2}{2} \eta \, d\eta \,, \tag{15}$$

obtained by integrating the differential Eq. (4). The η-axis in this approximation is the locus of maximum stream-line slope; the corresponding locus of the basic solution is the curve $\sigma = 4/5$. This suggests the picture of Fig. 3, with the implication that the point η_0 at which the curve $\sigma = 4/5$ crosses the η-axis is located in the gap between the outer and the inner region.

In order to determine η_0, it is convenient to introduce the quantity

$$H(\eta) = \left[\frac{\eta^{3/2} \varphi_\eta}{\left(\int_{-\infty}^x \eta \varphi_\eta \, dx \right)^{3/4}} \right]_{x=x_m(\eta)} , \tag{16}$$

where x_m is the location of the maximum stream-line slope (for a given value of η). In fact, for the basic solution, with x_m corresponding to $\sigma = 4/5$, H is a constant independant of the parameters C and x_0, namely

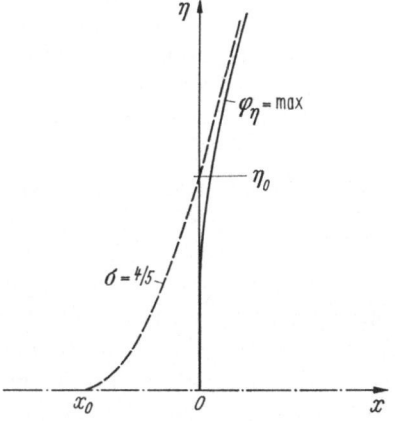

Fig. 3. Fitting in the basic solution.

$$H = 4/\sqrt{21} \,. \tag{17}$$

On the other hand, we obtain from the slender-body approximation $H = \sqrt{\eta}$ along the η-axis. Equating the two values at $\eta = \eta_0$ gives

$$\eta_0 = 16/21 (= 0.762) \,. \tag{18}$$

We can now obtain a rough estimate of C and x_0 by evaluating the integrals of (15) at $x = 0$, $\eta = \eta_0$. For this purpose we introduce the slender-body-type approximation

$$\varphi(x, y) = s'(x) \ln \frac{\eta}{\eta_0} - \left(1 - \frac{\eta}{\eta_0} \right) [\eta_0 \varphi_\eta(x, \eta_0) - s'(x)] + \varphi(x, \eta_0) \,, \tag{19}$$

18*

which correctly gives $\eta\varphi_\eta$ and the dominating term of φ_x for $\eta \to 0$, as well as both $\eta\varphi_\eta$ and φ_x for $\eta = \eta_0$. Using the basic solution (10) to determine $\varphi(x, \eta_0)$ and $\varphi_\eta(x, \eta_0)$, we obtain

$$\left.\begin{aligned}
C &= \frac{5}{4}\left(\frac{49}{48}\right)^{1/4}\left(\frac{324}{8575}\right)^{1/7} (= 0.787), \\
x_0 &= -\frac{3}{7}\left(\frac{49}{48}\right)^{1/4} (= -0.431).
\end{aligned}\right\} \tag{20}$$

It is worth noting that these values would have been only a half percent smaller (corresponding to a factor of unity instead of 49/48) if we had neglected the right-hand integral of (15) completely.

MILES [12] fitted in the basic solution in a somewhat different manner. Postulating that the singularity is situated at the centroid of $s(x)$ (for $x < 0$), he required the stream-line slope of the basic solution to coincide with that of the slender-body approximation at that point where the sonic line of the basic solution crosses the η-axis. For a parabolic-arc body this gives $C = 0.75$, $x_0 = -0.43$, with the sonic line crossing the η-axis at $\eta = 0.38$. The fact that the values obtained for the parameters are so close to our values, in spite of the implied assumption that the basic solution is a reasonable approximation as far in as $\eta = \eta_0/2$, seems to indicate that the determination of the parameters of the basic solution does not depend critically upon the method used.

5. An attempt at the Guderley expansion

Assuming, then, that the gap between the slender-body approximation and the Guderley expansion is located around η_0, the simplest way to find out how wide is the gap is perhaps to investigate the consequences of postulating zero width. Thus, our next step will be to try to connect directly the slender-body approximation to the Guderley expansion at $\eta = \eta_0$.

Integrating (15) with the slender-body approximation and equating the resulting streamline displacement at $\eta = \eta_0$ with the value obtained from the Guderley expansion leads to an equation of the second degree for the coefficients α_n. By keeping a finite number of terms in the expansion and fullfilling this equation at equally many "control points" along $\eta = \eta_0$, one should be able to obtain approximate values of the coefficients.

This has been tried for the parabolic-arc body

$$s = \left(1 + \frac{1}{2}x - \frac{1}{8}x^2\right)^2, \quad x > -2\,(\sqrt{3}-1)\,(\simeq -1.46), \tag{21}$$

the case for which the most complete set of experimental data is available [24, 8, 9]. This type of body is not as smooth as one would like, s'' having a jump discontinuity at the leading tip, so we must expect irregularities in the tip region.

A great number of trial solutions have been obtained, in particular by Y. SEDIN [27]. Strongly erratic values were obtained when control points approaching the tip section of the body were used. Attempts to improve the situation by iterating the slender-body approximation failed. Consistent results were obtained, with up to five terms, when the control points were located downstream of $x = -0.6$. In most cases all coefficients α_n were rather small except α_2, which was negative and considerably larger; a typical result was

$$\left.\begin{aligned}
\alpha_0 &= 0.044 \\
\alpha_1 &= 0.076 \\
\alpha_2 &= -0.237 \\
\alpha_3 &= -0.027.
\end{aligned}\right\} \tag{22}$$

Although the coefficients were small enough for the perturbation of the basic solution to be small at $\eta = \eta_0$, it was found that the nonlinear terms in the equations for α_n, which strictly should be negligible for the Guderley expansion to be valid, did influence considerably the values of the coefficients. It seems that this was because the systems of equations were ill-conditioned in consequence of the fact that errors in the tip region prohibited control points from being located there. Still, to some extent it might be the symptom of a gap beyond η_0, so that the Guderley expansion should be started at a somewhat larger value of η, $\eta = 1$ say, or perhaps even larger. This is entirely consistent with Yoshihara's [13] experience in treating the cone-cylinder body. Certainly one cannot go as close to the axis as $\eta_0/2$, the value suggested by MILES [12].

It is instructive at this stage to evaluate the pressure distributions measured by TAYLOR and McDEVITT [8, 9] outside parabolic-arc bodies. The corresponding distributions of $g' = \varphi_x - s'' \ln \eta$ for a number of sections with $x = $ constant have been plotted in Fig. 4. In spite of considerable scatter (partly obscured by the unintentional smoothing necessarily produced in interpolating experimental data) the data seem to be largely consistent with the

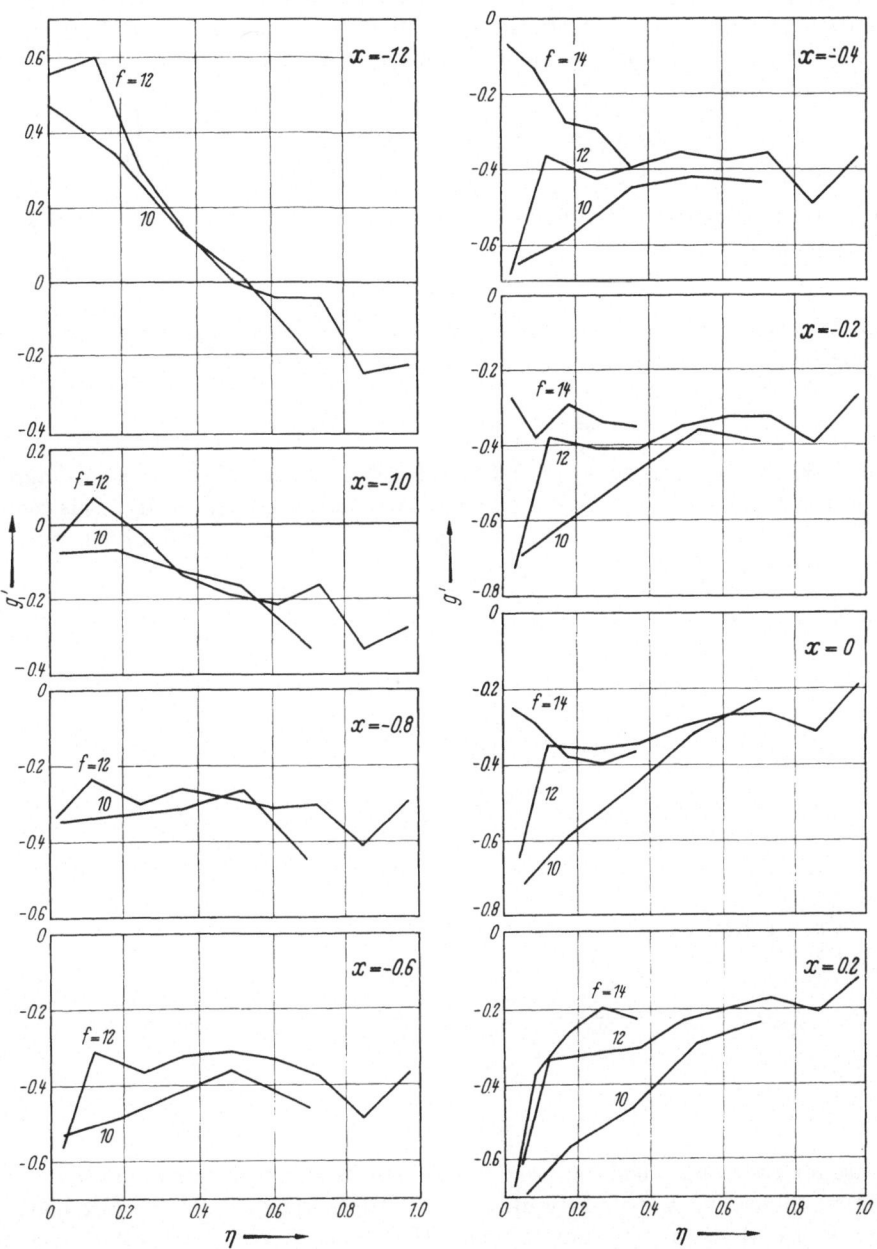

Fig. 4. The function $g' = \varphi_x - s'' \ln \eta$ for parabolic-arc bodies (f = fineness ratio) as determined from the experiments of [8] and [9].

contention of the slender-body approximation that g' is independant of η. However, when the tip region is approached, perhaps already for $x < -0.8$, the variation of g' becomes appreciable, except possibly when η is close to one (or η_0). This confirms our conclusion that the errors of the slender-body approximation are large in this region. This conclusion is also in agreement with the computations of [13].

6. The drag of the leading part of the body

Before turning to a concluding discussion of the problem of finding a more efficient integration procedure than the slender-body approximation, we shall calculate the drag of the body up to the section at which $s'' = 0$. This can be done by an outer integration [12, 25], and so is not directly influenced by local errors of approximation at the body surface.

The integration path extends from the body along the η-axis as far out as $\eta = 1$, where it turns and goes to upstream infinity along the line $\eta = 1$. The resulting drag formula, strictly valid to the first order, is

$$C_D = \tau^2 \left[\int_{\eta_b}^{1} \left(\varphi_\eta^2 + \frac{2}{3} \varphi_x^3 \right)_{x=0} \eta \, d\eta - 2 \int_{-\infty}^{0} (\varphi_\eta \varphi_x)_{\eta=1} \, dx \right], \tag{23}$$

with the cross-sectional area of the body at $x = 0$ as the reference area. The surface coordinate at $x = 0$ is given by

$$\eta_b = \tau^2 \sqrt{2(\gamma + 1)}. \tag{24}$$

If the influence of the perturbation terms in the Guderley expansion is small, the second integral may be evaluated by using the solution (22). The result is

$$\frac{C_D}{\tau^2} = \ln \frac{1}{\tau^2 \sqrt{2(\gamma + 1)}} + 0.95. \tag{25}$$

This is compared with experiments [8, 9, 24] in Fig. 5 and is seen to give as good agreement as the scatter would permit, not only for parabolic-arc bodies but generally. This tends to confirm that the basic solution is a reasonable approximation at $\eta = 1$ and adds to our confidence in the estimates of the parameters C and x_0.

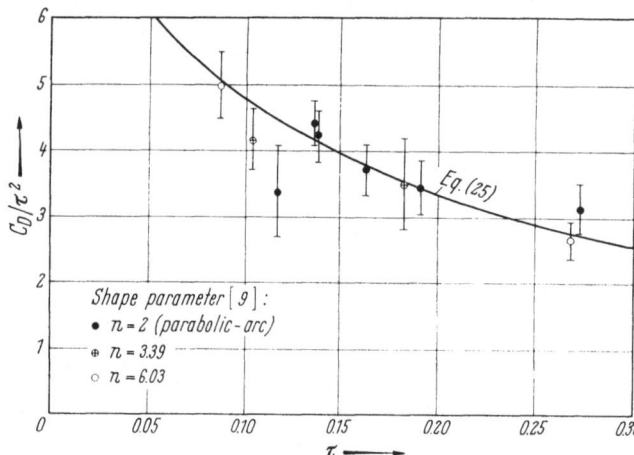

Fig. 5. The pressure drag upstream of the section where $s'' = 0$.

7. Concluding discussion

The case for the slender-body approximation can be summed up as follows:

(i)   In the region around and slightly upstream of the sonic line, the slender-body approximation is useful all the way out to where the Guderley expansion takes over, at $\eta = 1$ say.

(ii) If the body does not have a smoothly cusped tip, and perhaps even if it does, there is a region around the tip, extending perhaps as far out as $\eta = 1$, where the errors of (19) are considerable and preclude the approximation from being improved by iteration.

This must apply to the parabolic approximation as well, since that is essentially the slender-body approximation improved by including an estimate of the nonlinear term valid around the sonic line but not in the tip region.

This strongly suggests that one can improve the parabolic approximation by making the substitution for φ_{xx} a function of x (as proposed already in [1]) and using the Guderley expansion as a boundary condition. One then can account for the fact that φ_{xx} is negative in the larger part of the tip region, and that there will be a section, roughly halfway between the tip and the sonic line, where the coefficient of φ_x in the parabolic equation vanishes in the mean. The integration must proceed upstream and downstream from initial values in this section (they are easily obtained!). The author has worked through the analysis, with φ_{xx} derived from the slender-body approximation and the Guderley expansion, without finding any essential difficulties. However, there obviously is a question whether it would not be better to take the estimate of φ_{xx} for the tip region closer to the axis, and then perhaps to let it vary with η as well.

It was therefore thought advisable to postpone the computations based on this approach, and instead to start looking for a numerical method to provide accurate test solutions for the approximate ones. It seems that such a method perhaps can be established on the basis of the slender-body approximation and the Guderley expansion in the following way.

The first step would be to describe the field φ, for $\eta < 1$ say, by two functions, $F(x, \eta)$ and $G(x, \eta)$, generalizing the η-independent functions of the slender-body approximation:

$$\left.\begin{aligned}
F &\equiv \eta\varphi_\eta \\
G &\equiv \varphi - \eta\varphi_\eta \ln \eta\,.
\end{aligned}\right\} \tag{26}$$

From (4) then follows a system of equations for F and G:

$$\left.\begin{aligned}
F_\eta &= \eta \ln \eta \, (\ln \eta \, F_x + G_x) \, (F_{xx} + G_{xx}/\ln \eta) \\
G_\eta &= -\eta \ln \eta \, (\ln \eta \, F_x + G_x) \, (G_{xx} + \ln \eta \, F_{xx}) \\
\eta &= 0: F = s'(x) \\
\eta &= 1: \ G = \varphi(x, 1) = G[F(x, 1)],
\end{aligned}\right\} \tag{27}$$

with $G[F]$ denoting the functional relationship between F and G at $\eta = 1$ as defined by the Guderley expansion.

The differential equation for F if G is given, and for G if F is given, are both parabolic equations which presumably can be integrated stably with respect to η. Upstream of the sonic line the stable marching direction is *outwards* for F and *inwards* for G. So, briefly, this is what we propose to do. Using some reasonable estimates for G, the Guderley expansion, and the sonic line, F is integrated from the axis with the initial condition $F = s'(x)$. A boundary condition to the left (far upstream) is provided by the Guderley expansion, while a boundary condition to the right is obtained at the sonic line when integrating, by the method of characteristics, the flow equations in the region between the sonic line and the limiting characteristic. Reaching the outer boundary $\eta = 1$, we can determine a new Guderley expansion and hence initial values for G, so that we can march back to the axis with the equation for G (keeping F fixed). A boundary condition to the left (and an improved solution between the sonic line and the limiting characteristic) is obtained by reconstructing the sonic line by the method of characteristics along similar lines as in [13]. If the iteration of this procedure converges we have found a numerical method of solution.

Acknowledgement

The author wishes to express his thanks to Mr. Y. SEDIN of the SAAB Aircraft Co, Linköping, Sweden for valuable discussions and helpful numerical computations.

References

1. OSWATITSCH, K., and F. KEUNE: The flow around bodies of revolution at Mach number one. Proc. of Conf. on High-Speed Aeronautics, Polytechnich Inst. of Brooklyn, 1955, p. 113.

2. SPREITER, J. R., and ALBERTA Y. ALKSNE: Thin airfoil theory based on approximate solution of the transonic flow equation. NACA TN 3970 (1957).

3. — —: Slender-body theory based on approximate solution of the transonic flow equation, NASA Rep. 2 (1959).

4. COLE, J. D., and W. W. ROYCE: An approximate theory for the pressure distribution and wave drag of bodies of revolution at Mach number one. Proc. 6th Midwestern Conf. on Fluid Mechanics, 1959, p. 254.

5. EVANS, T.: An approximate solution for two-dimensional transonic flow past thin airfoils. Proc. Camb. Phil. Soc. 61, 573 (1965).

6. RUBBERT, P. E.: Analysis of transonic flow by means of parametric differentiation. Ph. D. dissertation, MIT (1965).

7. —, and M. T. LANDAHL: Solution of the transonic airfoil problem through parametric differentiation. AIAA J. 5, 470 (1967).

8. TAYLOR, R. A., and J. B. MCDEVITT: Pressure distributions at transonic speeds for parabolic-arc bodies of revolution having fineness ratios of 10, 12 and 14. NACA TN 4234 (1958).

9. MCDEVITT, J. B., and R. A. TAYLOR: Pressure distributions at transonic speeds for slender bodies having various axial locations of maximum diameter. NACA TN 4280 (1958).

10. MILES, J. W.: On linearized transonic flow theory for slender bodies. J. Aero. Sci. 23, 704 (1956).

11. GUDERLEY, K. G., and H. YOSHIHARA: An axial-symmetric transonic flow pattern. Qu. Appl. Math. 8, 333 (1951).

12. MILES, J. W.: On the sonic drag of a slender body, J. Aero. Sci. 23, 146 (1956).

13. YOSHIHARA, H.: On the flow over a cone-cylinder body at Mach number one. WADC Tech. Rep. 52—295 (1952).

14. MÜLLER, U.: Profile bei Schallanströmung im unendlich ausgedehnten Stromfeld. ZAMM 47, T 158 (1967).

15. GUDERLEY, K. G., and M. C. BREITER: The development of infinity of axisymmetric flow patterns with a free stream Mach number one. Aerospace Res. Lab., US Air Force, ARL 66-0066 (1966).

16. PERL, W., and M. M. KLEIN: Theoretical investigation of transonic similarity for bodies of revolution. NACA TN 2239 (1950).

17. COLE, J. D., and A. F. MESSITER: Expansion procedures and similarity laws for transonic flow. Z. angew. Math. Phys. 8, 1 (1957).

18. GUDERLEY, K. G.: Theorie Schallnaher Strömungen. Berlin/Göttingen/Heidelberg: Springer 1957. English translation: Theory of transonic flow. Pergamon Press 1962.

19. HAYES, W. D.: La seconde approximation pour les écoulements transsoniques non visqueux. J. de Méchanique 5, 163 (1966).

20. MÜLLER, E., und K. MATSCHAT: Ähnlichkeitslösungen der transsonischen Gleichungen bei der Anström-Machzahl 1. Proc. 11th Int. Congr. Appl. Mech. München 1964; Berlin/Heidelberg/New York: Springer 1966, p. 1061.

21. FALKOVICH, S. V., and I. A. CHERNOV: Flow of a sonic gas stream past a body of revolution. PMM 28, 342 (1964).

22. RANDALL, D. G.: Some results in the theory of almost axisymmetric flow at transonic speed. AIAA J. 3, 2339 (1965).

23. EUVRARD, D.: Écoulement transsonique a grande distance d'un corps de révolution. C.R. Acad. Sc. Paris 260 (Gr. 2), 5691 (1965).

24. DROUGGE, G.: An experimental investigation of the interference between bodies of revolution at transonic speeds with special reference to the sonic and supersonic area rules. Aero. Res. Inst. of Sweden (FFA), Rep. 83 (1959).

25. BERNDT, S. B.: On the drag of slender bodies at sonic speed. Aero. Res. Inst. of Sweden (FFA), Rep. 70 (1956).

26. RANDALL, D. G.: A marching procedure for the determination of inviscid two-dimensional sonic flow past a blunt symmetrical body. Aero. Res. Council (Great Britain) C.P. No. 992 (1968).

27. SEDIN, Y.: private communication (1968).

Similarity properties of the laminar or turbulent separation phenomena in a non-uniform supersonic flow

By

P. Carrière, M. Sirieix, and J.-L. Solignac

Office National d'Études et de Recherches Aérospatiales, Chatillon-sous-Bagneux (Seine)

C_f	Wall friction coefficient	v	$0y$ velocity component
D	Nozzle-throat diameter	$0x$	Reference axis (see section 3.1)
D_z	Nozzle-out diameter	X	Nozzle axis
f_1	Similitude function; Eqs.(3) and (15)	$0y$	Axis normal to X
f_2	Similitude function; Eqs.(6) and (15)	Y	Radial distance of a point in the flow
F	Pressure correlation function; Eqs.(7) and (16)		

Greek letters

g	$= F/\tilde{F}$	α	Mach angle
h	$= K/\tilde{K}$	γ	Specific-heat ratio
k	$= (l/\delta_0^*)/[F_R/(\theta_R - \theta_R)]$; Eq.(8)	δ^*	Displacement thickness of the boundary layer
K	$= k \cdot F_R$; Eq.(8)	η, ξ	Characteristic coordinates
l	Reference length; $l = x - x_0$	θ	Velocity inclination
M	Mach number	$\bar{\omega}(M)$	Function $p/p_i(M)$ (isentropic flow)
p	Pressure	ϱ	Specific mass
p_i	Stagnation pressure	τ	Shear stress

p' Non-dimensional pressure gradient $= \dfrac{\delta_0^*}{q_0} \cdot \dfrac{\partial p}{\partial x}$

Subscripts

P Büsemann pressure number $P(M) =$

0	at origin of interaction
R	at selected value of the F function
S	at separation point
W	at wall

$$\int -\frac{1}{M} \cdot \frac{\sqrt{M^2 - 1}}{1 + \dfrac{\gamma - 1}{2} M^2} \, dM$$

q Dynamic pressure $= \dfrac{1}{2} p\gamma M^2$

R	Nozzle longitudinal radius		
Re	Reynolds number		
s	Reduced abcissa $= \dfrac{x - x_0}{l}$		
T	Temperature		
u	$0x$ velocity component		

Superscripts

$\sim$	Case of uniform adiabatic flow
$-$	Case of non-separated flow

1. Introduction

A detailed and systematic experimental analysis of the separation phenomena in a plane and uniform supersonic flow, clearly emphasizing the influence of the laminar, turbulent or transitional nature of the boundary layer, has been presented first by CHAPMAN, KUEHN, and LARSON [1]. It has led these authors to point out a theoretical approach based on the existence of similarity properties in the development of the boundary layer around the separation point. This very simple method, applicable to laminar as well as turbulent flow, leads to a rather complete description of separation. It is particularly well adapted to practical applications, and has been used in that way by ERDOS and PALLONE [2]. More recently LEWIS, KUBOTA, and LEES [3] have considered an extension of Chapman's work to the case of uniform and non-adiabatic hypersonic flows.

Similar work has recently been undertaken at ONERA, within the framework of a critical study of theoretical methods for predicting the circumstances of separation. In particular, experimental research has been carried out to determine to what degree the similarity properties proposed by Chapman for uniform flow upstream of separation can be extended to separation phenomena in non-uniform supersonic flow, a case much more frequent in practice. The first results of this research are presented in this paper.

2. The free-separation theory: a brief survey

2.1 Theoretical basis

The analysis presented in Ref. [1] concerns the case of separation caused by the interaction between the boundary layer established in a plane, uniform supersonic flow, and a shock wave of high intensity generated by an obstacle or a pressure step. The uniform state upstream is defined by the Mach number M_0, the pressure p_0, and the evolution of the local characteristics of the boundary layer ($C_f(x)$, $\delta^*(x)$, ...). Without separation the effect of the variation $\frac{d\delta^*}{dx}$ on the flow uniformity is considered as negligible. (For notation, see list at beginning of paper.)

The interaction domain is characterized (Fig. 1), from its origin 0, by a rapid increase of the boundary-layer thickness leading to a progressive deviation of the external non-viscous flow,

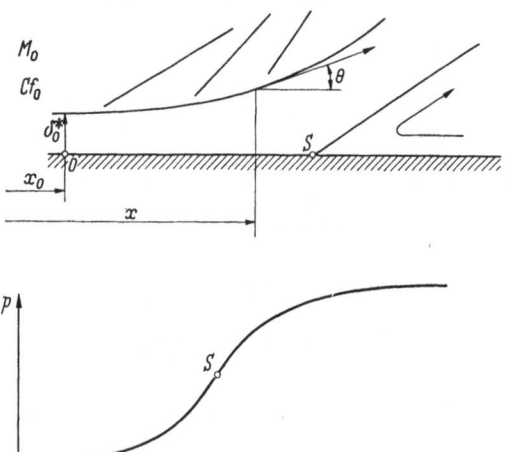

Fig. 1. Separation in a uniform flow. Notation.

and concurrently to a pressure rise at the wall. Chapman assumes that on the one hand the flow structure within this domain follows a law of similarity; on the other hand the deviation θ of the external non-viscous flow corresponds precisely to the displacement effect of the boundary layer, i.e.,

$$\frac{d\delta^*}{dx} = \theta - \theta_0. \qquad (1)$$

If we normalize the abscissas with an appropriate reference length l characterizing the extent of the domain, and the displacement thicknesses δ^* with a value δ_0^* at the origin, we obtain thus, with $s = \frac{x - x_0}{l}$,

$$\theta - \theta_0 = \frac{\delta_0^*}{l} \left[\frac{d\left(\frac{\delta^*}{\delta_0^*}\right)}{ds} \right]. \qquad (2)$$

Owing to the similarity assumption, the term between square brackets may be considered as an universal function $\tilde{f}_1(s)$, and Eq. (2) can be written

$$\theta - \theta_0 = \frac{\delta_0^*}{l} \tilde{f}_1(s). \qquad (3)$$

On the other hand, the boundary-layer momentum equation takes at the wall, where the velocity is nil ($u = v = 0$), the simplified form

$$\frac{dp}{dx} = \left(\frac{\partial \tau}{\partial y}\right)_w. \qquad (4)$$

If we integrate Eq. (4) from $x = x_0$, after making it non-dimensional by introducing the wall friction $\tau_{w_0} = \frac{1}{2}\varrho_0 u_0^2 C_{f_0}$ at $x = x_0$, we have

$$\frac{p - p_0}{q_0} = \frac{l}{\delta_0^*} C_{f_0} \int_0^s \frac{\partial(\tau_w/\tau_{w_0})}{\partial(y/\delta_0^*)} ds \qquad (5)$$

where $q_0 = \frac{1}{2}\varrho_0 u_0^2$. The integral on the right-hand side must, according to the assumption made, depend only on s, so that

$$\frac{p - p_0}{q_0} = \frac{l}{\delta_0^*} C_{f_0} \tilde{f}_2(s). \qquad (6)$$

By multiplying Eq. (3) by Eq. (6), $\frac{l}{\delta_0^*}$ is eliminated, and one obtains $\left(\frac{p-p_0}{q_0}\right) \cdot \left(\frac{\theta - \theta_0}{C_{f_0}}\right) = \tilde{f}_1 \cdot \tilde{f}_2$

or $\qquad \sqrt{\dfrac{\omega(M) - \bar{\omega}(M_0)}{\frac{1}{2}\gamma M_0^2 \bar{\omega}(M_0)} \cdot \dfrac{P(M) - P(M_0)}{C_f}} = \sqrt{\tilde{f}_1 \cdot \tilde{f}_2} = \tilde{F}(s). \qquad (7)$

Let us remark that, according to the Prandtl-Meyer law, $\theta - \theta_0 = P(M) - P(M_0)$ and that the isentropic recompression is defined by $\frac{p}{p_0} = \frac{\bar{\omega}(M)}{\bar{\omega}(M_0)}$, $P(M)$ and $\bar{\omega}(M)$ being known functions of the Mach number. $F(s)$ is thus by assumption an universal similarity function defining the Mach-number law (that is the pressure law within the interaction domain) and depending only on the laminar or turbulent nature of the boundary layer.

If the variation of $P(M) - P(M_0)$ as a function of $\frac{p-p_0}{q_0} = \frac{\omega(M) - \bar{\omega}(M_0)}{\frac{1}{2}\gamma M_0^2\bar{\omega}(M_0)}$ is expressed in linearized form, the simplified relation initially proposed by CHAPMAN is

$$\sqrt{\frac{M_0^2 - 1}{2c_{f_0}}} \cdot \frac{p - p_0}{q_0} = \tilde{F}(s). \qquad (7')$$

The length l which determines the physical scale of the domain may be obtained by dividing Eq. (6) by Eq. (3) which gives $\frac{l}{\delta_0^*} = \frac{\tilde{F}(s)}{P - P_0}\sqrt{\frac{\tilde{f}_1(s)}{\tilde{f}_2(s)}}$. This relation independent of s can be written

$$\frac{l}{\delta_0^*} = k\,\frac{\tilde{F}(1)}{P(1) - P_0}. \qquad (8)$$

2.2 Comparison with experiment

2.2.1 Pressure Evolution. In the case where the boundary layer is laminar and the wall adiabatic the existence of a quasi-universal function $\tilde{F}(s)$ for the pressure correlation is rather well verified experimentally, as shown in Fig. 2. The reduced abscissa s is defined here by taking

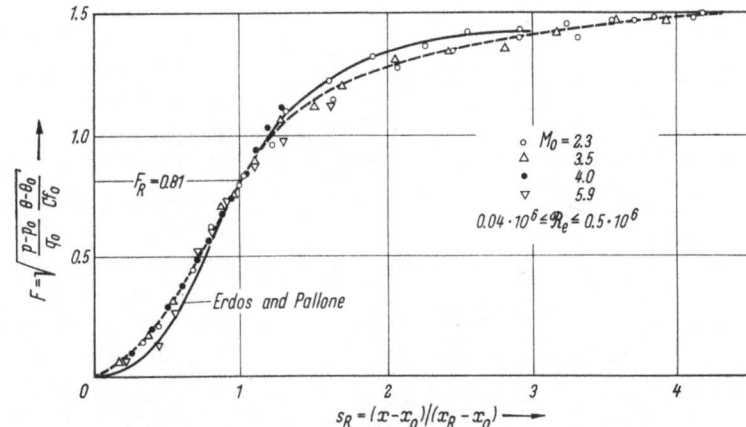

Fig. 2. Separation of a laminar boundary layer. Uniform flow.

as reference length the value $l = x_R - x_0$, x_R being the abscissa at which the function F takes the value $\tilde{F}_R = 0.81$. This reference value corresponds roughly to the separation point. The variation $\theta - \theta_0$ of the local direction of the non-viscous flow, which enters into the calculation of $\tilde{F}(s)$, has been deduced from the pressure with the assumption of a Prandtl-Meyer isentropic recompression.

19*

148   P. Carrière, M. Sirieix, and J.-L. Solignac

In these conditions one obtains actually an excellent grouping of experimental points from widely different experiments (see also Ref. [1]) in a broad range of Mach and Reynolds numbers. The correlation curve proposed by Erdos and Pallone [2] from a single experiment at

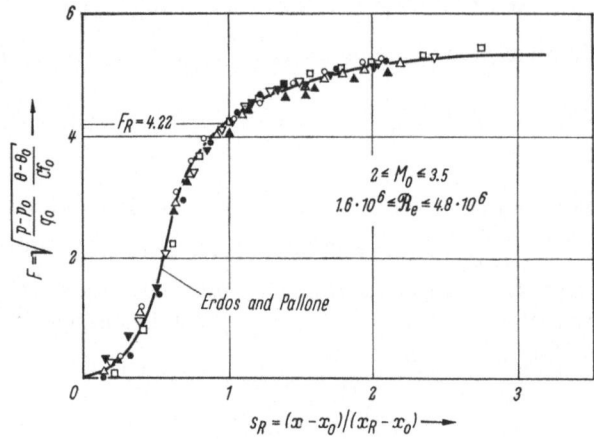

Fig. 3 Separation of a turbulent boundary layer. Uniform flow.

a moderate Mach number is quite near the average of the other experimental points. We shall, however, retain as a reference curve the dotted line of Fig. 2, which better represents this average.

When the boundary layer is turbulent (Fig. 3), the undisturbed flow being still uniform and the wall adiabatic, the correlation obtained is also satisfactory (the reference value of function $\widetilde{F}(s)$ being in this case taken as 4.22). The empirical function deduced from Erdos and Pallone well represents the results from very different experiments.

2.2.2 Interaction lengths. Thanks to the precise determination of the reference length $l_R = x_R - x_0$, it is possible to represent the evolution of pressure in the physical plane from the functions M_0, p_0, C_{f_0}, δ_0^*, so this constitutes an essential element for the practical exploitation of the method.

Theoretically the reduced reference length $\dfrac{l_R}{\delta_0^*}$ is a linear function of the parameter $\dfrac{\widetilde{F}_R}{\theta_R - \theta_0}$ [Eq. (8)] of slope $\widetilde{k}$. As seen in Fig. 4, where are assembled the various results of Figs. 2 and 3, the slope $\widetilde{k}_t = 0.52$ proposed by Erdos and Pallone is well verified in the turbulent case, but must be brought back in the laminar case to the value $\widetilde{k}_L = 0.68$, which is much closer to the experimental average. We shall then write $\widetilde{K}_t = \widetilde{k}_t \widetilde{F}_t(1)$, $\widetilde{K}_L = \widetilde{k}_L \widetilde{F}_L(1)$, so that Eq. (8) is

reduced, according to the case in question, to

$$\frac{l}{\delta_0^*} = \frac{\widetilde{K}_t}{P(1) - P_0} \qquad (8a)$$

$$\frac{l}{\delta_0^*} = \frac{\widetilde{K}_L}{P(1) - P_0}. \qquad (8b)$$

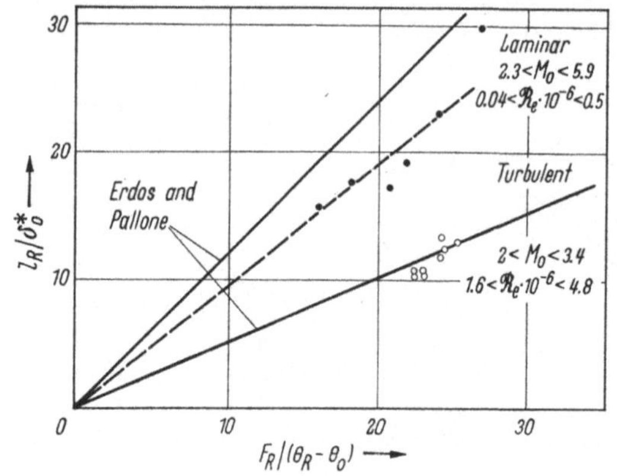

These results taken together confirm the existence of correlation functions which allow the prediction, with satisfactory precision when the flow is plane and uniform and the wall adiabatic, of the evolution of pressure in the so-called "free-interaction" separation domain. Let us now examine to what extent these results can be generalized when separation takes place in a non-uniform, plane or axisymmetrical flow.

Fig. 4. Interaction length.

3. Generalization of the Chapman theory
3.1 Description of the method [4]

Let us consider (Fig. 5) a separation zone originating at 0, appearing in a non-uniform supersonic flow with a X axis of symmetry. The non-uniformity of this flow may be characterized by the longitudinal component of the pressure gradient, for which we shall use the non-dimensional expression

$$p' = \frac{1}{q_0} \frac{\partial p}{\partial(x/\delta_0^*)} = \left(\frac{\partial p}{\partial s}\right) \cdot \frac{1}{q_0} \cdot \frac{\delta_0^*}{l} \tag{9}$$

which we shall assume to be almost constant within the domain $\frac{x-x_0}{l} = 0(1)$ where separation is liable to occur. The normal non-dimensional component $\frac{1}{q_0} \cdot \frac{\partial p}{\partial(y/\delta_0^*)} = -\frac{2\delta_0^*}{R}$ will be assumed negligible, δ_0^* remaining small as compared with R, the radius of curvature of the wall.

In the absence of separation, the evolution of the boundary layer leads to a "displacement" $\overline{\delta}^*(x)$ represented in the figure by $0x$. With separation, the frontier of the non-viscous flow is modified so that at a point Q, of abscissa x, the new direction $\theta(Q)$ of the velocity deviates from the direction that it would have had at point $\overline{Q}$ of the same abscissa without separation, the distance $\overline{Q}Q$ being equal to $\delta^* - \overline{\delta}^*$. The Mach-line η_0 originating at 0 gives the upstream limit of the perturbation zone due to separation. The Mach line $\overline{\xi}$ of the second family ending in $\overline{Q}$ intersects η_0 at $\overline{Q}_0$. In the

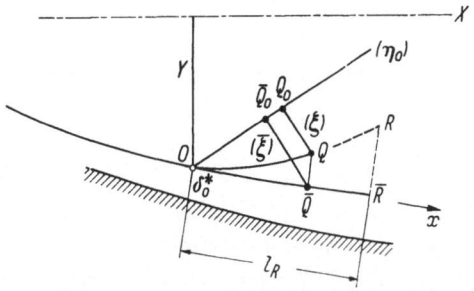

Fig. 5. Separation in a non-uniform flow. Notation.

same way, ξ ending in Q intersects η_0 at Q_0.

Let us call $P(M)$ the classical Prandtl-Meyer function, sometimes called the Busemann pressure number, defined by the relation $dP = 2\frac{\sqrt{M^2-1}}{q} dp$.

Let us apply the characteristic conditions to the two triangles $0\overline{Q}_0\overline{Q}$ and $0Q_0Q$ in their respective flows, assumed to be isentropic. One obtains in the undisturbed flow (triangle $0\overline{Q}_0\overline{Q}$)

$$P(\overline{Q}_0) - P(0) + \theta(\overline{Q}_0) - \theta(0) = -\int_0^{\overline{Q}} \frac{\sin\alpha\sin\theta}{Y} d\eta = \overline{A},$$

$$P(\overline{Q}) - P(\overline{Q}_0) - \theta(\overline{Q}) + \theta(\overline{Q}_0) = -\int_{Q_0}^{\overline{Q}_0} \frac{\sin\alpha\sin\theta}{Y} d\xi = \overline{B} \tag{10}$$

from which, after addition member by member,

$$P(\overline{Q}) - P(0) - \theta(\overline{Q}) - \theta(0) = \overline{A} + \overline{B} - 2\theta(\overline{Q}_0). \tag{11}$$

One finds likewise for the separation-perturbed flow (triangle $0Q_0Q$)

$$P(Q) - P(0) - \theta(Q) - \theta(0) = A + B - 2\theta(Q_0). \tag{11^1}$$

By taking the difference between these two equations one obtains a relation between the variations of P and θ due to separation for any abscissa x as follows:

$$P(Q) - P(\overline{Q}) = \theta(Q) - \theta(\overline{Q}) + [A - \overline{A} + B - \overline{B}]. \tag{12}$$

It is easily shown (see Ref. [5]) that the term in square brackets is of second order in $\frac{\delta^*}{Y}$. In other words, away from the vicinity of the axis the effect of the second member of Eq. (10) and similar terms is negligible to the second order of $\frac{\delta_0^*}{Y}$, which defines the limits of application of the method $\left(\frac{\delta_0^*}{Y} \ll 1\right)$.

150   P. Carrière, M. Sirieix, and J.-L. Solignac

Within the framework of such an approximation, the displacement effect due to separation, which is expressed by a relation similar to (1)

$$\theta - \bar{\theta} = \frac{d}{dx}\left(\delta* - \bar{\delta}*\right),$$

is then equivalent to

$$P - \bar{P} = \frac{d}{dx}\left(\delta* - \bar{\delta}*\right).$$

This can be written, by an operation similar to that of Section 2.1, as

$$P(s) - \bar{P}(s) = \frac{\delta_0^*}{l} \cdot \left[\frac{d}{ds}\left(\frac{\delta* - \bar{\delta}*}{\delta_0^*}\right)\right]. \tag{13}$$

In the same manner, we obtain by integration of the momentum equation at the wall

$$\frac{p(s) - p_0}{q_0} = \frac{\bar{\omega}(M) - \bar{\omega}_0}{\frac{1}{2}\gamma M_0^2 \bar{\omega}_0} = \frac{l}{\delta_0^*} C_{f_0}\left[\int_0^s \frac{\partial(\tau/\tau_{w_0})}{\partial(\partial/\delta_0^*)}\, ds\right]. \tag{14}$$

If we assume that there is still a similarity law for the separation in a non-uniform flow, the expressions in square brackets in Eqs. (13) and (14) will then be represented by functions of s and of the single perturbation parameter p' defined in (9), that is,

$$\left.\begin{aligned} P(s) - \bar{P}(s) &= \frac{\delta_0^*}{l} f_1(s, p')\\ \frac{\bar{\omega}(M) - \bar{\omega}(M_0)}{\frac{1}{2}\gamma M_0^2\bar{\omega}(M_0)} &= \frac{l}{\delta_0^*} C_{f_0}f_2(s, p') \end{aligned}\right\}. \tag{15}$$

Multiplying these two equations together one obtains

$$\frac{P - \bar{P}}{C_{f_0}}\frac{\bar{\omega} - \bar{\omega}_0}{\frac{1}{2}\gamma M_0^2\bar{\omega}_0} = f_1 \cdot f_2$$

which can also be written by analogy with (7)

$$F(s, p') = \sqrt{\frac{P - \bar{P}}{C_{f_0}}\frac{\bar{\omega} - \bar{\omega}_0}{\frac{1}{2}\gamma M_0^2\bar{\omega}_0}} = \frac{1}{\sqrt{C_{f_0}}}\Pi(M, M_0) \tag{16}$$

on the condition that only the positive values of $(\bar{\omega} - \bar{\omega}_0)$ are retained. Equation (16) generalizes Eq. (7) of the uniform case. In particular, we have

$$F(s, p') = \tilde{F}(s) \cdot g(s, p'), \tag{17}$$

where

$$\tilde{F}(s) = F(s, 0).$$

If we evaluate the reference length l from Eqs. (15), we obtain

$$\frac{l}{\delta_0^*} = \frac{F}{P - \bar{P}}\sqrt{\frac{f_1(s, p')}{f_2(s, p')}}.$$

Such an expression, which must be independent of s. can be written

$$\frac{l}{\delta_0^*} = \frac{K}{P(1) - \bar{P}(1)},$$

where

$$K = F(1)\sqrt{\frac{f_1(1, p')}{f_2(1, p')}}$$

and may be put in the form

$$\frac{l}{\delta_0^*} = \frac{\tilde{K}}{P(1) - \bar{P}(1)} h(p') \tag{18}$$

which reduces to (8) if $p' = 0$.

The determination of the functions $g(s, p')$ and $h(p')$ in Eqs. (17) and (18) from experiment should lead to the definition of the pressure distribution in the free interaction domain. In order to discuss the validity of the assumptions used in these purely formal considerations, an experimental study has been undertaken at ONERA concerning both laminar and turbulent separation.

3.2 Experimental study of laminar separation in a non-uniform supersonic flow

3.2.1 In order to outline the essential effects associated with thick boundary layers, and to avoid the initiation of transition phenomena during separation, the experiments have been carried out at high Mach-number in a wind tunnel designed especially for research at low Reynolds numbers. For practical reasons and to make the exploitation easier, the separation study has been made in a nozzle opening into a tank and submitted to a variable back pressure.

In Fig. 6 are shown the pressure distributions obtained in these conditions, the ratio $\frac{T_w}{T_f}$ being 0.64. The magnification of one of these curves around the origin 0 of the phenomenon under study shows the very gradual evolution of the interaction due to separation. This entails the existence, in the interaction domain, of a minimum value of the pressure smaller than p_0. This magnification also emphasizes the great importance of this upstream influence.

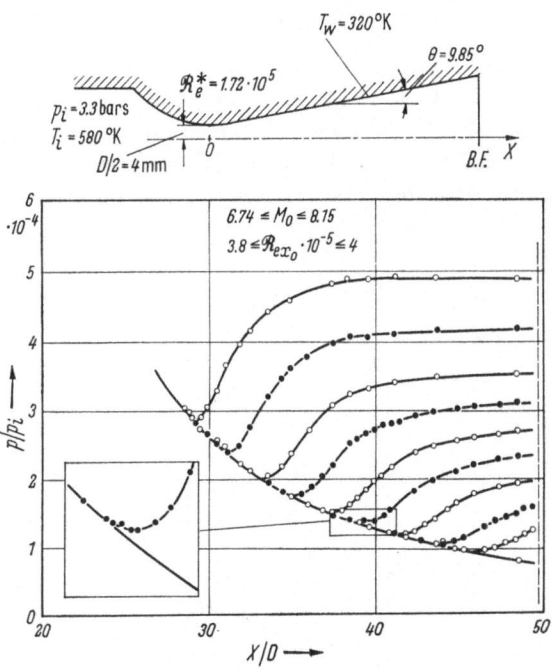

Fig. 6. Wall-pressure distribution along a Mach-number-8.5 conical nozzle. Laminar separation.

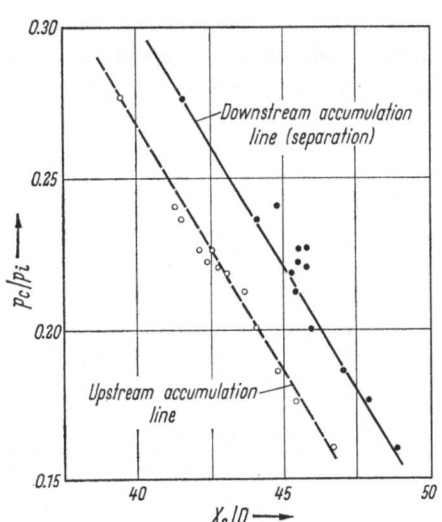

Fig. 7. Evolution of the laminar separation point (liquid film).

3.2.2 Evolution of the separation point. Jointly with the pressure measurements, the liquid-film technique has been used to localize the separation point. In accordance with CHAPMAN's (Ref. [1]) and other authors' remarks, two lines of accumulation have been obtained (Fig. 7). It has been verified by different means (wires, pressure sounding) that the trace furthest from the nozzle throat (downstream accumulation line) represents the locus of the separation points, of abscissa x_s, the evolution of which as a function of the tank pressure is given Fig. 7. This visualization technique, quite precise considering the extent of the interaction domain, has also provided verification that the separations obtained were reasonably axisymmetrical.

3.2.3 Pressures correlation. The abcissas x_s of the separation point being known, the reduced variable $s = \dfrac{x - x_0}{l_s}$ has been defined from the reference length $l_s = x_s - x_0$. The determination of the correlation function F involves the experimental values of $(P - \bar{P})$ and $(p - p_0)/q_0$, and also the friction coefficient C_{f_0} at the origin of the interaction. The latter has been deduced from the calculation of the laminar boundary layer at the nozzle wall [5].

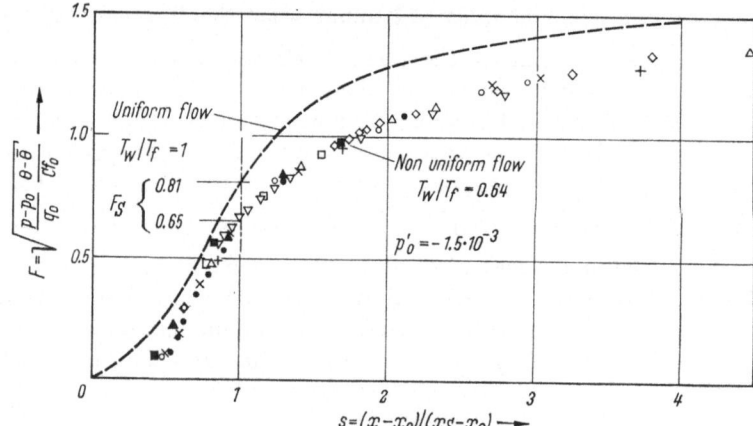

Fig. 8a. Pressure correlation.

The graph of $F(s)$ thus obtained (Fig. 8a) regroups in a quite remarkable fashion the various pressure distributions of Fig. 6, which correspond to Mach-numbers M_0 at the origin of interaction varying from 6.74 to 7.74 while the parameter p' moves very little around the average value -0.0015. The appreciable discrepancy should be noted between the correlation curve obtained and that corresponding to uniform adiabatic flow, the values of F_s at point $s = 1$ being respectively equal to 0.65 and 0.81.

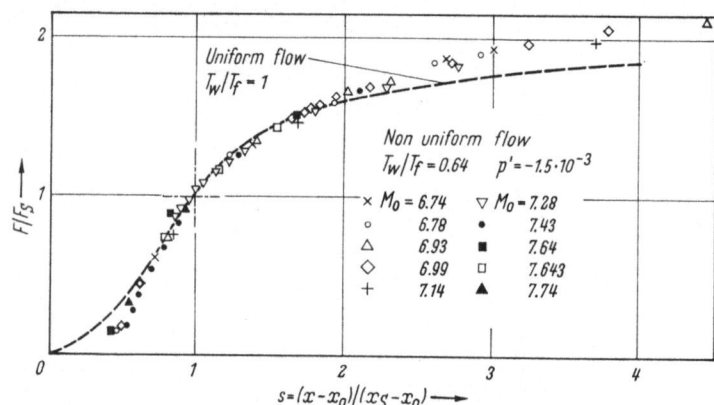

Fig. 8b. Laminar separation.

On the other hand, the whole of these results, presented in the normalized graph $\dfrac{F}{F_s} = F(s)$ (Fig. 8b), leads to a rather satisfactory regrouping of the pressure distributions for uniform and non-uniform flows, except in the immediate vicinity of the origin 0 of the interaction domain. This property indicates that $\dfrac{F(s, p')}{\tilde{F}(s)} = g(s, p')$ is independant from s within a large part of the interaction domain, and is reduced to a function $g(p')$ except in the initial region. The evolution of g as a function of p' will have to be made precise by additional experiments in which large variations of p' will have to be considered.

3.2.4 Interaction lengths. As Eq. (18) shows, the reference length l_s which defines the similitude variable s is in principle, for a given p', a function of the parameter $\dfrac{F_s}{P_s - \bar{P}_s}$, where subscript s indicates the separation point.

The experimental evolution of $\dfrac{l_s}{\delta_0^*}$ as a function of this parameter is represented Fig. 9, and compared with the results corresponding to a uniform and adiabatic upstream flow. The interaction lengths obtained in this latter case ($p' = 0, T_w/T_f = 1$) are larger. To understand the

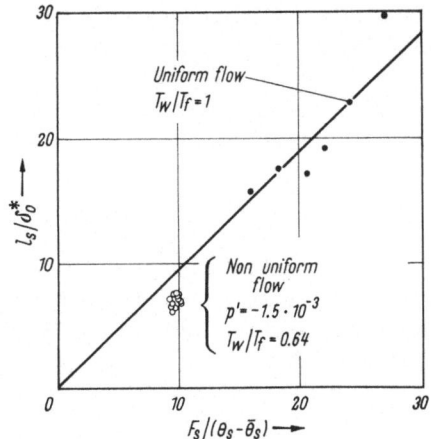

Fig. 9. Interaction length. Laminar separation.

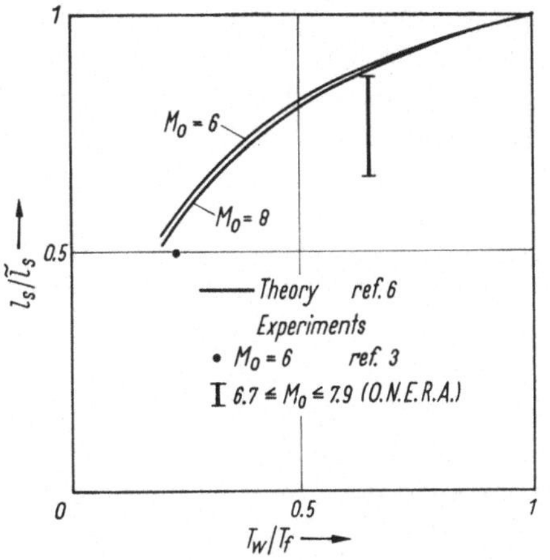

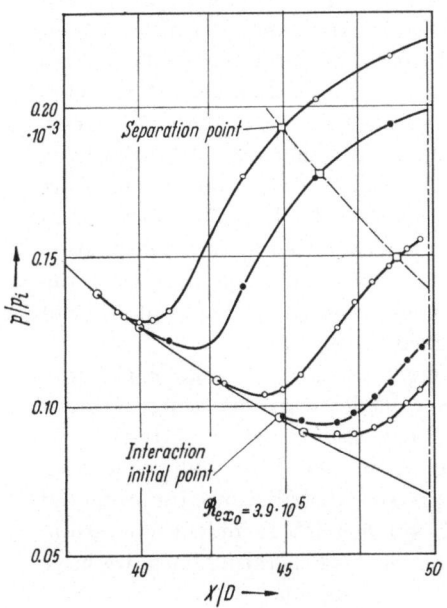

reasons of this discrepancy it is interesting to go back to the study made by LEWIS, KUBOTA, and LEES (Ref. [3]), which shows in particular that cooling of the walls plays an essential role in the extension of the interaction domain. The simple method recommended by these authors to take this influence into account, based on the work of CURLE [6], leads to a prediction of the cooling effect, the tendency of which is confirmed (Fig. 10) by our tests ($T_w/T_f = 0.64$).

3.2.5 Boundary-layer shock-wave interaction at the nozzle trailing edge. When the back-pressure decreases, the separation point moves along the nozzle towards the trailing edge and becomes fixed there for a certain value p_c of the tank pressure (Fig. 11). As the tank pressure continues to decrease, the shock wave that then sets near the trailing edge as long as the tank pressure is higher than the pressure at the nozzle outlet, initiates an interaction with the bound-

Fig. 10. Wall-temperature effect on the interaction length.

Fig. 11. Interaction at the trailing edge (laminar boundary layer). Wall-pressure distribution.

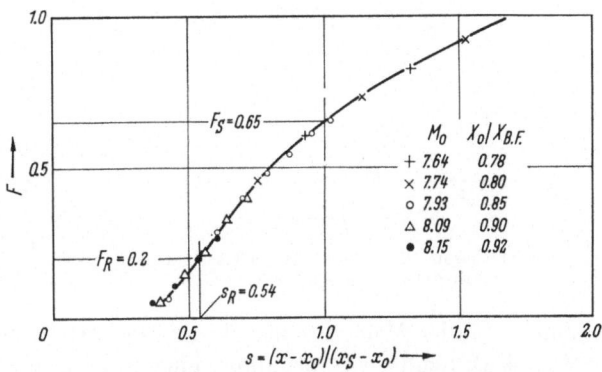

Fig. 12. Interaction at the trailing edge (laminar boundary layer). Wall-pressure correlation.

ary layer. The upstream extent of this interaction decreases as the intensity of this shock wave decreases. This is shown by the pressure evolution in Fig. 11.

It has seemed interesting (1) to examine if there exists, in the recompression zone, a critical point up to which the perturbation affecting the flow near the trailing edge modifies the upstream recompression curve, and (2) to study, when the separation point reaches the trailing edge and is fixed there, if the recompression curves follow a law of similarity different or not from that established in the case of free interaction.

The experimental results of Fig. 11, presented in Fig. 12 in the universal correlation graph of $F(s)$, give the following definite findings: On the one hand there appears no alteration in the evolution of the pressure curves, when the separation point approaches the trailing edge and is fixed there. On the other hand, the boundary-layer shock-wave interaction which develops in these conditions has the same similitude properties as does the free separation. The pressure then existing at the trailing edge, roughly equal to that of the tank, defines both the extent of the interaction and the pressure-evolution law in a form identical to that which would establish itself, with free separation, if the wall was extended beyond the trailing edge. Thanks to this remark it is legitimate to treat the boundary-layer shock-wave interaction problem, without separation, in a manner similar to that of the free separation (Lees-Reeves method).

3.3 Experimental study of turbulent separation in a non-uniform flow

3.3.1 Experimental set-up. The study of turbulent separation has been carried out, in the same way as for the laminar case, in various axi-symmetrical nozzles submitted to variable back-pressure. A first series of test has been undertaken of small-sized models ($D_E = 30$ mm), with the nominal Mach number in the outlet section adjusted at 3. Various configurations have been tested, including in particular a contoured nozzle and conical nozzles of 5°, 10°, and 17.5° half-angle. For each it has been verified that the boundary layer was turbulent from the throat, the pressure distributions $\frac{p}{p_i} = f\left(\frac{x}{D}\right)$ being practically independent of the variation, however large, of the Reynolds number, even if it is much higher than the critical value proposed by Herbert and Herdt [7]. An example of distributions obtained with the contoured nozzle is given Fig. 13. Note should be taken of the characteristic aspect of the recompression curves which, after a sharp rise followed by a plateau, show an important positive pressure gradient [8] (see also Ref. [7]). This phenomenon is typically tied to the recirculation conditions in the separated zone, and depends on the shape of the walls; it will not be examined in this paper.

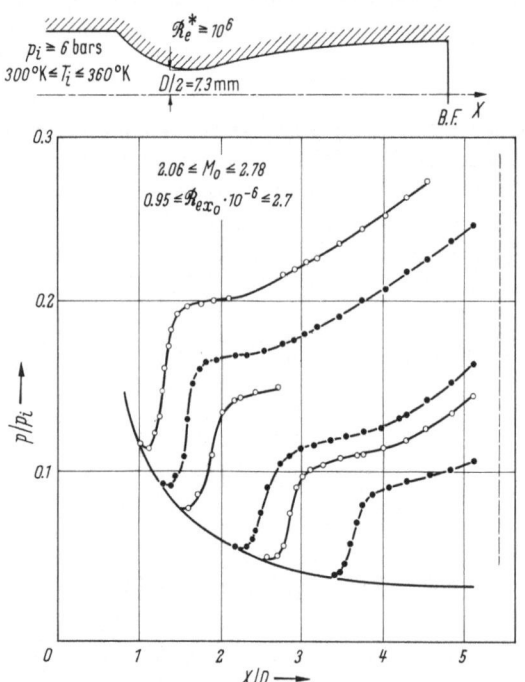

Fig. 13. Wall-pressure distribution along a Mach-number-3 contoured nozzle. Turbulent separation.

To complete and extend the Mach-number range considered in this study, we have carried out a second series of tests in a larger conical nozzle of 10° half-angle with a nominal outlet Mach number of 5.5. The corresponding pressure distributions are presented in Fig. 14. Difficulty was encountered in obtaining a perfectly symmetrical turbulent separation. By checking the transverse uniformity of the pressures along various generating lines, we have eliminated the cases showing too large discrepancies after data reduction.

3.3.2 Exploitation of pressure measurements. The number of measurements made in the case of a turbulent boundary layer was larger than in the laminar case, but less detail was obtained; in particular, the experimental determination of the separation point could not be assured. Under these conditions, the reduced abcissas s have been defined by taking as a reference the

length $l_R = x_R - x_0$, ending at the point, of abscissa x_R, at which the function reaches the value $F_R = 4.22$ characteristic of the separation point in uniform flow. The calculation of F uses the experimental pressure curves with and without separation, from which it is easy to obtain the values of $\theta_R - \bar{\theta}_R = P_R - \bar{P}_R$ and $(p - p_0)/q_0$, while the initial data C_{f_0} and δ_0^* were deduced from the calculation of the turbulent boundary layer [5].

The graph of $F(s)$ (Fig. 15) shows that the whole of the results obtained with various nozzles fall, with a moderate dispersion, on a single curve which is not much different from that proposed by ERDOS and PALLONE for the case of a uniform upstream flow. To attribute meaning to the observed discrepancies it will be necessary (1) to determine precisely the position of the separation point, and (2) to vary p' more widely so that we can define the perturbation function g associated with this parameter. During the present tests the value of p' stayed in the vicinity of the average value -0.001.

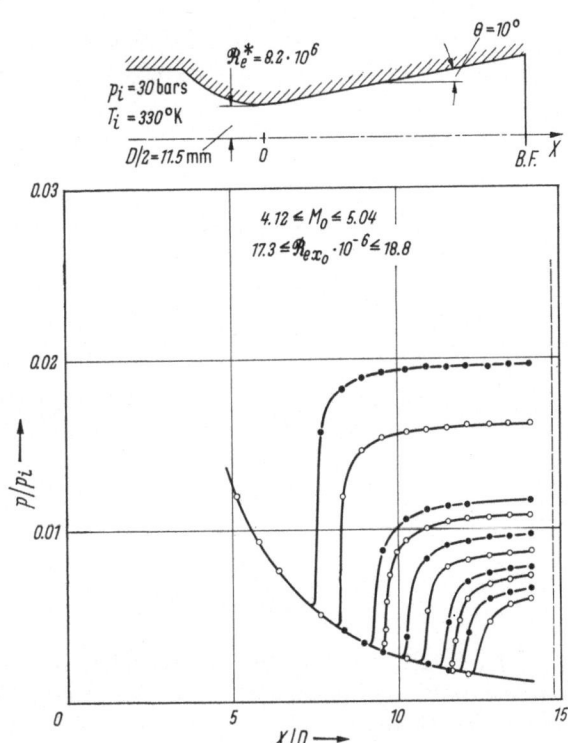

Fig.14. Wall-pressure distribution along a Mach-number-5.5 conical nozzle. Turbulent separation.

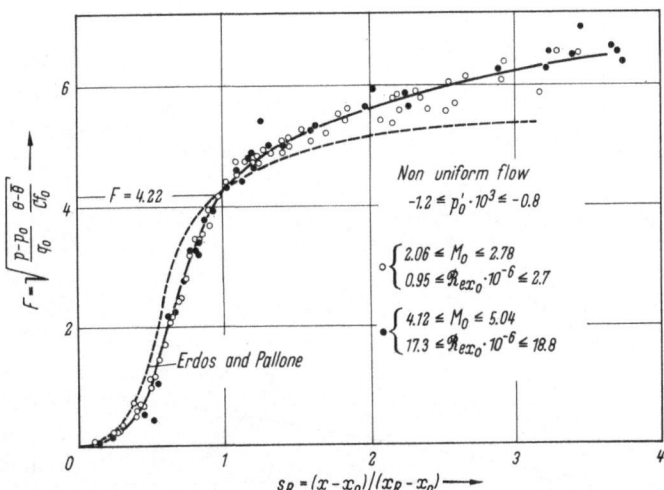

Fig.15. Pressure correlation. Turbulent separation.

3.3.3 Interaction length of turbulent separation in a non-uniform flow. While the pressure evolution within the interaction domain follows a similarity law foreseen in the analysis of Section 3.1, this is not the case for the interaction length.

The example given in Fig. 16 is significant in this respect. It concerns the separation conditions which establish themselves in a nozzle shaped for a nominal Mach of 3, whether it is or is not extended by a cylindrical duct of the same diameter D_E, and of length $0.8 D_E$. On the left-

156 P. Carrière, M. Sirieix, and J.-L. Solignac

hand graph are the recompression curves obtained from a quasi-identical origin x_0. Even though the initial conditions are very near to each other, it can be seen that the slopes of the two curves, and hence the interaction lengths relative to each, are quite different.

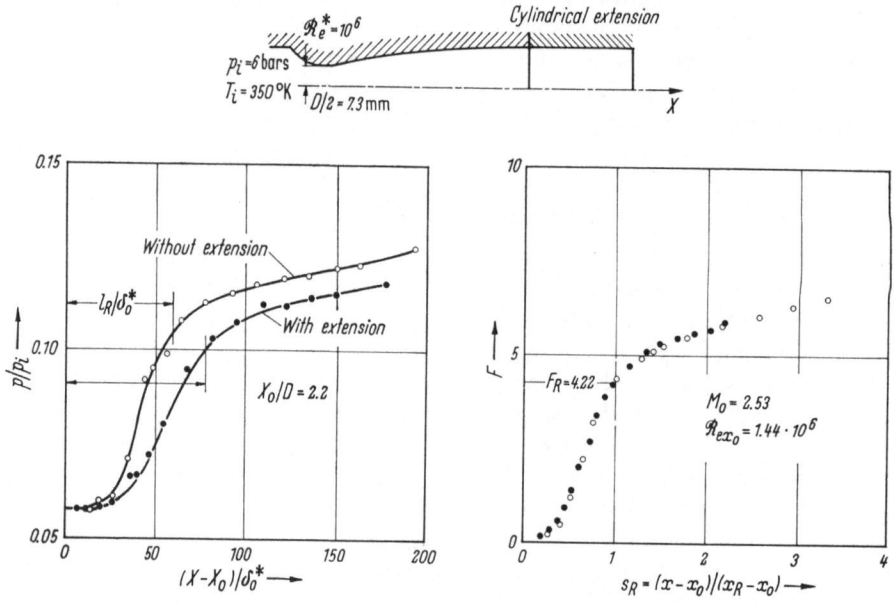

Fig. 16. Effect of downstream conditions on the turbulent separation.

This result, which seems to contradict the very notion of free interaction, has not been satisfactorily to date explained. Perhaps it is possible to connect these variations of the interaction length to three-dimensional effects. It should be noted, however (right-hand graph),

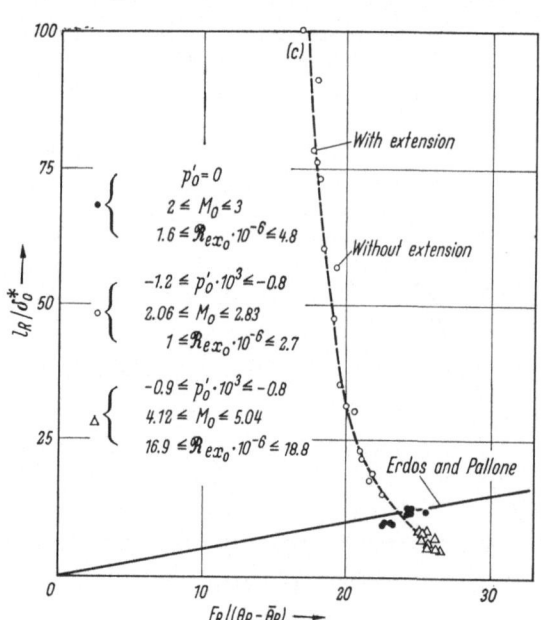

Fig. 17. Interaction length. Turbulent separation.

that the corresponding perturbation factors do not modify in the least the similarity law $F(s, p')$ which governs the pressures.

The incoherence of the evolution of the interaction lengths appears again in Fig. 17, where most of the results obtained are presented in the graph

$$\frac{l_R}{\delta_0^*} = f[F_R/(P_R - \bar{P}_R)].$$

The corresponding experimental points, including those of the preceeding example, define a very large range of variation of l_R/δ_0^*, which can not be attributed solely to the influence of the parameter p', the value of which stays almost constant at -0.001. The turbulent state of the boundary layer within the interaction domain also cannot be doubted. Failing a strict explanation, we can however make the following remarks concerning the functions $f_1(s, p')$ and $f_2(s, p')$, defined in Section 3.1 and considered by assumption as universal functions of s and p'. The product $f_1 f_2$, precisely equal to F^2, is not very sensitive to the effects of the perturbations that influence the interaction lengths. This can be seen in Figs. 15 and 16. On the other hand, the universality of the ratio f_1/f_2 found in Eq. (18), which defines l_R/δ_0^*, is made doubtful by the possible peturbations (three-dimensional effects, unstationary phenomena, etc.).

Lastly it seems worth-while to mention the following with regard to the unexpected regrouping of all results on the same curve (Fig. 17): On the one hand, the sharp decrease of l_R/δ_0^*, from very large values ($l_R/\delta_0^* = 100$) has been noted during tests of small nozzles ($D_E = 30$ mm and $10^6 \leq R_{ex_0} \leq 2.7 \cdot 10^6$). On the other hand, in this evolution the values of l_R/δ_0^* increase with increasing distance of the separation point from the nozzle-outlet section.

4. Conclusion

A detailed experimental analysis of the separation conditions in axi-symmetrical nozzles submitted to variable back-pressure has revealed the following results: (1) In the case of a laminar boundary layer, the similarity properties shown by the interaction domain due to separation when the upstream flow is uniform remain when it is non-uniform. The near-similarity parameters to be taken into account in this case are deduced from a generalization of the CHAPMAN-KUEHN-LARSON theoretical approach.

(2) The effect of wall cooling appears, in a first analysis, as a modification of the interaction lengths, rather well foreseen by the method proposed by LEWIS, KUBOTA, and LEES, and deduced from the work of CURLE.

(3) When the separation point becomes fixed at the nozzle trailing edge, the phenomena of boundary-layer shock-wave interaction are governed by the same similarity laws, whatever the intensity of the shock wave.

(4) In the case of a turbulent boundary layer the pressure evolution still follows a similarity law of the Chapman type. This does not seem to be true for the interaction lengths. It has not been possible to explain this experimental fact.

References

1. CHAPMAN, D. R., D. M. KUEHN and H. K. LARSON: Investigation of separated flows in supersonic and subsonic streams with emphazis on the effect of transition. NACA-TR 1356 (1968).
2. ERDOS, J., and A. PALLONE: Shock boundary layer interaction and flow separation. R.A.D. — TR 61-23, AVCO Corp., Aug. 15 (1961).
3. LEWIS, J. E., T. KUBOTA and L. LEES: Experimental investigation of supersonic laminar, two-dimensional boundary layer separation in a compression corner with and without cooling. AIAA J. 6, 7—14 (1968).
4. CARRIERE, P.: Recherches sur les décollements dans les tuyères propulsives. Rev. roumaine des

Sciences Techniques — Mécanique appliquée. Tome 13, No. 3 (1968) p. 339—415.
5. MICHEL, R.: Cours del' Ecole Nationale Supérieure de l'Aéronautique. Paris 1967.
6. CURLE, N.: The effect of heat transfer on laminar boundary layer separation in supersonic flow. Aero. Quart. 12, 309—336 (1961).
7. HERBERT, M. V., and R. J. HERDT: Boundary layer separation in supersonic propelling nozzles, in the presence of external flow. NGTE, Report 260, Aug. 1964.
8. CARRIERE, P.: Exhaust nozzles. In AGARD Lecture Series on supersonic Turbomachinery. Varenne (Italy), May 1967 (to be published).

Torque and flow patterns in supercritical circular Couette flow

By

W. Debler, E. Füner*, and B. Schaaf

University of Michigan

1. Introduction

The experiments described in this paper were undertaken to provide new and useful torque data in the supercritical regime of Taylor instability. The current interest in the non-linear mathematical problem associated with the fluid motion after secondary flow commences has demonstrated the desirability for additional experimental data for comparison with analytical results. The existing torque data is sparse (cf. Donnelly and Simon (1960)). Furthermore, previous information does not give any direct correlation with the multiple mode possibilities discussed by Coles (1965). Although Schultz-Grunow and Hein (1956) have related their own visual observations of azimuthal waves in the secondary flow to the torque observations of Wendt (1933), it was thought that simultaneous torque measurements and visual observations would establish more completely the torque-speed relationship for the particular modes which occurred. In addition, some of the transitions noted by Coles (1965) might be detectable in the torque data if the apparatus were sensitive enough.

There was a great deal of optimism that the desired sensitivity could be achieved with some modifications to the design used by Donnelly (1958). These changes were intended to facilitate the acquisition of data and thereby permit additional system parameters to be examined. Instead of allowing the dynamometer section of the outer cylinder to be suspended only by a torsion wire, an air bearing was installed to support this piece. Consequently there were no serious problems in maintaining alignment of the elements of the apparatus after the initial installation was carefully carried out.

The experimental results fulfilled some of the anticipations that the authors had at the outset of the work. The change in the torque-speed relationship after the onset of Taylor instability was found for a wide range of annulus gaps. Photographs were obtained which show the flow patterns that existed during the experiments. Azimuthal waves (i.e., secondary Taylor instability) were seen to occur only over a portion of the range of cylinder geometries examined. The onset of these waves was observed in the torque data. The Taylor numbers for primary and secondary instability were obtained. The features of the torque-speed relationship beyond the point of secondary instability were examined.

2. Apparatus

Figures 1 and 2 show an overall photograph of the apparatus and a sectional drawing that display some of the design features. In the tests which are described in this paper the inner cylinders were made of aluminium. These cylinders are described later. The outer cylinder is, in effect, stationary and the inner one is driven by a motor from below. It is immediately clear from the figure that the apparatus is not excessively long. The effect of the test-section length on the experimental results will be discussed subsequently.

---

* Present address: Siemens AG, Berlin, Germany.

The air bearing provided for axial alignment of the central portion of the outer cylinder and maintained an end clearance of .015 cm between this part and the fixed upper and lower sections of the three-part outer cylinder. Lucite was used in making this assembly to allow the flow to be viewed. Glass would have provided superior dimensional stability but its use would have

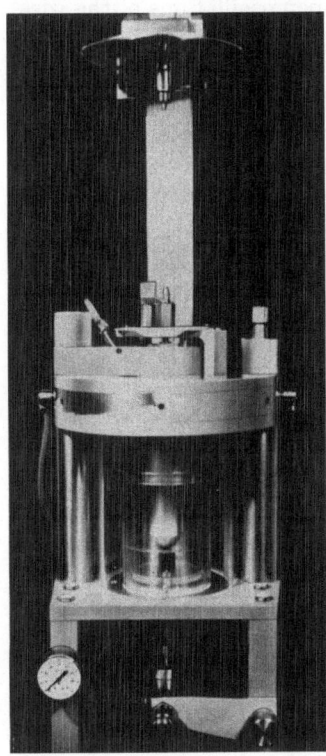

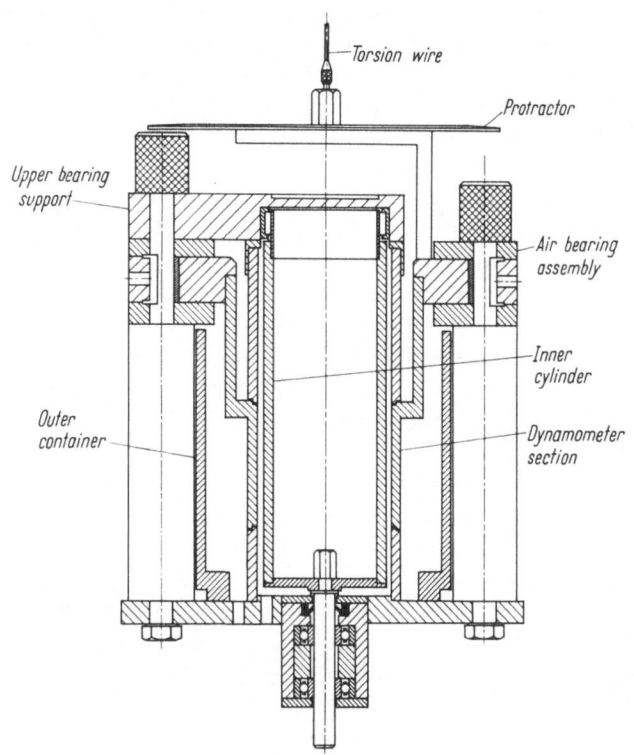

Fig. 1.
Photograph of experimental apparatus.

Fig. 2.
Sectional drawing of apparatus showing design details.

caused greatly increased difficulties in fabrication. The inside diameter of the outer cylinder was machined to $7.620 \pm .003$ cm, after which it was polished until it was transparent. Subsequent alignment checks showed the inner surface of this cylinder to be at $3.840 \pm .009$ cm with respect to the axis of rotation. Fabrication and assembly stresses may account for this situation.

The torque-sensitive portion of the apparatus was constrained by a steel torsion wire. The physical properties of the wires that were used are given in Table 1. The torsion constants of the wires were measured at the outset and conclusion of the experiments by measuring the frequency of torsional oscillations. Three different cylindrical masses were used for this calibration. The upper support for the torsion wire, a drill chuck, could be moved vertically to slightly pretension the wire, a procedure that was necessary because the air bearing, not the wire,

Table 1. *Torsional properties of steel wires used in experiments*

Diameter* inches	(cm)	Young's Modulus** (dynes/cm²)/(cm/cm)	Length cm	Spring constant dyne-cm/radian
0.0203 ± 0.0002	(0.0516)	$7.85 \ (10^{11})$	101.5	5382.85
0.0252 ± 0.0001	(0.0640)	$7.40 \ (10^{11})$	101.5	12008.42
0.0298 ± 0.0001	(0.0757)	$8.04 \ (10^{11})$	101.5	25537.32
0.0401 ± 0.0001	(0.1019)	$8.16 \ (10^{11})$	101.5	85098.12

* Average of ten measurements taken along the length of the wire.
** Average of ten measurements with each of three different cylindrical masses in torsional oscillation.

supported the weight of the test section. The angular deflection of the wire was read with a circular protractor that connected the wire to the air bearing by means of an ell-shaped arm. A vernier attachment permitted the rotation of the wire to be read to one-tenth of a degree. The determination of the deflection of the torsion wire was not a null method, in contrast to that of DONNELLY (1958). A permanent magnet was added to dampen oscillations of the wire, protractor, journal, and test-section assembly which occurred when the inner cylinder rotated at high speeds, i.e., in excess of 200 revolutions per minute (rpm). The aluminium disk on which the protractor was mounted provided the necessary conductor to effect the magnetic damping.

The inner cylinders were machined from aluminium stock. The final surface was smooth but no quantitative measurements of surface finish were made. The cylinders were sprayed with a dull black paint to improve the color-tone contrast in the photographs that were taken. The diameters of the eleven cylinders used in the experiments are 3.810, 4.953, 5.144, 5.334, 5.601, 5.715, 6.477, 6.655, 7.041, 7.125, and 7.239 cm. The diameters that are listed are the average of measurements taken at ten different locations along the length of the cylinders and have a tolerance of 0.003 cm. These cylinders allowed experiments to be conducted at values of η, the ratio of the inner-cylinder diameter to the outer-cylinder diameter, of 0.50, 0.65, 0.675, 0.70, 0.735, 0.75, 0.85, 0.875, 0.925, 0.935, and 0.95. This range of cylinders allowed for possible comparison with the work of COLES (1965), DONNELLY (1958), WENDT (1933) and TAYLOR (1936).

A variable-speed Graham transmission and motor assembly provided the power to the inner cylinder's drive shaft. A rubber coupling was used to isolate power-source vibrations from the fluid in the annulus. The speed of rotation was sensed by a magnetic transducer that was activated by the moving teeth of a 120-tooth gear mounted on the drive shaft. An electronic counter gave a convenient read-out of the speed. The power source gave good speed regulation, the speed did not vary more than 0.1 per cent over long periods of time. When low rotational speeds were used, i.e., below 40 rpm, the speed was measured by timing a number of rotations with a stop-watch.

Because the outer cylinder was segmented, a continuous outer wall was necessary to contain the fluid in the apparatus. Figure 1 shows a circular plastic cylinder that was used for this purpose. Subsequently a plastic box was installed. The flat front of the box, along with the fluid between the box and the segmented outer cylinder, gave an undistorted, glare-free view of the motion in the test annulus.

Silicone oil, Dow-Corning 210, was used in the experiments. Oils with a nominal viscosity of 5 and 50 centistokes were originally planned for the tests. After some initial trials an oil was blended from those at hand which had a nominal viscosity of 30 centistokes. The physical properties of these 5 and 30 centistoke oils are given in Table 2. The kinematic viscosities were determined with a certified Cannon-Fenske viscometer over the range from 21 to 40°C.

There was no temperature control of the oil in the apparatus during the test. A copper-constantan thermocouple was mounted in the upper third of the outside cylinder and projected slightly into the annulus. The voltage generated by this thermocouple in series with one in an ice bath was read on a digital microvoltmeter. With these temperature measurements and the

Table 2. *Physical properties of silicone oils used*

Temperature	Nominal 5 Centistokes		Nominal 30 Centistokes	
°C	ν^* Centistokes	ϱ^{**} g/cm³	ν^* Centistokes	ϱ^{**} g/cm³
22.5	5.75	0.922	35.50	0.935
25.0	5.50	0.920	33.35	0.934
30.0	5.04	—	30.85	—

* Values taken from calibration curves determined between 22.2 and 40°C with a Cannon-Fenske viscometer.

** Values taken from calibration curves determined between 21 and 27.2°C with a hydrometer.

viscosity-temperature calibration it was possible to specify the viscosity in the Taylor- and Reynolds-number calculations which were necessary.

3. Experimental procedure

Prior to beginning the experiments the apparatus was aligned with respect to the vertical and the centerline of the apparatus. Although the three elements of the outer cylinder were made of plastic and independently mounted, all were within 0.010 cm of concentricity with the axis of rotation of the inner cylinder.

The oil which was to be used for the particular experiment was divided into two batches. The first was mixed with a small amount of aluminium pigment [cf. COLES (1965)] and added to the annulus between the inner and outer cylinder. Simultaneously, clear oil was added to the space between the outer cylinder and the plastic box which formed the outside container. By carefully keeping the fluid levels in the two spaces equal, none of the pigment-laden oil seeped into the outer cavity via the narrow spaces between the segments of the outer cylinder. This procedure was necessary to obtain good photographs. The annulus was filled to a depth of 15.24 cm, the height permitted by the plastic box. Thus the ratio of the length of the annulus to the gap, the aspect ratio, encountered in the experiments ranged from 8 to 80. These values are intermediate to those of DONNELLY (1958) and WENDT (1933) for comparable η and less than those of TAYLOR (1936).

The motor was started with the transmission set to deliver zero output. The speed of rotation was increased in small steps. After each increment in speed had been completed, several minutes were allowed to elapse until it was certain that the new speed was constant and the torque had reached its new value. Whenever the critical Taylor number for axisymmetrical disturbances was known, either from theory or previous test, the increments in speed were reduced in the neighbourhood of this critical value. This was done to obtain the most information near the point of instability as well as to approach the instability point nearly quasi-statically. The experiment was continued to the speed of rotation that was deemed to be the upper limit for the test in question. The speed of rotation was then slowly reduced in a stepwise fashion, and the appropriate data recorded.

Photographs were usually taken whenever transitions were observed. The test section was illuminated by electronic flash units that shone through a diffusor to reduce the glare from bright surfaces. When tests were conducted at speeds above those at which waves first occurred, it was the practice to take photographs at regular intervals or whenever there was an occurrence which warranted a picture. At these higher speeds it was not possible to determine by visual ovservation whether or not a new wave was present.

A factor which determined when a photograph was to be taken was a slight but sudden drop in the torque when the unit was running at constant speed. The possibility existed that these changes in torque were associated with the development of a new wave mode. Photographs of this transition would be valuable. At the higher speeds associated with azimuthal waves there was also a tendency for the torque to be unsteady and cause the torsion wire to oscillate slightly. This occurred even though magnetic damping was used. If these oscillations changed magnitude, a photograph also was taken.

Lastly, photographs were made whenever the output of a thermistor became different or erratic. This thermistor was installed in the lowest segment of the outer cylinder and had a glass-bead encapsulation that was 0.051 cm in diameter. The rise time of the thermistor was given as 0.11 seconds by the manufacturer. The plug in which the thermistor was imbedded can be seen in the photographs which will be described later. The thermistor projected slightly into the flow and was one element in a Wheatstone bridge. An oscilloscope was included in the circuit to give a visual display of the response to the flow. Only qualitative results were obtained because the thermistor was not calibrated as an anemometer. When the oscilloscope was operated in the alternating-current mode, there was no change on the screen when the first instability occurred. When azimuthal waves commenced, the oscilloscope trace began a small periodic oscillation. As the speed was reduced, it was possible to ascertain the speed Ω_c at which the transition occurs by the absence of an alternating output from the thermistor.

Occasionally the thermistor signal would begin slowly to change its pattern. For example, a signal that was similar to a sine wave would develop a slight indentation near the point of maximum amplitude. This perturbation would slowly grow and move to the zero-amplitude position of the original signal. The new pattern that had evolved from the old was one of about twice the frequency and half the amplitude of the original one. Such a development also occasioned a photograph to be taken. At the time when this phenomenon first occurred it was thought that a change in the flow's wave modes may have caused the change in the thermistor output. No ultimate conclusions were drawn about these mutating wave forms.

There were also sudden changes in the regular patterns that were displayed on the oscilloscope. A pattern would disappear after a random disturbance appeared. The pattern that followed may or may not have been the same as the original one. After such abrupt changes became to be expected, every effort was made to photograph the flow during the time that the random disturbance was occurring on the oscilloscope screen. At speeds well above those at which the azimuthal waves first occurred, the signal from the thermistor began to include more and more harmonic components, and a speed could usually be reached where the output appeared random. The significance of this occurrence could not be assessed, and it did not influence the experimentation.

After a large number of tests had been conducted a removable disk was attached to the bottom of the rotating, inner cylinder. This phase of the experimentation was patterned after that of Wendt (1933) with the intention of showing the effect of the experimental boundary conditions at the bottom of the annulus on the results. In this way it was hoped that the role of the aspect ratio on the data could be assessed.

At the conclusion of an experiment the protractor was read while the inner cylinder was at rest. If this reading did not come within 0.2 degrees of the initial reading, the entire set of data was discarded. With the completion of a satisfactory experiment the torque and speed readings were reviewed and the data divided into groups. A least-squares, linear regression analysis was done on the data taken prior to any instability. A second group consisted of the data that were collected between the first and second Taylor instabilities. In the case of the wide-gap experiments in which a second Taylor instability did not occur, the data set was restricted to speeds of rotation relatively near that at which the first Taylor instability was initiated. These data were also given a linear regression analysis. On the basis of these calculations the intersection of the two curves was determined and the critical speed for the first Taylor instability was evaluated. The slopes and intercepts of these curves were also used in subsequent calculations and conclusions. Whenever it was unclear whether an observation belonged in the first or second group, it was put into neither one. The value of the speed at which the second instability occurred was inferred from the discontinuity in the slope of the speed-torque curve that was plotted for the experiment.

4. Results

4.1 Speed-torque relationships

A segment of the principal results of the research are shown in Fig. 3. Briefly, one can see that the data and the transitions in the flow are reproducable. The ordinate that was selected coalesces all of the data given for a given geometry, in the regime of primary flow, onto one straight line that is independent of the viscosity. Because the viscosity will affect the point of transition, one can expect that the data for different experiments, with their slightly different viscosities, will lie upon different curves after the speed for flow transition has been exceeded. One other feature of the figure is also evident. For this narrow gap, $\eta = 0.935$, a second discontinuity in the slope occurs soon after the first one. This discontinuity coincides with the occurrence of asymmetric Taylor cells, i.e., azimuthal waves.

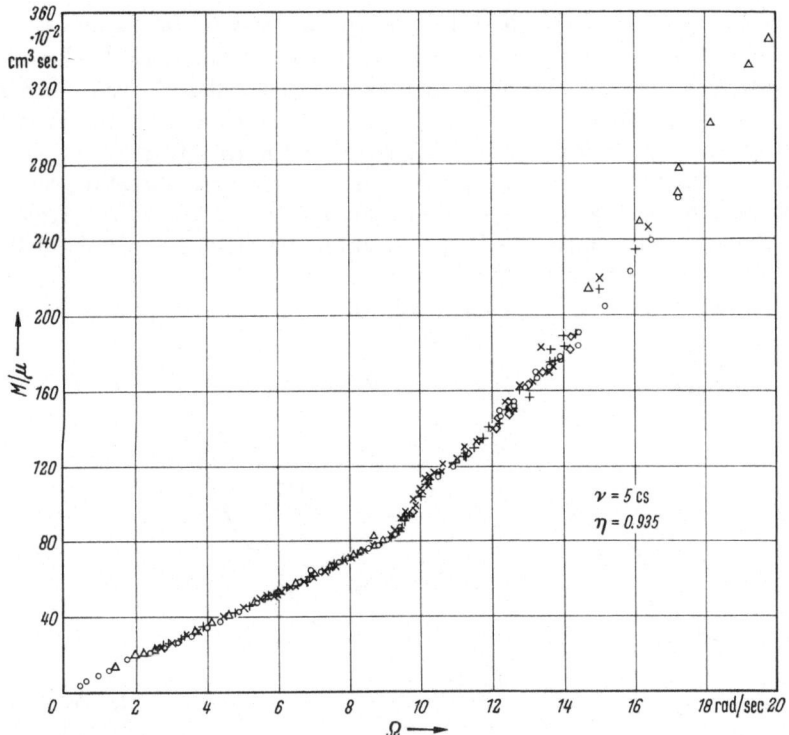

Fig. 3. Normalized torque-speed data for $\eta = 0.935$, $\nu = 5$ centistokes. ◯ Experiment No. 71; △ Experiment No. 72; + Experiment No. 73; × Experiment No. 74; ◇ Experiment No. 75; △ Experiment No. 76; ✕ Experiment No. 78.

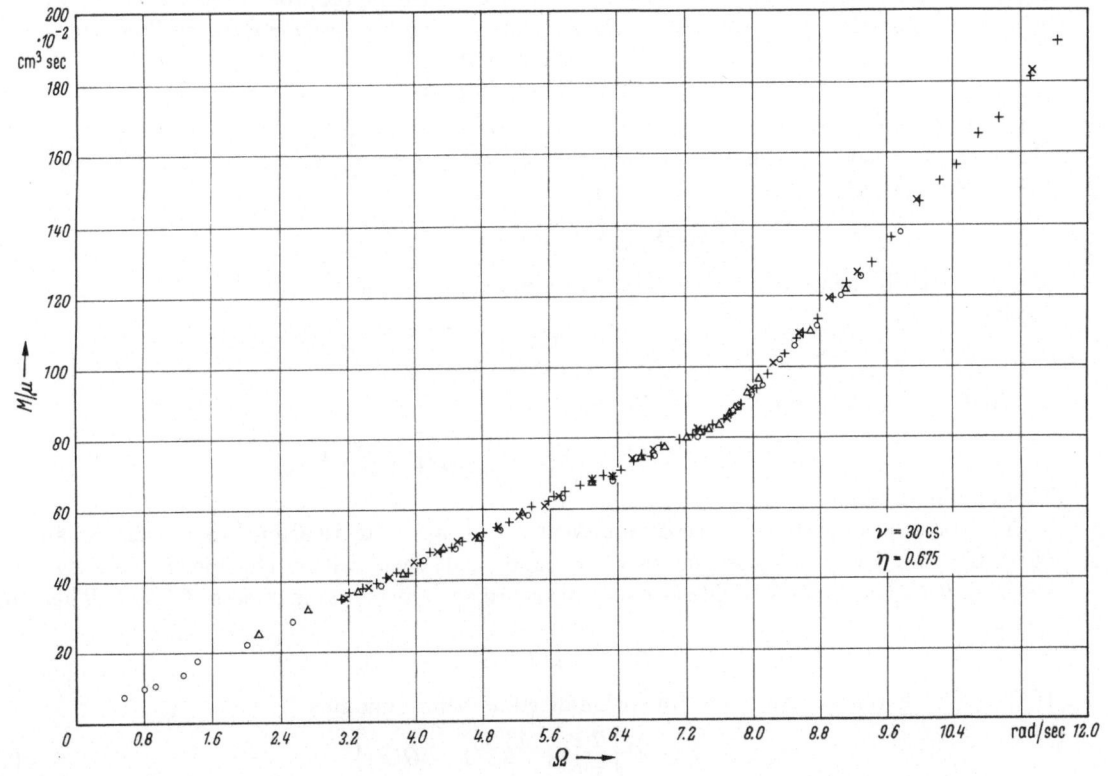

Fig. 4. Normalized torque-speed data for $\eta = 0.675$, $\nu = 30$ centistokes. ◯ Experiment No. 33; △ Experiment No. 34; + Experiment No. 35; × Experiment No. 36.

21*

Figure 4 shows a portion of the speed-torque data for $\eta = 0.675$. Again one has a discontinuity in the curve that is associated with the formation of axisymmetric Taylor cells; however, a second discontinuity does not appear in the figure, and would not, even if the speed of the inner cylinder were eight times that required for the first instability The propensity of the axisymmetric Taylor cells to become asymmetric after the first transitional speed has been exceeded by a small amount will be used by the authors to qualitatively catalog the experiments. Those cases for which the second discontinuity, occurring at Ω_c', does not take place within 50 percent of the speed at which the first transition occurred, Ω_c, will be called wide-gap experiments. If Ω_c' was less than $1.5\ \Omega_c$ the experiment will be designated as narrow-gap. In this sense the experiments for $\eta \geq 0.85$ were classified as narrow-gap, and for $\eta \leq 0.75$ were wide-gap.

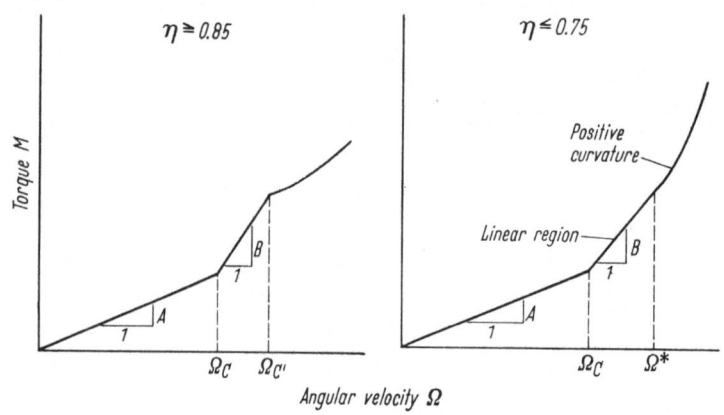

Fig. 5. Definition drawing for torque-speed characteristics.

During the experiment the data were plotted in the manner shown in Fig. 5. Clearly, in this figure the moment, M, is proportional to the angular velocity, Ω, for speeds $\Omega \leq \Omega_c$, with A being the proportionality constant. A theoretical solution for circular Couette flow has

$$M = \mu \left[\frac{4\pi R_1^2 R_2^2 h}{(R_2^2 - R_1^2)} \right] \Omega \tag{1}$$

in which R_1 and R_2 are the inner and outer radii, respectively, μ the viscosity, and h the length of the dynamometer section. The moment in Eq. (1) can be nondimensionalized by using the factor $\varrho \Omega^2 R_1^2 (2\pi R_2^2 h)/2$, with the result that

$$M/(\varrho \Omega^2 R_1^2 R_2^2 h\pi) = \lambda = 4(\nu/\Omega)\left(\frac{1}{R_2^2 - R_1^2}\right) \tag{2}$$

in which ϱ is the density and ν the kinematic viscosity. If one rewrites Eq. (2) in terms of the Taylor number,

$$T = 4\left(\frac{\Omega}{\nu}\right)^2 d^4\left(\frac{\eta^2}{1 - \eta^2}\right), \tag{3}$$

in which $d = R_2 - R_1$ and $\eta = R_1/R_2$, one has

$$\lambda = \frac{1}{T^{1/2}}\left[8\left(\frac{1 - \eta}{1 + \eta}\right)\left(\frac{\eta}{(1 - \eta^2)^{1/2}}\right)\right] = \frac{1}{T^{1/2}}\ F(\eta). \tag{4}$$

This equation defines F.

The present experimental results indicate that a linear relationship exists for speeds $\Omega_c \leq \Omega \leq \Omega_c'$. In those experiments where Ω_c' did not occur within the speed range studied, i.e., $\eta \leq 0.75$, there was a slight positive curvature at values of Ω in excess of $1.5\ \Omega_c$. Thus one can write

$$M = M_c + B(\Omega - \Omega_c) = A\Omega_c + B(\Omega - \Omega_c). \tag{5}$$

If Eq. (5) is rearranged and non-dimensionalized as before one has

$$\lambda = F\left[\frac{G/F}{T^{1/2}} + \frac{T_c^{1/2}}{T}\ (1 - G/F)\right] \tag{6}$$

in which F and G are both functions of η and T_c is the critical Taylor number.

In what follows the authors shall cite the values of G/F which they obtained for various η and thus provide reference values for the non-linear analytic work which is being done by other researchers.

Despite the fact that Eq. (4) gives the value for F, this function will be evaluated from the experimental data. For $\Omega \leq \Omega_c$, we have

$$\lambda = \frac{1}{T^{1/2}}\,F \quad \text{and} \quad \lambda = \frac{A\Omega}{\varrho\Omega^2 R_1^2 R_2^2 h\pi}.$$

Equating the last two equations one has after appropriate factoring

$$\frac{1}{T^{1/2}}\,F = \frac{A}{\mu}\left[\frac{1}{R_1^2 R_2^2 h}\,\frac{2}{\pi}\,\frac{d^2\eta}{(1-\eta^2)^{1/2}}\right]\frac{1}{T^{1/2}},$$

and

$$F = \frac{A}{\mu}\left[\frac{1}{R_1^2 R_2^2 h}\,\frac{2}{\pi}\,\frac{d^2\eta}{(1-\eta^2)^{1/2}}\right].$$

In the same way one has

$$G = \frac{B}{\mu}\left[\frac{1}{R_1^2 R_2^2 h}\,\frac{2}{\pi}\,\frac{d^2\eta}{(1-\eta^2)^{1/2}}\right];$$

so that

$$\frac{G}{F} = \frac{B}{A} = \frac{B'}{A'}$$

in which B' and A' are the appropriate slopes in Figs. 3 and 4.

Table 3 lists the values of F and G/F that were obtained in this research. As stated before, a least-squares regression analysis was used to obtain these values. The number of data points used in the analysis is also listed in the table. The fact that the experimental values of F are not in complete agreement with those calculated from Eq. (4) may be due to the end effects, a matter which will be discussed subsequently.

The experimental results are also presented in Fig. 6, which shows the trend of the results more clearly than the table. The scatter in the results may reflect upon the experimental technique; however, the accuracy of the individual data points is certainly satisfactory. Rather, the number of data points that were taken while the cells were axisymmetric may have contributed to the spread of the final results. In some cases the slope B (cf. Fig. 5) may have been

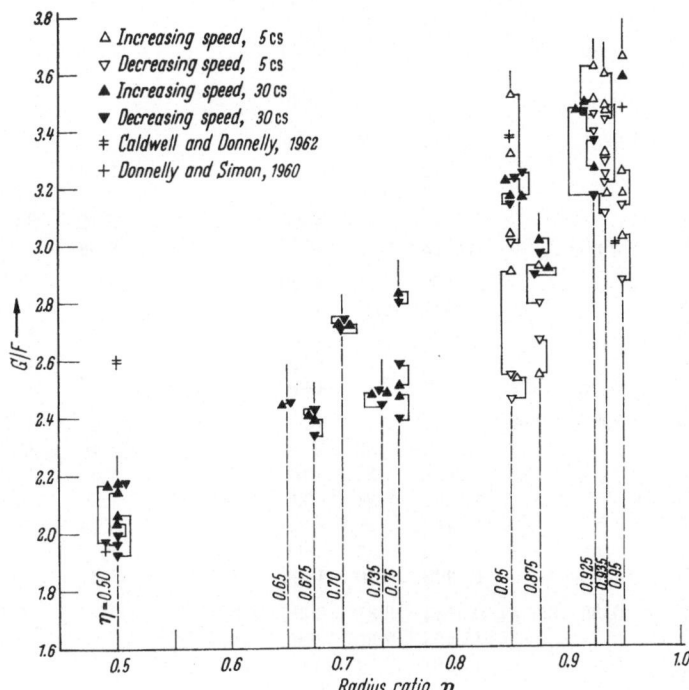

Fig. 6. Summary of G/F results for various η.

Table 3. *Summary of experimental results*

Radius Ratio	Theoretical Values		Experimental Results						Miscellaneous Data			
η	F	T_c	F	G/F	T_c	T'_c	Ω'_c/Ω_c	Ω/Ω_o*	ν (centistokes)	Wire Diameter (inches)	Data Points**	Exp. No.***
0.50	1.5396	6199.1	1.4810	2.06	5720	—	—	1.79	33.2	0.030	13, 10	U 3
0.50			1.4532	1.92	6350			1.63	33.2	0.030	7, 6	D 4
0.50			1.3185	2.14	5930			1.90	33.7	0.030	22, 18	U 5
0.50			1.3453	1.96	5930			2.02	33.7	0.030	11, 8	D 6
0.50			1.2174	2.16	5980			—	33.3	0.020	19, 10	U 7
0.50			1.2178	2.17	5840			—	33.3	0.020	19, 8	D 8
0.50			1.5422	2.03	6240			1.68	33.5	0.025	22, 23	U 11
0.50			1.5312	1.99	6300			1.67	33.5	0.025	30, 16	D 12
0.50			1.5160	2.16	6160			1.70	33.4	0.025	14, 6	U 13
0.50			1.5432	1.98	5920			1.71	33.4	0.025	24, 6	D 14
0.50			—	—	—			—	5.2	0.020		U 93
0.50	Average Value		**1.4166**	**2.06**	**6040**			**1.76**				
0.50	with disk		1.6539	1.97	6100				33.4	0.030	11, 7	U 1
0.50	with disk		1.6439	1.95	6280				33.4	0.030	10, 7	D 2
0.65	1.4535	4767.9	—	—	—	—	—	—	5.39	0.020		U 16
0.65			—	—	—			—	5.23	0.025		D 17
0.65			1.468	2.42	4710			2.64	32.3	0.030	15, 22	U 21
0.65			1.457	2.42	4620			2.67	32.3	0.030	7, 8	D 22
0.65	Average Value		**1.463**	**2.42**	**4660**			**2.59**				
0.65	with disk		1.4621	2.10	4840				32.4	0.030		U 19
	with disk		1.4309	2.09	4770				32.4	0.030		U 20
	with disk		1.4276	2.24	4800				32.3	0.030		U 23
	with disk		1.7222	1.99	4270				32.3	0.030		D 24
0.675	1.4225		—	—	—	—	—	—	5.18	0.025		U 29
0.675			1.4831	2.38	4440			1.25	32.4	0.030	17, 11	U 33
0.675			1.4846	2.33	4430			1.26	32.4	0.030	16, 8	D 34
0.675			1.4606	2.40	4600			1.45	32.4	0.030	34, 12	U 35
			1.4627	2.42	4490			1.46	32.4	0.030	17, 6	D 36
0.675	Average Value		**1.4725**	**2.38**	**4490**			**1.36**				
0.675	with disk		1.4414	2.33	4690				32.7	0.030	20, 11	U 31
0.675	with disk		1.4740	2.24	4640				32.7	0.030	12, 6	D 32
0.70	1.3867		1.3774	2.72	4530	—		2.03	33.8	0.030	10, 12	U 101
0.70			1.3970	2.73	4460			2.04	33.8	0.030	6, 6	D 102
0.70			1.4327	2.71	4510			2.11	33.4	0.020	10, 13	U 103
0.70			1.4377	2.70	4460			2.12	33.4	0.020	11, 10	D 104
0.70	Average Value		**1.4112**	**2.72**	**4450**			**2.08**				
0.735	1.3275		1.3470	2.48	4520	—		1.59	33.9	0.030	13, 10	U 105
0.735			1.3413	2.48	4460			1.61	33.9	0.030	16, 11	D 106
0.735			1.3683	2.47	4420			1.69	34.1	0.020	12, 15	U 107
0.735			1.3819	2.44	4370			1.71	34.1	0.020	14, 14	D 108
0.735	Average Value		**1.3596**	**2.46**	**4440**			**1.65**				
0.75	1.2998	4204.3	1.2627	2.50	4340	—		1.45	32.9	0.030	17, 11	U 37
0.75			1.2122	2.58	4140			1.48	32.9	0.030	11, 14	D 38
0.75			1.2656	2.82	4040			1.90	35.2	0.030	9, 15	U 109
0.75			1.2642	2.81	3930			1.90	35.2	0.030	14, 6	D 111
0.75			1.3749	2.48	4760			1.54	32.5	0.025	5, 5	U 110
0.75			1.3774	2.39	4560			1.57	32.5	0.025	9, 6	D 112
0.75	Average Value		**1.2928**	**2.60**	**4300**			**1.64**				
0.85	1.0536	3804.8	0.9844	3.31	3630	6460		—	5.44	0.030	3, 4	U 39
0.85			1.0424	2.90	3500	6360	1.35		5.55	0.030	4, 7	U 41
0.85			1.0101	2.54	3210	6310	1.40		5.55	0.020	13, 8	D 42
0.85			0.9862	3.52	3810	5610	1.21		5.55	0.020	10, 14	U 43

Table 3 (continued)

Radius Ratio	Theoretical Values		Experimental Results						Miscellaneous Data			
η	F	T_c	F	G/F	T_c	T'_c	Ω'_c/Ω_c	Ω/Ω_c*	ν (centistokes)	Wire Diameter (inches)	Data Points**	Exp. No.***
0.85			1.2431	3.05	3790	8240	1.47		5.55	0.020	5, 6	D 44
0.85			—	2.46	3890				5.29	0.025	17, 14	U 46
0.85			—	2.53	3790				5.29	0.025	7, 5	D 47
0.85			0.9681	3.16	3670				32.3	0.030	31, 11	U 50
0.85			0.9800	3.25	3670				32.3	0.030	19, 11	D 51
0.85			1.0305	3.21	3590	5170			35.7	0.030	16, 13	U 113
0.85			1.0159	3.21	3500	5130			35.7	0.030	5, 5	D 114
0.85			1.0295	3.16	3570	5230			34.3	0.030	17, 12	U 115
0.85			1.0127	3.15	3460	5230			34.3	0.030	19, 14	D 116
0.85	Average Value		**1.0275**	**3.03**	**3620**	**5970**	**1.28**					
0.85	with disk		1.3662	2.74	4310				5.33	0.030		U 45
0.875	0.9779	3723.0	0.9935	2.74	3670	6490	1.33	—	5.34	0.030		U 52
0.875			1.0646	2.92	4080	6660	1.28		5.88	0.030	19, 6	U 55
0.875			1.0976	2.79	3750	6260	1.29		5.88	0.030	8, 3	D 56
0.875			1.0621	2.54	4090	9580	1.58		5.88	0.020	22, 5	U 57
0.875			1.1112	2.66	4040	10100	1.58		5.88	0.020	11, 4	D 58
0.875			0.9188	2.97	3730	—	—		32.6	0.030	22, 8	U 61
0.875			0.8943	3.01	3670	—	—		32.6	0.030	14, 6	D 62
0.875			1.0096	2.91	3960	5150	1.14		32.5	0.030	23, 10	U 117
0.875			1.0051	2.90	3980	5200	1.15		32.5	0.030	18, 19	D 118
0.875	Average Value		**1.0158**	**2.83**	**3880**	**6980**	**1.34**					
0.925	0.7753	3575.9	0.8035	3.42	3840	4640	1.10	—	5.44	0.030		U 63
0.925			0.8057	3.50	3820	4600	1.10		5.44	0.030	21, 7	U 65
0.925			0.8349	3.39	4050	4840	1.09		5.50	0.030	19, 3	U 66
0.925			0.8173	3.62	3930	4840	1.11		5.50	0.030	12, 2	D 67
0.925			0.8214	3.46	3770	4910	1.14		5.50	0.030	11, 3	U 68
0.925			0.7892	3.26	3910	4840	1.11		32.4	0.040	31, 3	U 69
0.925			0.7455	3.36	3620	4890	1.16		32.4	0.040	21, 9	D 70
0.925			0.7740	3.50	3550	4300	1.10		35.0	0.030	43, 8	U 127
0.925			0.7591	3.17	3380	4190	1.12		33.0	0.030	8, 16	D 128
0.925			0.8248	3.49	3390		—		35.0	0.030	4, 7	U 129
0.925			0.7446	3.48	3340		—		35.0	0.030	22, 6	D 130
0.925	Average Value		**0.7927**	**3.42**	**3690**	**4670**	**1.12**					
0.935	0.7214		0.7311	3.48	3380	3940	1.08	—	5.44	0.020	26, 3	U 71
0.935			0.7298	3.43	3380	4030	1.09		5.44	0.020	43, 2	D 72
0.935			0.6965	3.59	3430	4170	1.10		5.44	0.030	16, 5	U 73
0.935			0.6957	3.21	3380	4090	1.10		5.44	0.030	12, 5	D 74
0.935			0.7018	3.48	3480	4140	1.09		5.44	0.030	34, 5	U 75
0.935			0.6852	3.24	3350	4170	1.12		5.44	0.030	17, 2	D 76
0.935			—	—	—	—	—		5.39	0.030		U 77
0.935			0.7624	3.19	3650	4580	1.12		5.11	0.030	8, 6	U 78
0.935			—	3.11	3770	—	—		—		7, 12	D 79
0.935			0.7647	3.31	3750	4620	1.11		5.54	0.020	13, 5	U 119
0.935			0.7611	3.29	3720	4680	1.12		5.54	0.020	11, 8	D 120
0.935	Average Value		**0.7254**	**3.33**	**3530**	**4270**	**1.10**					
0.935	with disk		0.7672	3.64	3700				5.34	0.030	21, 10	U 80
0.935	with disk		0.8445	2.99	3470				5.34	0.030	8, 15	D 81
0.935	with disk		0.7792	3.59	3740				5.19	0.030	20, 4	D 83
0.95	0.6392	3509.5	0.6168	3.65	3360	4070	1.10	—	5.50	0.030	4,5	U 84
0.95			0.6157	—	—	—	—		5.50	0.030		D 85
0.95			0.6259	3.25	3370	4040	1.10		5.49	0.030	20, 4	U 86
0.95			0.6352	3.13	3340	4090	1.11		5.49	0.030	30, 4	D 87
0.95			0.6058	—	—	—	—		32.8	0.030		U 88

determined by the least-squares linear regression analysis with too few points to eliminate any doubt about the numerical result. The slope of a line can appear to change little, especially when only a short segment is viewed, and yet the numerical value will be noticeably altered. It is believed that the average value of G/F for all of the experiments at a particular η will have been determined by a large enough sample so as to be a reasonable estimate of the true value. Table 3 lists the number of data points that was used in each determination of F and G.

Table 3 (continued

Radius Ratio	Theoretical Values		Experimental Results						Miscellaneous Data			
η	F	T_c	F	G/F	T_c	T'_c	Ω'_c/Ω	Ω/Ω_c*	v (centi-stokes)	Wire Diameter (inches)	Data Points **	Exp. No. ***
0.95			0.6067	—	—	—	—		32.8	0.030		D 89
0.95			0.6110	3.58	3440	—	—		32.4	0.040	30, 2	U 90
0.95			0.6225	—	—	—	—		32.4	0.040		D 91
0.95			0.6757	3.18	3780	4490	1.09		5.64	0.020		U 121
0.95			0.6607	3.16	3600	4510	1.12		5.64	0.020		D 122
0.95			0.6499	3.19	3490	4380	1.12		5.64	0.030	11, 7	U 123
0.95			0.6479	2.87	3700	4300	1.08		5.64	0.030	11, 7	D 124
0.95			0.7386	3.17	3670	4630	1.12		5.50	0.025	9, 9	U 125
0.95	Average Value		**0.6410**	**3.24**	**3530**	**4310**	**1.10**					

* Speed ratio over which the torque appears to vary linearly with speed.
** Number of data points prior to instability, number of data points after instability.
*** U = increasing speed, D = decreasing speed.

Some results from Donnelly and Simon (1960) and Caldwell and Donnelly (1962) have been included in Fig. 6. The data in the former paper were plotted by the present authors, and the resulting slopes were measured. Certain graphs in the latter paper contained informa-

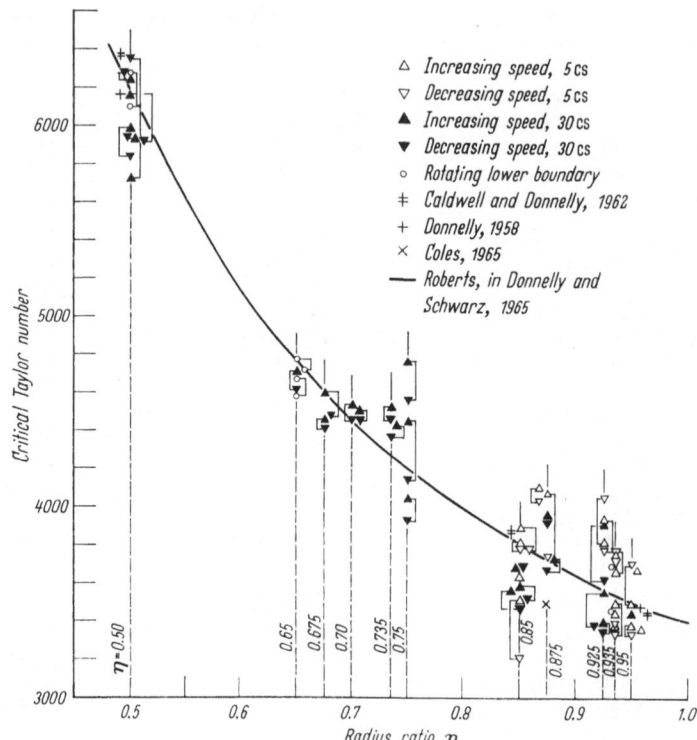

Fig. 7. Summary of critical Taylor numbers for various η.

tion from which G/F could be extracted. These graphs were scaled by the present authors and the needed values calculated. The present research results agree favorably with those of DON-NELLY and his co-workers.

4.2 Critical Taylor numbers

Table 3 also summarizes the values of T_c and T'_c that were obtained for various values of η. The latter parameter is the Taylor number at which the azimuthal waves commence. This information is also presented in Fig. 7 along with the theoretical values determined by ROBERTS in his appendix to DONNELLY and SCHWARZ (1965). The results of other workers are also presented for comparison. The critical Taylor number, T_c was obtained from the intersection of the two-speed torque lines that were found by statistical means. T'_c was calculated by using the speed at which it was judged from the speed-torque data that a discontinuity in the slope had occurred.

The scatter in the results shown in Fig. 7 can be ascribed to normal measurement errors, insufficient number of data points to determine unequivocally the value of Ω_c by statistical means, and an imprecise value of the viscosity of the fluid even though temperature measurements were made. The thermocouple was undoubtedly biased by the outside-wall temperature and may have lagged behind the temperature of the fluid in the annulus. The agreement with the analytical results may have been influenced by some slight eccentricity of the inner cylinder with respect to the outer one. Also, because of the finite length of the apparatus the primary

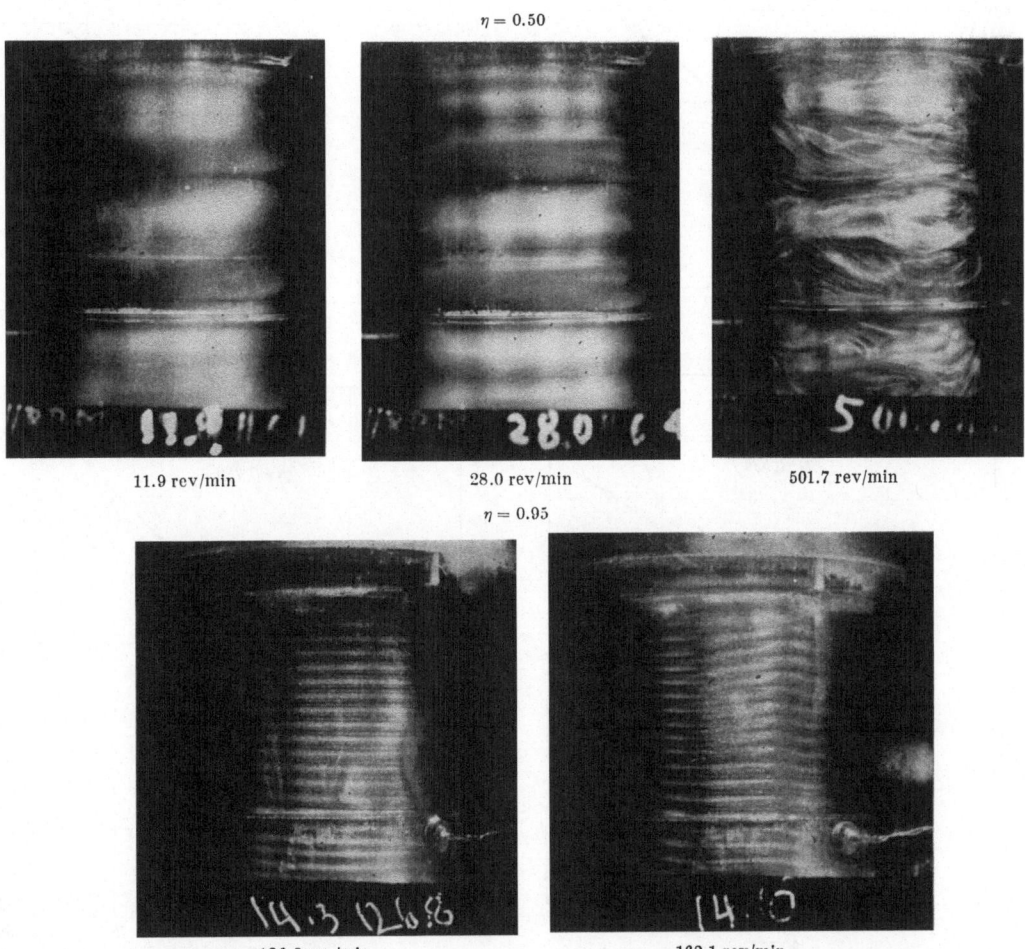

$\eta = 0.50$

| 11.9 rev/min | 28.0 rev/min | 501.7 rev/min |

$\eta = 0.95$

| 126.8 rev/min | 162.1 rev/min |

Fig. 8. Flow patterns for $\eta = 0.50$ and 0.95. For the former. the critical Taylor number occurred at 11.9 rpm. For the latter the first critical Taylor number occurred at 123.6 rpm. Azimuthal waves appeared at 136.2 rpm.

flow in the experiment may not have been the one used in the theoretical calculations. The effect of the experimental boundary conditions on the primary flow will be discussed in Section 4.5. Nevertheless, the results are within 5 percent of the theoretically predicted values.

It should be recalled that for $\eta \leq 0.75$ a second slope discontinuity does not readily occur. In the present experiments the speed of the inner cylinder was increased to about 8 Ω_c for $\eta = 0.75$ before azimuthal waves occurred; with $\eta = 0.65$ axisymmetric waves broke down near 12 Ω_c. In the case of $\eta = 0.50$ the flow appeared to a retain certain axisymmetric characteristics to the speed of 90 Ω_c, the highest value achieved. Fig. 8 presents pictures of the first occurrence of Taylor cells for wide- and narrow-gap configurations. For $\eta = 0.95$ the occurrence of azimuthal waves is also presented.

The visual observations which specified the speeds of rotation at which T_c and T_c' occurred agreed with the Taylor numbers calculated subsequently from the torque data. Additional agreement was found with the output from the thermistor probe. While no characteristic signal was found when the cells first formed, a gradual sinusoidal trace on the oscilloscope was produced by this sensor when azimuthal waves began to form. This signal changed in amplitude, frequency, and shape as the speed of the apparatus was subsequently increased.

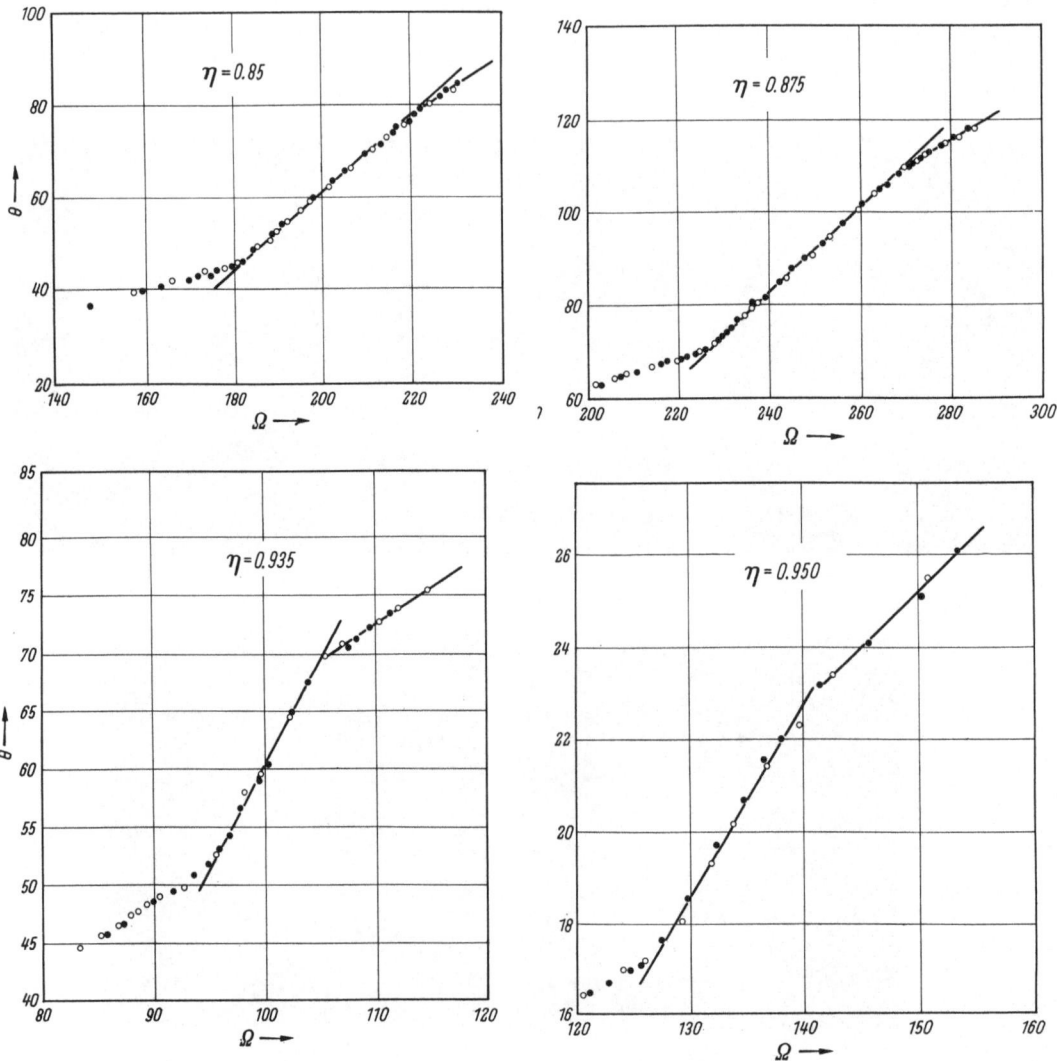

Fig. 9. Angle of wire deflection in degrees θ versus angular velocity in rpm, Ω, for various η near the transition points. Circles are for increasing speed and squares for decreasing speed.

4.3 Hysteresis

The results of the present research agree with the findings of CALDWELL and DONNELLY (1962). At Ω_c there is no hysteresis. Furthermore, the authors are convinced on the basis of Fig. 9 and other graphs like it, that the transition to azimuthal waves at Ω_c' is also reversible. At speeds beyond Ω_c' observations were made which indicated that in the present apparatus there was not a unique torque-speed relationship. More will be said concerning this in the next section.

4.4 Large Taylor number

Figures 10 and 11 are presentations of λ versus $R = \dfrac{\Omega R_1 d}{\nu}$ in logarithmic form. The same or similar parameters have been used by DONNELLY (1958), TAYLOR (1936), and WENDT (1933). Again these curves are typical of the wide- and narrow-gap experiments. Fig. 10 shows that after Taylor cells occur the curve gently changes slope over a range of R values and approaches a new negative value. In contrast to this, Fig. 11 has a sharp maximum in λ after the first minimum that occurred at Ω_c. This maximum occurs at Ω_c'.

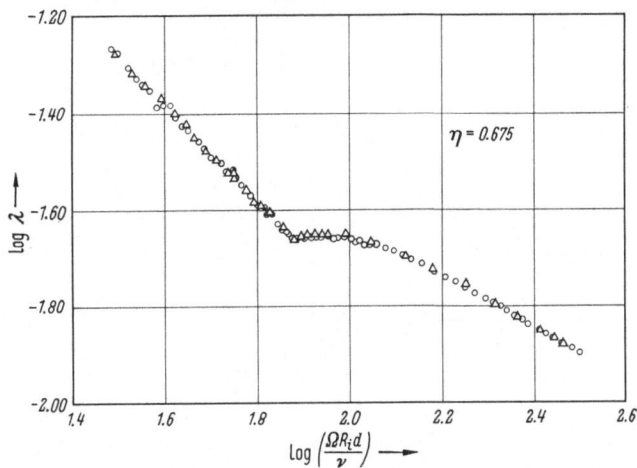

Fig. 10. Moment coefficient, λ, versus Reynolds number for $\eta = 0.675$. ◯ Experiment No. 35; △ Experiment No. 36.

As the Taylor number, proportional to R^2, is increased far beyond T_c', some non-uniqueness in the torque occured. This situation was often experienced, and the data in Fig. 11 exemplify this. At $\log R = 2.8$ the value of λ suddenly drops, and a new curve is traced to $\log R = 3.0$. As R is now reduced (i.e., speed is reduced) the values of λ coincide with those just previously measured. Below $\log R = 2.8$, λ is less for decreasing speed than for increasing speed. This lack of uniqueness could be associated with some of the admissible states that COLES (1965) presents. During the experiments it was not always possible convincingly to relate wave numbers with the torque.

The speed-torque curves given in Figs. 3 and 4 are not practical for displaying the relationship over a wide speed range. The logarithmic plots in Figs. 10 and 11 effectively compress the scale. Figure 12 is a presentation in logarithmic form of λ versus R for $\eta = .5$, .65, .675, .85, .875, .925, .935, and .95. Each of these curves is for a single experiment (i.e., either increasing or decreasing speed) and was selected on the basis of the wide speed range which it entailed. The values of R in Fig. 12 are based upon the viscosity that was thought to be present within the range $\Omega_c' - \Omega_c$ by using the thermocouple reading near the outside wall as a guide. Thus the Reynolds numbers may be slightly incorrect. Nevertheless, some interesting information is contained in this figure.

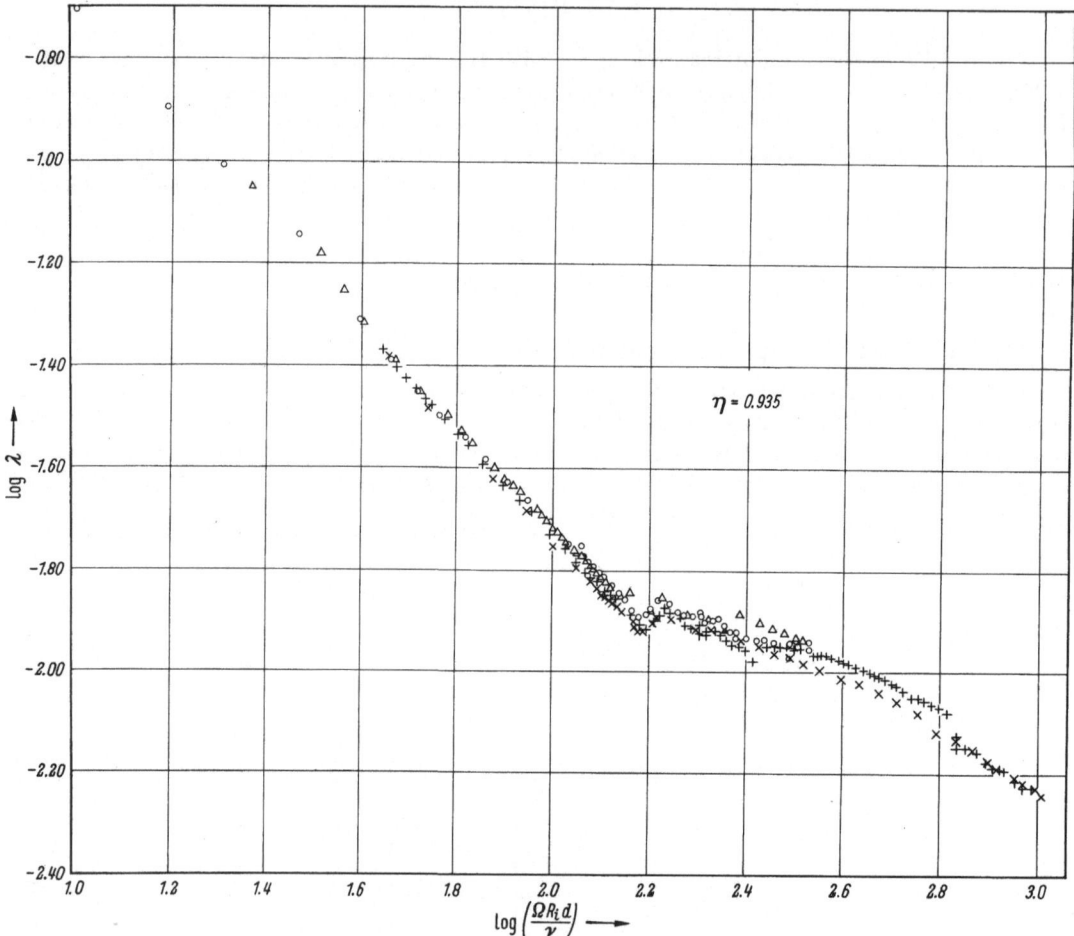

Fig. 11. Moment coefficient, λ, versus Reynolds number for $\eta = 0.935$. ⊙ Experiment No. 71; △ Experiment No. 72; + Experiment No. 75; × Experiment No. 76.

For $\eta = 0.5$, 0.65, and 0.675 the slope is -0.571, whereas for $\eta = 0.95$ it is -0.695. These numbers differ from that determined by Batchelor (appendix to Donnelly and Simon, 1960) as an asymptotic estimate and bracket the empirical conclusion of Donnelly and Simon (1960). For wide-gap experiments where only the axisymmetric mode seems to be present for a wide range or R, a $\lambda - R$ correlation can describe that particular mode. The authors believe that the multiple mode possibility for large η lessens the significance of the $\lambda - R$ slope for large R until more is understood about the problem.

The curves for $\eta = 0.85$, 0.875, and 0.925 appear to be composed of segments between which there is almost a jump discontinuity in magnitude, but not in slope. This discontinuity could occur (i.e., a sudden increase or decrease in torque) if the mode of the fluid motion were to suddenly change from one to another. Thus a logarithmic plot may well be a useful means of detecting mode changes. The photographs in Fig. 12 support this contention.

When operating cylinders with $\eta \geq 0.85$ at speeds well beyond Ω_c', the hysteresis mentioned in Section 4.3 was observed in addition to the sudden jumps in applied torque. The photographs which were taken were not always discriminating enough to tell if different azimuthal modes were present at the same speed when different torques were observed. More than likely, an optical technique similar to Coles' (1965) would be valuable to relate wave mode and torque. Such an investigation may show that some modes are more probable than others on the basis of a minimum torque.

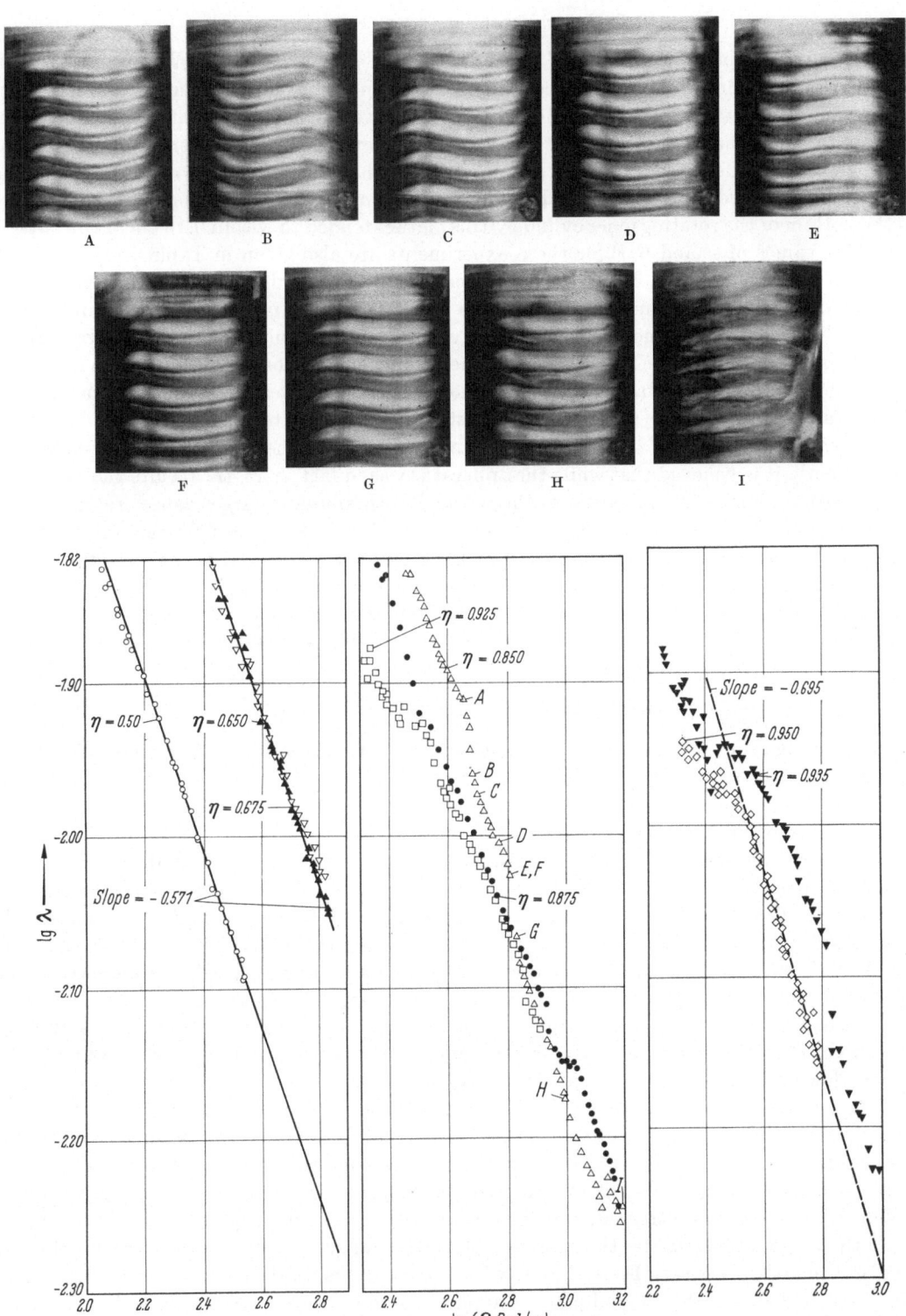

Fig. 12. λ vs. R plots for various values of η.

○ $\eta = 0.50$, Experiment No. 5; ▲ $\eta = 0.650$, Experiment No. 17; ▽ $\eta = 0.675$, Experiment No. 29; △ $\eta = 0.85$, Experiment No. 46;
● $\eta = 0.875$, Experiment No. 59; □ $\eta = 0.925$, Experiment No. 65; ▼ $\eta = 0.935$, Experiment No. 75; ◇ $\eta = 0.95$, Experiment No. 86.

— — — slope of -0.571; — · — · — slope of -0.695.

Insert A, $h/d = 1.148$, 4 waves; B, $h/d = 1.231$, 4 waves; C, $h/d = 1.223$, 4 waves; D, $h/d = 1.288$, 4 waves; E, $h/d = 1.293$, 4 waves;
F, $h/d = 1.323$, 4 waves; G, $h/d = 1.380$, 3 waves; H, $h/d = 1.410$, 2 waves; I: Turbulent.

174   W. Debler, E. Füner, and B. Schaaf

4.5 Experimental boundary conditions

Certainly the work of Coles and Van Atta (1966), Taylor (1935), and Wendt (1933) indicate that the experimental boundary conditions at the end of the annulus have a marked effect on the moment that is measured. The differences between the experimental and theoretical values of F given in Table 3 also raise the question as to the appropriateness of the present experiments, or indeed any other, as a comparison with the analytical problem for which no axial boundary condition is prescribed. Following Wendt (1933) the authors attached a plate to the bottom of the rotating inner cylinder. This plate extended to within 0.16 cm of the outside wall. The values of F and G/F for these experiments are also given in Table 3 for $\eta = 0.50$, 0.65, 0.675, 0.85, and 0.935. The values obtained are not greatly different from those obtained without the plate. This is in sharp contrast to the data of Wendt (1933)[1] who found a 300-percent difference in the amount for $\eta = 0.68$ when the lower boundary condition was varied. By comparing in Table 3 the values of F obtained from a one-dimensional calculation and those values obtained experimentally with and without a rotating lower boundary, one can deduce that the actual flow, prior to the first Taylor instability, is not quite one-dimensional. However, major changes in the lower boundary do not materially affect the torque that was measured. Consequently it is believed that while the apparatus was in fact short, the results that were obtained with it should be equivalent to those found with annuli of larger aspect ratio.

One effect of the lower boundary can be seen in Fig. 8. For $\eta = 0.95$ the axisymmetric Taylor cells near the bottom have boundaries that are not perpendicular to the axis of the cylinder. This is not believed to be a spiral mode since the effect is not present in the cells that are above the center of the dynamometer section of the outer cylinder. For $\eta \leq 0.925$ the axisymmetric cells were normal to the axis of the cylinder over the entire length of the annulus.

4.6 Photographic and visual observations

The photographs which were taken in the course of the experimentation were a source of pleasure for the authors. They displayed in a much more vivid way than the torque data what was occurring in the apparatus. The fact that there was not a second discontinuity in the torque-speed curve (cf. Fig. 4) is given support by photographs such as given in Fig. 8. This shows, for $\eta = 0.50$, the flow pattern at the speed where Taylor cells first appeared and at a speed nearly 90 times the critical. Despite the turbulent nature of the flow there are still vestiges of Taylor cells. This is similar to the observations of Pai (1943). At speeds of $5\,\Omega_c$ the photographs did not show azimuthal waves for $\eta = 0.75$. In this case, as well as for $\eta = 0.675$ and 0.70, the Taylor cells appeared to oscillate about their core axes as the speed was increased. This oscillation was observable by the pulsating reflection from the cells and occurred even though the cell boundaries remained horizontal. As the oscillation became more and more active the Taylor cells broke down to an asymmetric mode. This transition did not occur suddenly but would slowly mature while the apparatus was running at constant speed.

The authors' experimentation was originally motivated in part by a desire to extend the illuminating presentation of Schultz-Grunow and Hein (1956) by having simultaneous torque measurements and visual observations. Figure 13 contains a non-dimensional plot for $\eta = 0.935$ that is similar to Fig. 11. Here the black points indicate where photographs were taken as the speed was increased, and the open circles denote other data points. These figures show the manifold flow modes that are possible in this range of Reynolds numbers for the particular value of η selected. These figures with their various flow patterns lead one to expect the peculiar shapes of these non-dimensional plots which have been realized by the authors and Donnelly (1958). The features of such figures cannot be explained in detail unless concurrent visual and torque measurements are made and these in turn are correlated with the modes of flow that are possible in the particular Reynolds-number range.

---

[1] Wendt's results (1933) were for a stationary inner cylinder, which is not the case treated by the experiments of this paper.

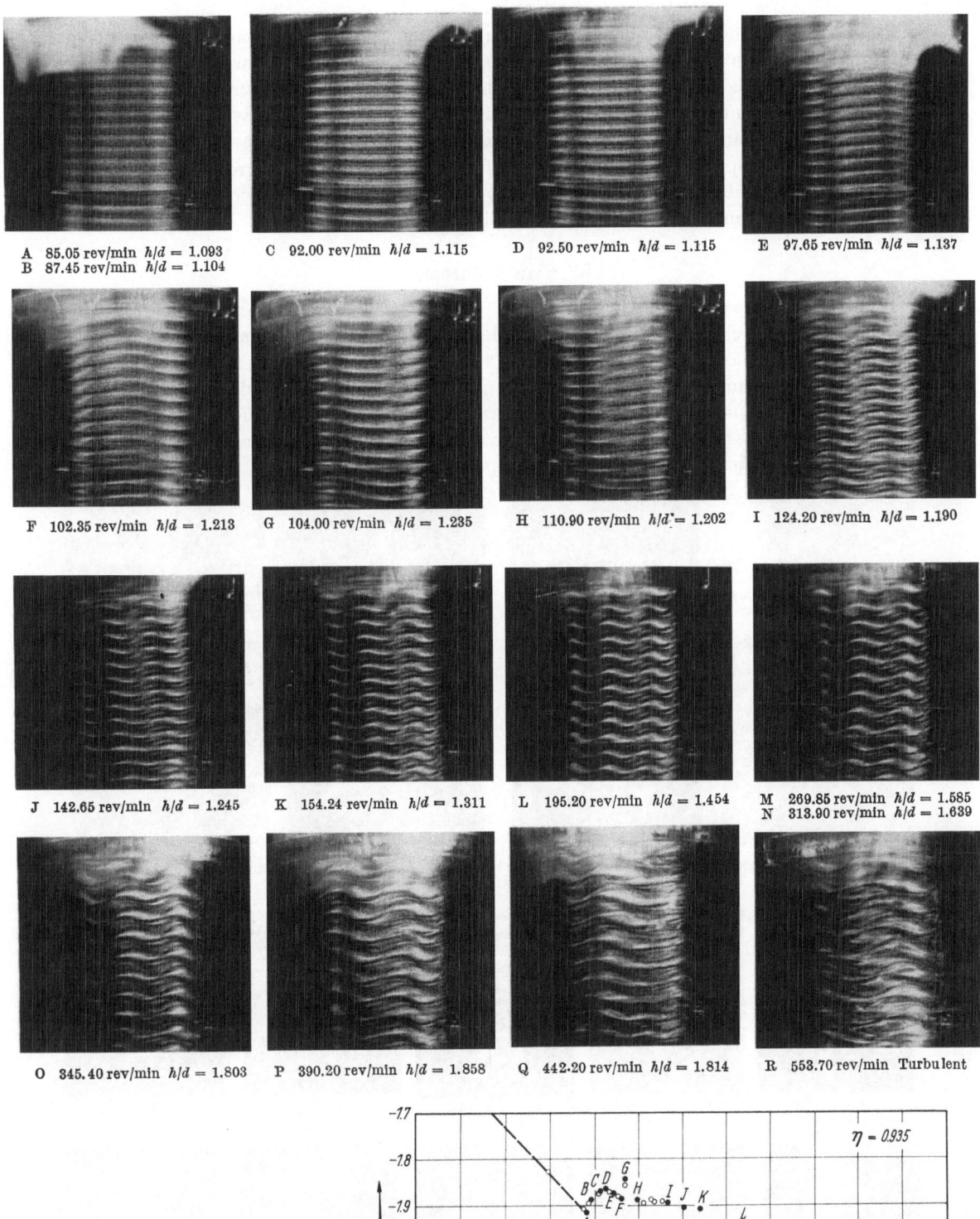

A   85.05 rev/min $h/d = 1.093$ C   92.00 rev/min $h/d = 1.115$ D   92.50 rev/min $h/d = 1.115$ E   97.65 rev/min $h/d = 1.137$
B   87.45 rev/min $h/d = 1.104$

F   102.35 rev/min $h/d = 1.213$ G   104.00 rev/min $h/d = 1.235$ H   110.90 rev/min $h/d = 1.202$ I   124.20 rev/min $h/d = 1.190$

J   142.65 rev/min $h/d = 1.245$ K   154.24 rev/min $h/d = 1.311$ L   195.20 rev/min $h/d = 1.454$ M   269.85 rev/min $h/d = 1.585$
N   313.90 rev/min $h/d = 1.639$

O   345.40 rev/min $h/d = 1.803$ P   390.20 rev/min $h/d = 1.858$ Q   442.20 rev/min $h/d = 1.814$ R   553.70 rev/min Turbulent

Fig. 13. Sequence of flow patterns in an annulus with $\eta = 0.935$. Experiment No. 77.

Figure 13 does show that there is a local maximum for λ at the point where azimuthal waves appear. The ratio of the cell height to annulus gap, h/d, shows some increase in the size of the axisymmetric cell as the speed is increased. However, no general conclusions can be drawn from these experiments concerning the rate of axisymmetric cell growth as a function of speed because an insufficient number of photographs was taken. Also, too few experiments were completely photographed, and some of the few results concerning cell growth are contradictory. Nevertheless, all the sets of photographs strongly suggest that when azimuthal waves appear their wave number is less than π (cf. Figs. 8 and 13).

4.7 Flow disturbances

In the course of experimentation at speeds in excess of Ω_c' it was often the case that there would be an abrupt change in the torque or a fleeting random signal from the thermistor that was inserted in the flow (cf. Section 3). Figure 14 shows a picture of the oscilloscope trace displaying such a random signal and photographs of the test section before, during, and after this random thermistor output. The cause of this disturbance, whether it be from the boundary or from within the flow, is unknown; but such transient disturbances undobtedly contribute to the unsteadiness in the torque at high rotational speeds.

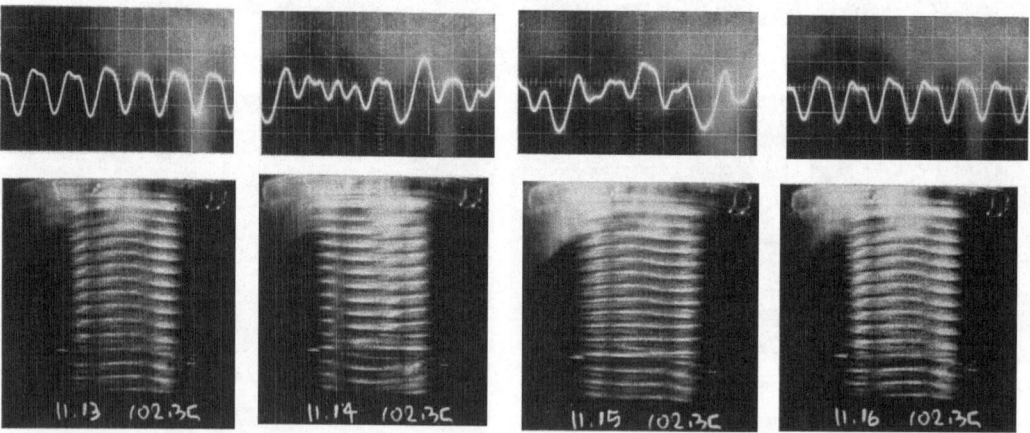

Fig. 14. Oscilloscope trace and flow patterns during a disturbance in the flow. Experiment No. 77. Oscilloscope trace is originated by a thermistor bead in the flow. All pictures taken at 102.35 rpm.

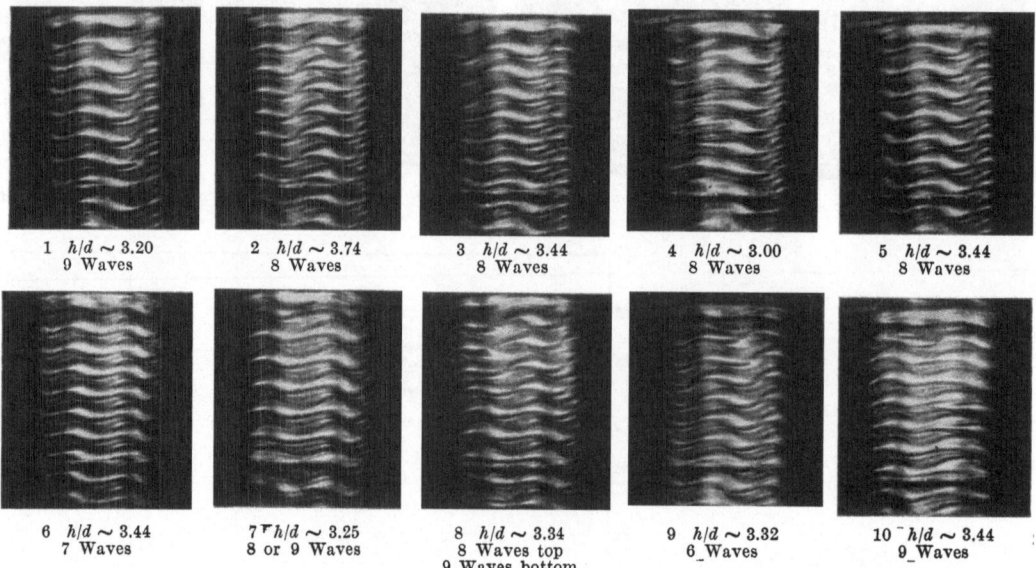

1　$h/d \sim 3.20$　　2　$h/d \sim 3.74$　　3　$h/d \sim 3.44$　　4　$h/d \sim 3.00$　　5　$h/d \sim 3.44$
9 Waves　　　　8 Waves　　　　8 Waves　　　　8 Waves　　　　8 Waves

6　$h/d \sim 3.44$　　7　$h/d \sim 3.25$　　8　$h/d \sim 3.34$　　9　$h/d \sim 3.32$　　10　$h/d \sim 3.44$
7 Waves　　　　8 or 9 Waves　　8 Waves top　　6 Waves　　　9 Waves
　　　　　　　　　　　　　9 Waves bottom

Fig. 15. Sequence of flows occurring at constant speed (276 rpm) subsequent to a 6-rpm increase in speed.

The multiple combinations of modes that are possible at certain operating speeds also can cause oscillations of the torque that is sensed. Figure 15 shows a sequence of 9 stages which the fluid passed through before arriving at the final flow pattern. All this was at constant inner cylinder speed. Most probably, the successive stages were influenced by those immediately before.

5. Discussion

The values of G are not explicitly given in Table 3 because it was thought that the ratio of G/F would provide for better correlation with the analytical and experimental results of others. The fact that the values of F do not agree completely with the theoretical calculations implies that the experimental configuration could also affect the value of the slope B' in Figs. 3 and 4.

In the case of the narrow-gap experiments, on the basis of the definition of Section 4.1, the value of G is appropriate for the entire small speed interval over which axisymmetric cells occur. The extent of the speed range in which this flow was found to exist is given in Table 3 by the column headed Ω'_c/Ω_c. These data point up the fact that when comparisons are made with analytical results, an appropriate upper speed limit must be selected. Furthermore, experimenters must be mindful of the variety of flow possibilities and properly restrict their empirical torque-speed formulas in order not to oversimplify the phenomena.

The wide-gap results for G/F (i.e., $\eta \leq 0.75$) should be interpreted as the slope ratio which is present for small $\Omega - \Omega_c$. Certainly, Fig. 4 shows that the extent of the constant-slope region for $\Omega > \Omega_c$ is quite large. Table 3, column 9, gives the authors' estimate of the maximum speed range over which the speed-torque relationship *appears* to be linear. Beyond the values quoted in the table there is a small positive curvature to the speed-torque curve for wide-gap experiments (cf. Fig. 3). The authors' experiments indicate that formulas such as

$$M = \frac{a}{\Omega} + b\Omega; \quad \Omega \geq \Omega_c, \quad a < 0, \quad b > 0,$$

are deficient for describing the phenomenon since their second derivatives are negative.

A physical dimension of the apparatus which could be a factor in the values of F and G/F that were obtained is the length of the middle section of the outer cylinder. The torque on this piece was transferred to the torsion wire and read from the protractor. Its length was 7.33 cm. For $\eta = 0.5$ there were approximately four Taylor cells in contact with this section, and for $\eta = 0.95$ there were about 38 cells. The former figure may be too small for a good statistical average.

Certainly, the results of others such as DONNELLY (1958), TAYLOR (1936), and WENDT (1933) can also be used to assess the merits of new work. The small-Reynolds-number comparison can be done conveniently only for Donnelly's work as was done in Figs. 6 and 7.

The high-Reynolds-number correlation can be implemented by using the ratio M/M_{LC}, in which M_{LC} is the moment calculated using the initial slope of the torque-speed data (i.e., laminar Couette flow). WENDT (1933) presents such a plot, and DONNELLY (1958) and Taylor's results (1936) can be converted to this ratio by scaling their logarithmic plots. This comparison is given in Table 4 and appears to be satisfactory.

The shape of the curves in Wendt's Fig. 13 (1933) near Ω_c is basically different than those which the authors obtained. Figures 10 and 11 show that for small η the change of slope after transition is gradual, and for large η the change is abrupt. DONNELLY (1958) makes this same observation.

The critical Taylor numbers that were found in the experiments could have been influenced, especially for the smallest gaps that were used, by the small amount of eccentricity in the apparatus. The recent work of RITCHIE (1968) suggests that for an eccentricity of 5 percent there should be a reduction in the critical Taylor number of the same percentage. Thus one might expect on this basis that the experimental results should be consistently lower than the theoretical ones. Figure 7 does not show this to be the case.

Finally, it should be noted that COLES (1965) has made the suggestion that azimuthal waves may not form for η less than about 0.715. The experiments show that waves do indeed appear for η less than this value but at speeds greatly in excess of Ω_c. It has been mentioned that these waves appear to form differently from those for narrow gaps, and the transition to them is not sudden, as for narrow gaps. Furthermore, the transition appears to be associated with some hystereses when increasing and decreasing the speed. Thus there may be additional factors when considering the formation of waves for η less than 0.715.

Table 4. *Comparison of ratio of actual moment to moment based on equation 1 for results of* Donnelly (1958), Taylor (1936), Wendt (1933) *and authors' experiments*

$R = \dfrac{\Omega_1 R_1 d}{\nu}$	η = 0.5		η = 0.675		η = 0.85			η = 0.925		η = 0.935		η = 0.95		
	Donnelly	Experiment No.6	Wendt	Experiment No.31	Taylor	Wendt	Experiment No.41	Taylor	Experiment No.65	Wendt	Experiment No.75	Donnelly	Taylor	Experiment No.84
120	1.36	1.54	1.60	1.46										
200	1.70	1.90	2.20	1.94						1.45	1.30			
300	1.97	2.13	2.63	2.40								1.70	1.90	1.66
500						3.05	2.90			2.60	2.48	2.22	2.34	2.18
700														
850										2.96		2.30	2.58	2.42
1000					4.20	4.50	4.20			3.40	3.06			
1222									3.39					
1500					5.20	5.35	5.15							
2000					6.0	5.95	6.15							

6. Conclusions

The experiments which have been discussed in the previous paragraphs showed that for $\eta \leq 0.75$ the Taylor cells that were formed were axisymmetric and remained so up to many times Ω_c. The experimental value of the critical Taylor number was found for several radius ratios. The ratio G/F was found for these cases. For $\eta = 0.5$, G/F is approximately 2.05. As Ω is increased well beyond the critical speed, the speed-torque curve has a positive curvature.

For $\eta \geq 0.85$ the axisymmetric Taylor cells change into an asymmetric mode prior to $1.4\Omega_c$. The ratio of the speeds at which azimuthal waves and axisymmetric cells appear, Ω'_c/Ω_c, was determined for five values of η. For $\eta = 0.95$, G/F is approximately 3.25. Between Ω_c and Ω'_c the slope of the speed-torque curve is, within experimental accuracy, a straight line. At speeds just in excess of Ω'_c the slope of the speed-torque curve is less than it was between Ω_c and Ω'_c. The transition at Ω'_c is reversible but for higher speeds the speed-torque relationship does not appear to be unique. The wave number for the initiation of asymmetric waves is less than π.

Acknowledgment

The efforts of the authors in completing these experiments were materially aided by the deft work of Messrs. B. Bourland and W. Huizenga. They kept the various components in operating condition and were ever ready to make the minor modifications and additions that were necessary to continue the research. The authors were fortunate to be able to cooperate with Dr. L. W. Wolf on his Rackham Research grant. This grant had as its purpose the development of a close interplay between digital computation and experimentation.

References

Caldwell, D. R., and R. G. Donnelly 1962: Proc. Royal Soc. A 267, 197—205.

Coles, D. 1965: J. Fluid Mech. 21, 385.

—, and C. Van Atta 1966: J. Fluid Mech. 25, 513.

Donnelly, R. J. 1950: Proc. Roy. Soc. A 246, 312.

—, and K. Schwarz 1965: Proc. Roy. Soc. A 283, 531—556.

Donnelly, R. J., and N. J. Simon 1960: J. Fluid Mech. 7, 401.

Pai, S.-I. 1943: NACA TN 892.

Ritchie, G. S. 1968: J. Fluid Mech. 32, 137.

Schultz-Grunow, F., and H. Hein 1956: Z. Flugwiss. 4, 28—30.

Taylor, G. I. 1936: Proc. Roy. Soc. A 157, 546—564.

Wendt, F. 1933: Ing. Arch. 4, 577—595.

One class of nonlinear stochastic differential equations characterized by random excitation

J. Drexler and **O. Kropáč**

Aeronautical Research and Test Institute, Prague

Introduction

Some years ago, a considerable effort has been concentrated in the Aeronautical Research and Test Institute in Prague on the study of fatigue phenomena characterized by random nature of environmental forces acting on elements of a construction or on constructions as a whole. Good experiences obtained while realizing simple and reliable self-excited systems for constant amplitude fatigue tests, have given rise to attempts of designing the random amplitude tests in an essentially similar way. The basic idea of generating the random processes consists then in alternating two subsequent test system states in random time intervals by applying a pseudotelegraphic signal [1], [2]; in one state, the damping term of the system is positive, in the other state, the damping term is negative (i.e. exciting).

Stability and stationarity investigations of such kind of random processes in the theoretical field [3] as well as on an analog computer [4] resulted in introducing a nonlinear feed-back in the damping term, proportional to the instantaneous value of the process envelope. During the exploitation of a real testing system further improvements have been made which make it possible to realize random amplitude tests with a priori prescribed probability of alternating two different random stress processes or alternating one random and one harmonic processes as well.

The mathematical interpretation of the attempts just mentioned may be expressed by a second order nonlinear stochastic differential equation characterized by random parametric excitation, as follows

$$\ddot{y}(t) + 2\omega_n \cdot F\{\psi(f_0, p; t); \varphi(f_0^*, q; t); A[y]\} \cdot \dot{y}(t) + \omega_n^2 \cdot y(t) = 0 \qquad (1)$$

representing thus a special class of differential equations being of an important technical use. The symbols used denote:

$y(t)$ the random stress process in the critical point of the examined specimen,

ω_n the natural circular frequency of the linear part of the system,

$\psi(f_0, p; t), \varphi(f_0^*, q; t)$ random pseudotelegraphic signals consisting of rectangular pulses having unit amplitudes and binomial distributions in their time durations. Both signals are generated by a special digital generator of binomial pulses, the occurence probability of which may be changed by probability transformers. The sampling frequences f_0, f_0^* may be present as well,

$A[y]$ the envelope of the random process $y(t)$.

Problem statement

For further investigation of the class of Eqs. (1), let us now consider the simplest one, for which the functional F is expressed by a linear form among the system parameters, as follows

$$F\{\} = \zeta_0 + c \cdot \psi(p, f_0; t) + k \cdot A[y], \qquad (2)$$

where ζ_0 is a dimensionless damping parameter of the testing system as a whole (i.e. including the non-negative specimen damping term, too), c is the constant amplitude of the pseudo-telegraphic signal $\psi(t)$, the latter consisting of rectangular pulses $+1$ and -1 (see Fig. 1), k is the feedback gain. Hence, the equation under question may be written in the form

$$\ddot{y} + 2\omega_n \cdot \{\zeta_0 + c \cdot \psi(p, f_0; t) + k \cdot A[y]\} \cdot \dot{y} + \omega_n^2 \cdot y = 0. \tag{3}$$

The parameters ω_n, ζ_0, k describe the configuration of the testing device, including the specimen; the amplitude c may be chosen within some predetermined limits. Let us further note, that $\zeta_0 \ll 1$ in the problem under question; consequently, one may assume, that the solution of Eq. (3) will be given by a narrow-band process.

The aim of the outmost importance for fatigue tests applications is now to find out such values of adjustable parameters c, p, f_0 and k, for which the envelope of the solution of Eq. (3) is a stationary random process characterized by first probability density approximating the Rayleigh or generalized Rayleigh densities.

Defining the envelope of the solution of Eq. (3) to be [5]

$$A[y(t)] = + \sqrt{y^2 + \left(\frac{\dot{y}}{\omega_n}\right)^2}, \tag{4}$$

the problem changes in that of determining the solution of a nonlinear first order stochastic differential equation

$$\dot{A}(t) = -\omega_n \{\zeta_0 + c \cdot \psi(p, f_0; t) + k \cdot A(t)\} \cdot A(t). \tag{5}$$

Taking into account, that Rayleigh probability densities are at most two-parametric ones and that no further postulations are laid upon the envelope correlation function, the solution of Eq. (5) may be limited to estimating the first moment to origin and the second central moment of the envelope only.

Estimation of the mean value and the variance of the stationary envelope process

After piece-wise integration of Eq. (5) for the r-th realization of the process $A(t)$ and applying the mean-value operators in respect to time and samples, resp., we get the expression for the mean value of the stationary envelope in the form

$$M[A] = \frac{1}{k} \cdot [(1 - 2p) \cdot c - \zeta_0]. \tag{6}$$

For determination of the envelope variance let us first express the solution of the Eq. (5) in the form

$$A_{(+1)}(t) = A_0 \frac{(c + \zeta_0) \cdot \exp\{-(c + \zeta_0) \cdot \omega_n \cdot (t - t_0)\}}{(c + \zeta_0) + k \cdot A_0 \cdot [1 - \exp\{-(c + \zeta_0) \cdot \omega_n \cdot (t - t_0)\}]} \tag{7}$$

assumed, $\psi(t)$ takes the value $+1$ in the time interval $\langle t_0, t\rangle$ and/or

$$A_{(-1)}(t) = A_0 \frac{(c - \zeta_0) \cdot \exp\{+(c - \zeta_0) \cdot \omega_n \cdot (t - t_0)\}}{(c - \zeta_0) - k \cdot A_0 \cdot [1 - \exp\{+(c - \zeta_0) \cdot \omega_n \cdot (t - t_0)\}]} \tag{8}$$

assumed, $\psi(t)$ takes the value -1 in the time interval $\langle t_0, t\rangle$ respectively.

From the formal point of view, it will be advantageous to use linearized forms

$$\left.\begin{array}{l} A_{(+1)}(t) = A_0[1 - (c + \zeta_0 + kA_0)\,\omega_n(t - t_0)] = A_0[1 - \alpha\omega_n(t - t_0)] \\ A_{(-1)}(t) = A_0[1 + (c - \zeta_0 - kA_0)\,\omega_n(t - t_0)] = A_0[1 + \beta\omega_n(t - t_0)] \end{array}\right\} \tag{9}$$

approximating Eqs. (7) and (8) the better, the higher is the sampling frequency f_0.

Considering the sequence of time instants $t_0, t_1, \ldots, t_{2l}$, which correspond to $+1$, -1 amplitude changes of the function $\psi(t)$ and vice versa, one may easily find the relative amplitude sequence $A_0, A_1, \ldots, A_{2l}$ interpreted by the form

$$A_{2l}^{(r)} = A_0^{(r)} \cdot \prod_{i=1,3,}^{2l-1} (1 + \beta_i^{(r)}\omega_n\tau_i^{(r)}) \cdot \prod_{j=0,2,}^{2l-2} (1 - \alpha_j^{(r)} \cdot \omega_n \cdot \tau_j^{(r)}); \quad \tau_k = t_k - t_{k-1}; \quad k = i.\,j. \tag{10}$$

where the index (r) refers again to the r-th realization of the envelope process. Equation (10) may be simplified assuming to be $\omega_n \beta_i \tau_i \ll 1$, $\omega_n \alpha_j \tau_j \ll 1$ as follows (see Fig. 1)

$$A_{2l}^{(r)} = A_0^{(r)} \left[1 + \sum_{i=1,3}^{2l-1} \omega_n \beta_i^{(r)} \tau_i^{(r)} - \sum_{j=0,2}^{2l-2} \omega_n \alpha_j^{(r)} \tau_j^{(r)} \right]. \tag{11}$$

For the stationary part of the solution of Eq. (5), the condition

$$A_0^{(r)} = M[A] \tag{12}$$

may be derived.

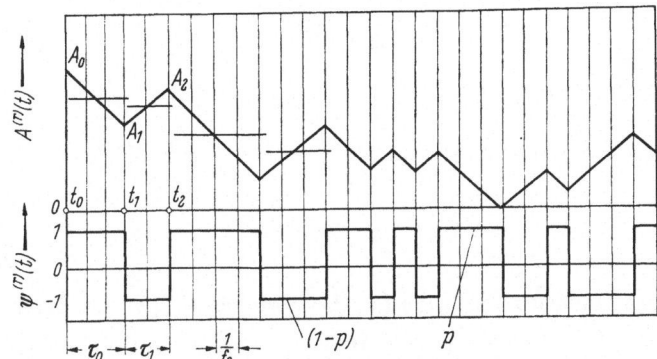

Fig. 1. The graphical interpretation of the simplified expressions (9) ÷ (11) for the random envelope process. Below the corresponding pseudotelegraphic signal.

Let us take now the centralized random quantities $A_{2l}^{0(r)} = A_{2l}^{(r)} - M[A]$, i.e. quantities

$$A_{2l}^{0(r)} = A_0^{(r)} \omega_n \left[\sum_{i=1,3}^{2l-1} \beta_i^{(r)} \tau_i^{(r)} - \sum_{j=0,2}^{2l-2} \alpha_j^{(r)} \cdot \tau_j^{(r)} \right] \tag{13}$$

and similarly

$$A_{2l+m}^{0(r)} = A_0^{(r)} \omega_n \left[\sum_{i=1,3}^{2l+m-1} \beta_i^{(r)} \cdot \tau_i^{(r)} - \sum_{j=0,2}^{2l+m-2} \alpha_j^{(r)} \cdot \tau_j^{(r)} \right]. \tag{14}$$

Then the product $A_{2l}^{0(r)} \cdot A_{2l+m}^{0(r)}$ may be expressed in the form

$$A_{2l}^{0(r)} \cdot A_{2l+m}^{0(r)} = M^2[A] \cdot \omega_n^2 \left\{ \sum_{i=1,3}^{2l-1} \beta_i^{(r)} \cdot \tau_i^{(r)} \cdot \sum_{i=1,3}^{2l+m-1} \beta_i^{(r)} \cdot \tau_i^{(r)} - \sum_{i=1,3}^{2l-1} \beta_i^{(r)} \cdot \tau_i^{(r)} \cdot \sum_{j=0,2}^{2l+m-2} \alpha_j^{(r)} \cdot \tau_j^{(r)} \right.$$

$$\left. - \sum_{i=1,3}^{2l+m-1} \beta_i^{(r)} \cdot \tau_i^{(r)} \cdot \sum_{j=0,2}^{2l-2} \alpha_j^{(r)} \cdot \tau_j^{(r)} + \sum_{j=0,2}^{2l-2} \alpha_j^{(r)} \cdot \tau_j^{(r)} \cdot \sum_{j=0,2}^{2l+m-2} \alpha_j^{(r)} \cdot \tau_j^{(r)} \right\}. \tag{15}$$

According to the definition, the variance $D_{2l}^{(r)}[A]$ relative to the r-th realization and the time interval $2l$ may be written as

$$D_{2l}^{(r)}[A] = M_{2l}^{(r)}[A_{2l}^0 \cdot A_{2l+m}^0]\big|_{m=0}. \tag{16}$$

We then get (bearing in mind, that the α's and β's are correlated)

$$D_{2l}^{(r)}[A] = M^2[A] \cdot \omega_n^2 \left\{ M_{2l}^{(r)}[\beta^2] \cdot M_{2l}^{(r)}[\vartheta_i^2] - 2 M_{2l}^{(r)}[\alpha\beta] \cdot M_{2l}^{(r)}[\vartheta_i] \cdot M_{2l}^{(r)}[\vartheta_j] \atop + M_{2l}^{(r)}[\alpha^2] \cdot M_{2l}^{(r)}[\vartheta_j^2] \right\}, \tag{17}$$

where

$$\vartheta_i = \sum_{i=1,3}^{2l-1} \tau_i; \quad \vartheta_j = \sum_{j=0,2}^{2l-2} \tau_j. \tag{18}$$

From the definition of the process $\psi(t)$,

$$M_{2l}^{(r)}[\vartheta_i] = \frac{N^{(r)}}{f_0}(1-p); \quad M_{2l}^{(r)}[\vartheta_j] = \frac{N^{(r)}}{f_0} \cdot p, \tag{19}$$

where $\dfrac{N^{(r)}}{f_0} = T^{(r)}$ is the observation time of the r-th realization. Let us write

$$M_{2l}^{(r)}[\vartheta_i^2] = M_{2l}^{(r)^s}[\vartheta_i] \cdot (1 + \varepsilon^2[\vartheta_i]),$$
$$M_{2l}^{(r)}[\vartheta_j^2] = M_{2l}^{(r)^s}[\vartheta_j] \cdot (1 + \varepsilon^2[\vartheta_j]), \tag{20}$$

where $\varepsilon[\,]$ is the operator of the variance ratio.

It may be shown that

$$\varepsilon_{2l}^2[\vartheta_i] = \frac{1}{N^{(r)}} \cdot \frac{1-p}{p}$$

$$\varepsilon_{2l}^2[\vartheta_j] = \frac{1}{N^{(r)}} \cdot \frac{p}{1-p} .$$

(21)

Hence, for $N^{(r)} \to \infty$

$$\left. \begin{aligned} M_{2l}^{(r)}[\vartheta_i^2] \doteq M_{2l}^{(r)^2}[\vartheta_i] = \frac{N^{(r)^2}}{f_0^2} \cdot (1-p)^2 \\ M_{2l}^{(r)}[\vartheta_i^2] \doteq M_{2l}^{(r)^2}[\vartheta_i] = \frac{N^{(r)^2}}{f_0^2} \cdot p^2 \end{aligned} \right\} .$$

(22)

Equation (17) may then be written in the form

$$D_{2l}^{(r)}[A] = M^2[A] \cdot 4\pi^2 \cdot \frac{f_n^2}{f_0^2} \cdot M_{2l}^{(r)}[\beta^2] \cdot (1-p)^2 \cdot N^{(r)^2} \cdot \left\{ 1 - 2 \cdot \frac{M_{2l}^{(r)}[\alpha\beta]}{M_{2l}^{(r)}[\beta^2]} \cdot \frac{p}{1-p} \right. $$
$$\left. + \frac{M_{2l}^{(r)}[\alpha^2]}{M_{2l}^{(r)}[\beta^2]} \cdot \frac{p^2}{(1-p)^2} \right\}$$

(23)

The stationarity condition for the variance $D_{2l}^{(r)}[A]$ has then to take the form

$$\left\{ 1 - 2\frac{M_{2l}^{(r)}[\alpha\beta]}{M_{2l}^{(r)}[\beta^2]} \cdot \frac{p}{1-p} + \frac{M_{2l}^{(r)}[\alpha^2]}{M_{2l}^{(r)}[\beta^2]} \cdot \frac{p^2}{(1-p^2)} \right\} = \frac{1}{N^{(r)^2}}$$

(24)

for arbitrary $N^{(r)}$. As $N^{(r)}$ increases, the value of the expression in brackets in Eq. (24) should approach zero.

In the problems investigated, approximate expressions may be taken in the form

$$M[\alpha^2] = M^2[\alpha] \cdot (1 + \varepsilon^2[\alpha]) \doteq M^2[\alpha]; \quad \varepsilon^2[\alpha] \ll 1$$
$$M[\beta^2] = M^2[\beta] \cdot (1 + \varepsilon^2[\beta]) \doteq M^2[\beta]; \quad \varepsilon^2[\beta] \ll 1$$
$$M[\alpha\beta] = M[\alpha] \cdot M[\beta] \cdot (1 + \varepsilon[\alpha] \cdot \varepsilon[\beta]) \doteq M[\alpha] \cdot M[\beta].$$

(25)

Hence, one obtains

$$\frac{M[\beta]}{M[\alpha]} = \frac{p}{1-p}$$

(26)

and after substitution for α and β respectively, the expression for the mean value of the stationary envelope

$$M[A] = -\frac{\zeta_0}{k}$$

is found, which is the same as that one obtained in the Eq. (6) for $p = 0.5$. Thus, for $p = 0.5$, one finds after corresponding substitution the final expression for the variance of the stationary envelope

$$D[A] = \left(\pi \cdot c \cdot \frac{f_n}{f_0} \right)^2 \cdot M^2[A].$$

(27)

Approximations to Rayleigh probability densities of the stationary envelope

From the theory of linear filtering of random signals [6] it is well known, that under specific assumptions when filtering the pseudotelegraphic signal for $p = 0.5$, the output signal probability density approaches the Gaussian law. Supposing k to be sufficiently small, one may assume, that it may approximately be the same case for the solution of the Eq. (5), too, the envelope being distributed according to Rayleigh as consequence.

The Rayleigh probability densities of interest may be derived from the generalized expression

$$W_1[A] = \begin{cases} \dfrac{A}{D[y]} \cdot \exp\left\{ -\dfrac{A^2 + A_0^2}{2D[y]} \right\} \cdot I_0 \left[\dfrac{A \cdot A_0}{D[y]} \right] & \text{for} \quad A > 0 \\ 0 & \text{for} \quad A < 0 \end{cases}$$

(29)

where I_0 is the Bessel function of the first kind of the imaginary argument, in the form

$$W_1[A] = \frac{A}{D[y]} \cdot \exp\left\{-\frac{A^2}{2D[y]}\right\} \tag{30}$$

assuming to be $\frac{A_0^2}{D[y]} \to 0$ or

$$W_1[A] \doteq \frac{1}{\sqrt{2\pi \cdot D[y]}} \cdot \exp\left\{-\frac{(A-A_0)^2}{2D[y]}\right\} \cdot \left(1 + \frac{D[y]}{8A \cdot A_0}\right) \cdot \sqrt{\frac{A}{A_0}}; \quad A > 0 \tag{31}$$

assuming $\frac{A_0^2}{D[y]} \gg 1$.

Let us examine the case of Eq. (30). The corresponding variance ratio

$$\varepsilon[A] = \frac{\sqrt{D[A]}}{M[A]} = \sqrt{\frac{4-\pi}{\pi}} = 0{,}523 \tag{32}$$

interpretes the RAYLEIGH density to be one-parametric one. The same variance coefficient may be expressed from Eqs. (27) and (28) in the form

$$\varepsilon[A] = \pi \cdot c \cdot \frac{f_n}{f_0}. \tag{33}$$

Comparing (32) with (33), one obtains the condition for the stationary envelope to be of the Rayleigh type, i.e. the condition

$$c \cdot \frac{f_n}{f_0} = 0{,}166 \tag{34}$$

which relates the parameters c and f_0 respectively of the investigated system.

The question may now arise about the condition for which the density probability (30) is not censored, the variance width being obviously determined by the maximum obtainable value A_{max}. For $\dot{A}(t) = 0$, one may find from Eq. (3) the variance width of $A(t)$ to be

$$\left\langle 0; \frac{c-\zeta_0}{k}\right\rangle. \tag{35}$$

It is useful to express the determined width by a multiple n_A of the root mean square value of the envelope $A(t)$; from Eqs. (28) and (35) one arrives to the last condition for the parameter c in the form

$$c = (0{,}523 n_A + 1) \cdot \zeta_0. \tag{36}$$

In the most technical problems, n_A is sufficient to be from 3 to 4. In such cases, which after all form the limit of technical possibilities of the generators of random processes, the following conditions are to be satisfied:

$$\left.\begin{array}{l} c = (2{,}57 \div 3{,}1) \cdot \zeta_0 \\ k = (0{,}51 \div 0{,}677) \end{array}\right\}. \tag{37}$$

For simulating the generalized RAYLEIGH probability density (31) the corresponding variance width of the envelope may be controlled by setting the parameters ζ_0 and c in such way, that inequalities $\zeta_0 < 0$, $|\zeta_0| > c$ are satisfied (see Fig. 5 in the next paragraph).

Some remarks to the investigated class of stochastic differential equations

The class of differential equations discussed in this paper is characterized by random parametric excitation, whereby the damping term is changing with the amplitude. This term can be graphically interpreted in the damping term—amplitude diagram. Such a diagram for the system described by Eq. (3) is shown on Fig. 2, where the geometric relations of the parameters of the damping term with the characteristic values of the envelope such as $M[A]$ and A_{max} are evident. (Because a linearized model was used for the estimation od $D[A]$, no similar graphical interpretation of $D[A]$ or $\sigma[A]$ can be shown.)

When choosing other shapes of the damping term—amplitude diagram, a set of modifications of a given type of system can be formulated. E.g. it is evident, that for $k = 0$, a linear system is established working just on the stability limits (see Fig. 3). The stabilization by the addend $k \cdot A$ in Eq. (3) is performed in forming a maximum value A_{max} on the intersection of the exciting branch with zero damping term line. As a more generalized case, the following system can be demonstrated having the exciting branch of the form $\{\zeta_0 - c + k_e \cdot A\}$ with

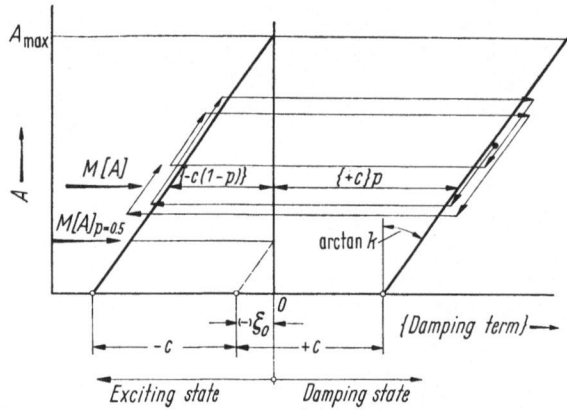

Fig. 2. Damping term — envelope diagram: Graphical representation of the variable damping term and the relation of its parameters to the significant values of the envelope A_{max} and $M[A]$, with a sample of the random process A (p = occurence probability of the damping state).

the occurence probability $(1 - p)$ and the damping branch described by the expression $\{\zeta_0 + c + k_d A\}$ with occurence probability p, where generally $k_d \neq k_e$ and k_d may be $k_d \gtreqless 0$ (see Fig. 4).

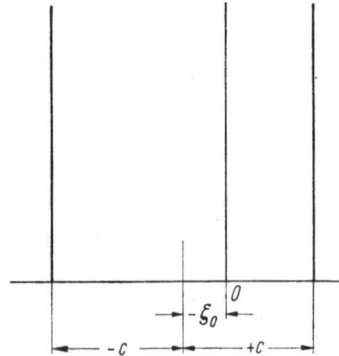

Fig. 3. Damping term — envelope diagram for the linear model.

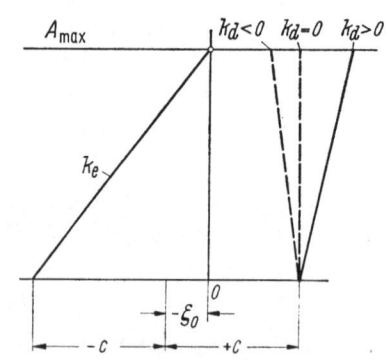

Fig. 4. Generalized damping term — envelope diagram.

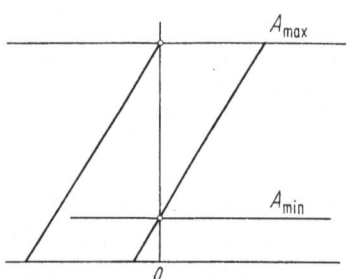

Fig. 5. Diagram alternative giving $A_{min} > 0$.

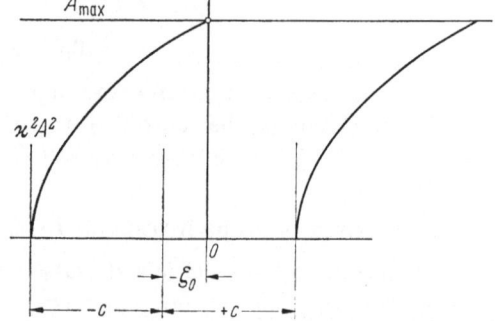

Fig. 6. Damping term — envelope diagram for the $\varkappa^2 A^2$ feed-back.

If the damping branch intersects the zero damping term line, a minimum value $A_{min} > 0$ can be reached (see Fig. 5) giving for $p = 0.5$ the probability density approaching the generalized Rayleigh function. The dependence of the damping term on A need not be linear. As an example, the damping term of the form $\{\zeta_0 + c \cdot \psi + \varkappa^2 A^2\}$ may be shown (see Fig. 6). This case for small values of A approaches the linear system while for large values of A has a character given in Fig. 2.

Some results of the analog study of the problem

Besides the analytical investigations, an analog study of the problem has been performed giving as well the initial informations as some stimulations for the design of real systems.

The analog experiments were arranged to form a complete factorial scheme, the choise of the level combinations for the parameters ζ_0, k and c being in accordance with practical requirements.

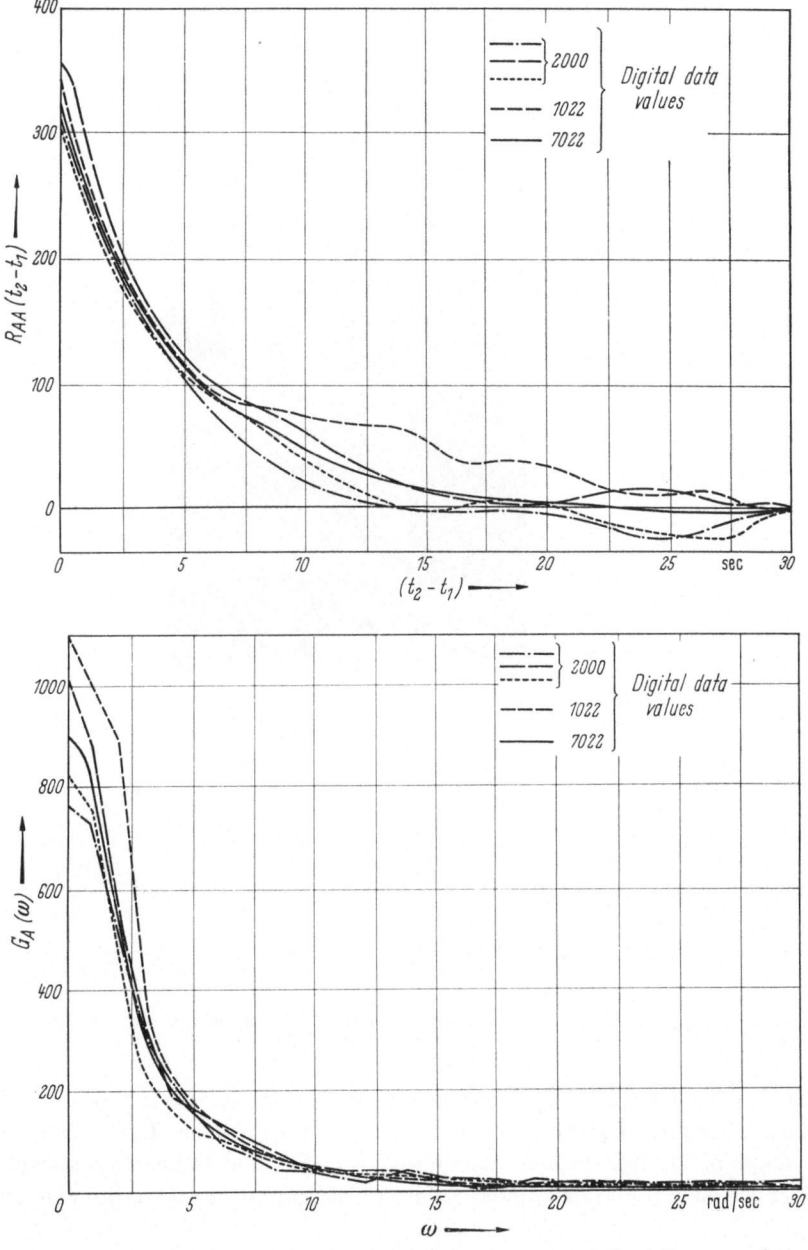

Fig. 7. Autocorrelation function (above) and power spectral density (below) of the stationary envelope process.

The description of the analog equipment and the results obtained have been summarized in a previous paper [4]. One could state, that the stationarity of the envelope process has been proved in respect to the first four moments of the first probability density function. An additional information to this statement is given on Fig. 7, where the autocorrelation functions of the individual samples as well as the corresponding power spectral densities are shown. The

186   J. Drexler and O. Kropáč

partial samples 1 to 4 contained at about 2000 digital values each, producing only random deviations as compared with the correlation function (power spectral density) calculated from the whole record available (7022 digital values).

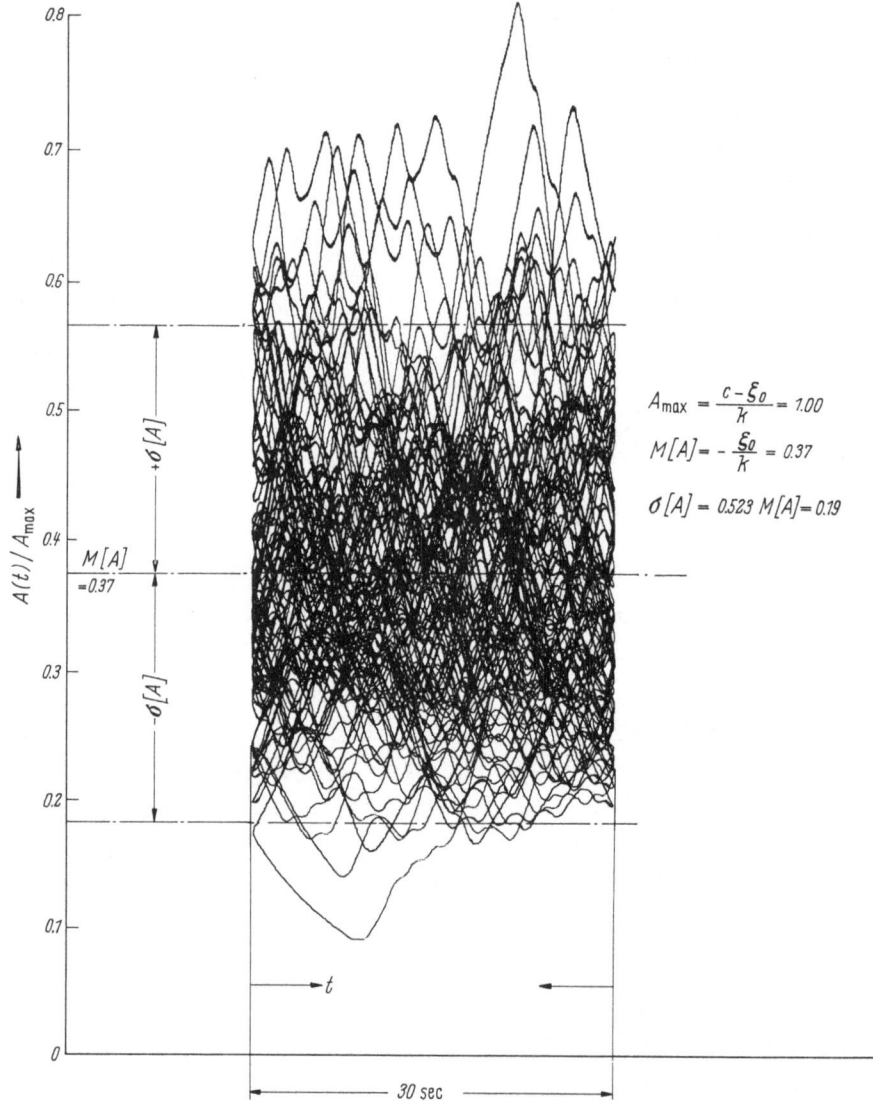

Fig. 8. A short-time realization of the envelope process as obtained on the analog computer. Observation time: $T = 60$ min; Over-lapping time: $T_0 = 30$ sec; Parameters: $p = 0.5$; $f_0 = 1.25$ Hz; $2\pi f_n = 20$ rad/sec; $c = 0.0628$; $\zeta_0 = -0.0372$; $k = 0.1$.

As an other example of interest, a graphical picture of the envelope process $A(t)$ is given on Fig. 8, obtained by manifold overlapping of the recorded realization. The picture presents some idea on the shape of the first probability density function and indicates a satisfactory agreement of the experimental and calculated values of mean and mean square characteristics.

Experimental results obtained on a real system

The conclusions of both analytical and analog investigations were proved on a real two-mass dynamic loading system; the results of this experimental activity are given in the following figures.

The block diagram of the testing device in question including the specimen, the electro-mechanical and electronic parts is shown in Fig. 9. The generator of random process producing rectangular pulses "0" and "1" actuates the integrated electro-mechanical system in a manner described by the equation

$$\dot{A} + \{\zeta_1 \cdot \omega_1 + c[\psi(p, f_0; t) - c_0] + k \cdot A\} \cdot A \doteq 0; \quad A \doteq -\frac{m_2}{m_1} \cdot B. \tag{38}$$

On Fig. 10, a record sample is shown illustrating the transition part of the envelop stress process $B(t)$ starting from an initial value considerably differring from the stationary mean value ($B_0 < B_{\min}$). Bellow, the record of the random pseudotelegraphic signal consisting of pulses "0" and "1" may be seen, the sampling frequency of which being $f_0 = 3.1$ Hz.

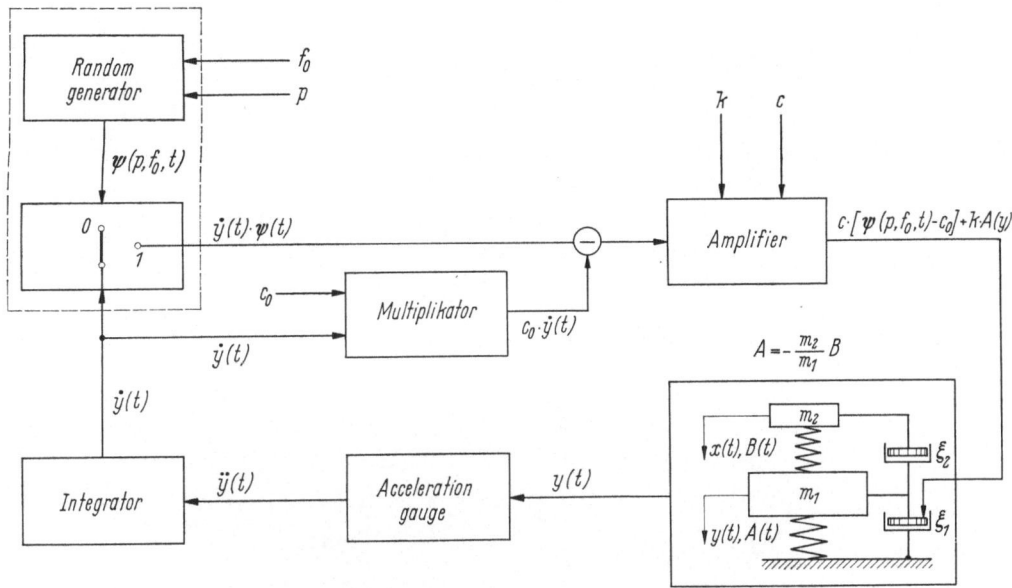

Fig. 9. Block-diagram of a real dynamic testing system.

On Fig. 11, the stress record samples of the same specimen are given, differring in sampling frequencies only ($f_0 = 3.1$ Hz and $f_0 = 50$ Hz respectively). In all mentioned cases, the value of probability occurence $p = 0.5$ was chosen. The influence of the occurence probabilities p relating to the "1" pulses on the resulting process is demonstrated on Fig. 12, these probabilities being $p = 0.25$ and $p = 0.75$ respectively.

The distribution functions of the envelope $B(t)$ of some stress processes are plotted in the Rayleigh probability paper on Fig. 13 showing acceptable agreement with the generalized Rayleigh distribution.

An example of the random alternating of two independent stationary random processes, thus producing a piece-wise stationary (i.e. with a prescribed occurence probability) random process, is shown in Fig. 14. Special reference is made to transition processes occuring after switching over the two given stationary states of the specimen. The corresponding occurence probability q was equal to 0.5, the amplitude frequency $f_0^* = 0.025$ Hz. In this case, the damping term functional F (see Eq. (1)) has the form

$$F\{\} = \{\zeta_0 + [\psi(p, f_0; t) + c_0] \cdot [(\varphi(q, f_0^*; t) + k_0) \cdot k \cdot B[x(t)] - c]\}. \tag{39}$$

In the last Fig. 15, an example of a stress process $B(t)$ is given in which two different types of processes are randomly alternated with an a priori described occurence probability. One of the processes is a stationary random process, the envelope of which has the generalized Rayleigh distribution, the second being a harmonic process with the mean square power corresponding to the equivalence $B_{\mathrm{harm}} = M[B]$. Note the presence of the transition zone in the upper part (passage from the random process to the harmonic one) and the absence of that zone in the lower part of the figure due to the mentioned power equivalence. A remark should be made

24*

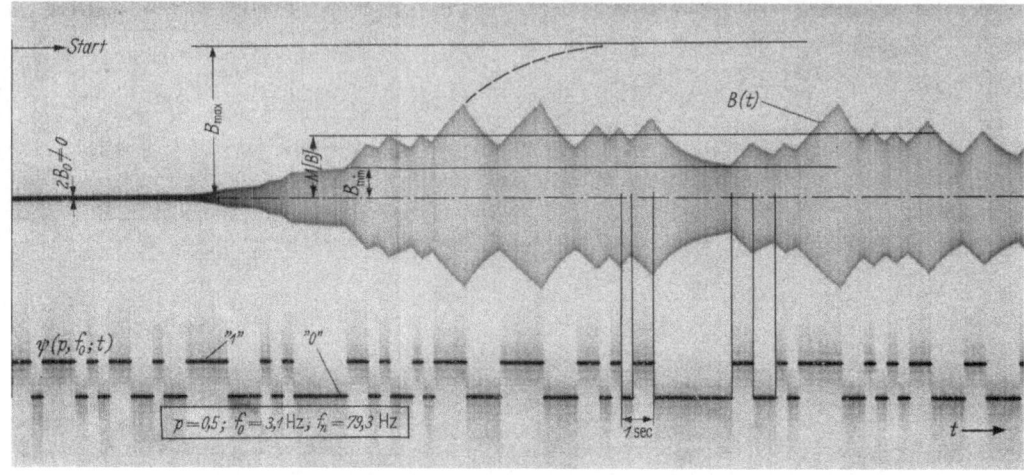

Fig. 10. Record sample of the random stress in the critical point of an investigated test specimen.

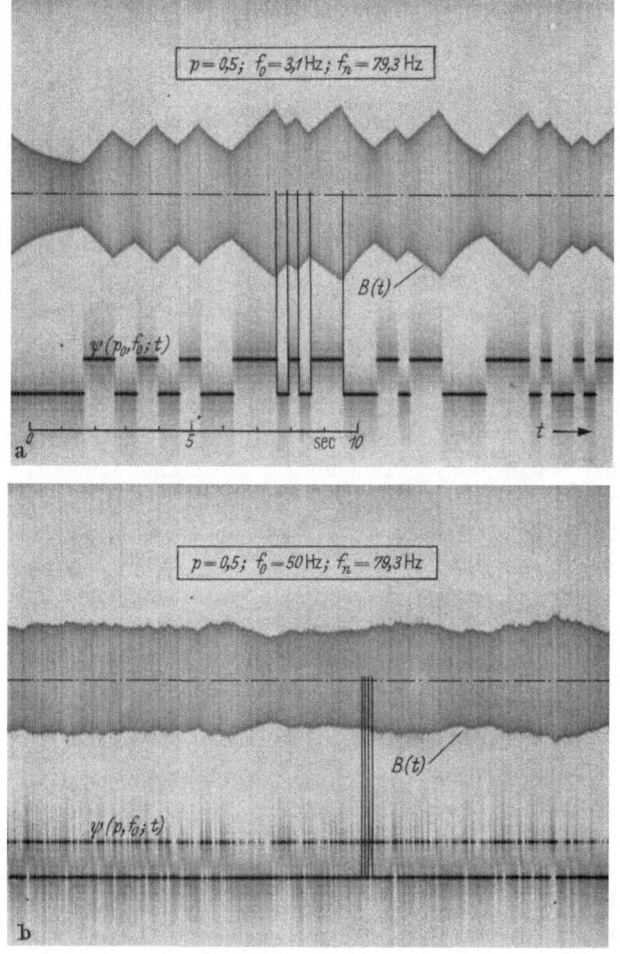

Fig. 11. Influence of the sampling frequency on the character of the envelope process.

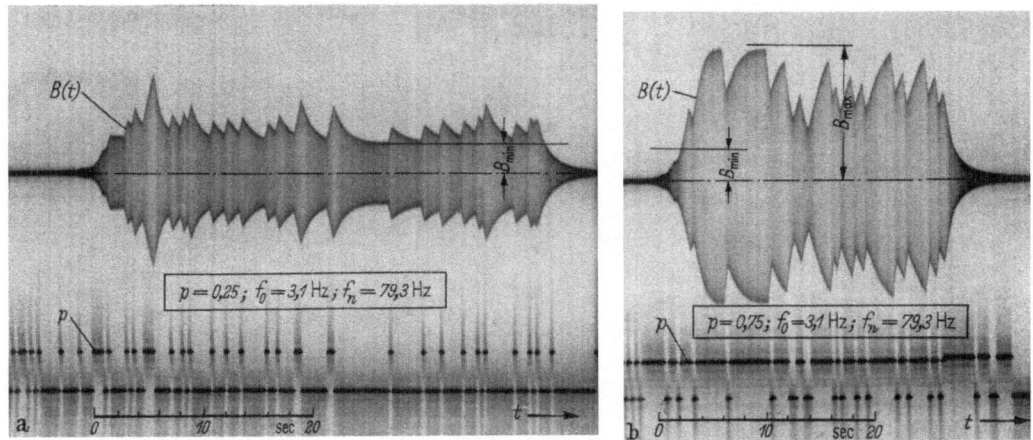

Fig. 12. Influence of the occurence probability of the "0" and "1" pulses on the character of the envelop process.

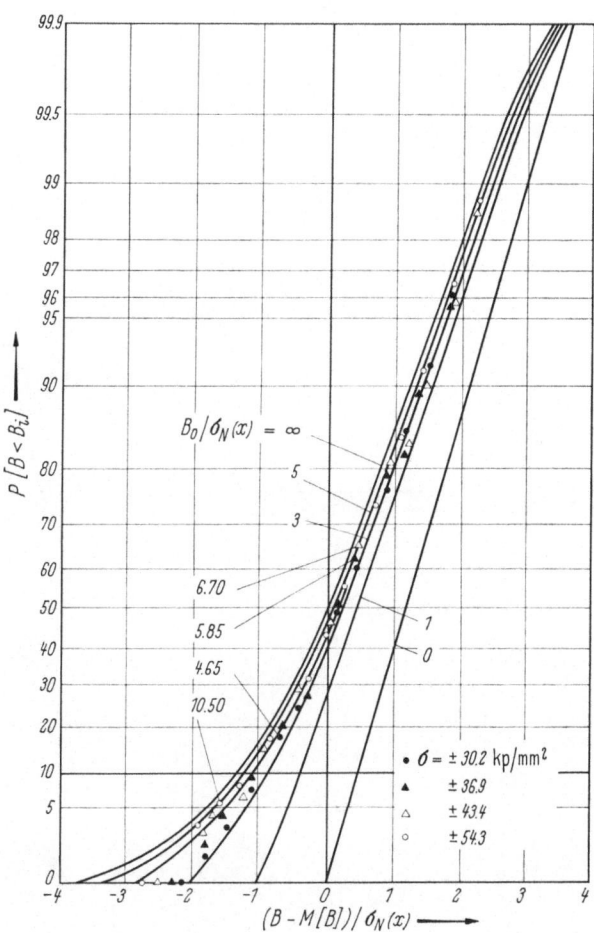

Fig. 13. The envelope distribution functions in the Rayleigh probability paper.

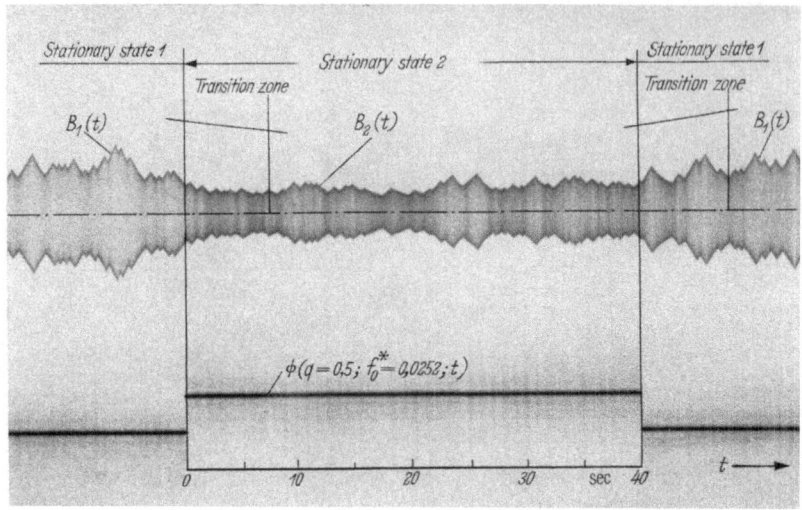

Fig. 14. Record sample of a piece-wise stationary random process (two randomly alternated independent random processes).

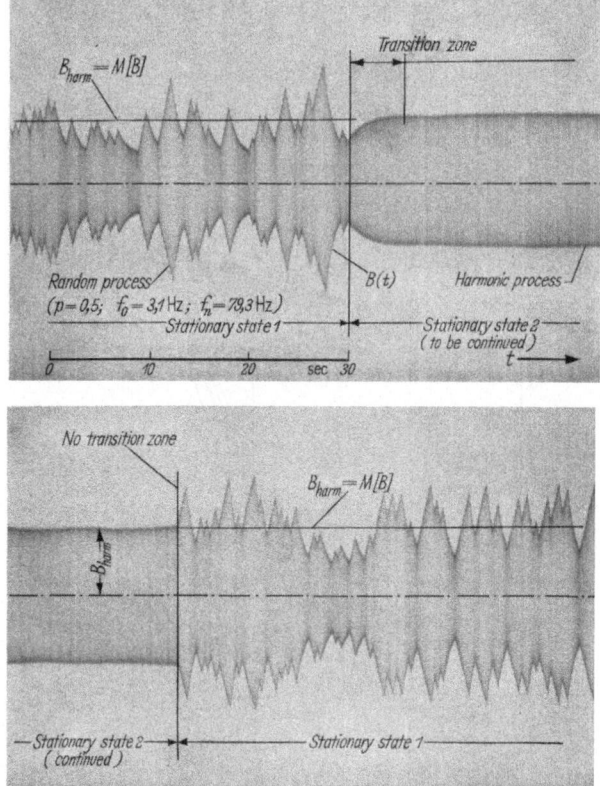

Fig 15. Record sample of a non-homogeneous random stress process (state 1: stationary random process; state 2: harmonic process).

on the importance of such kinds of stress processes when investigating fatigue damage phenomena. In one complex research program, the occurence probabilities p_{harm} relative to the harmonic process were set up according to the following values: 0; 0.0625; 0.125; 0.25; 0.50; 0.75; 1.00. The corresponding damping term functional takes here the form

$$F\{\} = \{\zeta_0 + \left([\psi(p, f_0; t) + c_0] \cdot \varphi(q, f_0^*; t) + c_1\right) \cdot [k \cdot \dot{B}[x] - c]\}. \tag{40}$$

Conclusion

The class of nonlinear stochastic differential equations some of which were discussed in this paper has been formulated while solving the problem of generating special random loading processes in a simple technical way. The simplicity as well as the corresponding reliability of the technical realizations had to be paid by an increased effort in the theoretical and experimental fields of the problem.

An exact mathematical solution could be achieved for the mean value (first moment to origine) of the envelope of the generated random process only, while for the estimations of the variance (second central moment) and of the probability density functions, some approximations based on linearization procedures had to be adopted giving technically acceptable compromises. It is obvious, that the discussed activities may be viewed as the first approximations to a problem of an outmost technical interest only. In this respect, further work especially in the theoretical domain is desirable concerning namely the estimations of the correlation functions and getting more informations on the transition behavior of the envelope stress process in the case of alternating the random processes differring in mean-square powers.

Nevertheless, the theoretical as well as experimental activities demonstrated in this paper have made it possible to design simple and versatile testing devices with stable performance parameters and a high reliability proved in more than 4000 operational hours up to date.

References

1. HAVEL, J.: An electronic generator of random sequences, Trans. Second Prague Conf. Inform. Theory, Prague, 1960, pp. 219—229. Devices for checking and further transformations of the output signal from the random process generator, Trans. Third Prague Conf. Inform. Theory, Prague: Academia 1964, pp. 265—276.
2. DREXLER, J., and J. HAVEL: Czechoslovak Patent Specification No. PV 1473/68.
3. BUNKE, H.: Über die fast sichere Stabilität linearer stochastischer Systeme. ZAMM 43, 12, 533—535 (1963).
4. KROPÁČ, O., and J. DREXLER: An analog study of random parametric vibration of a nonlinear dynamic second order system; Fourth Conf. on Nonlinear Oscillations, Prague, Sept. 1967, Proceedings, Prague; Academia 1968, pp. 349—359.
5. CRANDALL, S. H., and W. D. MARK: Random vibrations in mechanical systems, London: Academic Press 1963.
6. PANTELOPULOS, CH.: Processus aléatoires asymptotiquement stationnaires Laplaciennes produits par filtrage d'une suite périodique d'impulsions aléatoires, Trans. Second Prague Conf. Inform. Theory, Prague: Academia 1960, pp. 397—411.
7. TIMOSHENKO, S., and D. H. YOUNG: Vibration problems in engineering, Toronto-New York-London: Van Nostrand 1955.
8. MIDDLETON, D.: An introduction to statistical communication theory, New York: McGraw-Hill 1960.
9. KOZIN, F.: Stability of stochastic systems, IIIrd IFAC Congress, London, 1966, Paper 3A, 8 pp.
10. KOŽEŠNÍK, J.: Random vibration; Fifth Conf. on Dynamics of Machines, Liblice near Prague, October 1968, Proceedings, Prague: Academia 1968, pp. 41—57.

Célérités de fluctuations turbulentes de temperature et de vitesse dans une couche limite

Par

A. Favre, R. Dumas et E. Verollet

Institut de Mecanique Statistique de la Turbulence, Marseille

1. Introduction

Les mesures de corrélations spatio-temporelles doubles des composantes des fluctuations de vitesses ont conduit à définir une célérité dans une direction donnée [6, 7].

Les équations du mouvement peuvent s'exprimer en termes de célérités; elles expliquent les résultats expérimentaux d'après lesquels les célérités au long du mouvement moyen diffèrent en général de la vitesse de la matière, dans une couche limite [5, 6].

Notamment les célérités sont plus élevées que la vitesse de la matière près de la paroi, cet écart est encore plus grand si l'on ne considère, grâce à un filtrage en fréquence, que les fluctuations à grandes échelles dans l'espace. Par contre il semble que les célérités à petites échelles soient égales à celles de la matière.

Ces résultats ont été obtenus pour les composantes de la vitesse dans la direction du mouvement moyen; il est intéressant d'effectuer une comparaison avec ceux obtenus pour les fluctuations turbulentes de température en définissant les célérités correspondantes, en établissant leurs équations et en déterminant leurs valeurs à partir de mesures de corrélation spatio-temporelles.

Dans ce but des mesures sont effectuées dans une couche limite turbulente avec transfert thermique à la paroi. L'écart maximum de température de 25°C est suffisamment faible pour que la couche limite ne soit pas modifiée d'une façon notable du point de vue dynamique, le nombre de Richardson étant de 2×10^{-6} et celui de Grashof sur Reynolds au carré $\dfrac{Gr}{R_e^2} = \dfrac{g\beta(\Theta_0 - \Theta_e)\delta}{V_e^2}$ étant de $15 \ 10^{-5}$.

Les mesures ont été effectuées au long d'une ligne de courant à une distance de la paroi $y^+ = 66$, région où les gradients sont élevés mais où les tensions visqueuses sont négligeables. Des mesures de corrélations spatio-temporelles ont été aussi effectuées avec des décalages orthogonaux X_3 à la paroi.

2. Conditions experimentales

2.1. Dispositif de mesures

Une maquette (Fig. 1) de 3750 mm de longueur et de 800 mm d'envergure à paroi plane chauffée est disposée dans la soufflerie S.1. de l'IMST. [6]. La température Θ_0 de paroi est maintenue constante sur la surface avec un écart type de 0,5°C, sa valeur moyenne étant de 25,8°C en dessus de celle Θ_e de l'air de la soufflerie, dont la température ne varie pratiquement pas grâce à un échangeur thermique disposé en série dans le circuit de retour.

Les vitesses moyennes sont déterminées avec un tube de prise de pression totale de diamètre extérieur $\Phi = 0,8$ mm et les températures moyennes sont mesurées avec des thermocouples de 0,1 mm de diamètre.

Les mesures de fluctuations turbulentes sont effectuées avec deux fils chauds de 5μ de diamètre, dont l'un est maintenu à des distances du bord d'attaque $z_0 = 2780$ mm et de la paroi $y_0 = 1,7$ mm. Le support de ce fil est de forme annulaire pour réduire les effets du sillage sur le fil aval.

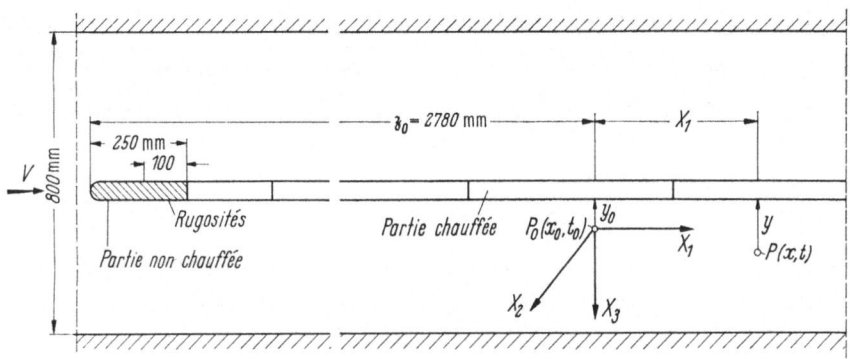

Fig. 1. Dispositif de mesures.

Des anémothermomètres à courant constant sont utilisés, les effets de non linéarités ne se manifestent pas en ce qui concerne les fluctuations de température qui de ce point de vue paraissent plus fiables que celles de vitesse.

2.2. Vitesses et temperatures moyennes

La Fig. 2 présente le profil de vitesse moyenne en fonction du logarithme du nombre $y^+ = \dfrac{yv^*}{\nu}$ à deux positions $z = 2775$ mm et 2855 mm.

Les points obtenus pour $y^+ < 58$ sont corrigés selon la méthode de MacMillan [11], et à l'aide des résultats obtenus à l'IMST [12]. La courbe tracée est le profil de vitesse proposé par van Driest [15].

A la position $z_0 = 2780$ mm les valeurs des paramètres sont:
vitesse hors de la couche limite $V_e = 16$ m/sec
épaisseur de la couche limite à $V = 0,99 V_e$, $\delta = 45,2$ mm

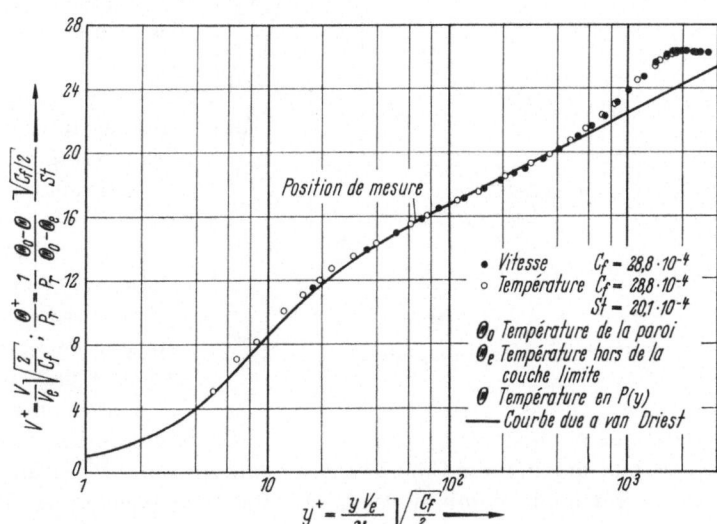

Fig. 2. Profils de vitesse et de température moyennes.

nombre de Reynolds $R_\delta = 48\,000$ (20°C)
coefficient de frottement $C_f = 28,8 \cdot 10^{-4}$
vitesse de frottement $V^* = 59$ cm/sec
nombre de Stanton $S_t = 20,1 \cdot 10^{-4}$.

Le gradient de pression statique longitudinal est négligeable.

Cette figure donne de même le profil de température moyenne $\dfrac{\Theta^+}{\mathrm{Pr}} = \dfrac{1}{\mathrm{Pr}} \dfrac{\Theta_0 - \Theta}{\Theta_0 - \Theta_e} \dfrac{\sqrt{C_f}}{S_t}$ qui coïncide pratiquement avec celui des vitesses dans le cas de ces expériences où le début de la partie chauffée est voisin du bord d'attaque.

2.3. Mesures de fluctuations turbulentes

Soient V_1' et Θ les fluctuations de la composante longitudinale de vitesse et de la température, la fluctuation de tension électrique aux bornes d'un fil chaud set selon la théorie "linéaire" [8]

$$e' = \alpha \frac{v_1'}{V_1} + \beta \frac{\theta'}{\Theta}.$$

Les coefficients α et β sont déterminés par un tarage expérimental; on définit le rapport $r = \dfrac{\alpha}{\beta}$ qui dépend de la température du fil. Au cours de ces mesures on a fait varier r toutes choses

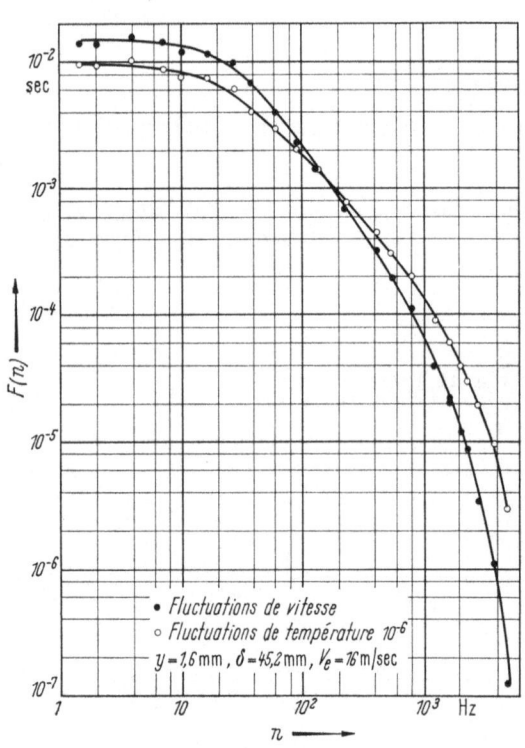

égales par ailleurs, de $0,25 \cdot 10^{-2}$ à $11,5 \cdot 10^{-2}$. Pour $r = 0,25 \cdot 10^{-2}$ le fil n'est pratiquement sensible qu'à la fluctuation thermique, la contamination en énergie de la fluctuation totale par les vitesses étant de l'ordre de 2%. Pour $r = 11,5 \cdot 10^{-2}$ la part dans la fluctuation totale de la fluctuation thermique paraît être inférieure à 35%. Toutefois des mesures de corrélation spatio-temporelle effectuées en écoulement isotherme pour un décalage $X_1 = 75$ mm ont donné des résultats voisins de ceux obtenus pour $r = 11,5 \cdot 10^{-2}$.

Dans ces conditions pour $r = 0,25 \cdot 10^{-2}$ les résultats concernent les fluctuations de température et lorsque r croît jusqu'à $11,5 \cdot 10^{-2}$ on interprêtera les résultats comme représentant principalement les effets des vitesses.

Les répartitions spectrales d'énergie normées $F(n)$ concernant les fluctuations de température et de vitesse (en écoulement isotherme) sont données dans la fig. 3.

On vérifie que les fluctuations thermiques ont une énergie relative plus forte aux fréquences moyennes et élevées que celles de la composante longitudinale de vitesse [3].

Fig. 3. Répartitions spectrales.

Dans une première série de mesure les coefficients de corrélation ont été déterminés avec l'enregistreur magnétique C.3 en utilisant un temps d'intégration de plusieurs minutes.

Des mesures plus nombreuses ont été effectuées avec l'appareil de Princeton Applied Research, qui est un calculateur à mémoire capacitive donnant 100 points de mesure et utilisé avec un temps d'intégration de 20 secondes.

La fig. 4 donne quelques exemples des 2000 courbes ainsi obtenues. Le temps optimum T_m correspondant au maximum maximorum de la corrélation est déterminé comme limite du milieu des cordes à corrélation constante.

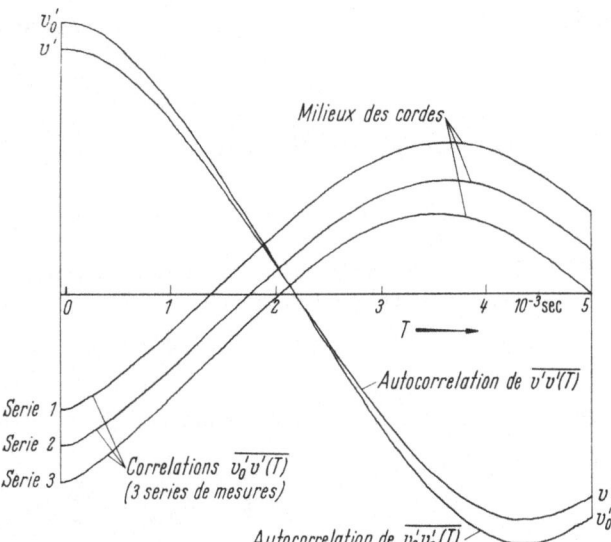

Fig. 4. Exemples de mesures de corrélations en fonction du décalage de temps T fréquence d'accord du filtre $n = 115$ hz.

3. Equations aux correlations spatio-temporelles de vitesse et de température — celerites

3.1. Conditions considérées

Considérons un écoulement d'air statistiquement stationnaire à vitesse faible (Mach $\sim 0{,}05$) et avec écart relatif de température modéré: $\dfrac{\Theta_w - \Theta_e}{\Theta_e} \approx 0{,}09$.

Les effets de compressibilité dûs aux variations de vitesse sont négligeables et les effets de dilatation dûs aux variations de température se réduisent en fait à la variation de la moyenne de la masse volumique. Ainsi dans les équations dynamiques $\varrho \approx \bar{\varrho}$.

Dans les équations de la chaleur l'expérience montrant que les fluctuations relatives de pression $\dfrac{p'}{p}$ sont négligeable (de l'ordre de 10^{-4}) il s'ensuit que d'après l'équation d'état: $\varrho\Theta \approx \overline{\varrho\Theta}$.

De plus les intensités de turbulence de température étant au maximum de l'ordre de $1{,}5 \cdot 10^{-2}$ on a $\overline{\varrho\Theta} \approx \bar{\varrho}\,\overline{\Theta}$ à $2 \cdot 10^{-4}$ près.

En ce qui concerne la viscosité μ et la conductivité k leurs fluctuations turbulentes relatives étant de l'ordre de 1% dans le cas examiné seront négligées: $k \approx \bar{k}$, $\mu \approx \bar{\mu}$. La chaleur massique C_p est pratiquement constante.

3.2. Correlations spatio-temporelles de vitesse et de température

Considérons les conditions de l'écoulement en un point $P(\underline{x})$ au temps t, et en un autre point $P_0(\underline{x}_0)$ au temps t_0. On pose, α étant un indice de direction:

$$v_\alpha = V_\alpha + v'_\alpha, \quad V_\alpha = \bar{v}_\alpha \Leftrightarrow \bar{v}'_\alpha = 0$$
$$\theta = \Theta + \theta, \qquad \Theta = \bar{\theta} \Leftrightarrow \bar{\theta}' = 0 \tag{1}$$
$$X_\alpha = x_\alpha - x_{0\alpha}, \quad T = t - t_0.$$

Les moyennes statistiques sont remplacées par les moyennes temporelles, selon l'hypothèse ergodique.

25*

3.2.1. Les équations dynamiques instantanées au point P s'écrivent:

$$\bar{\varrho}\frac{dv_\alpha}{dt} = -\frac{\partial p}{\partial X_\alpha} + \frac{\partial}{\partial X_\alpha}(f_{\alpha\gamma}) \tag{2}$$

le point $P_0(x_0)$ servant de référence des coordonnées avec

$$f_{\alpha\gamma} = -\frac{2}{3}\bar{\mu}\frac{\partial v_\alpha}{\partial X_\beta}\delta_{\alpha\beta} + \bar{\mu}\left(\frac{\partial v_\alpha}{\partial X_\gamma} + \frac{\partial v_\gamma}{\partial X_\alpha}\right). \tag{3}$$

On en déduit l'équation aux fluctuations:

$$\bar{\varrho}\frac{\partial v_\alpha'}{\partial t} + \bar{\varrho}V_\gamma\frac{\partial v_\alpha'}{\partial x_\gamma} + \bar{\varrho}v_\gamma'\frac{\partial V_\alpha}{\partial X_\gamma} + \bar{\varrho}v_\gamma'\frac{\partial v_\alpha'}{\partial X_\gamma} - \overline{\bar{\varrho}v_\gamma'\frac{\partial v_\alpha'}{\partial X_\gamma}} = -\frac{\partial p'}{\partial X_\alpha} + \frac{\partial f_{\alpha\gamma}'}{\partial X_\gamma}. \tag{4}$$

3.2.2. L'équation aux températures sera examinée, selon une suggestion de L. S. G. Ko-vasznay à partir de celle de l'entropie s, soit:

$$\varrho\frac{\theta\,ds}{dt} = \varphi + \frac{\partial h_\gamma}{\partial X_\gamma} \tag{5}$$

φ dissipation, $h_\gamma = k\dfrac{\partial\theta}{\partial X_\gamma}$ densité de flux de chaleur de conduction.

Pour un gaz parfait:

$$ds = C_p\frac{d\Theta}{\Theta} - \frac{Rdp}{p} \approx C_p\frac{d\theta}{\theta} \tag{6}$$

car dans le cas de ces expériences le terme de pression est négligeable compte tenu des fluctuations turbulentes et des gradients. De même pour les écoulements supersoniques les modes de turbulence d'entropie et de vitesse sont presque indépendants [10] au premier ordre. On remarque d'ailleurs que le nombre d'Eckert

$$E_C = \frac{V_e^2}{C_p(\Theta_0 - \Theta_e)} \approx 0{,}01$$

étant faible, on peut négliger en effet les termes de pression, mais aussi les termes de dissipation (4) dans l'équation aux températures. L'équation retenue est donc:

$$\bar{\varrho}C_p\frac{d\theta}{dt} = \varphi + \frac{\partial h_\gamma}{\partial X_\gamma} \approx \frac{\partial h_\gamma}{\partial X_\gamma} \tag{7}$$

et l'équation aux fluctuations de température s'écrit:

$$\bar{\varrho}C_p\frac{\partial\theta'}{\partial t} + \bar{\varrho}C_pV_\gamma\frac{\partial\theta'}{\partial X_\gamma} + \bar{\varrho}C_pv_\gamma'\frac{\partial\Theta}{\partial X_\gamma} + \bar{\varrho}C_pv_\gamma'\frac{\partial\theta'}{\partial X_\gamma} - \overline{\bar{\varrho}C_pv_\gamma'\frac{\partial\theta'}{\partial X_\gamma}} = \frac{\partial h_\gamma'}{\partial X_\gamma}. \tag{8}$$

L'équations aux fluctuations de vitesses longitudinales ($\alpha = 1$), et l'équation aux fluctuations de température sont donc de même forme quant aux termes de transport, et de variation locale.

En utilisant l'équation de continuité avec les approximations précitées, on peut admettre que:

$$\bar{\varrho}v_\gamma'\frac{\partial v_\alpha'}{\partial X_\gamma} \approx \frac{\partial(\bar{\varrho}v_\gamma'v_\alpha')}{\partial X_\gamma}$$

et

$$\bar{\varrho}C_pv_\gamma'\frac{\partial\theta'}{\partial X_\gamma} \approx \frac{\partial(\bar{\varrho}C_pv_\alpha'\theta')}{\partial X_\gamma}. \tag{9}$$

3.2.3. Pour obtenir des équations aux corrélations spatio-temporelles il suffit de multiplier les équations précédentes respectivement par les fluctuations $(v_\beta')_0$ et $(\theta')_0$ au point P_0 et au temps t_0 et d'en prendre les moyennes.

Compte tenu du fait que l'écoulement est statistiquement stationnaire on a, avec l'hypothèse ergodique (6)

$$\overline{()_0'\frac{\partial()'}{\partial t}} = \frac{\partial\overline{()_0'()'}}{\partial T} \tag{10}$$

les équations peuvent s'écrire alors:

$$\bar{\varrho}\,\frac{\partial \overline{v'_\alpha v'_{\beta 0}}}{\partial T} + \bar{\varrho}\,V_\gamma\,\frac{\partial \overline{v'_\alpha v'_{\beta 0}}}{\partial X_\gamma} + \overline{\varrho v'_\gamma v'_{\beta 0}}\,\frac{\partial V_\alpha}{\partial X_\gamma} + \frac{\partial(\overline{\varrho v'_\gamma v'_\alpha v'_{\beta 0}})}{\partial X_\gamma} = -\,\frac{\partial \overline{p' v'_{\beta 0}}}{\partial X_\alpha} + \frac{\partial \overline{f'_{\alpha\gamma} v'_{\beta 0}}}{\partial X_\gamma} \tag{11}$$

$$\bar{\varrho}C_p\,\frac{\partial \overline{\theta' \theta'_0}}{\partial T} + \bar{\varrho}C_p V_\gamma\,\frac{\partial \overline{\theta' \theta'_0}}{\partial X_\gamma} + \bar{\varrho}C_p\overline{v'_\gamma \theta'}_0\,\frac{\partial \Theta}{\partial X_\gamma} + \frac{\partial \overline{\varrho C_p \theta' v'_\gamma \theta'_0}}{\partial X_\gamma} = \frac{\partial \overline{h'_\gamma \theta'_0}}{\partial X_\gamma}. \tag{12}$$

3.2.4. Considérons des fluctuations de vitesse obtenus avec un filtre de fréquences défini par l'intégrale [1]:

$$\hat{v}'_\alpha(t) = \int\limits_{-\infty}^{+\infty} v'_\alpha(\tau)\,g(t-\tau)\,d\tau \tag{13}$$

$g(t)$ est la réponse impulsionnelle du filtre, avec les conditions

$$g(\infty) = g(-\infty) = 0\,.$$

Si nous appliquons l'intégration aux termes des équations aux fluctuations de vitesse, compte tenu des hypothèses, on a:

$$\widehat{\left(\frac{\partial v'}{\partial t}\right)} = \frac{\partial \hat{v}'}{\partial t} \quad \text{et} \quad \widehat{\overline{v'_\alpha v'_\gamma}} = \overline{\widehat{v'_\alpha v'_\gamma}} \tag{14}$$

et les équations aux fluctuations filtrées sont:

$$\bar{\varrho}\,\frac{\partial \hat{v}'_\alpha}{\partial t} + \bar{\varrho}\,V_\gamma\,\frac{\partial \hat{v}'_\alpha}{\partial X_\gamma} + \bar{\varrho}\hat{v}'_\gamma\,\frac{\partial V}{\partial X_\gamma} + \frac{\partial}{\partial X_\gamma}\left(\widehat{\varrho v'_\alpha v'_\gamma} - \overline{\widehat{\varrho v'_\alpha v'_\beta}}\right) = -\,\frac{\partial \hat{p}'}{\partial X_\alpha} + \frac{\partial \widehat{f'_{\alpha\gamma}}}{\partial X_\gamma} \tag{15}$$

de même forme que celles des fluctuations du champ total. Il en est de même pour les fluctuations de température.

Si donc on introduit les définitions des corrélations spatio-temporelles doubles et triples suivantes:

$$\widehat{Q}_{\alpha\beta}(\underline{x}_0, \underline{x}, T) = \widehat{v'(\hat{v}'_\beta)}_0, \qquad \widehat{Q}(\underline{x}_0, \underline{x}, T) = \overline{\widehat{\theta'}\,\widehat{\theta'}}_0,$$

$$\widehat{Q}_\gamma(\underline{x}_0, \underline{x}, T) = \widehat{v'_\gamma\,\hat{\theta}'}, \tag{16}$$

$$\widehat{K}_\beta(\underline{x}_0, \underline{x}, T) = \overline{\widehat{p'(\hat{v}'_\beta)}}_0,$$

$$\widehat{S}_{\alpha\gamma\beta}(\underline{x}_0, \underline{x}, T) = \overline{\varrho \widehat{v'_\alpha v'_\gamma}(\widehat{v'_\beta})}_0, \qquad \widehat{S}_\gamma(\underline{x}_0, x, T) = \bar{\varrho}C_p\,\overline{\widehat{\theta v'_\gamma}(\widehat{v'_\beta})}_0$$

les équations des corrélations spatio-temporelles avec filtrage en fréquences peuvent s'écrire:

$$\bar{\varrho}\,\frac{\partial \widehat{Q}_{\alpha\beta}}{\partial T} + \bar{\varrho}\,V_\gamma\,\frac{\partial \widehat{Q}_{\alpha\beta}}{\partial X_\gamma} + \bar{\varrho}\widehat{Q}_{\gamma\beta}\,\frac{\partial V_\alpha}{\partial X_\gamma} + \frac{\partial \widehat{S}_{\alpha\gamma\beta}}{\partial X_\gamma} = \frac{\partial \widehat{K}_\beta}{\partial X_\alpha} + \frac{\partial \overline{\widehat{f'_{\alpha\gamma}(\widehat{v'_\beta})}_0}}{\partial X_\gamma} \tag{17}$$

et

$$\bar{\varrho}C_p\,\frac{\partial \widehat{Q}}{\partial T} + \bar{\varrho}C_p V_\gamma\,\frac{\partial \widehat{Q}}{\partial X_\gamma} + \bar{\varrho}C_p\widehat{Q}_\gamma\,\frac{\partial \Theta}{\partial X_\gamma} + \frac{\partial \widehat{S}_\gamma}{\partial X_\gamma} = \frac{\partial \widehat{h'_\gamma \Theta'_0}}{\partial X_\gamma}. \tag{18}$$

Des équations correspondantes on été écrites en termes spectraux (9).

3.3. Célérités

3.3.1. Nous avons été conduits (6), (7) par des considérations expérimentales et théoriques, notamment celle de stationnarité statistique, à définir des célérites à partir des corrélations spatio-temporelles de vitesses:

$$\hat{C}_\gamma = -\left(\frac{\partial^2 \widehat{Q}_{\alpha\beta}}{\partial T^2}\,\middle/\,\frac{\partial^2 \widehat{Q}_{\alpha\beta}}{\partial T\,\partial X_\gamma}\right) \tag{19}$$

(sans sommation d'indice) et de températures :

$$\widehat{\Gamma_\gamma} = -\left(\frac{\partial^2 \hat{Q}}{\partial T^2} \middle/ \frac{\partial^2 \hat{Q}}{\partial T\, \partial X_\gamma}\right). \tag{20}$$

Les valeurs sont prises au décalage de temps T_m correspondant au maximum de la corrélation spatio-temporelle pour des positions données dans l'espace $\underline{x_0}$ et $\underline{x}$. Toutefois les formules restent valables indépendamment de ce choix de T.

Si $\underline{l}$ est le vecteur unitaire dans la direction de la vitesse moyenne au point P, les célérités selon cette direction sont :

$$\frac{1}{\hat{C}} = \frac{l_\gamma}{\hat{C}_\gamma} \quad \text{et} \quad \frac{1}{\hat{\Gamma}} = \frac{l_\gamma}{\hat{\Gamma}_\gamma}. \tag{21}$$

L'expérience montre que les termes $\dfrac{\partial^2 \hat{Q}_{\alpha\beta}}{\partial T^2}$ et $\dfrac{\partial^2 \hat{Q}}{\partial T^2}$ ne sont pas nuls. Des équations en termes de célérités sont ainsi obtenues en dérivant par rapport à T les Eqs. (17) et (18); soit pour les corrélations spatio-temporelles de vitesse :

$$\bar\varrho \,\frac{\partial^2 Q_{\alpha\beta}}{\partial T^2}\left(1 - \frac{|V|}{\hat{C}}\right) = -\bar\varrho\,\frac{\partial \hat{Q}_{\gamma\beta}}{\partial T}\frac{\partial V_\alpha}{\partial X_\gamma} - \frac{\partial^2 \hat{S}_{\alpha\gamma\beta}}{\partial T\,\partial X_\gamma} - \frac{\partial^2 \hat{K}_\beta}{\partial T\,\partial X_\alpha} + \frac{\partial \overline{(\hat{f}_{\alpha\gamma}(\widehat{v'_\beta})_0)}}{\partial X_\gamma} \tag{22}$$

et pour les corrélations spatio-temporelles de températures :

$$\bar\varrho C_p \,\frac{\partial^2 \hat{Q}}{\partial T^2}\left(1 - \frac{|V|}{\hat{\Gamma}}\right) = \bar\varrho C_p\,\frac{\partial \hat{Q}_\gamma}{\partial T}\frac{\partial \Theta}{\partial X_\gamma} - \frac{\partial^2 \hat{S}_\gamma}{\partial T\,\partial X_\gamma} + \frac{\partial \overline{(\hat{h}'\,\widehat{\theta'_0})}}{\partial X_\gamma}. \tag{23}$$

Dans le cas du champ turbulent total des équations précédentes sont formellement les mêmes.

Ces équations montrent que les célérités correspondant à la turbulence de vitesse ou de température ne sont pas d'une façon générale égales à la vitesse moyenne de la matière, les seconds membres des équations n'étant pas nuls à priori, notamment les termes de corrélations triples.

Dans chaque équation aux vitesses relatives à une composante, et dans l'équation aux températures, les termes de transport et une partie des termes moléculaires se présentent sous la même forme. Ceci peut expliquer une analogie des célérités qui n'est d'ailleurs que partielle en raison : de la présence de termes différents, de l'existence d'équations supplémentaires comme celle de continuité, ainsi que du fait que le système des équations statistiques est ouvert On. note d'ailleurs que les répartitions spectrales sont différentes.

3.3.2. Dans le cas des présentes expériences pour la couche limite $\alpha = \beta = 1$ l'écoulement est homogène en moyenne en X_2. De plus en admettant que le temps optimum T_m soit le même pour $\hat{Q}_{13}$ et $\hat{Q}_{11}$; avec les approximations habituelles de la couche limite et si $\dfrac{|V|}{\hat{C}} \approx \dfrac{V_1}{C_1}$ on a :

$$\bar\varrho \,\frac{\partial^2 Q_{11}}{\partial T^2}\left(\frac{\hat{C}_1 - V_1}{\hat{C}_1}\right) \approx -\frac{\partial^2}{\partial T\,\partial X_3}\left(\hat{S}_{13,1} - \overline{\hat{f}_{13}\,(\widehat{v'_1})_0}\right) - \frac{\partial^2 \hat{K}_1}{\partial T\,\partial X_1} \tag{24}$$

$$\text{à} \quad T = T_m.$$

De même pour la température avec des hypothèses similaires :

$$\bar\varrho C_p \,\frac{\partial^2 \hat{Q}}{\partial T^2}\left(\frac{\hat{\Gamma}_1 - V_1}{\hat{\Gamma}_1}\right) \approx -\frac{\partial^2}{\partial T\,\partial X_3}\left(\hat{S}_3 - \overline{\hat{h}'_3\,\widehat{\theta'_0}}\right) \tag{25}$$

$$\text{à} \quad T = T_m.$$

On peut penser qu'à la distance à la paroi considérée ($y/\delta = 0{,}038$) et pour les fluctuations à grandes échelles, les termes de viscosité et de conductivité peuvent être négligés.

4. Resultats experimentaux

4.1. Correlations spatio-temporelles avec décalage longidutinal d'espace

4.1.1. Temps optimums. La fig. 5 présente les résultats de mesures concernant les temps optimums T_m, correspondant au maximum de la corrélation, pour quatre décalages longitudinaux $X_1 = 17, 35, 75, 150$ mm et pour divers rapports de sensibilité r de $0{,}25 \cdot 10^{-2}$ à $11{,}5 \cdot 10^{-2}$, en fonction de différentes fréquences n. Les valeurs de T_m sont données aussi pour le champ turbulent total. On remarque que les temps T_m sont croissants avec la fréquence, jusqu'à une certaine valeur de celle-ci, fonction de X_1. Au delà de cette fréquence, T_m peut décroitre mais le coefficient de corrélation est alors faible, et on peut penser que les fréquences élevées ne correspondent pas uniquement à des fluctuations à petites échelles spatiales mais aussi à des harmoniques spectraux des fluctuations à grandes échelles.

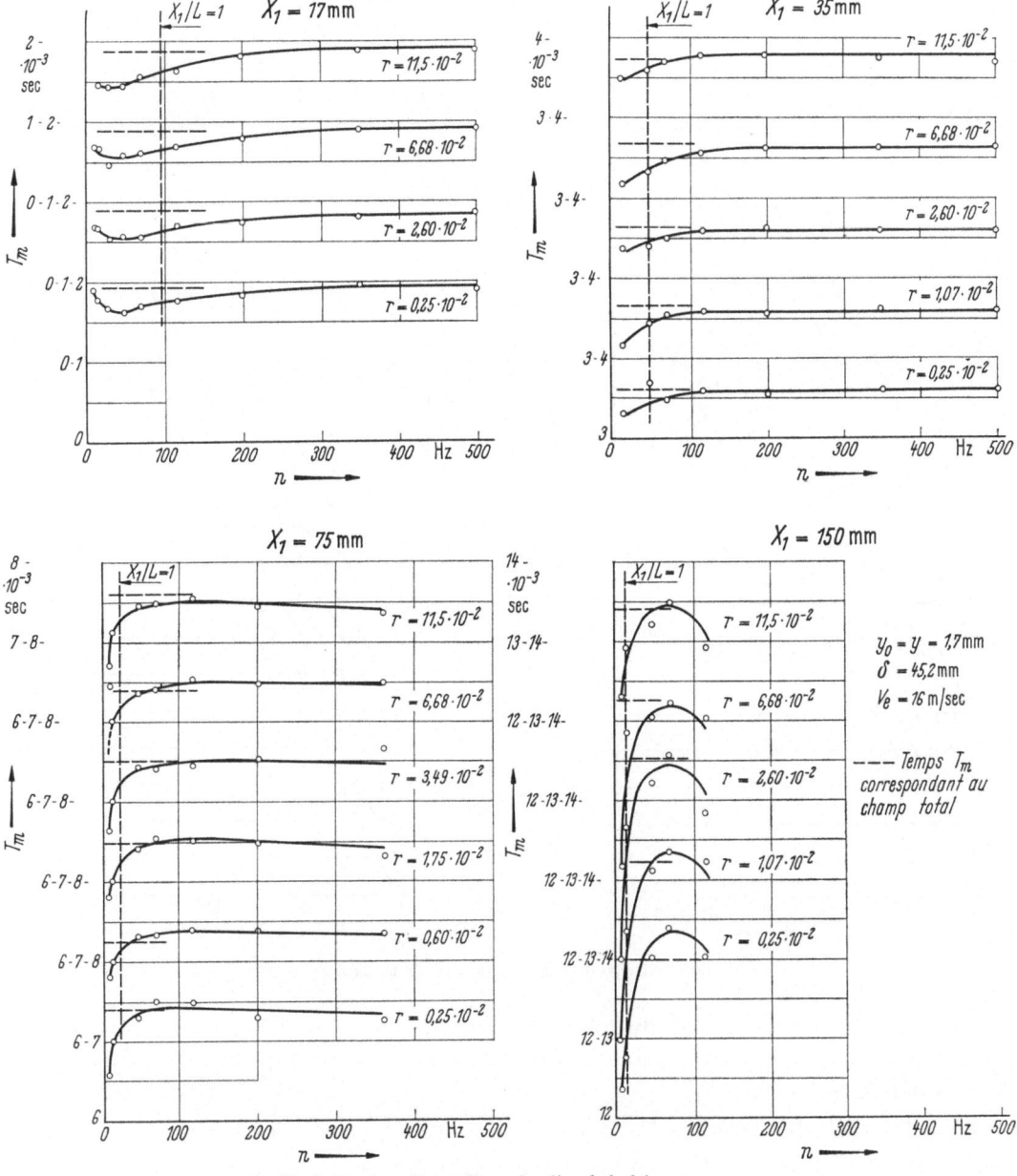

Fig. 5. Temps optimum T_m en fonction de la fréquence n.

A la distance $X_1 = 17$ mm, on constate au contraire une décroissance de T_m en fonction de n aux valeurs les plus faibles de la fréquence. Le coefficient de corrélation dans ce cas est élevé même pour les fluctuations à petites échelles, et on peut penser aussi que ces fréquences basses ne correspondent pas uniquement à des fluctuations à grandes échelles spatiales.

On définit une longeeur caractéristique de la fluctuation turbulente avec filtrage de fréquence $L = \dfrac{\hat{C}_1}{2\pi n}$, $\hat{C}_1$ étant la célérité considérée aux paragraphes 3.3.1 et 4.1.2.

La fig. 10 montre que pour les valeurs $\dfrac{X_1}{L} = 1$ le coefficient de corrélation reste compris entre 0,75 et 0,8 ce qui paraît satisfaisant pour suivre des fluctuations au long du mouvement moyen, compte tenu des remarques précédentes.

Pour ces raisons, et en accord avec les résultats antérieurs (6) on choisit pour mesurer la célérité une longueur relative pour laquelle l'analyse fréquentielle paraît significative du point de vue de l'échelle spatiale des fluctuations.

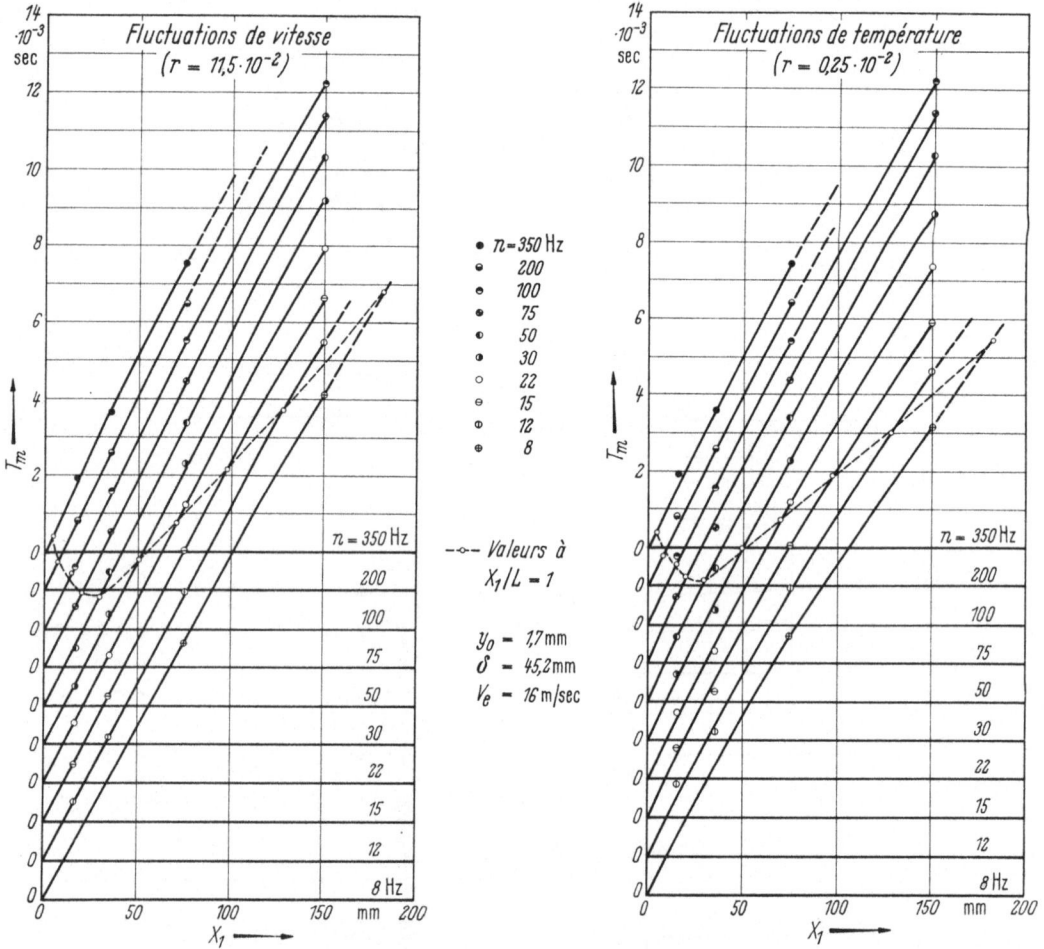

Fig. 6. Temps optimum en fonction du décalage X_1.

La fig. 6 présente les valeurs du temps T_m en fonction de la distance X_1, pour diverses fréquences n, mais ceci seulement dans le cas de fluctuations de température ($r = 0{,}25 \cdot 10^{-2}$) et de vitesse ($r = 11{,}5 \cdot 10^{-2}$).

4.1.2. Célérités. Les célérités $\hat{C}_\gamma$ et $\hat{\Gamma}_\gamma$ définies au paragraphe 3.3.1. sont les pentes des courbes $T_m(X_\gamma)$; si on appelle $\hat{C}_1(r)$ la célérité correspondant à un coefficient de sensibilité donné r, on a vu que (7):

$$\frac{1}{\widehat{C}_1} = \left(\frac{\partial T_m}{\partial X_1}\right)_n \tag{26}$$

pour $r \leq 0,25 \cdot 10^{-2}$, $\widehat{C}_1(r) \# \widehat{\Gamma}_1$ et pour $r \geq 11,5 \cdot 10^{-2}$, $\widehat{C}_1(r) \sim \widehat{C}_1$.

Dans la mesure où les courbes $T_m(X_1)$ sont des droites et où les bandes passantes sont étroites, ces célérités sont équivalentes aux vitesses de phase utilisées par d'autres auteurs [2].

Les principaux résultats sont résumés sur la fig. 7 donnant l'évolution des écarts relatifs de célérités qui apparaissent dans les Eqs. (24) et (25) en fonction de l'échelle $\frac{L}{\delta}$ des fluctuations à $\frac{X_1}{L} = 1$, et ceci pour les valeurs de r correspondant au cas des températures ($r = 0,25 \cdot 10^{-2}$) et des vitesses ($r = 11,5 \cdot 10^{-2}$). Les points de petites dimensions correspondent à des résultats directs et ceux de dimensions plus grandes à divers lissage de courbes.

On constate que les célérités des fluctuations de température ou de vitesse tendent vers la vitesse moyenne de la matière pour les fluctuations correspondant aux petites échelles.

Si l'on considère des échelles croissantes on voit que les célérités deviennent supérieures à la vitesse moyenne du fluide et que l'écart est moins accentué pour les vitesses que pour les températures.

Dans un schéma d'ondes les célérités précédemment définies correspondraient à des célérités de phase, et l'on pourrait définir aussi des célérités de groupe.

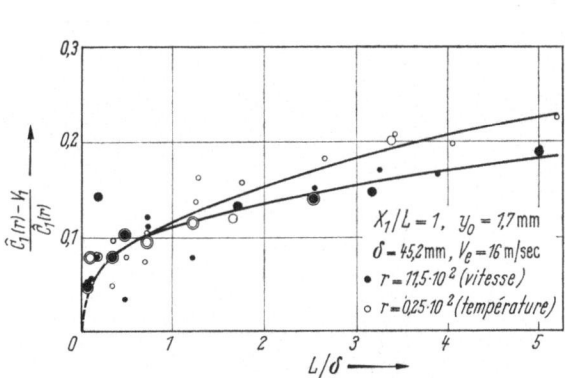

Fig. 7. Ecarts relatifs des célérités en fonction de l'échelle des fluctuations L/δ.

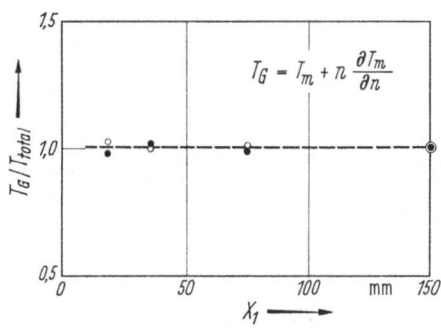

Fig. 8. Temps optimum T_G rapporté au temps optimum du champ total.

Soit ω la pulsation fonction du nombre d'onde k la célérité de phase est $C = \frac{\omega}{k}$ et celle de groupe $G = \frac{d\omega}{dk}$ (16). Si l'on admet que les célérités $C_1 \approx \frac{X_1}{T_m}$, on établit alors la relation:

$$T_G = T_m + n\frac{\partial T_m}{\partial n} \tag{27}$$

où T_G est le temps compensateur défini par $T_G = \frac{X_1}{G}$.

Les temps T_G ont été déterminés pour les valeurs $\frac{X_1}{L} = 1$, et on constate sur la fig. 8 que T_G reste voisin des temps optimums T_{tot} du champ turbulent total. Ceci paraît confirmer la signification physique des mesures effectuées sur des longeeurs $\frac{X_1}{L}$ adaptées à l'échelle spatiale des fluctuations.

4.1.3. Coefficients de correlation optimum. Les valeurs du coefficient de corrélation optimum $\widehat{R}(T_m, X_1)$ au long de la ligne de courant $y/\delta = 0,038$ sont données sur la fig. 9, en fonction de la fréquence n, pour les divers décalages X_1 et coefficients de sensibilité r.

A. Favre, R. Dumas et E. Verollet

Les mêmes résultats sont présentés dans le cas des fluctuations de vitesse et de température (Fig. 10) en fonction du rapport $\frac{X_1}{L}$. Comme trouvé précedemment en ce qui concerne les vitesses la corrélation dépend principalement du décalage longitudinal X_1 compté en nombre de longueurs caractéristiques L. Un résultat semblable avait été établi par Corcos pour les corré-

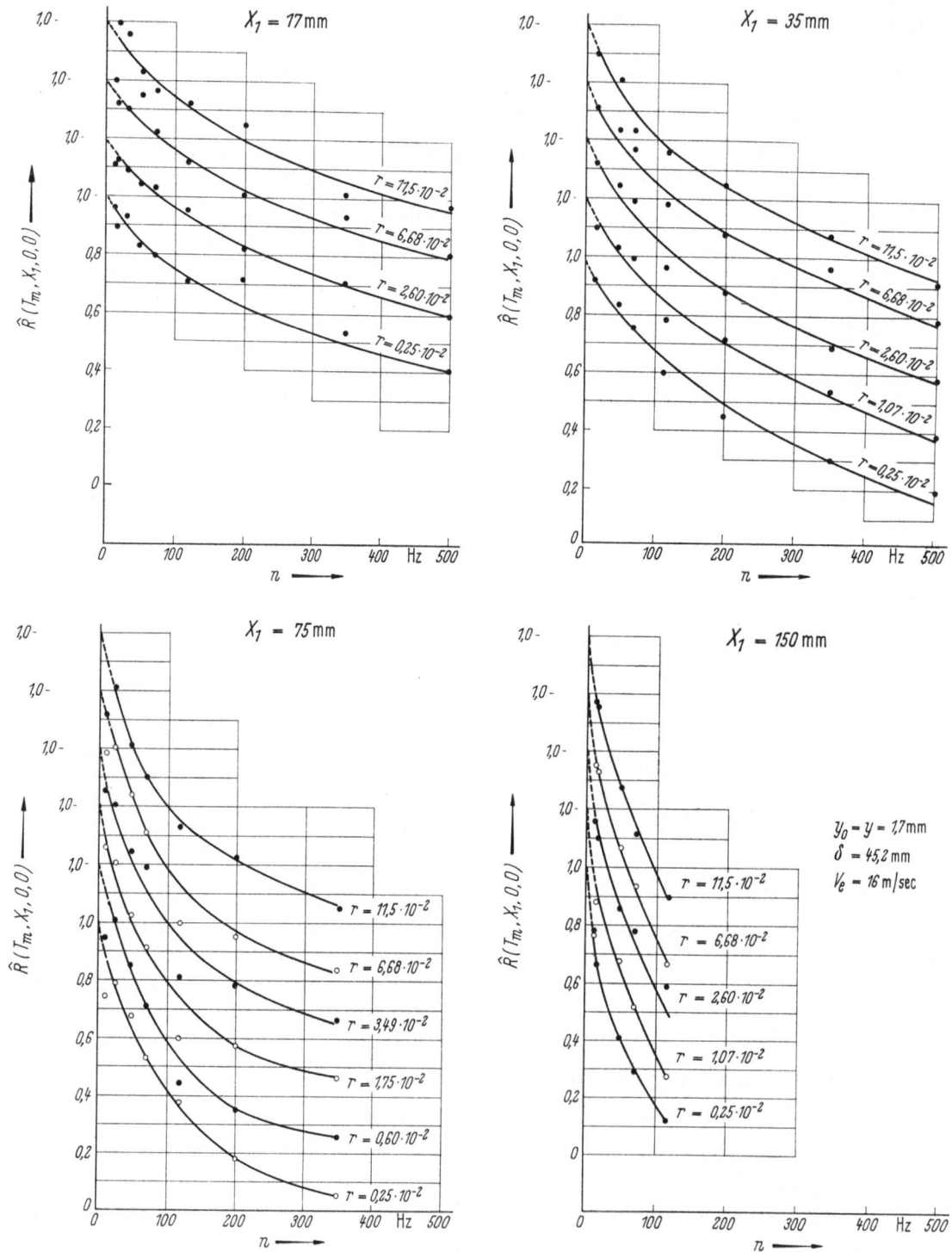

Fig. 9. Coefficients de corrélation optima en fonction de la fréquence n.

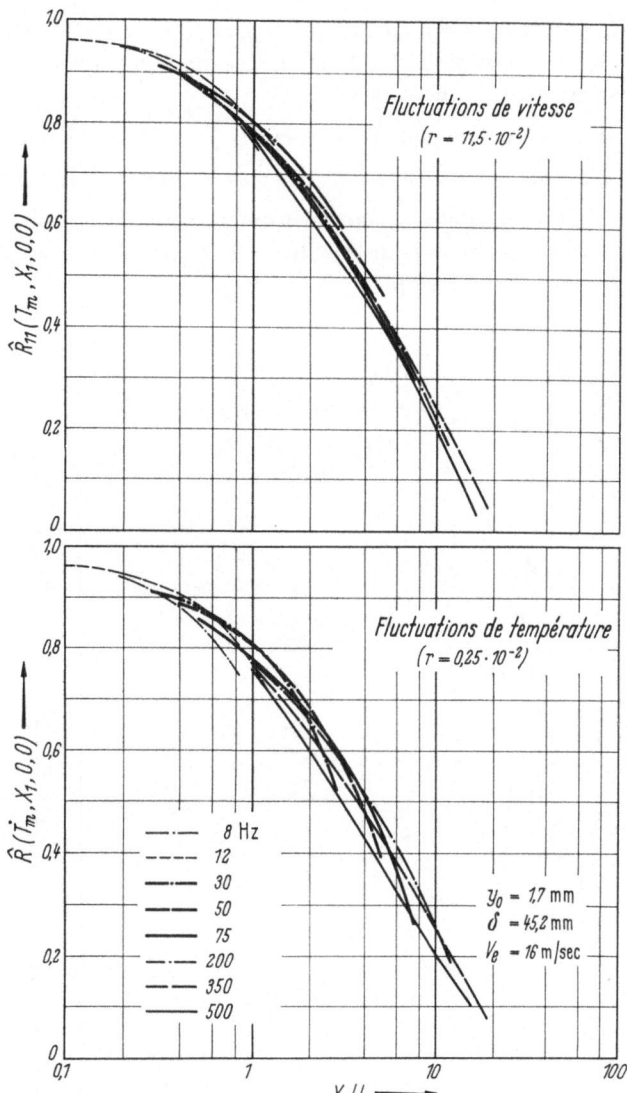

Fig. 10. Coefficient de corrélation avec temps optimum au long du mouvement.

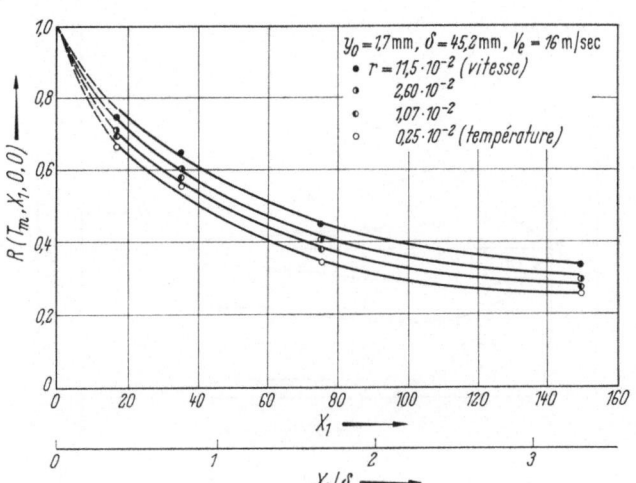

Fig. 11. Coefficient de corrélation optimum en fonction de l'écart X_1 (champ total).

204   A. Favre, R. Dumas et E. Verollet

lations spatio-temporelles de pression à la paroi [2]. Ce résultat apparaît aussi en ce qui concerne les corrélations pour les fluctuations de température, qui ont des valeurs voisines de celles des corrélations de vitesse.

En particulier pour les décalages $\frac{X_1}{L} = 1$ choisis pour déterminer les célérités, on constate un regroupement satisfaisant des corrélations aux diverses fréquences.

Si l'on considère le champ turbulent total les corrélations optima présentées en fonction de X_1 sont plus elevées pour les vitesses que pour les températures (Fig. 11), ce qui peut s'expliquer par l'énergie relative plus grande des vitesses aux basses fréquences (Fig. 3).

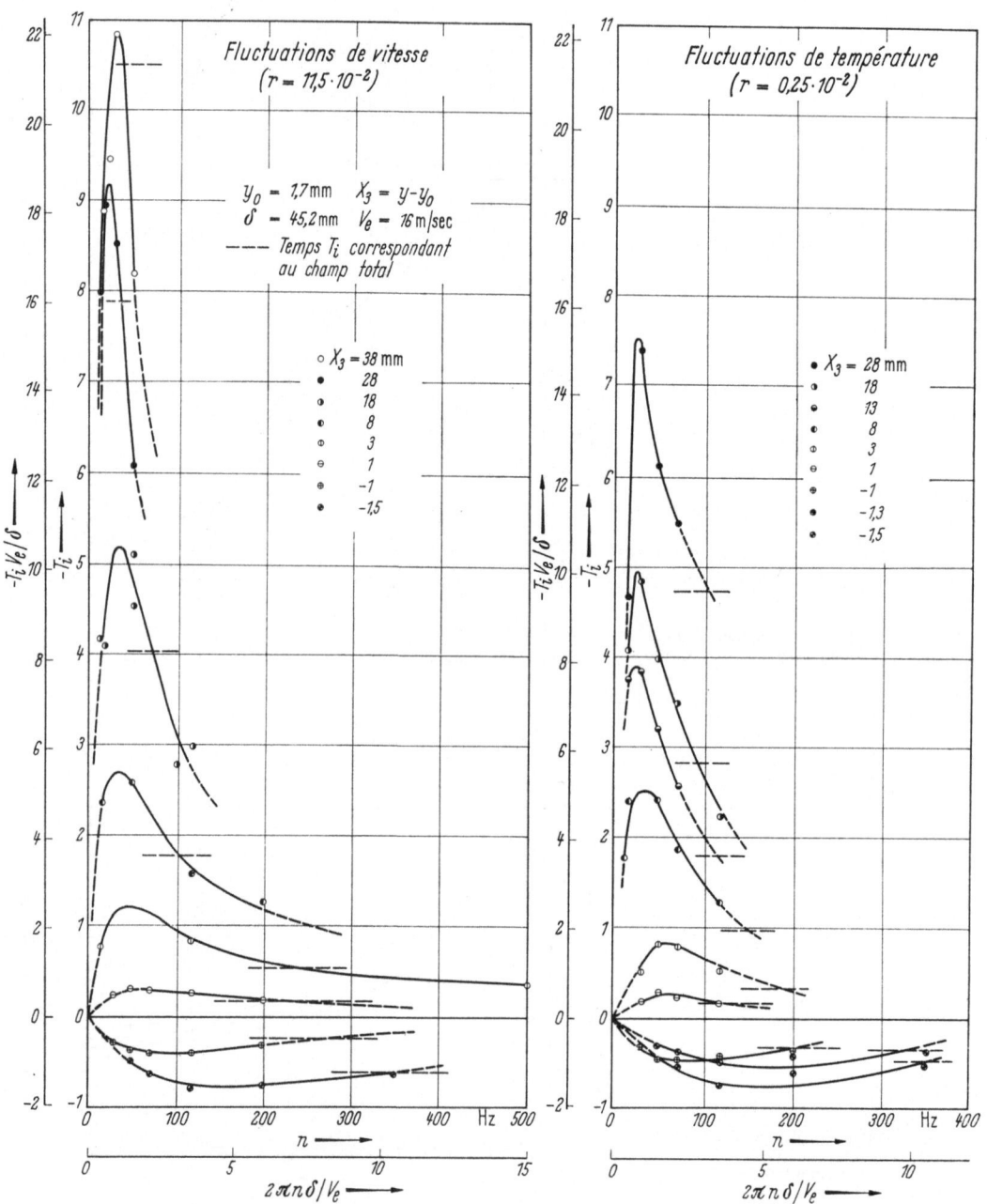

Fig. 12. Temps optimum T_i en fonction de la fréquence n.

4.2. Corrélations spatio-temporelles avec décalage orthogonal à la paroi

4.2.1. Les valeurs du temps optimum T_i, pour des décalages d'espace X_3 orthogonaux à la paroi, sont portés en fonction des fréquences n sur la fig. 12 pour les fluctuations de température ($r = 0.25 \cdot 10^{-2}$) et pour les fluctuations de vitesse ($r = 11.5 \cdot 10^{-2}$). Les valeurs de T_i correspondant au champ total sont également indiquées sur la figure.

Les temps T_i sont négatifs lorsque le point P est plus éloigné de la paroi que P_0 et positifs dans le cas contraire.

Les résultats sont en accord avec les précédents (5), (6) concernant les vitesses dans une couche limite isotherme, mais ils ont été complétés en déplaçant le fil chaud mobile jusqu'à une distance de 0,2 mm de la paroi. Pour les températures les résultats sont analogues, le temps T_i croît avec le décalage X_3, passe par un maximum, d'ailleurs plus faible que pour les vitesses, et paraît tendre vers zéro aux fréquences les plus basses et aux fréquences les plus élevées.

La fig. 13 récapitule les temps T_i relatifs au champ turbulent total.

Dans le cas des décalages X_3 positifs, les temps T_i en valeur absolue pour la température sont de l'ordre de 62% de ceux concernant les vitesses.

Si on exprimait ce résultat en termes de célérités C_3 ou Γ_3 du champ total pour les déplacements orthogonaux à la paroi, on trouverait donc des valeurs négatives plus élevées, en valeur absolue, pour les températures que pour les vitesses, que l'on pourrait estimer en moyenne à $C_3 \sim -3,8 \text{ ms}^{-1}$, $\Gamma_3 \sim -6,1 \text{ ms}^{-1}$.

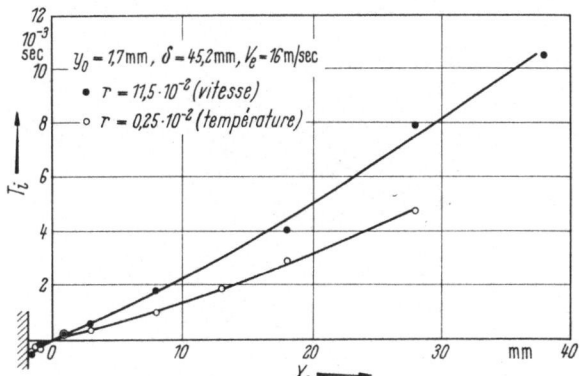

Fig. 13. Temps optimum T_i en fonction du décalage X_3 pour le champ total.

Rapporté à la vitesse $(V_1)_0$ la célérité de vitesse C_3 correspond à un angle de 22° par rapport à la paroi que l'on peut comparer à l'inclinaison des tourbillons donnés dans l'interprêtation de STERNBERG [14], l'angle correspondant serait de 33° pour les températures.

RUNDSTADLER, KLINE et REYNOLDS [13] considèrent que la structure de la couche limite comporte d'une part des filaments tourbillonnaires émis (ejected eddy) au voisinage de la paroi, et d'autre part des tourbillons plus diffus retournant vers la paroi (inflow eddy).

D'après ces auteurs, ces derniers tourbillons auraient un effet prépondérant sur la corrélation de vitesse prise orthogonalement à la paroi, déterminant aussi les temps T_i négatifs trouvés.

4.2.2. La fig. 14 donne les valeurs du coefficient de corrélation $\widehat{R}(T_i, X_3)$ avec le temps optimum T_i en fonction de la distance X_3 et pour diverses fréquences n, pour les vitesses et pour les températures.

Les résultats sont analogues à ceux trouvés précédemment pour les vitesses (6). En ce qui concerne les températures on peut noter qu'à fréquence n et décalage X_3 égaux, la corrélation conserve des valeurs plus élevées. Par contre la corrélation est un peu moindre pour le champ total des températures que pour celui des vitesses au même décalage X_3, en raison des différences de répartition spectrale.

4.2.3. Quelques mesures de corrélations au temps T_m ont été effectuées pour le champ total avec les décalages longitudinaux $X_1 = 75$ mm et 150 mm et des décalages orthogonaux X_3 variables, pour les températures et pour les vitesses.

Les résultats précédents (5), (6) relatifs aux fluctuations de vitesse sont confirmés: la ligne de corrélation maximum maximorum est une courbe dont la convexité est **tournée vers la paroi**. En ce qui concerne les fluctuations de température les résultats sont analogues mais cette convexité semble un peu plus marquée.

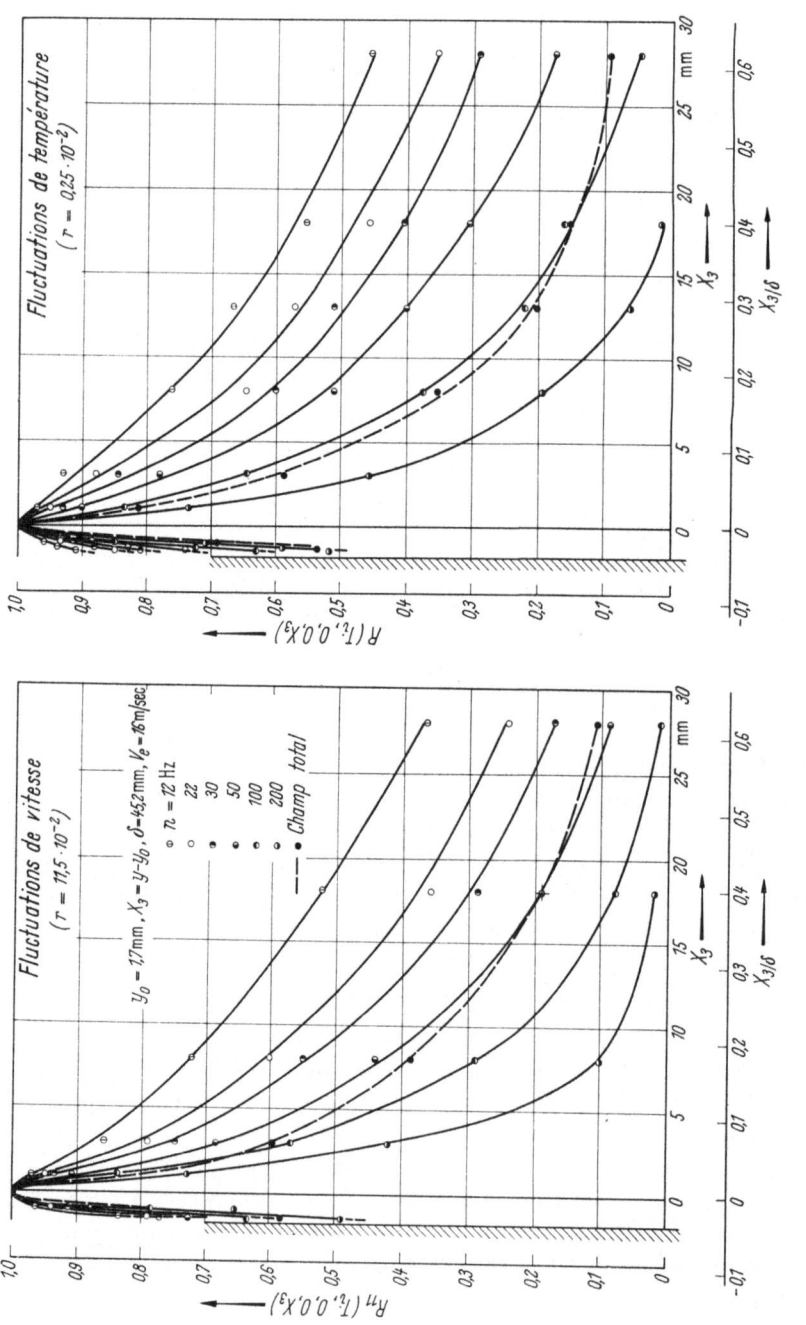

Fig. 14. Coefficients de corrélation optima en fonction de l'écart X_3.

5. Conclusions

5.1.

Les équations aux corrélations spatio-temporelles de vitesse, et de température, ont été écrites en termes de célérités pour le champ turbulent total et pour les fluctuations turbulentes filtrées en fréquences.

Dans le cas d'un filtrage parfaitement sélectif et dans un schéma d'ondes, ces célérités correspondraient aux célérités de phase.

Les équations montrent que dans le cas général les célérités diffèrent de la vitesse moyenne de la matière.

Dans le cas de nos expériences ces différences sont dues principalement aux corrélations spatio-temporelles triples de vitesse et de température, et peut-être pour les équations aux vitesses, aux corrélations pression-vitesse. L'équation aux températures et l'équation aux composantes longitudinales de vitesses présentent des termes de mêmes formes à l'exception du terme de corrélation pression-vitesse, ce qui conduirait à une analogie partielle, mais le système restant ouvert des différences peuvent se manifester, ce que l'on constate notamment dans la répartition spectrale d'énergie.

5.2.

Les mesures ont été effectuées dans une couche limite avec transfert thermique sans gradient de pression, la station de mesure fixe étant à $y/\delta = 0{,}038$.

Les résultats paraissent particulièrement significatifs, et leur interpretation physique plus simple, lorsqu'on choisit pour effectuer des mesures des distances adaptées à l'échelle des fluctuations. Pour les décalages longitudinaux X_1 on a été conduit à retenir les résultats obtenus pour $\frac{X_1}{L} = 1$. L étant une longueur caractéristique dépendant de la fréquence, le coefficient de corrélation conserve alors des valeurs comprise entre 0,75 et 0,8, et la célérité de groupe est égale à celle du champ total. Les fluctuations se comporteraient presque dans la direction longitudinale comme un système d'ondes. Dans la direction orthogonale les tentatives effectuées n'ont pas permis encore un choix assez satisfaisant des longueurs X_3 adaptées aux échelles des fluctuations. Les interprêtations portent donc sur le champ turbulent total.

5.3.

5.3.1. Dans la direction de l'écoulement les célérités correspondantes au champ de température et de vitesse tendent vers la vitesse moyenne du fluide quand l'échelle des fluctuations décroît. Par contre les célérités dépassent d'autant plus la vitesse moyenne que l'échelle des fluctuations est plus grande et cet effet est plus marqué pour les températures que pour les vitesses. Par exemple pour la distance $\frac{L}{\delta} = 5$ l'écart relatif atteint 19% pour les vitesses et 23% pour les températures.

Comme trouvé précédemment pour le champ de vitesse, les valeurs de coefficients de corrélations spatio-temporelles avec retard optimum relatifs aux températures se regroupent lorsque les décalages longitudinaux sont comptés en nombre $\frac{X_1}{L}$ de longueurs caractéristiques L des fluctuations.

5.3.2. Dans la direction orthogonale à la paroi, on trouve pour les températures, comme pour les vitesses, des temps optimums T_i négatifs pour le point le plus éloigné de la paroi mais avec des valeurs absolues plus faibles. Ces temps paraissent tendre vers zéro aux fréquences les plus basses, et présentent des maximums pour des fréquences correspondant à des longueurs caractéristiques de l'ordre de l'épaisseur de la couche limite.

Dans le cas du champ turbulent total ce temps est pour les températures de 62% de celui correspondant aux vitesses.

Le rapport des célérités orthogonales à la vitesse moyenne du fluide correspond à des inclinaisons par rapport à la paroi de l'ordre de 22° pour les fluctuations de vitesses, et de 33° pour les fluctuations de températures. Les valeurs du coefficient de corrélation spatio-temporelle avec retard optimum relatif aux températures, comme celui relatif aux vitesses, augmentent lorsque diminue la fréquence des fluctuations, mais les valeurs sont plus élevées à fréquence égale pour la température que pour la vitesse.

5.4.

Une analogie partielle se manifeste aussi bien d'après les équations que d'après les résultats expérimentaux entre les caractéristiques considérées des champs turbulents de température et de vitesse.

Les effets sont généralement plus marqués pour les températures.

Remerciements

The research reported in this document has been sponsored in part by the Air Force Office of Scientific Research under Contract AF 61 052 67 C 0025 through the European Office of Aerospace Research (E.O.A.R.) United States Air Force.

Bibliographie

1. Blanc-Lapierre, A., et B. Picinbono: Propriétés statistiques du bruit de fond. Masson Edit. 1961.
2. Corcos, G. M.: The structure of the turbulent pressure field in boundary layers flows. J. Fluid Mech. 18, 3, 353—378 (1964).
3. Corrsin, S., and M. S. Uberoi: Spectrums and diffusion in a round turbulent jet. NACA TN 2.124, Washington, July 1950.
4. Eckert, E. R. G., and R. M. Drake: Heat and Mass transfer, PP. A 3 Sec. Ed. New York: McGraw-Hill.
5. Favre, A.: Review on space time correlations in turbulent fluids. J. Appl. Mech. June 1965.
6. Favre, A., J. Gaviglio and R. Dumas: Space-time correlation in boundary layer. Int. Symp. on Boundary Layer and Turb including geo. Appl. KYOTO (Japan), September 1966. Phy. of Fluid Vol. 10, Part II, No. 9, September 1967.
7. Favre, A., and R. Dumas: Analytical expressions of space-time correlation celerities. AFOSR — ASR No. 3, Marseille, January 1967.
8. Favre, A., R. Dumas and E. Verollet: Facilities and methods of measurements for temperature space-time correlations in a boundary layer. AFOSR — ASR No. 3, Marseille, January 1967.
9. Favre, A., R. Dumas and E. Verollet: Boundary layer: expressions for celerities of fluctuations: first measurements for temperature fluctuations. AFOSR — FSR No. 4, January 1968.
10. Kovasznay, L. S. J.: Turbulence in supersonic flow. JHU — 5 Ae. Vol. 17, No. 9, Baltimore, September 1950.
11. MacMillan, F. A.: Experiments on pitot tubes in shear flow. ARC — R. and M. No. 3.028 — 1956.
12. Marcillat, J.: Fonctions de répartition des vitesses turbulentes dans une couche limite. Effets de la réponse de l'anémomètre à fil chaud. Thèse de Doctorat 3e Cycle, Marseille 1962.
13. Runstadler, P. W., S. J. Kline and W. C. Reynolds: An experimental investigation of the flow structure of the turbulent boundary layer. Report MD-8 Dpt. Mech. Eng. Stanford Univ. June 1963.
14. Sternberg, J.: On the interpretation of space-time correlation measurements in shear flow. Int. Symp. on Boundary Layer and Turb, including geo. Appl. KYOTO (Japan), September 1966. Phy. of Fluid Vol. 10, Part II, No. 9, September 1967.
15. van Driest, E. R.: On turbulent flow near a wall. J. Aero. Sci. Vol. 23, 1.007—1.011 and 1.036 (1956).
16. Whitham, G. B.: A note on group velocity. J. Fluid Mech. Vol. 9, Part 3, November 1960.

Axisymmetric plastic collapse of shells of revolution according to the Nakamura yield criterion

By

W. Flügge   and   **J. C. Gerdeen**

Stanford University,   Michigan Technological University,
Stanford   Houghton

Notations

Capital letters denote physical quantities and lower case letters denote the corresponding dimensionless quantities.

$\varepsilon_s, \varepsilon_\theta$	Plastic strain-rates of the middle surface of a shell
f_i, g_i, h_i	Yield functions
H	Thickness of a shell
H_i	A constituent regular face of a piecewise regular yield hypersurface for which $h_i = 1$
$H_1 \cap H_2$	An intersection of two regular faces
$H_1 \to H_2$	A transition between two regular faces corresponding to either a weak or a strong discontinuity
k	Geometrical parameter: $\dfrac{M_0}{LN_0} = \dfrac{H}{4L}$
K_s, K_θ	Plastic rates of curvature of the middle surface $\varkappa_s = K_s M_0/N_0$, $\varkappa_\theta = K_\theta M_0/N_0$
L	A typical shell length
λ, λ_i	Deformation parameters in Mises flow law
M_0	Yield bending moment/length, $\dfrac{\sigma_0 H^2}{4}$
M_s, M_θ	Bending moments/length, $m_s = M_s/M_0$, $m_\theta = M_\theta/M_0$
N_0	Yield normal force/length, $\sigma_0 H$
N_s, N_θ	Normal forces/length, $n_s = N_s/N_0$, $n_\theta = N_\theta/N_0$
Q_s	Transverse shear force/length, $q_s = Q_s/N_0$
Q	Generalized stress vector
q	Generalized strain-rate vector
R	Radius of a parallel circle, $r = R/L$
R_1, R_2	Principal radii of curvature of a meridian and parallel circle, respectively, $r_1 = R_1/L$, $r_2 = R_2/L$
S	Length measured along a meridian, $s = S/L$
σ_0	Yield stress in simple tension
θ	Coordinate defining the position of a meridian
V, W	Velocities in the S-direction and the direction normal to the shell middle surface, respectively, $v = V/L$, $w = W/L$
$(\ )'$	$\dfrac{d(\ )}{ds}$

Introduction

The plastic collapse of thin shells of revolution under axisymmetric loads is considered. A rigid perfectly-plastic material is assumed. Methods of limit analysis are applied to find collapse loads. The Tresca-Mohr yield criterion, as closely approximated by Nakamura [1, 2] is used.

In terms of the two principal stresses for a plane stress state, the Tresca-Mohr yield criterion is piecewise regular, and linear. In terms of the four stress resultants for a shell, the yield criterion is piecewise regular, but is nonlinear. Because of the nonlinearity, it has been necessary to make approximations, in order to obtain solutions. The approximations have been generally of three kinds: (1) for specific problems the stress and velocity fields have been assumed independently to find lower and upper bounds respectively, [3, p. 47; 4], (2) the uniform thickness shell has been approximated by a sandwich shell, [3, pp. 22–26] and (3) linear approximations have been made to the yield hypersurface, [3, pp. 30–32; 5].

Approximating the stress and velocity fields is a trial and error approach which must be repeated for each new problem. The other approximations result in wide bounds on the collapse loads. For this reason a closer quadratic approximation to the Tresca hypersurface has been made by Nakamura [1, 2]. In addition to the close bounds, closed form solutions for trajectories[1] within the parabolic faces can be obtained by simple quadrature. Also, the reduction of constituent faces from twelve to six is an advantage of the Nakamura hypersurface.

Although some solutions have been obtained for the Tresca criterion [3, pp. 90—93; 1, pp. 74, 96; 2], these are for one-parameter loadings of shallow shells having trajectories entirely within one face of the hypersurface. Physical intuition and experience with flat circular plates has helped to locate the trajectories for shallow shells. Solutions are lacking for non-shallow shells and for shells with multi-parameter loadings.

There are three major difficulties in more complicated problems: the trajectory may lie partly within constituent faces and partly in the intersections of faces, and its location is not known a priori; for the parts of a trajectory in intersections, e.g., in the edges of the Nakamura hypersurface, nonlinear differential equations must be solved; and the piecewise regular parts of the trajectory need to be matched together. The present paper is concerned with these difficulties.

Part I. Theory of piecewise smooth stress trajectories, constitutive law and yield criterion

The constitutive law for a shell is described in terms of generalized stress and strain [6, 7]. The state of stress at any point s, θ is represented by a four-dimensional vector Q with the components

$$Q_1 = n_s, \quad Q_2 = n_\theta, \quad Q_3 = m_s, \quad Q_4 = m_\theta. \tag{1}$$

The strain rates are described by the strain rate vector q, which is so chosen that the dot product $Q_i q_i$ represents the internal energy dissipation per unit area of the middle surface. Thus,

$$q_1 = \varepsilon_s, \quad q_2 = \varepsilon_\theta, \quad q_3 = \varkappa_s, \quad q_4 = \varkappa_\theta. \tag{2}$$

For stress states at yield, the end points of the vector Q lie on a hypersurface, which is convex [8, 9]. If this yield surface is derived from the Tresca-Mohr yield criterion, it consists of twelve regular pieces. Six of them are parabolic hypercylinders and the other six are 4th-degree noncylindric hypersurfaces. Nakamura [1, 2] has shown that an excellent approximation to the yield surface is obtained by retaining only the six hypercylinders H_i, described by the equations

$$f_i = \xi(m_\theta - \delta m_s) + \gamma m_s + [(\delta n_s - n_\theta)^2 + \gamma n_s^2] = 1, \quad i = 1, 2, \ldots, 6, \tag{3}$$

where

$$
\begin{aligned}
\xi &= 1, & \delta &= 0, & \gamma &= 0, & \text{for } H_\theta^+&: f_1 = h_\theta^+, \\
\xi &= -1 & \delta &= 0, & \gamma &= 0, & \text{for } H_\theta^-&: f_2 = h_\theta^-, \\
\xi &= 1, & \delta &= 1, & \gamma &= 0, & \text{for } H_{s\theta}^-&: f_3 = h_{s\theta}^-, \\
\xi &= -1, & \delta &= 1, & \gamma &= 0, & \text{for } H_{s\theta}^+&: f_4 = h_{s\theta}^+, \\
\xi &= 0, & \delta &= 0, & \gamma &= 1, & \text{for } H_s^+&: f_5 = h_s^+, \\
\xi &= 0, & \delta &= 0, & \gamma &= -1, & \text{for } H_s^-&: f_6 = h_s^-.
\end{aligned}
\tag{4}
$$

H_i denotes the regular part of the hypersurface for which $h_i = 1$.

The strain rates are given by the Mises flow law [8—10]. For all points within one face $f_k = 1$ of the yield surface, the flow law is

$$q_i = \lambda \frac{\partial f_k}{\partial Q_i} = \lambda f_{k,i}, \tag{5}$$

while for a point on the intersection of m faces it reads

$$q_i = \sum_{k=1}^{m} \lambda_k f_{k,i} \tag{6}$$

---

[1] A trajectory, as referred to in this text, is a locus of points on the yield hypersurface. This locus is a one-to-one map of points along the shell generator.

and the following conditions hold:

$$\lambda_k = 0 \text{ for } f_k < 1 \text{ and for } f_k = 1, \ \textstyle\sum_i dQ_i f_{k,i} < 0;$$

$$\lambda_k \geq 0 \text{ for } f_k = 1, \ \textstyle\sum_i dQ_i f_{k,i} = 0. \tag{7}$$

Stress compatibility equations, trajectories within faces

The strain compatibility equations are [12]

$$\varepsilon_s = \left(\frac{r\varepsilon_\theta}{r'}\right)' - \left(\frac{r}{r'}\right)^2 \frac{\varepsilon_\theta}{r_1} - \left(\frac{r}{r'}\right)^2 \frac{\varkappa_\theta}{kr_2}, \qquad \varkappa_s = \left(\frac{r\varkappa_\theta}{r'}\right)'. \tag{8a, b}$$

Substitution for the strain rates from the flow law (5) into (8a, b) gives two equations homogeneous in the deformation parameter:

$$\lambda f_{,1} = \left(\lambda \frac{r}{r'} f_{,2}\right)' - \left(\frac{r}{r'}\right)^2 \frac{\lambda}{r_1} f_{,2} - \left(\frac{r}{r'}\right)^2 \frac{\lambda}{kr_2} f_{,4}, \qquad \lambda f_{,3} = \left(\lambda \frac{r}{r'} f_{,4}\right)'. \tag{9a,b}$$

For $\lambda \neq 0$ (plastic flow), elimination of λ'/λ gives a stress compatibility equation

$$f_{,1} f_{,4} - f_{,2} f_{,3} = \left(\frac{r}{r'}\right) \left[f_{,4}(f_{,2})' - f_{,2}(f_{,4})' - \frac{1}{r_1}\left(\frac{r}{r'}\right) f_{,2} f_{,4} - \frac{1}{kr_2}\left(\frac{r}{r'}\right)(f_{,4})^2 \right], \tag{10}$$

for a stress trajectory within a face H.

For the faces $H_\theta^\pm$ and $H_{s\theta}^\pm$, substitution of $f = h_\theta^\pm$ and $f = h_{s\theta}^\pm$ from (3) into (10) gives differential equations for n_θ and for $(n_s - n_\theta)$ respectively. This differential equation, together with three equilibrium equations and one yield equation gives five equations for the five unknowns n_s, n_θ, m_s, m_θ and q_s. Thus, the stress solutions for trajectories within $H_\theta^\pm$ and $H_{s\theta}^\pm$ are fully prescribed.

However, for $H_s^\pm$ a different result is obtained. Substitution of $f = h_s^\pm$ into (9a, b) gives $\lambda = \lambda(s) = 0$ for $H_s^\pm$. Therefore, only a rigid body motion is permitted for stress trajectories in $H_s^\pm$. Also, another stress equation has not been found, i.e. there are only four equations for five unknowns and the stress solution for a trajectory within $H_s^\pm$ cannot be fully prescribed.

Stress trajectories in intersections

In an axisymmetrical shell problem, only two-dimensional intersections need be considered. The special case of more than two yield equations is coincidental and corresponds to trivial problems.

For an intersection of two faces, $H_1 \cap H_2$, substitution from the flow law (6) into (8a, b) gives

$$\lambda_i f_{i,1} = \left(\frac{r}{r'} \lambda_i f_{i,2}\right)' - \frac{1}{r_1}\left(\frac{r}{r'}\right)^2 \lambda_i f_{i,2} - \frac{1}{r_2 k}\left(\frac{r}{r'}\right)^2 \lambda_i f_{i,4}, \qquad \lambda_i f_{i,3} = \left(\frac{r}{r'} \lambda_i f_{i,4}\right)' \tag{11a, b}$$

where i is summed over 1, 2.

Because a stress solution can be obtained independently, these compatibility equations become a system of two linear homogeneous differential equations in the deformation parameters λ_1 and λ_2. Proof of the existence of solutions to such equations is given in [14, 15, 12]. It can be verified that a solution exists for the ratio λ_1/λ_2 and therefore, the solutions for λ_1 and λ_2 are only known within a common factor.

Strong discontinuity conditions

A stress trajectory on the yield surface may lie partly within several faces and intersections. Before solutions can be completed, boundary conditions at strong discontinuities, and transition conditions at weak discontinuities on the hypersurface need to be specified.

At a strong discontinuity ε_θ and $\varkappa_\theta$ are discontinuous. According to (8a, b), ε_s and $\varkappa_s$ must have discontinuities of higher order. As shown in [12, p. 36; 1, pp. 64—65], the strain-rates have the following ratios

$$\frac{\varepsilon_\theta}{\varkappa_s} = 0, \qquad \frac{\varkappa_\theta}{\varkappa_s} = 0, \qquad \frac{\varepsilon_s}{\varkappa_s} = \frac{\Delta\varepsilon_\theta}{\Delta\varkappa_\theta} \qquad (12\,\mathrm{a-c})$$

at a strong discontinuity, $s = s_0$, where $\Delta\varepsilon_\theta \equiv \varepsilon_\theta(s_0^+) - \varepsilon_\theta(s_0^-)$.

An important result of (12a—c) for the Nakamura (and Tresca) yield criterion is the following: A necessary condition for a strong discontinuity at $s = s_0$ is that the corresponding stress point lie in one of the hypersurfaces $H_s^\pm$ or in one of their boundaries.

From the flow-law (5), for $H_s^\pm$, $\frac{\varepsilon_s}{\varkappa_s} = \pm 2n_s$. Thus, the three boundary conditions for a strong discontinuity at $s = s_0$, a "yield-hinge" circle are

$$\pm m_s + n_s^2 = 1, \qquad w = 0, \qquad \frac{\Delta\varepsilon_\theta}{\Delta\varkappa_\theta} = \pm 2n_s \qquad (13\,\mathrm{a-c})$$

$\Delta w = 0$ because no shear deformation has been allowed. m_s, n_s and also q_s must always be continuous for equilibrium.

In addition other conditions governing the location of strong discontinuities can be derived. These conditions are listed in Table 1, and are derived from (8b), (13c) and the inequalities $\lambda_k > 0$, and $f_i \le 1$. For example, the possibility $H_\theta^+ \to H_s^-$ for $r' > 0$ must be ruled out because it violates the yield inequalities.

The conditions in Table 1 determine the existence of yield hinge circles and help locate the stress trajectory on the hypersurface adjacent to yield hinge circles, for discontinuities of various types. A rigid region is implied for $s \ge s_0$ and $s \le s_0$ for $\to H_s^\pm$ and $H_s^\pm \to$, respectively. Discontinuities of the type $H_i \to H_s^\pm \to H_j$, etc. can also exist, but the conditions may be different. They are not considered here for brevity.

The conditions of Table 1 do not apply in the special case of a cylindrical shell. These rules assume that r' and $\varkappa_\theta$ are not identically zero. However, the condition $\varkappa_\theta = 0$ for a cylindrical shell locates the possible stress points on the yield hypersurface to those where $\lambda_k f_{k,4} = 0$, from (6). The possibilities are $H_{s\theta}^+ \cap H_\theta^+$, $H_{s\theta}^- \cap H_\theta^-$, $H_\theta^+ \cap H_\theta^-$, $H_{s\theta}^+ \cap H_{s\theta}^-$.

Table 1. *Strong discontinuity locations and conditions*[1]

	Type	Geometry	Deformation	Stress
(a)	$H_{s\theta}^\pm \to H_s^\pm$	$r' > 0$		$n_\theta = 0$
	$H_\theta^\pm \to H_s^\pm$	$r' < 0$		$n_\theta = n_s$
(b)	$H_s^\pm \to H_\theta^\pm$	$r' > 0$		$n_\theta = n_s$
	$H_s^\pm \to H_{s\theta}^\pm$	$r' < 0$		$n_\theta = 0$
(c)	$H_{s\theta}^\pm \cap H_{s\theta}^\mp \to H_s^\pm$	$r' > 0$	$\lambda_2 = 0$	$n_s = \pm 1$
	$H_\theta^\pm \cap H_\theta^\mp \to H_s^\pm$	$r' < 0$	$\lambda_2 = 0$	$n_s = n_\theta = \pm 1$
	$H_{s\theta}^\pm \cap H_s^\pm \to H_s^\pm$	$r' > 0$		$n_s = n_\theta = 0$
	$H_\theta^\pm \cap H_s^\pm \to H_s^\pm$	$r' < 0$		$n_\theta = n_s$
(d)	$H_s^\pm \to H_\theta^\pm \cap H_\theta^\mp$	$r' > 0$	$\lambda_2 = 0$	$n_s = n_\theta = \pm 1$
	$H_s^\pm \to H_{s\theta}^\pm \cap H_{s\theta}^\mp$	$r' < 0$	$\lambda_2 = 0$	$n_s = \pm 1$
	$H_s^\pm \to H_\theta^\pm \cap H_s^\pm$	$r' > 0$		$n_\theta = n_s$
	$H_s^\pm \to H_{s\theta}^\pm \cap H_s^\pm$	$r' < 0$		$n_s = n_\theta = 0$

[1] The conditions apply at s_0^- for types $\to H_s^\pm$ and at s_0^+ for $H_s^\pm \to$.

Weak discontinuity conditions

Weak discontinuities can exist when the stress trajectory on the yield hypersurface makes a transition between faces $H_i \to H_j$, or between a face and an intersection $H_i \to H_i \cap H_j$, or between intersections. A weak discontinuity is defined as one for which ε_s and $\varkappa_s$ are discontinuous while ε_θ and $\varkappa_\theta$ are continuous. Continuity of ε_θ and $\varkappa_\theta$ can be used in the flow law (6), for both sides of the discontinuity $s = s_1$ to obtain two transition conditions. These are

$$[\lambda_1 f_{1,2} + \lambda_2 f_{2,2}]_{s_1^-} = [\mu_1 g_{1,2} + \mu_2 g_{2,2}]_{s_1^+}, \qquad [\lambda_1 f_{1,4} + \lambda_2 f_{2,4}]_{s_1^-} = [\mu_1 g_{1,4} + \mu_2 g_{2,4}]_{s_1^+} \qquad (14)$$

where g_i and μ_i are used to denote the yield functions and deformation parameter respectively, for $s \geq s_1$. The conditions noted for the flow law (6) must also be enforced. For example, if the stress trajectory is in $H_1 \cap H_2$ for $s \leq s_1$ and within H_1 for $s \geq s_1$, then it is required that $\mu_2 = 0$.

Sometimes, only trivial types of weak discontinuities are permitted, e.g. $\lambda_1 = \lambda_2 = \mu_1 = \mu_2 = 0$. If $\lambda_1 = \lambda_2 = 0$ at a point $s = s_1$, then (9a, b) give a solution for λ_1, $\lambda_2(s)$ both zero for the trajectory to one side of the point $s \leq s_1$. A similar statement applies for $s \geq s_1$. Then all the strain-rates are zero and therefore, the neighborhood about s_1 can move only as a rigid body.

Equilibrium requires that n_s, m_s and q_s are always continuous, but the possibility of discontinuous n_θ and m_θ exists. This leads to the following theorem:

Weak discontinuity theorem: At a weak discontinuity, discontinuous n_θ and m_θ correspond to continuous ε_θ and $\varkappa_\theta$ only if $\varepsilon_\theta = \varkappa_\theta = 0$. To prove the theorem the following lemma is used:

Lemma: The yield hypersurface is convex [8, 9]. For a stress point Q_i on the hypersurface, Fig. 1, and for any other statically admissible point Q_i^* (on or within the hypersurface) convexity is defined by

$$(Q_i^* - Q_i)\, q_i \leq 0 \qquad (15)$$

where q_i is given by (5) or (6).

Assume an orthonormal basis n_i for Q and q, such that $Q = Q_i n_i$. Define another "strain-rate vector" e as $e = \varepsilon_\theta n_2 + \varkappa_\theta n_4$, where at a weak discontinuity e must be continuous. Let $s = s_1$ be a weak discontinuity where Q is discontinuous, i.e.

$$Q = n_s n_1 + n_\theta^- n_2 + m_s n_3 + m_\theta^- n_4, \qquad \text{at } s = s_1^-$$

$$Q^* = n_s n_1 + n_\theta^+ n_2 + m_s n_3 + m_\theta^+ n_4, \qquad \text{at } s = s_1^+ .$$

(16a, b)

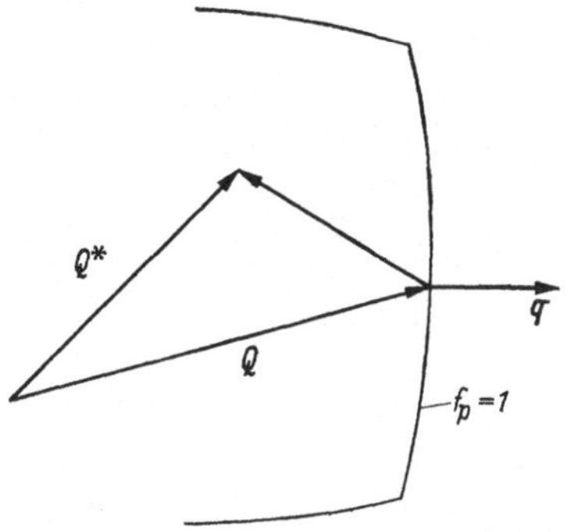

Fig. 1. Yield point Q and statically admissible point Q^*.

Case (1): Q^* also on the yield hypersurface. From (16a, b) and (15), $(Q_i^* - Q_i)\, q_i = (Q_i^* - Q_i)\, e_i \leq 0$. Because e is defined as continuous, $e_i = e_i^*$, this implies

$$(Q_i^* - Q_i)\, e_i^* \leq 0. \qquad (17)$$

$Q_i - Q_i^*$ and q_i^* must also satisfy (15), i.e.

$$(Q_i - Q_i^*)\, q_i^* = (Q_i - Q_i^*)\, e_i^* \leq 0. \qquad (18)$$

In order for both (17) and (18) to hold, it follows that

$$(Q_i^* - Q_i)\, e_i^* \equiv 0. \qquad (19)$$

If $Q_i^* - Q_i \neq 0$, then $e_i = e_i^* = 0$, and the theorem is proved for Case (1).

Discussion: This conclusion presupposes that $Q^* - Q$ and e^* ar not orthogonal. Suppose they are. Because q^* must be normal to a face f_k, then $Q^* - Q$ must lie on the tangent to $F_k(n_\theta, m_\theta) = 1$ (projection of $f_k(n_s, n_\theta, m_s, m_\theta) = 1$), at Q^*. From (15), all points Q in a neighborhood of Q^* must lie on the tangent to F at Q^* and consequently F must be linear in m_θ and n_θ at Q^*. The Nakamura faces (4), and also the Tresca faces [1, 2] are quadratic in n_θ. This is a contradiction. Therefore, the theorem holds. The exception is a two-dimensional state of stress, membrane state n_s, n_θ, $(m_s \equiv m_\theta \equiv 0)$ or bending state m_s, m_θ, $(n_s \equiv n_\theta \equiv 0)$, where either $\varkappa_\theta = 0$ or $\varepsilon_\theta = 0$ respectively, and the hypersurface degenerates to the linear 6-sided surface.

Comment: The theorem does not hold for piecewise linear approximations [3, pp. 23—25]. Such an approximation may result in associated velocity profiles quite different from those the original nonlinear hypersurface permits.

Finally, orthogonality must be considered for Q^* at an intersection. It can be shown that the same arguments hold.

Case (2): Q^* is within the yield hypersurface. From (5) and (6), $q_i^* \equiv 0$, $e_i^* \equiv 0$; hence $e_i \equiv 0$, and this holds for continuous Q_i as well as discontinuous Q_i. Thus, the *corollary*: At a weak discontinuity $s = s_1$, ε_θ and $\varkappa_\theta$ must be identically zero if the stress trajectory for $s > s_1$ enters the rigid domain bounded by the yield hypersurface whether Q_i is discontinuous or not.

Transition conditions. From the requirements on continuity of n_s, m_s, and q_s, the transition Eqs. (14a, b), the theorem and the yield criteria, conditions can be derived for the existence of all nontrivial weak discontinuities for piecewise smooth stress trajectories on the Nakamura yield hypersurface[1]. These conditions together with the strong discontinuity conditions enable systematic location of the stress trajectory on the hypersurface for any particular problem.

There are a great number of possible weak discontinuities to consider. The number of discontinuities of Type (a), $H_i \rightarrow H_j$, is 21, for the six faces H_i of the Nakamura hypersurface. There are 90 possible cases of Type (b), $H_i \cap H_j \rightarrow H_k$ and 120 of Type (c), $H_i \cap H_j \rightarrow H_k \cap H_l$. These 231 cases have been investigated and the conditions derived for existence of the nontrivial weak discontinuities are given in [12]. For rexample, $H_\theta^+ \cap H_\theta^- \rightarrow H_{s\theta}^+$ exists under the conditions that $\lambda_1 = 0$, $\lambda_2 = \mu$ and $n_s = 0$.

There are some cases of weak discontinuities which are nonexistent. For example, $H_i \rightarrow H_i$ cannot exist, because (14) requires $\lambda = \mu$. However, some cases of $H_i \cap H_j \rightarrow H_i \cap H_j$ do exist because at an intersection the strain-rate vector has more freedom than it has at a point within a face.

Part II. Computer algorithm solution to a collapse problem of conical shells

The theory of Part I forms the basis of decision making devices constructed into a computer algorithm. This algorithm is combined with a numerical integration scheme for testing conditions and inequalities at each step of integration. Conical shells are particularly chosen here as an illustration, but the same method has been used for other axisymmetric shell problems, and a general computer program has been written for any shell of revolution made up of parts, e.g. torispherical head pressure vessels, etc., [16].

Equation system

The equilibrium equations and kinetic relations are taken from [17], and are specialized here for conical shells with edge loading only as follows:

$$(sn_s)' - n_\theta = 0, \quad (sq_s)' + n_\theta \cot \alpha = 0, \quad (sm_s)' - m_\theta = \frac{sq}{k}, \qquad (20\,\text{a—c})$$

$$\varepsilon_s = v', \quad \varepsilon_\theta = \frac{v + w \cot \alpha}{s}, \quad \varkappa_s = kw'', \quad \varkappa_\theta = \frac{kw'}{s}. \qquad (21\,\text{a—d})$$

Additional equations are the compatibility Eqs. (9a, b) or (11a, b). These differ for a piece of a stress trajectory in H_i, (Case A) or in $H_i \cap H_j$, (Case B). For the Nakamura hypersurface, Eqs. (3) and (4), it can be shown that (9a, b) and (11a, b) reduce to the following:
Case A, H_i:

$$\lambda' = -\frac{\delta + 1}{s} \lambda, \quad n_\theta' = \delta n_s' + \xi \frac{\cot \alpha}{2k}. \qquad (22\,\text{a, b})$$

---

[1] A piecewise smooth trajectory is defined to be a continuous locus of stress points which is piecewise differentiable.

Case B, $H_i \cap H_j$:

$$\lambda_2' = \frac{\xi_1}{n_\theta\,(\xi_1 - \xi_2) + n_s\,(\xi_2\,\delta_1 - \xi_1\,\delta_2)} \left\{ - \left[n_\theta' - \delta_1 n_s' + \frac{\cot\alpha}{2k}\,\xi_1 \right] \lambda_1 \right.$$

$$+ \left[(2\delta_2\xi_1 - (1 + \delta_2)\,\delta_1\xi_2)\,\frac{n_s}{s\xi_1} + (\xi_2 - \xi_1)\,(1 + \delta_2)\,\frac{n_\theta}{s\xi_1} - n_\theta' + \delta_2 n_s' \right.$$

$$\left. \left. + \frac{\cot\alpha}{2k}\,\xi_2 \right] \lambda_2 \right\}, \qquad \lambda_1' = -\,(1 + \delta_1)\,\frac{\lambda_1}{s} - (1 + \delta_2)\,\frac{\xi_2}{\xi_1}\,\frac{\lambda_2}{s} - \frac{\xi_2}{\xi_1}\,\lambda_2'. \qquad \text{(23a, b)}$$

The remaining equations are the yield Eqs. (3), which can be solved for m_θ, and for m_θ and n_θ respectively, for Cases A and B. $H_s^\pm$ are not considered because a stress trajectory in $H_s^\pm$ corresponds to a rigid region.

The stress solution is now completely specified. The velocity solution can be found from (21 a–d), using the flow law (5) or (6) as follows:

Case A, H_i:

$$w' = \frac{\xi s}{k}\,\lambda, \qquad v = 2s\lambda(n_\theta - \delta n_s) - w\cot\alpha. \qquad \text{(24a, b)}$$

Case B, $H_i \cap H_j$:

$$w' = \frac{s}{k}\,(\xi_1\lambda_1 + \xi_2\lambda_2), \qquad v = 2s[(n_\theta - \delta_1 n_s)\,\lambda_1 + (n_\theta - \delta_2 n_s)\,\lambda_2] - w\cot\alpha. \qquad \text{(25a, b)}$$

The equation system can be set up for numerical integration. There are five differential equations to integrate. For Case A, these are Eqs. (20a, c), (22a, b), and (24a)[1], and for Case B, Eqs. (20a, c), (23a, b), and (25a). The equations for Case A can be integrated directly in closed form. However, they are set up for numerical integration on the computer, in order that the procedure be the same for both Case A and B.

Integration method and decisions at transition points

Plastic stress distributions are slowly varying in contrast to rapidly varying elastic solutions for the edge zone of a shell. Therefore, a second-order Adams integration procedure [18] was found to be sufficiently adequate when checked against a fourth-order Adams procedure [19]. For the second-order procedure, a Taylor series starting method was used. The integration has to be restarted each time the stress trajectory makes a transition at a weak discontinuity into a new region of different yield behavior.

The integration for each region of different yield behavior is governed by the inequalities, $h_i \le 1$ and $\lambda_i \ge 0$. For example, integration for Case A for $h_1 = 1$ proceeds until another of the $h_i = 1$. At this point a decision is made whether to proceed into H_i or into $H_1 \cap H_i$, by checking the weak discontinuity conditions [12]. As a second example of recognizing the end of an integration region, consider integration for Case B for $h_1 = h_2 = 1$. At a point $s = s_1$ during the integration, if λ_1 becomes zero then the trajectory may enter into H_2 for $s \ge s_1$. Otherwise, if another $h_i = 1$ at $s = s_1$, then the trajectory may enter H_i, $H_i \cap H_1$ or $H_2 \cap H_i$ for $s \ge s_1$. The location for $s \ge s_1$ can again be determined by testing the conditions [12]. Such decision making devices were built into a computer algorithm for automatic location of piecewise smooth trajectories on the Nakamura yield hypersurface. Stanford's B 5500 Burroughs computer, employing the ALGOL language, was used. At transition points between regions of Case A and Case B integrations, parabolic interpolation was used [20].

Three-parameter collapse of a conical shell

The conical shell problem shown in Fig. 2 is considered. The three loadings at the inner edge are N, M, and Q as shown, and

$$n_s = n, \qquad m_s = -m, \qquad q_s = q \text{ at } s = 1. \qquad \text{(26a–c)}$$

---

[1] Equation (20b) can be integrated directly by combination with (20a) to give $sq_s = -s\cot\alpha\,n_s +$ constant.

The outer edge of the collapse region is either the clamped edge $s_0 = \beta$ or it may be a yield hinge circle $s_0 = \beta_0 \leq \beta$. The boundary conditions at s_0 are given by Eqs. (13a—c).

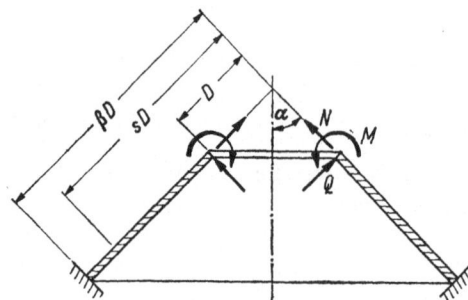

Fig. 2. Three-parameter loading of a cone.

An immediate difficulty arises for numerically integrating from the inner edge, because it is not known a priori the corresponding point on the yield hypersurface. However, the yield conditions are known at the outer edge—the stress trajectory will most generally lie in $H_{s\theta}^{\pm}$ for a strong discontinuity $H_{s\theta}^{\pm} \rightarrow H_s^{\pm}$ with $r' > 0$ and $0 < |n_s| < 1$, Table 1. (The cases $n_s = 0$ and $n_s = \pm 1$ lead to other trajectories, however, these solutions may be considered the limit cases for $n_s \rightarrow 0$ or $n_s \rightarrow \pm 1$). Because the location of the stress trajectory is known at the outer edge, it is decided to integrate backward from there.

The boundary condition (13a) is an equation between n_s and m_s. Consequently, either n_s or m_s must be assumed. Also, q_s must be assumed and the integration iterated to satisfy boundary conditions (26a—c). Alternatively, as chosen here, n_s and q_s can be assumed at $s = \beta_0$ and whatever set of n, m, q that result at $s = 1$ are taken as a point on the collapse load surface. A sufficient number of such integratons then gives enough data to construct the three-parameter collapse load surface in (n, m, q) space. n_s and q_s at $s = \beta_0$, however, must be selected so that $(h_s^{\pm})' \geq 0$ at $s = \beta_0$, in order that the inequality $h_s^{\pm} \leq 1$ be not violated for $s < \beta_0$. More details concerning the boundary conditions are given in [12].

The particular geometrical parameters chosen for the problem were

$$\cot \alpha = 1, \qquad k = 0.025\,^1, \qquad \beta = 1.35. \qquad (27\,\text{a—c})$$

With $\beta = 1.35$, all yield-hinge circles, $s_0 = \beta_0$ were encompassed, i.e. a complete (n, m, q) collapse load surface was found for $\beta_0 \leq 1.35$. (However, for shallower cones, larger β_0 occur, [12, p. 103].)

The second-order Adams procedure used employed a constant step size. It was found that a step size of $h = 0.005\,(\beta - 1)$ gave accuracy to four significant figures, when the results were compared for smaller step sizes. The computation procedure was found to be efficient. The run time was $10-15$ seconds for one entire integration on the B 5500 with testing performed at each step.

Results

A wide variety of yield behavior is exhibited in the collapse of the conical shell under three-parameter loading. In fact, 26 different classes of stress trajectories were found: 13 each for $q > 0$ and $q < 0$, respectively. The various trajectories and their corresponding yield conditions are tabulated in Table 2 for $q > 0$. The remaining 13 for $q < 0$ have similar yield conditions with the $+$ and $-$ signs interchanged. As evident from Table 2, there are as many as seven different yield criteria for one trajectory—class no. (13) (including H_s^+ at the yield hinge at $s = \beta_0$). Obviously, an intuitive approach is impracticable in a problem of this complexity and the necessity of a systematic method of solution is realized.

Two collapse load curves are plotted in Figs. 3 and 4. The first is a level curve of the (n, m, q) collapse load surface for $n \equiv 0$, the second is a level curve for $m \equiv 0$. Portions of these curves correspond to the different classes of yield behavior in Table 2. Two other degenerate classes of yield behavior are also exhibited by the curves. These occur when the yield-hinge circle coincides with the inner edge, i.e. $\beta_0 = 1$. The collapse region degenerates to the line $s = 1$ and the stress trajectory degenerates to a point in the surface H_s^- or H_s^+, for which either $n = 0$, $m = \pm 1$, or $n = \pm 1$, $m = 0$.

---

[1] $L = D \sin \alpha$, Fig. 2, is used in definition of k.

Table 2. *Classes of trajectories for a cone under three-parameter loading*

Classification Number	Yield Conditions in Regions: $(1 \le s \le s_1), \ldots, (s_k \le s \le \beta_0)$
(1)	$H_{s\theta}^+$
(2)	$H_{s\theta}^+ \wedge H_\theta^+,\; H_{s\theta}^+$
(3)	$H_{s\theta}^+ \wedge H_\theta^-,\; H_{s\theta}^+$
(4)	$H_{s\theta}^+ \wedge H_{s\theta}^-,\; H_{s\theta}^+$
(5)	$H_\theta^- \wedge H_\theta^+,\; H_{s\theta}^+ \wedge H_\theta^+,\; H_{s\theta}^+$
(6)	$H_{s\theta}^+ \wedge H_{s\theta}^-,\; H_{s\theta}^+ \wedge H_\theta^+,\; H_{s\theta}^+$
(7)	$H_\theta^-,\; H_{s\theta}^+ \wedge H_\theta^-,\; H_{s\theta}^+$
(8)	$H_\theta^- \wedge H_{s\theta}^-,\; H_\theta^- \wedge H_\theta^+,\; H_{s\theta}^+ \wedge H_\theta^+,\; H_{s\theta}^+$
(9)	$H_\theta^- \wedge H_{s\theta}^-,\; H_{s\theta}^+ \wedge H_{s\theta}^-,\; H_{s\theta}^+ \wedge H_\theta^+,\; H_{s\theta}^+$
(10)	$H_\theta^- \wedge H_\theta^+,\; H_\theta^-,\; H_{s\theta}^+ \wedge H_\theta^-,\; H_{s\theta}^+$
(11)	$H_\theta^- \wedge H_\theta^+,\; H_\theta^-,\; H_{s\theta}^+ \wedge H_\theta^-,\; H_{s\theta}^+ \wedge H_\theta^+,\; H_{s\theta}^+$
(12)	$H_\theta^- \wedge H_{s\theta}^-,\; H_\theta^- \wedge H_\theta^+,\; H_\theta^-,\; H_{s\theta}^+ \wedge H_\theta^-,\; H_{s\theta}^+$
(13)	$H_\theta^- \wedge H_{s\theta}^-,\; H_\theta^- \wedge H_\theta^+,\; H_\theta^-,\; H_{s\theta}^+ \wedge H_\theta^-,\; H_{s\theta}^+ \wedge H_\theta^+,\; H_{s\theta}^+$

In order to portray the yield behavior more completely, the entire collapse load surface was constructed by building up level curves of constant n. Two quadrants of the three-parameter surface are shown in Figs. 5 and 6 for $n > 0$ and $n < 0$ respectively. The collapse load q is positive in both figures. The two other quadrants for $q < 0$ are similar and can be identified by symmetry about the origin: $(n, m, q) \to (-n, -m, -q)$.

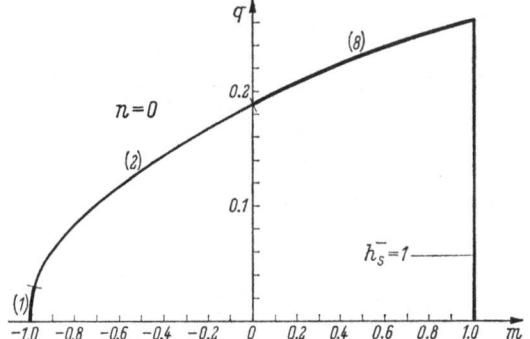

Fig. 3. Level curve of the collapse-load surface for $n = 0$, $k = 0.025$, $\cot \alpha = 1$.

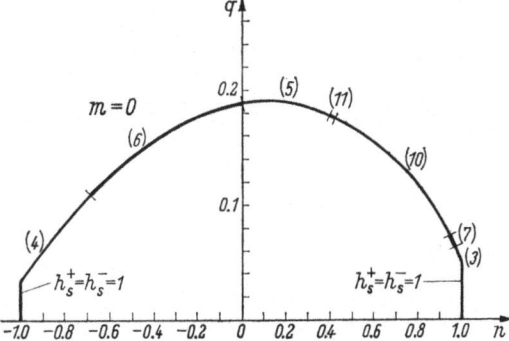

Fig. 4. Level curve of the collapse-load surface for $m = 0$, $k = 0.025$, $\cot \alpha = 1$.

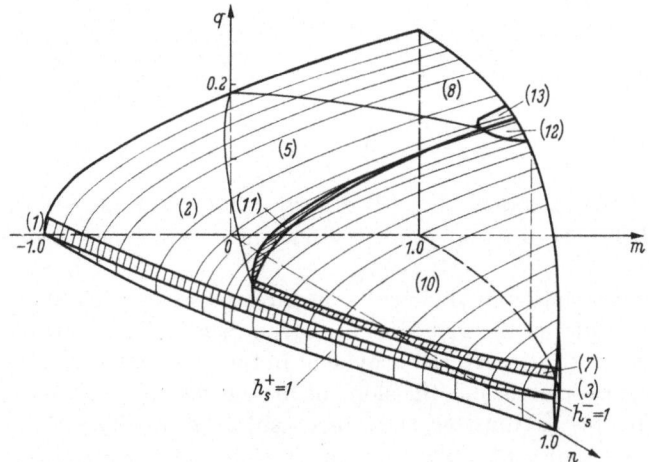

Fig. 5. Quadrant of the collapse-load surface for positiv n and q, $k = 0.025$, $\cot \alpha = 1$.

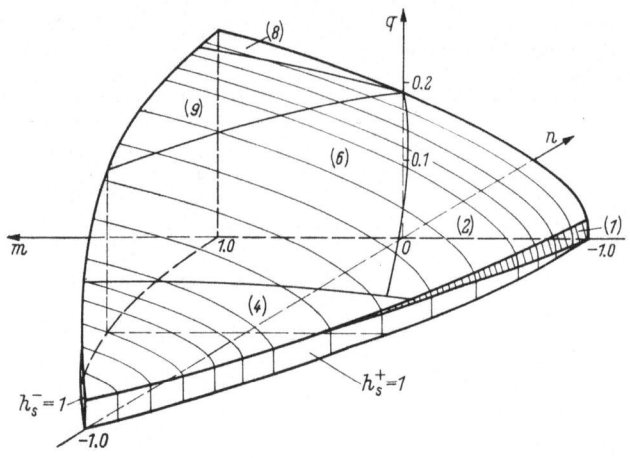

Fig. 6. Quadrant of the collapse-load surface for negativ n and positiv q,
$k = 0.025$, cot $\alpha = 1$.

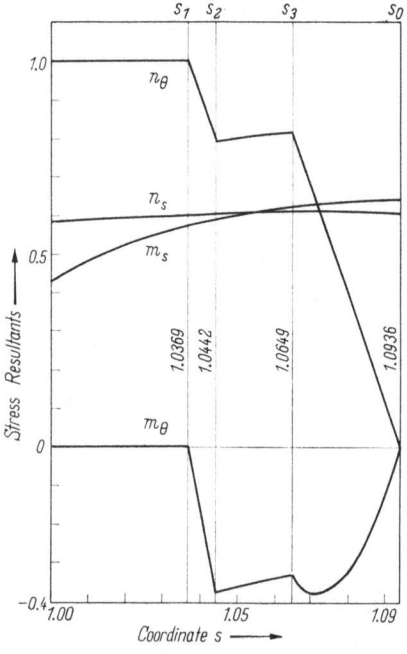

Fig. 7. Stress resultants for a No. (10) trajectory $k = 0.025$, cot $\alpha = 1$, $n = 0.58$,
$m = -0.43$, $q = 0.096$.

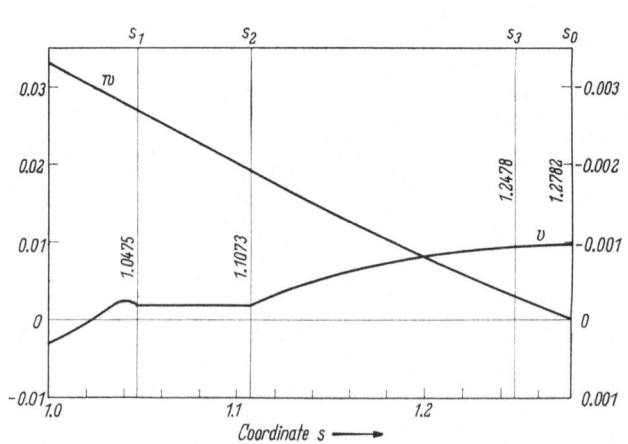

Fig. 8. Incipient velocities for a No. (8)
trajectory $k = 0.025$, cot $\alpha = 1$, $n = 0.1$,
$m = 0.91$, $q = 0.26$.

From Figs. 5 and 6 it is evident that Classes (2), (4), (5), (6), (8), (9), and (10) dominate. However, a conical shell of different geometry may have a different distribution of yield behavior.

To conclude the description of the results on three-parameter loading, an example plot of stress resultants and a plot of the velocities are given in Figs. 7 and 8 respectively. The piecewise smooth character of the trajectories is evident in the plots in Fig. 7. The curves of the stress resultants n_s and m_s are smooth, but the slope of the curves of n_θ and m_θ changes abruptly at the transition coordinates s_i. However, the stress resultants are all continuous. This continuity is required by the weak discontinuity theorem in order that the associated deformation at collapse be nontrivial.

The incipient velocities, v and w at collapse are shown in Fig. 8 for a Class (8) trajectory. It is noted that, at the transition coordinates s_i, both v and w' are continuous, which verifies the existence of weak discontinuities at these points, i.e., ε_θ and $\varkappa_\theta$ are continuous. The strain-rate $\varepsilon_s = v'$ is discontinuous at the weak discontinuity points, but the changes in curvature $(\varkappa_s)$ of the w plots are difficult to see (the w plot is linear in the regions $s_1 \leq s \leq s_2$ corresponding to $H_\theta^- \cap H_\theta^+$).

Conclusions

A theorem states that a weak discontinuity can exist with discontinuous stresses only if the strain rate ε_θ and the rate of curvature $\varkappa_\theta$ are identically zero. Such a discontinuity line can have only rigid body motion. The theorem is true for nonlinear hypersurfaces, but it does not hold for linear approximations. Therefore, nontrivial strain states correspond to piecewise smooth trajectories on the Tresca and Nakamura hypersurfaces, but may correspond to discontinuous stresses for linear approximations.

Rules for yield-hinge circles, weak discontinuity conditions, compatibility, and flow law requirements form the basis for construction of a computer algorithm for automatically locating stress trajectories on the yield hypersurface. This algorithm, combined with a numerical integration scheme for testing conditions and inequalities at each step of integration, make up a computer program for finding collapse loads in complicated problems.

The computer program was used to solve the problem of a conical shell under three-parameter loading—a direct stress, a bending moment, and a transverse shear load at the inner edge. Complete collapse load surfaces were found. Complicated yield behavior, twenty-six different classes of trajctories, were found. The most complicated trajectory exhibited seven regions of yield behavior—the trajectory lay partly in two faces of the yield hypersurface, partly in four intersections, and made a strong discontinuity at another hypersurface. Thus, the advantage of a systematic method of solution was realized.

References

1. NAKAMURA, T.: Plastic analysis of shells of revolution under axisymmetric loads, Diss. Stanford University, 1961, see also Dissertation Abstracts 27, 4316 (1962)[1].

2. FLÜGGE, W., and T. NAKAMURA: Plastic analysis of shells of revolution under axisymmetric loads. Ing. Archiv, 34, 238—247 (1965).

3. HODGE, P. G., Jr.: Limit analysis of rotationally symmetric plates and shells, Englewood Cliffs, N.J.: Prentice-Hall 1963.

4. KOOPMAN, D. C. A., and R. H. LANCE: On linear programming and plastic limit analysis. J. Mech. Phys. Solids, 13, 77—87 (1965).

5. DINNO, K. S., and S. S. GILL: Limit pressure of symmetrically loaded shells of revolution. Int. J. Mech. Sci. 7, 15—19 (1965).

6. PRAGER, W.: The general theory of limit design, Proc. 8th Int. Congr. Appl. Mech., 2, 65—72, Istanbul, (1952).

7. GVOZDEV, A. A.: The determination of the value of the collapse load for statically indeterminate systems undergoing plastic deformations. Int.

J. Mech. Sci. 1, 322—335 (1960). Translation from Proc. Conf. Plastic Deformations, 19 (1936), Akad. Nauk SSSR, Moscow-Leningrad 1938.

8. DRUCKER, D. C.: A more fundamental approach to plastic stress-strain relations, Proc. First U.S. Nat. Congr. Appl. Mech. 1951, pp. 487—491.

9. KOITER, W. T.: General theorems for elastic-plastic solids, Progress in Solid Mechanics, 1, Ed.: Sneddon, I. N., and R. Hill, 165—221, Tech. Univ. Delft, Netherlands, (1960).

10. — Stress-strain relations, uniqueness and variational theorems for elastic-plastic materials with a singular yield surface, Quart. Appl. Math. 11, 350 (1953).

11. ONAT, E. T., and W. PRAGER: Limit analysis of shells of revolution I, II, Proc. Roy. Netherlands Acad. Sci., B 57, 534—548 (1954).

12. GERDEEN, J. C.: Shell plasticity — piecewise smooth trajectories on the Nakamura yield surface, Diss. Division of Engineering Mechanics, Stanford University 1965.

---

[1] In the discussion, the following paper has been brought to the attention of the authors. Z. Mroz, B-Y. Xu; The load carrying capacities of symmetrically loaded spherical shells, Arch. Mech. Stos. 15, 245—266 (1963); this appears to be a second, independent presentation of Nakamura's yield surface.

13. Lance, R. H., and E. T. Onat: Analysis of plastic shallow conical shells, J. Appl. Mech. 30, 199—209 (1963).

14. Coddington, E., and N. Levinson: Theory of ordinary differential equations, New York: McGraw-Hill, Chapters 1, 3. (1955).

15. Ince, E. L.: Ordinary differential equations, Chapter III. Dover Publications 1956.

16. Gerdeen, J. C.: Theoretical analysis of the plastic collapse of thin shell structures, Battelle Special Report, Battelle Columbus Laboratories, Columbus, Ohio, 1966.

17. Flügge, W.: Stresses in shells, Berlin/Göttingen/Heidelberg: Springer 1962, pp. 320, 355—359.

18. Hildebrand, F. B.: Advanced calculus for applications, Prentice Hall 1963, pp. 96—98.

19. Stanford library program No. 76, "Fourth-order Adams Predictor—Corrector Method", Stanford Computation Center, Stanford, California.

20. Stanford library program No. 16, "Newton's forward and backward interpolation for equally spaced points", Stanford Computation Center, Stanford, California.

An elastic-plastic analysis of a crack in a plate of finite size

By

L. E. Hulbert, G. T. Hahn, A. R. Rosenfield, and **M. F. Kanninen**

Battelle Memorial Institute, Columbus

Introduction

Two great complexities are involved in determining the stress and strain fields in a body containing a crack. These arise in treating the plastically deformed regions at the ends of the crack and in accounting for the boundary conditions on the periphery of the body. Progress has been made in treating these complexities individually. A large number of completely elastic solutions for finite and semi-finite cracked plates (see, for example, the compilation by FEDDERSEN [1]) and of elastic-plastic solutions for infinite cracked plates are available. Few elastic-plastic solutions for finite regions have been reported, however. Of these the anti-plane-strain solutions of BILBY et al [2] and KOSKINEN [3] are particularly noteworthy.

In this paper a combined theoretical and experimental investigation of precracked thin sheets under uniaxial loading is described. The theoretical analysis employs the Dugdale model [4] and, simultaneously, accounts for the finite outer dimensions of the sheet containing the crack. Dugdale's crack model provides a simple yet effective way to account for the plastic deformation which occurs at the ends of a crack under conditions which approximate plane stress. In essence, the plastic regions are taken to be thin extensions of the crack itself. The length of these zones is such that the stress singularity, an inevitable feature of the elastic solution, is abolished leaving the stresses in the body everywhere finite.

In the experimental portion of the program rectangular center-notched foil coupons of annealed mild steel — a material which exhibits plastic zones in advance of the crack which are narrow and wedge-shaped like those of the Dugdale model — were used. Previous measurements on sheets and foils in which the plastic zone size was small compared to the overall dimensions, were found to be in good agreement with Dugdale's predictions of plastic zone size [4, 5] and of crack-tip displacement [5]. The present experiments extend to conditions under which the plastic zone size is comparable to the foil width.

The theoretical analysis will be described in the following section. Subsequently, the experimental program will be described and the results compared with those of the theory. In particular, it will be shown that the observed plastic zone sizes for a given applied stress are in good agreement with the theoretical values. The existence of an elastic-plastic instability predicted by the theory is also confirmed. In addition, the effect of imposing different kinds of boundary conditions is examined and is shown to have a substantial effect in many instances.

The theoretical analysis

The general method

The analysis of a Dugdale crack, even though plastic zones are considered to exist at its ends, lies completely within the framework of linear elasticity theory. That is, the elastic solution for a uniform pressure acting on a part of the edge of a flattened elliptic hole is used;

the surface tractions on the hole are attributed to the presence of a region of plastic deformation. The region occupied by this hypothetical material is, however, outside the region under consideration which remains completely elastic.

The analysis reported in this paper was performed with an extension of the numerical scheme known as the boundary point least squares method. This method, which has wide applicability to boundary value problems in general, has been applied extensively by HULBERT [6] for the solution of plane elasticity problems in multiply connected regions. Briefly, the method requires a series solution to the partial differential equation governing the problem. Each term of the series contains an arbitrary constant whose value is determined by satisfying the boundary conditions, not along the entire boundary, but only at a discrete number of boundary points. In contrast to the boundary point method known as collocation[1], the number of boundary value equations is greater than the number of free constants. The boundary data are then satisfied only in the least squares sense.

A major difficulty encountered in simple collocation is that the solution usually depends on the number and the location of the selected boundary points. The consequent tedious trial and error process required to find the optimum set of points (the set which minimizes the errors on the "unmatched" boundary segments) is avoided here. It will be shown below that the number and location of the boundary points and, to a certain extent, the number of stress function coefficients is not critical in the present problem.

In order to remove the stress singularity in accord with Dugdale's hypothesis, a relation involving the unknown stress function coefficients must be satisfied. However, unlike the boundary value equations, this relation must be satisfied exactly. This is accomplished by imposing a constraint upon the least squares solution through the introduction of a Lagrange multiplier. Hence, the name "boundary point least squares with constraints" can be given to the technique used in this analysis.

The stress function series

The problem to be considered here is shown in Fig. 1. A Dugdale crack of length $2c$ is centrally located in a rectangular sheet of width $2w$ and height $2h$. The sheet is stretched by a force $2wT$ (per unit thickness) acting on the edges parallel to the crack line. The yield stress of the material is denoted as Y.

Because of the symmetry, only the portion of the sheet in the first quadrant of the $x - y$ plane need be considered. The appropriate boundary conditions for three different types of loading are as shown in Fig. 2.

The boundary point least squares technique is based on the complex variable formulation of the Airy stress function as two analytic functions φ and ψ. The expressions for the stresses and displacements in terms of these two functions and their derivatives are given, for example, by MUSKHELISHVILI [8] as

$$\sigma_x + \sigma_y = 2[\varphi'(z) + \overline{\varphi'(z)}],$$

$$\sigma_y - \sigma_x + 2i\tau_{xy} = 2[\bar{z}\varphi''(z) + \psi'(z)], \tag{1}$$

$$2\mu(u + iv) = \varkappa\varphi(z) - z\overline{\varphi'(z)} - \overline{\psi(z)}$$

where $z = x + iy$ and $\varkappa = (3 - \nu)/(1 + \nu)$ for plane stress.

In the present problem functions satisfying the required boundary conditions on the x and y axes are employed. This, in effect, determines one stress function in terms of the other which remains arbitrary. A series form for this stress function can be adopted and the free constants contained in the series used to satisfy the conditions along the periphery of the plate (the lines $y = h$, $x = w$ in Fig. 2). The selected stress function will be composed of terms for which,

---

[1] This technique was used, for example, by KOBAYASHI, CHERPY, and KINSEL [7] in a strictly elastic analysis of a crack in a finite sized plate.

on the x axis,

$$\text{(A)} \quad \sigma_y = \tau_{xy} = 0$$

$$\text{(B)} \quad \begin{cases} \sigma_y = \tau_{xy} = 0, & |x| < a \\ v = \tau_{xy} = 0, & |x| \geq a \end{cases}$$

$$\text{(C)} \quad \begin{cases} \sigma_y = \tau_{xy} = 0, & |x| < c \\ \sigma_y = Y, \ \tau_{xy} = 0, & c \leq |x| \leq a \\ v = \tau_{xy} = 0, & a \leq |x|. \end{cases}$$

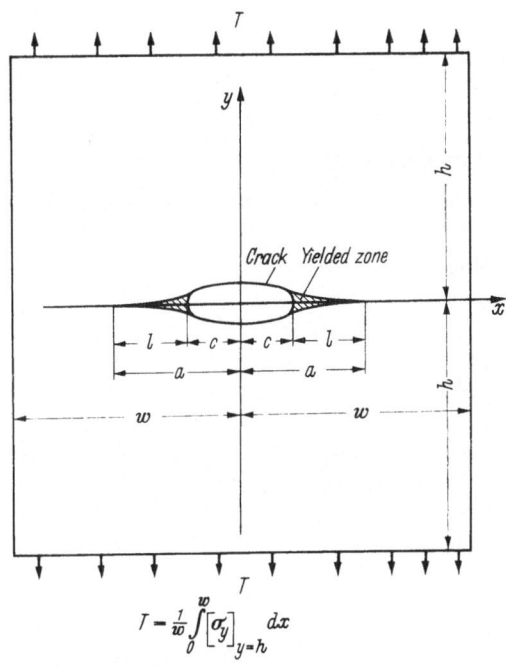

$$T = \frac{1}{w} \int_0^w [\sigma_y]_{y=h} \, dx$$

Fig. 1. The Dugdale model for a centrally located crack in a rectangular plate.

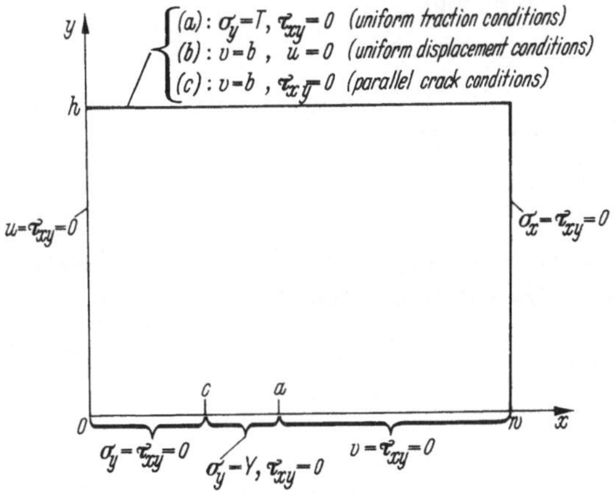

Fig. 2. The boundary value problem.

The boundary conditions on the y axis are satisfied by the symmetry properties of the selected functions.

Stress functions φ_1 and ψ_1 satisfying conditions of the type (A) are obtained by setting

$$\psi_1(z) = -\bar{\varphi}_1(z) - z\varphi_1'(z) \tag{2}$$

and taking

$$\varphi_1(z) = \sum_{n=1}^{N} A_n z^{2n-1}. \tag{3}$$

Conditions (B) are satisfied by functions φ_2 and ψ_2 where

$$\psi_2(z) = \bar{\varphi}_2(z) - z\varphi_2'(z) \tag{4}$$

and

$$\varphi_2(z) = \sqrt{z^2 - a^2} \sum_{n=0}^{N} B_n z^{2n}. \tag{5}$$

Notice that although the number N of terms in the series given by Eqs. (3) and (5) need not be the same, it has been found to be computationally advantageous to have it so.

The stress functions φ_3 and ψ_3 satisfying conditions of the type (C) are obtained by specializing Muskhelishvili's [9] solution for an elliptic hole subjected to a uniform pressure on a part of its edge. These are

$$\varphi_3(z) = -\frac{Y}{2\pi i} \left[(z+c) \log Z_1 + (z-c) \log Z_2 - i\alpha \left(z - \sqrt{z^2-a^2}\right) \right], \tag{6}$$

$$\psi_3(z) = -\frac{Y}{2\pi i} \left[c \log Z_1 - c \log Z_2 - i\alpha \frac{a^2}{\sqrt{z^2-a^2}} \right] \tag{7}$$

where

$$Z_1 = \frac{z + \sqrt{z^2-a^2} + c + i\sqrt{a^2-c^2}}{z + \sqrt{z^2-a^2} + c - i\sqrt{a^2-c^2}}$$

$$Z_2 = \frac{z + \sqrt{z^2-a^2} - c - i\sqrt{a^2-c^2}}{z + \sqrt{z^2-a^2} - c + i\sqrt{a^2-c^2}}$$

and

$$\alpha = i \log \frac{c - i\sqrt{a^2-c^2}}{c + i\sqrt{a^2-c^2}} = 2 \cos^{-1} \frac{c}{a}. \tag{8}$$

The singularity cancelling equation

Ignoring for the moment nonsingular terms, the stresses given by substituting the combined stress functions

$$\varphi(z) = \varphi_1(z) + \varphi_2(z) + \varphi_3(z) \tag{9}$$

$$\psi(z) = \psi_1(z) + \psi_2(z) + \psi_3(z) \tag{10}$$

into Eqs. (1) are

$$\sigma_x = \sigma_y = 2 \operatorname{Re} \left\{ \frac{z}{\sqrt{z^2-a^2}} \sum_{n=0}^{N} B_n z^{2n} - \frac{Y\alpha}{2\pi} \frac{z}{\sqrt{z^2-a^2}} \right\}.$$

The shear stress is finite everywhere. Therefore, in order to remove the singularity in accord with Dugdale's hypothesis, set

$$\sum_{n=0}^{N} B_n a^{2n} - \frac{Y\alpha}{2\pi} = 0$$

or, using Eq. (8)

$$\frac{Y}{\pi} \cos^{-1}\left(\frac{c}{a}\right) = \sum_{n=0}^{N} B_n a^{2n}. \tag{11}$$

Notice that for an infinite region, $B_0 = \frac{T}{2}$, $B_1 = B_2 = \cdots = B_N = 0$ and Eq. (11) becomes

$$\frac{c}{a} = \cos \frac{\pi}{2} \frac{T}{Y} \tag{12}$$

as given by Dugdale.

Introduction of the Lagrange multiplier

Equation (11) is a relation between the unknown coefficients which must be exactly satisfied. This cannot be accomplished by simply appending it to the set of boundary value equations in the boundary point least squares approach since none of the boundary value equations are satisfied exactly. Hence, the Lagrange multiplier technique has been incorporated to accomplish this.

The stresses and displacements given by (A) and (B) contain N and $N + 1$ arbitrary constants respectively. Type (C) contributes a single term. Thus, a number $M > 2N + 2$ points on the periphery of the sheet must be selected at which to set the boundary conditions. Omitting the details, the set of M equations in the $2N + 2$ stress function coefficients $A_1, A_2, \ldots, A_N, B_0, B_1, \ldots, B_N$, and Y, represented by the vector X, will then have the form

$$QX = R \tag{13}$$

where Q is an $M \times (2N + 2)$ matrix and R is an M component vector. The matrix Q contains the values of the stresses (or displacements) given by each term of the stress function evaluated at each given boundary point. R contains the specified values of the stresses (or displacements) at each boundary point. Now, because there are more equations to be solved than there are unknowns to be determined, the left hand sides of Eqs. (13) will not in general be equal to their right hand sides. Call these differences the δ_i's, $i = 1, 2, \ldots, M$. A solution is achieved in the least squares sense when the sum

$$\Delta = \sum_{i=1}^{M} \delta_i^2 = (QX - R)^T (QX - R) \tag{14}$$

is a minimum. (The superscript T is used here to indicate the transpose operation.)

In the boundary point least squares method with constraints a solution is sought in which Eq. (14) is minimized while Eq. (11) is satisfied exactly. A method to accomplish this employing Lagrange multipliers has been given by KORDA [10]. In applying this method in the present problem the augmented function $\Delta^\dagger$ defined as

$$\Delta^\dagger = \Delta + 2 \left(\frac{Y}{\pi} \cos^{-1}(c/a) - \sum_{n=0}^{N} B_n a^{2n} \right) \lambda \tag{15}$$

is minimized. Here λ denotes a single Lagrange multiplier.

More generally, if a matrix L having as many rows as there are equations of constraint is defined, the constraint equations[1] can be written as

$$LX = S. \tag{16}$$

Then, substituting Eqs. (14) and (16) into Eq. (15), gives

$$\Delta^\dagger = (QX - R)^T (QX - R) + 2(LX - S)^T \Lambda$$

where Λ is a column vector. Differentiating $\Delta^\dagger$ with respect to X and equating the result to zero gives

$$0 = (Q^T QX - Q^T R) + L^T \Lambda$$

or

$$[Q^T Q \mid L^T] \begin{bmatrix} X \\ \Lambda \end{bmatrix} = [Q^T R].$$

By appending Eq. (16) to this set the final form

$$\begin{bmatrix} Q^T Q & L^T \\ L & 0 \end{bmatrix} \begin{bmatrix} X \\ \Lambda \end{bmatrix} = \begin{bmatrix} Q^T R \\ S \end{bmatrix} \tag{17}$$

is obtained. This represents in the present problem a set of $2N + 2$ linear equations in an equal number of unknowns. The λ's, however, are of no particular interest and, having served their purpose, can be discarded once the solution has been obtained.

Notice that in order to solve the problem in this way, the tractions (or displacements) on the edge of the plate must be specified in advance. Then, for a given crack length, plastic zone size, and plate dimensions the value of the yield stress Y is calculated as a part of the solution. Because the problem is linear, the ratio T/Y will be correct for the given values of c and a.

The load to yield stress ratio

It was of considerable interest in this investigation to employ various kinds of boundary conditions on the loaded edges of the sheet. When a uniform tensile stress $(\sigma_y)_{y=h} = T$ is imposed the ratio T/Y, the quantity of greatest interest, can be obtained directly. However, when displacement or mixed boundary conditions are employed the additional computation

$$T = \frac{1}{w} \int_0^w (\sigma_y)_{y=h} \, dx \tag{18}$$

---

[1] In the present problem there is, of course, only one constraint equation and L consists of a single row. The derivation given is more general because in subsequent applications it is likely that additional constraints may be imposed.

must be performed. For all of the contemplated boundary conditions in the present work (as shown in Fig. 2) Eq. (18) can be rewritten as

$$T = \frac{1}{w} \left\{ Y(a - c) + \int_a^w (\sigma_y)_{y=0} \, dx \right\}. \tag{19}$$

Evaluating σ_y on the x axis for the combined stress function, substituting the results into Eq. (19), and using Eq. (11) gives

$$\frac{T}{Y} = \frac{2}{\pi} \left\{ \tan^{-1} \left(\frac{w}{c} \sqrt{\frac{a^2 - c^2}{w^2 - a^2}} \right) + \frac{c}{w} \tan^{-1} \sqrt{\frac{w^2 - a^2}{a^2 - c^2}} \right\}$$

$$- \frac{c}{w} + \frac{2}{wY} \sqrt{w^2 - a^2} \sum_{n=1}^{N} B_n (w^{2n} - a^{2n}) \tag{20}$$

whereupon the ratio T/Y can be obtained routinely as a part of the solution.

A comment on the further applicability of the analysis

The Dugdale crack is, in fact, a degenerate case of a more general class of problems now being considered by the present authors. As is well known, the main difficulty encountered in solving elastic-plastic problems is to determine the contour which separates the elastic and plastic regions. A general method is not available. But, in plane problems when the stress components in the plastic region are known (e.g., from the slip line field), a relatively straight forward approach employing the boundary point least squares technique can be used. The method rests on the fact that once an estimate of the contour is made the stresses along it can be evaluated from the plastic stress distribution. Consequently, by taking these stresses as the surface tractions on a "hole" in an elastic region, the problem is reduced to an elastic boundary value problem.

In the present problem, of course, the boundary is known a priori and the solution can be obtained directly. In general, however, an iterative process must be followed in which successively improved estimates of the elastic-plastic contour are made. The boundary point least squares technique can then be used to solve the elastic problem for the specified boundaries and surface tractions.

The experimental procedure

Previous experiments [5, 11] in this laboratory have shown that under conditions approximating plane stress, the width of the plastic zone is on the order of the sheet thickness. Accordingly, the idealized Dugdale zone of Fig. 1 is closely approximated in mild steel foil coupons $\sim 0.002''$ thick. These foils contained 0.08% C and were fully annealed. The lower yield stress was 32,000 psi and the ultimate tensile strength 45,000 psi. Four different geometries were used with heights of $3-1/4''$ and $4''$ and half crack lengths of $0.110''$, $0.220''$, and $0.440''$. All specimens were $3-1/2''$ wide.

In order to delineate the plastic zone clearly, the foils were first electropolished to give a highly smooth surface. Centrally located notches ($0.003''$ root radius) were introduced by an electrical discharge spark machine apparatus which produces an almost completely strain-free crack rim. The notched and polished foil specimens were then bonded between flat rectangular stainless steel grips using epoxy cement and incrementally loaded in a creep machine. The zone on the surface was photographed after each measurement and the zone length measured from the photographs.

A typical set of experimental results is presented in Fig. 3 where the plastic zones at each of five stages in the loading program are shown. It is evident that, with the exception of $T/Y = 0.91$ where unstable growth is occurring, the zone shape is in good accord with Dugdale's model (Fig. 1). The complete results of this experiment, designated as specimen No. 39, are given in Fig. 4.

It should be noted that the large amounts of plastic deformation observed in this work is a result of the foils being fully annealed. In contrast to the present work, as-rolled sheet and foil specimens will likely break before the plastic zones have spread completely across the specimen.

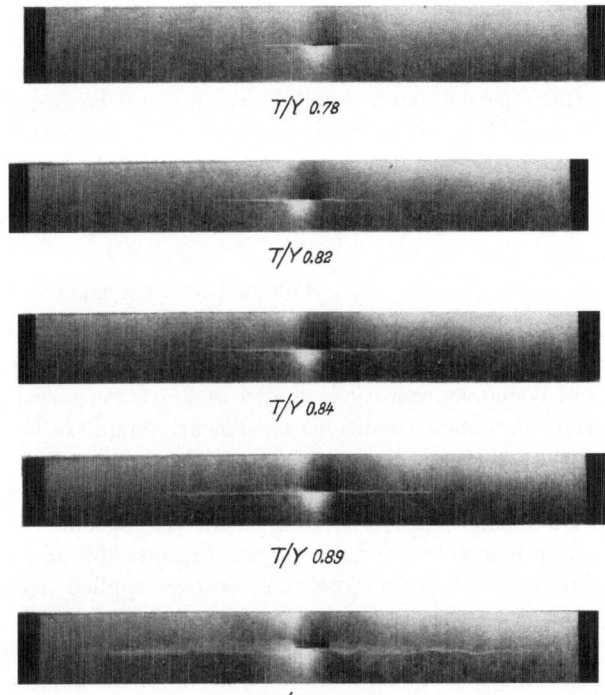

T/Y 0.78

T/Y 0.82

T/Y 0.84

T/Y 0.89

Fig.3. Plastic zones observed on the surface of steel foil at varying stress levels; $c = 0.110''$, $h = 1.75''$, $w = 1.625''$.

T/Y 0.91

Results and discussion

The Battelle boundary point least squares computer program for the CDC 6400 Computer was expanded to include the stress functions and the Lagrange multiplier technique described above. With the revised program a large number of solutions reflecting the effects of varying the overall dimensions of the plate and of imposing different sets of boundary conditions were obtained. In particular, the plastic zone lengths observed under given conditions were compared with the corresponding predicted values.

Preliminary computations

A cut and try procedure is usually required to find a suitable set of boundary points and stress function terms when a simple collocation method is employed. This is largely eliminated by using the boundary point least squares approach. A few trial computations may still be necessary to establish an adequate number of boundary points and stress function terms. However, because the solution is fairly insensitive to these values, the selections made at the outset are applicable to an entire set of computations. In the present work, for example, it was necessary to employ only two different combinations of boundary point distributions and stress function terms for all of the different plate and crack sizes considered.

The procedure followed was to first determine a satisfactory boundary point distribution for a fixed number of stress function coefficients. Then, an optimum number of coefficients was determined for the selected distribution. Two criteria were used to judge the results; the change in the applied load T and the change in the crack tip displacement v_c. The results of tentative computations made for each of the three different sets of boundary conditions are presented in the Appendix.

29*

On the basis of Tables A I and A III it was concluded that the same choices could be made for both the uniform traction boundary conditions and the parallel crack conditions. In particular, the values $2N = 30$ and $M = 72$ were selected[1]. For the uniform displacement conditions shown in Table A II the values $2N = 40$ and $M = 106$ (entry 12 in Table A IIa) were selected.

An examination of Tables A I through A III shows that for a given boundary point set there is an optimum number of stress function terms; a number which may be considerably less than the maximum number which could be employed. The maximum error for the considered problems occurs at the corners of the plates and is the largest for the uniform displacement boundary conditions. Further, it is readily apparent from these tables that the error in meeting the boundary conditions at the plate corners has little effect on the values of T/Y and Ev_c/Yc. Consequently, the values calculated for these quantities are considered to be accurate.

Comparison with experimental results

The plate dimensions and the crack lengths selected for the computations were either the same as or were integral multiples of those actually used in the experimental program. All three sets of boundary conditions shown in Fig. 2 were used in the computations. Because the uniform displacement conditions most nearly simulate the actual test conditions, however, these received special emphasis.

Computed values of T/Y for three sets of boundary conditions (for the plate dimensions and crack lengths corresponding to the test specimens) are presented in Table 1. Comparisons between these values and the observed results of four different tests are shown in Figs. 4 through 7. Notice that in these figures the average applied stress is plotted as a function of the plastic zone length.

Of greatest interest is the fact that the theoretical solutions predict the existence of a point of instability. As shown in Table 1, for all of the boundary condition types considered, the load required to produce a given extent of deformation increases to a maximum and decreases thereafter. For comparison with the experimental results, the theoretical curves shown in Figs. 4—7 are based on the computations only up to the maximum load. A horizontal extension, representing unstable growth at the critical load, is used thereafter.

It can be seen that the experimental results do in fact exhibit this kind of behavior. In the experiments, increasing the load by small increments produced instantaneous incremental changes in the zone lengths until a critical load was reached. An additional load increment then caused the deformed zone to grow slowly until it had spread completely across the specimen.

The significant effect obtained by changing the boundary conditions is also revealed by the figures. The infinite plate solution does not, of course, reflect either of these conclusions.

Finally, it can be seen that several entries in Table 1 exceed unity. This obviously requires stresses away from the neighborhood of the crack tips to exceed the yield stress[2]. No allowance was made in this investigation for plastic deformation other than at the crack tip. However, these calculations will still be realistic for materials which manifest an upper and a lower yield stress for then, because of the high strain concentration at the crack tips, the material there will deform at the lower yield stress Y while the remainder of the body remains elastic. In particular, in fully annealed steel foil the upper yield stress is roughly 11% greater than the lower yield stress.

Comparison with other theoretical results

In addition to the applied load, the principal results of the computations are the normal displacements at the center v_0 and the ends v_c of the crack. The corresponding values given by

---

[1] While the same type of distribution of boundary points was maintained for all plate sizes the actual number of boundary points varied slightly. The values cited here are for the 3-1/4″ × 3′1/2″ plates.

[2] This behavior occurs only for the uniform displacement conditions but is possible even when $T < Y$.

Table 1. *Computed T/Y values for $w = 1.75''$*

				Uniform Displacement b.c.		Uniform Traction b.c.			Parallel Crack b.c.	
			Infinite Plate							
c	a	l/c		$h{=}1.625''$	$h{=}2.00''$	$h{=}1.00''$	$h{=}1.625''$	$h{=}2.00''$	$h{=}1.625''$	$h{=}2.00''$
.110	.165	0.5	.535	.611	.574	.520	.528	.531	.533	.533
.110	.220	1.0	.667	.760	.714	.634	.652	.657	.660	.661
.110	.275	1.5	.738	.840	.789	.685	.713	.722	.727	.729
.110	.330	2.0	.784	.892	.837	.707	.747	.760	.766	.770
.110	.440	3.0	.839	.953	.892	.712	.775	.797	.806	.815
.110	.550	4.0	.872	.987	.922	.691	.777	.807	.821	.836
.110	.660	5.0	.893	1.007	.940	.662	.770	.802	.824	.845
.110	.770	6.0	.909	1.017	.949	.631	.760	.789	.821	.848
.110	.880	7.0	.920	1.020	.954	.602	.755	.774	.818	.850
.110	.990	8.0	.929	1.017	.954	.579	.756	.749	.816	.852
.110	1.100	9.0	.936	1.008	.953	.568	.767	.716	.818	.856
.220	.275	0.25	.410	.462	.435	.378	.395	.400		
.220	.330	0.5	.535	.603	.567	.480	.509	.518		
.220	.440	1.0	.667	.748	.702	.560	.613	.631		
.220	.550	1.5	.738	.824	.772	.579	.654	.680		
.220	.660	2.0	.784	.869	.813	.573	.671	.701		
.220	.770	2.5	.816	.896	.839	.557	.678	.706		
.220	.880	3.0	.839	.912	.855	.539	.681	.700		
.220	.990	3.5	.857	.919	.865	.523	.689	.685		
.220	1.100	4.0	.872	.919	.870	.517	.704	.661		
.440	.495	0.125	.303	.328	.309	.240	.270	.281		
.440	.550	0.25	.410	.441	.416	.312	.358	.373		
.440	.660	0.50	.535	.571	.537	.379	.450	.473		
.440	.770	0.75	.613	.645	.608	.403	.498	.522		
.440	.880	1.00	.667	.691	.653	.409	.527	.546		
.440	.990	1.25	.707	.720	.683	.411	.551	.553		
.440	1.100	1.50	.738	.738	.703	.414	.576	.546		

Dugdale's solution for the infinite plate have been derived by GOODIER and FIELD [12]. For plane stress their result can be reduced to:

$$\frac{Ev_0}{Yc} = \frac{2}{\pi} \log \left(\frac{a + \sqrt{a^2 - c^2}}{a - \sqrt{a^2 - c^2}} \right) \tag{21}$$

and

$$\frac{Ev_c}{Yc} = \frac{4}{\pi} \log \frac{a}{c} . \tag{22}$$

A sample comparison between the infinite plate values given by Eqs. (12), (21), and (22), and the computational results are presented in Table 2.

The data in Table 2 are for a single set of boundary conditions and plate size, and are limited to moderate plastic zone lengths. The comparison is, however, more or less typical of those obtained in all of the calculations. In particular, the data show that the relations between displacements and zone sizes for infinite plates, Eqs. (21) and (22), give an excellent approximation to the finite plate values. The T/Y values which correspond to a given crack-tip or crack-center displacement, on the other hand, can be quite different from the infinite plate values.

Also shown in Table 2 are the ratios of the applied stress in an infinite plate to that in a plate of finite dimensions for a given crack length. These ratios have been computed in two different ways: for the same plastic zone length and for the same crack-tip displacement. These have been denoted as $\left(\frac{T_\infty}{T}\right)_l$ and $\left(\frac{T_\infty}{T}\right)_{v_c}$, respectively. KANNINEN, et al [13] have shown that the condition for crack propagation under conditions approximating plate stress may be related

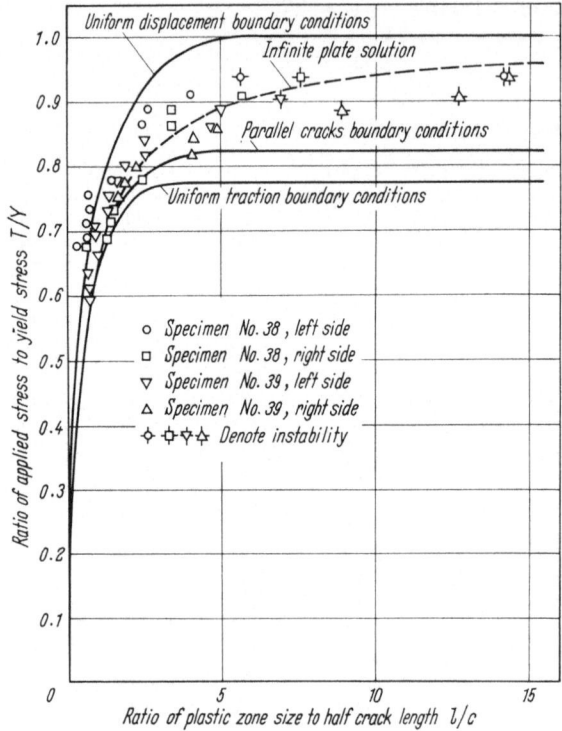

Fig. 4. Comparison of theoretical and experimental plastic zone sizes in mild steel foil; $c = 0.110''$, $w = 1.75''$, and $h = 1.625''$.

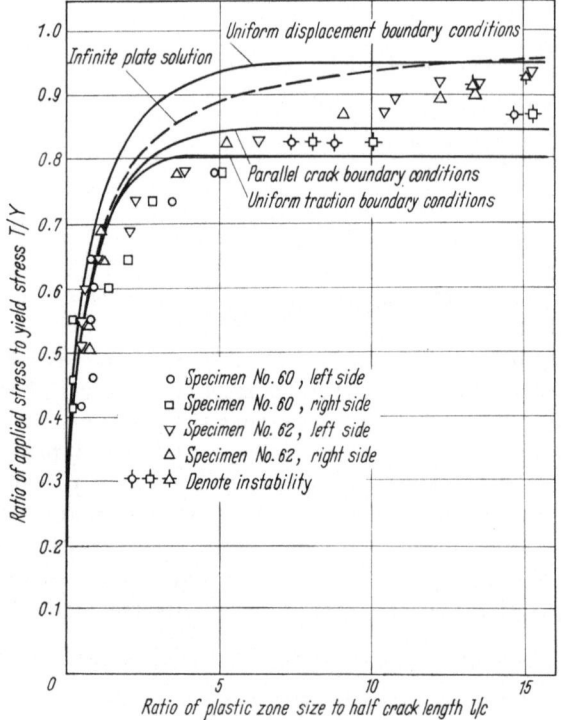

Fig. 5. Comparison of theoretical and experimental plastic zone sizes in mild steel foil; $c = 0.110''$, $w = 1.75''$, and $h = 2.00''$.

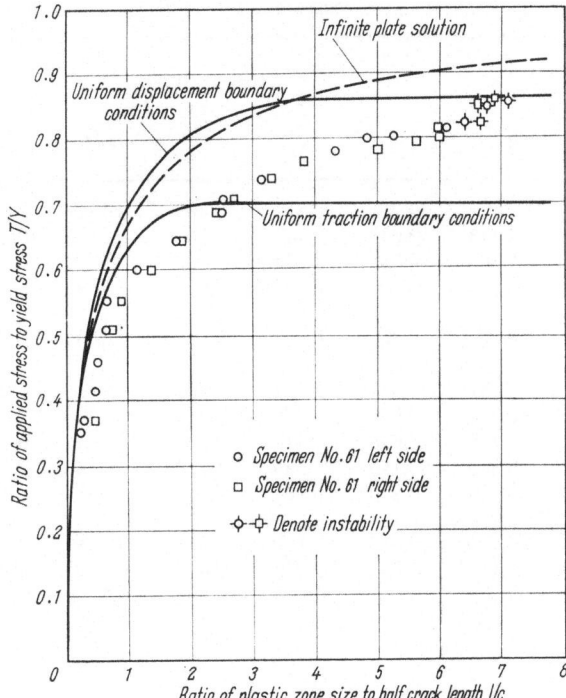

Fig. 6. Comparison of theoretical and experimental plastic zone sizes in mild steel foil; $c = 0.220''$, $w = 1.75''$, and $h = 2.00''$.

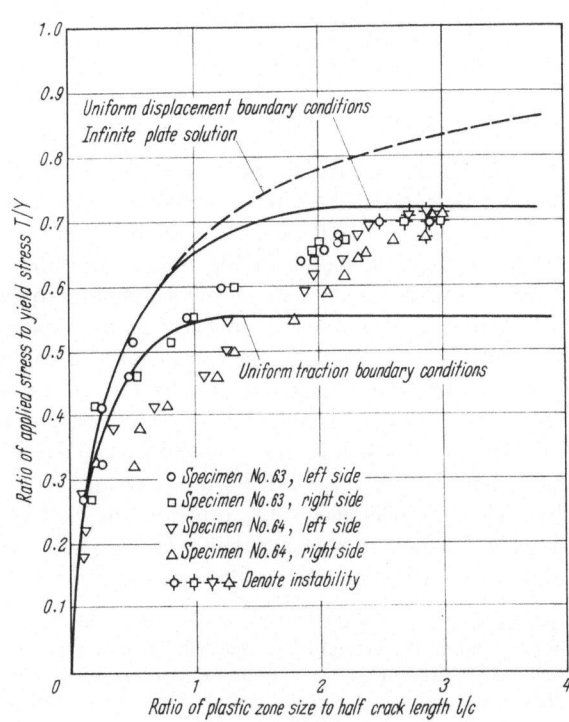

Fig. 7. Comparison of theoretical and experimental plastic zone sizes in mild steel foil; $c = 0.440''$, $w = 1.75''$, and $h = 2.00''$.

Table 2. *Some computational results for the uniform displacement boundary conditions*

c	a	l	Infinite Plate			$w = 1.75$ in., $h = 2.00$ in.			$\left(\dfrac{T_\infty}{T}\right)_l$	$\left(\dfrac{T_\infty}{T}\right)_{v_c}$
			$\dfrac{T_\infty}{Y}$	$\dfrac{Ev_0}{Yc}$	$\dfrac{Ev_c}{Yc}$	$\dfrac{T}{Y}$	$\dfrac{Ev_0}{Yc}$	$\dfrac{Ev_c}{Yc}$		
0.110	0.121	0.011	0.274	0.565	0.121	0.294	0.565	0.122	0.93	0.93
0.110	0.143	0.033	0.441	0.963	0.334	0.473	0.963	0.335	0.93	0.93
0.110	0.165	0.055	0.535	1.225	0.516	0.574	1.224	0.516	0.93	0.93
0.110	0.220	0.110	0.667	1.677	0.883	0.714	1.674	0.881	0.94	0.93
0.110	0.330	0.220	0.784	2.244	1.399	0.837	2.232	1.389	0.94	0.93
0.110	0.440	0.330	0.839	2.627	1.765	0.892	2.593	1.734	0.94	0.94
0.220	0.231	0.011	0.197	0.401	0.062	0.210	0.400	0.062	0.94	0.94
0.220	0.253	0.033	0.329	0.689	0.178	0.349	0.688	0.178	0.94	0.94
0.220	0.275	0.055	0.410	0.883	0.284	0.435	0.881	0.284	0.94	0.94
0.220	0.330	0.110	0.535	1.225	0.516	0.567	1.221	0.515	0.94	0.94
0.220	0.440	0.220	0.667	1.677	0.883	0.702	1.664	0.874	0.95	0.95
0.440	0.451	0.011	0.141	0.284	0.031	0.144	0.283	0.032	0.98	0.98
0.440	0.473	0.033	0.239	0.490	0.092	0.244	0.487	0.092	0.98	0.98
0.440	0.495	0.055	0.303	0.630	0.150	0.309	0.626	0.150	0.98	0.98
0.440	0.550	0.110	0.410	0.883	0.284	0.416	0.875	0.283	0.99	0.98
0.770	0.781	0.011	0.107	0.215	0.018	0.098	0.212	0.018	1.09	1.09
0.770	0.803	0.033	0.183	0.371	0.053	0.167	0.366	0.053	1.09	1.09
0.770	0.825	0.055	0.234	0.478	0.088	0.213	0.470	0.088	1.10	1.10
0.770	0.880	0.110	0.322	0.673	0.170	0.290	0.658	0.169	1.11	1.10
1.100	1.111	0.011	0.090	0.180	0.013	0.068	0.175	0.013	1.33	1.33
1.100	1.133	0.033	0.154	0.311	0.038	0.115	0.301	0.038	1.33	1.33
1.100	1.155	0.055	0.197	0.401	0.062	0.147	0.386	0.062	1.34	1.34
1.100	1.210	0.110	0.274	0.565	0.121	0.202	0.538	0.120	1.36	1.35
1.320	1.331	0.011	0.082	0.164	0.011	0.050	0.156	0.011	1.64	1.64
1.320	1.353	0.033	0.141	0.284	0.031	0.086	0.268	0.031	1.64	1.64
1.320	1.375	0.055	0.181	0.366	0.052	0.109	0.342	0.052	1.65	1.65
1.540	1.551	0.011	0.076	0.152	0.009	0.033	0.136	0.009	2.32	2.33

to a critical crack tip displacement. Consequently, it is more appropriate to make such comparisons on the basis of the same v_c value. However, as the data of Table 2 indicate, the distinction is negligible provided the plastic zone lengths are not too great.

The ratio T_∞/T is shown in Fig. 8 as a function of the crack length for uniform traction boundary conditions and in Fig. 9 for uniform displacement conditions[1]. For comparison, the elastic solution is also shown[2]. In the elastic crack development, no account is taken of the finiteness of the plate in the direction normal to the crack line. In the present work, however, there is obviously considerable dependence on h, particularly when $h < w$.

Figs. 8 and 9 suggest that under the conditions examined in the present paper (plane stress deformation — foil width = 3.5″), use of the purely elastic correction factor proposed by Isida results in errors $< 10\%$ provided the foil height is greater than its width. However, the extent to which this result is generally applicable needs to be established by more extensive elastic-plastic calculations. In the meantime, the apparent agreement between two results should be considered fortuitous and to result from the particular choice of geometries.

[1] It can be seen in Fig. 9 that when $c = 0$ the ratio of the infinite plate loading to that of the finite plate does not become unity as might be expected. The reason is that the boundary conditions in the two cases are fundamentally different irrespective of the presence of a crack.

[2] The results of all of the relevant elastic calculations are in good quantitative agreement. Isida's is generally considered most reliable [1].

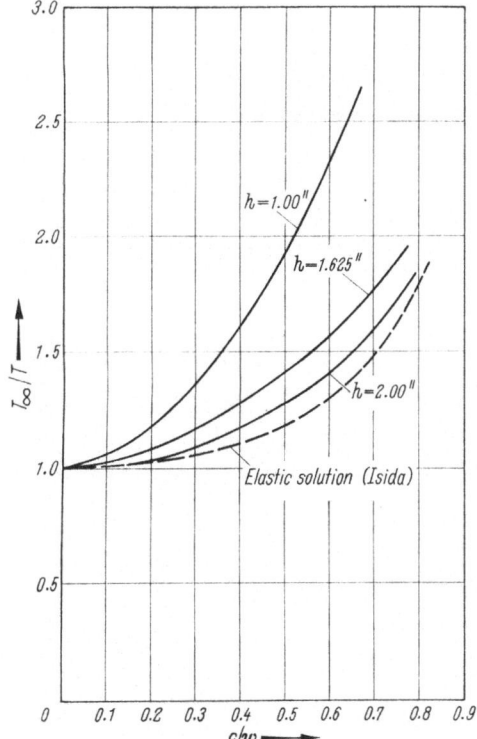

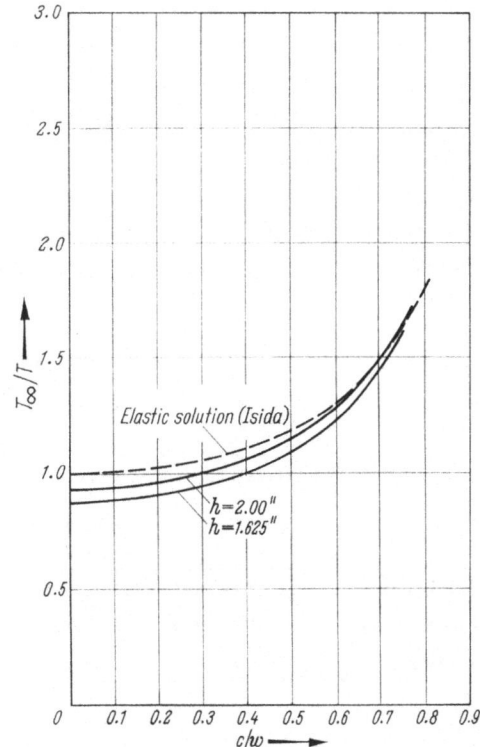

Fig. 8. Ratio of applied stress in an infinite plate to that in a plate of finite dimensions for uniform traction boundary conditions; $w = 1.75''$.

Fig. 9. Ratio of applied stress in an infinite plate to that in a plate of finite dimensions for uniform displacement boundary conditions; $w = 1.75''$.

Conclusions

1. The computational method, designated here as Boundary Point Least Squares with Constraints, provides an effective method for elastic-plastic stress analysis in situations where closed-form analytical solutions are unavailable.

2. The Dugdale model, modified to include finite plate effects, is in quantitative agreement with experimentally measured plastic zone sizes under conditions which approximate plane stress.

Acknowledgment

The theoretical portion of this study was supported by the Air Force Materials Laboratory, Wright-Patterson Air Force Base and the experimental portion by the Army Research Office, Durham. The authors are grateful to these two agencies and to C. R. BARNES for his expert technical assistance and to Mrs. PATRICIA BARE for her excellent work on the manuscript. The authors also wish to thank Dr. E. T. STEPHENSON of the Bethlehem Steel Corporation for providing the foil used in the experimental program.

References

1. FEDDERSEN, C. E.: ASTM-STP, 410, 77, 1966.
2. BILBY, B. A., A. H. COTTRELL, E. SMITH and K. H. SWINDEN: Proc. Roy. Soc. A 279, 1, 1964.
3. KOSKINEN, M. F.: J. Basic Eng. 85, 585, 1963.
4. DUGDALE, D. S.: J. Mech. Phys. Solids, 8, 100, 1960.
5. HAHN, G. T., and A. R. ROSENFIELD: ASTM-STP 432, 5, 1968.
6. HULBERT, L. E.: The numerical solution of two-dimensional problems of the theory of elasticity, Bulletin 198, Engineering Experiment Station, The Ohio State University, Columbus, 1963.
7. KOBAYASHI, A. S., R. D. CHERPY and W. C. KINSEL: J. Basic Eng., 86, 681, 1964.
8. MUSHKELISHVILI, N. I.: Some basic problems in the mathematical theory of elasticity, Fourth Edition. Groningen: Noordhoff 1963, p. 114.
9. MUSKHELISHVILI, N. I.: ibid. p. 346.
10. KORDA, P. E.: Report No. RF-1699-1, Department of Engineering Mechanics, The Ohio State University, Columbus, 1964.
11. HAHN, G. T., and A. R. ROSENFIELD: Acta Met. 13, 293, 1965.
12. GOODIER, J. N., and F. A. FIELD: in Fracture of solids, Ed.: Drucker and Gilman, New York: Interscience, 1963, p. 103.
13. KANNINEN, M. F., A. K. MUKHERJEE, A. R. ROSENFIELD and G. T. HAHN: in Mechanical behavior of materials under dynamic loads, Ed.; U. S. Lindholm, New York: Springer, 1968, p. 96.

Appendix

Results of preliminary computations

Table A1. *Trial computations for uniform traction boundary conditions*

(a) Number and Location of Boundary Points Varied—Fixed Number of Stress Function Coefficients

$$2N = 30$$

$$c = 0.110''　　　　w = 1.750''$$
$$a = 0.220''　　　　h = 1.625''$$

Number of Boundary Points	Number of Boundary Conditions	Description of Boundary Point Distribution	$\dfrac{T}{Y}$	$\dfrac{Ev_c}{Yc}$
28	56	Uniform spacing	.65162	.8838
55	110	Uniform spacing	.65162	.8837
36	72	High density near corner	.65162	.8837
40	80	Very high density near corner	.65162	.8837

(b) Number of Stress Function Coefficients Varied—Number and Location of Boundary Points Fixed

$$M = 72$$

$$c = 0.110''　　　　w = 1.750''$$
$$a = 0.220''　　　　h = 1.625''$$

Number of Stress Function Coefficients	Maximum Percent Error in Assigned Boundary Conditions[1]				$\dfrac{T}{Y}$	$\dfrac{Ev_c}{Yc}$
	Stresses on Free Edge		Stresses on Loaded Edge			
	Normal	Shear	Normal	Shear		
10	.39	.21	.27	.47	.65150	.8837
20	.004	.009	.014	.013	.65163	.8838
30	.004	.002	.004	.002	.65162	.8837
40	.10	.11	.09	.05	.65163	.8833
50	8.23	6.39	5.34	6.39	.65162	.8854

[1] Based on the difference between the boundary point least squares solution evaluated at the given point and the nonzero assigned value at the point.

Table A2. *Trial computations for uniform displacement boundary conditions*

(a) Number and Location of Boundary Points Varied—Fixed Number of Stress Function Coefficients

$$2N = 30$$

$$c = 0.110''　　　　w = 1.750''$$
$$a = 0.220''　　　　h = 1.625''$$

	Number of Boundary Points	Number of Boundary Conditions	Description of Boundary Point Distribution	$\dfrac{T}{Y}$	$\dfrac{Ev_c}{Yc}$
1.	22	42	Uniform spacing—Stress b.c. at corner point	.758	.8801
2.	29	56	Uniform spacing—Stress b.c. at corner point	.757	.8801
3.	25	48	High density near corner—Stress b.c. at corner point	.758	.8801
4.	21	41	Uniform spacing—Displacement b.c. at corner point	.774	.8797
5.	36	71	High density near corner—Displacement b.c. at corner point	.761	.8800
6.	35	69	Uniform spacing—Corner point excluded	.760	.8800
7.	35	69	High density near corner—Corner point excluded	.759	.8800
8.	51	101	Very high density near corner—Corner point excluded	.759	.8801
9.	30	58	Uniform spacing—Stress and displacement b.c. at corner point	.761	.8800
10.	37	73	Uniform spacing—Stress and displacement b.c. at corner point	.761	.8800
11.	37	73	High density near corner—Stress and displacement b.c. at corner point	.759	.8800
12.	53	106	Very high density near corner—Stress and displacement b.c. at corner point	.761	.8800

(b) Number of Stress Function Coefficients Varied—Number and Location of Boundary Points Fixed
$$M = 106$$

	$c = 0.110''$	$w = 1.750''$
	$a = 0.220''$	$h = 1.625''$

Number of Stress Function Coefficients	Maximum Percent Error in Assigned Boundary Conditions[1]				$\dfrac{T}{Y}$	$\dfrac{Ev_c}{Yc}$
	Stresses on Free Edge		Displacements on Loaded Edge			
	Normal	Shear	Normal	Tangential		
10	10.8	8.4	5.4	5.5	.802	.8796
20	4.5	5.5	3.6	2.0	.771	.8798
30	1.3	1.1	2.3	0.9	.761	.8800
40	1.1	1.1	2.3	0.4	.760	.8800
50	8.1	22.7	2.8	1.2	.759	.8800
60	8.7	22.1	1.7	0.8	.759	.8800

Table A 3. *Trial computations for parallel cracks boundary conditions*

Number of Stress Function Coefficients Varied—Number and Location of Boundary Points Fixed
$$M = 72$$

	$c = 0.110''$	$w = 1.750''$
	$a = 0.220''$	$h = 1.625''$

Number of Stress Function Coefficients	Maximum Percent Error in Assigned Boundary Conditions[1]				$\dfrac{T}{Y}$	$\dfrac{Ev_c}{Yc}$
	Stresses on Free Edge		Loaded Edge			
	Normal	Shear	Shear Stress	Normal Displacement		
10	.242	.171	.179	0.77	.6606	.97155
20	.005	.013	.008	.009	.6605	.97158
30	.004	.007	.004	.003	.6604	.97158
40	.440	.226	.024	.014	.6604	.97164
50	5.735	4.999	2.970	.092	.6604	.97153

[1] See footnote Table A 1.

The intermittently turbulent region of the boundary layer

By

R. E. Kaplan and J. Laufer

University of Southern California, Los Angeles

The intermittently turbulent region of the boundary layer

There is increasing evidence that fully developed turbulent shear flows exhibit a velocity structure that is more coherent than generally expected. It is possible to detect an ordered, relatively coherent structure that moves randomly in space and time and has characteristic lifetimes and length scales which are, in general, much larger than the dissipation scale of the flow.

Perhaps the most dramatic evidence of the existence of this type of quasi-ordered motion in the boundary layer is found in the work of KIM, KLINE, and REYNOLDS [Ref. 1], in which such a structure is detected near the sublayer by measurement of instantaneous velocity distributions by means of the hydrogen-bubble technique. These measurements showed the repeated appearance of this characteristic structure at random intervals of time. It is clear that without a set of "instantaneous" velocity profiles, conventional time averaging of the velocities would have overlooked such a phenomenon. The interest, of course, is not in the phenomenon itself, but on any light that it may cast on the production of turbulent energy by the shear flow, in this case through a possible instability of the basic flow.

The phenomenon to which we address ourselves is the structure associated with the motion of the interface between the turbulent and non-turbulent fluid at the outer edge of the turbulent boundary layer, although similar phenomena occur in turbulent jets, wakes, and free shear flows. It is known that for these phenomena conventional time averaging techniques yield little evidence of such structure, since they weigh equally contributions from the turbulent and non-turbulent regions of the flow. However, investigators such as KOVASZNAY and KIBENS [Ref. 2], and the authors, have demonstrated the existence of such a structure and made detailed quantitative measurements of its properties. The question that is relevant in this region of the flow is the relationship between this detectable structure and the entrainment process of non-turbulent fluid.

For the purpose of analyzing such phenomena, a facility for the digital processing of turbulent-boundary-layer data was established at the University of Southern California. The following is a description of the initial measurements processed by this facility, some interpretation of the results, and a brief account of the techniques used.

Conditional sampling and detector functions

There are two basic requirements that must be met before the probing of randomly occurring coherent structures can proceed.

i) The first of these is the ability to measure velocities at several points in the flow field simualtaneously. This necessitates the use of an array of velocity sensors (hot-wire anemometers) with the size of the array of the order of the characteristic scale of the structure. This technique was in fact successfully used by KOVASZNAY, KOMODA, and VASUDEVA [Ref. 3] in their study of the formation of turbulent spots during transition.

ii) The second requirement is the capability of performing a conditional sampling of these signals, that is, the ability to sample only during the period when the structure passes the array of tensors. We can compare this process to conventional time averages by performing the appropriate statistics not on $U(x, t)$, but on the product of U with a detector function $D(t)$ which has the value 1 during the time passage of the structure and 0 at all other times.

It is our contention that these two requirements are common to all problems of the type under discussion, with the major difference from problem to problem being in the generation of the detector function. In the intermittent zone, the question of what to detect is quite straightforward, namely, the presence or absence of turbulence. While it is clear what is to be detected, there are many possible techniques for mechanizing the detection. These are, in rough chronological order, the sensing of the streamwise component of vorticity [Ref. 4], the sensing of the spanwise component of vorticity [Ref. 2], the sensing of the transport of a marked species (such as smoke) [Ref. 5], and the sensing of the amplitude in a frequency band of the u' fluctuations [Ref. 6].

In brief, the detector evaluation is as follows. The desired criterion function (enumerated above) is significantly larger in the turbulent region than in the non-turbulent region of the flow. The criterion is conditioned, amplified, and rectified, and a threshold level is set. When the criterion function exceeds this threshold, the detector function is set to 1, otherwise it is 0. The major problem in the detection is that the criterion function is not explicitly zero in the non-turbulent portion and quite often crosses zero in the turbulent portion. Hence some sort of logic must be used to ascertain whether the indication given by the detector function is a true indication of the state, or a transient. This problem has been solved to one degree or another by the investigators listed above, but only in the case of the last two has the question really been crucial. When the detector function is used for conditional samling, it is "differentiated" to compute certain types of conditional averages. Any spurious indications of interface position can bias certain types of averages with an excess of signal not belonging to that sample set. Hence it is important that great care be given in the generation of the detector function. In this regard it is most fortunate that the work of KOVASZNAY and KIBENS, cited earlier, used an entirely separate technique for the generation of the detector function, for it demonstrated that the conditional averages described herein are independent of the form of the criterion function. The details of our criterion-function generation are described later.

Part of the studies in the intermittent zone have been based solely on the detector function. The intermittency factor γ is an appropriate time average of the detector function, that is,

$$\gamma(y) = \frac{1}{2T} \int_{-T}^{+T} D(y, t) \, dt$$

where we now explicitly note that D (and γ) are functions of y, the coordinate normal to the wall. In the same sense, we can define averages of some property Q, under some function f of D, as

$$\overline{Q}_f(y) = \frac{1}{2T\gamma_f} \int_{-T}^{+T} f\big(D(y, t)\big) Q(t) \, dt.$$

For example, turbulent zone averages use $f(D) = D$, while non-turbulent zone averages use $f(D) = 1 - D$. The function γ_f is generalized to be an integral over $f(D)$, and hence when $f(D)$ is a positive derivative (a Dirac delta), γ_f measures the number of discrete non-zero values of $f(D)$, with a similar result for the negative derivatives. The latter two averages are associated with the passage of the interface at a detector location and are formally equivalent to summations over appropriate interface crossings. It is clear that, by proper adjustment of the threshold levels at the disposal of the experimentalist, the intermittency factors can be brought into reasonable experimental agreement, because the resultant function D is integrated. In the interface averages, however, when the integral is in fact replaced by a summation, differences in the detector function cause significant differences in the averages. It is for this reason that we shall discuss formation of the detector function in some detail.

Generation of the detector function

The technique described in the following paragraphs is well suited to digital processing but is significantly more difficult to perform in the laboratory. We have mentioned a criterion function, which is the explicit representation of the physical phenomenon from which the detector function is derived. In general, the criterion function should be small in the non-turbulent regions and of appreciable magnitude in the turbulent regions. The most physical quantity available for this use is the velocity fluctuation itself, but a judgment based on this, or even its time derivative, is quite difficult because there are still appreciable velocity fluctuations in the non-turbulent portions of the flow. The "best" criterion is based on the local vorticity, which is zero in the non-turbulent regions and quite large in the turbulent regions. The vorticity was used by CORRSIN and KISTLER [Ref. 4] and a component of vorticity by KOVASZNAY and KIBENS [Ref. 2]. The problem in using vorticity as a criterion, in our view, is (i) that a multi-wire probe must be used merely to detect the interface position, and (ii) that the outputs of the wires in general must be linearized and carefully balanced to prevent the dominance of the u' fluctuation from one of the sensors. A more serious drawback of an instantaneous measurement is the fact that it is desired to base the detection on the emergence of the criterion function from a window in some phase plane (KOVASZNAY and KIBENS used as their criterion that both $\partial^2 u/\partial y\,\partial t$ and $\partial^3 u/\partial y\,\partial t^2$ must be larger than a set value). But the criterion function often crosses zero; and, while filtering can help remove some of these spurious zero crossings, and not seriously affect the transition in which the criterion function goes from a small to a large value, the other interface crossing (when the criterion function goes from a large to a small value) is made more ambiguous by the discharge of the filter capacitor, and the detailed setting of the threshold can have a profound influence on the results. It became apparent to the authors that an important requirement for good detection was some knowledge of the future. The same conclusion was reached independently by KOVASZNAY and KIBENS, who take some time to generate the detector function, and consequently must delay the signal from other wires to account for this delay. In the present study signals were stored in the computer for a small time interval so that various amounts of time delay (and lead) were attainable.

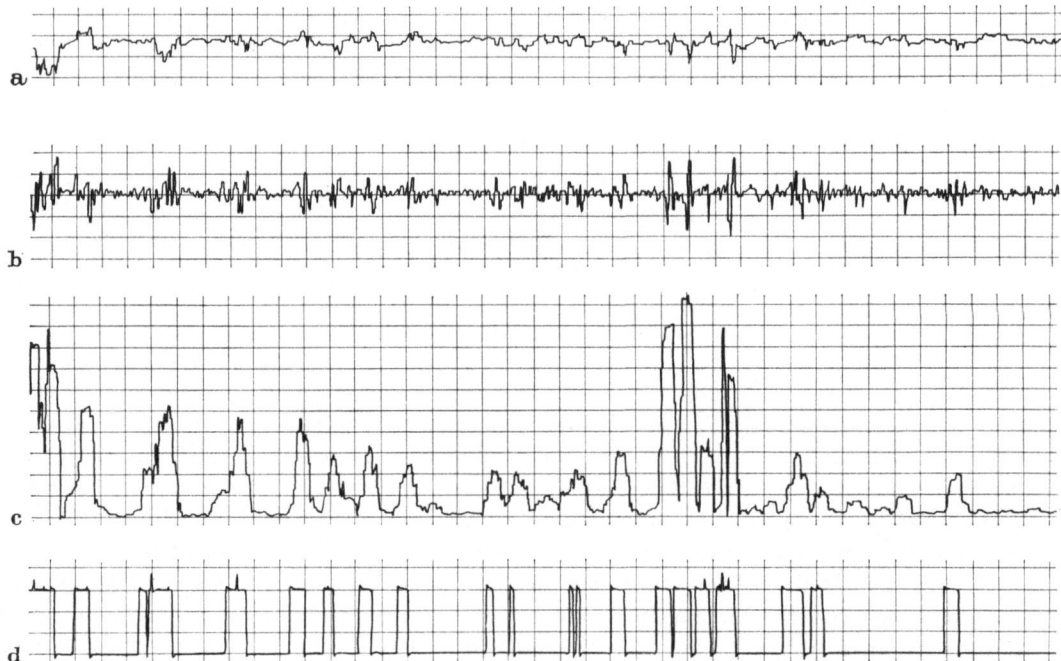

Fig. 1. Formation of the detector function. a Hot-wire voltage; b Time derivitive of a); c Short-time variance of b) (criterion function); d Detector function.

The detection technique used universally in this study was based on a detector consisting of a single hot wire. The variance of the time derivative of the hot-wire voltage about the mean was computed over a relatively short time interval (the time interval varied from 2.5 to 8 msec). It can be observed from Fig. 1 that this variance was small when the wire was in a non-turbulent region of the flow and large when it was in a turbulent region. In the figure, trace (a) is the hot-wire voltage, (b) its derivative, (c) the short-time variance, and (d) the detector function. The threshold for the detector function was derived by taking a fraction of the local maxima of the criterion function over a reasonable time interval. Since the short-time variance exhibited no "zero-crossings", the filtration effect of the trailing edge of the criterion function was avoided, although a type of filtration was inherent in the length of time interval over which the variance was evaluated. The variance was ascribed to some point in the time interval, and the velocities corresponding to that point were used in evaluating the statistics. Hence the filtering process was roughly centered about the time in question, with information from the near past and future equally weighed. It can be shown that this is a low-pass filter that is lacking in phase shift, but with terminal-slope asymptote the same as an analog "exponential" filter. While this is in no sense an optimal filter, it appears to be much more suited to the problem at hand than the type of linear filter with continuous analytic describing functions.

The structure of the intermittent region

The first computations performed in the intermittently turbulent region of the boundary layer involved processing signals coming from the ten-wire vertical rake illustrated in Fig. 6. (A brief description of the rake, the data-gathering system, and the digital analysis system is given in the Appendix.) The 0.0001-in.-diameter platium wires were arranged parallel to the plate and normal to the flow, with a nominal 0.1-inch spacing, and functioned as U wires. The rake was mounted to traverse the intermittent zone. The results reported in the following paragraphs were all made at a nominal 20-ft/sec free-stream velocity and in a zero-pressure-gradient boundary layer, 3.05-in. thick, artificially tripped 15 ft. ahead of the test station, with Reynolds number of 3600 based on the momentum thickness. Previous experiments performed in the tunnel established that the boundary layer was a representative one, and that its intermittency distribution followed the familiar scaling laws.

The recorded data for this configuration were initially programmed to reproduce the results of KOVASZNAY and KIBENS for a similar flow situation. The recorded signals did not provide vertical-velocity information (V'), and hence those measurements could not be checked.

The analysis of the signals proceeded as follows. Each wire in turn was used to generate a detector function. This detector function was then used to compute intermittency, turbulent and non-turbulent zone averages and variances, and conditional point averages associated with the passage of the interface. The familiar curve of intermittency distribution $\gamma(y)$ was found to agree with those of other observers, lying within the usual scatter band. The conditional averages measured in the present study are in agreement with those of [Ref. 2], as illustrated in Fig. 2. (The notation is that of [Ref. 2]). The conditional point averages associated with the leading interface encounter (leading-edge averages) corresponding to the transition from non-turbulent to turbulent fluid were also substantially the same as those found in [Ref. 2]. The geometry of the point averages is illustrated in Fig. 5.

The only area of disagreement with [Ref. 2] came in the evaluation of the trailing-edge averages. In [Ref. 2] these were essentially the same as the leading-edge averages, and in fact this similarity was used as a means of evaluating the time delay inherent in their detector-function generation. Use of the present detector function indicated a consistent difference in the trailing-edge averages, incapable of being resolved by varying the appropriate time delay. The present results indicate a small but definite velocity defect for trailing-edge point averages as compared to the leading-edge averages. Fig. 3 illustrates these point-averaged velocities at the interface location. This suggests that the particle velocity is measurably smaller in the vicinity of the trailing edge of the turbulent bump, and in particular at the interface. While no conclusion can be made about the velocity of the interface at the present time, one can anticipate a corresponding difference in interface velocity.

If the latter conjecture is true, there should be an observable difference in the interface geometry. With this goal, an attempt was made to provide a visual display of the intermittent

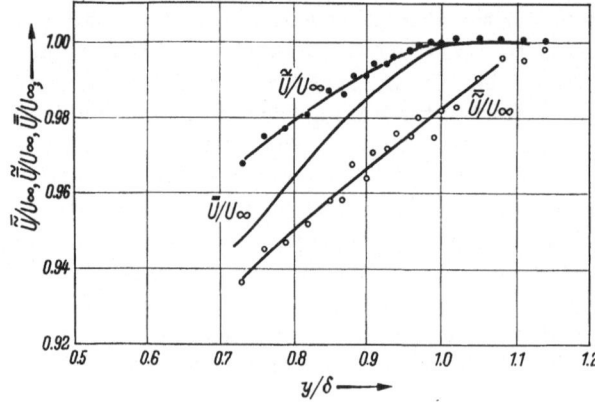

Fig. 2.
Conditional zone averages of the velocity.

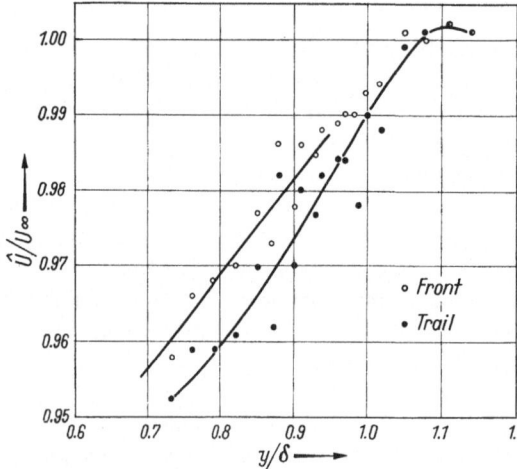

Fig. 3. Conditional point averages of the veclocity at
the interface.

zone. The result of such an experiment is presented in Fig. 4. At each time increment, a decision was made as to whether a given hot wire was in a turbulent or non-turbulent region of the flow. These decisions were presented on the line printer as asterisks (*) if the detector function was 1, and a blank otherwise. Each horizontal line of asterisks corresponds to a given wire output. To conserve space, ten successive time intervals were presented on the same figure. The time sequence in the figure proceeds from the lower right-hand corner, left to right, continuing in the left side of the next higher row. The time between successive presentations is approximately 1 msec, and the wire locations varied from y/δ of 0.7 to 1, corresponding to intermittency factor γ of 0.6 to 1. The apparent slopes on the figure are approximately 1.5 times the physical slopes to be expected if the pattern were frozen in space.

Several results are apparent from figures such as this. The first is that there is a regular difference between the leading and trailing edges of the interface, with the former being on the average steeper corresponding to its higher velocity. The second is the apparent absence of regions of non-turbulent fluid imbedded in a mass of turbulent fluid, and the converse, even though it is now known that the bumps are highly three-dimensional, and the two-dimensional rake could easily "graze the side" of a protruding turbulent lump. Thirdly, it is noted that the indication of turbulence (an asterisk) rarely appears isolated in the non-turbulent (blank) region, and vice versa. This quality is not programmed into the logic of forming the detector function, but is a consequence of the criterion function used. Finally, the extent of the lager-scale motions is apparent and complements the conditionally sampled structures presented here and in [Ref. 2].

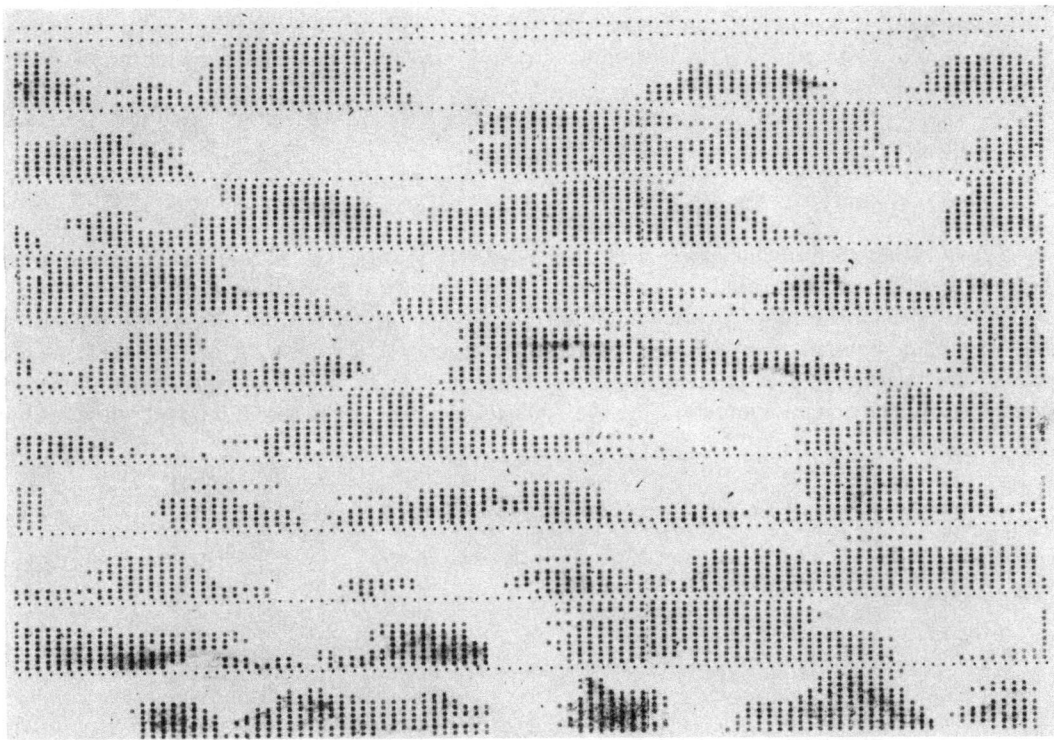

Fig. 4. Representation of the extent of the turbulent zone.

While the indication of the interface, strictly speaking, yields its elevation at a given stream-wise location as a function of time, its spatial structure is believed not to be significantly different from the situation as pictured. The tentative interpretation (see Fig. 5) is that the entrainment of non-turbulent fluid occurs predominantly along the relatively diffuse trailing edge of the protruding turbulent zone. Further measurements of the propagation velocity of the turbulent front are needed to verify this interpretation.

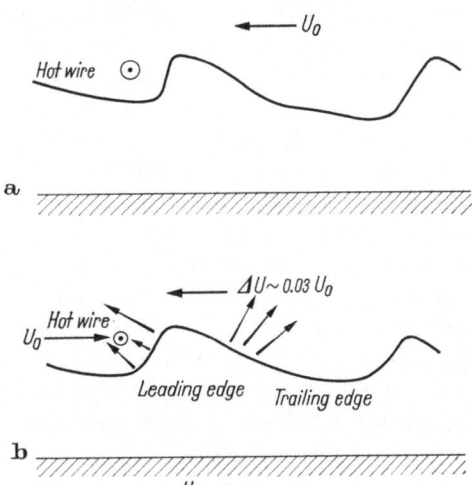

Fig. 5. Geometry of the intermittent region.
a) Tunnel coordinate system; b) Interface coordinate system.

Conclusion

In summary, it is noted that these initial results from the present facility shed further light on the processes associated with the entrainment of non-turbulent fluid. The presence of the large-scale motions of the interface and the local propagation of the fronts through small-scale turbulent diffusion are in fact consistent with Townsend's ideas [Ref. 7]. Whether the large-scale motion is in fact due to a Kelvin-Helmholtz type of instability as suggested by him, cannot be ascertained at this stage.

Appendix

Turbulence data-processing system

The data reported in this paper were processed in a unique digital facility. The system has two parts, which are separated both physically and logically. The data-gathering portion of the facility resided in the Low-Turbulence Wind Tunnel and consisted first of a probe with ten hot-wire anemometers arranged in a vertical two-dimensional rake with 0.1-inch spacing. All wires were inserted in a bank of constant-resistance bridges, so that they operated as constant-temperature hot-wire anemometers. At the typical velocities of the test (20 ft/sec) all wires had

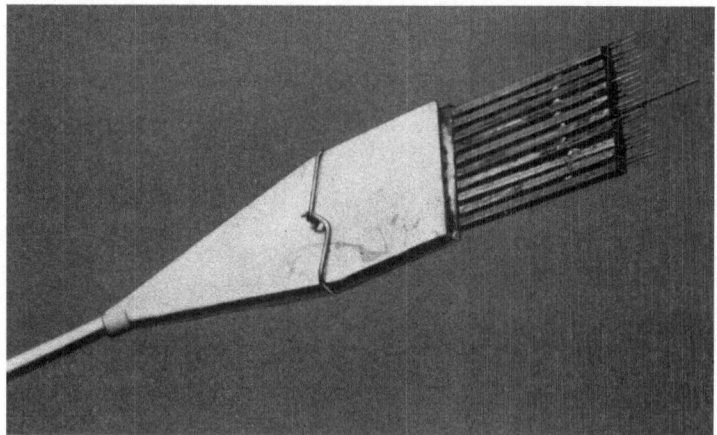

Fig. 6. The hot-wire rake.

frequency response above 30 kHz as determined from a pulse test. The output from these bridges was recorded as channels 1—10 on a 14-channel FM tape recorder at 60 ips (in/sec). At this speed, the signal-to-noise ratio (re 40% modulation) was 48 db. To counteract losing the small velocity fluctuations in the tape noise, signal conditioners consisting of a bucking DC amplifier, which subtracted most of the DC component of the wire output, and a 20-db amplification of the difference voltage were recorded. The frequency response of these components was in excess of 25 kHz for the signal levels typical of those from the hot wires. The frequency response of the tape recorder was DC-20 kHz. From the system gain, it is estimated that the signal-to-noise ratio of the fluctuations alone was in excess of 36 db, while for the reconstructed hot-wire voltages it exceeded 70 db.

Simultaneously, the voltage output from a pressure transducer monitoring the tunnel dynamic pressure was recorded FM, and a 20-kHz square wave was recorded direct. The recording procedure was as follows. First, five short runs were made to calibrate the hot-wire anemometers. The first two of these consisted of dialing indicated pressures into the pressure-transducer system to calibrate that portion of the electronics, placing the hot-wire array in the free stream of the tunnel, and recording the dynamic pressure and hot-wire voltages at three separate tunnel velocities, thus providing data for a linearization program which fitted King's law to the wire outputs over the small velocity range of interest. The calibration was completed in less than 5 minutes and was followed immediately by from two to four data runs of one-minute duration with the rake at various elevations from the surface. Generally, calibration and data gathering occupied from 10 to 12 minutes elapsed time in the tunnel, although setup of the ten probes, determination of the rake position relative to the wall, balancing and optimization of the frequency response of the ten bridges, and adjustment of record levels could consume an hour's time. The net result of this phase was the production of an analog tape with several experimental arrangements recorded on it.

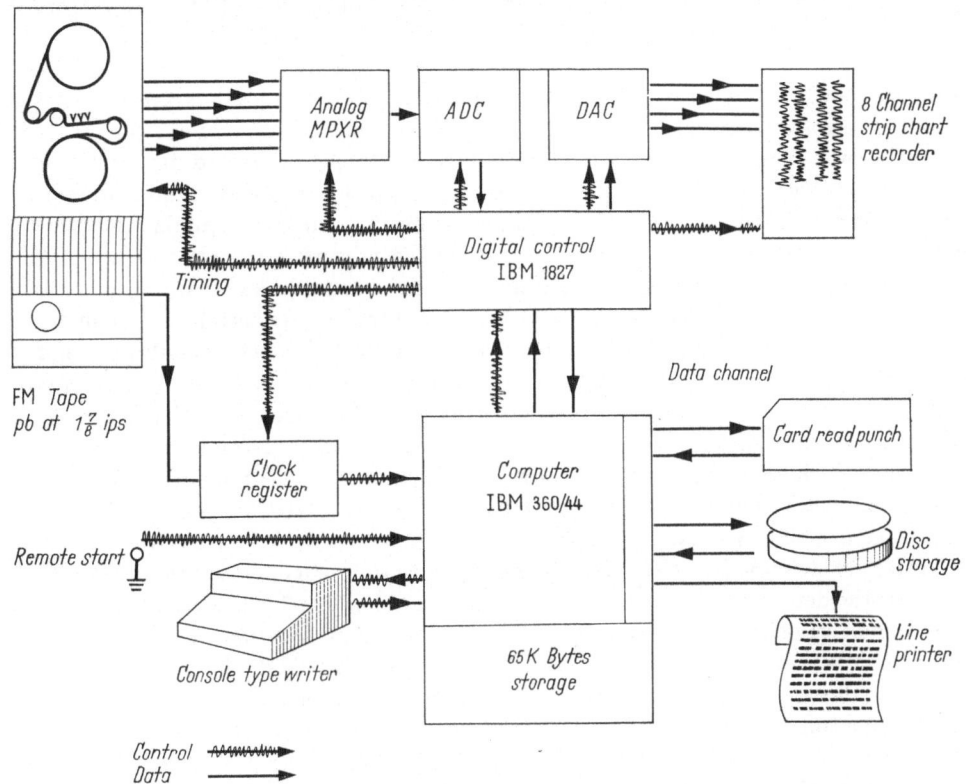

Fig. 7. Flow diagram of the analysis facility.

This tape was then carried to the analysis facility, physically located in the USC School of Engineering Systems Simulation Laboratory. In this laboratory were located the playback recorder (identical to the recording unit at the wind tunnel), an IBM 360/44 digital computer with conventional input/output devices, and an IBM 1827 digital input/output unit. The last device communicated with the analog portion of the laboratory through digital in (DI) and digital out (DO) lines to sense and transmit voltage levels and operate reaysl. The portion of the interface of particular interest was the Adage analog-to-digital converter (ADC) the voltage multiplexor (VMX), and the digital-to-analog converters (DAC). These devices were capable of being operated by the IBM 1827 and transmitting (or receiving) control signals and data. In our case, the ADC conversion provided a 14-bit conversion (including sign) in 2's-complement notation. Additionally, the computer has as a standard feature, 6 uncommitted external interrupt lines, the use of which will be described later.

A unique feature of the facility was the absence of an intermediate storage medium for the conversions, owing initially to the lack of digital tape units in the laboratory and the small capacity of the system's random-access discs. As development of the system progressed, the lack of this intermediate storage was not missed and actually simplified the tape accounting.

The programming system consists of a series of FORTRAN and assembly-language subroutines compatable with the 360/44 Programming System provided by IBM. The major feature

31*

of these routines was that they usurped many of the functions generally reserved to the Supervisor program, enabling the support of hybrid computation and the external interrupt feature. Only these system's programs were assembly-language programs; the computation was performed by FORTRAN programs, with a consequent loss of computing speed. Without going into the details of the programming, several general points of interest will be noted.

The first phase of the computation was the calibration of the hot-wire channels and the pressure transducer. The assumptions made in this calibration were (i) that the recorders, playback amplifiers, and converters were linear (experimentally verified), (ii) that both the gains and offsets of the recorded signals were unknown, and (iii) that King's law held for the constant-temperature hot-wire outputs. The calibration proceeded as follows. From the two pressure-transducer calibration signals, the dynamic-pressure calibration factors were recorded. This information was then used in the fitting of King's law to the converted numbers from the hot-wire channels, with the velocity determined from the measured dynamic pressure. Several thousand conversions were made during these calibration runs to minimize the effects of tape noise on the calibration (three velocities provide the minimum information to fit King's law when the offset for hot-wire voltage is not known). Since the problem of interest involved only a narrow range of velocities, the details of how the calibration was performed are of small importance. While there are still some small discrepancies in the static (DC) calibrations of the wires for the last runs in the groups, tne slope of the calibration is still sufficiently accurate to provide velocity differences for a given wire of 0.1% (the limits of the system). This was felt essential in view of the small differences between the conditional averages and the time averages.

The second phase of the computation was the actual evaluation of the desired experimental results. A detector function was generated from the converted hot-wire voltages for all ten channels, and the hot-wire voltages were linearized to give a velocity array and stored in a small ring buffer. The computation of conditional averages proceeded as follows. First, a decision as to whether the detector function for a given wire was either 0 or 1 was made, and the appropriate summation of velocity and squares of velocity for that wire was updated. Then a decision was made as to whether the detector function had changed. If it had, the appropriate conditional average for all wires (and the squares) was updated. Since each wire of the ten could generate a detector function for its position, this process was continued for all wires.

It is obvious from even thus brief description, that the time to update all the appropriate summations would vary appreciably, from a minimum when no detector function changed to a maximum when all detector functions changed. Since large amounts of bulk storage were not available at the computer, the conversions were occurring at the same time as the computation. If the conversion and calculation proceeded serially, the interval between conversions would have to be long enough to insure that the longest calculation could be completed. This would have resulted in an unduly long time interval between conversions. However, if it is recognized that the maximum calculation time occurs only rarely, the conversions and calculation can be overlapped. This overlapping requires a small buffer to keep active information from the recent past, and in practice conversions from only 16 time slices in the past were retained.

The programming feature that enabled the overlapping of conversion and calculation was essentially a small time-sharing system, operating internally under control of an external clock. Upon receipt of a pulse from the external clock (at 16-msec intervals) on one of the six available external interrupt lines mentioned previously, the computer interrupts its current task, stores the results of a previous conversion in locations accessible to the calculating program, updates the current conversion count, initiates a new conversion, and returns. The calculating program is entirely unaware of the interruption and resumes operation with the next instruction, and for all practical purposes is unaware of the source of data it is reducing. Two checks are built in; namely, that the calculating program does not overtake and pass the conversion routine, and because of the relatively small buffer size, fall too far behind.

It should be stressed that the calculation programs were written in IBM-supplied FORTRAN IV, and computation times could have been reduced by at least a factor of 4 if assembly-language programs had been used. It was felt that the added programming nuisance of assembly programming was unwarranted at this stage of the system development, but will certainly become necessary for future timing-sensitive calculations.

As a final note on the versatility of the programming scheme, it was generally true that the calculations program had at its disposal velocities at 11 past-time segments and 5 future-time segments of the time assigned to the detector function. This allowed the investigation of the importance of detector-time delay on the evaluated conditional averages. The buffer would have to be enlarged only slightly to enable space-time correlation to be made over reasonable time intervals, although the generation of conventional correlations were not a goal in this experiment.

Acknowledgement

This work was supported in part by the Office of Naval Research under Contract Nonr-228(33) and the National Science Foundation under Grant No. GK-1256.

References

1. KIM, H., S. KLINE and W. REYNOLDS: An experimental study of turbulence production near a smooth wall in a turbulent boundary layer with zero pressure gradient, AFOSR Sci. Rept. AFOSR-68-0383 January 1968.
2. KOVASZNAY, L. S. G., and V. KIBENS: To be published Refer to Kibens, V.: The intermittent region of a turbulent boundary layer, Dissertation, Johns Hopkins Univ., February 1968.
3. KOVASZNAY, L. S. G., H. KOMODA and B. R. VASUDEVA: Detailed flow field in transition, Proc. 1962 Heat Transfer and Fluid Mech. Inst. Stanford Univ. Press, 1962.
4. CORRSIN, S., and A. L. KISTLER: Free stream boundaries of turbulent flows. NACA Report 1244, 1955.
5. FIEDLER, H., and M. R. HEAD: Intermittency measurements in the turbulent boundary layer. JFM 24, 4 August 1966.
6. LAUFER, J., and R. E. KAPLAN: Concerning the large scale motion in the turbulent boundary layer. Fourth Euromech. Colloquium The Structure of Turbulence. Southampton, March 1967.
7. TOWNSEND, A. A.: The mechanism of entrainment in free turbulent flows. J. Fl. Mech. 26, 4, 689—715 (1966).

Postbuckling of an axially compressed oval cylindrical shell

J. Kempner and Y.-N. Chen

Polytechnic Institute of Brooklyn

Introduction

The present study represents an extension of the work on the problem of the nonlinear (postbuckling) behavior of a noncircular (oval) cylindrical shell under axial compression reported on earlier by the authors [1, 2]. In these reports attention was focused on the "far" postbuckling region, and not on the initial postbuckling region. In [2] it was found that, for a sufficiently out-of-round cylinder, the postbuckling region exhibited a relative maximum load which could exceed the classical buckling load. Thus, it was suggested that, for practical purposes, oval cylinders with sufficiently eccentric cross sections, might be less sensitive to initial imperfections than circular cylinders.

Before the authors could complete their present work on the extension of the results of [2] into the initial postbuckling region, HUTCHINSON [3] applied Koiter's theory to explore the sensitivity of the oval cylinder to initial imperfections. These results, which relate only to the neighborhood of the bifurcation point, showed that the initial buckling load was quite sensitive to initial imperfections. Inspite of such results, the present study shows, as was conjectured in [2], that for moderate to large eccentricities, the elastic collapse loads of the oval cylinder are likely to be far less sensitive to imperfections than the classical buckling load.

In the present work the approximate energy analysis proposed in [2] is extended to include the initial postbuckling region for the oval cylindrical shell. Due to the difficulties involved with maintaining accuracy in the numerical computations, a consistent-asymptotic solution was developed, which facilitated the determination of load-deformation curves in the initial postbuckling region. Furthermore, it is shown that the analytical results obtained are in reasonably good agreement with the results of laboratory tests.

The energy functional

The basic relations to be applied to determine the postbuckling behavior of homogeneous, isotropic, noncircular cylinders correspond to those of the Donnell-von Kármán type theory of thin-walled shells [2]. The analysis is based upon the principle of stationary potential energy, together with an appropriately assumed deflection function and a related Airy stress function, such that postbuckling configurations corresponding to stationary values of the total potential $W = W_1 + W_2$ are sought, where

$$W_1 = (1/2Eh) \int_0^{L_1} \int_0^{L_2} \{(\nabla^2 F)^2 + 2(1 + \nu)\,[(F_{,xs})^2 - F_{,xx}F_{,ss}]\}\,dx\,ds$$

$$W_2 = [Eh^3/24(1 - \nu^2)] \int_0^{L_1} \int_0^{L_2} \{(\nabla^2 w)^2 + 2(1 - \nu)\,[(w_{,xs})^2 - w_{,xx}w_{,ss}]\}\,dx\,ds$$

(1)

respectively, are the membrane energy and flexural energy of the shell. Inasmuch as controlled end-displacement loading is to be considered, the potential energy of applied end loads is iomtted without loss of generality, since its variation is identically zero. In (1), w denotes the geometrically admissible inward normal displacement component at an arbitrary point on the middle surface of the shell (Fig. 1); x and s, respectively, represent the axial and circumferential coordinates of the same point; L_1 and L_2, respectively, are the axial and circumferential lengths of the middle surface of the shell of wall-thickness h; E and ν are the Young's modulus and Poisson's ratio; $\nabla^2(\ ) = (\ )_{,xx} + (\ )_{,ss}$ and F, the Airy stress function, satisfies the compatibility equation

$$\nabla^4 F = Eh[(w_{,xs})^2 - w_{,xx} w_{,ss} - (1/r)\, w_{,xx}] \tag{2}$$

in which $\nabla^4 = \nabla^2\nabla^2$, and F is such that $F_{,ss} = N_x$, $F_{,xx} = N_s$ and $F_{,xs} = -N_{xs}$, where N_x, N_s and N_{xs}, respectively, are the axial, circumferential and shear membrane stress resultants. In the present study, the local radius of curvature of the noncircular cross section is taken to correspond to the family of doubly-symmetric, oval cross section cylinders considered in [2]. Thus, in (2)

$$r = r_0/[1 - \xi \cos (2s/r_0)] \tag{3}$$

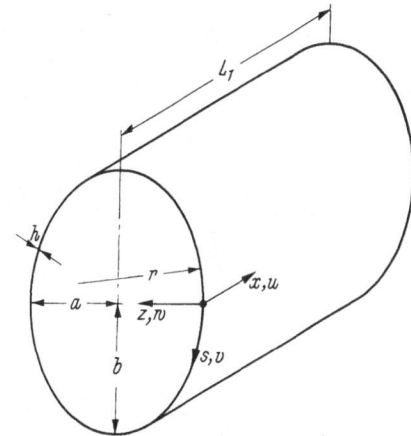

Fig. 1. Sign convention and geometry.

where $r_0 = L_2/2\pi$ is the average radius of curvature, and ξ, the eccentricity parameter, is restricted to range from zero to unity, corresponding to the major-to-minor axes ratio (b/a) varying from 1 to 2.06 (see Fig. 1).

In the ensuing work the admissible displacement w is chosen to be

$$
\begin{aligned}
w = (r_0\alpha/\sqrt{K}) \bigg[&(\sqrt{K}/\alpha)\, e + 2A \cos (2k_0 x/\alpha) + C \cos (4k_0 x/\alpha) \\
&+ 4\sqrt{2} \cos (k_0 x/\alpha) \sum_{n=0}^{N} B_n \cos nk_0 s + 2\sqrt{2} \cos (2k_0 x/\alpha) \sum_{p=1}^{P} D_p \cos 2pk_0 s \\
&+ 4 \sum_{m=1}^{M} E_m \cos mk_0 s \bigg]
\end{aligned} \tag{4}
$$

in which $k_0 = 1/r_0$ and $K = [12(1 - \nu^2)]^{1/2}(r_0/h)$. With the exception of the last sum, the radial deflection function (4) is identical to the deflection function used in [2]. The x-independent terms in the last sum are believed to be unimportant for the postbuckling patterns established in [2] for the far postbuckling region [4, 5], but play a significant role immediately after buckling, as pointed out by HUTCHINSON [3].

With the aid of (1) to (4), the nondimensional energy functional $V = (K^2/4EhL_1L_2)(W_1 + W_2)$ can be obtained in terms of the displacement parameters A, B_n, C, D_p, E_m and α, and the parameter e can be eliminated from V by utilizing the relation $V_{,e} = 0$, which is one of the conditions of stationary energy. The parameter α represents the nondimensional form of the axial wave length λ_x such that $\alpha = \lambda_x/L_2$.

The lengthy expression for V will not be shown; reference should be made to the appendices of [2] to which additional terms, due to the inclusion of the E_m-terms can be readily added. However, the complete relation for V can be expressed by the following formal representation:

$$
\begin{aligned}
V &= B_{ijkl} x_i x_j x_k x_l + C_{ijk} x_i x_j x_k + (Q_{ij} - \bar\varepsilon\delta_{ij})\, x_i x_j, \\
&\quad i, j, k, l = 1, 2, \ldots, N + P + M + 3
\end{aligned} \tag{5}
$$

where x_1 to x_{N+1}, x_{N+2} and x_{N+3}, x_{N+4} to x_{N+P+3}, and x_{N+P+4} to $x_{N+P+M+3}$, respectively, denote B_0 to B_N, A and C, D_1 to D_P, and E_1 to E_M [see (4)]; B_{ijkl}, C_{ijk}, and $(Q_{ij} - \bar\varepsilon\, \delta_{ij})$, respectively, are the coefficients of the biquadratic, the cubic, and the quadratic terms in V. Unless specified otherwise, repeated subscripts carry the usual meaning of summation in the $(N + P + M + 3)$-dimensional manifold, and δ_{ij} is the Kronecker delta. The end-shortening parameter $\bar{\varepsilon}$ is equal to twice the ratio of the unit end-shortening of the oval cylinder to the unit end-shortening of the equivalent circular cylinder of radius r_0 and wall-thickness h. Since it can be demonstrated that $\bar{\varepsilon}$ does not depend upon the E_m-terms in (4), the expression for this quantity can be taken from [2].

Thus,

$$\bar{\varepsilon} = \theta + 4\left[A^2 + \sum_{n=0}^{N}(1 + \delta_n^0)\, B_n^2 + C^2 + \sum_{p=1}^{P} D_p^2 \right]$$

in which $\theta = \sigma K/E$, where σ is the applied average compressive stress.

Furthermore, it can be shown that Q_{ij} forms a symmetric square matrix Q, which can be expressed in the partitioned form [2]

$$Q = \begin{bmatrix} Q^* & 0 & 0 \\ 0 & Q^{**} & 0 \\ 0 & 0 & Q^{***} \end{bmatrix} = \begin{bmatrix} Q^* & 0 & 0 \\ 0 & Q^{**} & 0 \\ 0 & 0 & R^{***} + \bar{\varepsilon} \end{bmatrix} \tag{6}$$

in which the sub-matrices Q^*, Q^{**}, and $Q^{***} \equiv R^{***} + \bar{\varepsilon}$, respectively, are symmetric with ranks equal to $N + 1$, $P + 2$, and M. In addition, R^{***} is diagonal and, therefore, non-singular.

Also,

$$C_{ijk} = 0, \quad \text{if} \quad i, j, k \leq N + 1 \tag{7a}$$

In other words, terms of the form $B_i B_j B_k$ are absent in V. This fact can be readily deduced by setting $A = C = D_1 = \cdots = D_P = E_1 = \cdots = E_M = 0$ in w and F, and then by observing that the cubic terms in V vanish, i.e., substitution of

$$w = BY(s) \cos (k_0 x/\alpha)$$

$$F = (1/2)\, fs^2 + BR(s) \cos (k_0 x/\alpha) + B^2 P(s) + B^2 T(s) \cos (2k_0 x/\alpha)$$

into (1) yields no terms proportional to B^3 by virtue of the orthogonality conditions:

$$\int_0^{L_1} \cos (k_0 x/\alpha)\, dx = 0, \qquad \int_0^{L_1} \cos (k_0 x/\alpha) \cos (2k_0 x/\alpha)\, dx = 0.$$

A less obvious implication of the orthogonality conditions is that

$$B_{ijkl} = 0 \tag{7b}$$

if one of the indices of this term is greater than $N + 1$ and the others are smaller than or equal to $N + 1$. Thus, terms of the form $B_m B_n B_r A$, $B_m B_n B_r C$, $B_m B_n B_r D_p$, and $B_m B_n B_r E_s$ are absent in V.

The properties of Q_{ij}, C_{ijk}, and B_{ijkl} given in (6) and (7) prove to be essential in the subsequent analysis.

The parametric expansion

In accordance with the principle of stationary potential energy, the reduced potential $V(x_i, \alpha)$ is stationary for any equilibrium configuration. This condition is equivalent to a set of displacement parameters x_i and α satisfying the simultaneous equations $f_n = 0$, $V_{,\alpha} = 0$, where f_n are the partial derivatives of V with respect to x_n. In the subsequent analysis only stable equilibrium configurations will be sought, i.e., states corresponding to a minimization of V.

With the aid of (5), f_n can be expressed in the form

$$\begin{aligned} f_n = {} & (B_{ijkn} + B_{ijnk} + B_{injk} + B_{nijk})\, x_i x_j x_k + (C_{ijn} + C_{inj} + C_{nij})\, x_i x_j \\ & + 2Q_{nj} x_j - 2\bar{\varepsilon} x_n, \qquad n, i, j, k = 1, 2, \ldots, N + P + M + 3. \end{aligned} \tag{8}$$

When $f_n = 0$ and $V_{,\alpha} = 0$, the roots x_i characterize an equilibrium configuration for any given value of $\bar{\varepsilon}$. In the procedure adopted in [2] the algebraic equations $f_n = 0$ were solved directly for fixed values of $\bar{\varepsilon}$ and α until a minimum of V with respect to α was found for each value of $\bar{\varepsilon}$. However, when an attempt was made to extend the procedure of [2] to the initial post-buckling region, numerical difficulties arose due to the smallness of the vector $\boldsymbol{x}$. Consequently, an alternative approach to the solution of the equations $f_n = 0$ is adopted in the analysis that follows.

The essence of this analysis lies in the interchange of the roles of $\bar{\varepsilon}$ and one of the unknowns, say, x_λ, which is now considered as given, while $\bar{\varepsilon}$ joins the ranks of the unknowns, all of which now depend upon x_λ. The selection of λ is arbitrary; but for reasons which become apparent later, x_λ is chosen to be that deflection parameter which is the amplitude of the dominant harmonic at the bifurcation point.

Next, a new set of functions $\varphi_n = x_n f_\lambda - x_\lambda f_n = 0$ is constructed, and a relation for the end-shortening parameter $\bar{\varepsilon}$ is obtained from the equation $f_\lambda = 0$. Obviously, the equations $\varphi_n = 0$ and $f_\lambda = 0$ can replace $f_n = 0$. Thus, with the aid of (8)

(no sum on λ)

$$\begin{aligned}
\bar{\varepsilon} = {} & (1/2)\,(B_{ijk\lambda} + B_{ij\lambda k} + B_{i\lambda jk} + B_{\lambda ijk})\,x_i x_j x_k / x_\lambda \\
& + (1/2)\,(C_{ij\lambda} + C_{i\lambda j} + C_{\lambda ij})\,x_i x_j / x_\lambda + Q_{\lambda j} x_j / x_\lambda
\end{aligned} \tag{9}$$

$$\begin{aligned}
\varphi_n = {} & (B_{ijk\lambda} + B_{ij\lambda k} + B_{i\lambda jk} + B_{\lambda ijk})\,x_i x_j x_k x_n \\
& - (B_{ijkn} + B_{ijnk} + B_{injk} + B_{nijk})\,x_i x_j x_k x_\lambda \\
& + (C_{ij\lambda} + C_{i\lambda j} + C_{\lambda ij})\,x_i x_j x_n + 2Q_{\lambda j} x_j x_n \\
& - (C_{ijn} + C_{inj} + C_{nij})\,x_i x_j x_\lambda - 2Q_{nj} x_j x_\lambda
\end{aligned} \tag{10}$$

$$(i, j, k, n = 1, 2, \ldots, N + P + M + 3).$$

In accordance with the definition of x_λ, the classical buckling solution clearly corresponds to the limiting case represented by x_λ approaching zero. In view of (9), if $\bar{\varepsilon}$ is to remain finite, the variational parameters x_i must also approach zero with x_λ. Thus, in the vicinity of the bifurcation point where x_i is expected to be very small, a small quantity $\delta \geq 0$ may be introduced such that

$$x_\lambda = \delta$$

and

$$x_i = \delta(a_i + b_i \delta + c_i \delta^2 + \cdots) \tag{11}$$

provided that it is understood that $a_\lambda = 1$, $b_\lambda = c_\lambda = \cdots = 0$.

Upon substitution of (11) into (9) and (10), parametric expansions for $\bar{\varepsilon}$ and φ_n are obtained in terms of powers of δ. The equations $\varphi_n = 0$ can then be replaced by a series of sets of equations, determined by enforcing the requirement that the multipliers of the powers of δ in φ_n must vanish. The expansion of $\bar{\varepsilon}$, and the first three sets of the sequential equations are

$$\begin{aligned}
\bar{\varepsilon} = {} & \bar{\varepsilon}_0 + \delta[Q_{\lambda j} b_j + (1/2)\,(C_{ij\lambda} + C_{i\lambda j} + C_{\lambda ij})\,a_i a_j] \\
& + \delta^2[Q_{\lambda j} c_j + (1/2)\,(C_{ij\lambda} + C_{i\lambda j} + C_{\lambda ij})\,(a_i b_j + a_j b_i) \\
& + (1/2)\,(B_{ijk\lambda} + B_{ij\lambda k} + B_{i\lambda jk} + B_{\lambda ijk})\,a_i a_j a_k] + 0(\delta^3)
\end{aligned} \tag{12}$$

and

$$Q_{nj} a_j - \bar{\varepsilon}_0 a_n = 0, \tag{13}$$

$$\begin{aligned}
D_{nj} b_j = {} & (1/2)\,(C_{ij\lambda} + C_{i\lambda j} + C_{\lambda ij})\,a_i a_j a_n \\
& - (1/2)\,(C_{ijn} + C_{inj} + C_{nij})\,a_i a_j
\end{aligned} \tag{14}$$

$$\begin{aligned}
D_{nj} c_j = {} & Q_{\lambda j} b_n b_j - (1/2)\,(C_{ijn} + C_{inj} + C_{nij})\,(a_i b_j + a_j b_i) \\
& + (1/2)\,(C_{ij\lambda} + C_{i\lambda j} + C_{\lambda ij})\,(a_i a_j b_n + a_i a_n b_j + a_n a_j b_i) \\
& + (1/2)\,(B_{ijk\lambda} + B_{ij\lambda k} + B_{i\lambda jk} + B_{\lambda ijk})\,a_i a_j a_k a_n \\
& - (1/2)\,(B_{ijkn} + B_{ijnk} + B_{injk} + B_{nijk})\,a_i a_j a_k
\end{aligned} \tag{15}$$

in which
$$\bar{\varepsilon}_0 = \lim_{\delta \to 0} \bar{\varepsilon} = Q_{\lambda j} a_j, \tag{16}$$

$$D_{nj} = Q_{nj} - Q_{\lambda j} a_n - \bar{\varepsilon}_0 \delta_{nj}. \tag{17}$$

Equations (13), (14), and (15), solved in that order, determine the unknown parameters a_i, b_i, and c_i, respectively. When these equations are satisfied, the remainders in the equations $\varphi_n = 0$ are of the order of δ^5.

Solution of the recurrence equations

In the following analysis, the recurrence equations shown in (13) to (15) are solved in the order in which they are listed. In this connection, let the components of any $N + P + M + 3$-dimensional vector, say t be partitioned in such a way that

$$t^* = (t_i^*) \qquad i = 1, 2, \ldots, N + 1,$$

$$t^{**} = (t_i^{**}) \qquad i = N + 2, N + 3, \ldots, N + P + 3,$$

$$t^{***} = (t_i^{***}) \qquad i = N + P + 4, N + P + 5, \ldots, N + P + M + 3.$$

Of course, the totality of these three vectors in proper order, as indicated by their subscripts, is the vector t itself, i.e.,

$$t = (t^*, t^{**}, t^{***})$$

Aided by the form of the square matrix Q shown in (6), together with the convention introduced above, (13) yields three sets of homogeneous equations, viz.,

$$(Q_{nj}^* - \bar{\varepsilon}_0 \delta_{nj})\, a_n^* = 0, \tag{18a}$$

$$(Q_{nj}^{**} - \bar{\varepsilon}_0 \delta_{nj})\, a_n^{**} = 0, \tag{18b}$$

$$R_{nj}^{***}\, a_n^{***} = 0. \tag{18c}$$

Hence, $\bar{\varepsilon}_0$ is an eigenvalue of either the matrix Q^* or the matrix Q^{**}. For the linear buckling problem, only the lowest eigenvalue is of interest for a given value of α. Moreover, for the same value of α, the lowest eigenvalues calculated with (18a) and (18b) are different. Thus, without loss of generality, henceforth, $\bar{\varepsilon}_0$ will be defined as the lowest eigenvalue of the matrix Q^* for a given value of α. Consequently, if the same α is used in both (18a) and (18b), $\bar{\varepsilon}_0$ is not an eigenvalue of the matrix Q^{**}. Of course, the end shortening parameter at the bifurcation point, denoted by $\bar{\varepsilon}_{cr}$, is obtained by minimizing $\bar{\varepsilon}_0$ with respect to α, which, in turn, corresponds to a minimum of the total potential.

In view of the above, a_n^* constitutes the eigenvector of the matrix Q^*, associated with the lowest eigenvalue $\bar{\varepsilon}_0$ of Q^*, with $a_\lambda = 1$. Since a_λ belongs to the group a_i^*, λ takes on integral values between 1 and $N + 1$; i.e., $1 \leq \lambda \leq N + 1$. The fact that $\bar{\varepsilon}_0$ is not an eigenvalue of the matrix Q^{**} implies that the matrix $[Q^{**} - \bar{\varepsilon}_0]$ is non-singular. Therefore, (18b) and (18c) only yield the trivial solutions $a_i^{**} = 0$ and $a_i^{***} = 0$. The latter follows from the observation that R^{***} is non-singular [see (6)].

Thus, from (18)

$$(Q_{nj}^* - \bar{\varepsilon}_0 \delta_{nj})\, a_n^* = 0, \qquad |Q_{nj}^* - \bar{\varepsilon}_0 \delta_{nj}| = 0, \tag{19a, b}$$

$$1 \leq \lambda \leq N + 1, \qquad a_i^{**} = 0, \qquad a_i^{***} = 0 \tag{19c, d, e}$$

represent the solution of (13).

Furthermore, (6), (7) and (19) lead to the separation of (14) into the following three distinct sets of linear equations:

$$D_{nj}^* b_j^* = 0, \qquad D_{nj}^{**} b_j^{**} = -g_n^{**}, \qquad R_{nj}^{***} b_j^{***} = -g_n^{***} \tag{20a, b, c}$$

in which

$$g_n^{**} = (1/2)\, (C_{ijn} + C_{inj} + C_{nij})\, a_i^* a_j^*, \tag{21a}$$

$$g_n^{***} = (1/2)\, (C_{ijn} + C_{inj} + C_{nij})\, a_i^* a_j^* \tag{21b}$$

and [see (16) and (17)]

$$D_{nj}^* = Q_{nj}^* - Q_{\lambda j}^* a_n^* - \bar{\varepsilon}_0 \delta_{nj}, \qquad D_{nj}^{**} = Q_{nj}^{**} - \bar{\varepsilon}_0 \delta_{nj}. \qquad (22\,\text{a, b})$$

Although the right-hand sides of (21) are identical in form, they are not the same, since the subscript n ranges over different values in these expressions.

As stated previously, the square matrix $R^{***} = [R_{nj}^{***}]$ is non-singular. Also, $D^{**} = [D_{nj}^{**}]$ is obviously not singular, since $\bar{\varepsilon}_0$ is not an eigenvalue of Q^{**}. The fact that the square matrix $D^* = [D_{ij}^*]$ is also non-singular follows from the relation

$$|D^*| = -\bar{\varepsilon}_0 \tilde{H}_{\lambda\lambda} a_i^* a_i^*$$

which can be obtained from the repeated application of the Addition Theorem [6], together with (19) and (22a), where $\tilde{H}_{ij}$ denotes the cofactor of $H_{ij} = Q_{ij}^* - \bar{\varepsilon}_0 \delta_{ij}$, and $\tilde{H}_{\lambda\lambda}$ is different from zero. It follows immediately from (20) that

$$\boldsymbol{b}^* = 0 \qquad \boldsymbol{b}^{**} = -D^{**-1}\boldsymbol{g}^{**}, \qquad \boldsymbol{b}^{***} = -R^{***-1}\boldsymbol{g}^{***}. \qquad (23\,\text{a, b, c})$$

Finally, (15) is separable into three distinct groups involving $\boldsymbol{c}^*$, $\boldsymbol{c}^{**}$, $\boldsymbol{c}^{***}$, which can be determined in the same manner as the vector $\boldsymbol{b}$. However, the complete solution for the vector $\boldsymbol{c}^*$ is not necessary, since (12) reveals that only the sum $Q_{\lambda j}^* c_j^*$ is required. Substitution of (19), (22), and (23) into (15) leads to

$$Q_{\lambda j}^* c_j^* = -(1/2)\,(C_{ij\lambda} + C_{i\lambda j} + C_{\lambda ij})\,(a_i b_j + a_j b_i) - g_\lambda^* + g_k(a_k + 2b_k)/a_n a_n \qquad (24)$$

where

$$g_n^* = (1/2)\,(B_{ijkn} + B_{ijnk} + B_{injk} + B_{nijk})\,o_i^* a_j^* a_k^* \qquad (25)$$

and, of course, $\boldsymbol{g} = (g_n) = (\boldsymbol{g}^*, \boldsymbol{g}^{**}, \boldsymbol{g}^{***})$. In addition, it can be shown that $\boldsymbol{c}^{**} = \boldsymbol{c}^{***} = 0$.

Equations (19), (23), and (24) provide sufficient information for the ensuing asymptotic analysis.

Initial postbuckling behavior of the oval cylinder

The behavior of a cylindrical shell under axial compression is usually characterized by the load deformation response; i.e., by a relation between the parameters $\bar{\varepsilon}$ and θ, where, as defined previously, the former is the end-shortening parameter, while the latter, defined previously as $\sigma K/E$, is twice the ratio of the applied average compressive stress to the classical buckling stress of the equivalent circular cylinder. As shown in [2], the load and deformation are related by

$$\theta = \bar{\varepsilon} - 4x_i x_i, \qquad i = 1, 2, \ldots, N + P + 3. \qquad (26)$$

The results established in the previous sections, as shown in (6), (7), (16), (19) and (23) to (25), can be combined to yield

$$\bar{\varepsilon} = \bar{\varepsilon}_0 + \delta^2[g_k(a_k + 2b_k)/a_n a_n] + 0(\delta^3), \qquad (27)$$

$$\theta = \bar{\varepsilon} - 4a_i a_i \delta^2 + 0(\delta^3). \qquad (28)$$

Similarly, the corresponding expansion for the energy functional $V(\boldsymbol{x}, \boldsymbol{a})$ can be obtained by combining (5) to (7), (12), (16), (19), (23) to (25) and (27). In the process, terms of order δ^3 in (27) are omitted. Thus,

$$V = -(1/2)\,(\bar{\varepsilon} - \bar{\varepsilon}_0)^2 a_i a_i a_j a_j/g_k(a_k + 2b_k) + 0(\delta^5). \qquad (29)$$

It should be noted that contributions of the components of the vector $\boldsymbol{c}$ have been implicitly incorporated in (27) to (29) with the aid of (24). These relations are valid asymptotically in the vicinity of the bifurcation point.

A useful expression can be readily deduced by combining (27) and (28), after omitting terms of order δ^3, to yield

$$\theta = \bar{\varepsilon} - 4(\bar{\varepsilon} - \bar{\varepsilon}_0)a_i a_i a_j a_j/g_k(a_k + 2b_k). \qquad (30)$$

32*

Upon the substitution of $\bar{\varepsilon}_0 = \bar{\varepsilon}_{cr}$, (30) yields the asymptote of the load-deformation response curve. Of course, $\alpha = \alpha_{cr}$, which is the value of the wave length parameter at the critical point.

The slope T_a of the asymptote can be easily obtained from (30) by differentiating θ with respect to $\bar{\varepsilon}$. Hence,

$$T_a = 1 - 4a_i a_i a_j a_j / g_k(a_k + 2b_k). \tag{31}$$

Thus, the nonlinear postbuckling curve θ vs $\bar{\varepsilon}$ branches away from the prebuckling membrane solution, corresponding to which the slope is obviously $T_m = 1$, at the critical point with a slope $T_a \neq 1$.

It is interesting to note that the partial sum $g_k^{**}b_k^{**}$ in the denominator of (31) is negative at the bifurcation point where $\bar{\varepsilon}_0 = \bar{\varepsilon}_{cr}$. Recalling (20b) and (22b), when $\bar{\varepsilon}_0$ is set equal to $\bar{\varepsilon}_{cr}$,

$$g_k^{**}b_k^{**} = -(Q_{ij}^{**} - \bar{\varepsilon}_{cr}\delta_{ij})\, b_i^{**}b_j^{**}.$$

Due to the fact that the matrix Q^{**} is real and symmetric, the canonical form of Q^{**} is purely diagonal. Thus,

$$g_k^{**}b_k^{**} = -\sum_{i=N+2}^{N+P+3}(q_i^{**} - \bar{\varepsilon}_{cr})\, \hat{b}_i^{**2}$$

in which q_i^{**} are the eigenvalues of Q^{**} when the wave length parameter $\alpha = \alpha_{cr}$, and $\hat{b}_i^{**}$ represents the components of the vector b^{**} projected along the principal axes of Q^{**}. The fact that $g_k^{**}b_k^{**}$ is negative follows from the observation that $q_i^{**} - \bar{\varepsilon}_{cr} > 0$, since $\bar{\varepsilon}_{cr}$ is the smallest possible value of all eigenvalues of Q for all values of α.

In view of the fact that $g_k^{**}b_k^{**}$ is negative, it can be shown that the approximate asymptotic slope

$$T_b = 1 - 4a_i a_i a_j a_j / (g_k a_k + 2g_k^{***}b_k^{***}) \tag{32}$$

which requires considerably less computation than does T^a, provides an approximation for T_a such that, if ψ denotes the angle measured counterclockwise from the prebuckling solution $(\theta = \bar{\varepsilon} \leq \bar{\varepsilon}_{cr})$ to the asymptote at the bifurcation point, then

$$0 < \psi_a < \psi_b$$

where ψ_a and ψ_b are computed from (31) and (32), respectively.

The numerical results which will be discussed in detail in the following section indicate that ψ_b is a close upper bound of ψ_a.

Results and discussion

In the present study computations based upon both the asymptotic solution and the earlier solution of [2] for the far postbuckling region were carried out.

For the asymptotic solution the determination of the vector a and b follows from the solution of two sets of linear algebraic equations, viz., (19) and (20). In the program written for a CDC-6600 computer, the coefficients B_{ijkl}, C_{ijk}, and Q_{ij} [see (5)] are first compiled for a given value of the end-shortening parameter $\bar{\varepsilon}$ and a selected value of the wave length parameter α, with the aid of an automatic process built into the computing program. After this preliminary step, the remaining computations are readily accomplished in several distinct phases. The eigenvector a, together with the lowest eigenvalue $\bar{\varepsilon}_0$, is obtained through the solution of the homogeneous system (19), with $a_\lambda = 1$. The vector g is then compiled and, subsequently, the vector b is determined [see (20), (21), (25)]. The correct α for the fixed value of $\bar{\varepsilon}$ follows by minimizing the total potential V, obtained from (29), with respect to α. Finally, the load-deformation relation θ vs $\bar{\varepsilon}$ and the approximate asymptotic slope T_b are calculated from (30) and (32), respectively.

The direction of bifurcation, which is measured by the angle ψ defined earlier, can be estimated from the relation

$$\psi_b = \tan^{-1}[(T_b - 1)/(T_b + 1)].$$

Table 1 shows that, in all but one case, the direction of bifurcation is toward the third quadrant in the manner displayed by the sketch. The exception, corresponding to the extreme case $K = 100$, $\xi = 1.0$, corresponds to a branching towards the fourth quadrant. In any case, the result indicates a drop in the load carried by the oval cylinder once the classical buckling load has been reached. This behavior is similar to that of the circular cylindrical shell and, thus, the present results confirm those of HUTCHINSON [3], who analyzed this aspect of the oval cylinder with the aid of a Koiter-type theory. Furthermore, the observations that, for a wide range of geometric parameters, $\psi_b \ll 45°$ also indicates that the x-independent terms in the deflection function [i.e., E_m-terms in (4)] are significant at the bifurcation point. The importance of such terms was also pointed out in [3]. In the present work the need for these terms in the determination of ψ_b follows from the fact that, if the partial vector b^{***}, which corresponds to E_m, is ignored, T_b reduces to

$$T_c = 1 - 4a_i a_i a_j a_j / g_k a_k$$

which must be less than 1, since $g_k a_k > 0$. This conclusion implies that $\psi_c > 45°$, which differs substantially from the results for ψ_b. However, for the stable region of the asymptotic solution it was not found necessary to include the E_m-terms.

The load-deformation response in the initial postbuckling region determined from the present asymptotic analysis is shown in Figs. 2 and 3. The response curves corresponding to

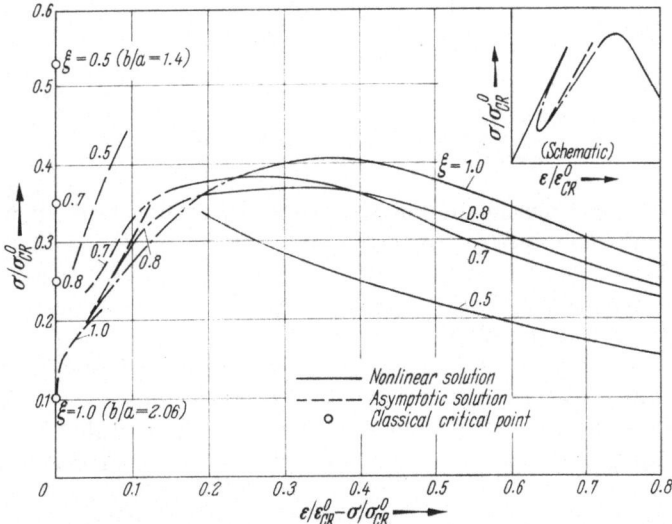

Fig. 2. Theoretical response curves ($r_0/h = 100$, $K = 330$).

the fully developed postbuckling configurations [2] are also displayed to facilitate comparison with the present analysis. Except for the additional case, $\xi = 0.7$, the results related to the fully developed configurations for $K = 330$ are taken from [2], while the data for $K = 1000$ were computed during the course of the present study. Clearly, the asymptotic curves and those obtained from the direct nonlinear solution match reasonably well, although not precisely. The smallness of the discrepancy between the two analyses indicates that the asymptotic curves are quite reliable, especially when $\bar{\varepsilon}$ is small.

The results presented thus far suggest that, as in the case of the circular cylinder, the initial postbuckling curve starts with an unstable branch which exhibits a decreasing end shortening until a vertical tangent corresponding to a minimum value of $\bar{\varepsilon}$ on a plot of θ vs $\bar{\varepsilon}$ is reached (see sketch on Fig. 2). Subsequently, a first relative minimum for the load θ is expected, followed by stable equilibrium states corresponding to loads increasing with increasing end shortening until the maximum postbuckling load is attained. Thus, after the first minimum load is reached, the behavior of the oval cylinder can be markedly different from that of the circular cylinder, since the latter can hardly, if ever, (see [7]), support a load much higher than the minimum postbuckling load once the cylinder has buckled. Of course, this difference in behavior is best exemplified by the extreme cases, $\xi = 1.0$ and $\xi = 0$. For in-between cases, the postbuckling behavior exhibits characteristics of both the extremely oval cylinder and the circular cylinder.

Since the required numerical accuracy for the determination of points lying too close to the prebuckling curve is prohibitive, a reasonable estimate for the first minimum load must be assumed in order to complete the response curve. In [3] Hutchinson obtained a load reduction

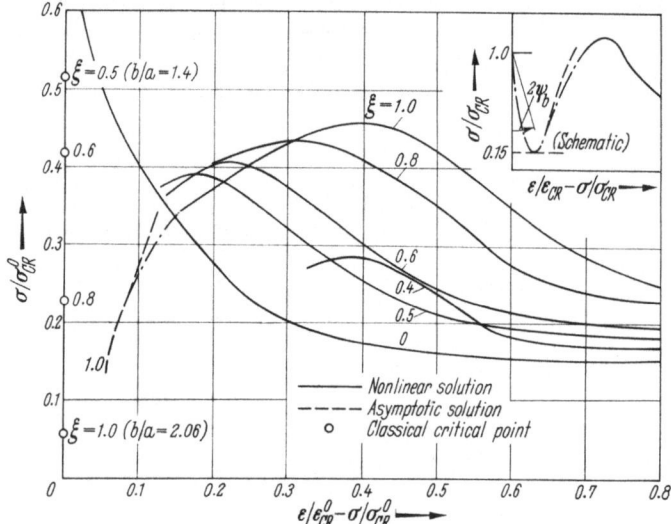

Fig. 3. Theoretical response curves ($r_0/h = 303$, $K = 1000$).

parameter with which the dependency of the classical buckling load upon the amplitude of imperfections for initially imperfect oval cylinders can be plotted with the aid of Koiter's well-known formula. Although such curves were not displayed in [3], it can be shown numerically that the reduction of the classical buckling load due to the presence of imperfections is practically of the same order for all ovals in the range $0 \leq \xi \leq 0.9$. Hence, it may be assumed that the load corresponding to the first minimum of the oval cylinder is approximately 15% of its classical buckling load, as in the case of a circular cylinder for the deflection functions assumed in (4). With this assumption and the data shown in Table 1 and Figs. 2 and 3, the entire postbuckling curve can be constructed as illustrated schematically on Fig. 3.

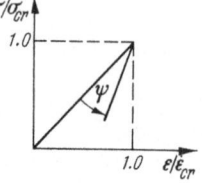

Table 1. *Bifurcation Angle* ψ_b (Degrees)

ξ	$K = 100$	$K = 330$	$K = 500$	$K = 1,000$	$K = 3,000$	$K = 5,000$
0.1	4.7	2.4	2.2	1.7	.77	.90
0.3	4.1	2.3	1.9	1.4	1.0	.63
0.5	4.3	2.2	1.7	1.3	.81	.63
0.7	4.9	2.6	2.1	1.5	.91	.74
0.9	11.2	4.6	3.5	2.3	1.4	1.0
1.0	57.9	34.7	27.4	21.1	13.6	11.3

Despite the aforementioned agreement with the results in [3] for the region immediately adjacent to the classical buckling point, it should be noted that the first stable branch determined by the present theory for an oval cylinder is located much closer to the prebuckling curve

$(\theta - \bar{\varepsilon} = 0)$ than that for a circular cylinder. Hence, the snap-through phenomenon for an oval cylinder of even moderate eccentricity is substantially less drastic than that for a circular cylinder, while the snap-through action of an oval cylinder with large eccentricity is not even noticeable. This behavior is observed in Figs. 4 and 5, respectively, which exhibit the experi-

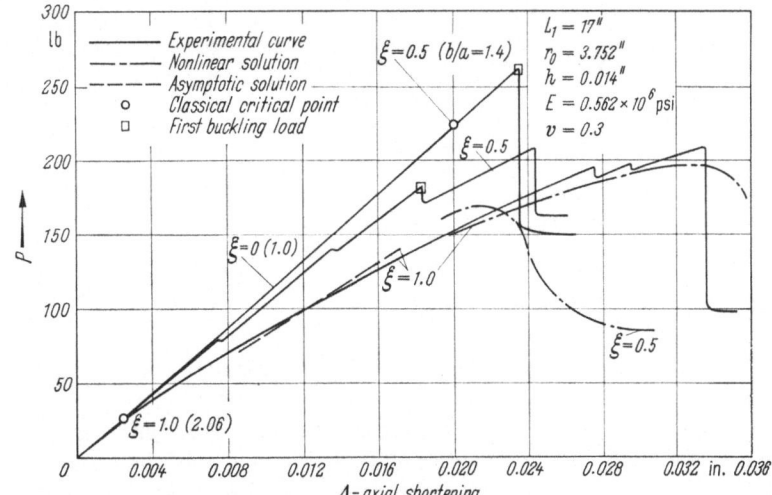

Fig. 4. Experimental and theoretical response curves ($r_0/h = 268$, $K = 884$).

mentally determined response curves for oval cylinders with thickness ratios of $r_0/h = 268$ and 375. More extensive test results for these Mylar cylinders are summarized in Table 2. It is interesting to note that even the moderately eccentric oval cylinder $\xi = 0.5$ (i.e., $b/a = 1.4$) does not show a drastic drop in load after the first buckles are developed (at the ends of the

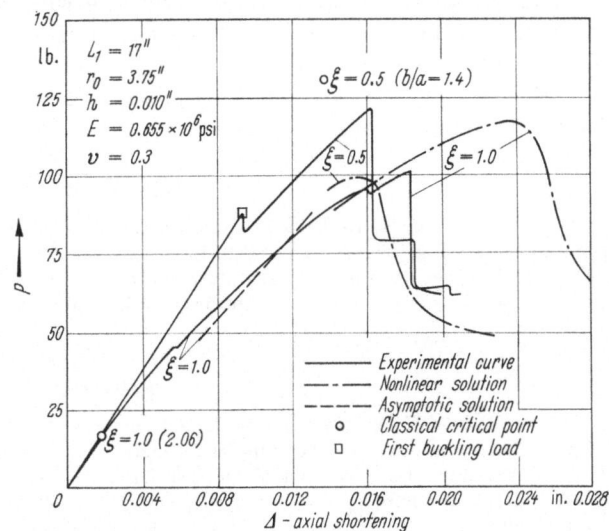

Fig. 5. Experimental and theoretical response curves ($r_0/h = 375$, $K = 1237$).

minor axis). After this point is reached, the cylinder continues to carry a load appreciably higher than the "first" buckling load. For $\xi = 1$ (i.e., $b/a = 2.06$), the experimental response curves extend through the region of the bifurcation point and the first minimum load without any noticeable drop in load and, ultimately, reach a point corresponding to 7 to 8 times the classical buckling load. After such a maximum load is attained, the collapse is as catastrophic as that of a circular cylinder. As predicted by the theory, it was observed during the experiments that buckles initiate in the regions at the ends of the minor axis and progress to regions of greater curvature, until final collapse occurs (also see [2] and [3]).

Table 2. *Test Data*

Spec. No.	b/a	r_0/h	P_{cr}	$(P_{cr})_{exp.}$	P_{max}^*	$(P_{max})_{exp.}$	$(P_{max})_{exp}/P_{cr}$
1, 2	1.0	375**	255	144—159	255	144—159	.57—.62
3—6	1.0	268***	431	258—295	431	258—295	.60—.69
7—10	1.4	375	131	64—118	99	120—125	.92—.96
11—14	1.4	268	223	165—183	168	204—225	.92—1.0
15—17	2.06	375	13.5	—	117	93—102	6.9—7.6
18—21	2.06	268	26.0	—	197	180—217	6.9—8.3

* Calculated for $r_0/h = 303$ and modified for proper thickness.
Axial length $= 17''$, Average inner radius $= 3.745''$.
P_{cr} = classical buckling load.
P_{max} = theoretical maximum postbuckling load.
All loads in pounds.
** $E = .655 \times 10^6$ psi, $\nu = 0.3$.
*** $E = .562 \times 10^6$ psi, $\nu = 0.3$.

The theoretical curves shown for the oval cylinders are in surprisingly good agreement with the test records.

Thus, the test results strongly suggest that for oval cylinders with moderate to large eccentricities, for which the maximum postbuckling load is of the order of or higher than the classical buckling load, the maximum load serves as a more meaningful definition of the buckling (collapse) load than does the classical buckling load. On the other hand, for oval cylinders exhibiting small eccentricity, stability may be better characterized by the classical buckling load, which is now higher than the maximum postbuckling load. Of course, while the results of the sensitivity analysis of HUTCHINSON clearly do not apply to the former case, they are applicable in the latter case.

No analytical results are currently available with regard to the sensitivity of the oval in connection with the maximum postbuckling load. However, since the test data shown in Table 2 agree reasonably well with the theoretical predictions for the maximum load, it is suggested that oval cylinders with moderate to large eccentricities are not very sensitive to initial imperfections.

Acknowledgment

The authors would like to thank Professor B. ERICKSON and Mr. G. FEINSTEIN for their invaluable contributions to the testing program referred to in this paper.

The research was sponsored by Air Force Office of Scientific Research, Office of Aerospace Research, United States Air Force, and was monitored under the technical supervision of Dr. JACOB POMERANTZ, AFOSR.

References

1. KEMPNER, J., and Y.-N. CHEN: Large deflections of an axially compressed oval cylindrical shell. Proc. Eleventh Int. Cong. of Appl. Mech., Munich, 1964, Berlin/Heidelberg/New York: Springer 1966, p. 299.
2. — — Buckling and postbuckling of an axially compressed oval cylindrical shell. Proc. Symp. on the Theory of Shells to Honor Lloyd H. Donnell, McCutchan Pub. Co., May 1967, p. 141.
3. HUTCHINSON, J. W.: Buckling and initial postbuckling behavior of oval cylindrical shells under axial compression. J. Appl. Mech. 35, 1, 66 (1968).
4. KEMPNER, J.: Postbuckling behavior of axially compressed circular cylindrical shells. J. Aero. Sci. 21, 5, 329 (1954).
5. ALMROTH, B. O.: Postbuckling behavior of axially compressed circular cylinders. AIAA J. 1, 3, 630 (1963).
6. MUIR, T.: A treatise on the theory of determinants. New York: Dover 1960, p. 31.
7. HOFF, N. J.: Thin shells in aerospace structures. Astronautics and Aeronautics, 5, 2, 26 (1967).

Stability analysis of branching solutions of the Navier-Stokes equations

By

K. Kirchgässner and P. Sorger

Institut für Angewandte Mathematik und Mechanik der DVLR, Freiburg

Introduction

This paper describes a method for the investigation of some properties of branching solutions of the Navier-Stokes boundary-value problem. Questions arising in this context are closely connected to problems in hydrodynamic stability. Many of the nonlinear aspects of this theory have been investigated thoroughly (cf. J.T. STUART [11]). However, it was only in recent times that a mathematically rigorous basis was layed for these results, beginning with the use of topological methods by VELTE [13] and IUDOVICH [3] and the application of the method of SCHMIDT-LYAPUNOV by KIRCHGÄSSNER [7] and IUDOVICH [4].

The main emphasis of this paper lies on the mathematical aspects of hydrodynamic stability theory. Wes hall prove some new results for the celebrated Taylor problem and indicate the close connection between the "principle of exchange of stabilities" (P.E.S.) and the selection of a distinct period by nature as the only stable, axisymmetric, and periodic solution of the problem. The method can be applied quite analogously to the Bénard problem; the implications are indicated in this paper.

1. Formulation of the problem and summary of results

The initial boundary-value problem is considered which describes the flow of a viscous incompressible fluid between two coaxial cylinders of infinite length, the inner cylinder rotating with constant angular velocity while the outer cylinder is at rest.

Let R_1, R_2 be the radii of the inner and the outer cylinder respectively and Ω_1 ($\Omega_1 > 0$) the angular velocity of the inner cylinder. We choose $(R_2 - R_1)^2/\nu$, $R_2 - R_1$, $R_1\Omega_1$ and $\varrho(R_1\Omega_1)^2$ as quantities of reference for time, length, velocity, and pressure (ν kinematic viscosity, ϱ density). The similarity parameter thus introduced into the problem is the Reynolds number

$$\lambda = \frac{(R_2 - R_1)\, R_1\Omega_1}{\nu}.$$

We choose an appropriate system of dimensionless cylindrical coordinates r, θ, z and restrict the analysis to flows which are independent of θ and periodic in z-direction. Let $v(r, z, t) = \{(v)_1, (v)_2, (v)_3\}$ and $q(r, z, t)$ denote the velocity vector and the pressure function respectively. Then the initial- and boundary-value problem describing motions of this restricted class reads as follows:

$$\frac{\partial v}{\partial t} - Dv = \lambda \nabla q - \lambda(v \cdot \nabla v + B(v)\, v),$$

$$\nabla \cdot v = 0,\ v|_{t=0} = v_0, \tag{1.1}$$

$$v = \{0, 1, 0\}\ \text{for}\ r = r_1,\ v = 0\ \text{for}\ \ r = r_2,$$

$$v(r, z) = v(r, z + 2\pi/\alpha),$$

where r_i denote the dimensionless radii of the cylinders, $\alpha = 2\pi/l$ the wave number, and l the period of the solutions, which can be chosen as an arbitrary positive real. Furthermore, the following notations have been used:

$$D = \begin{pmatrix} \Delta - 1/r^2 & 0 & 0 \\ 0 & \Delta - 1/r^2 & 0 \\ 0 & 0 & \Delta \end{pmatrix}, \qquad B(r) = \begin{pmatrix} 0 & -(v)_2/r & 0 \\ (v)_1/r & 0 & 0 \\ 0 & 0 & 0 \end{pmatrix},$$

$$\Delta = \frac{1}{r}\frac{\partial}{\partial r}\left(r\frac{\partial}{\partial r}\right) + \frac{\partial^2}{\partial z^2}; \qquad \nabla = \left\{\frac{\partial}{\partial r}, 0, \frac{\partial}{\partial z}\right\}.$$

In experiments for a wide range of values of λ, steady motions are observed. For values of λ less than a critical value λ_c, all unstationary motions seem to approach arbitrarily close to the so-called Couette flow. For values of λ greater than λ_c and less than a second critical value, all unstationary motions tend to the so-called Taylor vortices, which are steady toroidal motions periodic in z-direction. Moreover, in experiments one distinct mode is always observed.

Thus there are, besides others, three main problems linked with the theoretical investigation of (1.1).

I. Nonuniqueness of the stationary problem. The stationary problem corresponding to (1.1) is defined in a natural way. Now, for this stationary problem, does there exist, besides the Couette solution, a solution of the Taylor vortex type? How many exist and how can they be calculated?

II. Stability of stationary solutions. Given a stationary solution $v*$ of (1.1). For which class of initial conditions and for which region of the parameter λ do the solutions of (1.1) tend to $v*$?

III. Selection of a distinct wave number. The results under I. ensure that a continuum of steady solutions of (1.1) exists, where the solutions are distinguished by their wave number. What is the reason that nature selects a distinct wave number within the class of admissible wave numbers?

In this first section the results are outlined without making use of the more precise notions in function spaces that will be applied later. For a more precise formulation of our results, compare theorem 1—3 in the following sections.

I. Nonuniqueness of the stationary problem. The Couette solution is given by

$$v^0(r) = \{0, V(r), 0\}, \qquad q^0(r) = \int_{r_1}^{r} \frac{V^2(\varrho)}{\varrho}\,d\varrho + \text{const.},$$

$$V(r) = -\frac{r_1 r}{r_2^2 - r_1^2} + \frac{r_1 r_2^2}{(r_2^2 - r_1^2)\,r}, \qquad r_i = \frac{R_i}{R_2 - R_1}, \qquad i = 1, 2.$$

By setting

$$v(r, z) = v^0(r) + u(r, z),$$

$$q(r, z) = q^0(r) + p(r, z),$$

the stationary problem has the following form:

$$Du = -\lambda\nabla p + \lambda(u \cdot \nabla v^0 + v^0 \cdot \nabla u + B(v^0)\,u + B(u)\,v^0) + \lambda(u \cdot \nabla u + B(u)\,u),$$

$$u = 0 \quad \text{for} \quad r = r_1, r_2, \tag{1.2}$$

$$u(r, z + 2\pi/\alpha) = u(r, z).$$

For this nonlinear boundary-value problem the following results are derived concerning the existence and uniqueness of nontrivial solutions:

Let $\lambda_1(\alpha)$ be an algebraically simple eigenvalue of least modulus of the linearized problem of (1.2). If τ_0, $\tau_0 > 0$, is sufficiently small, then there exists for each $\tau = (\lambda - \lambda_1(\alpha)) \in (0, \tau_0]$ a non-trivial real solution $u^\alpha(r, z; \sqrt{\tau})$ which is unique up to translations in z-direction and has the following properties:

a) u^α is a holomorphic function in $\sqrt{\tau}$

b) $\lim_{\tau \to 0} \|u^\alpha(r, z; \sqrt{\tau})\| = 0$.

Property b) guarantees that the solution u^α bifurcates from the trivial solution; it is therefore called a branching solution. For $\tau < 0$ no real branching solutions exist.

These results are obtained by applying SCHMIDT-LYAPUNOV theory to the functional equation which is a weak form of (1.2) in a suitable Hilbert space. The results will be formulated in theorem 1 in section 2 and are submitted for publication elsewhere by the authors [8]. The continuum of branching solutions obtained is shown qualitatively in Fig. 2. The linearized eigenvalue problem of (1.2) has been studied thoroughly in linear stability analysis (cf. [1]). One of the authors has proven in [6] that for each α, $\alpha > 0$, there exists a positive eigenvalue $\lambda_1(\alpha)$ of least modulus. The dependance of λ_1 on α is still an open problem. However, all numerically obtained values suggest the form of the curve $\lambda_1(\alpha)$ shown qualitatively in Fig. 1. The most important property of this graph is the fact that it has a unique minimum at some α_0. It will turn out, that as long as $\lambda_1(n\alpha) \neq \lambda_1(\alpha)$ for $n = 2, 3, \ldots$, the eigenvalue $\lambda_1(\alpha)$ is simple in the underlying Hilbert space. Under the assumption that the function $\lambda_1(\alpha)$ has a unique minimum the simplicity follows for almost all α. VELTE [13] has proven that at least a continuous subset of the positive real axis exists for which the condition stated above is satisfied.

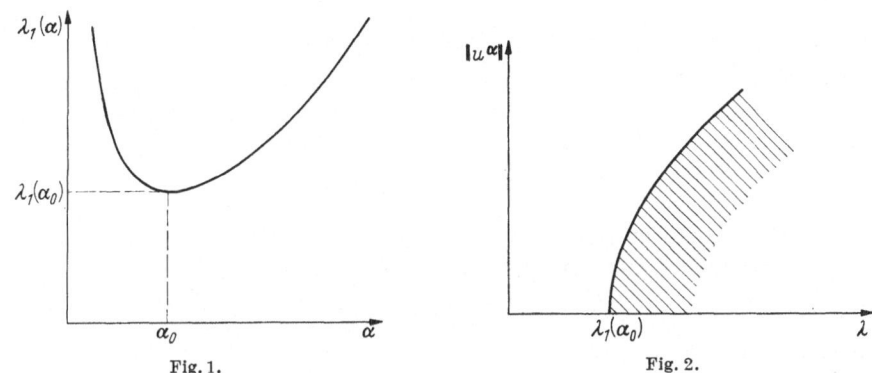

Fig. 1.   Fig. 2.

II. Stability of stationary solutions. Under the assumption that the P.E.S. holds, it can be shown that the Couette solution is stable for $\lambda < \lambda_1(\alpha)$ and unstable for $\lambda \in (\lambda_1(\alpha), \lambda_1(\alpha) + \tau)$, whereas each branching solution α is stable within the class of functions having the same wave number α for $\lambda \in (\lambda_1(\alpha), \lambda_1(\alpha) + \tau)$, τ sufficiently small. Here stability is understood in the following local sense: if for a steady solution v^* and a function v_0 defining the initial condition for a time dependent solution $v(t)$, $||v^* - v_0||$ is sufficiently small, then $\lim_{t \to \infty} ||v(t) - v^*|| = 0$. If for all time dependent solutions with v_0 in any small neighborhood of v^*, $\limsup_{t \to \infty} ||v(t) - v^*|| > 0$, then v^* is called unstable. The analysis strongly depends on a theorem given by PRODI [10]. The proofs are given in section 3 (cf. Theorem 2).

III. Selection of a distinct wave number. Under the assumption that $\lambda_1(\alpha)$ has a unique minimum for some value α_0, it is possible to show that all branching solutions u^α, $\alpha \neq \alpha_0$ are unstable with respect to a broader class of admissible time-dependent solutions of (1.1). Thus, the only possibly stable branching solution is u^{α_0} (cf. theorem 3 in section 4). All results analogously apply to the Bénard problem. Rolls, rectangular and hexagonal cells are branching solutions of the stationary problem (cf. [2—4]). Since the P.E.S. holds in this case if the bounding planes are both rigid or both free, the results on the stability of these solutions are proven without any assumption. The distinction of a certain wave number implies in this connection that only those solutions of a certain cell pattern are stable for which $\beta^2 + \gamma^2 = \alpha_0^2$, where β and γ are the wave numbers in two orthogonal directions of the bounding plane. The shape of the eigenvalue curve $\lambda_1(\alpha)$ is known explicitly in the case of two free boundaries (cf. [1]). Therefore, in this case no assumption whatsoever has to be made to prove the results. The theory outlined in this paper, however, does not given any explanation for the preference of cell patterns of a special structure such as rolls or hexagons.

2. Nonuniqueness of the stationary problem

To begin we have to introduce some notation. Let Ω be the strip $(r_1, r_2) \times (-\infty, +\infty)$ and Ω_α any domain of periodicity $(r_1, r_2) \times (z_0, z_0 + 2\pi/\alpha)$, z_0 arbitrary. $C_{\alpha,\sigma}^\infty$ denotes the class of real, infinitely often differentiable, solenoidal vector functions $u(r, z)$, which vanish in a boundary strip of Ω and are periodic in z with $2\pi/\alpha$.

We define N_α as the completion of $C_{\alpha,\sigma}^\infty$ with respect to the norm $|\ |$ in $L_2(\Omega_\alpha)$ and N_α^1 as the completion of $C_{\alpha,\sigma}^\infty$ with respect to the Dirichlet norm $\|\ \|$ given by:

$$\|u\|^2 := \int\limits_{\Omega_\alpha} \int \left\{ (\nabla u)^2 + \frac{(u)_1^2 + (u)_2^2}{r^2} \right\} r \, dr \, dz.$$

The corresponding scalar product in N_α is denoted by $(\cdot, \cdot)$, in N_α^1 by $((\cdot, \cdot))$, and N_α^2 is defined as $N_\alpha^2 := N_\alpha^1 \cap W_2^2(\Omega_\alpha)$, where $W_2^2(\Omega_\alpha)$ is a Sobolev space in usual notation.

The boundary-value problem will now be investigated in a weak form in the Hilbert space N_α^1. We define the operators L and T by the following identities:

$$
\begin{aligned}
((Lu, \Phi)) &= -(v^0 \cdot \nabla u + u \cdot \nabla v^0 + B(v^0)\, u + B(u)\, v^0, \Phi), \\
((Tu, \Phi)) &= -(u \cdot \nabla u + B(u)\, u, \Phi), \quad \Phi \in N_\alpha^1.
\end{aligned}
\tag{2.1}
$$

The right-hand sides of these equations define linear continuous functionals on N_α^1. For the first equation this follows from v^0 bounded in Ω_α, Schwarz' and Friedrichs' inequality. For the second equation this is implied by Sobolev's embedding theorem (cf. [9]):

$$\left| (u \cdot \nabla u + B(u)\, u, \Phi) \right| \leq c\, |u|_{L_4}^2\, \|\Phi\| \leq \bar{c}\, \|u\|^2\, \|\Phi\|.$$

By the theorem of F. Riesz unique elements Lu, Tu exist in N_α^1 so that (2.1) holds.

With (2.1) the boundary-value problem (1.2) can be written as

$$u = \lambda Lu + \lambda Tu, \quad u \in N_\alpha^1. \tag{2.2}$$

For the operators L and T it follows:

Lemma 1: L is compact in N_α^1, $\|L\| \leq l$, T is a bounded quadratic operator with

$$\|Tu_1 - Tu_2\| \leq t\, \|u_1 + u_2\| \cdot \|u_1 - u_2\|, \quad u_1, u_2 \in N_\alpha^1,$$

where l and t only depend on r_1 and α.

The proof is given in [8].

Now, α is chosen in such a way that $\lambda_1(\alpha)$ is a simple[1] eigenvalue of L, which is always possible according to Velte [13]. We denote by φ_0 the corresponding eigenfunction of L and by ψ_0 the eigenfunction of the adjoint operator L^*, where φ_0 and ψ_0 are normalized by $\|\varphi_0\| = 1$, $((\varphi_0, \psi_0)) = 1$. The last condition can always be satisfied, since $((\varphi_0, \psi_0)) = 0$ would imply that $\lambda_1(\alpha)$ has higher multiplicity than one. The theory of Schmidt-Lyapunov is applicable to (2.2).

Lemma 2: (cf. Vainberg and Trenogin [12]). Let the operator $\hat{L}$ be defined by $\hat{L} := L - \frac{1}{\lambda_1}((\psi_0, \cdot))\, \varphi_0$, $\|\hat{L}\| \leq \hat{l}$, then $R(\lambda_1) = (I - \lambda_1 \hat{L})^{-1}$ exists and is bounded, $\|R(\lambda_1)\| \leq \varrho$. For a proof we again refer to [8].

Now we define real quantities ζ and τ by $\zeta = \frac{\lambda}{\lambda_1(\alpha)}((u, \psi_0))$, $\tau = \lambda - \lambda_1(\alpha)$. As can be easily seen, the Eq. (2.2) is equivalent to the system

$$
\begin{aligned}
\text{a)} \quad & u = \zeta \varphi_0 + \tau R(\lambda_1)\, \hat{L}u + (\lambda_1 + \tau)\, R(\lambda_1)\, Tu, \\
\text{b)} \quad & 0 = \frac{\tau \zeta}{\lambda_1 + \tau} + (\lambda_1 + \tau)\, ((\psi_0, Tu)).
\end{aligned}
\tag{2.3}
$$

For small values of $|\zeta|$ and $|\tau|$, (2.3 a) can be solved by Banach's contraction mapping principle. The sphere $K(0, \delta_0)$ is mapped into itself and in $K(0, \delta_0)$ a Lipschitz condition with constant

---

[1] By "simple eigenvalue" we always understand an algebraically simple eigenvalue.

$q < 1$ is satisfied for $|\zeta| \leq \zeta_0$, $|\tau| \leq \tau_0$, if ζ_0 and τ_0 fulfill the following conditions:

$$\zeta_0 + \tau_0 \tilde{l} \delta_0 + (\lambda_1 + \tau_0)\, \tilde{t}\, \delta_0^2 = \delta_0,$$

$$\tau_0 + 2\tilde{t}(\lambda_1 + \tau_0)\, \delta_0 = q,$$

$$\tilde{l} = \varrho \hat{l}, \quad \tilde{t} = \varrho t.$$

Therefore the elements of the sequence u_k, defined by

$$u_{k+1} = \zeta \varphi_0 + \tau R(\lambda_1)\, \hat{L} u_k + (\lambda_1 + \tau)\, R(\lambda_1)\, T u_k, \tag{2.4}$$

$$u_0 = \zeta \varphi_0$$

converge strongly in N_α^1 to a solution $u^\alpha(r, z; \zeta, \tau)$ of (2.3a). A slight modification of the proof shows that u^α depends holomorphically on ζ and τ. Moreover the Lipschitz condition applied to 0 and u^α implies

$$\|u^\alpha\| \leq |\zeta| + q\,\|u^\alpha\|,$$

from which $\lim\limits_{|\zeta| \to 0} \|u^\alpha\| = 0$ follows.

Inserting the power series for u^α into (2.3b) leads to a transcendental equation in ζ and τ of the form

$$0 = \frac{\tau \zeta}{\lambda_1 + \tau} + \sum_{\nu=2}^{\infty} a_\nu \zeta^\nu + \sum_{\substack{\mu=1 \\ \nu=2}}^{\infty} a_{\nu,\mu} \zeta^\nu \tau^\mu. \tag{2.5}$$

The number of roots of this equation for which $\zeta(0) = 0$ determines the number of branching solutions of (2.2). For the coefficients we derived in [8] the following results:

Lemma 3: The coefficients a_{2n}, $a_{2n,\mu}$ in (2.5) vanish:

$$a_{2n} = 0, \quad a_{2n,\mu} = 0 \quad \text{for } n, \mu = 1, 2, \ldots$$

Remark: The eigenfunctions φ_0, ψ_0 are of the form $\varphi_0 = 2\,\mathrm{re}(\varphi_{01}(r)\, e^{i\alpha z})$, $\psi_0 = 2\,\mathrm{re}\,(\psi_{01}(r)\, e^{i\alpha z})$, where re denotes the real part, and it follows immediately that

$$(\psi_0, \varphi_0 \cdot \nabla \varphi_0 + B(\varphi_0)\, \varphi_0) = 0. \tag{2.6}$$

The number of solutions of (2.3b) for small $|\zeta|$, $|\tau|$ is determined, by use of Newton's diagram, by the terms of lowest order in (2.5):

$$0 = \frac{\tau \zeta}{\lambda_1} + a_3 \zeta^3 + \text{higher-order terms in } \zeta \text{ and } \tau.$$

The right-hand side of this equation is an odd function of ζ (cf. Lemma 3). Thus we get the solutions $\zeta = 0$ and, if $a_3 \neq 0$,

$$\zeta_+(\tau) = \sqrt{-\frac{\tau}{\lambda_1 a_3}} + \text{higher-order terms in } \sqrt{\tau},$$

$$\zeta_-(\tau) = -\zeta_+(\tau).$$

The solutions of (2.2) corresponding to these roots are the trivial solution $u = 0$ and the nontrivial solutions $u_+ = u^\alpha(r\ z; \zeta_+(\tau), \tau)$, $u_- = u^\alpha(r, z; -\zeta_+(\tau), \tau)$. But since u^α is an odd function of ζ, these nontrivial solutions coincide after a translation by π/α in the z-direction.

The fact that $a_3 < 0$ could only be established numerically. However, if L would be self-adjoint, as is the case in the Bénard problem, $a_3 < 0$ could be proven rigorously (cf. [4]). u^α is a classical solution of the problem (1.2), as was already pointed out by VELTE [13] and can also be established directly [7].

Theorem 1: Let $\lambda_1(\alpha)$ be a simple eigenvalue of L and $a_3 < 0$. Then there exists for each $\tau \in (0, \tau_0]$, $\tau_0 > 0$ sufficiently small, a nontrivial branching solution $u^\alpha(r, z; \sqrt{\tau})$ of the functional Eq. (2.2) in N_α^1. This solution is unique up to translations in the z-direction and is a holomorphic family of $\sqrt{\tau}$. There is no real, nontrivial solution in N_α^1 for $\tau < 0$ which continuously depends on τ and for which $u(r, z; 0) = 0$.

3. Stability of stationary solutions

Now we turn to the investigation of the initial boundary-value problem (1.1). Let $u^*(r, z)$, $u^* \in N_\alpha^2$, be any steady solution of (1.1). Then we consider solutions $v(r, z, t) = u^*(r, z) + u(r, z\ t)$ of (1.1) with the following properties:

a) $u(t)$ is a vectorvalued function, mapping $t \to u(t)$, where $t \in (0, \infty)$ and $u(t) \in N_\alpha^2$. Furthermore $u(t)$ is square integrable with respect to the time in the Bochner sense. We use the notation:

$$u \in L_2([0, \infty), N_\alpha^2)$$

b) $u(t) \in C([0, \infty), N_\alpha^1)$ ($\|u\|$ depends continuously on t),

c) $u(t)$ has a weak derivative

$$\frac{du}{dt} \in L_2([0, \infty), N_\alpha).$$

$L_2(\Omega_\alpha)$ can be decomposed into two orthogonal subspaces: $L_2(\Omega_\alpha) = N_\alpha \oplus G$, where G consists of all functions g which are gradients of a scalar function $p \in W_2^1(\Omega_\alpha)$, $g = \nabla p$. Let P be the orthogonal projection operator of $L_2(\Omega_\alpha)$ in N_α. Then (1.1) can be written as follows (cf. PRODI [10]):

$$\frac{du}{dt} + Au + \lambda M(u^*)\, u = \lambda Ru$$

$$u|_{t=0} = u_0, \qquad u_0 \in N_\alpha^1 \tag{3.1}$$

with

$$A = -PD$$

$$M(u^*)\, u = P\{u^* \cdot \nabla u + u \cdot \nabla u^* + B(u^*)\, u + B(u)\, u^*\}$$

$$Ru = -P\{u \cdot \nabla u + B(u)\, u\}.$$

Now we list some properties obtained by PRODI [10] of the operators occurring in (3.1).

Lemma 4:

i) M and R are operators acting from N_α^1, N_α^2 respectively in N_α satisfying the following inequalities:

$$|Rv|^2 \leq c_1\, \|v\|^3\, |Av|,$$

$$|M(u^*)\, v| \leq c_2\, \|v\|.$$

ii) The operator $\tilde{A}(\lambda, u^*) = A + \lambda M(u^*)$ with the domain of definition $D(\tilde{A}) = N_\alpha^2$ considered as operator in N_α is closable.

iii) The domain of the closed operator (again denoted by $\tilde{A}$) is independent of u^* as long as $u^* \in N_\alpha^2$.

iv) The spectrum of $\tilde{A}$, $\Sigma(\tilde{A})$, lies entirely in the parabola

$$\mathrm{re}\,(\mu) = \frac{1}{4 c_2^2\, |\lambda|^2}\, \mathrm{im}\,(\mu)^2 - c_2^2\, |\lambda|^2.$$

v) For $\mu < -\dfrac{c_2^2\, |\lambda|^2}{4}$, μ is in the resolvent set of $\tilde{A}(\lambda, u^*)$ and the following estimate is valid:

$$\|(\tilde{A} - \mu I)^{-1}\| \leq \frac{1}{-\mu - \dfrac{1}{4}\, c_2^2\, |\lambda|^2}.$$

As a further result on $\tilde{A}$ we need the following:

Lemma 5: For each real λ the spectrum of the operator $\tilde{A}(\lambda, u^*)$ consists of isolated eigenvalues with finite multiplicities.

Proof: By setting $\tilde{A}u - \mu u = g$ we obtain after scalar multiplication by u in N_α with Schwarz's inequality the estimate

$$\|u\|^2 - \mu\, |u|^2 \leq |g| \cdot |u| + |\lambda|\, c_2\, |u|\, \|u\|.$$

Using the inequality $2ab \leq \gamma^2 a^2 + b^2/\gamma^2$, we get

$$\frac{1}{2} \, ||u||^2 - \mu \, |u|^2 \leq |g| \cdot |u| + \frac{c_2^2 \, |\lambda|^2}{2} \, |u|^2. \tag{3.2}$$

Let $\mu < -c_2^2 \, |\lambda|^2/2$; then μ belongs to the resolvent set of $\tilde{A}$ (cf. Lemma 4, v). Now, take any bounded set S in N_α; then, for $\mu < -c_2^2 \, |\lambda|^2/2$, the set U, $U = (\tilde{A} - \mu I)^{-1} S$, is by (3.2) bounded in N_α^1. By Sobolev's embedding theorem U is compact in N_α. Thus $(\tilde{A} - \mu I)^{-1}$ is a compact operator for some values of the resolvent set and thus, by a well known theorem (cf. [5], p. 187), the assertion of the lemma is proven.

The properties ii) and v) of Lemma 4 ensure by the theorem of HILLE-YOSIDA (cf. [14], p. 246) that the closure of $\tilde{A}$ generates a strongly continuous semigroup $G(t)$ with

$$||G(t)||_{N_\alpha \to N_\alpha} \leq \exp\left(\frac{c_2^2 \, |\lambda|^2}{4} \, t\right).$$

With this semigroup $G(t)$ any solution of (3.1) can be represented by (cf. [10])

$$u(t) = G(t) \, u_0 + \int_0^t G(t - s) \, R u(s) \, ds. \tag{3.3}$$

The abstract integral Eq. (3.3) is the basis for the stability analysis. Starting from this equation the following theorem links the question of stability with the spectrum of $\tilde{A}$:

Lemma 6: (PRODI [10]). Let the following assumptions be valid:

i) For all eigenvalues μ, $\mu \in \Sigma(\tilde{A})$,

$$\mathrm{re} \, (\mu) \geq \gamma > 0,$$

ii) $||u_0||$ sufficiently small.

Then a solution $u(t)$ of (3.3) exists for all times t and is stable in the following sense:

$$||u(t)|| \leq c \exp (-\gamma' t), \text{ for some } \gamma' > 0.$$

Since it is our aim to prove also the instability of some u^*, we need an extension of this lemma to the case where $\mathrm{re}(\mu_k) \leq 0$ for a finite number of eigenvalues μ_k, $\mu_k \in \Sigma(\tilde{A})$.

Lemma 7: Let $u(t)$ be a solution of (3.1) for all $t \in [0, \infty)$ with $|Au| \leq C$, and let the following conditions be satisfied:

i) There exists a simple eigenvalue $\mu_0 \in \Sigma(\tilde{A})$, $\mu_0 < 0$,

ii) $P_0 u_0 \neq 0$, where P_0 is the projection of u_0 into the eigenspace belonging to μ_0; then

$$\limsup_{t \to \infty} ||u(t)|| > 0\,[1].$$

Proof: By the conditions of the lemma the spectrum of $\tilde{A}$ can be divided into two parts $\Sigma(\tilde{A}) = \Sigma_1 + \Sigma_2$, where Σ_1 consists of the eigenvalues μ_k, with $\mathrm{re}(\mu_k) \leq 0\,[2]$, $k = 0, 1, \ldots, n$, and Σ_2 is the rest. This implies a decomposition of $\tilde{A}$ according to a decomposition of space $N_\alpha = M \oplus N$ in such a way that the spectra of the parts $\tilde{A}P$ and $\tilde{A}(1 - P)$ coincide with Σ_1, Σ_2 respectively and $\tilde{A}P$ is a bounded operator on M (cf. [5, p. 178]). Here P denotes the projection operator of N_α in M along N.

Since for $\Sigma(\tilde{A})$ condition iv) of Lemma 4 holds, the imaginary axis cannot be an accumulation point of Σ_2. Thus we have for $\mu \in \Sigma(\tilde{A}(1 - P)) = \Sigma_2$, $\mathrm{re}(\mu) \geq \omega > 0$, for some ω. Now, for $\tilde{A}(1 - P)$ the conditions of the theorem of HILLE-YOSIDA are satisfied and thus $\tilde{A}(1 - P)$ generates a strongly continuous semigroup $V(t)$, for which the following estimate holds (cf. PRODI [10]):

$$||V||_{N \to N} \leq c \exp (-\omega' t), \quad 0 < \omega' < \omega.$$

---

[1] After submission of this paper it could be shown that the conditions $|Au| \leq C$, μ_0 real are obsolete (cf. KIRCHGÄSSNER, ZAMM, Sonderheft 1969, to appear).

[2] Because of Lemma 4, iv) and Lemma 5 there can be only a finite number of μ_k.

Introducing the notation $(1 - P)\, u = v$, $Pu = w$ we have to consider

$$\frac{dv}{dt} + \tilde{A}v = \lambda\, (1 - P)\, Ru = f_1(u),$$ (3.4a)

$$v|_{t=0} = (1 - P)\, u_0$$

in the Hilbert space N

and

$$\frac{dw}{dt} + \tilde{A}w = \lambda P\, Ru = f_2(u),$$ (3.4b)

$$w|_{t=0} = Pu_0$$

in the finite space M.

Property i) of Lemma 4 and the condition on $|Au|$ ensure that, assuming $||u|| \to 0$ for $t \to \infty$, we also have $|Ru|^2 \leq c_1\, ||u||^3\, |Au| \to 0$. Thus, in view of the continuity of $||u||$, $|f_1(u)|$ and $|f_2(u)|$ are bounded and tend to zero as $t \to \infty$.

Integration of (3.4a) with the aid of $V(t)$ yields

$$v(t) = V(t)\, (1 - P)\, u_0 + \int_0^t V(t - s)\, f_1(s)\, ds.$$

Using the estimate on $V(t)$ and taking into account that $|f_1|$ is bounded, we can conclude that $|v(t)|$ is bounded as well. The projection P is given by $P = \sum_{k=0}^n P_k$, where P_k are the projections into the eigenspaces belonging to μ_k, and $P_k P_l = \delta_{kl} \cdot P_k$, where δ_{kl} denotes the Kronecker symbol. Setting $P_0 u = (\psi, u)\, \varphi = z\varphi$, $\lambda P_0 Ru = f_{20}\varphi$, where ψ, φ are the eigenfunctions of $\tilde{A}^*$ and $\tilde{A}$ respectively belonging to μ_0 normed in an appropriate way, we obtain for z the ordinary differential equation

$$\frac{dz}{dt} + \mu_0 z = f_{20}(t)$$

$$z(0) = P_0 u_0,$$

where $f_{20}(t)$ is bounded and decreases with time. Integration yields, because of condition ii) in the lemma, $P_0 u \to \infty$ as $t \to \infty$. Since $||u|| \to 0$ implies $|u| \to 0$, and since P_0 is a bounded operator, we have $|P_0 u| \to 0$ contradicting $|P_0 u| \to \infty$ as $t \to \infty$.

Lemmas 6 and 7 show that the stability of a steady solution u^* of (1.1) depends essentially on the spectrum of the operator $\tilde{A}(\lambda, u^*) = A + \lambda M(u^*)$. Let $\lambda_1(\alpha)$ be a simple eigenvalue of L (cf. section 2). If $(\psi_0, \varphi_0) \neq 0$[1], then $\mu = 0$ is a simple eigenvalue of $\tilde{A}(\lambda_1(\alpha), v_0)$, since $\tilde{A}^m u = 0$, m integer, $m > 1$, would imply $\tilde{A}^{m-1} u = \varphi_0$ and thus $(\psi_0, \varphi_0) = 0$.

We are now interested in the behaviour of this eigenvalue $\mu = 0$ in a neighbourhood of $\lambda_1(\alpha)$. This behaviour depends on the steady solution u^* we have chosen. Thus we distinguish two eigenvalue problems for small values of $|\tau| = |\lambda - \lambda_1(\alpha)|$:

a) $\tilde{A}(\lambda_1(\alpha) + \tau, v^0)\, \varphi - \mu\varphi = 0,$ (3.5)

b) $\tilde{A}(\lambda_1(\alpha) + \tau, v^0 + u^\varkappa (\sqrt{\tau}))\, \varphi - \mu\varphi = 0.$

We recall that v^0 is the Couette solution, which is independent of τ and $u^\varkappa(\sqrt{\tau})$ is the branching solution for given α, which depends holomorphically on $\sqrt{\tau}$ (cf. theorem 1).

To determine μ in dependance of $\sqrt{\tau}$, analytic perturbation theory is applied. The case (3.5a) is rather trivial; thus we restrict the considerations to (3.5b). For this purpose it has to be ensured that $\tilde{A}$ is a holomorphic family of type (A) (cf. KATO [5], p.375), i.e.,

---

[1] This is shown in the proof of Lemma 9.

i) $D(\tilde{A}(\sqrt{\tau})) = D(\tilde{A}(0))^1$, ii) $(\tilde{A}(\sqrt{\tau})\,u, \Phi)$ is a holomorphic function of $\sqrt{\tau}$ for all $u \in D(\tilde{A}(0))$ and $\Phi \in N_\alpha$. For condition i) compare Lemma 4, iii). The second condition is satisfied because of the strong convergence of the power series for $u^\alpha(\sqrt{\tau})$ in N_α^1. Thus it is justified to use the results of analytic perturbation theory for operators of type (A) (cf. [5]).

Lemma 8: If $\tilde{A}(\sqrt{\tau})$ is of type (A) in $\sqrt{\tau}$ and if $\mu(0) \in \Sigma(\tilde{A}(0))$ is a simple eigenvalue, then, for sufficiently small $\sqrt{\tau}$, $\mu(\sqrt{\tau})$ is simple and μ together with the corresponding eigenprojection depend holomorphically on $\sqrt{\tau}$.

With the aid of Lemma 8 the eigenvalue $\mu(\sqrt{\tau})$ with $\mu(0) = 0$ of (3.5) is investigated.

Lemma 9: Let $\mu = \gamma_1 \sqrt{\tau} + \gamma_2 \tau + \cdots$ be the analytic branch of eigenvalues of (3.5) corresponding to $\mu(0) = 0$, and let φ_0, ψ_0 denote the eigenfunctions introduced in section 2, then

i) $(\psi_0, \varphi_0) \neq 0$,

ii) $\gamma_1 = 0$,
$$\gamma_2 = \frac{-1}{\lambda_1(\alpha)\,(\psi_0, \varphi_0)} \quad \text{for} \quad u^* = v^0,$$

iii) $\gamma_1 = 0$,
$$\gamma_2 = \frac{+1}{\lambda_1(\alpha)\,(\psi_0, \varphi_0)} \quad \text{for} \quad u^* = v^0 + u^\alpha(\sqrt{\tau}).$$

Proof: i) and ii): If we choose $u^* = v^0$ we have $\tilde{A}(\sqrt{\tau}) = A + (\lambda_1 + \tau)\,M(v^0)$. Setting

$$\mu = \gamma_1 \sqrt{\tau} + \gamma_2 \tau + \cdots,$$
$$\varphi = \varphi_0 + \varphi_1 \sqrt{\tau} + \varphi_2 \tau + \cdots,$$
$$\tilde{A} = \tilde{A}(0) + \tau M(v^0),$$

and evaluating (3.5a) by comparing coefficients of the same order in $\sqrt{\tau}$ we obtain:

$$\text{a)} \quad \tilde{A}(0)\,\varphi_0 = 0,$$

$$\text{b)} \quad \tilde{A}(0)\,\varphi_1 - \gamma_1 \varphi_0 = 0, \tag{3.6}$$

$$\text{c)} \quad \tilde{A}(0)\,\varphi_2 + M(v^0)\,\varphi_0 - \gamma_2 \varphi_0 = 0.$$

Multiplying (3.6c) by ψ_0 in N_α, we get

$$\gamma_2(\varphi_0, \psi_0) = (\psi_0, M(v^0)\,\varphi_0) = -\frac{((\psi_0, \varphi_0))}{\lambda_1(\alpha)} = -\frac{1}{\lambda_1(\alpha)}.$$

Thus $(\psi_0, \varphi_0) \neq 0$, which proves i), and

$$\gamma_2 = -\frac{1}{\lambda_1(\alpha)\,(\varphi_0, \psi_0)}.$$

iii) Here we have $\tilde{A}(\sqrt{\tau}) = A + (\lambda_1 + \tau)\,M(v^0 + u^\alpha(\sqrt{\tau}))$. Setting

$$\mu = \gamma_1 \sqrt{\tau} + \gamma_2 \tau + \cdots,$$
$$\varphi = \varphi_0 + \sqrt{\tau}\,\varphi_1 + \tau \varphi_2 + \cdots,$$
$$u^\alpha = \sqrt{\tau}\,u_{10} + \tau u_{20} + \tau^{3/2}\,u_{30} + \cdots,$$
$$\tilde{A}(\sqrt{\tau}) = \tilde{A}(0) + \sqrt{\tau}\,\lambda_1(\alpha)\,M(u_{10}) + \tau\big(M(v^0) + \lambda_1(\alpha)\,M(u_{20})\big) + \cdots$$

and comparing coefficients of the same order of $\sqrt{\tau}$ in (3.5b) yields

a)
$$\tilde{A}(0)\,\varphi_0 = 0,$$

b)
$$\tilde{A}(0)\,\varphi_1 + \lambda_1(\alpha)\,M(u_{10})\,\varphi_0 - \gamma_1 \varphi_0 = 0, \tag{3.7}$$

c)
$$\tilde{A}(0)\,\varphi_2 + \lambda_1(\alpha)\,M(u_{10})\,\varphi_1 + \lambda_1(\alpha)\,M(u_{20})\,\varphi_0 + M(v^0)\,\varphi_0 - \gamma_2 \varphi_0 = 0.$$

---

[1] For simplification we use the notation $\tilde{A}(\sqrt{\tau}) := \tilde{A}(\lambda_1(\alpha) + \tau, v^0 + u^\alpha(\sqrt{\tau}))$.

Since $v^0 + u^\alpha(\sqrt{\tau})$ is a stationary solution of (3.1) we have

a)
$$\tilde{A}(0)\, u_{10} = 0,$$

b)
$$\tilde{A}(0)\, u_{20} + \lambda_1 M(u_{10})\, u_{10} = 0, \qquad (3.8)$$

c)
$$\tilde{A}(0)\, u_{30} + \lambda_1 M(u_{20})\, u_{10} + M(v^0)\, u_{10} = 0,$$

which is obtained analogously to (3.7). From (3.8a) it follows that $u_{10} = c\varphi_0$, $c \neq 0$. Inserting this relation into (3.7b) and using (2.6) yields, after multiplication by ψ_0 in N_α, $\gamma_1 = 0$. Now, comparing (3.7b) and (3.8b) leads with $u_{10} = c\varphi_0$ to $\varphi_1 = u_{20}/c + \bar{c}\varphi_0$, where $\bar{c}$ is some constant. Inserting this relation into (3.7c) and again multiplying by ψ_0 gives:

$$\frac{\lambda_1(\alpha)}{c}\left(\psi_0,\, M(u_{10})\, u_{20}\right) + c\bar{c}\left(\psi_0,\, M(\varphi_0)\, \varphi_0\right) + \frac{\lambda_1(\alpha)}{c}\left(\psi_0,\, M(u_{20})\, u_{10}\right) + \left(\psi_0,\, M(v^0)\, \varphi_0\right) - \gamma_2(\psi_0, \varphi_0) = 0.$$
$$(3.9)$$

But the second term vanishes because of (2.6) and from (3.8c) it follows that

$$\left(\psi_0,\, M(u_{20})\, u_{10}\right) = -\frac{1}{\lambda_1(\alpha)}\left(\psi_0,\, M(v^0)\, u_{10}\right) = c/\lambda_1^2(\alpha)$$

and thus (3.9) leads to

$$\gamma_2 = \frac{1}{\lambda_1(\alpha)\,(\psi_0, \varphi_0)},$$

which proves the lemma.

We now turn to the formulation of our first stability theorem.

Assumption 1: ("Principle of exchange of stabilities") Let $\lambda \in [\lambda_1(\alpha) - \delta,\, \lambda_1(\alpha))$, $\delta > 0$, δ sufficiently small, then re $\Sigma(\tilde{A}(\lambda, v^0)) > 0$.

This is a somewhat weaker form of the classical P.E.S. It ensures that v_0 is stable for $\lambda \in (\lambda_1(\alpha) - \delta,\, \lambda_1(\alpha))$, and thus we obtain from Lemma 9, ii) that $(\psi_0, \varphi_0) > 0$.

Theorem 2: Let assumption 1 be satisfied and let $\lambda_1(\alpha)$ be a simple eigenvalue of L. Then we have in N_α^1:

i) a) If $||v^0 - u_0||$ is sufficiently small, then the Couette solution v^0 is stable for $\lambda < \lambda_1(\alpha)$ in the following sense: $||v^0 - u|| \leq c \exp(-\omega t)$, for some $\omega > 0$, where u is any solution of (3.1).

b) The Couette solution v^0 is unstable for $\lambda > \lambda_1(\alpha)$ and we have for any solution u of (3.1) which exists for all times t with $u|_{t=0} = u_0 \in N_\alpha^1$ and $P_0 u_0 \neq 0$:

$$\limsup_{t \to \infty} ||v^0 - u|| > 0.$$

ii) If $||v^0 + u^\alpha(\sqrt{\tau}) - u_0||$ is sufficiently small the branching solution is stable for $\lambda > \lambda_1(\alpha)$, $\lambda - \lambda_1(\alpha) = \tau$ sufficiently small, in the following sense:

$$||v^0 + u^\alpha - u|| \leq c \exp(-\omega t), \text{ for some } \omega > 0,$$

where u is any solution of (3.1).

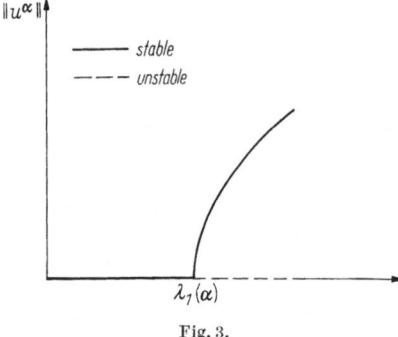

Fig. 3.

Proof: We observe that theorem 1 guarantees the existence of a branching solution $u^\alpha(\sqrt{\tau})$. Assumption 1 and the fact that $\tilde{A}(\sqrt{\tau})$ is of type (A) ensure with Lemma 9 that for v^0 the assumptions of the lemma of PRODI (Lemma 6) are satisfied as long as $\lambda < \lambda_1(\alpha)$, while for the branching solution $v^0 + u^\alpha(\sqrt{\tau})$ they are satisfied for $\lambda > \lambda_1(\alpha)$, $\lambda - \lambda_1(\alpha) = \tau$ sufficiently small. On the other hand the assumptions of Lemma 7 are fulfilled for v^0 if $\lambda > \lambda_1(\alpha)$, ensuring the instability of Couette flow.

Remarks

1. Up to now all considerations have been restricted to the spaces N_α^k. This means we had to specify the value of the wave number α. Theorem 2 is only valid in the class of functions u having the same wave number α.

2. The same considerations apply to the Bénard case, where the P.E.S. is valid for two rigid or two free bounding surfaces. The simplicity of $\lambda_1(\alpha)$ is guaranteed for cells of roll, rectangular or hexagonal pattern. Thus, theorem 2 can be analogously proven for these cell flows and yields therefore the stability of these flows within the same cell pattern (cf. [4]).

4. Selection of a distinct wave number

It is quite suggestive that the distinction of a specific wave number is closely related to the form of the eigenvalue curve $\lambda_1(\alpha)$. But results in this direction which are strong enough have not yet been rigorously established .Thus, we are forced to make an additional assumption.

Assumption 2: The function $\lambda_1(\alpha)$ has a unique minimum.

Let α_0 denote the value of α where this minimum is attained. As long as $\lambda_1(n\alpha) \neq \lambda_1(\alpha)$ for $n = 2, 3, \dots$ the eigenvalue $\lambda_1(\alpha)$ of L is simple and thus $\mu = 0$ is a simple eigenvalue of $\tilde{A}(\lambda_1(\alpha), v^0)$. Let us consider the eigenvalue problem

$$\tilde{A}\big(\lambda_1(\alpha) + \tau, v^0\big)\, \varphi - \mu(\alpha, \tau)\, \varphi = 0.$$

For each $\alpha_1 > 0$ and $|\alpha - \alpha_1| + |\tau|$ sufficiently small, μ depends holomorphically on α and τ. Since $\tilde{A}$ is real and $\mu(\alpha_1, 0)$ is simple, $\mu(\alpha, \tau)$ must be real and simple as well. With assumption 1 it follows that sign $\mu(\alpha, \tau) = -\text{sign } \tau$. Thus, for any interval $[\check{\alpha}, \hat{\alpha}]$, $\check{\alpha} \leq \alpha \leq \hat{\alpha}$, we can define a strip above the graph of $\lambda_1(\alpha)$:

$$S_\delta = \{(\alpha, \lambda);\quad \alpha \in [\check{\alpha}, \hat{\alpha}],\ \lambda_1(\alpha) < \lambda < \lambda_1(\alpha) + \delta\},$$

so that in S_δ we have $\mu < 0$.

Now, let $\alpha > \alpha_0$. We define α_1 by $n\alpha_1 = \alpha$, $(n-1)\,\alpha_1 = \beta$, n an appropriate positive integer, so that $(\beta, \lambda_1(\alpha)) \in S_\delta$, and we consider the spaces N_{α_1}, $N_{\alpha_1}^1$, $N_{\alpha_1}^2$ in the analysis of the preceding sections. The branching solution $v^0 + u^\alpha(\sqrt{\tau})$, the existence of which was established in theorem 1, is an element of $N_{\alpha_1}^1$. Moreover, the spectrum of

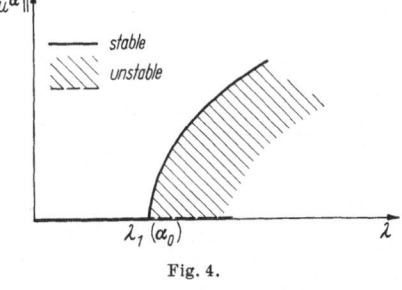

$\tilde{A}(\lambda_1(\alpha), v^0)$ contains the simple negative eigenvalue, $\mu(\beta, \lambda_1(\alpha) - \lambda_1(\beta))$. Possibly there are some other eigenvalues with negative real parts, but there can be only a finite number of them (cf. Lemma 5 and 6). Thus, Lemma 7 can be applied and it ensures that for small $\tau > 0$, $v^0 + u^\alpha(\sqrt{\tau})$ is unstable in $N_{\alpha_1}^1$.

If $\alpha < \alpha_0$ we proceed in an analogous way by defining $n\alpha_1 = \alpha$, $(n + 1)\,\alpha_1 = \beta$.

Fig. 4.

Theorem 3: Let assumptions 1 and 2 hold. Then,

i) If $\lambda_1(n\alpha) \neq \lambda_1(\alpha)$, $n = 2, 3, \dots$, and $\alpha \neq \alpha_0$, the branching solution $v^0 + u^\alpha(\sqrt{})$ is unstable.

ii) The Couette solution v_0 is stable for $\lambda < \lambda_1(\alpha_0)$ and unstable for $\lambda > \lambda_1(\alpha_0)$.

iii) The branching solution $v^0 + u^{\alpha_0}(\sqrt{\tau})$ is stable in all spaces N_α^1 for which α_0/α is integer.

References

1. CHANDRASEKHAR, S.: Hydrodynamic and hydromagnetic stability, Oxford (1961).
2. GÖRTLER, H., K. KIRCHGÄSSNER and P. SORGER: Branching solutions of the Bénard problem, (1967), to appear in Problems of Hydrodynamic and Continuum Mechanics, dedicated to L. I. Sedov.
3. IUDOVICH, V. I.: O vozniknovenii konvektsii (On the origin of convection), PMM 30, No. 6 (1966).
4a. — Svobodnaia Konvektsiia i vetvlenie (Free convection and bifurcation), PMM 31, No. 1 (1967).

4b. — Ustoichivost' Konvektsionnykh potokov (Stability of convection flows), PMM 31, No. 2 (1967).

5. KATO, T.: Perturbation theory for linear operators, Berlin/Heidelberg/New York: Springer 1966.

6. KIRCHGÄSSNER, K.: Die Instabilität der Strömung zwischen zwei rotierenden Zylindern gegenüber Taylor-Wirbeln für beliebige Spaltbreiten. ZAMP 12 (1961).

7. — Verzweigungslösungen eines stationären hydrodynamischen Randwertproblems, Habilitation, Freiburg, 1966.

8. —, and P. SORGER: Branching analysis for the Taylor problem, to appear in Quart. J. Mech. Appl. Math. (1969).

9. LADYSHENSKAJA, O. A.: Funktionalanalytische Untersuchungen der Navier-Stokesschen Gleichungen, Berlin 1965.

10. PRODI, G.: Teoremi di tipo locale, per il sistema di Navier-Stokes e stabilita delle soluzioni stazionarie, Rend. Sem. Mat. Univ. Padova 32 (1962).

11. DAVEY, A., R. C. DI PRIMA and J. T. STUART: On the instability of Taylor vortices, J. Fluid Mech. 31 (1968).

12. VAINBERG, M. M., and V. A. TRENOGIN: The methods of Lyapunov and Schmidt in the theory of non-linear equations, and their further development, Russ. Math. Surv. 17, 1—60 (1962).

13. VELTE, W.: Stabilität und Verzweigung stationärer Lösungen der Navier-Stokesschen Gleichungen beim Taylorproblem, Arch. Rat. Mech. Anal. 22, 1—14 (1966).

14. YOSIDA, K.: Functional analysis, Berlin/Heidelberg/New York: Springer 1965.

Photoviscoelastic analysis by use of polyurethane rubber

By

T. Kunio and Y. Miyano

Keio University, Koganei-shi, Tokyo

I. Introduction

Photoviscoelastic materials such as polyurethane rubber have a significant time- and temperature dependence mechanically as well as optically. Therefore, in order to perform the practical optical analysis of viscoelastic stress or strain, mechanical and optical characterizations of the materials are required. In other words, it is necessary to determine the time- and temperature dependent visco- and photoviscoelastic coefficients.

On the basis of the consideration that the "Correspondence Rule" concerning the mechanical behavior of viscoelastic materials originally proposed by ALFREY [1] may be also applicable to their optical behaviors if the materials are linear-viscoelastic, the research group at Caltech [2—4] in this field has given a fundamental theory for photoviscoelasticity and performed the characterizations of several viscoelastic materials by mathematical interconversion of the experimental data obtained from the constant strain rate testing.

In our experiments, polyurethane rubber [5] was employed specimens of which are linearly viscoelastic, thermorheologically simple, incompressible and fairly transparent.

Mechanical and optical characterizations of this material essential to the practical photoviscoelastic analyses which have already been carried out by two different kinds of experiments; i.e. constant strain rate- and creep testings [6]. It was found that the both results from these experiments agreed fairly well each other. This teaches us that the polyurethane rubber employed here has a linearly photoviscoelastic properties, because the mathematical interconversion from one to another can be allowed under the condition of existence of its linearity. Thus, the performance of the photoviscoelastic analysis is possible by use of this material after the experimental determination of its various characterization coefficients. Also, a new collocation method [6] somewhat different from that by SCHAPERY [7] is proposed here in the mathematical interconversion.

In this paper, the practical photoviscoelastic analyses were carried out on two problems, employing polyurethane rubber models. The first problem was a static two-dimensional strain analysis, that is, a square plate with a central hole that was loaded non-proportionally[1] on the adjacent sides under condition of constant temperature, at which the material showed fairly viscoelastic behavior. Generally speaking, the axes of principal stress, principal strain and polarization of light do not always coincide in the photoviscoelasticity [4, 8]. The variations of the fringe patterns of isoclinics and isochromatics with time were recorded by a camera used in the ordinary photoelastic techniques. Thus, the principal stress difference, the principal strain difference, and their axes were obtained from experimental data. It was found that these results agreed quite well with the theoretically calculated results.

The second problem was concerned with the analysis of stress wave propagation [9]. Dynamic stress concentration due to viscoelastic wave was analyzed for a prismatic bar with

---

[1] Non-proportional loading means that the distribution pattern of the load vary with time.

a hole. Also the analysis of the stress wave through a simple prismatic bar, with no hole, was performed for the purpose of comparison. The bars were loaded in impact by shock waves at various constant temperatures. The fringe patterns of the specimens were recorded by a camera at various time intervals from the impact, using the "Delayed Synchro Flash" and an ordinary photoelastic equipment. The time-varying stress was calculated from the experimental data and photoviscoelastic coefficients. From these results, it was found experimentally that the dynamic stress concentration due to a hole is smaller than the corresponding static stress concentration for a small initial period of time t immediately following impact, and the former tends to approach the latter as time t increases. Furthermore, such a tendency occurs more quickly as the temperature falls down.

II. The photoviscoelastic coefficients of polyurethane rubber

The polyurethane rubber employed in this investigation is the same as that which was used in previous papers. This material exhibits excellent light-transmission properties and has high birefringence sensitivity. The composition of the material are Poly-urethane-glycol (molecular

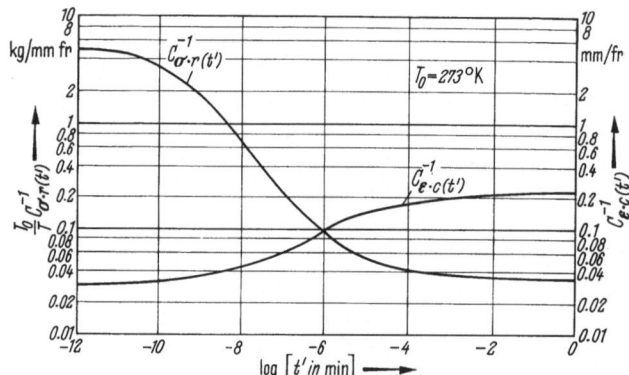

Fig. 1. Master curves of inverse relaxation stress-birefringence coefficient $C_{\sigma \cdot r}^{-1}(t')$ and inverse creep strain-birefringence coefficient $C_{\varepsilon \cdot c}^{-1}(t')$ for polyurethane rubber.

weight $2000-3000$), T.D.I. and Poly-propyrene-triol (molecular weight 1000) which are grouped into Tri-methylol-propane. The glass transition temperature of the material was measured as $T_g = -56°C$. Fig. 1 shows the experimental results of characterization concerning the inverse relaxation stress-birefringence coefficient $C_{\sigma \cdot r}^{-1}(t')$ and the inverse creep strain-birefringence coefficient $C_{\varepsilon \cdot c}^{-1}(t')$ for practical photoviscoelastic analysis [6].

III. A square plate having a central hole loaded non-proportionally on adjacent sides

In case of ordinary photoelasticity, the directions of principal stress, principal strain and polarization of light at any point in a two-dimensional model generally coincide completely with one another, while they do not always do this in photoviscoelasticity [8]. However, under the special conditions, for example for proportional loading[1] at constant temperature, all three of the above will have the same direction. In such a case, the photoviscoelastic situations at any time with respect to stress and strain are just similar to those in standard photoelasticity. Therefore, the time-dependent principal stress and strain differences can be analyzed photoelastically, using the characterization master curves of $C_{\sigma \cdot r}^{-1}(t)$ and $C_{\varepsilon \cdot c}^{-1}(t)$, after determination of the fringe order at any particular time.

The experiments demonstrated here are concerned with case of disagreement on these items. As an example for the case of non-proportional loading may be regarded. In this case the principal stress differences, principal strain differences and their directions at any time were analyzed from recording of the time-dependent photoviscoelastic isoclinics and isochromatics.

---

[1] Proportional loading means that the distribution pattern of the load does not vary but its magnitude may vary with time.

1. Experimental method of procedure

Figure 2 shows the geometry of the specimen employed and the loading conditions, in which the dead load over all the time was indicated by W_1 and the step load applied after $t = 0$ by W_2. In this way, the non-proportional loading could be realized. The constant temperature kept during the experiment was as low as $T = -41.3°C$, because the polyurethane rubber shows decisive viscoelastic behavior at this temperature.

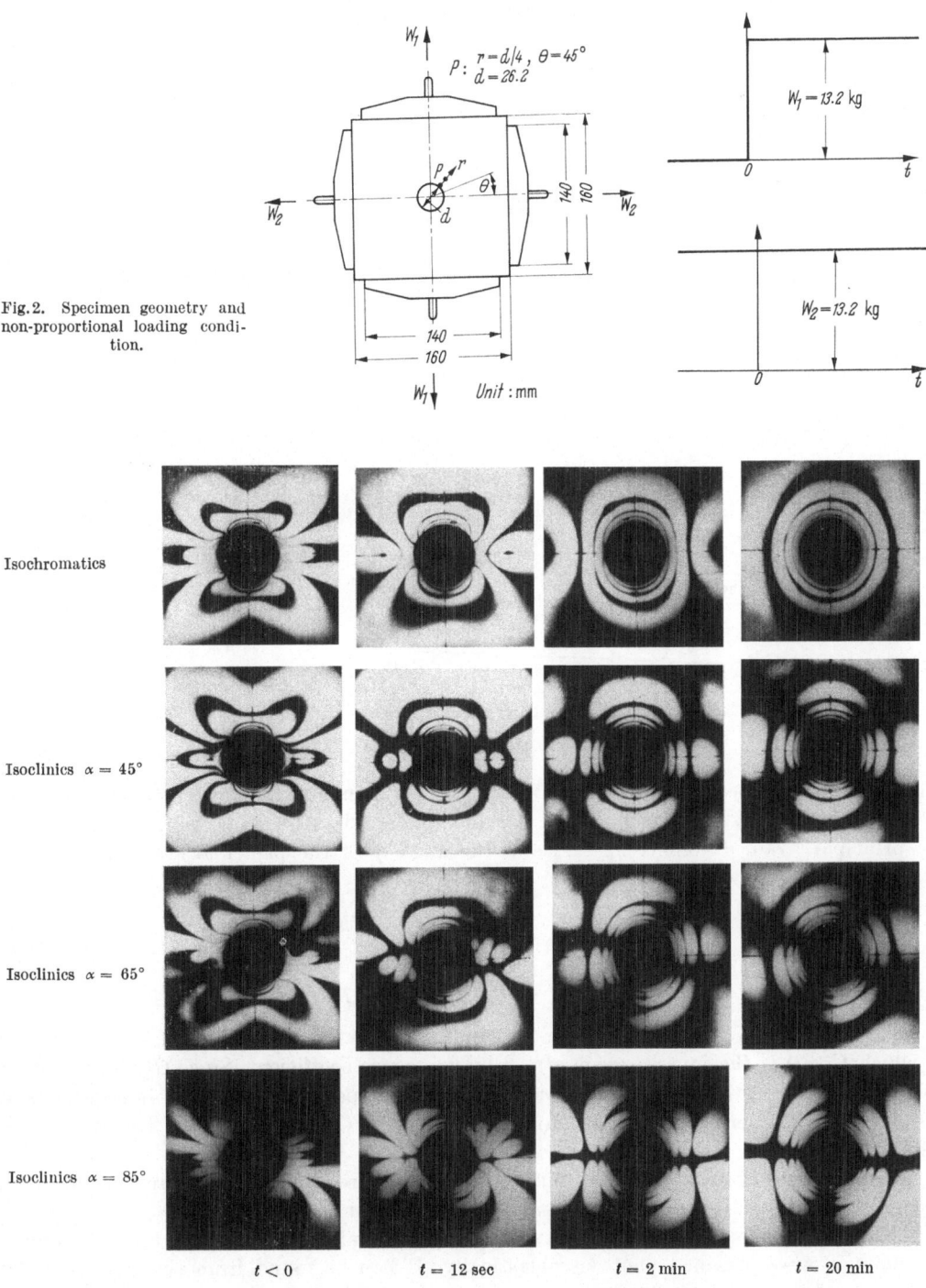

Fig. 2. Specimen geometry and non-proportional loading condition.

Fig. 3. Typical fringe patterns due to non-proportional loading at $T = -41.3°C$.

Standard linear- and circular polariscopes were used for taking photoviscoelastic pictures of isoclinic and isochromatic patterns, respectively. Monochromatic light of $\lambda = 5770-5790$ Å was obtained from a mercury lamp by aid of an orange filter.

Some typical pictures of isoclinic and isochromatic fringe patterns at different times, measured from the instant ($t = 0$) of W_1 loading, are demonstrated in Fig. 3. Isoclinic parameter α, i.e. the direction of polarization of light was measured as the angle to the horizontal direction.

2. Calculation of stress and strain

From Fig. 3, the variations of fringe order $N(t)$ and isoclinic parameter $\alpha(t)$ at a point P ($r = d/4$, $\theta = 45°$), in Fig. 2 with time can be read as shown in Fig. 4. The principal stress difference $\sigma_1 - \sigma_2$ and the angle β of the principal stress direction to horizontal line can be calculated by introducing the above data $N(t)$ and $\alpha(t)$ into the following equations:

$$(\sigma_1 - \sigma_2) \cos 2\beta = \frac{1}{h} \int_{-\infty}^{t} C_{\sigma \cdot r}^{-1}(t - \tau) \frac{dN(\tau) \cos 2\alpha(\tau)}{d\tau} d\tau$$

$$(\sigma_1 - \sigma_2) \sin 2\beta = \frac{1}{h} \int_{-\infty}^{t} C_{\sigma \cdot r}^{-1}(t - \tau) \frac{dN(\tau) \sin 2\alpha(\tau)}{d\tau} d\tau,$$

(1)

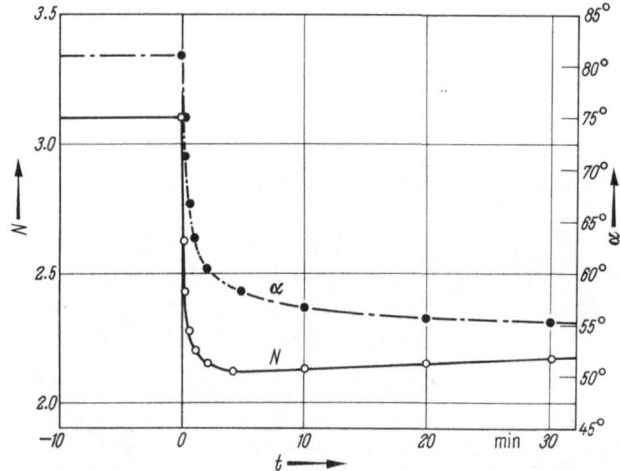

Fig. 4. Fringe orders N and angle α at point P at $T = -41.3°C$.

where h is the thickness of the specimen and $C_{\sigma \cdot r}^{-1}(t)$ is the inverse relaxation stress-birefringence coefficient which has been defined in a previous paper [6]. Also as far as the calculations of the principal strain difference $\varepsilon_1 - \varepsilon_2$ and the angle γ of the principal strain direction, the following equations were used:

$$(\varepsilon_1 - \varepsilon_2) \cos 2\gamma = \frac{1}{h} \int_{-\infty}^{t} C_{\varepsilon \cdot c}^{-1}(t - \tau) \frac{dN(\tau) \cos 2\alpha(\tau)}{d} d\tau$$

$$(\varepsilon_1 - \varepsilon_2) \sin 2\gamma = \frac{1}{h} \int_{-\infty}^{t} C_{\varepsilon \cdot c}^{-1}(t - \tau) \frac{dN(\tau) \sin 2\alpha(\tau)}{d} d\tau$$

(2)

in which $C_{\varepsilon \cdot c}^{-1}(t)$ is the inverse creep strain-birefringence coefficient. Since $C_{\sigma \cdot r}^{-1}(t')$ and $C_{\varepsilon \cdot c}^{-1}(t')$ given by Fig. 1 are the coefficients at the reference temperature $T = 0°C$, it is necessary to convert them to coefficients for the temperature $-41.3°C$ by utilizing the time-temperature shift factor [6, 10].

The variations of principal stress difference and its direction with time are shown in Fig. 5 and those of principal strain difference and its direction in Fig. 6.

Thus, the principal stress difference, the principal strain difference and their directions at the point P were obtained photoviscoelastically. It can be found from the figures that the experimental and theoretical results agree fairly well with each other, and also that, because the input to the specimen is given by loads at the boundary, the stress responds immediately after loading, while the strain response shows a remarkable time-dependent viscoelastic behavior.

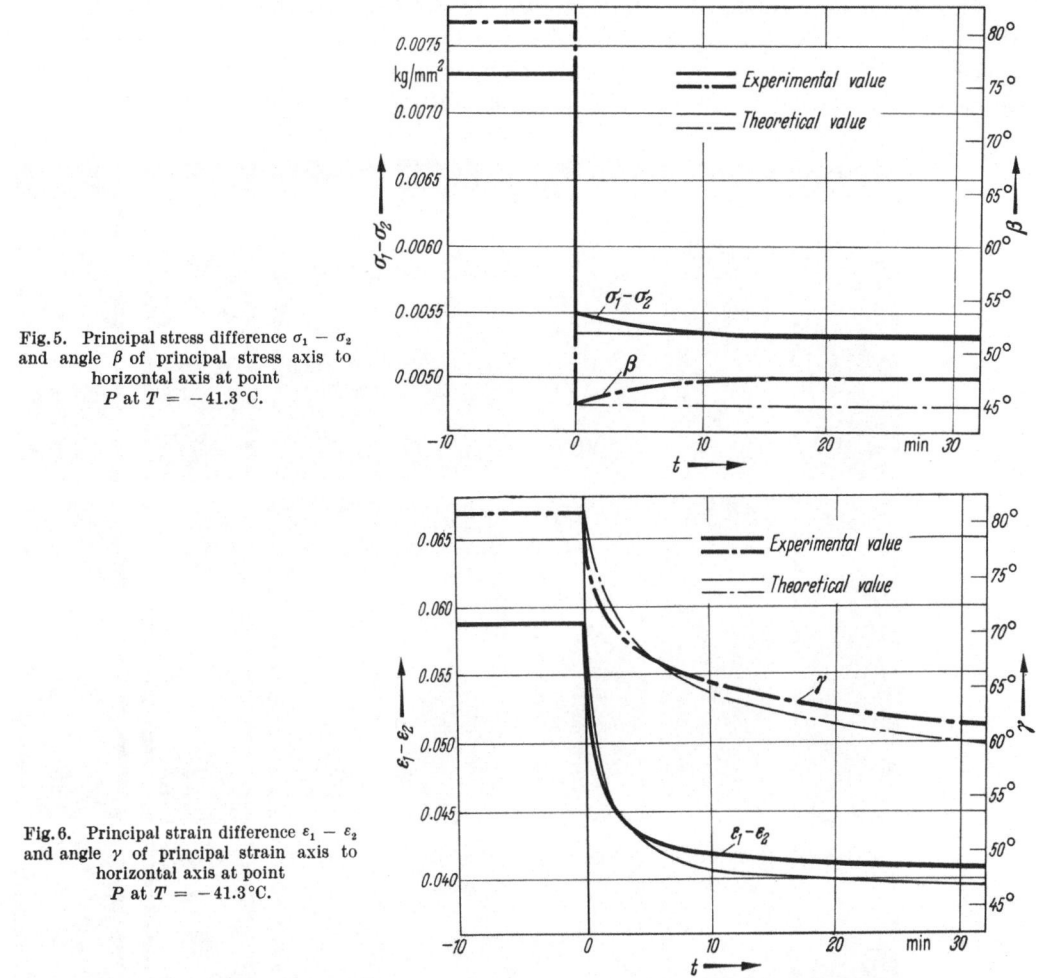

Fig. 5. Principal stress difference $\sigma_1 - \sigma_2$ and angle β of principal stress axis to horizontal axis at point P at $T = -41.3\,°\mathrm{C}$.

Fig. 6. Principal strain difference $\varepsilon_1 - \varepsilon_2$ and angle γ of principal strain axis to horizontal axis at point P at $T = -41.3\,°\mathrm{C}$.

IV. Stress wave propagation in a prismatic bar with a hole, loaded by impact

Another problem in practical photoviscoelastic stress analysis is that of stress-wave propagation due to an impact loading [9]. Although our polyurethane rubber can be regarded as the rubberlike-elastic material at room temperature in ordinary time scale, it behaves as a viscoelastic material for small initial periods of time following impact. Therefore, experiments were performed in which a compressive step-load was applied to the end of the prismatic bar having a circular hole, at various constant temperatures (near room temperature), and the variation of fringe patterns with time, which appeared due to the stress wave propagation in the bar, were observed.

Calculation of their viscoelastic stresses were carried out by use of $C_{\sigma\cdot r}^{-1}(t)$ at various temperatures.

Fig. 7. Specimen geometry and impact loading condition.

274 T. KUNIO and Y. MIYANO

1. Experimental methods and data

The specimens used in the experiments were machined in the form shown in Fig. 7 from a plate of polyurethane rubber. The impact loads $w = 0.0197\,\mathrm{kg/mm^2}$ were applied stepwise at the ends of the specimens by a shock tube, and the specimens were kept at four stages of temperatures 19°, 10°, 0° and -5°C, because the material of the specimens shows a remarkable viscoelastic behavior for impact load at these temperatures. The ordinary circular polariscope

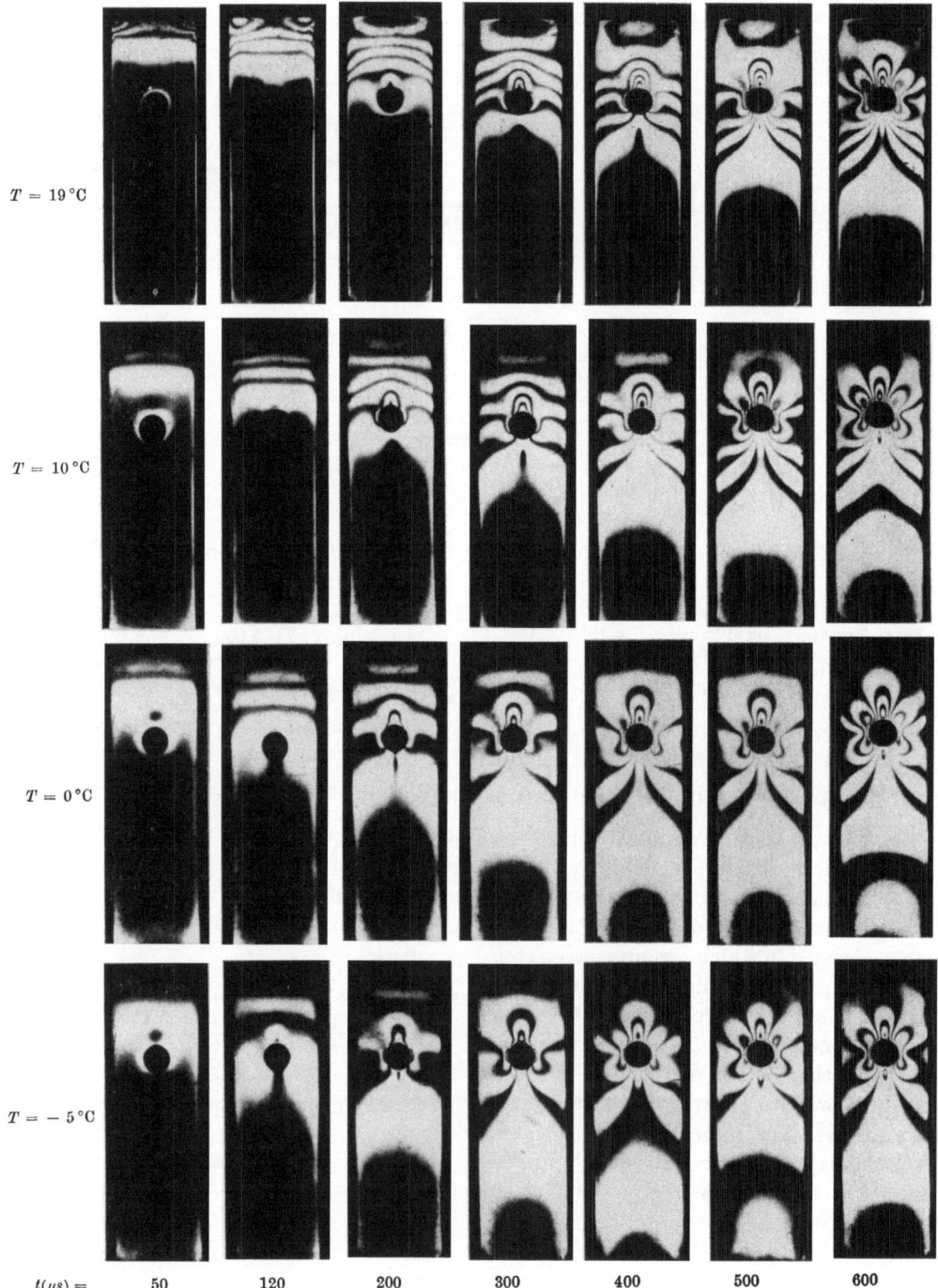

Fig. 8. Typical isochromatic fringe patterns of prismatic bar having a hole due to impact load at various constant temperatures.

was used for taking pictures of fringe patterns. The fringe patterns at any time ($t = t$) from the instant of impact ($t = 0$) were taken by a Xenon lamp with the Delayed Synchro Flash apparatus, operated by the signal of the trigger applied near the end of the specimen (see Fig. 7). The light had a wave length of between 6000 to 6500 Å.

Typical pictures of the fringe patterns obtained are shown in Fig. 8.

2. Stress concentration in a prismatic bar with a hole

To investigate the stress concentration near a circular hole in a prismatic bar due to impact load, the variation of stress at a point A in Fig. 9 was read from the fringe patterns shown in Fig. 8. Judging from the fact that the fringe patterns around a circular hole do not undergo

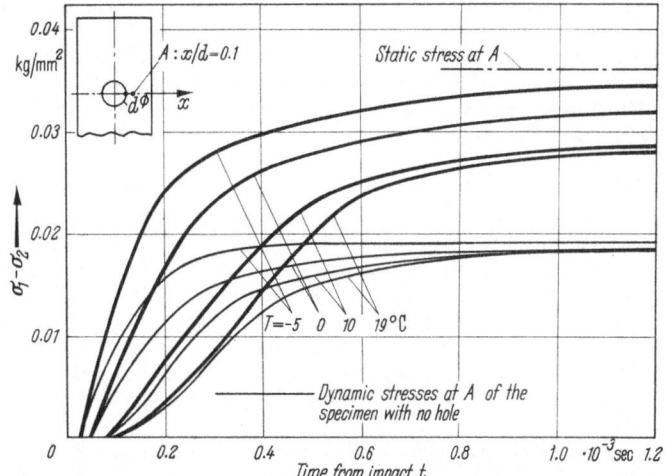

Fig. 9. Fringe orders N at point A of prismatic bar with a hole due to impact load at various constant temperatures.

such a remarkable change as seen in Fig. 8 after the arrival of the stress wave at the point A, it may be considered that the direction of the polarization of light agrees practically with that of the principal stress. Therefore, the following equation, which is reduced by putting $\alpha = \beta$ in eqs. (1), was employed for analysis of the stress.

$$\sigma_1 - \sigma_2 = \frac{1}{h} \int_{-\infty}^{t} C_{\sigma \cdot r}^{-1}(t - \tau) \frac{dN(\tau)}{d\tau} d\tau. \tag{3}$$

The variations of principal stress difference $\sigma_1 - \sigma_2$ at point A with time are drawn in Fig. 10, taking various temperatures as parameters.

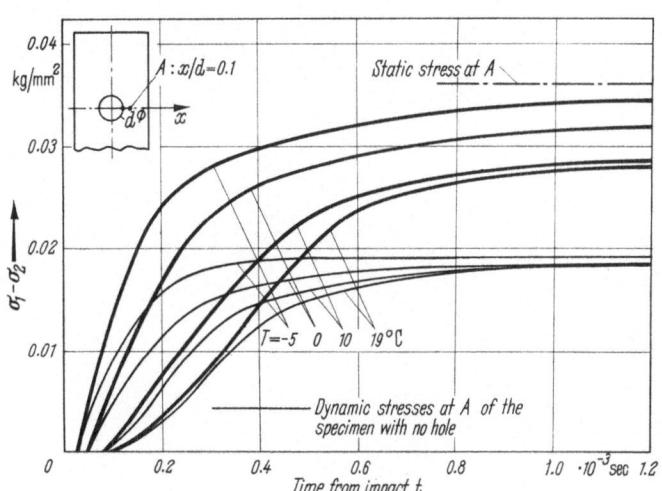

Fig. 10. Principal stress difference $\sigma_1 - \sigma_2$ at point A of prismatic bar with a hole due to impact load at various constant temperatures.

The stress at point A is still increasing even at the time when the stress at the corresponding point in the bar with no hole, having the same geometry as the specimen employed, has already been saturated to a certain stress level, and finally it will attain to the stress value due to static loading. Furthermore, it is possible that the time necessary to arrive at the final value becomes gradually shorter as the temperature falls. In other words, the dynamic stress concentration is smaller than the static one, and the former will approach the latter with time. In addition, the dynamic stress concentration approaches the static one more quickly as the temperature falls.

Acknowledgement

One of the authors have stayed for some time at California Institute of Technology. It is a great pleasure to express our appreciation to Prof. M. L. Williams, a former-professor at Caltech, for giving us the fundamental concepts of photoviscoelasticity.

The authors are extremely grateful to Prof. H. Okamura of Tokyo University for suggesting to us the use of the experimental apparatus, called Delayed Synchro-Flash.

Our thanks are also due to Messrs. M. Takashi, T. Morita, T. Tamura, A. Misawa and others for their aborious help in the experiments.

References

1. Alfrey, T.: Quart. Appl. Math. II, 113—119 (1944).

2. Williams, M. L., and R. J. Arenz: Exper. mechanics, 249—262 (September 1964).

3. Arenz, R. J., C. W. Ferguson and M. L. Williams: Exper. Mech. 183—188 (April 1967).

4. Arenz, R. J., C. W. Ferguson, T. Kunio and M. L. Williams: GALCIT SM 63-31 (October 1963).

5. San Miguel, A., and E. N. Duran: Exper. Mech. 84—88 (March 1964).

6. Kunio, T., Y. Miyano and T. Tamura: Bull. JSME, 6, No. 49 (1969) (to be published).

7. Schapery, R. A.: Proc. of the Fourth U.S. National Congress of Applied Mechanics, Vol. 2 (1962), pp. 1075.

8. Read, W. T.: J. Appl. Phys. 21, 671—674 (1950).

9. Ferguson, C. W.: J. SMPTE, 73, 782—787 (September 1964).

10. Williams, M. L., R. F. Landel and J. D. Ferry: J. Amer. Chem. Soc. 77, 3701—3707 (1955).

Pulsatile flows in arteries

By

S. C. Ling, H. B. Atabek and J. J. Carmody

The Catholic University of America, Washington

1. Introduction

Recent interest in the hydrodynamic events relating to biomedical problems of the arteries as well as the development of artificial heart-assist devices have prompted one to study the nature of pulsatile flows in arteries. In particular, the problem of atherosclerosis is suspected to be related to the shearing stress developed over the vascular wall. The endothelial cells which normally line the intima of blood vessels can be eroded under high wall shear. It was found by FRY [1] that these cells have an acute yield stress of the order of 350 to 400 dynes/cm². Shearing stresses exceeding this limit will cause marked endothelial proliferation and sub-endothelial lipid deposition. On the other hand, there are problems of keeping the shearing stresses within a range to permit the endothelial cells to establish and to stabilize the newly formed fibrin surfaces over prostheses such as heart valves and heart-assist devices. Basic understanding of these problems could be gained through detailed knowledge of the pulsatile flow fields.

Hot-film anemometry has been successfully developed for the mapping of flow fields in the arteries of living animals as well as simulated laboratory models. It was found that under normal conditions flow in the arteries can be classified as nonsteady laminar flow. That is, the flow profiles are developed locally and transiently during the passage of each pulsatile wave. The rise time of the pressure-gradient wave front is of the order of 0.02 sec, and the apparent propagating velocity of the wave front is of the order of 600 cm/sec. Therefore, the physical width of the traveling wave front is about 12 cm. This, in conjunction with the short development time for the boundary layer, causes the entrance effect to be limited to a short distance from any branching points or heart. Also, the thicknesses of the boundary layers during systole were found to have a uniform value of about 1.5 mm at all points along the aorta and arteries, except at zones close to the branching points where the boundary layers are correspondingly thinner due to entrance effect. In general, flow profiles were found to be blunt and axially symmetrical. The average shearing stress at the wall has a value of 20 to 40 dynes/cm², while the peak shearing stress may vary from 80 to 160 dynes/cm². Shearing stresses at the lips of arterial branches are correspondingly higher. Both turbulence and secondary flows were found to be inhibited and localized by the nature of pulsatile flow. Only weak turbulence was observed in the aortic arch of small canines. These turbulences were noted to be generated and dissipated within each heart cycle. By the same token, they were convected spatially for a limited distance of about 6 cm, representing the displacement of blood for one heart beat.

To test some of the in vivo observations we have conducted several analogous analytical and model experiments. In general, we have found that the existing theories based on the linearized long-wave approximation do not give the correct flow profiles and wall shears. However, the model experiments confirm the basic flow characteristics found in living animals.

2. Anemometry

Linear constant-temperature hot-film anemometry was used for the present investigation. Two types of micro probes were developed: the cone-type probe for measurement of the flow profiles, and the surface-type probe for the measurement of wall shears. The detailed structure of these probes is shown in Fig. 1. The cone-type probe was designed for insertion through the vascular wall at an angle of 70° with respect to the axis of the vessel. This configuration was selected to provide for good control in positioning the sensing element under pulsatile vibrations,

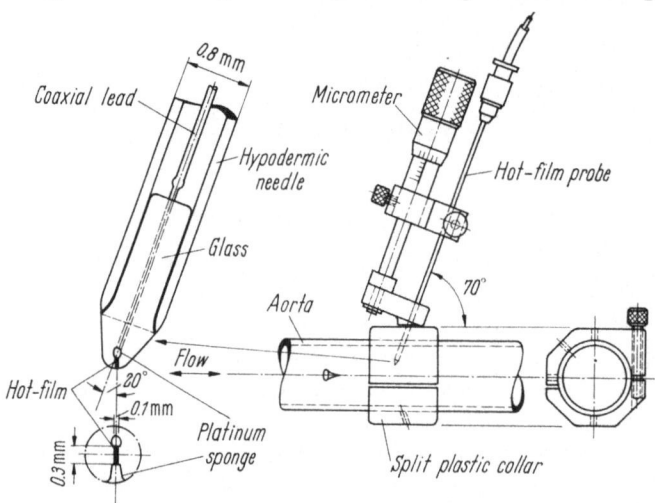

Cone-type velocity probe

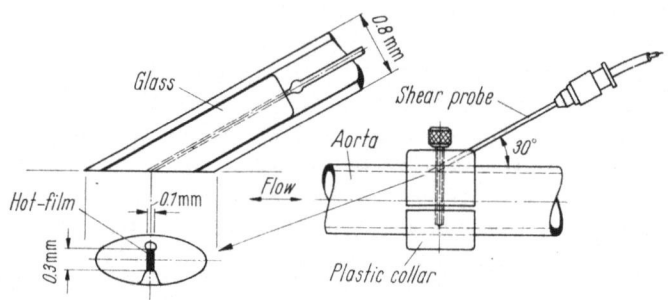

Surface-type shear probe

Fig. 1. Hot-film probes for measurement of velocity field and wall shear.

and at the same time to minimize the effect of blockage to the flow cross section. Since reverse flows are often encountered in the arteries, the probes have to be capable of detecting bidirectional flows. The probes shown in Fig. 1 represent the best design of many forms tested. It should be noted that blood vessels pulsate radially about 10% of their mean radius. To map the flow field it would be best to have the probe fixed with respect to the wall. However, this was not possible because the wall is too soft and fragile. An alternative solution was to use a split plastic collar of fitted size over the blood vessel to serve as a jig for the probes as well as a stable reference for controlling the radial position of the probe by a micrometer. The split plastic collar was made adjustable within a limited range so as to allow some radial motion of the artery, and thereby reduces error due to local constriction. It was found that the velocity probe will be inaccurate when operated closer than 0.4 mm from the vascular wall. This is due to the fact that the high velocity gradient found near the wall will effect the calibration of the probe, and also that both the wall effect and the radial wall motion will cause error in the detected signals. Hence, the flow field near the wall is best detected by the wall shear probe, since it is less sensitive to wall motion. In this paper all flow profiles were referred to the fixed mean radius, r_0, of the pulsating artery. This would introduce error in the flow profiles. However, by plotting the correct velocity gradient near the wall, using the data from direct wall shear measurements, this error is minimized.

Both types of flow probe were calibrated in known flow fields in order to ascertain their operating characteristics. It was found that an oscillating flow field in a rigid tube of limited length is highly reproducible and predictable. In contrast to the steady-flow case, it is also relatively unaffected by the entrance effect. That is, the velocity field approaches that of a long tube within 20 diameters from the ends of the tube if the oscillating excursion of the fluid is limited to less than 10 diameters. The measured flow profiles were found to be essentially identical to those predicted by theory [2] for an α parameter ranging from 1 to 20. The α parameter is expressed as $\alpha = r_0(\omega/\nu)^{1/2}$, where ω is the angular velocity of the oscillating field and ν the kinematic viscosity. Typical flow profiles for the cases $\alpha = 10$ and 1.0 are shown in Fig. 2. The theoretical profiles are depicted as solid curves against the experimental points. Both blood and a glycerin-water mixture having the same viscosity as blood give essentially the same flow profiles. Pure glycerin was used to obtain the flow profiles at $\alpha = 1.0$. Because the measured flow fields fit so well with the theory, both the absolute centerline velocity and

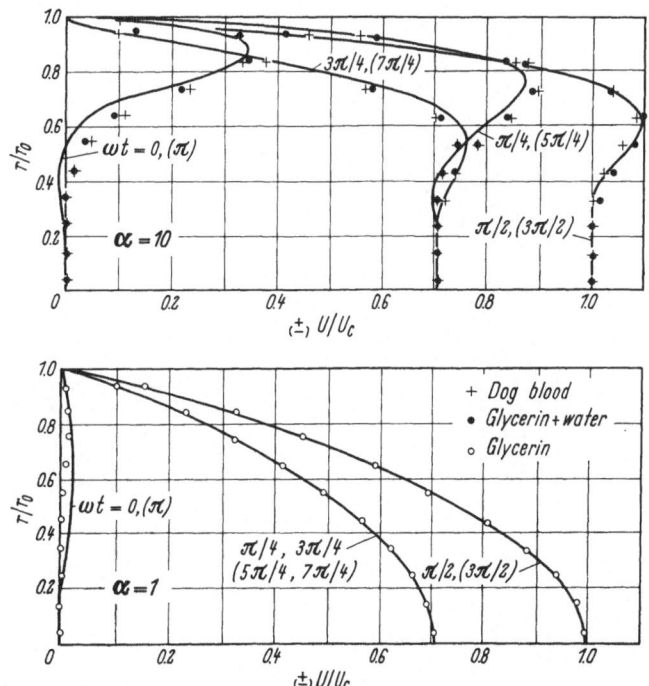

Fig. 2.
Oscillatory flow profiles in a rigid tube.

the velocity gradient at the wall may be related through the theory to the known piston (average) velocity. Hence, these fields can be used for calibrating both the velocity and the shear probes. A simple calibrator was constructed of a tube 100 cm in length and 1.27 cm in diameter. Blood in the tube was driven by a piston with a fixed sinusoidal displacement of 5 cm, while its angular velocity was made variable. Ports at midlength of the tube were provided for insertion of the test probes. A constant-temperature bath was also provided to keep the blood sample in the calibrator at the same temperature as the animal under study. Because the calibrator uses only 100 cc of blood, a blood sample can be taken from the animal under study without affecting its physical condition. In this manner, absolute calibrations can be obtained independent of the physical properties and temperature of the test fluid. To reduce blood clotting over the probe, the sensing elements were operated at a low temperature of 5° to 8°C above the ambient temperature.

3. Experimental measurements in living animals

Experimental measurements were made in canines and pigs of different sizes and weights. In general, a complete survey of the flow field representing a whole circulation system was found

280   S. C. Ling, H. B. Atabek and J. J. Carmody

to be extremely difficult to obtain. This is because major surgery was required to permit access to a limited segment of the aorta. In addition, the circulation systems were subjected to a host of physiological variables that were never steady for a duration of time. Flow profiles in this experiment were obtained through recordings of the pulsatile velocity waves taken at various radial positions in the arteries. During the period of recording there were unavoidable variations in the heart rate, heart output, and pulse wave form. Consequently two velocity probes had to be used to perform the measurements. One of these probes was used as a reference probe, while the other probed the velocity field. Based on the reference probe, data samples were selected in which the heart rate, pulse amplitude, and wave form were similar within a limited range. These selected data were then averaged to obtain the velocity profiles, from which the corresponding discharge waves were obtained.

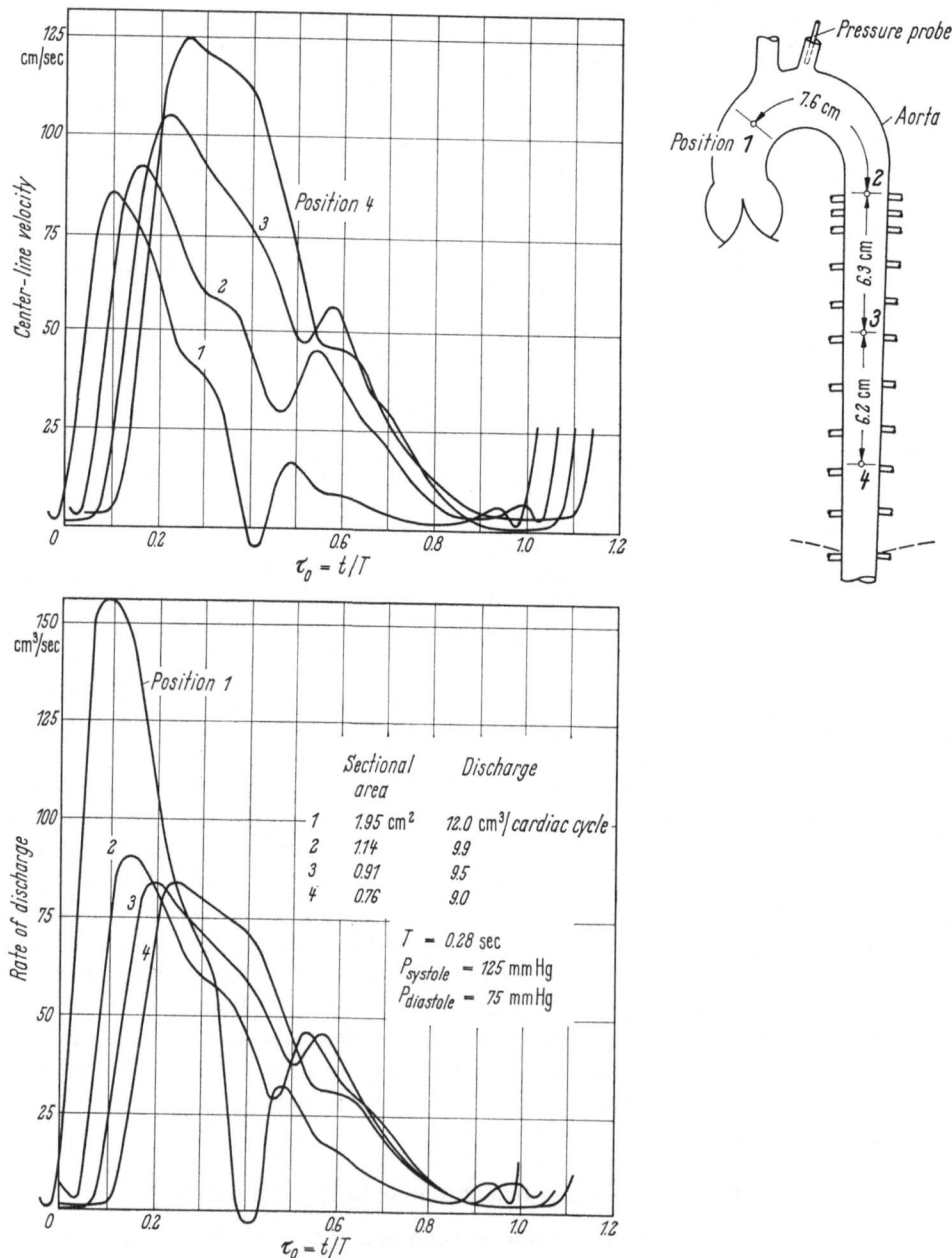

Fig.3. Discharge and centerline velocity waves along the aorta of a 31 kg canine.

A typical measurement representing clean flow waves along the main part of the aorta is shown in Fig. 3. The corresponding flow profiles are shown in Fig. 4. The time base, t, was nondimensionalized by the cardiac period, T, and plotted as τ_0. It is interesting to note that the peak centerline velocities increase with distance from the heart, while the reverse is true

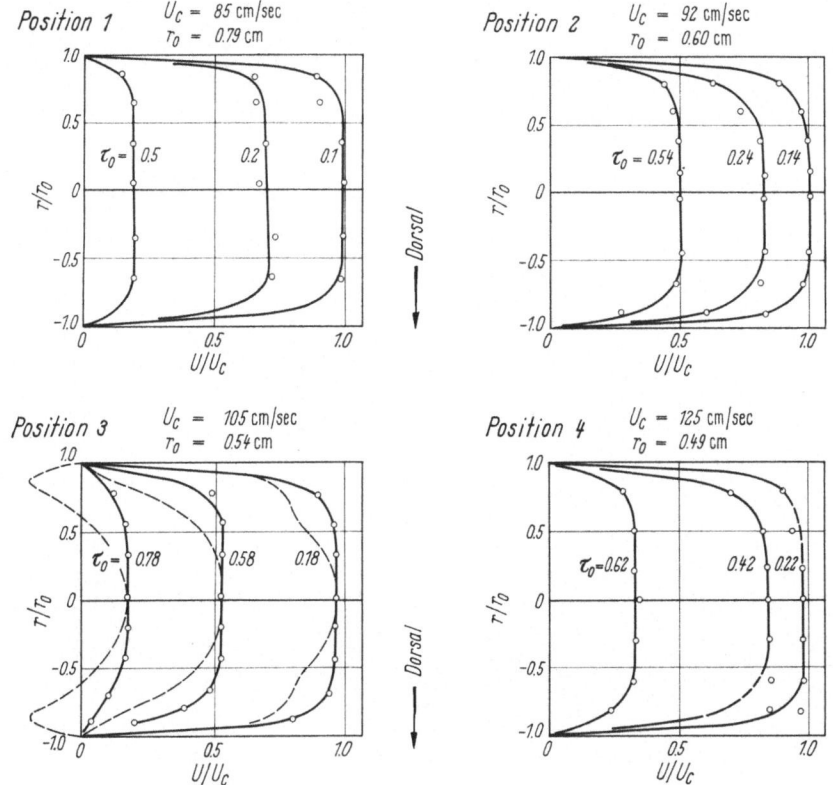

Fig. 4. Velocity profiles corresponding to the flow waves in Fig. 3.

for the peak flow discharge. These results are due mainly to the fact that the aorta is tapered and the flow cross-section decreases in the downstream direction, as indicated in Fig 3. The velocity profiles at the four indicated positions are normalized by their local peak centerline velocities, U_c. It is apparent that these velocity profiles are all blunt and similar in form. They are also quite independent of the entrance effect from either the heart or the main arteries. From these flow data, one can show also that the wall shear stresses tend to increase downstream.

By comparing the discharge waves between position 1 and 2 in Fig. 3, one notes that there is a very large dynamic flow storage in the ascending aorta and the aortic arch. However, some outflow through the brachiocephalic arteries must be taken into account. The effect of dynamic flow storage in the descending aorta is relatively small. At $\tau_0 > 0.6$ one notes that the discharge rates at positions 2, 3, and 4 are essentially the same, with the basic flow derived largely from both the ascending aorta and the aortic arch.

In general, the velocity wave forms in arteries are highly variable from one animal to the next, quite unlike the pressure wave forms which are related mainly to the charge and discharge characteristics of the overall circulation system. The velocity waves are sensitive to the local variation in wave transmission characteristics of the aorta. Two additional types of flow waves and their corresponding flow profiles observed in the descending aorta of pigs are shown in Fig. 5 and Fig. 6. In Fig. 5 the flow wave shows a large reverse flow. During the systolic period the usual blunt velocity profiles are noted. For the case in Fig. 6, there is a relatively large base flow. The velocity profiles are again blunt and similar in form throughout the cardiac cycle.

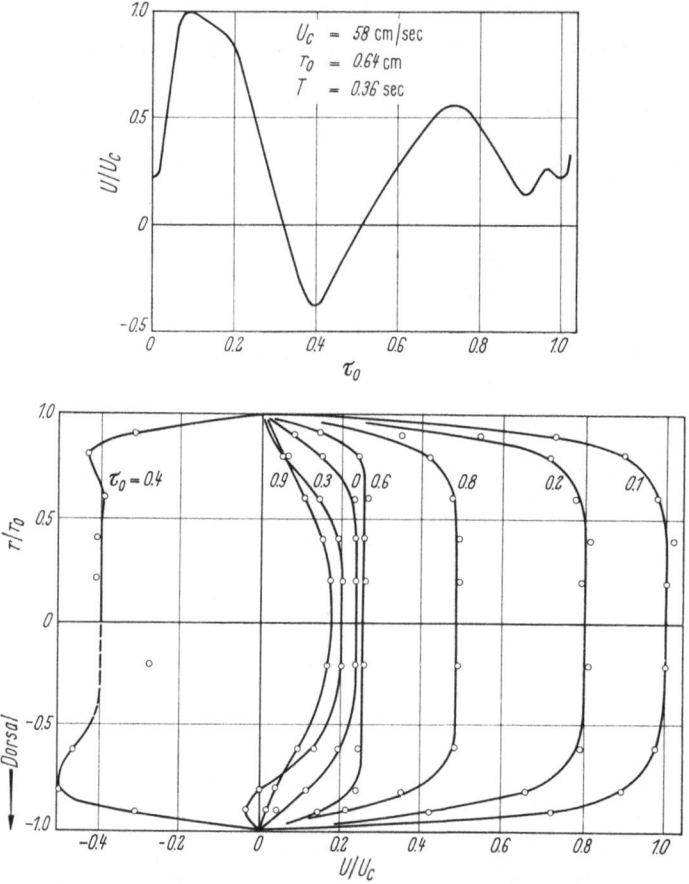

Fig. 5. Centerline velocity wave and velocity profiles in the descending aorta of a pig.

One notes that for cases with no net flow reversal, both the velocity profiles near the wall and the wall shears tend to remain positive during the deceleration period. Tentative measurements conducted in small arteries of a few millimeter size, such as the coronary arteries, were found to have the same blunt velocity profiles. The observed blunt flow profiles during systole can be explained by the fact that these flows are all accelerated locally and transiently, since the characteristic diffusion parameter [3] vt/r_0^2 is $\ll 1$, where t is the rise time of the flow wave. Hence, the velocity profiles in arteries should be essentially independent of the entrance effect, except at zones close to the branching points.

A recording of the centerline velocity waves taken in the aortic arch of a small dog is shown in Fig. 7. In this recording there is one missing heart beat. One notes that during this period there are no significant waves being contributed by the preceding cardiac cycles. This would indicate that pulsed waves are independent in nature. It is also noted that the flow is turbulent, and that the tubulence was created and dissipated within each heart beat.

The flow data at position 3 of Fig. 3 were selected to test the existing theories [4] to see how well they can predict the flow profiles. Under the assumption of developed flow, one need only prescribe either the local pressure gradient wave, the flow discharge wave, or the centerline velocity wave as the input function in order to obtain the flow profiles from the analytical solution. For the present case, it is most convenient to use the centerline velocity wave. On the assumption that the flow is periodic, the centerline velocity wave can be represented in terms of a Fourier series. The velocity profiles associated with each of the wave components can be calculated from the analytical solutions. The present linearized long-wave solution has been extended by Atabek [5] to include the physical properties of the aorta and its surrounding tethering materials, as well as the conditions of initial stress and strain. It was found that

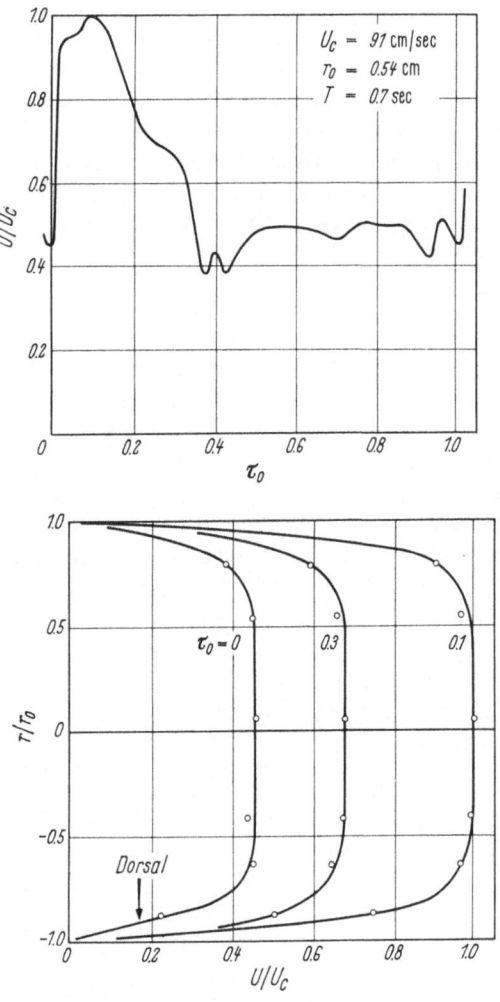

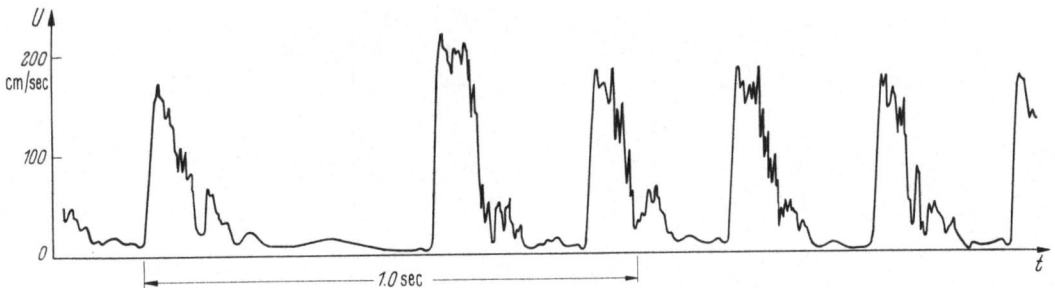

Fig. 6. Centerline velocity wave and velocity profiles in the descending aorta of a pig.

Fig. 7. Centerline velocity waves taken in the aortic arch of a 17 kg canine.

for a normal physical case, the improved solutions for the flow profiles do not differ much from those of a rigid tube. In line with the assumption of established flow, the steady flow component is taken to be parabolic in form. The corresponding integrated flow-profiles are plotted as dotted curves in Fig. 4. It is apparent that at $\tau_0 = 0.18$ the flow profile is less than blunt. Also at $\tau_0 = 0.78$ the theory indicates a strong negative annular flow which was not indicated by the experiment. However, when the parabolic steady-flow component was replaced by a blunter profile, such that the profile at $\tau_0 = 0.18$ is made identical with the measured one, then the profiles at subsequent times give a better fit. At $\tau_0 = 0.78$, however, the profile still shows some negative flow near the wall. Similar results were obtained when the flow wave was treated

36*

as a single pulse by means of the Fourier integral analysis. In view of these results, one faces several basic questions in the application of the linearized theory: What type of flow profile should be assigned to the steady-flow component, what are the errors due to neglecting the tapering geometry of the arteries, and what are the effects due to neglecting the convective terms in the Navier-Stokes equations? To settle some of these basic questions we have conducted several model and analytical experiments in which both the geometry and the physical properties of the flow model were known and less complex.

4. Experimental measurements in models

Precision mappings of the dynamic flow fields in living animals are often difficult to perform, owing to the fact that the circulation systems are subjected to constant changes. Hence, it is important to supplement the study by model experiments. Two types of model system were built and investigated. One is a physical model which was designed to simulate the system of a medium-sized canine, and the other is a model with a long uniform artery. The physical model was intended for precision study of the flow profiles in the aorta and its main arteries, while the second model was for checking the theories.

All model arteries were made of G.E. RTV-108 silicone rubber. This material at a strain of percent has an incremental Young's modulus of elasticity of about 10^7 dynes/cm^2, which is close to that of the arteries at the mean aortic pressure [6]. Both the nonisotropic and nonlinear elastic properties of the blood vessels were not simulated. The models were molded on Teflon mandrels, whose size was that of the arteries at their uninflated state. Silicone rubber was squeegeed on to the mandrel in thin layers. These layers were applied and cured one layer at a time until the desired thickness was obtained.

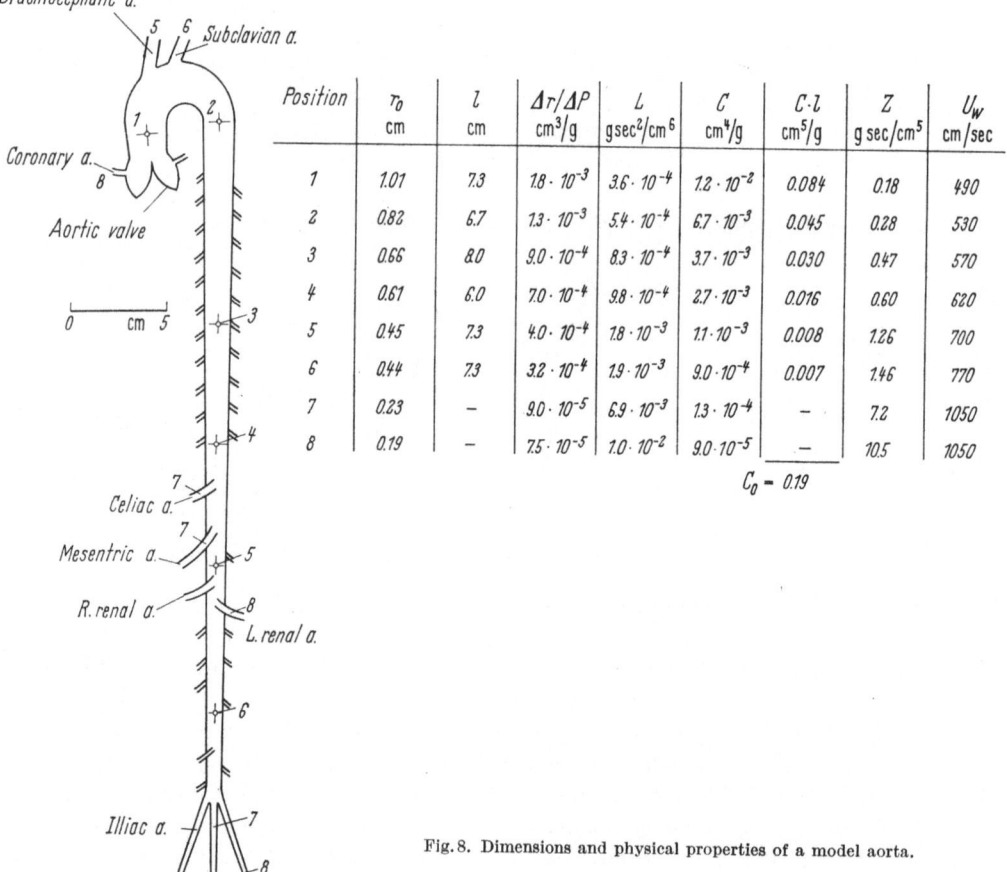

Position	r_0 cm	l cm	$\Delta r/\Delta P$ cm^3/g	L gsec^2/cm^6	C cm^4/g	$C\cdot l$ cm^5/g	Z g sec/cm^5	U_w cm/sec
1	1.01	7.3	$1.8\cdot10^{-3}$	$3.6\cdot10^{-4}$	$1.2\cdot10^{-2}$	0.084	0.18	490
2	0.82	6.7	$1.3\cdot10^{-3}$	$5.4\cdot10^{-4}$	$6.7\cdot10^{-3}$	0.045	0.28	530
3	0.66	8.0	$9.0\cdot10^{-4}$	$8.3\cdot10^{-4}$	$3.7\cdot10^{-3}$	0.030	0.47	570
4	0.61	6.0	$7.0\cdot10^{-4}$	$9.8\cdot10^{-4}$	$2.7\cdot10^{-3}$	0.016	0.60	620
5	0.45	7.3	$4.0\cdot10^{-4}$	$1.8\cdot10^{-3}$	$1.1\cdot10^{-3}$	0.008	1.26	700
6	0.44	7.3	$3.2\cdot10^{-4}$	$1.9\cdot10^{-3}$	$9.0\cdot10^{-4}$	0.007	1.46	770
7	0.23	–	$9.0\cdot10^{-5}$	$6.9\cdot10^{-3}$	$1.3\cdot10^{-4}$	–	7.2	1050
8	0.19	–	$7.5\cdot10^{-5}$	$1.0\cdot10^{-2}$	$9.0\cdot10^{-5}$	–	10.5	1050

$$C_0 = 0.19$$

Fig. 8. Dimensions and physical properties of a model aorta.

A Newtonian fluid of glycerin-water mixture having a kinematic viscosity of 0.04 cm²/sec at 26°C was used for all model experiments. Although blood had been used also in the model systems, the results were identical to those of the glycerin mixture.

4.1 Physical Model System

The design parameters for the model arteries are governed by the characteristic line impedance $Z = (L/C)^{1/2}$ and the propagating wave velocity $U_w = (1/LC)^{1/2}$, where $L = \varrho/\pi r_0^2$ is the inertance per unit length, ϱ the density of fluid, and $C = 2\pi r_0 \, \Delta r/\Delta P$ the capacitance per unit length. The value for U_w should be corrected for viscous effect [5], which amounts to a reduction of 2 to 5 percent for a wide range of α parameter. In model design this can be ignored. The line capacitance for arteries is generally obtained by the dynamic measurement of the change in radius of the vessel per unit change in internal pressure, $\Delta r/\Delta P$, under living condition [7]. At present the experimental data were found to be inaccurate. Fortunately, several experimental data are known to be reliable and can be used to obtain a more correct estimate for the distribution of the line capacitance. These are the line inertance, wave velocity, and the overall time constant for the aortic system, $R_0 C_0$, where C_0 is the total capacitance of the aorta, and R_0 is the total vascular bed load. A typical model for the aorta of a medium-sized canine is shown in Fig. 8. The distribution of U_w along the aorta is shown in the listed table. The corresponding line inertance, L, which is a function of the geometry, is also listed. From these the required distribution of line capacitance, C, was calculated. Also from the sum of the product of C and its corresponding line segment, l, one obtains the total capacitance of the system

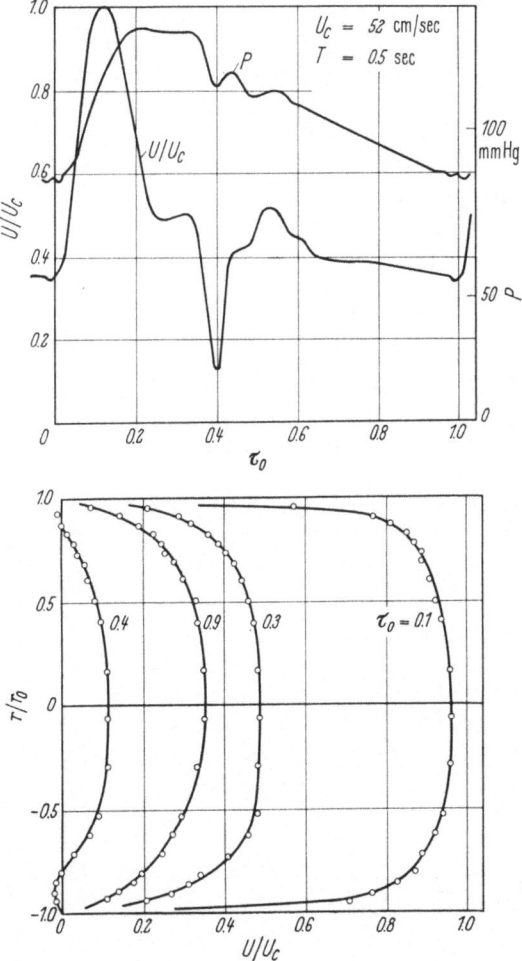

Fig. 9. Centerline velocity wave, pressure wave, and velocity profiles of a model aorta.

as $C_0 = 0.19$ cm⁵/g. The total vascular bed resistance, R_0, for the system, based on a mean flow rate of 30 cc/sec at a mean aortic pressure of 140 g/cm², is 4.7 g sec/cm⁵. Hence, the time constant for the system is $R_0C_0 = 0.9$ sec. This is close to the average value found in animals. The corresponding line impedance, Z, is also listed. It is noted to increase monotonically from the heart to the terminal arteries. The vascular bed resistance for each main artery is about 30 to 40 g sec/cm⁵. In the model, the terminal bed loads were simulated by 50 cm length of 3.2 mm I.D. tubes, having a characteristic line impedance of about 19 g sec/cm⁵, and an effective resistive loading of 35 g sec/cm⁵.

From the required distribution of line capacitance, the values of $\Delta r/\Delta P$ can be found. These are then used to calculate the required wall thicknesses for the model. Using the silicone-rubber material mentioned above, one finds that the required wall thicknesses for the entire model vary within a limited range of 0.7 to 0.4 mm.

The input function for the model system was controlled by the model heart system. In this case, it was simulated by the ventricular pressure wave form, which was produced by an electro-pneumatic system. The aortic valve of this model was made similar to the living counterpart, so as to create a realistic dicrotic flow wave in the aorta.

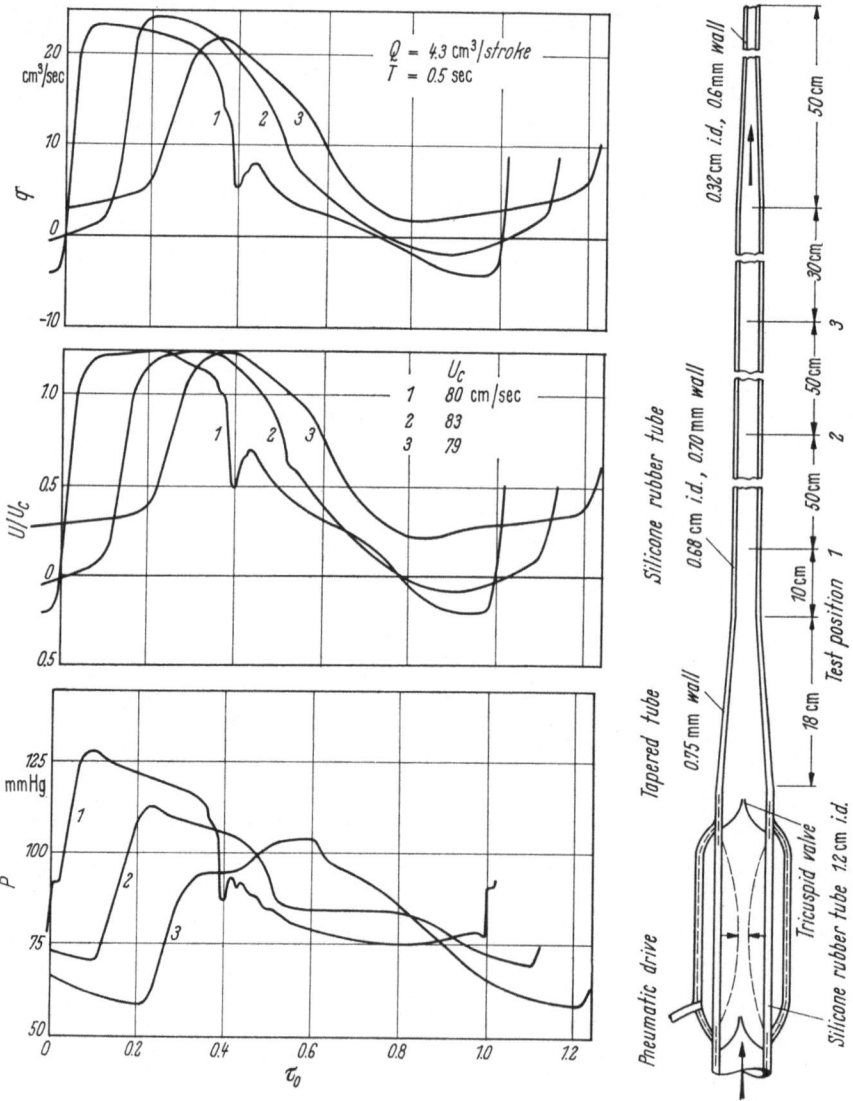

Fig. 10. Discharge waves, centerline velocity waves, and pressure waves in a uniform artery.

Further consideration for the model system is the tethering effect. It was found experimentally that a weak longitudinal tethering will essentially inhibit the longitudinal mode of wall motion. Also there was no detectable change in the velocity field whether the flow model was supported on a sheet of foam rubber or on a rigid surface. Hence, all model tests were conducted with the model laying on a rigid table. The observed effect of longitudinal tethering is in full accord with the theory [5].

The model system was found to behave like a real circulation system. The observed flow profiles in the aorta were in every respect similar to those found in the animals. A typical centerline flow wave and velocity profiles taken at position 3 of the model are shown in Fig. 9.

4.2 Pulsatile Flow in a Uniform Artery

The model system for observing pulsed flow through a uniform artery is shown in Fig. 10. The system consists of a heart pump and a tapered section to produce pulsed waves similar to those of the living counterpart. In order to achieve a fully developed flow condition for checking the existing theories, the uniform artery was made extra long, 140 cm, and small in diameter, 0.68 cm. That is, an inlet transition parameter [3] of $xv/\bar{U}r_0^2 > 0.16$ was obtained in the test section, where x is the distance from the inlet and $\bar{U}$ is the mean velocity. The line impedance for the uniform artery is 2.6 g sec/cm^5. The system was terminated by a tapered section and a smaller tube of 0.32 cm I.D. to simulate the resistive bed load as well as to provide a proper impedance matching.

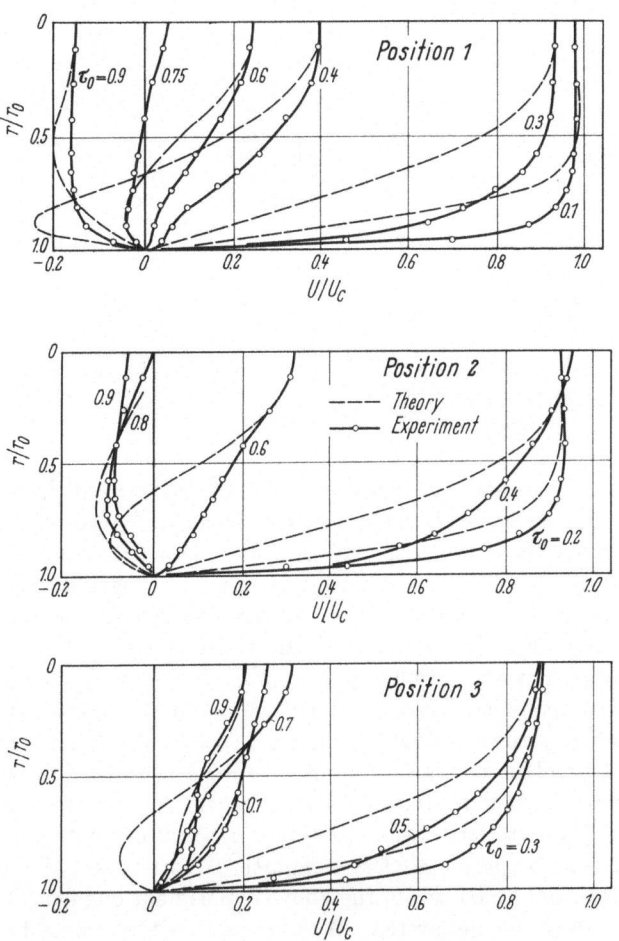

Fig. 11. Velocity profiles for the corresponding flow waves of Fig. 10.

The measured discharge waves, q; centerline velocity waves; and pressure waves, P, at the three marked test positions are shown in Fig. 10. The corresponding measured flow profiles at different pulse periods τ_0 are shown in Fig. 11. The related flow profiles predicted by the existing theories [5] are shown as dashed curves. It is clear that the basic differences between the measured and predicted flow profiles are similar to those found in the animal experiment. During the systolic period the measured profiles are blunter than those predicted by the theory. At the deceleration period the theory again indicated strong negative annular flows near the wall. This was absent in the flow model. However, at the quiescent period of $\tau_0 > 0.5$ the measured and predicted velocity profiles are nearly the same. The measured wall shears are shown in the upper part of Fig. 12. The shape of these waves is very similar to that of the velocity waves at the center of the artery. This would mean that the boundary layer tends to stay thin

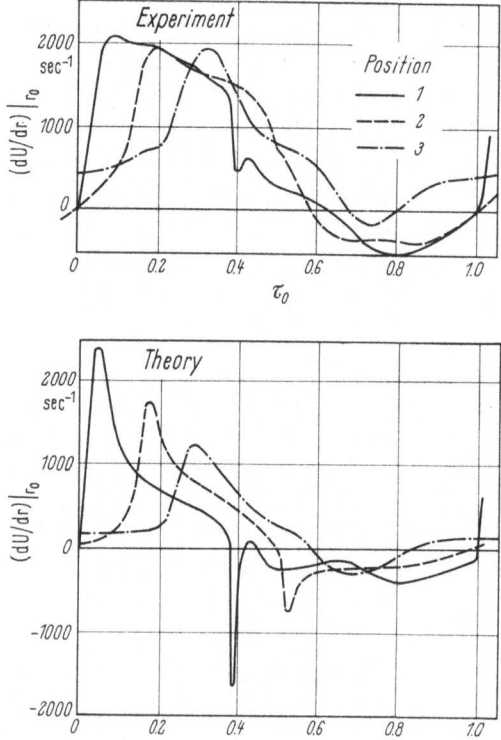

Fig. 12. Shear waves at the wall corresponding to the flow waves of Fig. 10.

with time, in contrast to the case of a rigid tube. The time-averaged shearing stress at the three indicated positions is approximately 28 dynes/cm². This could account for the observed mean-pressure drop of 12 mm Hg between positions 1 and 3, which are 100 cm apart. The shear waves predicted by the theory [5] are shown in the lower part of Fig. 12. The mean shear stresses for these waves were found to be 3 to 4 times lower than the measured values. Similar results were observed by Fry [8]. From the differences between the measured and predicted wave forms, one notes that the problem is unrelated to the proper selection of flow profiles for the steady flow components, since the steady component can modify the mean shear level but not the wave form. By the same token, one can not attribute the differences to the entrance effect, since the artery is long enough to eliminate such a problem. Furthermore, the observed differences in the flow profiles are definitely not due entirely to the tapered geometry of the aorta since the same characteristics were observed in a uniform artery.

The existing theory [5] predicted the existence of two types of waves: the radial (first kind) and the axial (second kind). From the above experiment one can find the relative magnitudes of these waves in the model system. We assume that there were two pairs of waves, one outgoing and one reflected, associated with each kind of wave component. From the data of the centerline velocity waves and the discharge waves taken at two of the test positions, and with the assumption of normal tethering characteristics for the artery, the relative magnitudes of these waves at test position 2 were found and are tabulated as follows:

	First-kind Waves		Second-kind Waves	
	Outgoing	Reflected	Outgoing	Reflected
1st harmonic	1.00	0.22	0.017	0.012
2nd harmonic	0.44	0.05	0.015	0.010

One notes that the components of the second-kind waves were negligible.

4.3 Pulsatile Flow Through Arterial Branch

An experimental study was conducted to map the velocity field in the vicinity of a model arterial branch representing that of a renal artery. The geometry of the model is shown in Fig. 13. It consists of a uniform artery of 0.42 cm I.D. connecting at an angle of 45° to a uniform main artery of 0.68 cm I.D. The arterial branch was installed between the tapered section of the heart pump and the uniform artery described in the preceding section. This arterial branch was terminated in a similar manner as the main branch, except a longer section of

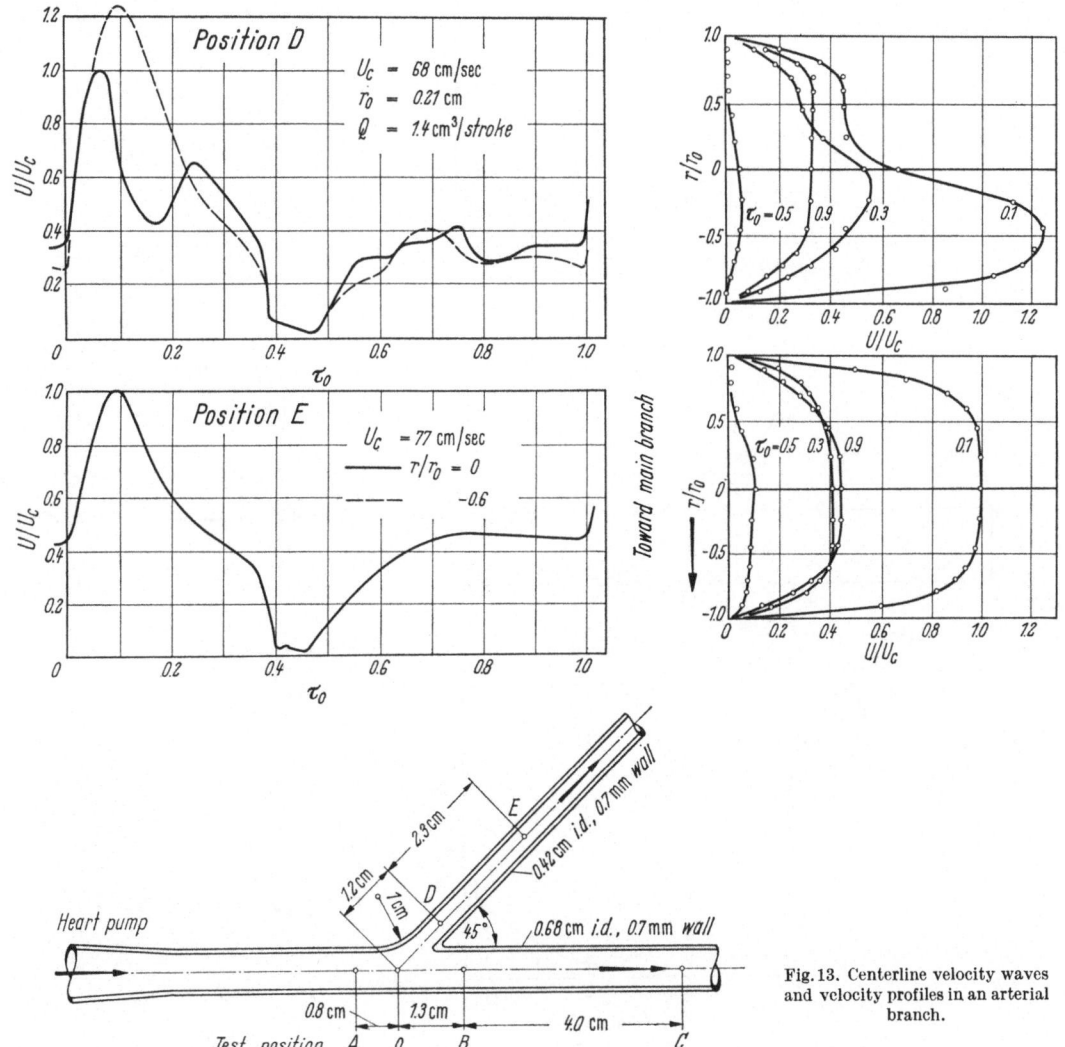

Fig. 13. Centerline velocity waves and velocity profiles in an arterial branch.

290   S. C. Ling, H. B. Atabek and J. J. Carmody

terminating line was used to simulate higher bed load. The flow rate in the side branch was set at about 1/3 of the total flow in the system. This division of flow was set on the high side so that some features of the onset of flow instability could be demonstrated. The velocity profiles of this experiment were taken in the plane of bifurcation and are shown in Figs. 13 and 14. At the inlet of the side branch, position D, one notes that there is a strong jet developed over the lip of bifurcation. The velocity wave at the center of the artery shows a dip at $\tau_0 = 0.15$. This was caused by the onset of flow instability occurring on the wall opposite to the bifurcation. The corresponding velocity wave at $r/r_0 = -0.6$ is shown to be stable. The peak shear stress occuring at the lip of bifurcation has a value of 140 dynes/cm². At position E, which is 2.3 cm downstream, the flow profiles are essentially symmetrical and show no significant influence of the entrance effect or of the unstable flow field.

The velocity profiles for the main branch are shown in Fig. 14. At position B, a steep velocity profile was developed at the lip of bifurcation. The measured peak shear stress at this point is about 260 dynes/cm². This is nearly equal to the yielding stress limit of the endothelial cells. An

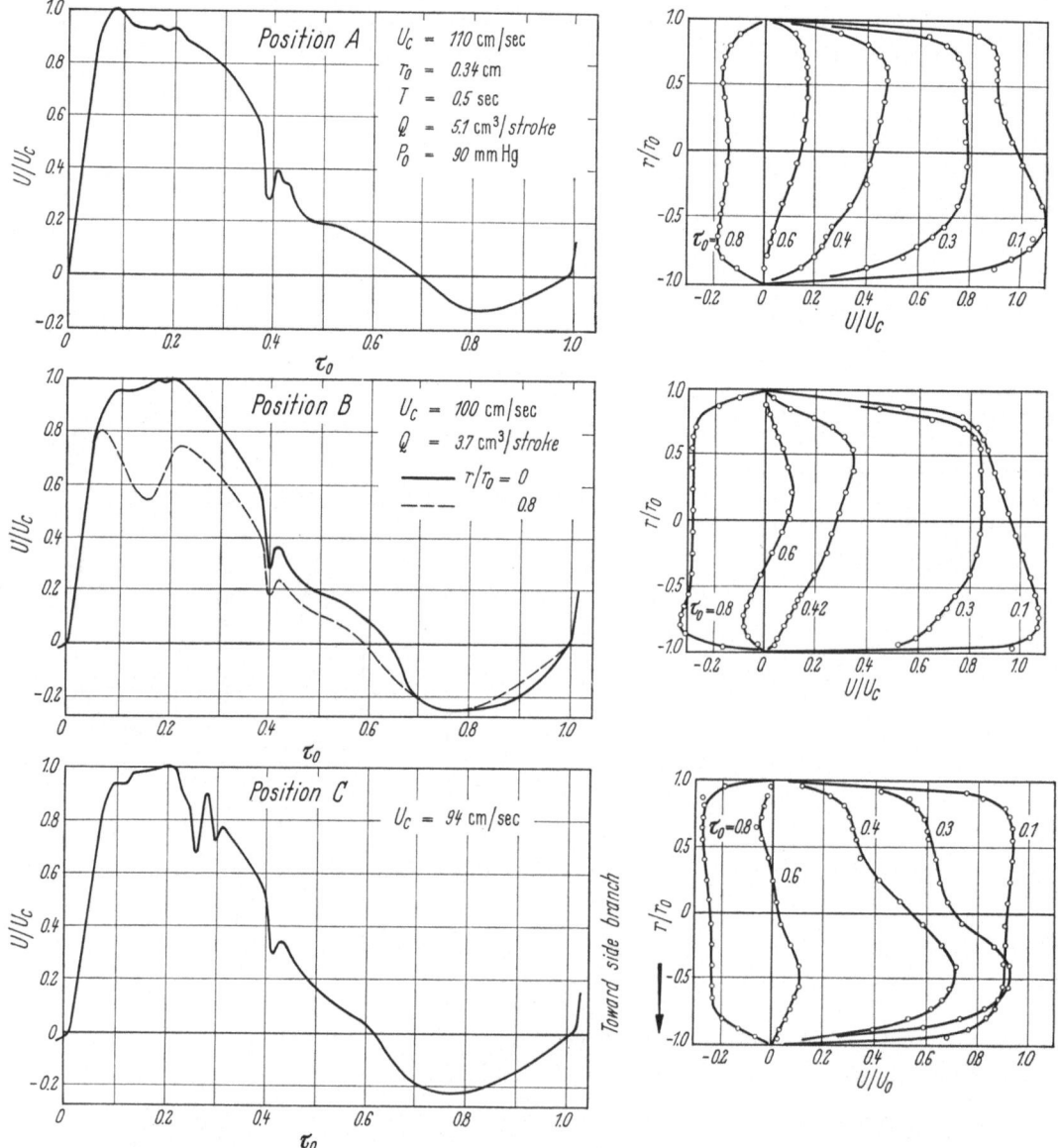

Fig. 14. Centerline velocity waves and velocity profiles in the main artery near an arterial branch.

incipient flow instability was also observed at $r/r_0 = 0.8$ and $\tau_0 = 0.15$ as indicated by the notch in the velocity wave at position B. This disturbance was convected downstream and is seen at position C as the eddies riding on the centerline velocity wave. The flow profile at $\tau_0 = 0.1$ for position C is clearly locally accelerated and is independent of the entrance effect. However, at $\tau_0 = 0.3$ and 0.4 the velocity profiles were strongly affected by the unstable field being convected downstream. At distances greater than 10 cm from position B the flows fields were found to be free of all distrubances created at the entrance.

5. Conclusion

From the above experimental studies of the pulsatile flows in both living animals and models one may conclude that:

1. Pulsatile flows through arteries are decidedly different from steady or steady oscillatory flows in rigid tubes. Theories based on the linearized long-wave approximation fail to provide correct flow profiles and wall shears.

2. Flow profiles were found to be developed locally and transiently during the passage of pressure-gradient waves. Both entrance effect and locally developed flow instability are confined to a short distance in the arteries.

3. Wall-shear stresses of the order of the yielding stress level of endothielal cells were found to exist on the lips of arterial bifurcations.

4. Quantitative flow data can be obtained from realistic model systems to supplement data which would be difficult to obtain in living systems. These data are useful for guiding the development of more advanced theories.

Acknowledgement

This work was supported by the Department of Health, Education, and Welfare, NIH, under Grant HE 12083-01.

References

1. FRY, D. L.: Acute endothelial changes associated with increased blood velocity gradients, Circ. Res. 22, 165 (1968).
2. SCHLICHTING, H.: Boundary layer theory, McGraw-Hill 1960, p. 229.
3. ATABEK, H. B.: Development of flow in the inlet length of a circular tube starting from Rest, J. Appl. Math. Phys. 13, 417 (1962).
4. WOMERSLEY, J. R.: An elastic tube theory of pulse transmission on oscillatory flow in mammalian arteries, Wright Air Dev. Ctr., TR 56—614, 1957.
5. ATABEK, H. B.: Wave propagation through a viscous fluid contained in a tethered, initially stressed, orthotropic elastic tube, Biophys. J., 8, 626 (1968).
6. BERGEL, D. H.: The static elastic properties of the arterial wall, J. Physiol. 156, 445, 458 (1961).
7. PATEL, D. J.: Relationship of radius to pressure along the aorta in living dogs, J. Appl. Physiol. 18, 1111 (1963).
8. FRY, D. L.: In vivo studies of pulsatile blood flow: the relationship of the pressure gradient to the blood velocity, Pulsatile Blood Flow, Ed. Attinger, McGraw-Hill 1964, p. 101.

An experimental study of combined longitudinal and torsional plastic waves in a thin-walled tube

J. Lipkin and **R. J. Clifton**

Brown University, Providence

1. Introduction

In recent years there has been considerable research done on the elastic-plastic behavior of materials at high rates of strain. With few exceptions the experimental investigations have dealt with relatively simple stress states such as occur in uni-axial stress, uni-axial strain, pure bending, or pure torsion. Experimental information from dynamic tests involving more complex stress states is needed in order to develop constitutive equations for describing material behavior under more general dynamic loading conditions.

There are essentially two types of tests that can be conducted to determine material behavior under dynamic loading: (a) nominally homogeneous deformation tests and (b) stress wave propagation tests. In tests of type (a) the specimens used are short enough that many waves traverse the specimen during the time of interest. Under these conditions it is assumed that stress and strain are nominally uniform throughout the specimen so that measurements of forces on the ends, and relative displacements of the ends, can be used to obtain directly relationships between stresses and strains. This type of test has been used by LINDHOLM [1, 2] for combined tension and torsion of thin-walled tubes. For equal biaxial tension, another type of nominally homogeneous deformation test is that of the expansion of thin spherical diaphragms by a sudden pressure pulse as studied by GERARD and PAPIRNO [3]. In tests of type (b) it is the nature of the stress wave propagation in the specimen that is of interest. In these tests, the time interval during which measurements are made is generally restricted to sufficiently short times that reflected waves do not have time to reach points where measurements are being made. Because it is not possible to measure stresses and strains at the same point, these tests do not give directly relationships between stresses and strains. Instead, it is necessary to assume a form for the constitutive relation describing the material behavior and then see whether or not any undetermined functions and/or parameters in these constitutive relations can be determined so that agreement between the experimental results and theoretical predictions is obtained. So far, although stress wave propagation tests have been used extensively for both one dimensional stress and one dimensional strain tests, no stress wave propagation tests appear to have been reported for biaxial stress states.

Preliminary results are presented in this paper for stress wave propagation tests involving combined tension and torsion of thin-walled tubes. The loading consists of a static torque, to a shear stress beyond the yield point of the material, followed by a longitudinal tensile impact. This loading was suggested by Clifton's analysis [4], which showed that for a rate independent, elastic-plastic, isotropic work-hardening material such a loading would produce two types of plastic waves, both involving combined longitudinal and torsional motion.

In the next section, a rate independent theory of combined longitudinal and torsional plastic wave propagation for quite general work-hardening materials is presented. This theory is helpful in understanding the type of information that can be obtained from wave propagation tests such as those reported here. The details of the experiment are discussed in Section 3. Experimental results are presented and compared with theoretical predictions in Section 4.

2. Theoretical Background

2.1 Strain-rate independent theory

In order to provide a theoretical basis for the design and interpretation of the experiment we consider an elementary theory of combined longitudinal and torsional plastic wave propagation in thin-walled tubes. This theory is analogous to that given in [4] for elastic-plastic,

isotropic work-hardening materials except that here we extend the theory to include more general work-hardening laws. For the analysis we neglect variations across the wall thickness and assume axial symmetry so that there are only two independent variables; namely, time, t, and distance along the tube axis, x. Furthermore, lateral inertia effects are neglected so that there are only two non-zero stress components, the axial stress σ and the torsional shear

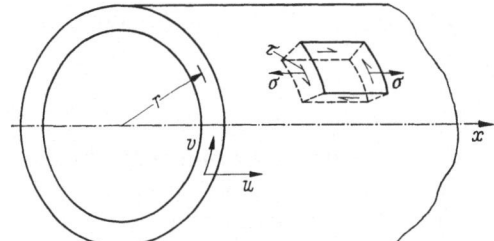

Fig. 1. Stresses and velocities in the tube.

stress τ. The motion is described by the longitudinal and circumferential particle velocities u, and v respectively (see Fig. 1). Small strains and rotations are assumed for which the elastic and plastic strain rates satisfy

$$\varepsilon_t^e + \varepsilon_t^p = u_x, \tag{1a}$$

$$\gamma_t^e + \gamma_t^p = v_x \tag{1b}$$

in which ε and γ are longitudinal and torsional strains respectively. In Eqs. (1) superscripts e and p denote respectively elastic and plastic parts of the total strain, and subscripts t and x denote partial differentiation with respect to the corresponding variable. The material is assumed to be a strain-rate independent elastic-plastic material for which the plastic behavior can be characterized by a loading surface

$$f(\sigma, \tau, \varepsilon^p, \gamma^p, \alpha_1, \ldots, \alpha_n) = 0 \tag{2}$$

and a flow rule

$$\varepsilon_t^p = G \frac{\partial f}{\partial \sigma} \left(\frac{\partial f}{\partial \sigma} \sigma_t + \frac{\partial f}{\partial \tau} \tau_t \right), \tag{3a}$$

$$\gamma_t^p = G \frac{\partial f}{\partial \tau} \left(\frac{\partial f}{\partial \sigma} \sigma_t + \frac{\partial f}{\partial \tau} \tau_t \right) \tag{3b}$$

where $\alpha_1, \ldots, \alpha_n$ are various parameters which are specified by a system of n ordinary differential equations relating the functions in the argument of f and their time derivatives. These equations are assumed to be homogeneous of degree one in the various time derivatives so that the material behavior is independent of the rate of strain. The scalar function G in Eqs. (3) is determined from the condition $f_t = 0$. Assuming that the material satisfies Drucker's stability postulate, the function G must be positive.

Two particularly simple cases of the material behavior characterized by Eqs. (2) and (3) are those of isotropic work-hardening and kinematic work-hardening. For isotropic work-hardening, the loading surface f and the scalar G, assuming the von Mises yield criterion, are

$$f \equiv \sigma^2/3 + \tau^2 - k^2 = 0, \tag{4a}$$

$$G = \frac{1}{4k^3 \left(\dfrac{dk}{dW_p} \right)}. \tag{4b}$$

In Eqs. (4), k is a single valued function of the plastic work W_p; the function $k = k(W_p)$ can be determined from a uni-axial stress-strain curve. For kinematic work-hardening, the loading surface f and the scalar G are

$$f \equiv \frac{(\sigma - \alpha_1)^2}{3} + (\tau - \alpha_2)^2 - k_0^2 = 0, \tag{5a}$$

$$G = \frac{1}{4h\left\{\left(\frac{\sigma - \alpha_1}{3}\right)^2 + (\tau - \alpha_2)^2\right\}}, \tag{5b}$$

where k_0 is a constant equal to the yield stress in pure shear. The parameters α_1 and α_2 are the coordinates in σ-τ space of the center of the ellipse which, following Prager's assumption, translates in the direction of the plastic strain-rate vector. Thus,

$$(\alpha_1)_t = h\varepsilon_t^p, \tag{6a}$$

$$(\alpha_2)_t = h\gamma_t^p. \tag{6b}$$

Work-hardening is accounted for by taking h in Eqs. (6) to be a function of the equivalent stress $(\alpha_1^2/3 + \alpha_2^2)^{1/2}$. This function can be determined from a uni-axial stress-strain curve.

The elastic strain rates are given by

$$\varepsilon_t^e = \frac{1}{E}\sigma_t, \tag{7a}$$

$$\gamma_t^e = \frac{1}{\mu}\tau_t \tag{7b}$$

where E is Young's modulus and μ is the shear modulus. The governing system of equations consists of Eqs. (1), (2), (3), (7) and the equations of motion

$$\sigma_x = \varrho u_t, \tag{8a}$$

$$\tau_x = \varrho v_t. \tag{8b}$$

Eliminating strains from Eqs. (1), (3), and (7), the governing system of partial differential equations can be written as

$$L(w) \equiv Aw_t + Bw_x = 0 \tag{9}$$

in which

$$w = \begin{bmatrix} u \\ \sigma \\ v \\ \tau \end{bmatrix}, A = \begin{bmatrix} \varrho & 0 & 0 & 0 \\ 0 & M & 0 & N \\ 0 & 0 & \varrho & 0 \\ 0 & N & 0 & P \end{bmatrix}, B = \begin{bmatrix} 0 & -1 & 0 & 0 \\ -1 & 0 & 0 & 0 \\ 0 & 0 & 0 & -1 \\ 0 & 0 & -1 & 0 \end{bmatrix}$$

where ϱ is the mass of the tube per unit volume and

$$M = 1/E + G\left(\frac{\partial f}{\partial \sigma}\right)^2,$$

$$N = G\frac{\partial f}{\partial \sigma}\frac{\partial f}{\partial \tau},$$

$$P = \frac{1}{\mu} + G\left(\frac{\partial f}{\partial \tau}\right)^2.$$

The preceding equations apply at any point where the loading condition $\frac{\partial f}{\partial \sigma}\sigma_t + \frac{\partial f}{\partial \tau}\tau_t \geq 0$ is satisfied. For unloading, the material behavior is assumed to be elastic; the corresponding equations are obtained by setting $G = 0$ in Eqs. (9).

2.2 Characteristic properties of the equations

Equations (9) constitute a quasi-linear symmetric hyperbolic system of first order partial differential equations. The characteristic speeds c, or equivalently the speeds of propagation of acceleration waves across which stresses and velocities are continuous but their derivatives are discontinuous, are the roots of the characteristic equation

$$\det(-Ac + B) = 0. \tag{10}$$

Equation (10) is a quadratic equation for ϱc^2 for which the roots are given by

$$\varrho c^2 = \frac{M + P \pm [(M - P)^2 + 4N^2]^{1/2}}{2[MP - N^2]}. \tag{11}$$

The wave speeds of Eq. (11) corresponding to the '+' and '−' signs will be denoted by c_f (fast waves) and c_s (slow waves) respectively. These wave speeds can be shown to satisfy the inequalities

$$c_2 \leq c_f \leq c_0, \tag{12a}$$

$$c_s \leq c_2 \tag{12b}$$

where $c_2 = (\mu/\varrho)^{1/2}$ is the elastic shear wave speed and $c_0 = (E/\varrho)^{1/2}$ is the elastic bar velocity.

The jump in the time derivative of w across an acceleration wave with velocity c is proportional to the null vector l for the characteristic matrix $(-Ac + B)$. The null vectors l_f^+ and l_s^+ corresponding to $(-Ac_s + B) l_s^+ = 0$ and $(-Ac_f + B) l_f^+ = 0$ respectively can be written as

$$l_f^+ = \begin{bmatrix} Nc_f \\ -N\varrho c_f^2 \\ \frac{1}{\varrho c_f} - Mc_f \\ M\varrho c_f^2 - 1 \end{bmatrix}, \quad l_s^+ = \begin{bmatrix} \frac{1}{\varrho c_s} - Pc_s \\ P\varrho c_s^2 - 1 \\ Nc_s \\ -N\varrho c_s^2 \end{bmatrix}. \tag{13}$$

The null vectors l_f^- and l_s^- corresponding to characteristics $\frac{dx}{dt} = -c_f$ and $\frac{dx}{dt} = -c_s$ respectively are obtained from Eqs. (13) by substituting $-c_f$ for c_f and $-c_s$ for c_s. Because of the symmetry of the matrices A and B, the vectors l_f^+, l_f^-, l_s^+, and l_s^- are mutually orthogonal with respect to the positive definite matrix A.

Since all components of the null vectors, and consequently the jump vectors, are generally non-zero it follows that acceleration waves generally involve jumps in the derivatives of both stresses and both velocities. Thus, acceleration waves show a coupling between longitudinal and torsional motion which arises because of the coupling between σ_t and γ_t^p, and between τ_t and ε_t^p in Eqs. (3). Another observation that can be made using Eqs. (13) is that for either fast or slow acceleration waves the direction of the two dimensional vector corresponding to the jump in the time derivatives of the stress vector is the same as that of the jump in the time derivative of the velocity vector. Furthermore, the directions of the jumps in the time derivative of the stress (or velocity) vector for fast and slow waves are mutually orthogonal. The latter result has been discussed by HILL [6] for a three-dimensional elastic-plastic continuum.

Through any point (x, t) there are four characteristic curves corresponding to the four slopes $\frac{dx}{dt} = \pm c_f$, and $\frac{dx}{dt} = \pm c_s$. Along these characteristic curves the partial differential equations (9) reduce to ordinary differential equations. The ordinary differential equation along a characteristic curve with null vector l is given by $l \cdot L(w) = 0$, where the dot denotes the inner product [5]. In incremental form these equations can be written as

$$(1/\varrho c_f^2 - M)(\varrho c_f \, dv - d\tau) + N(\varrho c_f \, du - d\sigma) = 0; \quad \frac{dx}{dt} = + c_f, \tag{14a}$$

$$(1/\varrho c_s^2 - P)(\varrho c_s \, du - d\sigma) + N(\varrho c_s \, dv - d\tau) = 0; \quad \frac{dx}{dt} = + c_s. \tag{14b}$$

The incremental relations along characteristics $\frac{dx}{dt} = -c_f$ and $\frac{dx}{dt} = -c_s$ are obtained by replacing c_f and c_s in Eqs. (14a) and (14b) by $-c_f$ and $-c_s$ respectively. Equations (14), with the coefficients that correspond to the particular case of isotropic work-hardening, are used for the numerical solutions that were obtained to compare theory with experiment.

2.3 Simple waves

Solutions of Eqs. (9) which are constant along straight lines in the x-t plane are called simple wave solutions. To study such solutions we consider a one-parameter family of straight lines

$$\beta = t - \frac{x}{c} , \tag{15}$$

where β is the parameter and c is the velocity $\frac{dx}{dt}$ for the line β. Assuming the vector w to be a function $w = w(\beta)$ we find from Eqs. (9) and (15) that $w' \equiv \frac{dw}{d\beta}$ must satisfy

$$(-Ac + B) w' = 0. \tag{16}$$

Comparing Eqs. (16) with Eqs. (10) we see that the straight lines (15) along which a simple wave solution is constant are characteristics of Eqs. (9). A simple wave solution for which the wave speeds of the straight line characteristics correspond to $c_f (c_s)$ will be called a fast (slow) simple wave. The vector $w'(\beta)$ associated with a characteristic c is proportional to the jump vector $[w_t]$ for an acceleration wave having the same velocity. (The coefficients M, N, and P in Eqs. (9) are assumed to have the same value on the characteristic as on the acceleration wave having the same velocity.) As a result, the stress trajectory for a fast (slow) simple wave has the same direction in σ-τ space as does the jump in the time derivative of the stress vector for a fast (slow) acceleration wave. From the discussion of acceleration waves it follows that stress (or velocity) trajectories for fast and slow simple waves are mutually orthogonal.

Stress trajectories corresponding to simple wave solutions are particularly helpful in understanding the nature of slow and fast simple waves. The slope of the stress trajectory for slow simple waves $\left(\frac{d\tau}{d\sigma}\right)_s$ can be expressed in several ways [4]. The form given in Eq. (17) is particularly convenient for establishing the relation between the normal to the current yield surface and the direction of the stress trajectory for slow simple waves.

$$\left(\frac{d\tau}{d\sigma}\right)_s = \frac{(\varrho c_s^2/E - 1)\left(\frac{\partial f}{\partial \tau}\right)}{(\varrho c_s^2/\mu - 1)\left(\frac{\partial f}{\partial \sigma}\right)} . \tag{17}$$

In σ-τ space let Ω and χ denote the counterclockwise angles from the σ-axis to the yield surface normal and to the direction of the stress trajectory for a slow simple wave respectively. From Eq. (17) these angles satisfy

$$(\cot \Omega + \tan \chi) \tan (\chi - \Omega) = \eta \tag{18}$$

where η is a non-negative quantity given by

$$\eta = \frac{(1 + 2\nu)}{2(1 + \nu)(c_2^2/c_s^2 - 1)}$$

and ν is Poisson's ratio.

In order for loading to occur for the slow simple wave solution, the angles Ω and χ cannot differ by more than $\pi/2$. If we consider the yield surface normal to lie in the first quadrant of the σ-τ plane, (i.e. $0 \leq \Omega \leq \pi/2$) then from Eq. (18) the angle χ must satisfy $\Omega \leq \chi \leq \pi/2$. The maximum value of the angle $(\chi - \Omega)$, for a given value of η, occurs for Ω satisfying

$$\cot \Omega = (\eta + 1)^{1/2} \tag{19}$$

and is given by

$$(\chi - \Omega)_{\max} = \tan^{-1} \frac{\eta}{2(\eta + 1)^{1/2}} . \tag{20}$$

From Eqs. (19) and (20) we find, for example, that if $c_2/c_s = 5$ and $\nu = 0.3$, then $\eta = .026$ and $(\chi - \Omega)_{\max} = 0.72$ degrees. These calculations indicate that the stress trajectories for fast and slow simple waves are respectively, essentially tangential and normal to the current yield surface. This result suggests the possibility of determing the shape and growth of subsequent yield surfaces, for high rates of strain, by studying stress trajectories for simple waves.

In stress wave propagation experiments, measurements are usually made of either strains or particle velocities rather than stresses. For simple waves, however, stress, particle velocity, and strain increments satisfy the following relations obtained from Eqs. (1) and (8).

$$d\sigma = -\varrho c\, du, \quad d\tau = -\varrho c\, dv, \tag{21a}$$

$$du = -c\, d\varepsilon, \quad dv = -c\, d\gamma. \tag{21b}$$

Eliminating du and dv in Eqs. (21) we obtain the following relationship between stress increments and strain increments for simple waves

$$d\sigma = \varrho c^2\, d\varepsilon, \quad d\tau = \varrho c^2\, d\gamma. \tag{22}$$

Thus, if the velocity of propagation c of each level of particle velocity or strain is obtained from experiments, then the two stress-time profiles can be obtained from either measured velocity-time profiles by use of Eqs. (21a) or measured strain-time profiles by use of Eqs. (22). Another consequence of Eq. (21) is that for simple waves, the slopes of the trajectories corresponding to a given state are the same in stress, velocity, and strain spaces. That is, for simple waves

$$\frac{d\tau}{d\sigma} = \frac{dv}{du} = \frac{d\gamma}{d\varepsilon}. \tag{23}$$

In order to illustrate the relationships between stresses, particle velocities, and strains for simple waves, solutions have been obtained for constant velocity longitudinal impact (zero angular velocity) of a pre-torqued tube assuming both an isotropic work-hardening material and a kinematic work-hardening material. For these calculations the following emperical uni-axial stress-strain relation was assumed

$$\varepsilon = \sigma/E + 0.437 \times 10^{-8}(\sigma - \sigma_y)^{1.767}, \quad \varepsilon \geq \sigma_y/E \tag{24}$$

where the yield stress σ_y and Young's modulus E were taken to be 1500 psi and 10×10^6 psi respectively. For $\varepsilon \leq 2\%$, Eq. (24) is a good representation of the quasi-static stress-strain curves obtained for coupon samples cut from the annealed 1100 aluminum tubes described in the next section.

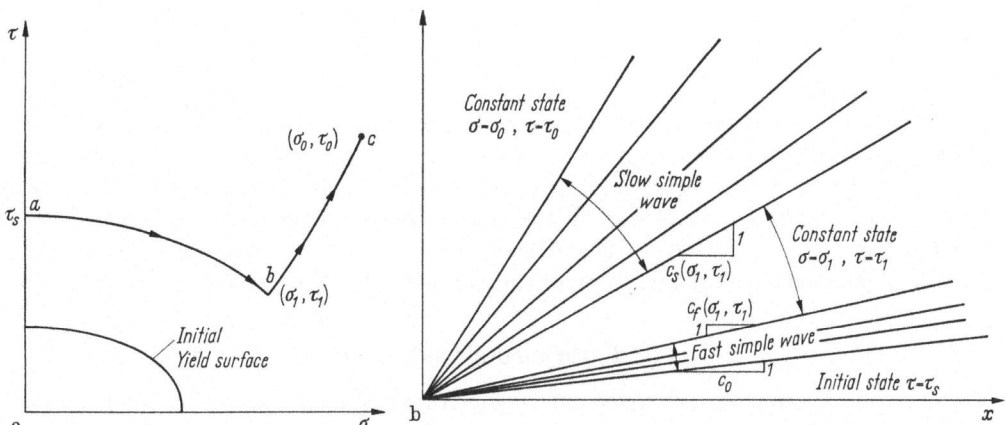

Fig. 2. Qualitative features of the solution for step-loading of a pre-torqued tube. a) stress path; b) solution in the t-x plane.

The qualitative features of the solutions are shown in Fig. 2. The solutions consist of an initial constant state region, a fast simple wave for which the stress trajectory ab essentially follows the current yield surface, a constant state region corresponding to the state at b, a slow simple wave for which the stress trajectory bc is essentially normal to the current yield surface, and a constant state region in which the particle velocities are those imposed at the end $x = 0$. The stress, velocity and strain trajectories for both the isotropic work-hardening case and the kinematic work-hardening case are shown in Fig. 3. In Fig. 3a the translation of the initial yield surface is shown for the kinematic work-hardening model. In Fig. 3c $\Delta\gamma$ is the change in shear strain from the value produced by the static torque. The values of the initial shear stress, $\tau_s = 3700$ psi, and the longitudinal impact velocity, $u_0 = 600$ in./sec., are comparable to the values used for the experiment described in detail in Section 4.

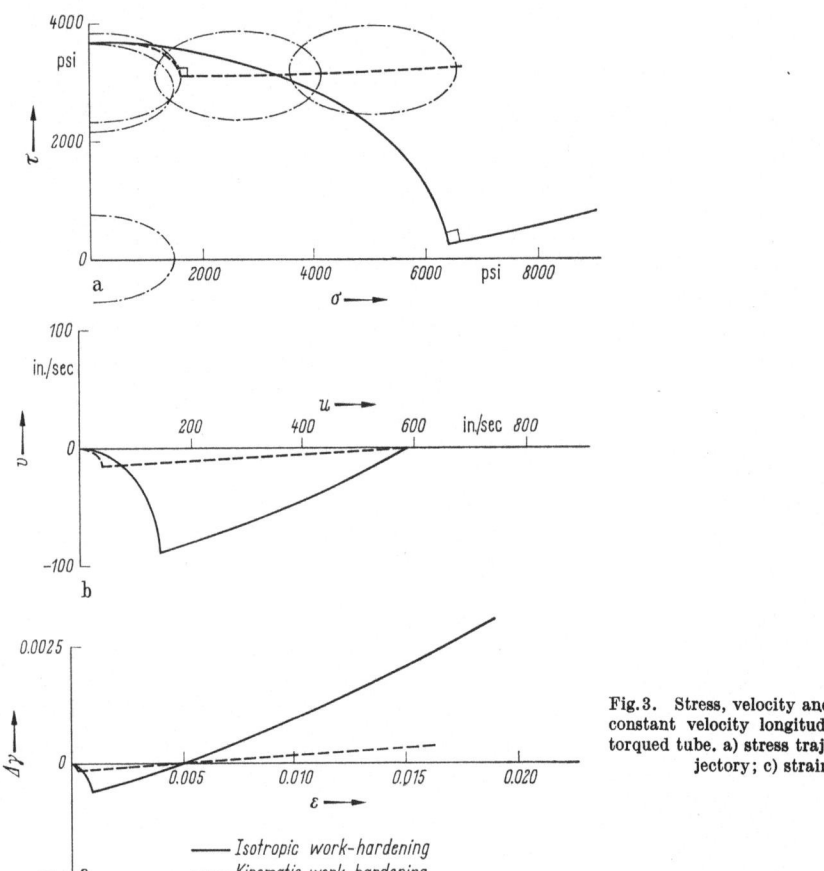

Fig. 3. Stress, velocity and strain trajectories for constant velocity longitudinal impact of a pre-torqued tube. a) stress trajectory; b) velocity trajectory; c) strain trajectory.

——— Isotropic work-hardening
– – – Kinematic work-hardening

From Fig. 3 it is clear that under these loading conditions the response of an isotropic work-hardening material differs significantly from that of a kinematic work-hardening material. This suggests that the stress wave propagation is sufficiently sensitive to the form of the loading surface that an experiment employing analogous loading conditions can yield information on the loading function f in Eq. (2). The strains corresponding to the fast simple wave are small so that their measurement by means of strain gages should be reliable.

3. Experimental Technique

The general experimental arrangement is shown in Fig. 4. The static torque is applied by thin cables *5* which are attached to the hardened steel impact-torsion plate *3*. This plate is bolted to a flange on the end of the specimen and rests on the central guide rail *13*. The cables are attached to the loading beam *7* which is forced down using a hand pump and a hydraulic jack *6*. Bending during application of the static torque does not occur because the cable loading arrangement is nominally symmetrical. In addition, if any unbalance between the forces in the loading cables does occur, it is resisted by the impact-torsion plate bearing on the central guide rail. After the static torque is applied, a Hyge shock tester *17* mounted horizontally is used to accelerate a hitter mass *15* to the velocity desired for the experiment. The hitter mass slides down the guide rail and through the specimen, striking the impact-torsion plate and accelerating it to a nominally constant velocity while extending the end of the specimen. Thus the impact-torsion plate, the specimen, and the torsion loading cables move together as a result of the impact; the cables are thin enough that they offer negligible resistance to the longitudinal motion of the end of the specimen. The impact-torsion plate, together with the flange at the impact end of the specimen, have a large enough polar moment of inertia so that the boundary condition of zero angular velocity is satisfied with good accuracy during the early times that are of interest in the experiment.

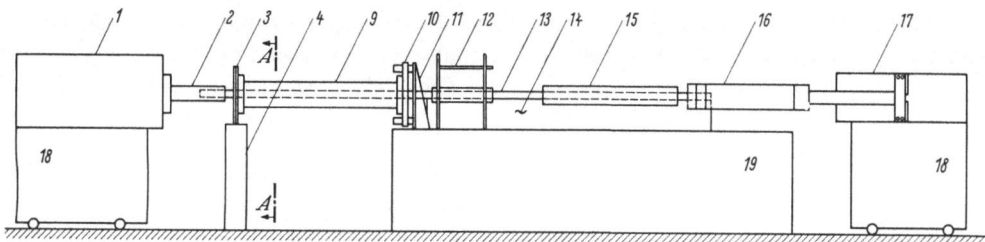

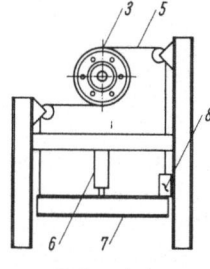

Section A–A
Detail of loading frame

Fig. 4. General arrangement of test set-up.
1 Anvil mass; *2* Anvil extension; *3* Impact-torsion plate; *4* Torsion loading frame; *5* Loading cable; *6* Hydraulic Jack, Blackhawk Model RC 251; *7* Loading beam; *8* Load cell, BLH model U-1; *9* Thin-walled tube test specimen; *10* Rear mounting plate; *11* Support bracket; *12* Catcher; *13* Central guide rail; *14* Central guide rail support; *15* Hitter mass; *16* Hyge ram extension; *17* Hyge shock tester, CVC modified model HY-6134; *18* Supporting carriage; *19* Supporting table.

The entire apparatus shown in Fig. 4 is aligned with the axis of the Hyge. Thus the extension of the Hyge gun *16* is concentric with, and slides over, the fixed central guide rail. The guide rail is supported as a mono-rail until it enters the specimen. It is unsupported for the length of the specimen, and is "built-in" with a tight, sliding fit in the anvil extension. The sliding fit is needed because the anvil mass *1* and supporting carriage *18* move horizontally to absorb the impact.

The end of the specimen opposite from the impact-torsion plate is attached to the rear mounting plate *10*. This plate is restrained from rotation about the tube axis, but can move axially when the entire specimen moves as a result of the impact. In this way the large forces needed to deform the tube plastically in the longitudinal direction are not transmitted to the supporting system *11*.

The catcher *12* is a spring-loaded device which is tripped as the hitter enters the specimen. It then closes so that the hitter mass rebounding from the impact at the anvil end of the specimen cannot get through and damage the extended Hyge ram.

The test specimen is cut from extruded tube stock of commercially pure aluminium (1100-0). The nominal dimensions of the specimen are: outside diameter = 3.320 inches, wall thickness = .200 inches, length $\simeq$ 36 inches. Using these dimensions it is estimated that the maximum shear stress that can be obtained from the static torque without causing inelastic buckling is approximately 15,000 psi.

The use of extruded tube stock for specimens means that a great deal of machining can be avoided. However, an important detail of such a specimen is the design of a suitable method for gripping the ends so that both static shear and tensile impact can be applied with limited local deformation. A workable design was achieved by welding stiff flanges to both ends of the specimen. The material used for these flanges is a 5456-H321 aluminium-magnesium alloy, readily weldable to the 1100-0 specimen with a tungsten-inert-gas process. A filler metal of

5356 alloy is added to the joint for increased strength. Fig. 5 is a drawing of the finished specimen which also shows the welding detail.

Since the experimental technique was developed, a series of 8 tests have been conducted, 7 with the static shear stress $\tau_s \simeq 1800$ psi and one with $\tau_s = 3670$ psi. The maximum longitudinal velocity of the impact end has been between 600 and 700 in./sec. for each of these tests. The specimens for these tests are prepared by first welding the flanges to the ends and then annealing before the final machining operations and the strain gage application. In the annealing process, the temperature is raised to 1000°F in about eight hours and held at this level for two hours, after which the specimen is oven cooled. The yield strength of the 5456-H321 alloy used for the specimen flanges is reduced by this annealing, however it remains significantly higher than that of the specimen.

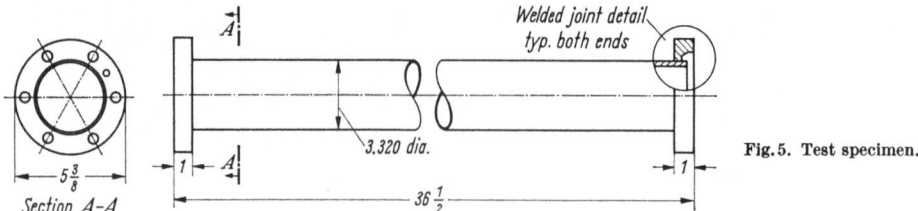

Fig. 5. Test specimen.

The final machining operations include squaring and facing the ends of the specimen which mate with the impact-torsion plate and the rear mounting plate. In the 7 tests where τ_s was less than 2000 psi, the strain gages were applied immediately after this machining operation. For the test with $\tau_s = 3670$ psi it was necessary to pre-twist the specimen before application of the strain gages. This pre-straining made it possible to reach a relatively large shear stress in the specimen without encountering shear strains that are too large for the strain gages being used in the experiment.

Three element, 45° etched foil strain gage rosettes are used for the experiments. During the initial quasi-static torque loading, the load-strain response of the specimen is monitored by using the shear gages in one of the strain gage rosettes and a load cell connected in series with one of the torsion loading cables.

Three strain gage rosettes spaced at approximately 1.5 inch intervals are used for each test. The first rosette is located approximately 1.5 inches from the bead weld around the outside of the specimen flange at the impact end. The gages are located near the impact end in order to obtain the complete strain-time profiles during the propagation of the plastic loading wave of combined stress. For the time scale of interest in the experiment, the length of the hitter mass, rather than the length of the specimen, determines the duration of loading at the impact end. After the tension wave in the hitter (reflected from its free end) reaches the impact-torsion plate, the velocity of the impact end is somewhat reduced, initiating an unloading wave that propagates into the specimen. By using a 30 inch long hitter, unloading at the impact end is delayed until 300 microseconds after impact.

Three oscilloscopes equipped with Polaroid cameras are used to record the various strain gage signals associated with the wave propagation phenomenon in the specimen. The longitudinal velocity-time profile of the impact end is obtained from the strain-time profile of an elastic wave propagating in a long, thin titanium rod attached to the impact end. All of the oscilloscope traces are triggered simultaneously before the beginning of the strain gage signals by using the output of a semi-conductor gage attached to a short titanium rod on the back of the impact-torsion plate.

Electrically isolated constant current strain gage circuits are used for the dynamic strain measurements. Each circuit-gage combination is calibrated just prior to a test by switching a large resistor in parallel with the gage, and using the manufacturer's gage factor to convert the resulting change in voltage observed on the oscilloscope to an equivalent strain.

4. Experimental results and comparison with theory

Oscilloscope traces corresponding to the velocity-time profile at the impact end and strain-time profiles at distances of 3 and 4.5 inches from the impact end are shown in Fig. 6 for the test in which $\tau_s = 3670$ psi. The strain-time profile in the titanium "velocity bar" is shown as the bottom trace of Fig. 6a. The time scale is 50 μsec./division; the maximum strain corresponds to a velocity of 630 inches/sec. The rise time for the velocity of the impact end of approximately 150 μsec. is a consequence of the relatively large mass at the end of the specimen. The oscillations at early times are presumably due to wave reflections in the specimen flange and the

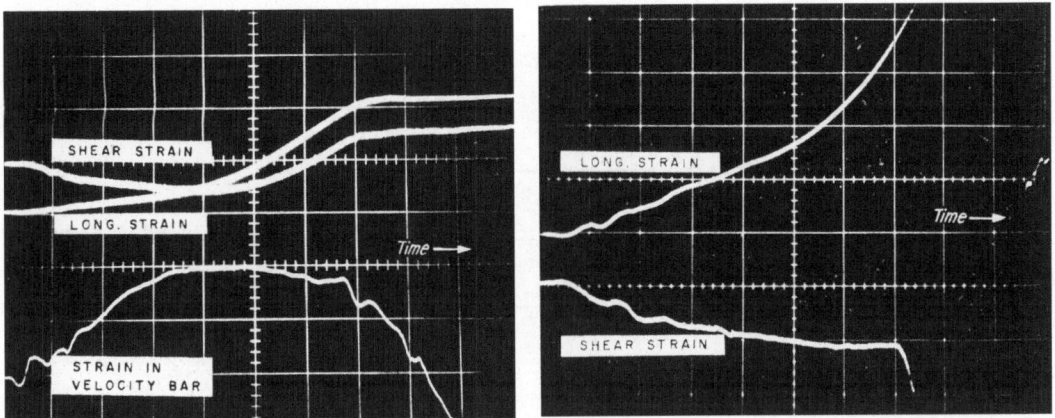

Fig. 6. Oscilloscope traces ($\tau_s = 3670$ psi, $u_0 = 630$ inches/sec.). a) Longitudinal and shear strains at $x = 4.5$ inches and strain in velocity bar, sweep rate $= 50$ μsec./cm; b) Longitudinal and shear strains at $x = 3.0$ inches, sweep rate $= 25$ μsec./cm.

impact-torsion plate. From 150 to 280 μsec. after impact the velocity remains nearly constant, and then decreases as unloading begins from the free end of the hitter mass. The sharp drop in the trace at approximately 325 μsec. after impact does not correspond to a sudden velocity decrease at the impact end; instead, it corresponds to the arrival of the front of the elastic wave in the titanium bar after being reflected from its free end.

The strain-time profiles for longitudinal and torsional strains at $x = 4.5$ inches are also shown in Fig. 6a. For a given level of strain, the displacement of the trace corresponding to the shear strain in Fig. 6 is approximately 4.5 times that of the trace for the tensile strain. The essential features of these traces are (a) tensile strain is monotonically increasing and reaches a plateau value of approximately 1.55%, (b) shear strain decreases (i.e. shear strain is reduced from the level produced by the initial torque), remains relatively uniform for awhile, and then increases and becomes nearly constant at approximately the same time that the tensile strain becomes constant. Strain-time profiles at $x = 3.0$ inches are shown in Fig. 6b. For this record an expanded time scale and greater sensitivity were used in order to clarify the behavior near the wave front. The interruption of the shear strain trace resulted because of the failure of a strain gage lead. The qualitative features of the strain-time profiles in Fig. 6 are also typical of those obtained in the tests with $\tau_s \simeq 1800$ psi.

Using the strain records from successive strain gage stations it can be seen that the front of the disturbance propagates at a velocity approximately equal to the elastic bar velocity. It is also clear that changes in both longitudinal and torsional strains occur across this front; the change in the longitudinal strain corresponds to tension and the change in the torsional strain corresponds to a reduction of the pre-strain.

For the velocity boundary condition imposed in the experiment, the solution corresponding to the theory described in section 2 does not consist of simple wave regions and constant state regions as shown in Fig. 2. It is therefore necessary to use difference solutions of the governing equations to compare theory and experiment. Such solutions have been obtained based on the rate independent theory for the case of isotropic work-hardening. The difference equations used are obtained by integration along characteristics as done by COURANT, ISAACSON, and REES

38 B*

[7]. A detailed discussion of the difference method is given in a report by VITIELLO and CLIFTON [8]. The prescribed longitudinal velocity of the impact end is taken to be a non-decreasing function of time obtained by passing a smooth curve through the oscilloscope trace for the strain in the velocity bar. The tangential velocity of the impact end is taken to be zero. For these

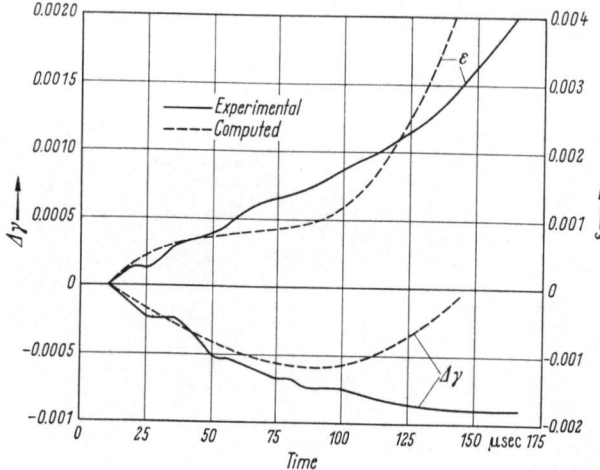

Fig. 7. Comparison of computed and measured strain-time profiles at $x = 3.0$ inches. ($\tau_s = 3670$ psi, $u_0 = 630$ in./sec.).

boundary conditions unloading does not occur so that the problem of locating elastic-plastic boundaries does not arise. The uni-axial stress-strain curve for the material is taken to be that of Eqs. (24).

Computed and measured strain-time profiles for the test corresponding to the oscilloscope traces of Fig. 6 are shown in Figs. 7 and 8. For the computations the space increment Δx was

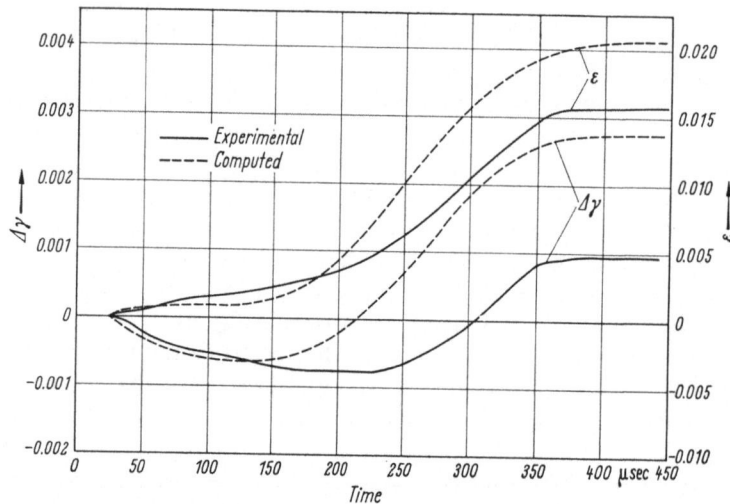

Fig. 8. Comparison of computed and measured strain-time profiles at $x = 4.5$ inches. ($\tau_s = 3670$ psi, $u_0 = 630$ in./sec.).

taken to be 0.25 inches and the time increment was $\Delta x/c_0$. By comparing numerical calculations for different mesh spacings the errors in the computed values of strain are estimated to be less than 2%. Reflections from the end $x = L$ did not influence either the computed or measured strain-time profiles during the time shown in Figs. 7 and 8. There is an uncertainty of a few microseconds with regard to correlating the zero time for computed and measured strain-time profiles. This matter was handled by making the arrival of the first disturbance at $x = 3.0$ inches to be the same for the experiment and the numerical solution.

The qualitative features of the experimental and computed strain-time profiles are in good agreement, however, significant quantitative differences do occur. The most notable difference is in the shear strain-time profile at $x = 4.5$ inches where the experimental curve, after decreasing initially, does not begin increasing as soon as the computed curve and finally approaches a plateau strain that is only about one-third the computed plateau value. The computed value of the longitudinal strain plateau also exceeds that recorded from the experiment. Such a discrepancy in the longitudinal strain plateau value could be explained in terms of a dynamic stress-strain curve that is above the static curve used in the computations, however computations using the same static stress-strain curve showed good agreement in the plateau values of longitudinal strain for an experiment with $\tau_s = 1830$ psi.

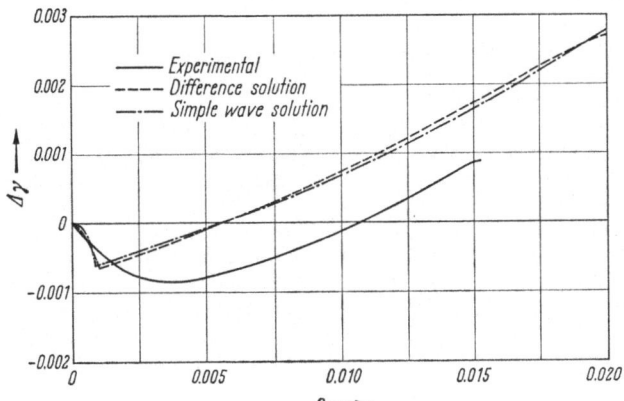

Fig. 9. Comparison of computed and experimental strain trajectories at $x = 4.5$ inches. ($\tau_s = 3670$ psi, $u_0 = 630$ in./sec.).

Computed and measured strain trajectories at $x = 4.5$ inches are shown in Fig. 9. The strain trajectory for the numerical solution is observed to agree remarkably well with that predicted by a simple wave solution corresponding to constant velocity impact at a velocity equal to the maximum velocity observed in the experiment. This is an indication that strain trajectories are not very sensitive to the details of the imposed velocity-time profile. The experimental strain trajectory has the same general features as the computed ones except in the transition region where the slope changes sign. Interpreted in terms of a change in shape of the yield surface the experimental strain trajectory corresponds to the yield surface being pushed farther out along the positive σ-axis as the stress trajectory comes around the yield surface than is predicted by the assumption of isotropic work-hardening.

5. Concluding remarks

Experimental results on the propagation of plastic waves of combined stress in a thin-walled tube have been presented. For the initial and boundary conditions imposed in the experiment two types of waves result, both of which involve a coupling of torsional and longitudinal motion. The qualitative features of the experimental strain-time profiles are shown to be predictable by means of strain-rate independent theory for an elastic-plastic isotropic work-hardening material. There are, however, systematic discrepancies between the predicted and the observed response. These discrepancies are believed to be due primarily to differences between the actual and the assumed material behavior. More work needs to be done before an appropriate representation of the actual material behavior can be deduced from the experiments.

At this point the authors are impressed more by the degree of agreement obtained by use of such an elementary theory than by the observed discrepancies. The results would seem to indicate that the one-dimensional idealization of the wave propagation phenomena is satisfactory, and that the representation of the material behavior by an elastic-plastic work-hardening model is substantially correct.

38 A*

Acknowledgment

This work was supported in its early stages by the Department of Defense, Advanced Research Projects Agency through Contract SD-86 with Brown University and subsequently by the National Science Foundation under Contract NSF GK 1013, and the Ballistic Research Laboratories, Aberdeen Proving Ground, Maryland, under Contract DAADOS-68 C-0083.

Support of the first author as a NASA trainee is also gratefully acknowledged.

References

1. Lindholm, U. S., and L. M. Yeakley: A dynamic biaxial testing machine, Experimental Mechanics, Vol. 7, 1967, pp. 1—7.

2. Lindholm, U. S.: Some experiments in dynamic plasticity under combined stress, Mechanical Behavior of Materials Under Dynamic Loads, New York: Springer 1968.

3. Gerard, G., and R. Papirno: Dynamic biaxial stress-strain characteristics of aluminium and mild steel, Trans. Amer. Soc. Metals, Vol. 49, 132—148 (1957).

4. Clifton, R. J.: An analysis of combined longitudinal and torsional plastic waves in a thin-walled tube, Proc. of the Fifth U.S. Natl. Cong. of Appl. Mech., Univ. of Minnesota, 1966, pp. 465—480.

5. Courant, R., and D. Hilbert: Methods of mathematical physics, Vol. II: Partial Differential Equations, New York: Interscience Publishers 1962.

6. Hill, R.: Acceleration waves in solids, J. Mech. Phys. Sol. 10, 1—16 (1962).

7. Courant, R., E. Isaacson and M. Rees: On the solution of nonlinear hyperbolic differential equations by finite differences, Comm. Pure Appl. Math. 5, 243—255 (1952).

8. Vitiello, D. L., and R. J. Clifton: The numerical solution of a problem in the propagation of plastic waves of combined stress, Brown University, Division of Applied Mathematics, Report No. 4, under Contract DA-31-124-ARO-D-358, July, 1967.

Tension field theory,
a new approach which shows its duality with inextensional theory

By

E. H. Mansfield

Royal Aircraft Establishment, Farnborough

1. Introduction

Tension field theory describes the highly buckled (wrinkled) state of membranes or very thin plates whose boundaries are subjected to certain planar displacements well in excess of those necessary to initiate buckling. The present interest in tension field theory is because lightweight structures with stretched membrane components have potential applications in space. In addition, membrane structures which are pre-tensioned by internal pressure have application to lightweight portable bridges, protective coverings and various air cushion devices. The theory was conceived by WAGNER (1929) [1] whose primary concern was to explain the behaviour of thin metal webs in beams and spars carrying a shear load well in excess of the initial buckling value. Such webs offer little resistance to the compressive strain component of the shear and the spar flanges must be held apart by struts to prevent collapse. In the simple case of rigid spar flanges and rigid perpendicular struts the stress field in the web in the highly buckled state is primarily that of tension at 45°. As the shear load increases so does the magnitude of this tensile stress field and, just as a taut string resists a kinking action, so too does this tensile stress field resist the out-of-plane displacements engendered by the buckling action of the compressive stresses; these opposing actions result in a decreasing wavelength along the compressive buckles which form at right angles to the tension field. Strictly speaking such problems are non-linear and their exact analysis presents formidable difficulties. However, within the framework of large-deflexion plate theory it may be shown that for large values of the ratio (applied shear strain)/(shear strain at initial buckling) the relation between applied loads and planar displacements and stresses again approaches linearity, and it is this asymptotic regime for which tension field theory is applicable. In this regime the flexural stresses and the planar compressive (post-buckling) stresses are negligible compared with the tensile stresses; the assumption that their magnitude is zero is physically equivalent to the assumption of zero flexural membrane stiffness, and it is this which characterises tension field theory: the membrane is envisaged as being finely wrinkled at right angles to the lines of tension. In general these "tension rays" are not necessarily parallel and the boundary conditions need not be those of pure shear, as in our previous example, but shear must play a dominant role in the boundary deformation because of the requirement that the principal strains at any point are of opposite sign. This requirement will be considered in greater detail later but it is clear that if the principal strains are both positive so too are the principal stresses, and if the principal strains are both negative the membrane is ineffective in carrying load.

In his original paper Wagner presented a method for determining the distribution of tension rays in the general case, but this was based on lengthy geometrical considerations. REISSNER (1938) [2] achieved a markedly simpler analysis based on straightforward calculus and presented the first solution of a tension field problem — the torsion of an annular membrane — in which the rays are not parallel; of course, this problem has the compensating advantage of rotational symmetry. More recently STEIN and HEDGEPETH (1961) [3] introduced the concept of a variable Poisson's ratio with particular reference to the partly wrinkled membrane. In the specific problems which they solve, however, the tension rays are either parallel or possess rotational symmetry.

The complexity of all these analyses can be traced to the fact that they are based essentially on the appropriate field equations in which equal and simultaneous attention is devoted to the displacement components along and normal to the (unknown) tension rays. The present analysis, however, is based on a remarkably simple variational principle which enables us to determine the distribution of tension rays by focussing attention solely on the displacement components along the tension rays. The comparative simplicity of the resulting equations has enabled us to solve a variety of problems which do not exhibit any simplifying degree of symmetry. Further, the analysis can readily be modified to include anisotropic and/or variable membrane stiffness while, of greater importance perhaps, the theory lends itself to the logical determination of *approximate* solutions where an exact solution is intractable. Finally it is shown that an analogy exists between tension field theory and the inextensional theory for thin plates, the tension rays corresponding under certain specified conditions to the generators of the deflected plate. [Inextensional theory was conceived by the author (1955) [4] to describe the large-deflexion behaviour of a very thin plate which tends to deflect into a developable surface if this is kinematically possible. Then, the boundary conditions preclude the formation of significant middle surface forces and the simplifying assumption is made that the middle surface strains are zero so that the plate is constrained to deform into a developable surface.]

2. List of symbols

a	width of membrane	x, y	Cartesian coordinates
c, C	constants	α	angle that tension ray/generator makes with x-axis
D	flexural rigidity of plate $Et^3/12(1 - \nu^2)$		
E	Young's modulus	β	defined in section 5
F	defined by equation (3.8)	$\gamma_{\zeta\alpha}$	shear strain
H	(point on) edge of regression	Δ_α	total elongation of tension ray
m_s	moment/unit length applied to plate boundary	ε_α	"projected" strain normal to tension ray
		ε_s	strain along boundary
M	moment applied to plate, about y-axis	$\varepsilon_\eta, \varepsilon_\zeta$	strain along tension ray
M_η	moment/unit length about a generator	ζ	distance along tension ray measured from an involute to edge of regression
$\mathcal{M}_\alpha$	total moment about an α-generator		
n	normal distance from boundary point to tension ray/generator	η	distance along tension ray measured from edge of regression
p_s	load/unit length applied to plate boundary	η_1, η_2	values of η at boundaries
P	concentrated load	η_x	value of η at x-axis
r_1, r_2	radii of annular membrane/plate	θ	angular coordinate
s	distance along boundary, measured clockwise	λ	angle between tension ray and boundary
t	thickness of membrane/plate	μ	η_2/η_1 or η_1/η_2
T	torque applied to plate, about x-axis	ν	Poisson's ratio
u	displacement along tension ray	ϱ	radius of curvature of edge of regression
U	displacement of boundary in x-direction	σ	non-dimensional stress
$\mathcal{U}$	strain energy of membrane/plate	σ_η	tensile stress along tension ray
v	displacement normal to tension ray	ψ	defined in section 3.4
V	displacement of boundary in y-direction	ω, Ω	rotations

3. Analysis

We consider an initially flat membrane of arbitrary shape and variable stiffness. The membrane is loaded only at its boundaries which are subjected to given planar displacements or, if a boundary is straight, either the displacements are given or the edge is free. [In inextensional theory we consider a normally loaded thin plate of arbitrary shape and variable rigidity whose boundaries are free or, if straight, either free or clamped.] We assume that the boundary conditions generate a tension field over the entire membrane, an assumption which, for topological reasons, restricts the number of free edges that can be considered to two or less. The case of the partly wrinkled membrane is not considered although it can, in theory, be analysed by matching displacements and stresses along the (unknown) boundaries separating the wrinkled and unwrinkled regions.

3.1 The tension ray

The lines of principal stress necessarily form an orthogonal curvilinear network and it is convenient to consider the equilibrium of an infinitesimal curvilinear rectangle bounded by such lines. Equilibrium in the direction of the zero principal stress gives

$$t\sigma_\eta \varkappa = 0 \qquad (3.1)$$

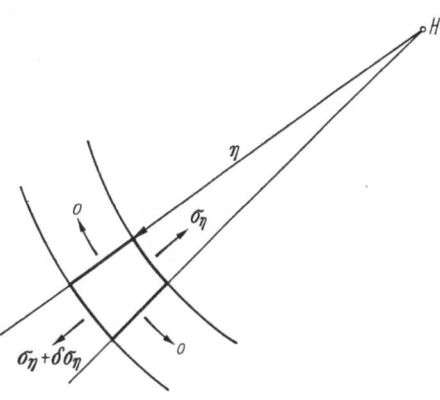

Fig. 1.

where σ_η is the non-zero (tensile) principal stress and $\varkappa$ is the curvature of this principal stress trajectory. Accordingly $\varkappa$ is zero and the trajectories of the tensile principal stresses are straight lines, hereafter referred to as *tension rays*. See Fig. 1.

Equilibrium in the direction of a tension ray demands continuity of the tensile load carried between adjacent rays, and hence along any ray

$$\eta t\sigma_\eta = \text{constant}, \qquad (3.2)$$

where η is the distance from the point of intersection of adjacent rays. Of course, if the tension rays are parallel Eq. (3.2) reduces to

$$t\sigma_\eta = \text{constant}. \qquad (3.3)$$

3.2 The principle of maximum strain energy

The fundamental problem of tension field theory lies in determining the orientation of the tension rays. In this connection we note again that there are no direct and shear stresses across adjacent rays; accordingly the strain energy of the membrane is solely due to tensile stresses directed along the rays and, furthermore, it would be unaffected by the introduction of cuts in the membrane along these rays. However, if the strained membrane were to be cut along closely spaced straight lines which do *not* coincide with the tension rays, two features are apparent. First, the resultant strain energy would be solely due to tensile stresses directed along these lines and, second, this strain energy would be *less* than in the uncut membrane because, with no transfer of energy from the boundaries, successive cuts could only release energy. It follows that the true distribution of tension rays maximises this "tensile" strain energy; it is this principle which forms the basis of our variational technique and which highlights the analogy with inextensional theory. [According to inextensional theory the plate deforms into a developable surface whose generators can be determined by maximising the strain energy associated with an arbitrary distribution of generators, an erroneous distribution being tantamount to the introduction of constraints.]

3.3 Variational analysis and the analogy

The analogy between tension field theory and inextensional theory is best demonstrated by presenting the respective analyses in parallel. Fig. 2 is appropriate to both theories; it shows a membrane/plate with a reference axis Ox; the

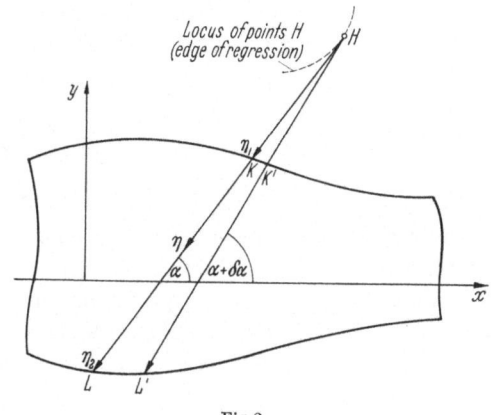

Fig. 2.

39*

lines LK, $L'K'$ are arbitrary adjacent cuts/generators which intersect the x-axis at angles α and $(\alpha + \delta\alpha)$ and meet at the point H. A typical point in the membrane/plate is specified by coordinates α, η where η is the distance along the α-line from the point H - itself a function of α and x.

In tension field theory the elemental slice $KLL'K'$ undergoes *radial tension* such that

$$\eta E t \varepsilon_\eta = c, \text{ a constant}, \qquad (3.4\,\text{a})$$

where ε_η is the direct strain *along* the line LK. The constant c is determined by equating the line-integral of the strain ε_η to the known change in length of LK, denoted by Δ_α:

$$\int_{\eta_1}^{\eta_2} \varepsilon_\eta \, d\eta = \Delta_\alpha, \qquad (3.5\,\text{a})$$

whence,

$$c = \frac{\Delta_\alpha}{\int_{\eta_1}^{\eta_2} \frac{1}{\eta E t} \, d\eta}. \qquad (3.6\,\text{a})$$

In inextensional theory the elemental slice $KLL'K'$ undergoes *conical bending* such that

$$\eta M_\eta / D = c, \text{ a constant}, \qquad (3.4\,\text{b})$$

where M_η is the moment per unit length *about* the line LK. The constant c is determined by equating the line-integral of the moment M_η to the known moment of the applied loads about LK, denoted by $\mathcal{M}_\alpha$:

$$\int_{\eta_1}^{\eta_2} M_\eta \, d\eta = \mathcal{M}_\alpha, \qquad (3.5\,\text{b})$$

whence,

$$c = \frac{\mathcal{M}_\alpha}{\int_{\eta_1}^{\eta_2} \frac{D}{\eta} \, d\eta}. \qquad (3.6\,\text{b})$$

The strain energy of the membrane/plate will now be determined, and in doing so it is convenient to integrate first over an elemental slice bounded by adjacent cuts/generators. Thus

$$\mathcal{U} = \frac{1}{2} \iint E t \varepsilon_\eta^2 \, dA$$

$$= \frac{1}{2} \iint_{\eta_1}^{\eta_2} E t \, \varepsilon_\eta^2 \eta \, d\eta \, d\alpha$$

$$= \frac{1}{2} \int F \, d\alpha \qquad (3.7\,\text{a})$$

where

$$F = \frac{\Delta_\alpha^2}{\int_{\eta_1}^{\eta_2} \frac{1}{\eta E t} \, d\eta}. \qquad (3.8\,\text{a})$$

$$\mathcal{U} = \frac{1}{2} \iint \frac{M_\eta^2}{D} \, dA$$

$$= \frac{1}{2} \iint_{\eta_1}^{\eta_2} \frac{M_\eta^2}{D} \eta \, d\eta \, d\alpha$$

$$= \frac{1}{2} \int F \, d\alpha \qquad (3.7\,\text{b})$$

where

$$F = \frac{\mathcal{M}_\alpha^2}{\int_{\eta_1}^{\eta_2} \frac{D}{\eta} \, d\eta}. \qquad (3.8\,\text{b})$$

Now from geometrical considerations the value of η at the x-axis is given by

$$\eta_x = \pm \frac{dx}{d\alpha} \sin \alpha, \qquad (3.9)$$

(the sign depending on whether H is above or below the x-axis) and it follows that F is a known function of x, α, $dx/d\alpha$ and the boundary conditions/applied loading. The relation between α and x is determined by maximising $\mathcal{U}$, whence

$$x'' F_{x'x'} + x' F_{xx'} + F_{\alpha x'} - F_x = 0 \qquad (3.10)$$

where $x' = \frac{dx}{d\alpha}$, $x'' = \frac{d^2x}{d\alpha^2}$ and F_x, for example, stands for $\frac{\partial F}{\partial x}$ where F is now formally regarded as a function of independent variables α, x, x'. Equation (3.10) is a non-linear differential equation of the second order and it determines the (α, x) relationship. The solution of Eq. (3.10) for a number of particular problems will be given shortly, but first we draw attention to a simple but important class of problems in which F does not contain x explicitly: the equation may then be integrated once to give

$$F_{x'} = \text{constant}. \qquad (3.11)$$

Also we note that for the practically important case in which Et is constant

$$F \propto \frac{\Delta_\alpha^2}{\ln(\eta_2/\eta_1)} . \tag{3.12}$$

The analogy between tension field theory and inextensional theory is immediately apparent from the preceding analyses. It can be expressed formally as follows:

	tension field theory	inextensional theory
membrane/plate geometry:	$Et \leftrightarrow 1/D$	
boundary conditions: $\{$	free edge $\leftrightarrow$ supported edge, $\Delta_\alpha \leftrightarrow \mathcal{M}_\alpha$	
orientation of rays/generators:	$(\alpha, x) \leftrightarrow (\alpha, x)$	
strains/moments:	$\varepsilon_\eta \leftrightarrow M_\eta$	

Some care must be exercised in interpreting this analogy. First, the boundary conditions require a detailed elaboration which is given in Section 3.4. Second, a given inextensional solution does not necessarily imply an analogous tension field solution unless it can be verified that

$$\Delta_\alpha > 0 \tag{3.13}$$

and

$$\varepsilon_\alpha + \nu \varepsilon_\eta \lessgtr 0 \tag{3.1 4}$$

everywhere. The second of these inequalities involves the transverse strain ε_α whose determination is given in the Appendix. Finally, we note that in inextensional theory the α, x relation is unaffected by a reversal of the applied loads, whereas in tension field theory a reversal of the boundary displacements will cause a tension field (if at all) with quite different characteristics. This naturally raises the question in inextensional theory as to the analogy of this further tension field. The answer is simply that *both* analogous systems of generators correspond to *local* stable states. In general, of course, these states contain different strain energies and there will be a preference for the state which contains the greater energy. A simple case in which there is no preference is that of the corner loaded square plate with free edges; a deflexion pattern with generators parallel to either diagonal is equally probable, while the two tension field analogies correspond to the positive and negative shear distortion of a square membrane.

3.4 The analogous boundary conditions

Consider first the statement *free edge* $\leftrightarrow$ *supported edge* which expresses the fact that these (straight) edges necessarily coincide with the tension rays/generators. The term *supported* is used here in a deliberately loose manner to mean simply that the edge is maintained straight by some external agency or structure which can apply, or resist, a bending moment, torque and shear force; the term does not necessarily imply any restriction on the rotations or vertical displacement of the edge. When there is only one such edge, as in a cantilever plate, the term *clamped* is more appropriate. But if there are two such edges they may undergo relative rotations and a vertical displacement and the term *supported* covers a variety of conditions including, for example, *clamped*, *freely clamped* and *simply supported*. Indeed, the problem now falls naturally into two classes depending on whether the moments, torques and forces acting at each supported edge are statically determinable or not. (The reader will recognise a parallel here with the clamped-clamped beam). In either case it is to be noted that the forces acting on a supported edge can be resolved into a moment $\mathcal{M}_{\alpha,0}$ and concentrated loads P_1, P_2 acting at the corners as shown overleaf. In the statically determinable case $\mathcal{M}_{\alpha,0}$, P_1, P_2 are known; in the statically indeterminate case the unknown values of $\mathcal{M}_{\alpha,0}$, P_1, P_2 are to be treated as given constants in the derivation of the governing equations and then determined by the displacement conditions (cf. the clamped-clamped beam). The analogue of $\mathscr{M}_{\alpha,0}$ is $\varDelta_{\alpha,0}$ while the analogue of the corner loads P_1, P_2 is given below.

Consider now the analogous boundary conditions associated with the statement $\varDelta_{\alpha} \leftrightarrow \mathscr{M}_{\alpha}$. First, it must be recognised that $\varDelta_{\alpha}$ depends only on the displacements *at the boundary* and a

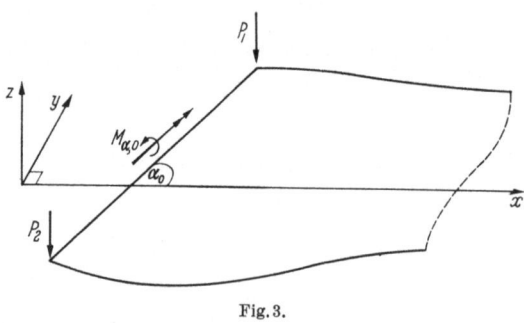

corollary to this is that the analogy is restricted to cases in inextensional theory in which the loads and moments are applied *at the boundary*. The basic component types of boundary condition are listed below in analogous pairs. In the associated tension field diagrams the shaded lines indicate that the boundary undergoes a rigid body displacement only, while the corresponding full lines in the diagrams appropriate to inextensional theory indicate that the boundary is unsup-

Fig. 3.

ported; the broken lines indicate an unspecified continuation of the membrane/plate. Attention must also be drawn to the sign conventions adopted, for their consistency only becomes apparent when they are related to s which is measured along the boundary in a clockwise direction; we adopt a left hand screw convention for describing positive rotations and moments.

Tension field theory	*Inextensional theory*

(i) Membrane with opposing boundaries undergoing rigid body displacements:

$$\varDelta_{\alpha} = (U_1 + U_2) \cos\alpha + (V_1 + V_2) \sin\alpha. \qquad (3.15\,\mathrm{a})$$

Plate under combined torsion and bending moment, opposing boundaries unsupported:

$$\mathscr{M}_{\alpha} = T \cos\alpha + M \sin\alpha. \qquad (3.15\,\mathrm{b})$$

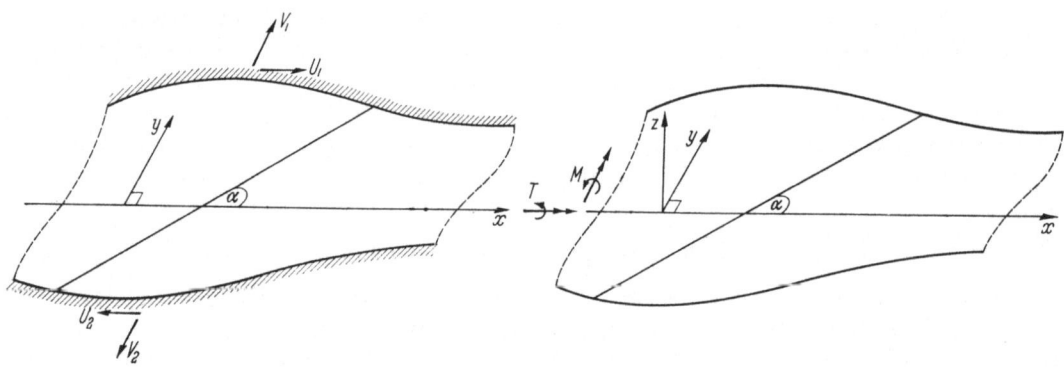

Fig. 4.   Fig. 5.

[The apparent lack of symmetry here can be reconciled by noting that, because of its assumed rigidity, the moments and torques can be assumed to be applied in arbitrary proportions at each end of a generator, i.e. we could write $T = T_1 + T_2$, $M = M_1 + M_2$.]

(ii) Membrane with boundaries undergoing rotation about corner point or partial rotation (kinking) about boundary point:

$$\varDelta_{\alpha} = n_1 \omega_1 + n_2 \omega_2. \qquad (3.16\,\mathrm{a})$$

Plate with unsupported boundaries carrying concentrated loads at corner or elsewhere on boundary:

$$\mathscr{M}_{\alpha} = n_1 P_1 + n_2 P_2. \qquad (3.16\,\mathrm{b})$$

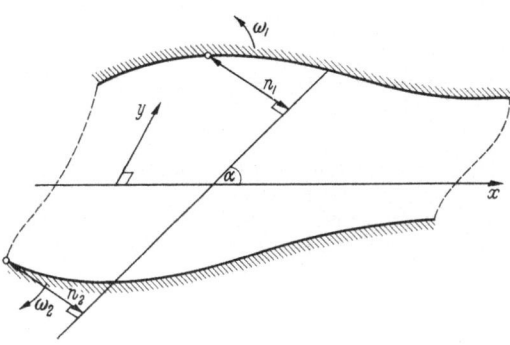

Fig. 6.

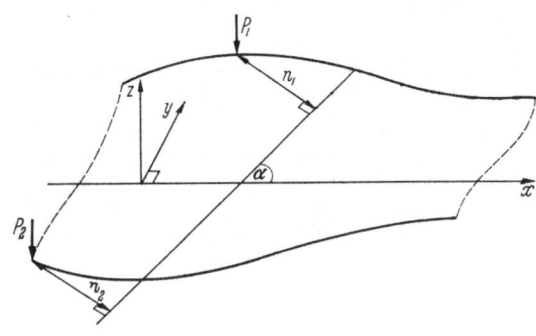

(iii) Membrane with boundaries subjected to varying strain ε_s, $d\omega/ds$ zero:

$$\Delta_\alpha = \int \varepsilon_{s,1} \cos \psi_1 \, ds$$
$$- \int \varepsilon_{s,2} \cos \psi_2 \, ds. \quad (3.17\text{a})$$

Plate with unsupported boundaries subjected to varying moment/unit length m_s:

$$\mathscr{M}_\alpha = \int m_{s,1} \cos \psi_1 \, ds$$
$$- \int m_{s,2} \cos \psi_2 \, ds. \quad (3.17\text{b})$$

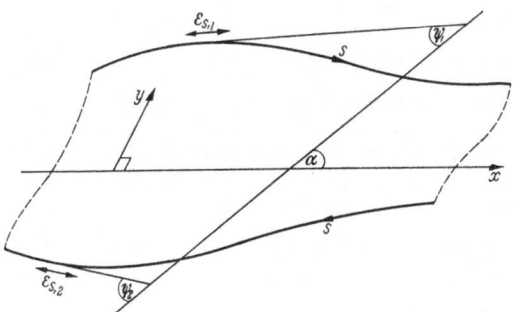

Fig. 8.

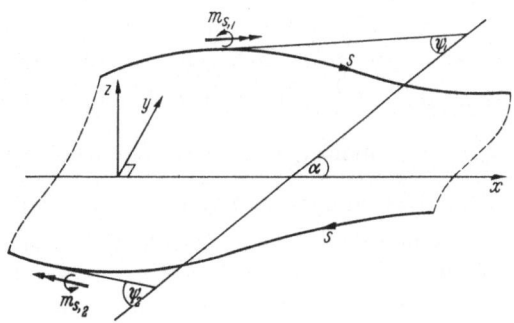

Fig. 9.

(iv) Membrane with boundaries subjected to varying rate of rotation $d\omega/ds$, ε_s zero:

$$\Delta_\alpha = \int n_1 \left(\frac{d\omega}{ds}\right)_1 ds$$
$$+ \int n_2 \left(\frac{d\omega}{ds}\right)_2 ds. \quad (3.18\text{a})$$

Plate with boundaries subjected to varying loads/unit length p_s:

$$\mathscr{M}_\alpha = \int n_1 p_{s,1} \, ds$$
$$+ \int n_2 p_{s,2} \, ds. \quad (3.18\text{b})$$

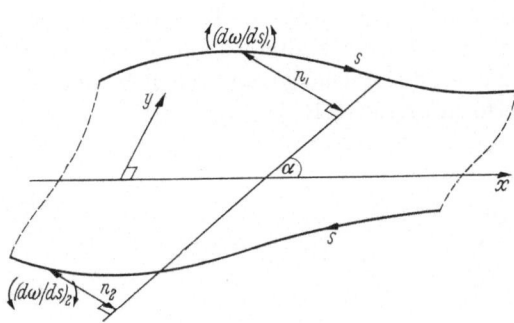

Fig. 10.

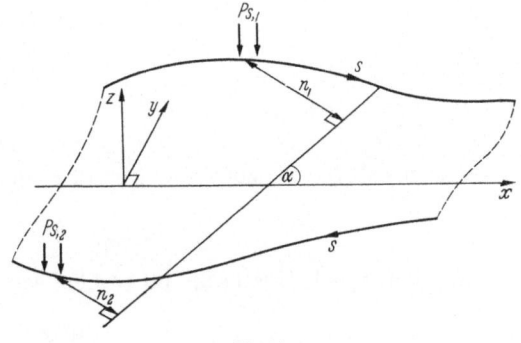

Fig. 11.

312   E. H. MANSFIELD

The analogous terms in the boundary equations above are summarised in the following table.

Tension field theory		Inextensional theory	
displacement	U, V	T, M	torque, bending moment
rotation	ω	P	concentrated load
strain	ε	m	moment/unit length
change in curvature	$d\omega/ds$	p	load/unit length

Uniform strain field

If the displacements along the complete boundary are consistent with a uniform strain field the tension rays are parallel and coincide with the direction of the positive principal strain; this direction may be determined by maximising $\Delta_\alpha/(\eta_2 - \eta_1)$.

4. Shearing of membrane strip

In this section we consider a number of problems associated with the shearing of a parallel strip with free edges.

4.1 Shearing of semi-infinite strip

Consider first the semi-infinite parallel strip whose opposite edges undergo a constant shear displacement U_0 as shown in Fig. 12. We will obtain the solution for the case of a free edge at $\alpha_0 = \frac{1}{2}\pi$, but it will be realised that this necessarily contains the solution for $\alpha_\infty < \alpha_0 < \frac{1}{2}\pi$; indeed, it also contains the solution for $\alpha_0 > \frac{1}{2}\pi$, for the additional triangular region beyond $\alpha_0 = \frac{1}{2}\pi$ is unstressed. [Strictly speaking there are limitations on the value of α_0 arising from the inequality (3.14). Thus it is shown in the Appendix that when $\tan \alpha_0 > \nu^{-1/2}$ the membrane will not be completely wrinkled so that tension field theory is not valid everywhere. From a practical standpoint, however, this restriction—which also applies in Sections 4.2 4.3 — is not of much significance.]

With the notation of Fig. 12 it is seen that $d\alpha/dx$ is negative so that the locus of points H lies below the x-axis and accordingly $\eta_1 > \eta_2$. Thus

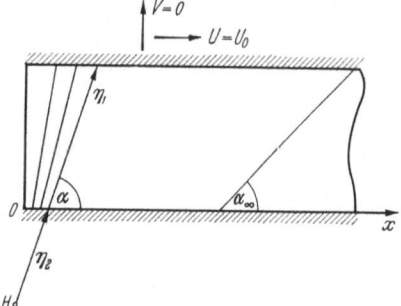

Fig. 12.

$$\left.\begin{aligned} \Delta_\alpha &= U_0 \cos \alpha, \\ \eta_1 - \eta_2 &= a \operatorname{cosec} \alpha, \\ \eta_x = \eta_2 &= -x' \sin \alpha, \end{aligned}\right\} \tag{4.1}$$

so that

$$F \propto \frac{\cos^2 \alpha}{\ln\left(1 - \dfrac{a \operatorname{cosec}^2 \alpha}{x'}\right)}. \tag{4.2}$$

[Note, however, that a pre-knowledge of the sign of x' is not essential to the analysis. Thus if x' had been assumed positive the only alteration in the expression for F is an overall sign change, and this itself does not affect the form of Eq. (3.10)].

At this stage it is convenient to introduce

$$\mu = \eta_1/\eta_2 = 1 - \frac{a \operatorname{cosec}^2 \alpha}{x'}, \quad \text{(i.e. } \mu \geq 1) \tag{4.3}$$

in terms of which Eqs. (4.2) and (3.11) become

$$\sin^2 2\lambda = C\mu \left(\frac{\ln \mu}{\mu - 1}\right)^2. \tag{4.4}$$

The constant of integration is determined from the condition that as $x \to \infty$ the tension rays become parallel and $\mu \to 1$, while $\alpha \to \frac{1}{4}\pi$, the value which maximises $\Delta_\alpha/(\eta_2 - \eta_1)$. Hence $C = 1$ and the relation between α and x' is readily determined numerically from Eqs. (4.3) and (4.4); the α, x relation is obtained by integration.

The stresses are given by Eqs. (3.4 a) and (3.6 a). As $x \to \infty$ it may be shown that

$$\sigma_{\eta,\infty} = EU_0/2a \tag{4.5}$$

and it is convenient to introduce

$$\sigma = \sigma_\eta/\sigma_{\eta,\infty}, \tag{4.6}$$

which is now a non-dimensional measure of the stress and the equivalent of a stress concentration factor. The peak values of σ, denoted by σ^*, occur along the x-axis ($\eta = \eta_2$) where it may be shown that they satisfy the relation

$$\sin 2\alpha = \frac{2\sigma^* \ln \sigma^*}{\sigma^{*2} - 1}. \tag{4.7}$$

Contours of constant σ and tension ray lines at $5°$ intervals are shown in Fig. 13, together with a shaded region which indicates the unstressed triangular zone which occurs when $\alpha_0 > \frac{1}{2}\pi$. We have included this because in the analogous inextensional problem of the cantilever strip under torsion precisely the same unstressed zone occurs *provided* α_∞ *is constrained to be* $\frac{1}{4}\pi$.

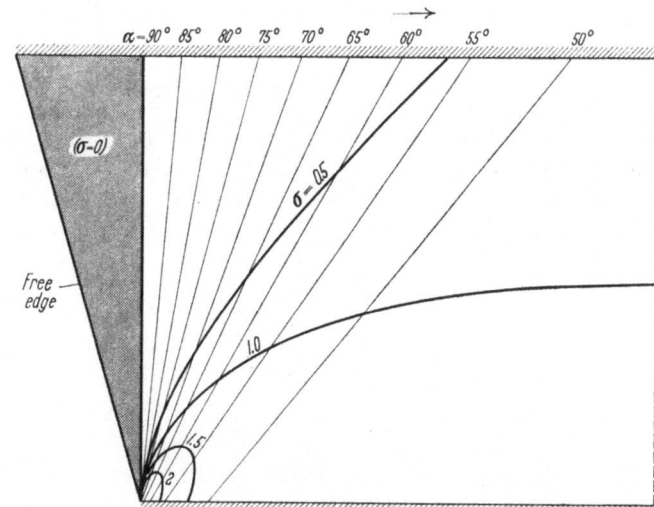

Fig. 13. Tension ray lines and stresses in semi-infinite membrane strip under shear ($\alpha_0 > \pi/4$).

If it is *not* so constrained its preferred value is $\frac{3}{4}\pi$ and the solution is then analogous to the tension field case with $U = -U_0$, which can immediately be obtained from the present solution from arguments of symmetry. Finally we note that the σ-contours of Fig. 13 retain their validity in the analogous inextensional problems if we re-define

$$\sigma = \frac{\text{bending stress normal to generator}}{[\text{bending stress normal to generator}]_{x \to \infty}}.$$

Case when $\alpha_0 < \alpha_\infty$.

When $\alpha_0 < \alpha_\infty$ it is seen that $d\alpha/dx$ is positive and hence the locus of points H lies *above* the x-axis and

$$\left. \begin{array}{l} \eta_2 - \eta_1 = a \operatorname{cosec} \alpha, \\ \eta_x = \eta_2 = x' \sin \alpha \end{array} \right\} \tag{4.8}$$

so that

$$F \propto \frac{\cos^2 \alpha}{-\ln \left(1 - \dfrac{a \operatorname{cosec}^2 \alpha}{x'} \right)}. \tag{4.9}$$

This expression for F differs only in sign from that of Eq.(4.2), and accordingly Eq.(4.4) retains its validity with the proviso that $\mu \leq 1$. The detailed solution now follows that of Section 4.1; the peak values of σ, which again satisfy Eq.(4.7), now occur along the edge opposite to the x-axis. Contours of constant σ and tension ray lines at 5° intervals are shown in Fig.14.

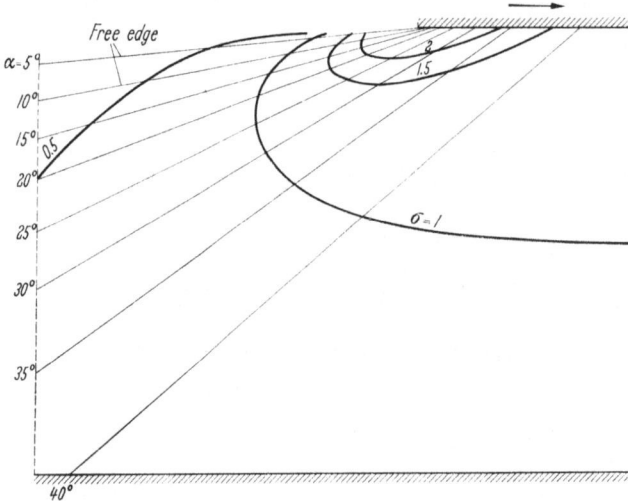

Fig.14. Tension ray lines and stresses in semi-infinite membrane strip under shear $(\alpha_0 < \pi/4)$.

4.2 Strip of finite length

The solution for a strip of finite length is given by Eqs.(4.3) and (4.4) with the constant C assuming values other than unity. There are three basic forms for the solution depending on the behaviour of the sign of $d\alpha/dx$, and these are illustrated in Figs. 15, 16 and 17. In

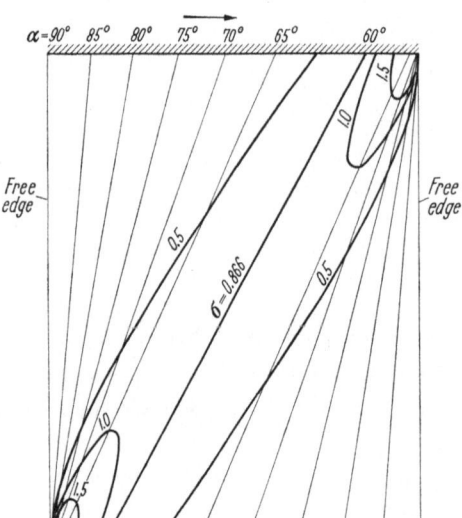

Fig.15. Tension ray lines and stresses in finite membrane strip under shear $C = 3/4$, $\alpha > \pi/4$.

Figs.15 and 16 we have taken $C = \dfrac{3}{4}$, but in Fig.15 attention is confined to the branch of Eq.(4.4) for which $\dfrac{\pi}{4} < \alpha < \dfrac{\pi}{2}$, while Fig.16 corresponds to the branch for which $0 < \alpha < \dfrac{\pi}{4}$. The solutions have point symmetry about the centre tension ray for which $\mu = 1$ and $\sin 2\alpha = \sqrt{C}$; to the left of these centre lines $\mu > 1$ in Fig.15 while $\mu < 1$ in Fig.16. In Fig.17 we have taken $C = \dfrac{7}{6}$ and confined attention to the case when $d\alpha/dx$ is negative which means that $\mu > 1$ everywhere. Equation (4.4) shows that there is a minimum value of μ, which occurs when $\alpha = \dfrac{\pi}{4}$, and this indicates a cusp in the locus of points H. (The same solution, apart from a change in the position of the x-axis, is obtained by assuming that $d\alpha/dx$ is positive and $\mu < 1$.)

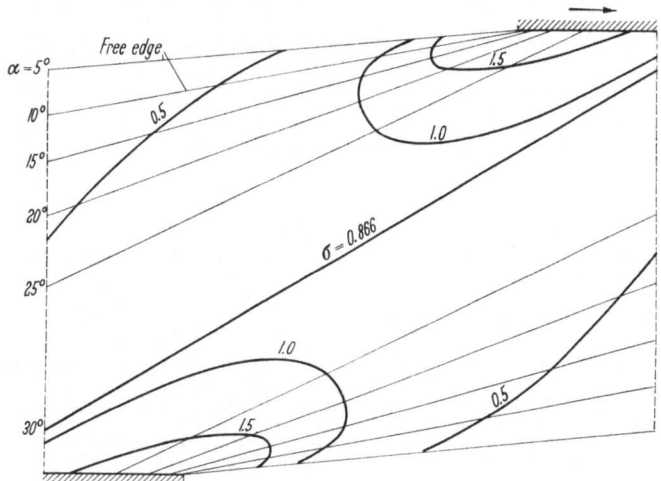

Fig. 16. Tension ray lines and stresses in finite membrane strip under shear $C = 3/4$, $\alpha < \pi/4$.

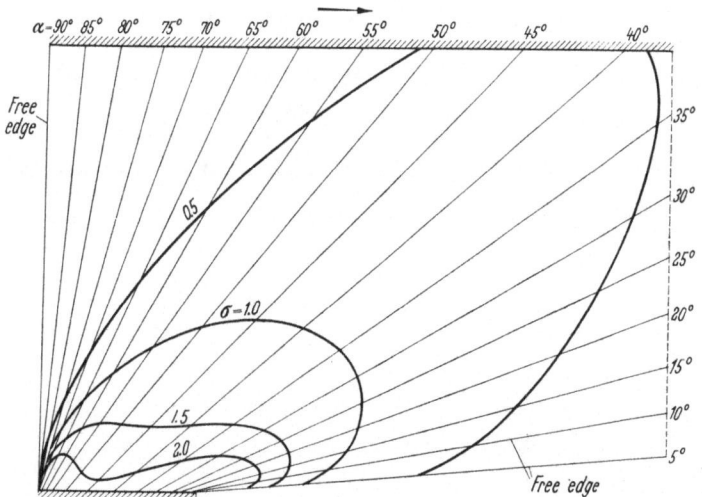

Fig. 17. Tension ray lines and stresses in membrane strip under shear ($C = 7/6$).

4.3 Shearing and lateral contraction of membrane strip

In the previous examples the opposite edges of the strip tend to be pulled together by the action of the membrane stresses, and the condition that $V = 0$ is only possible if these edges are held apart in a rigid manner; when they are held apart elastically some contraction occurs, and we consider now the simple case in which this contraction is constant with $V = -V_0$. Attention is confined to the case of the semi-infinite strip with a free edge at $\alpha_0 = \dfrac{1}{2}\pi$, but there is no intrinsic difficulty in obtaining solutions for the finite strip with arbitrary end values for α.

With the notation of Fig. 12, but with $V = -V_0$, we have

$$\Delta_\alpha = U_0 \cos \alpha - V_0 \sin \alpha, \tag{4.10}$$

and hence from equations (4.10), (4.1b, c) and (3.12)

$$F \propto \frac{(U_0 \cos \alpha - V_0 \sin \alpha)^2}{\ln\left(1 - \dfrac{a \csc^2 \alpha}{x'}\right)}. \tag{4.11}$$

40*

At this stage it is convenient to introduce some additional notation which is relevant to the boundary conditions. First, we note that an unstressed triangular region exists beyond a tension ray line for which $\Delta_\alpha = 0$. The limiting value of α is accordingly given by

$$\alpha^* = \tan^{-1}(U_0/V_0). \tag{4.12}$$

Second, we note that α_∞ is determined by maximising $\Delta_\alpha/(\eta_1 - \eta_2)$, whence

$$\alpha_\infty = \frac{1}{2}\alpha^*. \tag{4.13}$$

Equation (3.11) can now be expressed in the form

$$\{\sin\alpha \sin(\alpha^* - \alpha)\}^2 = C\mu\left(\frac{\ln\mu}{\mu - 1}\right)^2 \tag{4.14}$$

where μ is given by Eq. (4.3) and, for the semi-infinite strip,

$$C = \sin^4\frac{1}{2}\alpha^*. \tag{4.15}$$

The α, x relation may now be determined by integration, while the stresses are given by Eqs. (3.4a) and (3.6a). As $x \to \infty$ it may be shown that

$$\sigma_{\eta,\infty} = \frac{E}{2a}\{(U_0^2 + V_0^2)^{1/2} - V_0\} \tag{4.16}$$

and it is convenient to reintroduce σ [see Eq. (4.6)] as a stress concentration factor. The peak values of σ, denoted by σ^*, occur along the x-axis where it may be shown that they satisfy the relation

$$\frac{\sin\alpha \sin(\alpha^* - \alpha)}{\sin^2\frac{1}{2}\alpha^*} = \frac{2\sigma^* \ln\sigma^*}{\sigma^{*2} - 1}. \tag{4.17}$$

Contours of constant σ and tension ray lines at $5°$ intervals are shown in Fig. 18 for the case in which $V_0 = \frac{1}{2}U_0$.

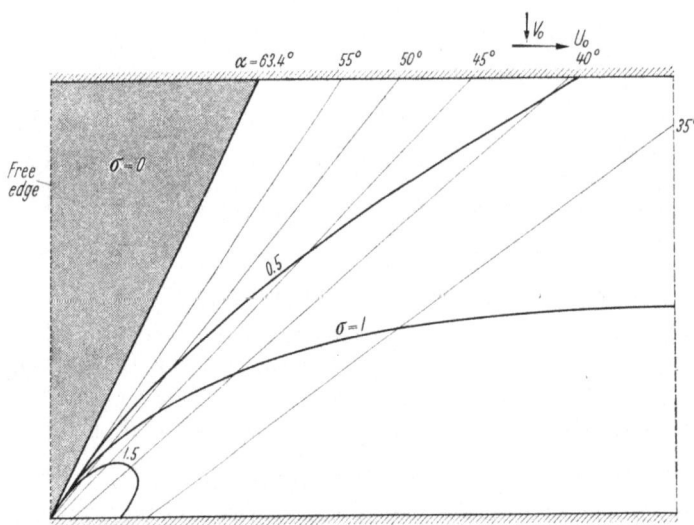

Fig. 18. Tension ray lines and stresses in semi-infinite membrane strip under shear and lateral contraction ($V_0 = 1/2\ U_0$).

5. An annular membrane

Throughout the preceding analysis the angle α was measured from some fixed direction, namely the x-axis, and this feature enabled us to use the area relation

$$\delta A = \eta\ \delta\alpha\ \delta\eta$$

in the derivation of Eq. (3.7). But the ordinate x itself is not an essential ingredient in the analysis and some other reference system may be preferable. To demonstrate this we now consider a membrane bounded by concentric circles and a radial cut, the circular boundaries undergoing a relative rotation about their common centre. We adopt the notation of Fig. 19 in which a typical tension ray is specified by the angle α and by the angular position θ of its intersection with the inner circular boundary. In the analysis it is preferable to use a hybrid coordinate system β, α where β is the angle that the tension ray makes with the radius at θ, i.e. $\beta = \alpha - \theta$. From geometrical considerations

$$\eta_\theta = HA = r_1 \cos \beta (1 - \beta'), \tag{5.1}$$

where a dash denotes differentiation with respect to α, and

where
$$\left.\begin{aligned} AB &= r_1\{(k^2 - \sin^2 \beta)^{1/2} - \cos \beta\} \\ k &= r_2/r_1 \end{aligned}\right\}. \tag{5.2}$$

Thus

where
$$\left.\begin{aligned} \eta_2/\eta_1 &= \mu, \text{ say} \\ &= 1 + \frac{J(\beta)}{1 - \beta'} \\ J(\beta) &= \sec \beta \, (k^2 - \sin^2 \beta)^{1/2} - 1 \end{aligned}\right\}. \tag{5.3}$$

Now

$$\Delta_\alpha = \Omega r_1 \sin \beta \tag{5.4}$$

and hence from Eq. (3.8a)

$$\frac{F}{Et\Omega^2 r_1^2} = \frac{\sin^2 \beta}{\ln\left(1 + \frac{J(\beta)}{1 - \beta'}\right)}. \tag{5.5}$$

This expression for F does not contain α explicitly and Eq. (3.10) can therefore be integrated once to give

$$F - \beta' F_{\beta'} = CEt\Omega^2 r_1^2, \tag{5.6}$$

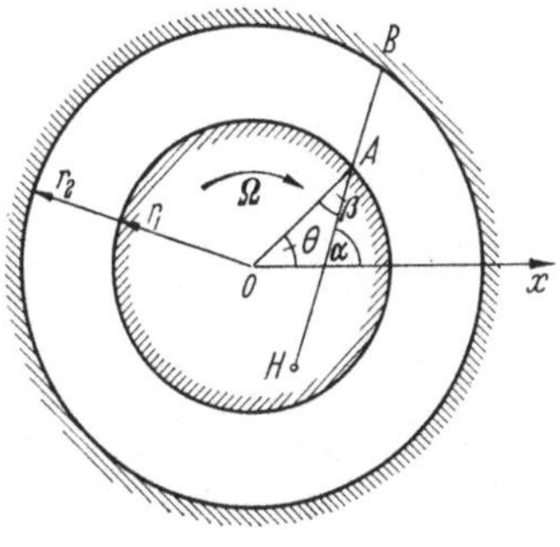

whence

$$C\mu \ln^2 \mu = \sin^2 \beta \left\{ \mu \ln \mu + (\mu - 1)\left(\frac{\mu - 1}{J(\beta)} - 1\right) \right\}. \tag{5.7}$$

Fig. 19.

By letting C assume various values we obtain from Eqs. (5.3) and (5.7) a relation between β and β' which may be integrated to yield the β, α or θ, α relation. In particular, with $C = 0.584$ we obtain the solution for the annulus with a radial cut.

Finally we note that if the annulus is continuous $\beta' = 0$ and β may be determined from the condition

$$F_\beta = 0, \tag{5.8}$$

whence

$$\ln\left(\frac{k^2 - \sin^2 \beta}{\cos^2 \beta}\right) = \frac{\tan^2 \beta \, (k^2 - 1)}{k^2 - \sin^2 \beta}, \tag{5.9}$$

which is in agreement with the solution obtained by REISSNER (1938) [2]. The reader will note that this also gives the inextensional solution to the axially loaded flat annular spring.

6. The anisotropic membrane

A membrane which exhibits a tensile stiffness which varies with direction can also be analysed by the method of section 3 simply by replacing E by $E(\alpha)$, where $E(\alpha)$ is the tensile modulus in the direction α. In particular, if the displacements along the complete boundary are consistent with a uniform strain field the tension rays are parallel and their direction is that which maximises $E(\alpha) \, \varepsilon^2(\alpha)$, where $\varepsilon(\alpha)$ is the (uniform) strain in the direction α.

As a simple example consider a symmetrical orthotropic membrane whose stiffness properties are specified by

$$E = E_0(1 + K \cos^2 2\alpha) \tag{6.1}$$

and whose boundaries undergo displacements consistent with a uniform shear strain referred
to the x-axis. The direct strain $\varepsilon(\alpha)$ is thus proportional to $\sin 2\alpha$ and α is determined by maximising

$$(1 + K \cos^2 2\alpha) \sin^2 2\alpha.$$

When $K \leqq 1$, so that the degree of anisotropy is small, it may be shown that $\alpha = \frac{1}{4}\pi$; but
if $K > 1$ it may be shown that α is *either* of the two roots (which are symmetrically disposed
about $\frac{1}{4}\pi$) of the equation

$$\sin 2\alpha = \left(\frac{1 + K}{2K}\right)^{1/2}. \tag{6.2}$$

For example, if $K = 2$ we find that $\alpha = 30°$ or $60°$. This problem demonstrates clearly that a
symmetry in the membrane and in its boundary displacements does not necessarily imply a
symmetry in the associated tension field.

7. Approximate solutions

As stated in the Introduction the present analysis lends itself to the logical determination
of approximate solutions. The method is the same as that used in inextensional theory: a
suitable relation is assumed between x and α which contains one or more arbitrary parameters
which are to be determined from the condition that the strain energy is a maximum.

Acknowledgement

The author wishes to thank E. T. Haynes for his enthusiastic assistance with the computation.

Appendix

Strains and displacements

In the following analysis it is convenient to use two systems of coordinates as shown in
Fig. 20. First, following Reissner we use an orthogonal system of curves ζ, α where ζ is measured
along a tension ray from some (arbitrary) curve Q which is an *involute* of the *edge of regression*
curve H. In addition we use a non-orthogonal system η, α where η is measured along a tension
ray from the edge of regression. When it is necessary to distinguish between these two systems,
as in partial differentiation, we use the notation $\left(\frac{\partial}{\partial\alpha}\right)_\eta$ for example to indicate that η is constant.
The displacement in the direction of increasing ζ (with α constant) is denoted by u, while that
in the transverse direction of increasing α (with ζ constant) is denoted by v.

The strain components in the orthogonal system are given by

$$\left.\begin{aligned}
\varepsilon_\zeta &= \frac{\partial u}{\partial \zeta}, \\
\varepsilon_\alpha &= \frac{1}{\eta}\left\{u + \left(\frac{\partial v}{\partial\alpha}\right)_\zeta\right\}, \\
\gamma_{\zeta\alpha} &= \frac{1}{\eta}\left(\frac{\partial u}{\partial\alpha}\right)_\zeta + \eta\frac{\partial}{\partial\zeta}\left(\frac{v}{\eta}\right)
\end{aligned}\right\}. \tag{A.1}$$

These can be expressed purely in terms of the η, α system by noting that

$$\left.\begin{aligned}
\left(\frac{\partial}{\partial\zeta}\right)_\alpha &\equiv \left(\frac{\partial}{\partial\eta}\right)_\alpha, \\
\left(\frac{\partial}{\partial\alpha}\right)_\zeta &\equiv \left(\frac{\partial}{\partial\alpha}\right)_\eta + \varrho\left(\frac{\partial}{\partial\eta}\right)_\alpha
\end{aligned}\right\} \tag{A.2}$$

where ϱ is the radius of curvature of the edge of regression, which can be expressed in the notation of the main text as follows,

$$\varrho = x'' \sin \alpha + 2x' \cos \alpha. \tag{A.3}$$

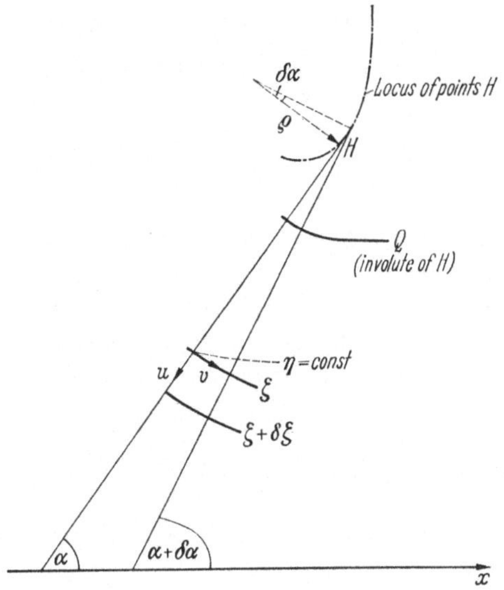

Fig. 20.

Substitution of Eq. (A.2) into Eq. (A.1), omitting the suffices and writing ε_η for ε_ζ, as in the main text, gives

$$\left.\begin{aligned} \varepsilon_\eta &= \frac{\partial u}{\partial \eta}, \\ \varepsilon_\alpha &= \frac{1}{\eta}\left(\frac{\partial v}{\partial \alpha} + \varrho\,\frac{\partial v}{\partial \eta} + u\right), \\ \gamma_{\zeta\alpha} &= \frac{1}{\eta}\left\{\frac{\partial u}{\partial \alpha} + \varrho\,\frac{\partial u}{\partial \eta} + \eta^2\,\frac{\partial}{\partial \eta}\left(\frac{v}{\eta}\right)\right\} \end{aligned}\right\}. \tag{A.4}$$

When the α, x relationship has been determined by the methods of the main text the strain ε_η is given by

$$\frac{\partial u}{\partial \eta} = \frac{\varDelta_\alpha}{\eta E t \displaystyle\int\limits_{\eta_1}^{\eta_2} \frac{d\eta}{\eta E t}}. \tag{A.5}$$

The value of u at the boundary is determined by the boundary displacements, for example

$$u_2 = U_2 \cos \alpha + V_2 \sin \alpha, \tag{A.6}$$

and accordingly equation (A.5) may be integrated to determine u throughout the membrane.

The displacement v can now be determined from the vanishing of $\gamma_{\zeta\alpha}$, whence

$$v = -\eta \int \frac{1}{\eta^2}\left(\frac{\partial u}{\partial \alpha} + \varrho\,\frac{\partial u}{\partial \eta}\right) d\eta \tag{A.7}$$

the constant of integration being determined from the known boundary values.

These equations enable ε_α to be found and hence the validity of the inequality (3.14) can be checked. A thorough check, however, is not only time-consuming but also frustrating because if the inequality is not satisfied — an indication that the membrane is not completely wrinkled — it is, nevertheless, virtually impossible to obtain a more meaningful solution. Indeed, the approximate nature of tension field theory itself does not justify too scrupulous an attention to detail. For these reasons we suggest that a check on the inequality (3.14) be confined to the boundaries where the critical regions often occur and where ε_α assumes the following particularly simple form:

$$\varepsilon_\alpha = \varepsilon_s - \varepsilon_\eta \cot^2 \lambda, \tag{A.8}$$

where ε_s is the strain along the boundary and λ is the angle that the tension ray makes with a tangent to the boundary. Thus at the boundary the inequality assumes the form

$$\varepsilon_s + (\nu - \cot^2 \lambda)\, \varepsilon_\eta \lessgtr 0. \tag{A.9}$$

Finally we note that if ε_s is zero the inequality requires that

$$\cot^2 \lambda \geq \nu,$$

i.e.

or
$$\left.\begin{array}{l} \lambda \leq \quad 63\tfrac{1}{2}^{\circ} \\[4pt] \lambda \geq 116\tfrac{1}{2}^{\circ} \end{array}\right\} \text{ if } \nu = \frac{1}{4}\,.$$

The form of the wrinkles. The assumption in tension field theory that the wavelength of the wrinkles is vanishingly small implies, of course, that the theory cannot predict the wavelengths or depth of wrinkles which occur in practice. But some information can be given on the maximum *slope* of the wrinkled surface. Thus if we consider a system of wrinkles parallel to the y-axis, say, in which the deflexion w is given by

$$w = w_0 \sin (2\pi x/l) \tag{A.10}$$

the maximum slope is given by

$$\begin{aligned} \left(\frac{\partial w}{\partial x}\right)_{\max} &= \Theta, \text{ say} \\[6pt] &= \frac{2\pi w_0}{l}\,. \end{aligned} \tag{A.11}$$

Further, the distance measured along the middle surface over a complete wavelength is

$$l\left(1 + \frac{1}{4}\,\Theta^2\right).$$

The average contractile strain in the x-direction due to wrinkling is accordingly $\dfrac{1}{4}\,\Theta^2$ and the total contractile strain is therefore $\left(\dfrac{1}{4}\,\Theta^2 + \nu\varepsilon_y\right)$ which may be equated to $-\varepsilon_x$. Thus in the notation of the main text

$$\Theta = 2\{-(\varepsilon_\alpha + \nu\varepsilon_\eta)\}^{1/2}. \tag{A.12}$$

The above simple analysis ignores the strains due to the (neglected) compressive stresses and should therefore be regarded as an upper limit.

References

1. WAGNER, H.: Z. Flugtechn. Motorluftschiffahrt (1929).
2. REISSNER, E.: On tension field theory. Proc. V Int. Cong. Appl. Mech. (1938).
3. STEIN, M., and J. M. HEDGEPETH: Analysis of partly wrinkled membranes. NASA TN D-813 (1961).
4. MANSFIELD, E. H.: The inextensional theory for thin flat plates. Quart. J. Mech. Appl. Math. VIII, 3 (1955).

Collapse of a homogeneous fluid mass in a stratified fluid

By

C. C. Mei

Massachusetts Institute of Technology, Cambridge

1. Introduction

In an experimental study of the turbulent wake behind an object moving horizontally in a stratified liquid, SCHOOLEY and STEWART [1] found that the growth of the wake is suppressed in the vertical direction as the well-mixed, nearly homogeneous fluid in the wake seeks to its own level of density. Thus, in the immediate vicinity of the body, the wake grows uniformly in all transverse directions, but after a certain distance behind the body, it begins to shrink in height and to spread in horizontal width at a rate faster than that in a homogeneous fluid. WU [2], utilizing the fact that the longitudinal variation of a wake boundary is relatively small compared with the transversel variation, conducted two-dimensional experiments on the gravitational collapse of a homogeneous fluid mass in a stratified fluid. The homogeneous fluid is originally confined in a hollow circular cylinder and has the same density as the ambient stratified fluid at the level of the horizontal axis of the cylinder. At the time $t = 0$, the cylinder is removed and the vertical collapse of the homogeneous fluid within is then recorded cinematically as a function of time. Related studies have been reported in [3], [4], and [5].

Experiments in [1] indicate that the collapse of a turbulent wake in a stratified fluid is accompanied by very effective excitation of internal waves. The initial potential energy excess in the system is transformed into kinetic energy of motion in both fluids. In his two-dimensional experiment, WU [6] also recorded internal waves in the "principal stage" defined by $3 < t(-g\varrho_0'/\varrho_0)^{1/2} < 25$.

It is possible that phenomena related to WU's studies in water also occur in the atmosphere in which an air mass of a more or less uniform density is dispersed horizontally within a stratified surrounding; for example, the eventual flattening of a buoyant plume is at least partially similar.

In the present paper, two aspects of the collapsing phenomenon relevant to WU's experiments will be investigated for an initially slender interface. In the initial period when the total time is less than the time scale characterized by the Brunt-Väisälä frequency, internal wave motion is ineffective in the stratified fluid. A theory for the spreading of the homogeneous mass is developed. In the second part, attention is foucsed on the internal wave generation in the stratified fluid. Viscosity is neglected.

2. The initial stage

2.a The approximate equations

With the coordinate system shown in Fig. 1, the initial density is assumed to be constant in the inner fluid (1) and linearly varying in the outer fluid (2), i.e.,

$$\varrho = \varrho_1 = \text{constant} \qquad (x, y) \text{ in } R_1 \tag{2.1}$$

$$\varrho = \varrho_2 = \varrho_1(1 - \varepsilon y) \qquad (x, y) \text{ in } R_2. \tag{2.2}$$

The linear assumption, Eq. (2.2), is an adequate approximation whenever the maximum height of the interior fluid is much smaller than the length scale characterizing the density gradient in the exterior fluid, i.e.,

$$-\eta_1 \varrho_2' / \varrho_1 \gg 1.$$

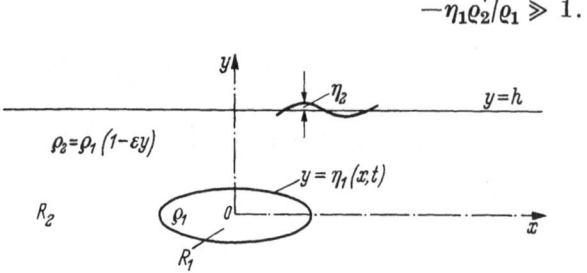

Fig. 1. Definition sketch.

In the initial interval before the total time reaches the time scale characterized by the Brunt-Väisälä frequency, internal waves in the stratified fluid may be ignored. Further analysis can be based on either (1) expansion in powers of t, or (2) the longwave approximation. Both approaches have been employed by PENNEY and THORNHILL [7] for the dispersion of a fluid column on a free surface, and agreement on gross features are found for squat columns. For analytical convenience the long-wave approximation will be used here, which is suitable for interfaces with a small height-to-width ratio. Accordingly, the pressure in both fluids may be assumed as hydrostatic:

$$p_1 = \varrho_1 g(\eta_1 - y) + p_2)_{2y = \eta_1} \tag{2.3}$$

$$p_2 = g \int_y^{2h + \eta_2} \varrho_1 (1 - \varepsilon y)\, dy. \tag{2.4}$$

In [7], the dispersion of a fluid column on the interface of a two-layered system is also studied, in which it is shown that, for long waves, the free-surface motion can be ignored if the depth of the upper layer is much greater than the amplitude of the interface. A similar argument leads to the same result here. We may therefore conclude that the outer fluid yields statically as the inner fluid disperses and merely provides the driving pressure gradient on the interface:

$$u_2, \quad \eta_2 \gg u_1, \quad \eta_1. \tag{2.5}$$

It follows from Eqs. (2.3) and (2.4) that

$$\frac{1}{\varrho_1} \frac{\partial p_1}{\partial x} = \varepsilon g \eta_1 \frac{\partial \eta_1}{\partial x}. \tag{2.6}$$

The quantity $\varepsilon g \eta_1$ may be viewed as the effective gravity. The subscript 1 is now unnecessary for fluid 1 and will be omitted henceforth. Assuming the interface to be symmetrical in both x and y as suggested by experiments [2], we may express the shallow-water equations for fluid 1 as

continuity: $\eta_t + (u\eta)_x = 0$ \tag{2.7}

x-momentum: $u_t + uu_x + \varepsilon g \eta \eta_x = 0.$ \tag{2.8}

With the initial half-width of fluid 1, L, as the characteristic length and $(g\varepsilon)^{-1/2}$ as the characteristic time, the following dimensionless variables and equations can be written:

$$(\eta, x) = L(\eta^*, x^*), \qquad t = Tt^*$$
$$u = (L/T)\, u^*, \qquad T = (g\varepsilon)^{-1\,2} \tag{2.9}$$

$$\eta_{t^*}^* + u^* \eta_{x^*}^* + \eta^* u_{x^*}^* = 0 \tag{2.10}$$

$$u_{t^*}^* + u^* u_x^* + \eta^* \eta_x^* = 0. \tag{2.11}$$

From now on the symbols * are dropped for convenience. The initial velocity will be assumed zero and the initial interface given, i.e.,

$$\text{at } t = 0: \quad u(x, 0) = 0, \quad \eta(x, 0) = \eta^0(x). \tag{2.12}$$

Equations $(2.10)-(2.11)$ are analogous to those describing a compressible gas in one-dimensional motion, with the ratio of specific heats $\gamma = 3$. Exact general solutions for this set of equations are well-known [8], [9]. To summarize the pertinent results, we note that the characteristics in the x-t plane are straight lines with the following relations:

$$C_+: \ u + \eta = \text{constant} = \alpha \qquad \frac{dx}{dt} = u + \eta$$
$$\text{along} \tag{2.13}$$
$$C_-: \ u - \eta = \text{constant} = -\beta \qquad \frac{dx}{dt} = u - \eta.$$

Hence, we have

$$u = (\alpha - \beta)/2, \quad \eta = (\alpha + \beta)/2 \tag{2.14}$$

$$x_\beta - \alpha t_\beta = 0, \quad x_\alpha + \beta t_\alpha = 0 \tag{2.15}$$

which can also be transformed to (u, η) variables as

$$x_\eta + \eta t_u - u t_\eta = 0, \quad x_u - u t_u + \eta t_\eta = 0. \tag{2.16}$$

Upon elimination of x, Eqs. (2.15) can be reduced to the Poisson-Euler-Darboux equation for t:

$$(\alpha + \beta)\, t_{\alpha\beta} + t_\alpha + t_\beta = 0. \tag{2.17}$$

This has the general solution

$$t = \frac{1}{\alpha + \beta}\, [F(\alpha) + G(\beta)] = \frac{1}{2\eta}\, [F(u + \eta) + G(\eta - u)] \tag{2.18}$$

where F and G are arbitrary functions.

2.b The explicit solution

To satisfy the initial conditions, $u(x, 0) = 0$, $\eta(x, 0) = \eta^0(x)$, we obtain from Eq. (2.18) that $F = -G$, hence

$$t = \frac{1}{2\eta}\, [F(u + \eta) - F(\eta - u)]. \tag{2.19}$$

From the first of Eqs. (2.16) it follows that

$$t_u = -(x_\eta/\eta), \quad u = 0, \quad \eta = \eta^0. \tag{2.20}$$

Now, in the transformed plane (u, η) the initial conditions can be stated as

$$t(0, \eta) = 0, \quad x(0, \eta) = x^0(\eta) \tag{2.21}$$

where $x^0(\eta)$ and $\eta^0(x)$ are the inverse of each other. Hence the righthand side of Eq. (2.20) is known, and Eqs. (2.12) and (2.20) provide two Cauchy initial conditions. By differentiating the general solution, Eq. (2.19), we obtain

$$t_u\big|_{u=0} = F'(\eta)/\eta = -x_\eta^0(\eta)/\eta.$$

Hence we have

$$F(\eta) = -x_0(\eta) + \text{constant}$$

and the solution for t is

$$t = \frac{1}{\alpha + \beta}\, [x^0(\beta) - x^0(\alpha)] = \frac{1}{2\eta}\, [-x^0(u + \eta) + x^0(\eta - u)]. \tag{2.22}$$

To return to (x, t) as the independent variables, it is convenient to use the integrals of Eq. (2.15):

$$x = \alpha t + x^0(\alpha), \quad x = -\beta t + x^0(\beta) \tag{2.23}$$

which can be verified with the help of Eq. (2.22). By solving from Eq. (2.23) for α and β in terms of x, t, u and η can immediately be obtained from Eq. (2.14).

41*

We shall give only the explicit result for the special case of an elliptical initial shape:

$$\eta(x, 0) = \eta^0(x) = b_0(1 - x^2)^{1/2} \qquad |x| < 1$$
$$= 0 \qquad\qquad\qquad |x| > 1. \tag{2.24}$$

Hence, considering $x > 0$ only, we have

$$x\big|_{u=0} = x^0(\eta) = (1 - \eta^2/b_0^2)^{1/2} \qquad 0 < \eta < b_0. \tag{2.25}$$

The final results are

$$u = b_0^2 xt/(1 + b^2 t^2) \tag{2.26}$$

$$\eta = b_0(1 - x^2 + b^2 t^2)^{1/2}/(1 + b^2 t^2). \tag{2.27}$$

Equation (2.27) can be expressed in the form

$$(x/a)^2 + (\eta/b)^2 = 1 \tag{2.28}$$

with

$$a = (1 + b^2 t^2)^{1/2} \tag{2.30}$$

$$b = b_0(1 + b^2 t^2)^{-1/2}. \tag{2.31}$$

Hence, the boundary of the homogeneous fluid will always be elliptical, with the major axis lengthening with time according to Eq. (2.30) and the minor axis shortening according to Eq. (2.31). The area of the ellipse above the line $x = 0$ is, of course, constant at $(\pi b_0/2)$. The tip of the homogeneous region is seen to trace out a hyperbola in the $x - t$ plane, as is indicated by Eq. (2.30).

Empirical results [2] for the rate of horizontal spread of an initially circular interface are given as follows:

$$a = 1 + 0.29\, t^{1.08}, \qquad 0 < t < 2.5, \text{ "initial stage"} \tag{2.32}$$

$$a = 1.03\, t^{0.55} \qquad\qquad 3 < t < 25, \text{ "principal stage"}. \tag{2.33}$$

The initial linear rate of spreading (based on rather limited records) does not conform with the expected quadratic rate which should be valid for all dynamical systems starting from rest.

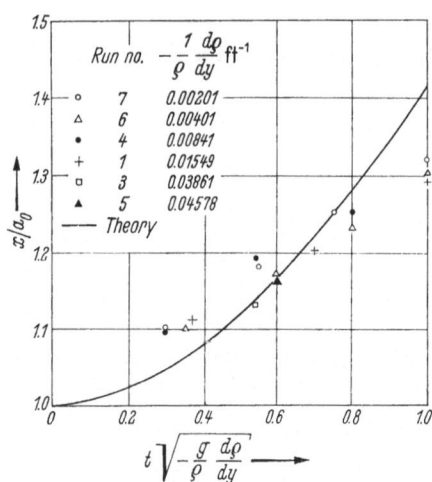

Fig. 2. Horizontal expansion of the interface tip during initial stage, experiments by Wu (1965).

It should be mentioned that, owing to the removal of the cylindrical partition, additional disturbance besides the normal interfacial mixing undoubtedly exists in those experiments. This and the fact that the initial interface is not squat in shape make comparison between experimental data and Eq. (2.30) not entirely appropriate. Nevertheless, some agreement in order of magnitude is found in the range $0 < t < 1$ (Fig. 2).

Finally, from Eq. (2.13) it follows that the two families of characteristics have slopes

$$\frac{dx}{dt} = \pm \eta(x, 0).$$

For an initial shape where $\eta(x, 0) = 0$ at the tip ($x = 1$), crossing of two neighboring C_+ characteristics occurs immediately. It is therefore necessary to inquire into the possibility of shocks. A proof is given in Appendix A that if the initial value of η, $\eta(x, 0) = \eta^0(x)$, is continuous, then multivaluedness in η will not develop. Hence the validity of the solution is ensured.

3. Internal waves in the stratified fluid

3.a Linearized theory

When the time after the initiation of collapse is comparable to the Brunt-Väisälä time scale $(g\varepsilon)^{-1\,2}$, internal waves are the most conspicuous feature in the outer fluid. Part of the potential energy of the system is gradually converted to kinetic energy of wave motion in the stratified fluid, thus the advance of the homogeneous fluid is slowed down.

A brief theoretical study related to this problem has been reported in [10], where an impulsive point source is used as a mathematical model. The results on the wave pattern do not, however, seem to conform with observations. For a complete theory one must treat the motion of the two fluids simultaneously, with the interaction properly accounted for. It is evident that the dynamics of the homogeneous fluid remains non-linear. However, the essential feature in the outer fluid is a small oscillation about an equilibrium state, and linearized equations may be expected to suffice. Available experimental observations for an initially circular interface [6] indicate that the collapse of the homogeneous fluid can be described by a rather simple empirical law, while the wave phenomena in the outer fluid exhibit much greater complexity. We shall therefore focus our attention on the outer fluid and make the simplifying assumption that the interface shape is given at all time. An important difficulty still remains if the initial interface is of finite height. Here the slender-body approximation will further be made in order to obtain some qualitative understanding.

With the Boussinesq approximation and the assumption of a linear density gradient, we may combine the linearized equations of motion into the following single equation [11]:

$$\frac{\partial^2}{\partial t^2} \nabla^2 \{\} + \frac{\partial^2}{\partial x^2} \{\} = 0. \tag{3.1}$$

The symbol $\{\}$ may represent any one of the unknowns ϱ, u, v, and p, which are defined here as small departures from equilibrium. Dimensionless variables are again used as defined in Eq. (2.9). In addition, ϱ and p are normalized by ϱ_1, and $\varrho_1 L^2/T^2$, respectively. The initial conditions are

$$\{\} = \frac{\partial}{\partial t} \{\} = 0, \quad t = 0. \tag{3.2}$$

The boundary condition at the interface is

$$v = \frac{\partial \eta}{\partial t}, \quad y = 0 \tag{3.3}$$

according to the slender body approximation, where $\eta(x, t)$ denotes the position of the interface. The presence of the channel boundaries and of the free surface is ignored. We note that in the dimensionless form, $\varrho(x, y, t)$ is identical to the vertical displacement of a fluid particle $\zeta(x, y, t)$; this follows from the definition $\zeta_t = v$ and the linearized incompressibility condition $\varrho_t = v$. In the sequel we shall concentrate on ϱ or ζ only. The solution can be easily obtained by the usual transform method:

$$\varrho = \zeta = \frac{1}{2\pi} \int_{-\infty}^{\infty} d\alpha \, e^{i\alpha x} \frac{2}{\pi} \int_{0}^{\infty} d\beta \sin \beta y \frac{\beta}{\alpha^2 + \beta^2} \frac{1}{2\pi i} \int_{c-i\infty}^{c+i\infty} dp \cdot e^{pt} \frac{M(\alpha, p)}{p^2 + \alpha^2(\alpha^2 + \beta^2)^{-1}} \tag{3.4}$$

with

$$M(\alpha, p) = p^2 \bar{\tilde{\eta}}(\alpha, p) - p \tilde{\eta}^0(\alpha), \quad \eta^0(x) \equiv \eta(x, 0) \tag{3.5}$$

where ($\tilde{}$) denotes the Fourier exponential transform for x and ($\bar{}$) denotes the Laplace transform for t.

As the explicit inversion of Eq. (3.4) does not seem feasible, only the large-time behavior of the integral will be studied here. Without loss of generality, attention will also be limited to $x > 0$, $y > 0$. If it is assumed for the time being that $M(\alpha, p)$ has no singular points for all Re $p \geq 0$, the Laplace inversion integral can be approximated by collecting the residues from the poles at $p = \pm i\alpha(\alpha^2 + \beta^2)^{-1/2}$. Thus we have

$$\varrho = \zeta = \int\!\!\!\int_{-\infty}^{\infty} d\alpha \, d\beta \, [K(\alpha, \beta) \, e^{itf_-(\alpha,\beta)} - K^*(\alpha, \beta) \, e^{itf_+(\alpha,\beta)}] \tag{3.6}$$

where the sine integral has been re-expressed in the exponential form, and

$$\begin{matrix} K(\alpha, \beta) \\ K^*(\alpha, \beta) \end{matrix} = \frac{1}{4\pi^2} \frac{\beta}{\alpha(\alpha^2 + \beta^2)^{1/2}} M\left(\alpha, \mp \frac{i\alpha}{(\alpha^2 + \beta^2)^{1/2}}\right), \tag{3.7}$$

$$f_{\pm}(\alpha, \beta) = \mu\alpha + \nu\beta \pm \frac{\alpha}{(\alpha^2 + \beta^2)^{1/2}}, \tag{3.8}$$

$$\mu = x/t, \quad \nu = y/t. \tag{3.9}$$

The extended method of stationary phase for double integrals, well-known in diffraction theory [13], [14], will now be applied to the integrals in Eq. (3.6). The critical points are found from

$$\frac{\partial f_{\pm}}{\partial \alpha} = 0, \quad \frac{\partial f_{\pm}}{\partial \beta} = 0 \tag{3.10}$$

which lead to

$$\mu \pm \frac{\beta^2}{(\alpha^2 + \beta^2)^{3/2}} = 0 \tag{3.11}$$

$$\text{for } f_{\pm}$$

$$\nu \mp \frac{\alpha\beta}{(\alpha^2 + \beta^2)^{3/2}} = 0. \tag{3.12}$$

Consider first the upper signs, i.e., f_+. For positive square roots and $\mu > 0$, no real solution exists for Eq. (3.11). Next consider te lower signs (f_-). Since $\nu > 0$, α and β must be of opposite sign for real solutions to Eq. (3.12). Hence, there are no critical points for f_+ and two for f_-:

$$\text{I: } (\alpha_0, -\beta_0), \quad \text{II: } (-\alpha_0, \beta_0) \tag{3.13}$$

with

$$\alpha_0 = \frac{\mu\nu}{(\mu^2 + \nu^2)^{3/2}} = \frac{txy}{r^3}, \quad \beta_0 = \frac{\mu^2}{(\mu^2 + \nu^2)^{3/2}} = \frac{tx^2}{r^3}. \tag{3.14}$$

At the critical points it can be easily calculated that

$$f = \mp \frac{y}{r}, \quad f_{\alpha\beta}^2 - f_{\alpha\alpha}f_{\beta\beta} = \frac{x}{t}, \quad \text{at} \quad \alpha = \pm\alpha_0, \quad \beta = \mp\beta_0. \tag{3.15}$$

Finally, the leading term of the asmyptotic expansion for large t and fixed x/t, y/t is

$$\varrho = \zeta \cong \frac{2\pi}{t} |f_{\alpha\beta}^2 - f_{\alpha\alpha}f_{\beta\beta}|_0^{-1/2} \left[K(\alpha_0, -\beta_0) e^{itf_-(\alpha_0, -\beta_0)} + K(-\alpha_0, \beta_0) e^{itf_-(-\alpha_0, \beta_0)} \right.$$

$$= -\frac{1}{2\pi} \frac{r^2}{t^{3/2} x^{1/2} y} \left[M_1 e^{\frac{-ity}{r}} + M_2 e^{\frac{ity}{r}} \right] \tag{3.16}$$

where

$$M_1 = M(\alpha, p)|_{\substack{\alpha = \alpha_0, \\ p = \frac{-iy}{r}}}, \quad M_2 = M(\alpha, p)|_{\substack{\alpha = -\alpha_0 \\ p = \frac{iy}{r}}}. \tag{3.17}$$

The next term in the asymptotic expansion is at least $0(t^{-1})$ higher [14].

It should be remarked that as $x \to 0$ the two critical points coalesce. However, the behavior of $f_-(\alpha, \beta)$ near that point ($\alpha = 0$, $\beta = 0$) ceases to be regular and Taylor's expansion is impossible (since $f_{\alpha\alpha}$, $f_{\beta\beta} \to \infty$). The usual asymptotic approximation about a caustic therefore fails. From now on the neighborhood of $x = 0$ is excluded from consideration.

We now assume specifically that the interface is an ellipse with constant area. The initial major and minor axes are 1 and b_0, respectively:

$$\eta(x, t) = \frac{b_0}{a} (1 - x^2/a^2)^{1/2}, \quad |x| < a, \quad a = a(t). \tag{3.18}$$

As suggested by the experimental investigation of an initially circular interface [2], we further postulate that the lengthening of the major axis can be described by a similar relation (cf. Eqs. (2.32) and (2.33)):

$$a(t) = (1 + t/T_0)^{1/2} \tag{3.19}$$

where T_0 is an empirical constant. The function $M(\alpha, p)$ is then

$$M(\alpha, p) = \pi b_0 \left[p^2 \int_0^\infty d\tau \, e^{-p\tau} \frac{J_1(\alpha\sqrt{1 + \tau/T_0})}{\alpha\sqrt{1 + \tau/T_0}} - \frac{p}{\alpha} J_1(\alpha) \right]. \tag{3.20}$$

In the bracket the Laplace integral represents the effect of continuing motion of the interface, whereas the remaining term pJ_1/α represents the effect of the initial configuration. It can be shown that the integral term has no singular point for Re $p \geq 0$. Substituting as indicated by Eq. (3.17), we have

$$\varrho = \zeta \cong -\frac{1}{2\pi} \frac{r^2}{t^{3/2}x^{1/2}y} 2 \, \text{Re} \left[M_1 e^{\frac{-ity}{r}} \right] \tag{3.21}$$

where

$$M_1 = \frac{\pi b_0 r^2}{tx} \left\{ -\frac{y}{r} \int_0^\infty d\tau \, e^{\frac{iy\tau}{r}} \frac{J_1(\alpha_0\sqrt{1 + \tau/T_0})}{\sqrt{1 + \tau/T_0}} + i J_1(\alpha_0) \right\}. \tag{3.22}$$

Now, for $\alpha_0 = \frac{txy}{r^3} \gg 1$, e.g., $x, y = 0(1)$, the integral term in the bracket is of order $0(\alpha_0^{-1})$ (Appendix B), while the term representing the initial shape dominates and gives

$$\varrho = \zeta \cong -\sqrt{\frac{2}{\pi}} b_0 \frac{r^{1^1/_2}}{t^3 x^2 y^{3/2}} \cos\left(\frac{txy}{r^3} - \frac{3\pi}{4}\right) \sin\frac{yt}{r}, \qquad \frac{txy}{r^3} \gg 1. \tag{3.23}$$

For either very small y or very large x, $\alpha_0 = txy/r^3$ can be much less than 1. Asymptotic evaluation of the integral in Eq. (3.22) for small α_0 leads to the following formula (see Appendix B):

$$\varrho = \zeta \cong \frac{b_0 T_0}{2} \frac{y}{t^{3/2}x^{1/2}} \cos\frac{yt}{r}, \qquad \frac{txy}{r^3} \gg 1. \tag{3.24}$$

3.b Discussion

The physical significance of the present theory and its relevance to the experimental results of Wᴜ [6, 12] will be discussed on the basis of Eqs. (3.23) and (3.24). The phase function is

$$\varphi = -\alpha_0 x + \beta_0 y + \alpha_0 t(\alpha_0^2 + \beta_0^2)^{-1/2} = yt/r \tag{3.25}$$

from which the following characteristics of the local wave pattern can be deduced:

constant phase line $\qquad x = (t^2 - \varphi^2)^{1/2} \, y$ $\qquad\qquad\qquad\qquad$ (3.26)

frequency $\qquad\qquad \omega = \alpha_0(\alpha_0^2 + \beta_0^2)^{-1/2} = y/r$ $\qquad\qquad$ (3.27)

wave number $\qquad \vec{k} = \alpha_0\vec{i} - \beta_0\vec{j} = \frac{tx}{r^2}\left(\frac{y}{r}\vec{i} - \frac{x}{r}\vec{j}\right) = \frac{\varphi}{r}\frac{x}{y}\left(\frac{y}{r}\vec{i} - \frac{x}{r}\vec{j}\right),$ $\quad$ (3.28)

wave length $\qquad l = \frac{2\pi r^2}{tx} = \frac{2\pi ry}{\varphi x}$ $\qquad\qquad\qquad\qquad$ (3.29)

phase velocity $\qquad \vec{C} = \frac{\omega\vec{k}}{k^2} = \frac{yr}{xt}\left(\frac{y}{r}\vec{i} - \frac{x}{r}\vec{j}\right) = \frac{y^2}{\varphi x}\left(\frac{y}{r}\vec{i} - \frac{x}{r}\vec{j}\right)$ $\qquad$ (3.30)

group velocity $\qquad \vec{C_g} = \frac{\partial\omega(\vec{k})}{\partial\vec{k}} = \frac{r}{t}\left(\frac{x}{r}\vec{i} + \frac{y}{r}\vec{j}\right).$ $\qquad\qquad$ (3.31)

Equation (3.26) indicates that a constant phase line is a straight line passing through the origin, with the slope $dy/dx = (t^2 - \varphi^2)^{-1/2}$. As time progresses the crests and the troughs turn from vertical towards horizontal. Since the square root is real only where $t > \varphi$, the numerical value of φ represents the threshold time at which a phase line first appears. In particular the successive crests and troughs have the threshold times

$$t_c = \left(n + \frac{1}{2}\right)\pi, \qquad n = 0, 1, 2, 3, \ldots \tag{3.32}$$

Equations (3.30) and (3.31) imply that the group velocity is everywhere perpendicular to the phase velocity, with their vector sum pointing in the positive x direction. This feature is already present in monochromatic waves and is a consequence of the continuity equation for an incompressible fluid (PHILLIPS [11]). As a check, a simple calculation produces the following expression for the fluid-velocity vector:

$$\vec{q} = u\vec{i} + v\vec{j} = -\frac{r^2}{t^{3/2}x^{1/2}y}\,\mathrm{Im}\,(M_1 e^{-ity/r})\,\vec{e}_r, \quad t \gg 1. \tag{3.33}$$

The energy flux $p\vec{q}$ is also radially outward along the phase lines.

To an observer fixed in space, the waves approach successively with decreasing wavelengths, since φ increases. The magnitude of the phase velocity is larger for smaller φ. Hence longer waves are always ahead of and outrun the shorter waves. If one follows a particular phase line, the wavelength is seen to increase with x; therefore all the waves are being drawn out as they propagate. These typical features of dispersion are quite similar to the transient wave propagation due to an initial disturbance on the surface of homogeneous water.

The amplitude variation in the region of moderate x, y and large t follows from Eq.(3.23) as

$$A(x, t) = \sqrt{\frac{2}{\pi}}\,\frac{b_0}{t^3}\,\frac{r^{5^1/_2}}{x^2 y^{3/2}}\cos\left(\frac{txy}{r^3} - \frac{3\pi}{4}\right) \tag{3.34a}$$

$$= \sqrt{\frac{2}{\pi}}\,\frac{y^{3/2}}{\varphi^3}\,\frac{r^{2^1/_2}}{x^2}\cos\left(\frac{x}{r^2}\varphi - \frac{3\pi}{4}\right). \tag{3.34b}$$

The cosine term evidently represents the interference effect due to the finite width of the homogeneous fluid. Being independent of T_0, Eq.(3.23) is entirely the result of the initial interface configuration. It follows that the wave pattern is largely caused by the initial impulse and not so much by the continual spreading of the interface. This was also stipulated by Wu [12]. From Eq.(3.34a) we see that the amplitude of the waves decreases very fast with t for a fixed observer. For a particular phase line, however, the amplitude increases slowly (as $x^{1/2}$) as the distance of travel increases.

For moderately large t but very small y or very large x such that $txy/r^3 \ll 1$, we have from Eq.(3.24) that

$$A(x, t) = \frac{b_0 T_0}{2}\,\frac{y}{t^{3/2}x^{1/2}} = \frac{b_0 T_0}{2}\,\frac{y^{5/2}}{\varphi^{3/2}x^2}. \tag{3.35}$$

Thus the amplitude of the constant phase line diminishes with increasing x as x^{-2}. In this region the effect of continual interface motion dominates over the initial impulse.

As mentioned before, existing experiments [6, 12] were performed for an initially circular interface, and quantitative comparison with the present theory is not possible. Nevertheless, several aspects such as the straight phase lines and their turning towards the x-axis, the variation of wave speeds and of wavelength, the interference effect, etc. appear to be in essential agreement in a qualitative sense. Two important differences still remain. First, the observed phase lines are propagated bodily downstream and do not remain radial. Second, for a fixed initial radius the observed maximum amplitude seems to occur around the ray $\theta = 60°$. Both features are most likely due to the finite initial height of the interface and are not predicted here.

Appendix A

On the possibility of shocks

The explicit equations for the straight characteristics are

$$C_\pm: x = \zeta \pm \eta^0(\xi)\,t$$

where ξ is a parameter denoting the initial position of the characteristic. Consider $x > 0$ only. For $\eta^0(x)$ positive and monotonically decreasing with x, only C_+ curves will intersect. The envelope of all C_+ characteristics is easily found to be a branch of a hyperbola

$$x = (1 + b^2 t^2)^{1/2}$$

in the first quadrant of the $x - t$ plane. Now, it would seem that doublevaluedness may occur on any point as an additional C_+ (coinciding with C_-) may arise from the undisturbed zone. Upon comparison with Eq. (2.30) it is seen that the envelope is precisely the path of the tip of the interior fluid (end of the major axis). Hence, the discontinuity, if it occurs, must occur at the wave front.

The shock conditions are derived from tne conservation Eq. (2.10) and the relation

$$(u\eta)_t + (u^2\eta + \eta^3/3)_x = 0,$$

which represents the conservation of momentum for the entire depth $0 < y < \eta$. It is obtained by multiplying Eq. (2.11) by η and using the continuity equation. Denoting the shock speed by C and the jump of a quantity across the shock by [ ], i.e., [ ] = ( )$_+$ − ( )$_-$, we than have

$$C[\eta] = [u\eta]$$

$$C[u\eta] = [u^2\eta + \eta^3/3].$$

If in particular the fluid is at rest in front of the shock, i.e., $u_- = \eta_- = 0$, the only possibility is that $\eta_+ = 0$ also. It follows that the interface must be continuous at the wave front, although the fluid velocity is not.

Appendix B

Asymptotic evaluation of an integral

We consider the integral

$$I = \int_0^\infty d\tau\, e^{\frac{iy\tau}{r}} \frac{J_1(\sqrt{1+\tau/T_0})}{\sqrt{1+\tau/T_0}} \tag{B.1}$$

$$\alpha_0 = txy/r^3. \tag{B.2}$$

By the transformation $1 + \tau/T_0 = \xi^2$, I can be written as

$$I = 2T e^{\frac{-iyT_0}{r}} \int_1^\infty d\xi\, J_1(\alpha_0\xi)\, e^{i\frac{y}{r}T_0\xi^2}. \tag{B.3}$$

Large α_0: Approximating the Bessel function for large argument and transforming again from ξ to z by $\xi = z(\alpha_0 r/yT_0)$, we have

$$I \cong T_0\sqrt{\frac{2}{\pi\alpha_0}}\, e^{\frac{-iyT_0}{r}} \left[N_1 e^{\frac{-i3\pi}{4}} + N_2 e^{\frac{i3\pi}{4}} \right]$$

where

$$N_{1,2} = \sqrt{\gamma} \int_{1/\gamma}^\infty dz\, z^{-1/2}\, e^{i\lambda(\pm z + z^2)}$$

with

$$\gamma = \alpha_0 r/yT$$

and

$$\lambda = \alpha_0\gamma = \alpha_0^2 r/yT_0 \gg 1.$$

The method of steepest descent can now be applied to the integrals N_1 and N_2 in a standard manner. It suffices to point out that for N_1, the steepest path in the complex z plane is a branch of hyperbola from $z = 1/\gamma$ to infinity ($\infty e^{i\pi/4}$). There is no saddle point, and the integral is of the order $0(\alpha_0^{-1})$. For N_2 the path can be deformed to a hyperbola going to $\infty e^{5i\pi/4}$ and returning along a straight path crossing the saddle point at $z = \frac{1}{2}$ at an angle $\frac{\pi}{4}$ with the real axis. The saddle-point contribution dominates and gives

$$N_2 = \sqrt{\frac{2\pi}{\alpha_0}} \exp\left\{ -\frac{i}{4}\left(\frac{t^2x^2y}{T_0 r^5} - \pi \right) \right\} \cdot [1 + 0(\alpha_0^{-1/2})]$$

$$I = -\frac{2T_0}{\alpha_0} \exp\left\{ -i\frac{yT_0}{r}\left[1 + \left(\frac{tx}{2T_0 r^2} \right)^2 \right] \right\} [1 + 0(\alpha_0^{-1/2})]. \tag{B.4}$$

Small α_0: Letting $\alpha_0\xi = z$ in Eq. (B.3), we have

$$I = \frac{2T_0}{\alpha_0} e^{\frac{-iyT_0}{r}} \int\limits_{\alpha_0}^{\infty} dz\, J_1(z)\, e^{i\lambda z^2}$$

with

$$\lambda = yT_0/r\alpha_0^2 \gg 1.$$

The path of steepest descent is from α_0 to $\infty\, e^{i\pi/4}$ along a hyperbola in the complex z plane. The leading term is

$$I = \frac{itx}{2r^2} e^{\frac{-iyT_0}{r}} [1 + 0(\alpha_0)]. \tag{B.5}$$

Acknowledgment

The author owes his initial interest in this problem to JOHN F. KENNEDY, State University of Iowa. Thanks are also due to RICHARD BARAKAT for an important reference [14]. The financial sponsorship of the U.S. Office of Naval Research under Contract NONR-1841-59 is gratefully acknowledged.

References

1. SCHOOLEY, A. H., and R. W. STEWART: Experiments with a self-propelled body submerged in a fluid with a vertical density gradient, J. Fluid Mechanics 9, 83—96 (1963).
2. WU, J.: Collapse of turbulent wakes in density stratified media, Report 231—214, Hydronautics, Inc., 1965.
3. STOCKHAUSEN, P. J., C. B. CLARK and J. F. KENNEDY: Three-dimensional momentumless wakes in density stratified liquids, Technical Report No. 93, Hydrodynamics Laboratory, Massachusetts Institute of Technology, 1966.
4. VAN DE WATERING, W. P. M.: The growth of turbulent wake in a density stratified fluid, Report 231—212, Hydronautics, Inc., 1966.
5. SCHOOLEY, A. H.: Wake collapse in a stratified fluid, Science 157, 150—155 (1967).
6. WU, J.: Flow phenomena caused by the collapse of a mixed region in a density stratified medium, Report 231—211, Hydronautics, Inc., 1966.
7. PENNEY, W. G., and C. K. THORNHILL: The dispersion, under gravity, of a column of fluid supported on a rigid horizontal plane, Phil. Trans. Royal Soc. London, Series A, 244, 1952, pp. 285—311.
8. COURANT, R., and K. O. FRIEDRICHS: Supersonic flows and shock waves, Interscience, 1949.
9. AMES, W.: Nonlinear Partial Differential Equatiots in Engineering, Academic Press 1965.
10. WONG, K. K.: On internal gravity waves generated by local disturbances, Report 231—235, Hydronautics, Inc., 1965.
11. PHILLIPS, O.; M.: The dynamics of the upper ocean, Cambridge University Press 1966.
12. WU, J.: Wave collapse and subsequent generation of internal waves in a density stratified medium, Report 231—217, Hydronautics Inc., 1968.
13. BORN, M., and E. WOLF: Principles of optics, Pergamon Press, 1959.
14. CHAKO, N.: Asymptotic expansions of double and multiple integrals occurring in diffraction theory. J. Institute of Mathematics and Applications 1, 372—422 (1965).

Stability of rotors with unsymmetrical flexible bearings

By

P. T. Pedersen

Technical University of Denmark, Copenhagen

Introduction

In classical studies of rotors supported by flexible bearings, isotropy of the bearings is assumed in most cases. F. M. DIMENTBERG [1], A. TONDL [2], and W. KELLENBERGER [3] have treated the problem of rotors with anisotropic flexible bearings with the principal axes of flexibility being in parallel. The present paper deals with the problem of rotors consisting of n discs on a massless shaft mounted in unsymmetrical flexible bearings, i.e. flexible bearings having their principal axes independently oriented. The flexural rigidity of the rotors is assumed to be isotropic.

The first part deals with free and forced vibrations when damping is neglected. It is shown that in free motion the centerline of the rotor is confined to surfaces containing the centerline of the undeflected rotor, so that points of the centerline are moving in straight lines. In forced motion points of the centerline move in ellipses the parameters of which depend on the distribution of rotor unbalance.

In the second part the influence of damping is treated. The following two types of linear damping forces are taken into account: external damping proportional to the velocity of the motion, and internal structural damping proportional to the rate of change of curvature of the shaft. It is found that internal damping causes self-excited vibrations at speeds of revolution in excess of certain speeds. The limits of stability is defined as the angular velocity at which the greatest real part of all the complex eigenvalues of the governing differential equation is zero. The corresponding imaginary part of the eigenvalue is called the whirling speed.

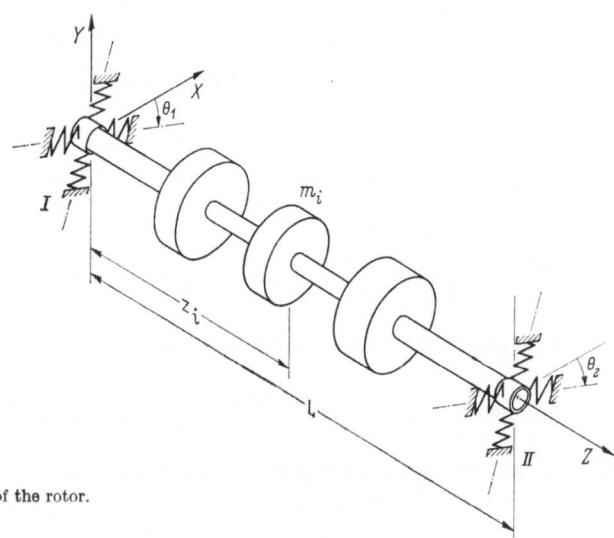

Fig. 1. Geometry of the rotor.

42*

If the rotor consists of a massless shaft with a single disc, the solution of the linear differential equation yields simple expressions for the limit of stability and the corresponding whirling frequency. Upper and lower limits are found for the whirling frequency expressed by the two critical frequencies for a system without damping. For vanishing external damping the whirling frequency is independent of the internal damping and also of the orientation of the principal axes of the bearings, and hence it depends only of the four principal flexibilities of the two bearings. In this case it is also shown that the limit of stability increases with the difference between the two critical frequencies without damping.

Figure 1 shows a simplified model of the rotor in a system of rectangular coordinates XYZ fixed in space. The model consists of n unbalanced discs on a massless isotropic shaft, supported by two anisotropic flexible bearings (I and II). The principal flexibilities of bearing I are C_{11}^{I} and C_{22}^{I}, and the angle between the principal direction of flexibility C_{11}^{I} and the XZ-plane is θ_1. Correspondingly for bearing II the principal flexibilities are C_{11}^{II} and C_{22}^{II}, and the angle between the principal direction of flexibility C_{11}^{II} and the XZ-plane is θ_2. The mass of disc number i is m_i.

Figure 2 shows the projection of dics number i on the XY-plane. C denotes the geometrical center of the deflected shaft and has the XY-coordinates (u_{2i-1}, u_{2i}). The distance between the center of the shaft C and the center of mass of the disc G is ε_i. The angular velocity of the rotor about the axis OZ is Ω. The angle between the line CG and a plane, which rotates with the rotor and contains the centerline of the undeflected shaft is denoted α_i. Torsional vibrations are not taken into account, which means that the angle α_i is independent of time t. The gyroscopic effect of the discs is also neglected.

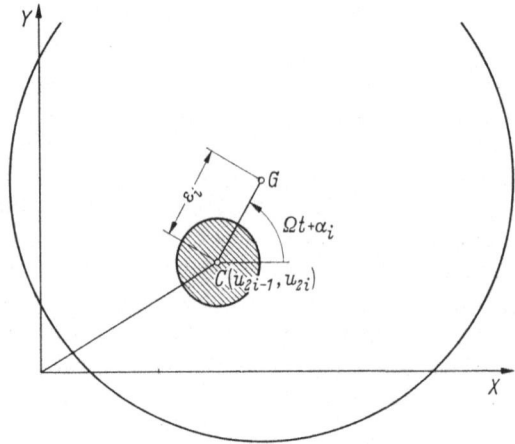

Fig. 2. Unbalance of disc number i.

Rotor without damping

Equations of motion

The equations of motion for the system without damping are derived by means of Lagrange's equations. The kinetic energy T becomes

$$T = \frac{1}{2} \sum_{i=1}^{n} m_i \{ r_i^2 \Omega^2 + [\dot{u}_{2i-1} - \varepsilon_i \Omega \sin (\Omega t + \alpha_i)]^2 + [\dot{u}_{2i} + \varepsilon_i \Omega \cos (\Omega t + \alpha_i)]^2 \} \qquad (1)$$

where r_i is the radius of inertia of disc number i. The potential energy takes the form

$$V = \frac{1}{2} \sum_{j=1}^{2n} \sum_{k=1}^{2n} s_{jk} u_j u_k \qquad (2)$$

where s_{jk} are influence coefficients of stiffness of the n masses. These influence coefficients are functions of the angles θ_1 and θ_2. As it is assumed, that the shaft has the same flexural rigidity in all directions perpendicular to the centerline the influence coefficients are independent of the time t and of the angular velocity Ω.

Employing Lagrange's equations

$$\frac{d}{dt}\left(\frac{\partial L}{\partial \dot{u}_j}\right) - \frac{\partial L}{\partial u_j} = 0 \qquad (j = 1, 2, \ldots, 2n) \tag{3}$$

where $L = T - V$, the equations of motion are obtained as

$$m_i \ddot{u}_{2i-1} + \sum_{j=1}^{2n} s_{ij} u_j = \Omega^2 m_i \varepsilon_i \cos (\Omega t + \alpha_i) \qquad (i = 1, 2, \ldots, n) \tag{4}$$

and

$$m_i \ddot{u}_{2i} + \sum_{j=1}^{2n} s_{ij} u_j = \Omega^2 m_i \varepsilon_i \sin (\Omega t + \alpha_i) \qquad (i = 1, 2, \ldots, n). \tag{5}$$

The equations of motion (4) and (5) can be expressed in matrix form as

$$[M]\{\ddot{U}\} + [S]\{U\} = \Omega^2[M]\{F\} \tag{6}$$

where $[M]$ is a diagonal matrix of order $2n$,

$$[M] = \begin{bmatrix} m_1 & 0 & 0 & 0 & 0 & \cdot & 0 \\ 0 & m_1 & 0 & 0 & 0 & \cdot & 0 \\ 0 & 0 & m_2 & 0 & 0 & \cdot & 0 \\ 0 & 0 & 0 & m_2 & 0 & \cdot & 0 \\ \cdot & \cdot & \cdot & \cdot & \cdot & \cdot & \cdot \\ 0 & \cdot & \cdot & \cdot & 0 & m_n & 0 \\ 0 & \cdot & \cdot & \cdot & 0 & 0 & m_n \end{bmatrix} \tag{7}$$

$[S]$ is the stiffness matrix the elements of which are the influence coefficients of stiffness s_{ij}, $\{U\}$ is a column matrix of order $2n$ representing the displacement coordinates u_j, and $\{F\}$ is a column matrix of order $2n$ with the elements

$$\begin{aligned} f_{2i-1} &= \varepsilon_i \cos (\Omega t + \alpha_i) \\ f_{2i} &= \varepsilon_i \sin (\Omega t + \alpha_i) \end{aligned} \qquad (i = 1, 2, \ldots, n). \tag{8}$$

Critical angular velocities

The solution of the inhomogeneous linear differential Eq. (6) consists of the general solution of the homogeneous equation representing the free vibrations plus a particular integral of the inhomogeneous equation representing the forced vibration. As the solution of the homogeneous part of (6) is of the form $\{A\} \cos \omega t$ or $\{B\} \sin \omega t$ angular velocities which equal ω will be critical; in this case there will be resonance between free and forced vibrations. Substituting $\{U\} = \{V\} e^{i\omega t}$ in the homogeneous part of (6) where i equals $\sqrt{-1}$ we obtain

$$(-\omega^2[M] + [S])\{V\} = \{0\}. \tag{9}$$

In order to have a nontrivial solution of (9), the determinant of the coefficient matrix must equal zero, i.e.

$$\det ([S] - \omega^2[M]) = 0. \tag{10}$$

As $[S]$ and $[M]$ are both symmetrical and positive definite the roots of (10)

$$\omega_1^2, \, \omega_2^2, \, \omega_3^2, \, \ldots, \, \omega_{2n}^2$$

are all real and positive. If the bearings of the rotor in consideration are anisotropic the $2n$ critical angular velocities are normally distinct.

Free critical motion

If the critical angular velocity ω_r is distinct from all the other critical angular velocities, then the corresponding mode $\{U\}_r$ can be expressed as

$$\{U\}_r = (s_1 \cos \omega_r\, t + s_2 \sin \omega_r t)\, \{V\}_r \tag{11}$$

where $\{V\}_r$ is the modal column, and s_1 and s_2 are arbitrary coefficients. Eq. (11) shows, that the ratios between X- and Y-coordinates of the discs are constant, and independent of time t. Consequently, free vibrations of centers of the discs are represented by straight lines. The angles between these straight lines and the XZ-plane are

$$\beta_{ir} = \mathrm{Arctan}\frac{v_{2i-1,r}}{v_{2i,r}} \tag{12}$$

and the ratio between the amplitudes of disc number i and disc number j is

$$\frac{a_{ir}}{a_{jr}} = \frac{\sqrt{v_{2i-1,r}^2 + v_{2i,r}^2}}{\sqrt{v_{2j-1,r}^2 + v_{2j,r}^2}} \; . \tag{13}$$

Forced vibrations

A particular solution of the inhomogeneous equation of motion (6) represents a forced vibration of the discs caused by the unbalance. For a rotor in stable rotation, this is the motion which will remain when the free vibrations have been attenuated. If the solution of (6) is written in the form

$$\{U\} = \{A\} \cos \Omega t + \{B\} \sin \Omega t \tag{14}$$

then the column matrices $\{A\}$ and $\{B\}$ of orders $2n$, after reduction of $\cos \Omega t$ and $\sin \Omega t$, can be calculated from the equations

$$[S]\{A\} - \Omega^2[M]\{A\} = \Omega^2[M]\{C\} \tag{15}$$

and

$$[S]\{B\} - \Omega^2[M]\{B\} = \Omega^2[M]\{D\} \tag{16}$$

where $\{C\}$ and $\{D\}$ are column matrices with the elements c_j and d_j given by formulas

$$\begin{aligned} c_{2i-1} &= \varepsilon_i \cos \alpha_i \\ c_{2i} &= \varepsilon_i \sin \alpha_i \end{aligned} \qquad (i = 1, 2, \ldots, n) \tag{17}$$

and

$$\begin{aligned} d_{2i-1} &= -\varepsilon_i \sin \alpha_i \\ d_{2i} &= \varepsilon_i \cos \alpha_i \end{aligned} \qquad (i = 1, 2, \ldots, n). \tag{18}$$

The solution (14) shows, that in forced vibration the centers of the discs move in elliptic orbits.

The forced motion becomes unbounded, when the imposed angular velocity Ω approaches any of the critical values

$$\omega_1, \omega_2, \ldots, \omega_{2n}.$$

In any such case the direction of the greatest principal axis of the ellipse will coincide with the directions of the lines along which the center moves in free vibrations of the corresponding critical angular velocities, and the eccentricity of the ellipse will tend to infinity. Thus the rotor behaves as if it looses the flexural rigidity in a certain twisted plane, when the angular velocity of the shaft approaches a critical value.

Figure 3 shows how the first four critical angular velocities of a rotor with uniformly distributed mass vary with the mutual orientation of the principal axes of the bearings. For the numerical calculations the uniform elastic shaft was approximated by a massless shaft with 35 equally distributed concentrated masses. For θ_1 equal $\pi/4$ the corresponding surfaces of motion are shown in Fig. 4.

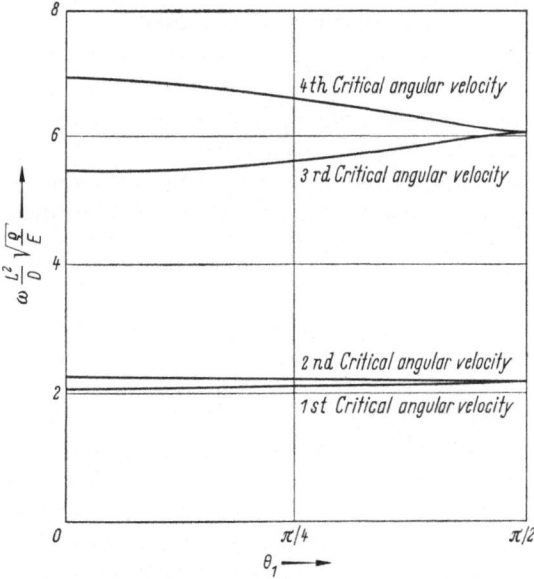

Fig.3. Critical angular velocities of a rotor with constant diameter D, density ϱ, and Young's modulus E, mounted in bearings with the flexibility matrices

$$\begin{bmatrix} \cos \theta_1 \sin \theta_1 \\ -\sin \theta_1 \cos \theta_1 \end{bmatrix} \times \begin{bmatrix} 0.2 & 0 \\ 0 & 0.1 \end{bmatrix} \times \begin{bmatrix} \cos \theta_1 & -\sin \theta_1 \\ \sin \theta_1 & \cos \theta_1 \end{bmatrix} \times \frac{L^3}{ED^4}$$

and

$$\begin{bmatrix} 0.2 & 0 \\ 0 & 0.1 \end{bmatrix} \times \frac{L^3}{ED^4}.$$

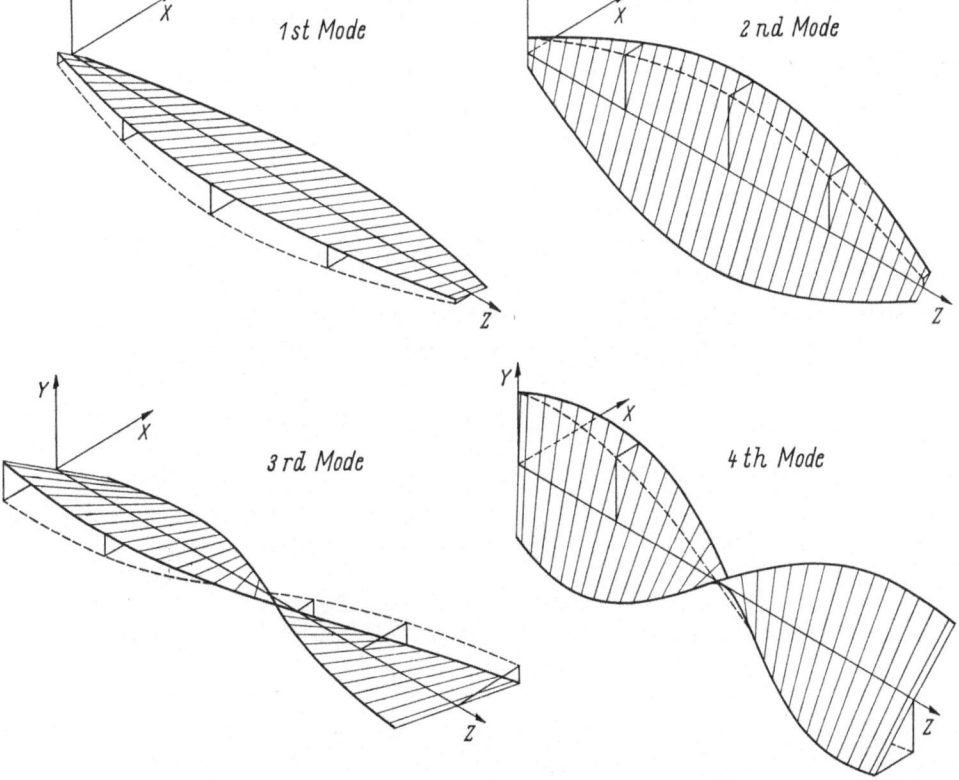

Fig.4. Vibration surfaces corresponding to the critical angular velocities shown in fig.3 for $\theta_1 = \pi/4$.

336   P. T. PEDERSEN

Rotor with damping

Damping assumptions

In our study of the influence of damping on the stability of the rotor, it is assumed that internal as well as external damping forces are acting.

The external damping is assumed to be proportional to the velocity of motion. If the damping coefficient is b, the X-component of the damping force on disc number i will then be

$$d_{ix}^e = -b\dot{u}_{2i-1} \tag{19}$$

and the Y-component

$$d_{iy}^e = -b\dot{u}_{2i}. \tag{20}$$

The internal structural damping is assumed to be proportional to the rate of change of curvature of the shaft and acting as a set of concentrated forces at the discs. Fig. 5 shows the projection of disc number i of the deflected system in the $\dot{X}Y$-plane. A system of axes ξ, η rotating

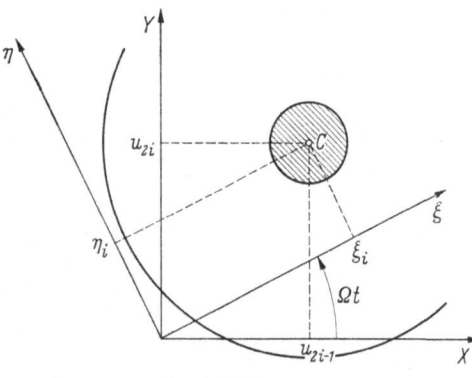

Fig. 5. Rotating axes.

with the imposed angular velocity Ω is introduced. Since torsional vibrations are not taken into account the angle between a diameter, fixed in the shaft, and the rotating coordinate system will be independent of time. The position of the center of disc number i is denoted by (ξ_i, η_i, z_i) in the rotating coordinate system. If the ξ-component of the curvature of the shaft is approximated by

$$\left(\frac{\Delta^2\xi}{\Delta z^2}\right)_i = \frac{\dfrac{\xi_{i+1}-\xi_i}{z_{i+1}-z_i}-\dfrac{\xi_i-\xi_{i-1}}{z_i-z_{i-1}}}{\dfrac{1}{2}(z_{i+1}-z_{i-1})} \tag{21}$$

and similarly for the η-component, the component of the equivalent load expressed in the rotating system in the ξ-direction becomes

$$d_{i\xi}^s = \frac{1}{2}c(z_{i+1}-z_{i-1})\frac{d}{dt}\left(\frac{\Delta^2\xi}{\Delta z^2}\right)_i. \tag{22}$$

Similarly the component in the η-direction becomes

$$d_{i\eta}^s = \frac{1}{2}c(z_{i+1}-z_{i-1})\frac{d}{dt}\left(\frac{\Delta^2\eta}{\Delta z^2}\right)_i. \tag{23}$$

Here c is a structural damping coefficient. If the components (22) and (23) are transformed to the fixed coordinate system, the component of the structural damping force in the X-direction becomes

$$d_{ix}^s = \frac{1}{2}c(z_{i+1}-z_{i-1})\left[\frac{d}{dt}\left(\frac{\Delta^2 x}{\Delta z^2}\right)_i + \Omega\left(\frac{\Delta^2 y}{\Delta z^2}\right)_i\right] \tag{24}$$

while the component in the Y-direction will be

$$d_{iy}^s = \frac{1}{2}c(z_{i+1}-z_{i-1})\left[\frac{d}{dt}\left(\frac{\Delta^2 y}{\Delta z^2}\right)_i - \Omega\left(\frac{\Delta^2 x}{\Delta z^2}\right)_i\right] \tag{25}$$

where

$$\left(\frac{\Delta^2 x}{\Delta z^2}\right)_i = \frac{\dfrac{u_{2i+1}-u_{2i-1}}{z_{i+1}-z_i}-\dfrac{u_{2i-1}-u_{2i-3}}{z_i-z_{i-1}}}{\dfrac{1}{2}(z_{i+1}-z_{i-1})} \tag{26}$$

and where

$$\left(\frac{\Delta^2 y}{\Delta z^2}\right)_i = \frac{\dfrac{u_{2i+2}-u_{2i}}{z_{i+1}-z_i}-\dfrac{u_{2i}-u_{2i-2}}{z_i-z_{i-1}}}{\dfrac{1}{2}(z_{i+1}-z_{i-1})}. \tag{27}$$

If the bearings of the rotor are massless the approximations of curvature (26) and (27) for $i = 1$ and $i = n$ contain the displacements of the bearings. In analogy to the notation of the coordinates of the displacements of the discs, the displacements of the bearings are denoted by (u_{-1}, u_0) and (u_{2n+1}, u_{2n+2}). In order to reduce the number of equations of motion to $2n$, the displacements of the bearings are expressed by the displacements of the n discs (u_i; $i = 1, 2, \dots, 2n$). This is obtained in two steps. First, note that

$$
\begin{Bmatrix} u_{-1} \\ u_0 \\ u_{2n+1} \\ u_{2n+2} \end{Bmatrix} = [F_B]\{P\} \tag{28}
$$

where the elements p_j of the column matrix $\{P\}$ denotes the forces at the discs. The rectangular matrix $[F_B]$ of order $4 \times 2n$ is the corresponding flexibility matrix, the elements of which are the displacements of the bearings caused by unit forces at the discs. However, the relation between the forces at the discs and the displacements of the discs can be expressed by means of the stiffness matrix $[S]$

$$
\{P\} = [S]\{U\}. \tag{29}
$$

Substituting (29) into (28) we finally find that the displacements of bearings can be written as

$$
\begin{Bmatrix} u_{-1} \\ u_0 \\ u_{2n+1} \\ u_{2n+2} \end{Bmatrix} = [F_B][S]\{U\}. \tag{30}
$$

Equations of motion

The equations of motion are obtained by use of Lagrange's equation in the form

$$
\frac{d}{dt}\left(\frac{\partial L}{\partial \dot{u}_j}\right) - \frac{\partial L}{\partial u_j} = Q_j \qquad (j = 1, 2, \dots, 2n) \tag{31}
$$

where L is the difference between the kinetic energy (1) and the potential energy (2) and where Q_j represents the generalized damping forces, i.e.

$$
\begin{aligned} Q_{2i-1} &= d_{ix}^e + d_{ix}^s \\ Q_{2i} &= d_{iy}^e + d_{iy}^s \end{aligned} \qquad (i = 1, 2, \dots, n). \tag{32}
$$

Substitution yields the following equations

$$
\begin{aligned} &m_i \ddot{u}_{2i-1} + b\dot{u}_{2i-1} - \frac{1}{2}c(z_{i+1} - z_{i-1})\left[\frac{d}{dt}\left(\frac{\Delta^2 x}{\Delta z^2}\right)_i + \Omega\left(\frac{\Delta^2 y}{\Delta z^2}\right)_i\right] \\ &+ \sum_{j=1}^{2n} s_{ij}u_j = \Omega^2 m_i \varepsilon_i \cos(\Omega t + \alpha_i) \qquad (i = 1, 2, \dots, n) \end{aligned} \tag{33}
$$

and

$$
\begin{aligned} &m_i \ddot{u}_{2i} + b\dot{u}_{2i} - \frac{1}{2}c(z_{i+1} - z_{i-1})\left[\frac{d}{dt}\left(\frac{\Delta^2 y}{\Delta z^2}\right)_i - \Omega\left(\frac{\Delta^2 x}{\Delta z^2}\right)_i\right] \\ &+ \sum_{j=1}^{2n} s_{ij}u_j = \Omega^2 m_i \varepsilon_i \sin(\Omega t + \alpha_i) \qquad (i = 1, 2, \dots, n). \end{aligned} \tag{34}
$$

In matrix form the Eqs. (33) and (34) can be written as

$$
[M]\{\ddot{U}\} + [C_1]\{\dot{U}\} + [B]\{\dot{U}\} + \Omega[C_2]\{U\} + [S]\{U\} = \Omega^2[M]\{F\}. \tag{35}
$$

The square matrices $[M]$ and $[S]$ and the column matrix $\{F\}$ represent the same quantities as those used in the equation of motion (6) for the rotor without damping. The square matrices $[C_1]$ and $[C_2]$ of orders $2n$ result from the internal damping and the matrix $[B]$ is a diagonal matrix of order $2n$ which is due to the external damping. It should be observed, that the matrix $[S]$ and, in the case of massless bearings, also the matrices $[C_1]$ and $[C_2]$ are functions of the orientation of the principal axes of the bearings.

Limits of stability

In order to find the limit of stability for the system, free vibrations are studied, i.e. the homogeneous part of (35) is examined.

$$[M]\{\ddot{U}\} + ([C_1] + [B])\{\dot{U}\} + (\Omega[C_2] + [S])\{U\} = \{0\}. \tag{36}$$

This second order differential equation can be written as the following differential equation of the first order

$$\{\dot{V}\} - [K]\{V\} = \{0\} \tag{37}$$

where $[K]$ is a square matrix of order $4n$ and defined by

$$[K] = -\begin{bmatrix} [0] & [M] \\ [M] & [C_1]+[B] \end{bmatrix}^{-1} \times \begin{bmatrix} -[M] & [0] \\ [0] & \Omega[C_2]+[S] \end{bmatrix} \tag{38}$$

and $\{V\}$ is a column matrix also of order $4n$ and defined by

$$\{V\} = \begin{Bmatrix} \{\dot{U}\} \\ \{U\} \end{Bmatrix}. \tag{39}$$

In order to solve Eq. (37) a solution of the form

$$\{V\} = e^{\nu t}\{R\} \tag{40}$$

is assumed, where ν is a parameter yet to be determined and where $\{R\}$ is a column matrix. If (40) is substituted into (37) and the exponential $e^{\nu t}$ divided out of the resulting equation, the following homogeneous equation is obtained

$$([K] - \nu[I])\{R\} = \{0\} \tag{41}$$

where $[I]$ denotes the unity matrix of order $4n$. In order to have a nontrivial solution of (41), the determinant of the coefficients must equal zero, i.e.

$$\det([K] - \nu[I]) = 0. \tag{42}$$

Equation (42) shows, that the parameter ν can be determined as the eigenvalues of the square matrix $[K]$. In general, the matrix $[K]$ will have $4n$ complex eigenvalues. These complex eigenvalues occur as complex conjugate pairs with corresponding conjugate eigenvectors.

If all the eigenvalues ν_r have non-positive real parts, then it is seen from (40) and (39) that the amplitudes will be bounded at all times, and hence the free motion is stable. But if one of the eigenvalues has a positive real part, the amplitude will increase exponentially with time and the motion becomes unstable. Hence the problem of determining the limit of stability for the rotor is reduced to finding the angular velocity of the shaft Ω_{st}, at which the greatest real part of all the complex eigenvalues ν_j equals zero. The corresponding imaginary part of the eigenvalue is the whirling velocity.

Numerical example

In numerical calculations of limits of stability and whirling velocities for rotors with several discs the complex eigenvalues of the matrix $[K]$ in Eq. (42) can be found by a power method used in conjunction with a deflation method removing from the matrix eigenvalues already found. Then by successive interpolations those angular velocities are found which make the greatest real part of all the eigenvalues equal to zero.

This method has been applied to a rotor consisting of four discs each of mass m placed on a massless shaft of length L, diameter D, and Young's modulus E. The discs are spaced in such a manner that the distance between each consecutive pair of discs is $L/5$. The distances between the outer discs and the bearings are also assumed to be $L/5$. The flexibility matrices of the bearings of the rotor are

$$\begin{bmatrix} \cos\theta_1 & \sin\theta_1 \\ -\sin\theta_1 & \cos\theta_1 \end{bmatrix} \times \begin{bmatrix} 0.8 & 0 \\ 0 & 0.4 \end{bmatrix} \times \begin{bmatrix} \cos\theta_1 & -\sin\theta_1 \\ \sin\theta_1 & \cos\theta_1 \end{bmatrix} \times \frac{L^3}{ED^4}$$

and

$$\begin{bmatrix} 0.8 & 0 \\ 0 & 0.4 \end{bmatrix} \times \frac{L^3}{ED^4}.$$

For this rotor the limit of stability Ω_{st} and the corresponding whirling velocity ω_w is shown in Fig. 6 as functions of the coefficient of external damping and of the mutual orientation of the principal axes of the bearings at a given value of the coefficient of internal structural damping.

At angular velocities in excess of Ω_{st} the amplitude of one of the two free motions with whirling velocities near the first critical angular velocities ω_1 and ω_2 increases exponentially with time. Then at a certain angular velocity a limit is reached, at which one of the two free motions with whirling velocities near ω_3 and ω_4 becomes unstable—and so on. Fig. 7 shows these limits and the corresponding whirling velocities in the case of no external damping.

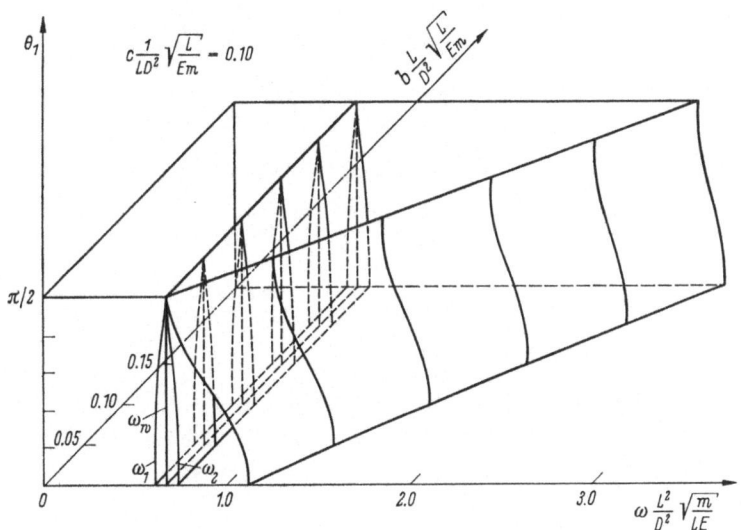

Fig. 6. Stability limit Ω_{st} and whirling velocity ω_w, for a rotor with four masses on a massless shaft, as functions of the mutual orientation of the principal axes of the bearings and of the external damping.

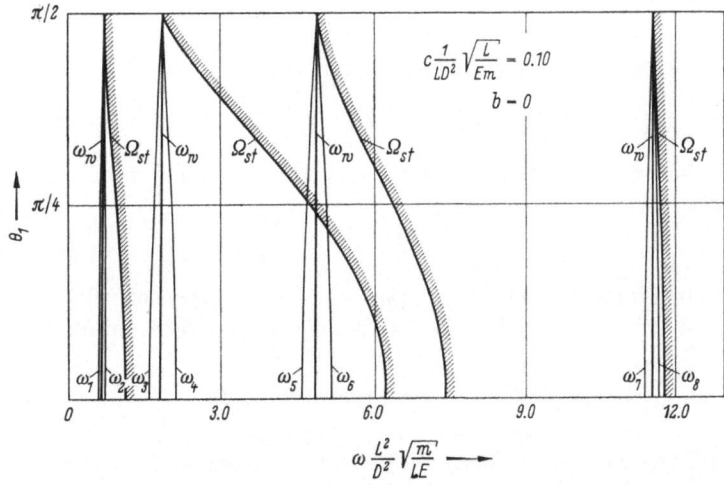

Fig. 7. Stability limits Ω_{st} and whirling velocities ω_w, for a rotor with four masses on a massless shaft, as functions of the mutual orientation of the principal axes of the bearings.

43*

A single disc on a massless shaft

In order to get a better understanding of the influence of the parameters involved, the special case of a rotor having a single disc on a massless shaft, will be considered. In this case, the simplified characteristic equation of (42) can be written in the following form

$$m^2 N \nu^4 + b^2 \nu^2 N + 2b\nu^3 mN + (m\nu^2 + b\nu)(1 + df\,\nu)(A + B + 2f)$$
$$+ (1 + df\,\nu)^2 + d^2 \Omega^2 f^2 = 0 \tag{43}$$

where

$$N = (A + f)(B + f) - C^2$$

f = flexibility of the shaft in rigid bearings

$$A = (C_{11}^{\mathrm{I}} \cos^2 \theta_1 + C_{22}^{\mathrm{I}} \sin^2 \theta_1)\left(\frac{L - z_1}{L}\right)^2$$
$$+ (C_{11}^{\mathrm{II}} \cos^2 \theta_2 + C_{22}^{\mathrm{II}} \sin^2 \theta_2)\left(\frac{z_1}{L}\right)^2$$

$$B = (C_{11}^{\mathrm{I}} \sin^2 \theta_1 + C_{22}^{\mathrm{I}} \cos^2 \theta_1)\left(\frac{L - z_1}{L}\right)^2$$
$$+ (C_{11}^{\mathrm{II}} \sin^2 \theta_2 + C_{22}^{\mathrm{II}} \cos^2 \theta_2)\left(\frac{z_1}{L}\right)^2$$

$$C = (C_{22}^{\mathrm{I}} - C_{11}^{\mathrm{I}})\left(\frac{L - z_1}{L}\right)^2 \cos \theta_1 \sin \theta_1$$
$$+ (C_{22}^{\mathrm{II}} - C_{11}^{\mathrm{II}})\left(\frac{z_1}{L}\right)^2 \cos \theta_2 \sin \theta_2$$

$$d = \frac{L}{z_1(L - z_1)}\,c.$$

At the limit of stability Ω_{st} the real part of ν is zero. In order to find Ω_{st}, ν in (43) is substituted by $i\omega_w$, where i equals $\sqrt{-1}$ and ω_w is the whirling velocity. The resulting equation is separated into its imaginary and real parts. The imaginary part becomes

$$-2bmN\omega_w^3 - m\,df(A + B + 2f)\,\omega_w^3 + b(A + B + 2f)\,\omega_w + 2df\,\omega_w = 0 \tag{44}$$

which gives the squared whirling velocity

$$\omega_w^2 = \frac{b(A + B + 2f) + 2df}{[2bN + df(A + B + 2f)]\,m}. \tag{45}$$

The real part of Eq. (43) becomes

$$m^2 N \omega_w^4 - b^2 N \omega_w^2 - m(A + B + 2f)\,\omega_w^2 - b\,df(A + B + 2f)\,\omega_w^2$$
$$+ 1 - d^2 f^2\,\omega_w^2 + d^2 \Omega_{st}^2 f^2 = 0 \tag{46}$$

which gives the following expression for the limit of stability

$$\Omega_{st}^2 = \frac{1}{d^2 f^2}\{[b^2 N + d^2 f^2 + (m + b\,df)(A + B + 2f)]\,\omega_w^2 - m^2 N \omega_w^4 - 1\}. \tag{47}$$

From the definitions of A, B, and C follows, that the flexibility matrix $[F]$ can be written as

$$[F] = \begin{bmatrix} A + f & C \\ C & B + f \end{bmatrix}. \tag{48}$$

The eigenvalues of (48) are equal to $1/(\omega_1^2 m)$ and $1/(\omega_2^2 m)$ respectively, where ω_1 and ω_2 are the critical angular velocities of the rotor without damping. Therefore

$$\left.\begin{matrix} \omega_1^2 \\ \omega_2^2 \end{matrix}\right\} = 2/\{m[(A + B + 2f) \pm \sqrt{A^2 + B^2 - 2AB + 4C^2}]\}. \tag{49}$$

By means of (49) the whirling velocity (45) can be written as

$$\omega_w^2 = \frac{b(\omega_1^2 + \omega_2^2) + 2m\,df\omega_1^2\omega_2^2}{2b + dfm(\omega_1^2 + \omega_2^2)} \tag{50}$$

and the limit of stability (47) is given by

$$\Omega_{st}^2 = \frac{1}{d^2 f^2} \left\{ \left[\frac{b^2}{m^2 \omega_1^2 \omega_2^2} + d^2 f^2 + (m + b\, df) \frac{\omega_1^2 + \omega_2^2}{m \omega_1^2 \omega_2^2} \right] \omega_w^2 - \frac{\omega_w^4}{\omega_1^2 \omega_2^2} - 1 \right\}. \tag{51}$$

If the ratio between the external and the internal coefficients of damping b/d increases, the whirling velocity ω_w tends asymptotically to the maximum value

$$\max(\omega_w^2) = \frac{1}{2}(\omega_1^2 + \omega_2^2). \tag{52}$$

At the same time, the limit of stability increases.

From Eq. (50) it is seen that the whirling speed for vanishing external damping has a minimum

$$\min(\omega_w^2) = \frac{2\omega_1^2 \omega_2^2}{(\omega_1^2 + \omega_2^2)}. \tag{53}$$

Introducing the flexibilities, this value can be written

$$\min(\omega_w^2) = 2 \Big/ \left\{ m \left[2f + \left(\frac{L - z_1}{L} \right)^2 (C_{11}^{I} + C_{22}^{I}) + \left(\frac{z_1}{L} \right)^2 (C_{11}^{II} + C_{22}^{II}) \right] \right\}. \tag{54}$$

In this case the whirling speed is independent of the orientation of the anisotropic bearings and of the structural damping. For vanishing external damping the limit of stability is given by

$$\Omega_{st}^2 = \omega_w^2 + \left(\frac{(\omega_1^2 - \omega_2^2)\, \omega_w^2}{2df\, \omega_1^2 \omega_2^2} \right)^2. \tag{55}$$

Equation (55) shows, that the limit of stability increases with the difference between the two critical angular speeds.

References

1. DIMENTBERG, F. M.: Flexural vibrations of rotating shafts, London 1961.
2. TONDL, A.: Some problems of rotor dynamics, Prague 1965.
3. KELLENBERGER, W.: Die Stabilität schnellaufen- der und anisotrop gelagerter Wellen mit äußerer und innerer Dämpfung, Brown Bovery Mitteilungen Nov./Dec. 1963.
4. WILKINSON, J. H.: The Algebraic Eigenvalue Problem, Oxford 1965.

Creep rupture

By

Y. N. Rabotnov

Moscow State University, Moscow

1.

At the present time numerous attempts were made to formulate the problem of rupture in terms of the mechanics of continuous media and that is the principal tendency of the modern development of the mechanics of solids bodies.

The solid bodies and the materials used for technical purposes are various and the character of their rupture is different depending on the conditions of loading, therefore the above mentioned general problem has many aspects and must be solved in different ways. Neverthless some general considerations may be formulated.

The traditional scheme of strength design of any structural element was the following one: using the methods of the theory of elasticity (in ideal) or the approximate means of the so called strength of materials (in practice) the stress distribution is found. Then a certain equivalent stress corresponding to the admitted strength criterion is calculated and compared with the admissible stress in every point of the body. The development of mathematical methods permits to obtain the solution of the first part of the problem, it is to find the stress distribution within any prescribed degree of exactitude. In contrast the considerations upon which the solution of the second part of the problem is based are rather uncertain and inaccurate.

The strength criterion used in the described scheme can be called the local strength criterion, we assume that the strength of a body as a whole is violated if a certain condition is not fulfilled through out the entire body.

A very different principle is used in the theory of limit design of ideally plastic structures. The formation of plastic zones in restricted volumes does not affect the bearing capacity of the structure. The problem is to find the combination of external forces under which the structure as a whole is no more rigid, but not to investigate the process of the spreading of plastic zones during the loading process. The corresponding strength criterion can be called the global criterion of strength.

Another example of the global approach is the theory of quasi-brittle fracturing of IRWIN-OROWAN.

In this theory the evaluation of the strength of a body containing a crack is reduced to the determination of the system of external forces under which the crack becomes unstable. A very high value of stresses near the end of the crack in this case is not limiting.

The quoted examples of the global approach have nevertheless a rather particular character, in reality the situations are much more various.

The theory of plasticity of hardening materials follows the line of development of the theory of elasticity, this theory permits to find the stresses and strains in a given body under the action of a given system of extermal forces but does not contain any elements permitting to judge whether the calculated state of stress and strain is admissible or not. The theory of brittle crack propagation permitted to propose certain practical methods of comparative estimation of materials destinated for the service under the condition when the brittle fracturing may occur. But, wehn we consider the strength of a structure, the determination of the ultimate load for a given body containing several cracks seems to have a rather restricted practical value.

In studying plastic properties of metals under normal temperature conditions the formulation of basic theoretical concepts preceded the accumulation of experimental results. In the field of fatigue strength and creep strength the situation is quite different. A very great amount of experimental data was obtained before any attempts of theoretical generalisation were made. This situation is easy to explain. Really, the fatigue strength and the creep strength define the true life-time of a number of structures and the elaboration of rational design methods is a vital necessity for the everyday engineering practice. That is the reason to fix our attention on these problems.

The mechanical theory of creep was created very recently. Nowadays we can calculate the accumulation of strain with the time and the stress redistribution in the processes of the duration, varying from few seconds up to about three years. And luckily, the first object for the application on of the new theory was a turbine disc. One can not raise the strength of a rotating disc by making it thicker, and therefore, a sufficiently accurate design is a real necessity. On the other hand the state of stress in a rotating disc is biaxial, which is the reason why even the first investigators were obliged to write down the basic equations of creep in a general form for the complex state of stress and not for simple tension or compression. One must add that, when manufacturing the turbine disc, the quality of the material is controlled and the service conditions are well determined.

In the field of fatigue strength the situation is different. Fatigue rupture usually occurs in machine parts having stress concentrators. The problem of stress distribution around holes and notches is difficult even if one supposes that the material is ideally elastic. Therefore even if we would have at our disposal a good theory of fatigue fracture, its application to real problems would meet serious difficulties.

As to the stress concentration in the elastic range, one should mention that in this field we have at our disposal a large amount of theoretical results, permitting to know the so called "theoretical stress concentration factor" for different cases. At the same time it is well known that numerous attempts to find a correlation between the theoretical stress concentration factor and the effective stress concentration factor, defining the real strength of a specimen in fatigue test or creep test gave no result. But it is only natural, since in fatigue and creep conditions the rupture criterion can not be a local one, the conditions of rupture are defined by the state of stress in certain finite volumes and this state of stress is changing during the process of rupture.

Using the term—"process of rupture", we must emphasize that the concept of "globality" is refered not only to space but also to time. It is well known indeed that the process of creep of a metal is accompanied by the formation of microscopic cracks on the grain boundaries. Phenomenologically one can say that damage-accumulation occurs. Under cyclic loading the situation is similar, at first the dissipated submicroscopic and microscopic cracks are formed and only at a certain moment appears a macroscopic crack, spreading over the cross-section of a body. Using the phenomenological approach we shall not enter into the details of the description of the process. The only one essential thing is that creep or cyclic loading is accompanied by a change of the structural state of the material which can be called damage accumulation.

2.

After the prelimenary remarks of a rather general character, we shall fix our attention on the problem of fracturing of metallic bodies at high temperature under the action of constant or slowly changing loads. The general point of view on the creep process can be formulated as follows [1, 2]. The creep strain rafe $\dot{\varepsilon}_p$ at any instant is determined by the actual stress σ, temperature T and depends on the structural state of the material which can be characterised by a number of structural parameters q_i. The change of these parameters with the time is described by a certain set of kinetic equations.

Supposing these kinetic equations to be linear with respect to the differentials of functions q_i, σ, ε_p, T, and of the time t, we obtain the following system of constitutive equations:

$$\dot{\varepsilon} = f(\sigma, \tau, q_s) \, dq_i = a_i \cdot d\varepsilon_p + b_i \cdot d\sigma + c_i \cdot dt + f_i \, dT. \qquad (2.1)$$

In previous papers by the author [1] it was shown that this point of view is sufficiently general, the different known theories of creep following from 2.1 as particular cases. For example, one of these structural parameters can be characterized as a measure of strain hardening, in the simplest case one can assume it to be equal to the accumulated creep strain.

The measure of damage of the material can be taken as one of structural parameters, in this case the equations will describe the process of creep rupture.

To make the idea clear, we assume that the parameter of damage, denoted as ω, is only one structural parameter entering the constitutive equations. We shall write these equations in the simplest form

$$\dot{\varepsilon} = f(\sigma, \tau, \omega), \quad \dot{\omega} = \varphi(\sigma, \tau, \omega). \qquad (2.2)$$

The equations of this type describe in a rather satisfactory way the short-time creep, when the temperature is so high that strain hardening practically does not occur. On the other hand, when considering a very long time creep, one can neglect the first stage of creep, or to add the corresponding part of the creep strain to the instanteneous creep strain, as it is done by Professor ODQVIST. Thus the Eqs. (2.2) describe the second and third stages of creep but not the first one.

The value of ω can be treated as the relative decrease of the cross-sectional area due to the formation of cracks. Then one can introduce the effective stress

$$s = \frac{\sigma}{1 - \omega}$$

and make a natural assumption that the creep strain rate and the rate of crack formation are depending only on the effective stress, when the temperature is fixed.

Assuming for instance the power law of creep and crack formation we have the following system of equations.

$$\dot{\varepsilon} = \varepsilon_0 \left(\frac{\sigma}{1 - \omega} \right)^n, \quad \dot{\omega} = \left(\frac{\sigma}{1 - \omega} \right)^k. \qquad (2.3)$$

The second Eqs. (2.3) describing the kinetics of crack formation was proposed by KACHANOV [3] (with the notation $\psi = 1 - \omega$) and independently by RABOTNOV [4] in a slightly different form. A number of investigations by KACHANOV-ODQVIST and others took into account the variation of the cross-sectional area at finite strain and the instantaneous plastic deformation (or the deformation on the first stage of creep).

These results are well known. Our point of view is that the accumulation of damage affects the creep strain rate. Actually, the first equation in (2.3) gives the acceleration of creep.

Equations (2.3) are sufficient for solving the simplest problems, for example the problem of finding the time to rupture of a girder. The traditional method of analysis is the following one. Using the creep equations, the stress distribution in the elements of the girder is to be found. When the external forces are constant and the steady state creep equations are used, the stresses do not vary with time. Then, using the stress-time to rupture diagram, or the second of Eq. (2.3), one finds the time to rupture of the most stressed bar. In reality the rupture is preceded by the acceleration of creep and that leads to the considerable stress redistribution. The numerical examples show that the stresses in the bars became almost equal and for the estimate of the life time one can assume within a certain degree of accuracy the stress distribution to be the same as in the case of an ideally plastic structure in the ultimate state.

The experimental investigation of short time creep, and creep rupture, showed that for a number of materials one can put $n = k$. Then the value of ε_0 is the uniform strain at the moment of rupture, when $\omega = 1$. The system (2.3) is then reduced to only one equation

$$\dot\varepsilon/\varepsilon_0 = \left(\frac{\sigma}{1 - \varepsilon/\varepsilon_0}\right)^n. \qquad (2.4)$$

The structure of this equation is similar to that of the strain hardening theory of creep. Now it is only natural to assume that the creep law

$$\dot\varepsilon = f(\sigma, \varepsilon)$$

can be used not only for the description of the first stage of creep, as in the strain hardening theory, but can be valid up to the moment of rupture. This idea was used by LEPIN [11] who showed that creep, relaxation and time-to-rupture of a number of technical alloys can be predicted by the aid of the following general equation

$$\dot\varepsilon = \varepsilon^a \exp\left(\frac{\varkappa\sigma}{1 - \varepsilon/\varepsilon_0}\right). \qquad (2.5)$$

The interpretation of the parameter ω as a relative reduction of the cross-sectional area is rather conventional and not necessary. Returning to the Eqs.(2.2), we can assume any form of the functions entering these equations, the parameters of them being determined from experiment. Some examples are given in [6]. The same considerations permit to investigate some more involved cases, for instance the processes of fracturing under cyclic loading. In this connection one should mention the recent paper by KOSTIUK [7]. He succeded to give a rather satisfactory theory of fatigue, creep strength and low cycle fatigue, using three structural parameters:

$$q_1 = \int \sigma \, d\varepsilon_p, \quad q_2 = \int \sigma \, d\varepsilon_0, \quad q_3 = \omega.$$

3.

Now we consider the general case of the tri-axial state of stress. The characteristic feature of modern heat resisting alloys is the brittle character of fracturing when the time to rupture is large enough. The experiments of JOHNSON [8] and others had shown that the maximum tensile stress $\sigma_{\max}$ can be taken as a characteristic stress, defining the life-time of the specimen. In our country this question was investigated by SDOBYREV [9]. The analysis of his own experiments and of all published experimental results obtained by others showed that the maximum tensile stress criterion is not exact. He proposed an empirical formula for the equivalent stress.

$$\sigma_\delta = \beta\sigma_{\max} + (1 - \beta)\,\sigma_0. \qquad (3.1)$$

Here σ_0 is the stress intensity. The value of β is approximately $1/2$.

The traditional scheme of the determination of life-time in creep conditions is based upon the use of a local strength cirterion. Using any creep theory the stress distribution is found, then one can calculate the equivalent stress from formula (3.1), for instance, and plot the corresponding point on the stress—time-to-rupture diagram. In reality the accumulation of damage leads to stress redistribution and at the instance when the first macroscopic crack appears the state of stress is essentially different from that calculated one with the aid of the usual creep equations. The idea of introducing the structural parameters seems to be useful even in this case. We introduce in a formal way the symmetric second range tensor ω_{ij} which we shall denote the tensor of damage. Then we put $\psi_{ij} = \delta_{ij} - \omega_{ij}$ where δ_{ij} is Kronecker's tensor. We define now the effective stress tensor s_{ij} by the aid of the following relation:

$$\sigma_{ij} = \Omega_{ijkl}s_{kl} \qquad (3.2)$$

where:

$$\Omega_{ijkl} = \frac{1}{4}\left(\psi_{ik}\,\delta_{jl} + \psi_{il}\,\delta_{jk} + \psi_{jk}\,\delta_{il} + \psi_{jl}\,\delta_{ik}\right).$$

Then

$$s_{ij} = \Omega_{ijkl}^{-1}\sigma_{kl}. \qquad (3.3)$$

The definition of the scalar parameter of damage and of the equivalent stress in the case of uniaxial tension given above, follows from here as a particular case.

Now we shall assume that the creep strain rate and the damage accumulation rate depend only upon the effective stresses s_{ij}:

$$\dot{\varepsilon}_{ij} = f_{ij}(s_{kl}), \quad \dot{\psi}_{ij} = \varphi_{ij}(s_{kl}). \tag{3.4}$$

If we admit that the steady creep rate obeys the potential law, and initially when $\omega_{ij} = 0$, $s_{ij} = \sigma_{ij}$, we must put:

$$\dot{\varepsilon}_{ij} = \varepsilon_0 \frac{\partial \psi}{\partial s_{ij}}. \tag{3.5}$$

As to the equations of the accumulation of damage, one can make different assumptions about their structure. Assuming that these equations are also of a potential type, we have:

$$\dot{\psi} = \frac{\partial \psi}{\partial s_{ij}}. \tag{3.6}$$

We admit then that the potential function ψ depends on a quadratic invariant of the tensor s_{ij} namely:

$$S = (1 - \mu)\, s_{ij} s_{ij} + \mu (s_{kk})^2. \tag{3.7}$$

Now we consides the case when the principal axes of the stress tensor are fixed and the principal stresses are σ_1, σ_2 and σ_3. The tensors ψ_{ij} and s_{ij} here have the same principal axes and

$$s_1 = \frac{\sigma_1}{\psi_1}, \quad s_2 = \frac{\sigma_2}{\psi_2}, \quad s_3 = \frac{\sigma_3}{\psi_3}.$$

The damage accumulation equations assume the form:

$$\dot{\psi}_1 = 2\psi^1(S) \left[\frac{\sigma_1}{\psi_1} + \mu \frac{\sigma_2}{\psi_2} + \mu \frac{\sigma_3}{\psi_3} \right],$$
$$S = \frac{\sigma_1^2}{\psi_1^2} + \cdots + 2\mu \left(\frac{\sigma_1 \sigma_2}{\psi_1 \psi_2} + \cdots \right). \tag{3.8}$$

Experiments show that the time to rupture is the same in both cases, at uniaxial tension with the stress σ, and at biaxial tension with the same stress σ. This result follows also from the empirical formula of SDOBYREV. This requirement permits to find the value of μ, once the function $\psi^1(S)$ is known. For example, if $\psi^1(S) = S$, the reciprocal value of the time to rupture is proportional to the third power of the stress and $\mu = 1/\sqrt{2} - 1$. In this particular case, considering the case of pure shear when $\sigma_1 = \sigma_2 = \tau$ we obtain the value of the equivalent stress $\sigma_e = 1{,}29\tau$, when the formula of SDOBYREV gives $\sigma_e = 1{,}36\tau$ at $\beta = 0{,}5$. Thus, in plane stress the equations (3.8) give the results which only slightly differ from these given by the formula of SDOBYREV.

4.

In constructing the variant of the creep rupture theory considered above, the basic concepts of the theory of plasticity were used without sufficient modification, namely the assumption of the damage potential depending on a certain quadratic invariant of the tensor s_{ij}. Moreover, it follows from the given equations that in compression in the planes orthogonal to the direction of compression, the value of ψ increases, that is the material becomes stronger. Even in this case when $\mu < 1$ the finite time to rupture exists, but the rupture occurs as a result of the accumulation of damage in the planes parallel to the direction of compression.

In the theory of plasticity are frequently used different piecewise linear approximations of the yield condition and of the corresponding flow rule. In a similar way, we can make different assumptions of this kind in the theory of creep rupture. These assumptions must follow from the direct experimental data and in the same time must permit the solution of particular problems.

The experiment shows that within a certain degree of accuracy the rupture criterion is the maximum tensile stress and the correction given by the formula of SDOBYREV, for instance, in many cases can be neglected. On the other hand the macroscopic crack is always orthogonal to the direction of the maximum principal stress. The dependence of the rate of damage on the effective stress is nonlinear. Therefore ω_1 grows much faster than ω_2 and ω_3 and at the moment when ω_1 reaches the critical value equal to unity, the values of ω_2 and ω_3 are small. Of course, it is true when $\sigma_1 > \sigma_2, \sigma_3$, and the difference is large enough.

Now the principal values of the tensor s_{ij} can be given in a following way:

$$s_1 = \frac{\sigma_1}{1 - \omega_1}, \quad s_2 = \sigma_2, \quad s_3 = \sigma_3.$$

In this approach the damage is characterised by a single component $\omega_1 = \omega$ (index omitted) which can be considered as a scalar.

The creep equations have the same form as above

$$\dot{\varepsilon}_i = \varepsilon_0 \frac{\partial \varphi}{\partial s_i} \tag{4.1}$$

at $s_i = \sigma_i$ the function Φ is the steady creep potential function.

The kinetics of the damage accumulation is described now by a single equation of the form:

$$\dot{\omega} = \varphi(\sigma_e) \tag{4.2}$$

where the equivalent stress σ_e is a function of s_i determined in accordance with the rupture criterion we have admitted.

Now we shall consider few examples of the application of this simplified theory.

a) *Pure bending of a bar having a rectangular cross-section.* The state of stress is uniaxial, the strain rate dependence on stress is given by

$$\dot{\varepsilon} = \varepsilon_0 \left(\frac{\sigma}{1 - \omega} \right)^n. \tag{4.3}$$

The creep damage accumulation rate obeys the following equations:

$$\dot{\omega} = \left(\frac{\sigma}{1 - \omega} \right)^k \quad \text{at} \quad \sigma > 0$$

and

$$\dot{\omega} = 0 \qquad \text{at} \quad \sigma < 0. \tag{4.4}$$

The cross-sections remaining plane, one can integrate these equations numerically, for instance in the case when the bending moment is constant. The neutral axis moves in the direction of the compressed part of the section, at a certain moment t_1 the value of ω becomes equal to unity at the extreme point of the section and the macrocrack aa appears. In a rather short time this crack spreads across the bar. Therefore the value of t_1 can be taken as a real estimate of the life-time.

b) *Fracture of the rotating disc having a central hole.* The solution of this problem is given in [1, 2] and will not be considered here. It is to be noticed that the method is applicable only when the circumferencial stress is much larger then the radial one. This assumption is valid for a disc having the central hole but is not true for a disc having no hole. In this case the damage cannot be characterized by a single scalar parameter.

c) *Fracture of a tube under internal pressure.* When the state of strain is plane, the creep rate is defined by the stress intensity

$$\sigma_0 = \frac{\sqrt{3}}{2} \left[\frac{\sigma_1}{1 - \omega} - \sigma_2 \right]. \tag{4.5}$$

Here σ_1 is the circumferential stress, and σ_2 is the radial stress. The creep strain rate distribution obeys the associated flow rule, the condition of incompressibility of course does not hold, the formation of microcracks is accompanied by the increase of volume. If we take the equivalent stress according to the criterion of SDOBYREV, we must put in (4.2)

$$\sigma_e = \left[1 - \beta \left(1 - \frac{\sqrt{3}}{2} \right) \right] \frac{\sigma_1}{1 - \omega} - \beta \frac{\sqrt{3}}{2} \sigma_2. \tag{4.6}$$

When $\beta = 0$ the equivalent stress is $\dfrac{\sigma_1}{1-\omega}$. The equations of equilibrium, compatibility and accumulation of damage are quasi-linear, their numerical integration permits to find the value of time t when $\omega = 1$ on the outer surface of the tube. On the next stage the cracked zone will have the form of a circular ring spreading over the whole thickness of the tube [5].

A further simplification of the proposed sheme will be based on the well-known fact that in creep conditions the distribution of stresses does not differ very much from that of an ideally plastic body in the limit state. So using the power law of creep, characterized by the exponent n we obtein in a formal way at $n = \infty$, the stress distribution, corresponding to the ideally plastic body. But, unlike the case of ideal plasticity this stress distribution is realized at any value of the external forces. Thus, the yield stress is supposed to be constant inside the body but will depend on the value of the acting load, in accordance with the equilibrium conditions.

Now we can propose the following approximate method for the solution of creep rupture problems. We assume that the effective stresses satisfy a certain condition of plasticity, for example the intensity of the effective stresses is constant. Under plane strain conditions we have

$$\frac{\sigma_1}{1-\omega} - \sigma_2 = \lambda\sigma_s. \tag{5.1}$$

Here σ_s is an arbitrary stress $\lambda(t)$ is a parameter which can be found from the equations of the equilibrium. The kinetics of damage accumulation is described, as formerly, by the Eq. (4.2), where σ_e is defined by (4.6).

The Eq. (5.1) is similar to that used in statics of soils. Joining the equilibrium equations we obtain a hyperbolic system and can use the standard method of characteristics. We shall not consider the details of its application. It is to be noticed only that the angle ψ between the directions of characteristics and the direction of the second principal stress axes is given by the formula

$$\mathrm{tg}\,\psi = \pm \sqrt{1-\omega}. \tag{5.2}$$

At the point where fracturing begins, ω becomes equal to unity, and both characteristics coincide, generating the crack.

In application of this method to problems of stress concentration we shall admit that inside the body appears a well restricted domain where the creep and damage accumulation are intense, while in the other part of the body the creep rates are small and can be neglected. In the proposed model the quasi-plastic domains are surrounded by the rigid parts of the body and the border line is the characteristic. As much as the damage parameter ω grows, the angle between the characteristics becomes smaller and the quasiplastic domain becomes more and more narrow.

The numerical procedure can be described as follows. The distribution of ω at any instant t being known, we construct the field of characteristics and calculate the stresses. Now, using Equation (4.2) we can calculate the distribution of ω at the moment $t + \Delta t$ and so on.

The results are rather simple when the problem can be reduced to one-dimensional cases, some of these will be considered here.

a) *Pure bending of a bar having a rectangular cross-section.* The plasticity condition (5.1) is replaced by:

$$\sigma = \lambda\sigma_s(1-\omega) \quad \text{in the upper part (tension)}$$

$$\sigma = -\lambda\sigma_s \qquad\;\; \text{in the lower part (compression)}.$$

The neutral axis being the axis of symmetry at $t = 0$ moves downwards. In the upper half of the section the stress distribution remains uniform and therefore ω reaches the ultimate value $\omega = 1$, at the same time in every point of the upper half of the section. Using the power law of creep and damage accumulation at the same value of exponent $n = 3$, we find the time the appearance of the crack at $t = 0.373$ (in a suitable scale). According to the exact solution the crack reaches the middle line of the section at $t = 0.428$.

b) *Stress concentration around a circular notch.* We consider a strip with two symmetric circular notches. The radius of the notch being a and the minimum width of the strip $2h$. Starting from the boundary of the notch we can construct only the axisymmetric stress field and therefore the problem is reduced to a one-dimensional one. The numerical integration in this case is rather simple. Taking $n = k = 3$, $a = h$ we obtained the following results. If $\beta = 0$ (maximum principal stress criterion) the supporting factor, it is the ratio of the stress in a notched specimen to the stress in an unnotched specimen, corresponding to the same life-time was found to be 1.024 (2.4% may be the error in computing). Taking $\beta = 1/2$ we obtained the value of supporting factor to be 1.25. This is very close to the experimental result for an aluminium alloy.

c) *Wedge under the action of a uniform load acting normally along one side.* At $t = 0$ the bisector separates the upper part of the wedge where the radial stress is positive, from the lower part where this stress is negative. This borderline turns downwards. As a result we have to consider three domains, in the first (upper) the radial stress is always positive in the second (middle) the sign of this stress is changed and in the third (lower) it is always negative. In the first and third domains the equations can be integrated in closed form, but in the second zone only numerical integration is possible. This problem was solved in the diploma work of Mr. LUBART, student at the Moscow University.

In conclusion we must notice that the introduction of the parameter of damage into the constitutive equations seems to be useful not only in the theory of strength of metals at high temperatures. Similar considerations are valid for high-polymers. Some features of their behaviour can be characterized as destructive plasticity. The internal mechanism of this destructive plasticity can be different, for the plastic material itself it is connected with the rupture of molecular bonds. For glass reinforced plastics this destructive plasticity is connected with the formation of internal cracks as a result of internal stress concentration.

The applications of the same ideas to the fatigue strength were already mentioned. But the most complicated problem is the one of rupture of plastic metals under normal temperature conditions, when the strains are large enough.

References

1. RABOTNOV, YU. N., Creep of Construction Elements, "Science", 1966, (in Russian).
2. RABOTNOV, YU. N., On the Equations of State for Creep, Progress in Mechanics, 1963, (in Russian).
3. KACHANOV, L. M., Theory of Creep, Physmatgis, 1960, (in Russian).
4. RABOTNOV, YU. N., On the Mechanism of Gradual Failure, Questions of Strength of Materials and Construction, Academy of Sciences, 1959, pp. 5—7, (in Russian).
5. MILEIKO, S. T., and YU. N. RABOTNOV, Creep and Ultimate Strength, Academy of Sciences, Novosibirsk, 1963, (in Russian).
6. RABOTNOV, YU. N., On Failure due to Creep, Applied Math. and Tech. Phys., 1963, No. 2, (in Russian).
7. KOSTIUK, A. G., On the Theory of Plastic Deformations of Polycrystalline Materials, Mechanics of Solid Bodies, 1967, No. I, pp. 95—101, (in Russian).
8. JOHNSON, A. E., J. HENDERSON and B. KHAN, Complex Stress Creep, Relaxation, and Fracture of Metallic Alloys. H. M. Stat. Office, NEL, Edinburg, 1962.
9. SDOBYREV, V. P., Criteria of Ultimate Strength for Various Heat-treated Alloys in Intricate States of Stress, Academy of Sciences, OTN, Mechanics and Mechanical Engineering, No. 6, 1959, pp. 93—99, (in Russian).
10. RABOTNOV, YU. N., Influence of Stress Concentrations on Ultimate Strength, MTT, No. 3, 1967, pp. 36—41, (in Russian).
11. LEPIN, G. F., On a Statistical Theory of Creep and Relaxation, Academy of Sciences, OTN, 1957, No. 9, pp. 130—134, (in Russian).

Deductions from profile-excited random vibration response

J. D. Robson

University of Glasgow, Glasgow

Synopsis. When certain points of a system are constrained to follow a randomly undulating profile, the spectral density of its response depends on the dynamic properties of the system, the spectral density of the profile, and the velocity of traversal. The relationship connecting response spectral density to profile spectral density is first developed for two important special cases: (a) where there is a single profile-imposed displacement, and (b) where two displacements are imposed by a single profile. It is then shown that the form of the profile spectrum is deducible in each case by consideration of response spectra alone.

1. Introduction

The response of a vehicle to the undulations of the roadway on which it runs, whether arising as motion of the vehicle as a whole, or as stresses in its components, is a matter of some practical interest. Suitably idealised, with the vehicle considered as a linear dynamical system and the road profile assumed to be an ergodic randomly-varying function of distance, the problem is of considerable academic interest also, having features which make it somewhat different from the usual random vibration response problem.

In such a system the displacement imposed on the vehicle at each of its four points of contact with the road is a randomly-varying function of time, and the spectral density of any response is related in the usual way to the spectral densities (direct and cross) of these imposed displacements. But these displacements, arising as they do from the traversal of a given surface, must possess a degree of correlation dictated by the cross-correlations of the road profile. Moreover their frequency content is directly influenced by the velocity at which the surface is traversed. It is of some interest to establish the manner in which these special factors influence the problem.

This paper will present certain results which arise from the special nature of the problem and which may be of practical interest; it will be shown in fact that certain characteristics of the profile may be predictable by suitable analysis of response records, without any detailed knowledge of the dynamic characteristics of the vehicle. In order to derive these results it will be necessary to set up basic response analysis for such a profile-excited system.

In this work we shall confine attention to two practically important special cases. In the response of certain components of a vehicle the effect of one of the four excitations will predominate: in such cases it will be sufficient to consider response to a single profile-imposed excitation. Where this approach is not adequate it will be more realistic to use a two-excitation analysis: we shall consider the case where both wheels on the same axle have a common imposed motion and where rear wheels follow the same profile as the front wheels but at a constant distance behind them. Four-input analysis can be developed by the same methods but is not needed for the purposes of this paper.

2. Response to profile excitation

a) Single excitation

Suppose that a single point of a given system is constrained to traverse a profile $\psi(x)$ with velocity v. (Here x is distance in the direction of motion; $\psi(x)$ is a member function of the ergodic random process $\{\psi(x)\}$ and represents displacement normal to x.) It will also be necessary to consider the imposed displacement at the point as a function of time, and this will be denoted by the different symbol $y(t)$. As $x = vt$, $\psi(x)$ and $y(t)$ are related by

$$\psi(vt) = y(t). \tag{1}$$

In considering response we shall wish to relate the velocity-dependent spectral density $S(f)$ of $y(t)$ to the fixed spectral density $\mathscr{S}(n)$ of $\psi(x)$, where f and n represent frequency and wavenumber respectively, so that $f = nv$. This relation is best approached through the corresponding autocorrelation functions.

The autocorrelation functions of $\psi(x)$ and $y(t)$ are defined by

$$\begin{aligned} \mathscr{R}(\delta) &= \langle \psi(x)\, \psi(x + \delta) \rangle, \\ R(\tau) &= \langle y(t)\ y(t + \tau) \rangle, \end{aligned} \tag{2}$$

and with $\delta = v\tau$, as $x = vt$, it follows that

$$R(\tau) = \mathscr{R}(\delta). \tag{3}$$

The spectral densities can then be related: making use of the Fourier transform relationship between spectral density and autocorrelation function,

$$\begin{aligned} S(f) &= \int\limits_{-\infty}^{\infty} 2R(\tau)\ e^{-i2\pi f\tau} d\tau \\ &= \int\limits_{-\infty}^{\infty} 2\mathscr{R}(\delta)\ e^{-i2\pi nv\,\delta/v}\ d(\delta/v) \\ &= \frac{1}{v} \int\limits_{-\infty}^{\infty} 2\mathscr{R}(\delta)\ e^{-i2\pi n\delta}\ d\delta = \frac{1}{v}\ \mathscr{S}(n). \end{aligned} \tag{4}$$

It is now possible to relate response spectral density at any point of the system to the spectral density of the profile. Let the response spectral density be denoted by $S^0(f)$ and the corresponding excitation spectral density by $S(f)$; then if the harmonic receptance connecting excitation at the two points is denoted by $\alpha(if)$, it follows that

$$S^0(f) = |\alpha(if)|^2\, S(f), \tag{5}$$

and so, by (4), that

$$S^0(f) = |\alpha(if)|^2 \frac{1}{v}\ \mathscr{S}(n). \tag{6}$$

b) Double excitation

Suppose now that two points A and B of a given system (corresponding to front and rear wheels of a vehicle) pass consecutively over the same profile. The profile may still be described by a single function $\psi(x)$ and its spectral density by $\mathscr{S}(n)$: the displacements of the two points, considered as functions of the same time variable will not be identical, however, and may be denoted by $y_A(t)$, $y_B(t)$. If the separation of the two points in the direction of x is a, and the traversal velocity is v, then

$$\begin{aligned} y_A(t) &= \psi(x) \\ y_B(t) &= \psi(x - a) \end{aligned} \tag{7}$$

with $x = vt$.

The direct spectral densities $S_A(f)$, $S_B(f)$ of the imposed displacements $y_A(t)$ and $y_B(t)$ will be given as before by

$$S_A(f) = S_B(f) = \frac{1}{v}\, \mathscr{S}(n). \tag{8}$$

But to determine the response in this case we shall also need to know their cross spectral density $S_{AB}(f)$. This also is more easily approached through the cross-correlation function.

The cross-correlation function $R_{AB}(\tau)$ is defined by

$$R_{AB}(\tau) = \langle y_A(t)\, y_B(t+\tau)\rangle: \tag{9}$$

then by (9) and (7)

$$
\begin{aligned}
R_{AB}(\tau) &= \langle y(vt)\, y(vt + v\tau - a)\rangle \\
&= \langle y(x)\, y(x + \delta - a)\rangle \\
&= \mathscr{R}\,(\delta - a).
\end{aligned}
\tag{10}
$$

The cross spectral density $S_{AB}(f)$ can now be determined, using the Fourier transform relationship as before:

$$
\begin{aligned}
S_{AB}(f) &= \int\limits_{-\infty}^{\infty} 2R_{AB}(\tau)\, e^{-2i\pi f\tau}\, d\tau \\
&= \frac{1}{v} \int\limits_{-\infty}^{\infty} 2\mathscr{R}(\delta - a)\, e^{-i2\pi n\delta}\, d\delta \\
&= \frac{1}{v} \int\limits_{-\infty}^{\infty} 2\mathscr{R}(\delta - a)\, e^{-i2\pi n(\delta - a)}\, d(\delta - a) \cdot e^{-i2\pi na} \\
&= \frac{1}{v}\, e^{-i2\pi na}\, \mathscr{S}(n).
\end{aligned}
\tag{11}
$$

The response spectral density at a point O can now be related to the profile spectral density. Denoting the relevant receptances (suitably defined to permit superposition) by $\alpha_{OA}(if)$, $\alpha_{OB}(if)$, the response spectral density $S^0(f)$ is given by

$$
\begin{aligned}
S^0(f) = {}&\alpha_{OA}^*(if)\, \alpha_{OA}(if)\, S_A(f) + \alpha_{OA}^*(if)\, \alpha_{OB}(if)\, S_{AB}(f) \\
&+ \alpha_{OB}^*(if)\, \alpha_{OA}(if)\, S_{BA}(f) + \alpha_{OB}^*(if)\, \alpha_{OB}(if)\, S_B(f),
\end{aligned}
\tag{12}
$$

where * indicates complex conjugate.

Using (8), (11), and (12) therefore

$$
\begin{aligned}
S^0(f) = \frac{1}{v}\, \big[&\alpha_{OA}^*(if)\, \alpha_{OA}(if) + \alpha_{OA}^*(if)\, \alpha_{OB}(if)\, e^{-i2\pi na} \\
&+ \alpha_{OB}^*(if)\, \alpha_{OA}(if)\, e^{i2\pi na} + \alpha_{OB}^*(if)\, \alpha_{OB}(if) \big]\, \mathscr{S}(n).
\end{aligned}
\tag{13}
$$

Equation (13) gives the required relationship but it is more conveniently written in the form

$$S^0(f) = \frac{1}{v}\, [A(f) + D(f)\cos 2\pi an + E(f)\sin 2\pi an]\, \mathscr{S}(n), \tag{14}$$

where

$$
\begin{aligned}
A(f) &= \alpha_{OA}^*\alpha_{OA} + \alpha_{OB}^*\alpha_{OB}, \\
D(f) &= 2\mathscr{R}e\,[\alpha_{OA}^*\alpha_{OB}], \\
E(f) &= 2\,\mathscr{I}m\,[\alpha_{OA}^*\alpha_{OB}].
\end{aligned}
$$

It is naturally implied that values of f, n, and v in (6), (13), and (14) will be compatible: if two of these variables are specified the third must follow. It may be noted that the response relationship embodied in (14) is one of considerable complexity: any change of frequency at constant v will affect all terms containing n as well as those containing f explicitly, and similarly for changes of velocity at constant f.

It is this interrelation between the effects of varying v and f which gives the profile-excitation problem its special features.

3. Deductions from observed responses

In the more usual form of random vibration response problem, where the excitation is specifically a randomly varying function of time, it is quite impossible to deduce any information about the excitation by consideration of the response alone: the dynamic characteristics of the system affect the response and there is no way of eliminating them from the analysis.

But in the case of profile-excitation the possibility of varying the traversal velocity offers the opportunity of generating a set of different but closely related excitations, from the responses of which the characteristics of the excitation can be deduced. This is true not only for the single-excitation case but for double excitation also.

a) Single excitation

Suppose that the response spectral densities $S_1^0(f)$, $S_2^0(f)$ are established when one point of a system having receptance $\alpha(if)$ traverses a given profile of (unknown) spectral density $\mathscr{S}(n)$ at velocities v_1 and v_2 respectively. Then by (6)

$$\left. \begin{aligned} S_1^0(f) &= |\alpha(if)|^2 \frac{1}{v_1} \mathscr{S}(f/v_1) \\ S_2^0(f) &= |\alpha(if)|^2 \frac{1}{v_2} \mathscr{S}(f/v_2) \end{aligned} \right\} \tag{15}$$

and hence

$$\frac{S_1^0(f)}{S_2^0(f)} = \frac{v_2 \mathscr{S}(f_1/v)}{v_1 \mathscr{S}(f/v_2)}.$$

Thus, writing $f/v_1 = n_0$

$$\mathscr{S}\left(\frac{v_1}{v_2} n_0\right) = \frac{S_2^0(n_0 v_1)}{S_1^0(n_0 v_1)} \frac{v_2}{v_1} \mathscr{S}(n_0). \tag{16}$$

If therefore we give $\mathscr{S}(n_0)$ any arbitrary value at any desired wave number n_0, the whole of the right hand side consists of known quantities, and the value of $\mathscr{S}\left(\frac{v_1}{v_2} n_0\right)$ can be determined.

Moreover recurrent use of this relationship — choosing the original $\frac{v_1}{v_2} n_0$ as the new n_0 — enables the form of the whole spectrum to be established. This result, along with certain others, was given in J. Sound Vib., 7, 156—158, (1968).

Absolute values of the $\mathscr{S}(n)$ spectrum cannot of course be determined without additional information. It would, for example, be possible for the profile spectral density to be doubled for all n without affecting any of the ratios considered.

b) Double excitation

Similar deductions can be achieved in the much more complicated case where two points of the system follow a given profile, so that response is described by Eq. (14). In the case of the single excitation it was possible to eliminate the system receptance $\alpha(if)$ by establishing response spectra at two different traversal velocities. But (14) is much more complicated than (6) containing as it does three separate receptance parameters, and consisting of four terms instead of one. Elimination will only be possible if responses are established at a number of velocities, and then only if some care is taken as to the choice of velocities.

It will be convenient to choose traversal velocities v_1, v_2, ... such that

$$v_2/v_1 = v_3/v_2 = \cdots = r, \text{ a constant}; \tag{17}$$

suppose that the corresponding response spectral densities $S_1^0(f)$, $S_2^0(f)$ are established. It will be convenient also to concentrate on frequencies f_1, f_2, ..., such that

$$f_2/f_1 = f_3/f_2 = \cdots = r. \tag{18}$$

We can then compile a set of equations of the form

$$\left.\begin{aligned}
S_1^0(f_1) &= \frac{1}{v_1}\,[A(f_1) + D(f_1)\cos 2\pi a n_0 + E(f_1)\sin 2\pi a n_0]\,\mathscr{S}(n_0) \\
S_2^0(f_1) &= \frac{1}{v_2}\,[A(f_1) + D(f_1)\cos 2\pi a n_0/r + E(f_1)\sin 2\pi a n_0/r]\,\mathscr{S}(n_0/r) \\
&\;\vdots \\
S_2^0(f_2) &= \frac{1}{v_2}\,[A(f_2) + D(f_2)\cos 2\pi a n_0 + E(f_2)\sin 2\pi a n_0]\,\mathscr{S}(n_0) \\
S_3^0(f_2) &= \frac{1}{v_3}\,[A(f_2) + D(f_2)\cos 2\pi a n_0/r + E(f_2)\sin 2\pi a n_0/r]\,\mathscr{S}(n_0/r) \\
&\;\vdots \\
S_3^0(f_3) &= \frac{1}{v_3}\,[A(f_3) + D(f_3)\cos 2\pi a n_0 + E(f_3)\sin 2\pi a n_0]\,\mathscr{S}(n_0) \\
S_4^0(f_3) &= \frac{1}{v_4}\,[A(f_3) + D(f_3)\cos 2\pi a n_0/r + E(f_3)\sin 2\pi a n_0/r]\,\mathscr{S}(n_0/r) \\
&\;\vdots \\
&\text{etc.,}
\end{aligned}\right\} \tag{19}$$

where at each frequency the velocities are chosen such that the same set of wave numbers is covered.

If these equations are to be solved for the unknown $\mathscr{S}(n)$, their number must be adequate to permit solution. As we shall not hope to determine absolute values of profile spectra, but only the ratios $\mathscr{S}(n_0/r)/\mathscr{S}(n_0)$, $\mathscr{S}(n_0/r^2)/\mathscr{S}(n_0)$ etc. we must ensure that the number of the unknowns exceeds the number of equations by one. The numbers of unknowns and equations will depend on the numbers of velocities and frequencies considered.

Let the number of frequencies considered be p and the number of velocities considered at each frequency be q. Then the number of equations is pq and the number of unknowns, $A(f)$, $D(f)$, $E(f)$, $\mathscr{S}(n)$, is $3p + q$. These numbers are shown in Table 1.

Table 1

p		$q =$ 1	2	3	4	5
1	Equations	1	2	3	4	5
	Unknowns	4	5	6	7	8
2	Equations	2	4	6	8	10
	Unknowns	7	8	9	10	11
3	Equations	3	6	9	12	15
	Unknowns	10	11	12	13	14
4	Equations	4	8	12	16	20
	Unknowns	13	14	15	16	17

Inspection shows that the combinations $p = 2$, $q = 5$ and $p = 3$, $q = 4$ are favourable. There is little to choose between these two possibilities, both requiring tests at six velocities, but the first involves the solution of fewer equations and adds an additional point to the spectrum.

For the two-frequency case the equations (19) become

$$\left.\begin{aligned}
S_1^0(f_1) &= \frac{1}{v_1}\,[A(f_1) + D(f_1)\cos 2\pi a n_0 + E(f_1)\sin 2\pi a n_0]\,\mathscr{S}(n_0) \\
&\;\vdots \\
S_5^0(f_1) &= \frac{1}{v_5}\,[A(f_1) + D(f_1)\cos 2\pi a n_0/r^4 + E(f_1)\sin 2\pi a n_0/r^4]\,\mathscr{S}(n_0/r^4) \\
S_2^0(f_2) &= \frac{1}{v_2}\,[A(f_2) + D(f_2)\cos 2\pi a n_0 + E(f_2)\sin 2\pi a n_0]\,\mathscr{S}(n_0) \\
&\;\vdots \\
S_6^0(f_2) &= \frac{1}{v_6}\,[A(f_2) + D(f_2)\cos 2\pi a n_0/r^4 + E(f_2)\sin 2\pi a n_0/r^4]\,\mathscr{S}(n_0/r^4)\,.
\end{aligned}\right\} \tag{20}$$

Here the response spectral densities are obtained by measurement; the velocities $v_1, v_2, \ldots, v_6$ and the frequencies f_1, f_2 are known; the unknowns are $A(f_1)$, etc., $A(f_2)$ etc., $\mathscr{S}(n_0)$ etc.

Equation (20) can be given the form usual to linear simultaneous equations by dividing through by the $\mathscr{S}(n)$ variables. They then become

$$
\begin{bmatrix}
a_{11} & 0 & b_{11} & 0 & c_{11} & 0 & 0 & 0 & 0 & 0 \\
a_{21} & 0 & b_{21} & 0 & c_{21} & 0 & d_{21} & 0 & 0 & 0 \\
a_{31} & 0 & b_{31} & 0 & c_{31} & 0 & 0 & d_{31} & 0 & 0 \\
a_{41} & 0 & b_{41} & 0 & c_{41} & 0 & 0 & 0 & d_{41} & 0 \\
a_{51} & 0 & b_{51} & 0 & c_{51} & 0 & 0 & 0 & 0 & d_{51} \\
0 & a_{12} & 0 & b_{12} & 0 & c_{12} & 0 & 0 & 0 & 0 \\
0 & a_{22} & 0 & b_{22} & 0 & c_{22} & d_{22} & 0 & 0 & 0 \\
0 & a_{32} & 0 & b_{32} & 0 & c_{32} & 0 & d_{32} & 0 & 0 \\
0 & a_{42} & 0 & b_{42} & 0 & c_{42} & 0 & 0 & d_{42} & 0 \\
0 & a_{52} & 0 & b_{52} & 0 & c_{52} & 0 & 0 & 0 & d_{52}
\end{bmatrix}
\begin{bmatrix}
x_1 \\ x_2 \\ y_1 \\ y_2 \\ z_1 \\ z_2 \\ w_1 \\ w_2 \\ w_3 \\ w_4
\end{bmatrix}
=
\begin{bmatrix}
e_{11} \\ 0 \\ 0 \\ 0 \\ 0 \\ e_{12} \\ 0 \\ 0 \\ 0 \\ 0
\end{bmatrix}
w_0
\tag{21}
$$

where $x_1 = A(f_1)$, $x_2 = A(f_2)$, $y_1 = D(f_1)$, $y_2 = D(f_2)$, $z_1 = E(f_1)$, $z_2 = E(f_2)$

$$
w_0 = \frac{1}{\mathscr{S}(n_0)}, \quad w_1 = \frac{1}{\mathscr{S}(n_0/r)}, \quad w_2 = \frac{1}{\mathscr{S}(n_0/r^2)}, \quad w_3 = \frac{1}{\mathscr{S}(n_0/r^3)}, \quad w_4 = \frac{1}{\mathscr{S}(n_0/r^4)}
$$

and
$$
a_{11} = a_{21} = a_{31} = a_{41} = a_{51} = 1,
$$
$$
a_{12} = a_{22} = a_{32} = a_{42} = a_{52} = 1,
$$

$$
b_{11} = b_{12} = \cos 2\pi a n_0, \qquad c_{11} = c_{12} = \sin 2\pi a n_0,
$$
$$
b_{21} = b_{22} = \cos(2\pi a n_0/r), \qquad c_{21} = c_{22} = \sin(2\pi a n_0/r),
$$
$$
b_{31} = b_{32} = \cos(2\pi a n_0/r^2), \qquad c_{31} = c_{32} = \sin(2\pi a n_0/r^2),
$$
$$
b_{41} = b_{42} = \cos(2\pi a n_0/r^3), \qquad c_{41} = c_{42} = \sin(2\pi a n_0/r^3),
$$
$$
b_{51} = b_{52} = \cos(2\pi a n_0/r^4), \qquad c_{51} = c_{52} = \sin(2\pi a n_0/r^4),
$$
$$
d_{21} = -v_2 S_2^0(f_1), \quad d_{31} = -v_3 S_3^0(f_1), \qquad d_{41} = -v_4 S_4^0(f_1), \quad d_{51} = -v_5 S_5^0(f_1),
$$
$$
d_{22} = -v_3 S_3^0(f_2), \quad d_{32} = -v_4 S_4^0(f_2), \qquad d_{42} = -v_5 S_5^0(f_2), \quad d_{52} = -v_6 S_6^0(f_2),,
$$

and
$$
e_{11} = v_1 S_1^0(f_1), \qquad e_{12} = v_2 S_2^0(f_2).
$$

Solution of (21) gives w_1, w_2, w_3, w_4 in terms of w_0, and so gives $w_0/w_1 = \mathscr{S}(n_0/r)/\mathscr{S}(n_0)$, $w_0/w_2 = \mathscr{S}(n_0/r^2)\mathscr{S}/(n_0)$, etc. as required. This permits four points on the $\mathscr{S}(n)$ curve to be fixed in addition to one arbitrarily fixed at n_0. If more points are required they can be provided by choosing a different value of n_0, but keeping the same v_1, v_2: different values of f must then be taken so that different values of $S^0(f)$ must be used.

Equation (21) has been used to establish the form of a profile, using response spectra computed from a known profile spectrum and from arbitrarily chosen receptances. The correct spectrum was recovered with high accuracy.

The remaining variables $x_1, x_2, \ldots$ etc. can also be defined in terms of w_0, so some description of system receptance is possible. The separate receptances $\alpha(if)$ cannot however be recovered.

45*

Studies of flow around blunt bodies by numerical methods

By

V. V. Rusanov and A. N. Lyubimov

Institute for Applied Mathematics, Moscow

The invention of the electronic computer and its use in scientific research have revealed new perspectives in the study of physical processes. In gas dynamics the use of computers has made it possible not only to obtain numerical solutions of complicated problems but also to study axisymmetric and spatial gas flows in detail. In this paper some general problems and special features connected with such applications of numerical methods are discussed.

Finite-difference methods developed recently by the authors of this paper and their colleagues [1, 2] have made it possible to obtain highly accurate numerical solutions to many problems connected with three-dimensional and axisymmetric gas flows around blunt bodies. On the basis of these methods, systematic and detailed studies of the flow fields between the shock wave and the body in both the subsonic and supersonic regions have been made. Some interesting flow details have been detected and studied, for example, the position and shape of the shock wave and sonic surface, the distribution and peculiarities in the behavior of various flow variables, etc.

Analytical, experimental, and numerical methods

The partial differential equations that describe the flow of a compressible gas have accurate analytical solutions in some special and often non-practical cases. Therefore, analytical studies of gas flow are limited. They include, in general, self-similar solutions, linear models, and various approximate methods, which as a rule do not provide full information on the whole flow field. The essential advantage of numerical methods is the practical possibility of solving the exact equations and determining the gas parameters at any point of the flow field with predetermined accuracy.

The analytical solution is represented generally by means of a formula with parameters. It is possible therefore to analyze a complete set of problems, which is often important for practical purposes. A numerical computation, on the contrary, is always made for a specific problem with fixed parameters, and for a comparative analysis one has to make a number of runs with the computer. These specific features of numerical methods determine both their fields of application and their connection with analytical and experimental studies.

The possibility of obtaining full information stimulates the use of numerical methods for the qualitative and quantitative study of the structure of complicated three-dimensional flows, as well as for the detection of new physical phenomena that can not be found with the aid of approximate methods. From the data obtained in many separate computations one can construct, by special processing, simple "empirical" formulas just as one can obtain them from experimental data. Finally, numerical solutions can be used for control and correction of the results obtained by various approximate methods. It is obvious from this that analytical and numerical methods do not oppose but supplement each other and must be developed in close relationship.

Let us now consider the relationship between numerical methods and experiment.

The above-mentioned analogy between a physical experiment and the solution of a mathematical problem on the computer — a "mathematical experiment" — becomes apparent in many ways. It can be said that the object of the mathematical experiment is the mathematical modeling of Nature. By comparing numerical results with those of the real experiment one can find how close the mathematical model is to the real physical process. It is clear that the evaluation of the accuracy of the numerical solution must be accomplished by the methods of numerical analysis only, independently of the physical experiment. Then, if the accuracy of both the numerical and the experimental data is sufficiently high, the differences between them show how close the mathematical model is to Nature.

In our opinion, the method of parallel and coordinated conduct of numerical and experimental studies is a most promising one. The experimental information about the qualitative structure of the problem is highly important for the development of the numerical algorithm. At the same time, the accurate solution for a mathematical model facilitates the planning of the experiment and increases its reliability.

As an example one can consider the work of J. XERIKOS and W. A. ANDERSON [3] on the determination of the flow field around a sphere. These authors state that in planning the experiment they used information on the flow field obtained by the numerical method of integral

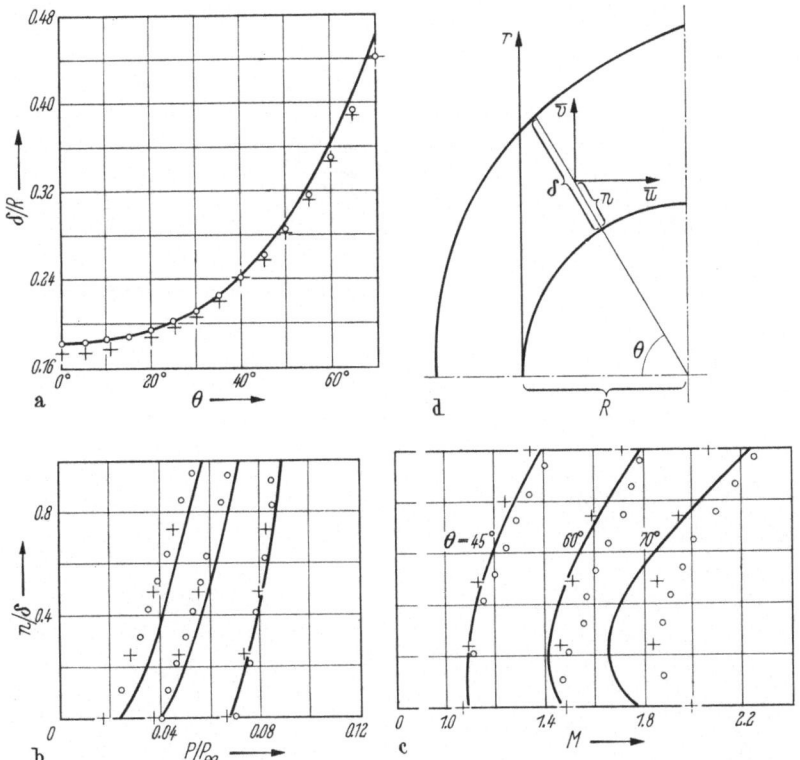

Fig.1. Comparison between calculations and experiment for flow around a sphere; $M_\infty = 3.975$.

relations. Figure 1 presents a comparison between the experimental data (circles) and the numerical results (solid lines) obtained in [3] and those obtained by us using the finite-difference method (crosses) for the same $M_\infty = 3.975$. As one can see, the shock-wave shapes (a) concide satisfactorily in all three cases, the experimental static-pressure values (b) match rather well with those found by us, and the Mach-number curves (c) diverge in all cases. To find the cause of this discrepancy one has to go through a more detailed comparison between the experimental and numerical data.

The requirements for the numerical algorithm

For the effective use of numerical methods the mathematical algorithm must satisfy some fairly rigid requirements. It seems to us that these requirements can be formulated briefly as follows:

1. The algorithm must be applicable to a certain accurately described class of problems, so that the computation of any problem of this class can be accomplished by standard techniques.

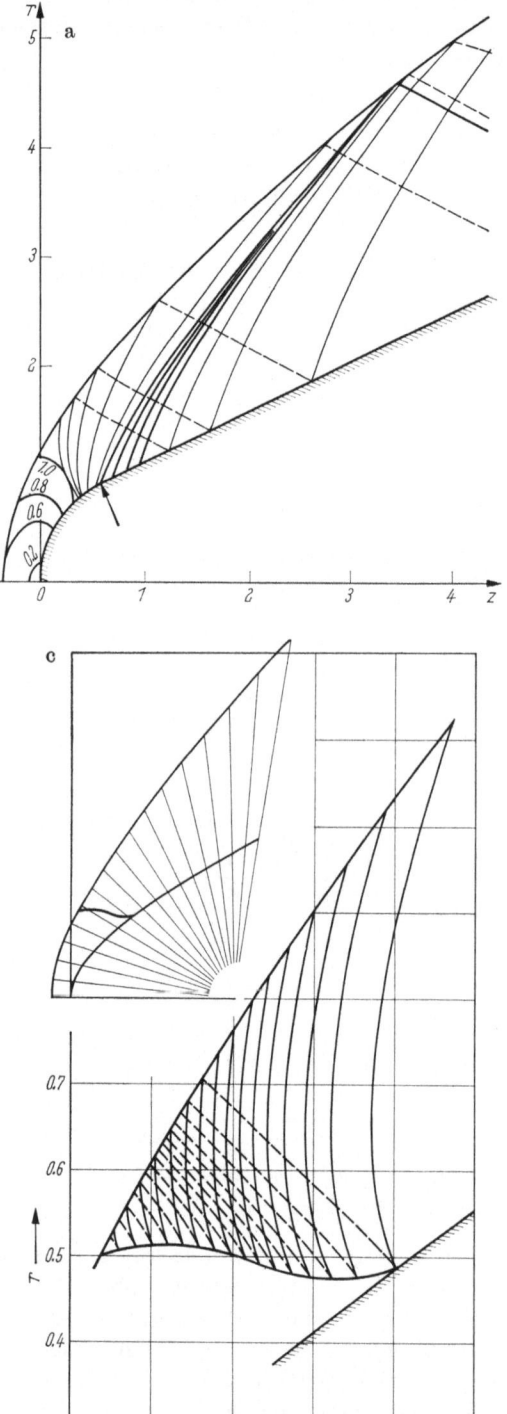

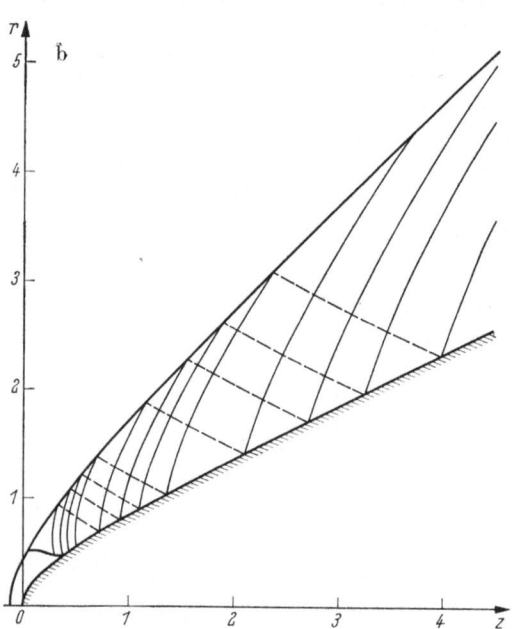

Fig. 2a. Characteristic mesh for flow over a blunted cone.

Fig. 2b. Characteristic mesh for flow over a hyperboloid.

Fig. 2c. Characteristic mesh for flow over a hyperboloid.

2. The algorithm must allow the computation of any problem of this class with given accuracy.

3. The algorithm must include an effective and flexible system of data processing and information representation in a convenient form.

Not having the time to consider here the theory and development of numerical algorithms, we shall only make some general remarks on the problem.

The set of algorithms that satisfies the first requirement depends critically on the class of problems. The extension of the latter can make the algorithm non-applicable, even if it was well developed and successfully used before. As an example, the well-known method of characteristics, which had been applied with success to many problems in gas dynamics, can be mentioned. It is well-known that the accuracy of computation in the supersonic region of the flow around a blunt body can decrease so much as to make it impossible to carry out the computation. This is due to the approach of a different characteristic of the same family. Such a situation can be seen in Fig. 2(a), where the characteristic mesh for such a case is shown. For comparison, the characteristic net for the flow around a hyperboloid with the same limiting half angle is shown in Figs. 2(b) and 2(c).

With respect to the second requirement, we note that the theoretical possibility of computation with an arbitrary accuracy is necessary but not yet sufficient. The order of the number of operations N that are needed for the solution of a problem with a solution error ε is not a minor matter. Moreover, not only is the way in which $N(\varepsilon)$ increases as $\varepsilon \to 0$ of practical importance, but also the actual value of N in the given interval of variation of ε. It seems that at present the so-called n "economical" finite-difference methods are most promising for gas-dynamic problems. In these methods, $N(\varepsilon)$ increases as $1/\varepsilon$ to the power of approximately one.

The third requirement is especially important for multidimensional problems, in which the primary output data consist of tens and hundreds of thousands of numbers. It is obvious that without preliminary processing the direct use of such an amount of information is practically impossible. The effective processing algorithm must ensure fast and convenient presentation of any information about the flow field, including computation of the gas parameters at given points, calculation of the extremal values, construction of level lines and surfaces, integration, differentiation, etc. An example of the use of the processing program is the construction of the characteristic mesh on the basis of results obtained by the finite-difference method (Fig. 2).

In conclusion we note that the construction of algorithms satisfying the above requirements is not a simple task and is based on both the theory of quasilinear multi-dimensional partial differential equations and the theory of finite-difference methods of numerical analysis.

Some calculated results

The systematic study of the gas flow around blunt bodies, some results of which are presented here, has been carried out by means of finite-difference algorithms for the full gas-dynamic equations. The description and investigation of the algorithms is contained in previous papers [1, 2]. In presenting these examples we do not attempt to give full information on the flows studied. Our principal purpose is to draw attention to the possibility of detailed investigation of these complicated three-dimensional flows by numerical methods and to show some interesting and new phenomena that were discovered.

All results are represented in the cylindrical co-ordinate system (z, r, θ). The z axis is directed along the intersection line of the symmetry planes of the body. The point $z = 0, r = 0$ lies on the surface. The angle θ is measured from the symmetry plane of the flow, with the value $\theta = 0$ on the weather side. The pressure p and density ϱ are ratioed to the values p_∞, ϱ_∞ in the undisturbed flow; the velocity-vector components u, v, w are ratioed to $(p_\infty/\varrho_\infty)^{1/2}$. The unit of length is the radius of curvature of the body at the nose point. The gas is perfect, with $k = 1.4$.

The equation of the shock wave is written in the form $r = F(z, \theta)$. Thus we have the derivative $F_z = dF/dz = \tan\psi$, where ψ is the angle between the z axis and the tangent to the shock in a meridional plane $\theta = \mathrm{const}$.

In Fig. 3, F_z is plotted as a function of z in the three meridional planes $\theta = 0$, $\pi/2$, and π for a spherically blunted cone of half angle $\beta = 10°$ at a free-stream Mach number $M_\infty = 10$ and an angle of attack $\alpha = 5°$. The interference of the compression and rarefaction waves with the shock wave determines the complicated from of the shock-wave surface. In the plane $\theta = 0$ the interference is the most strong; there are two maxima and two minima of F_z. For $\theta = \pi/2$ and π the interference it not so strong, and the extremal values of F_z are less marked.

In Fig. 4 the values of F are given as a function of θ in different planes $z = \mathrm{const}$ for the spherically blunted cone with $\beta = 25°$, $M_\infty = 6$, and $\alpha = 10°$. For $z \geq 18$ there occurs an

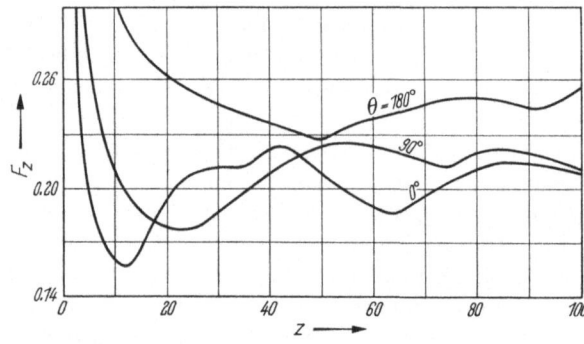

Fig. 3. F_z as a function of z for a spherically blunted cone; $\beta = 10°$, $M_\infty = 10$, $\alpha = 5°$.

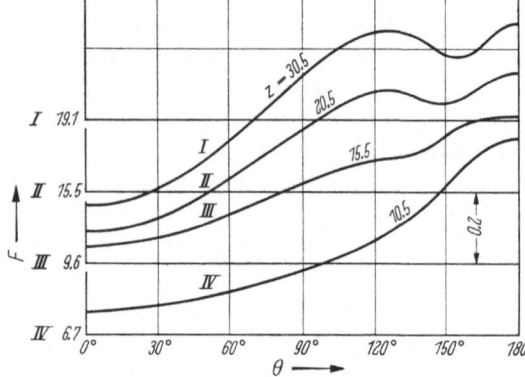

Fig. 4. F as a function of θ for a spherically blunted cone; $\beta = 25°$, $M_\infty = 6$, $\alpha = 10°$.

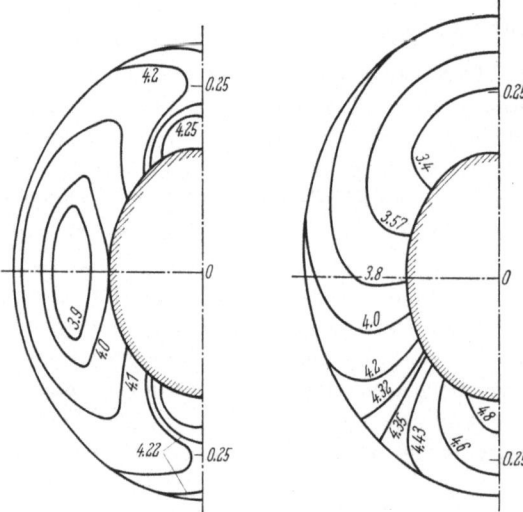

Fig. 5. Lines of constant density for an elliptical paraboloid; $M_\infty = 4$, $z = 0.04225$, $\alpha = 0°$ on the left, $\alpha = 15°$ on the right.

interesting new phenomenon. On the surface of the shock wave there appears a cavity which moves towards the plane $\theta = \pi$ as z increases. We see that in the three-dimensional case the shock can have a rather complicated shape even for a comparatively simple body.

No less sophisticated is the behavior of the gas-dynamic variables in the flow field between the body and the shock. In Fig. 5 the lines $\varrho = \mathrm{const}$ are given for the flow around the elliptical paraboloid.

$$r = [(1.5 \cos^2 \theta + 2.5 \sin^2 \theta) z]^{-1/2}$$

at $M_\infty = 4$ for $\alpha = 0°$ on the left and $\alpha = 15°$ on the right, both in the plane $z = 0.04225$. In Fig. 6 the lines $w = \mathrm{const}$ are given for the same paraboloid, again at $M_\infty = 4$, for $\alpha = 5°$ on the left and $\alpha = 15°$ on the right, both at $z = 0.1$. In the latter figure one can see a saddle-point singularity that moves towards the body surface as α increases.

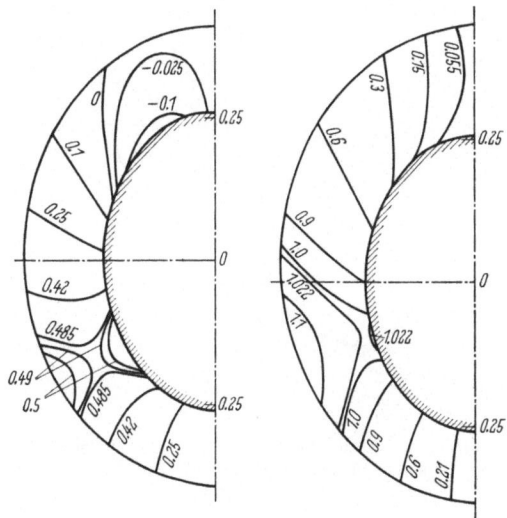

Fig. 6. Lines of constant w for an elliptical paraboloid; $M_\infty = 4$, $z = 0.1$, $\alpha = 5°$ on the left, $\alpha = 15°$ on the right.

In Fig. 7 the lines $p = \mathrm{const}$ are given for the spherically blunted cone for $\beta = 10°$, $M_\infty = 10$, $\alpha = 10°$, $z = 21$. Here there is on the lee side a saddle-point singularity that moves towards the shock as z increases. On the weather side the distribution of the isobars does not differ qualitatively from that for a sharp cone.

In Fig. 8 the lines of constant Mach number M are drawn for the circular paraboloid at $M_\infty = 4$, $\alpha = 10°$, $z = 20$. It is interesting that for this case there is a region in which $M > M_\infty$.

It is important for flow analysis as well as for the development of numerical methods to know the form and location of sonic surfaces. Figure 9 shows the intersection lines of the sonic surface and a plane $z = \mathrm{const}$ for a circular paraboloid at $z = 0.0751$ (on the left) and an elliptical paraboloid at $z = 0.1$ (on the right), both at $M_\infty = 4$ for $\alpha = 0°, 5°, 10°, 15°$. It is remarkable that the sonic surfaces for all the angles of attack intersect almost on the same line, which approaches the paraboloid surface as z increases.

The question of whether the stagnation streamline coincides with the line of maximum entropy in three-dimensional flow has been studied in recent years both from the experimental and analytical point of view. We have examined this topic numerically, and one of the results is shown in Fig. 10 for a circular paraboloid at $M_\infty = 4$. The stagnation streamlines for $\alpha = 10°$ (solid) and $\alpha = 15°$ (dashed) were obtained by means of numerical integration. Their deviations from the lines of maximum entropy do not exceed 1%, and the entropy value on the body surface in the vicinity of the stagnation point differs from the maximum by not more than 0.2%. The lines of maximum entropy were found by interpolation for the entropy function.

The distribution of entropy in the flow around the blunt body is represented in Fig. 11, where the lines $S = p/\varrho^k = \mathrm{const}$ are drawn for the spherically blunted cone for $\beta = 25°$,

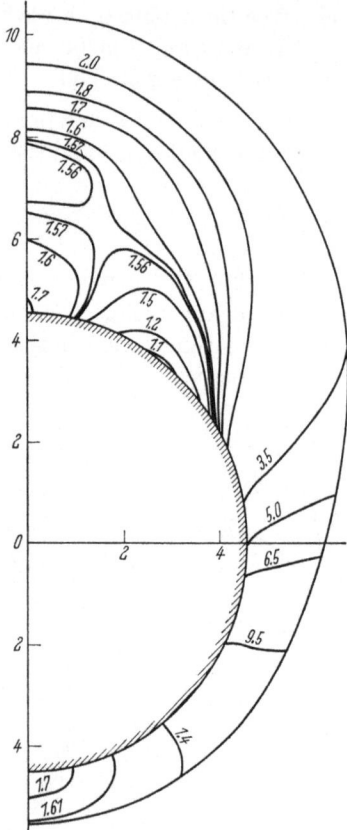

Fig. 7. Lines of constant pressure for a spherically blunted cone; $\beta = 10°$, $M_\infty = 10$, $\alpha = 10°$, $z = 21$.

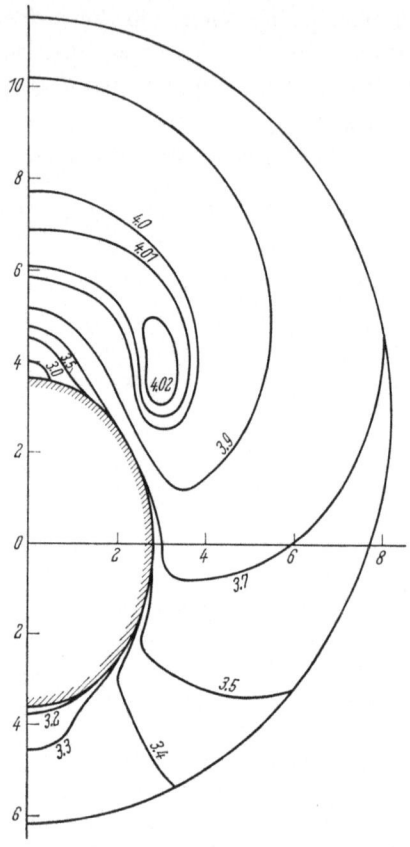

Fig. 8. Lines of constant Mach number for a circular paraboloid; $M_\infty = 4$, $\alpha = 10°$, $z = 20$.

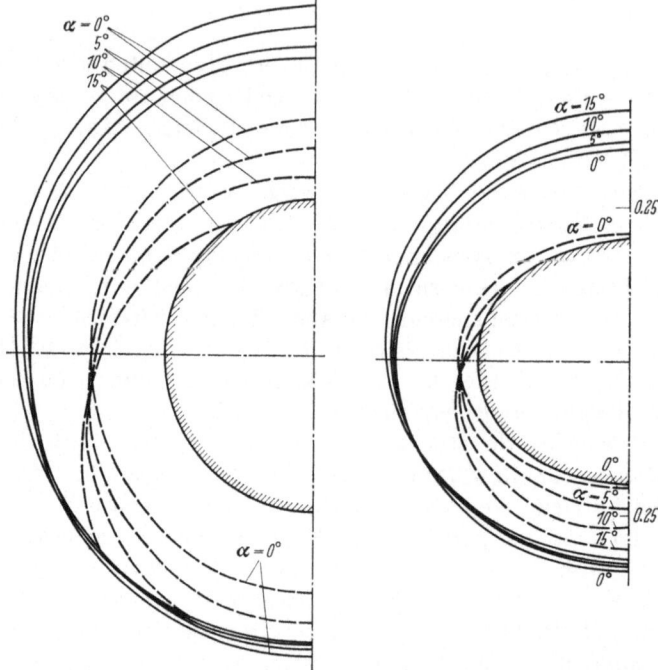

Fig. 9. Position of sonic surface for a circular paraboloid at $z = 0.0751$ on the left, for an elliptical paraboloid at $z = 0.1$ on the right; $M_\infty = 4$.

$M_\infty = 6$, $\alpha = 17.5°$, $z = 11$. For most of the flow, the lines $S = \text{const}$ have the same shape as for a sharp cone. An essential difference is observed only on the lee side and near the body surface, where the generation of the entropy layer is distinctly seen.

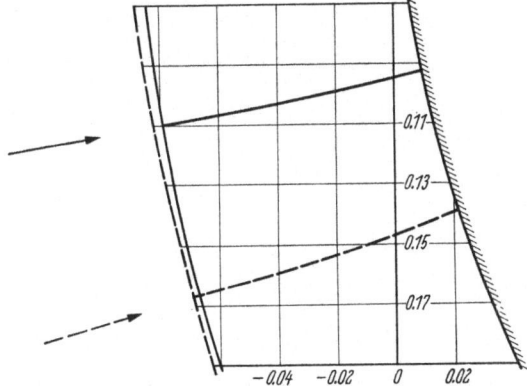

Fig. 10. Stagnation streamlines for a circular paraboloid at $\alpha = 10°$ (solid) and 15° (dashed); $M_\infty = 4$.

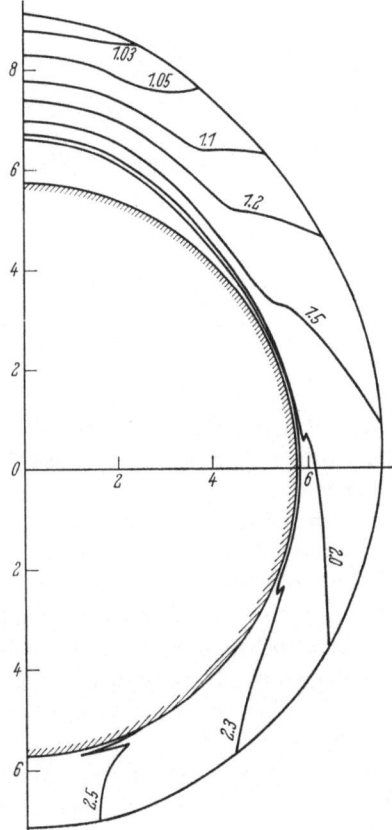

Fig. 11. Lines of constant entropy for a spherically blunted cone; $\beta = 25°$, $M_\infty = 6$, $\alpha = 17.5°$, $z = 11$.

References

1. Babenko, K. I., G. P. Voskresenskiĭ, A. N. Lyubimov and V. V. Rusanov, Spatial Flow past a Smooth Body in an Idealized Gas, "Science", 1964, (in Russian).
2. Rusanov, V. V., Spatial Flow past a Blunt Body in a Supersonic Stream of Gas, Sh. V. M. and M. F. T. 8, No. 3, 1968, (in Russian).
3. Xerikos, J., and W. A. Anderson, An Experimental Investigation of the Shock Layer Surrounding a Sphere in Supersonic Flow. AIAA, 3, N 3, (1965).

A generalization of Kepler orbits for large satellites

By

W. Schiehlen   and   **O. Kolbe**

Technische Hochschule, München   Universität Stuttgart

1. Introduction

In a central gravity field mass points and homogeneous spherical bodies are moving on Kepler orbits. For bodies of arbitrary shape, e.g. artificial satellites, Kepler orbits are only an approximation. The deviations from the Kepler orbits increase with the body extension a relative to the orbit radius r. The ratio a/r is small for hitherto existing satellites: $a/r \simeq 10^{-6}$. Anyhow, satellites with $a/r \simeq 10^{-4}$ are projected for special missions. Further, satellites are in discussion reaching from close to the earth surface up to altitudes far beyond the synchronous altitude [1]. Such satellites will yield ratios $0 < a/r < 1$ during erection.

Circular orbits have been found by DUBOSHIN [2] as particular solutions for the motion of beam-shaped satellites of any size. A special orientation of the beam is necessary: either along the orbit radius vector, or tangential, or normal to the orbit plane. The most striking result is the dependence of the orbit period with respect to the beam length. The attitude stability of beam satellites corresponding to the Duboshin solution has been investigated by MAGNUS [3]. Only the orientation along the orbit radius vector proves to be stable. In a more general paper, HOFER [4] presented further circular orbits as particular solutions for cylindrical satellites. The orbit period is calculated as a function of the geometrical body parameters. However, the orbit stability is not investigated.

In the present paper noncircular, so-called generalized Kepler orbits are derived for cylindrical satellites. In fact, in the general case apogee and perigee radii are functions of the geometrical body parameters and the initial conditions. Apogee and perigee points revolve on circles around the gravity centre: synonymously/counterwise to the orbit direction for satellites of the beam/disk type. Stable orientation to the gravitational attraction centre is shown to be feasible by a suitable flywheel control device. Thus, attitude stability has not to be treated. On the contrary, orbit stability investigations will be an essential part of this paper. New phenomena appear which even hold for the known circular orbits.

2. Equations of motion

The general equations for a fixed body in a central gravity field are in the presentation of MAGNUS [5]:

$$\dot{\boldsymbol{p}} = -\Gamma m_E \int\limits_{m_S} \frac{\boldsymbol{r}_D}{r_D^3} \, dm_S, \tag{1}$$

$$\dot{\boldsymbol{h}} = -\Gamma m_E \int\limits_{m_S} \frac{\boldsymbol{r}_S \times \boldsymbol{r}_D}{r_D^3} \, dm_S. \tag{2}$$

Here, $\boldsymbol{p}$ denotes the momentum and $\boldsymbol{h}$ the angular momentum of the satellite with mass m_S. The radius vectors $\boldsymbol{r}_D$ and $\boldsymbol{r}_S$ are defined by Fig. 1, Γ is the gravity constant and m_E is the mass of the gravitional attraction centre. Assuming a motion of the satellite centre of mass in the e_2, e_3-plane of the inertial frame, one gets for the momentum in the moving frame e_1, e_2, e_3:

$$\dot{p} = m_S \begin{pmatrix} 0 \\ -2\dot{r}_C\dot{\tau} - r_C\ddot{\tau} \\ \ddot{r}_C - r_C\dot{\tau}^2 \end{pmatrix} \qquad (3)$$

where τ denotes the true anomaly.

The angular momentum will be in the body-fixed principal axes frame $\bar{e}_1$, $\bar{e}_2$, $\bar{e}_3$:

$$\dot{h} = \begin{pmatrix} A\dot{\omega}_1 - (B - C)\,\omega_2\omega_3 + A_F\dot{w} \\ B\dot{\omega}_2 - (C - A)\,\omega_3\omega_1 + A_F\omega_3 w \\ C\dot{\omega}_3 - (A - B)\,\omega_1\omega_2 - A_F\omega_2 w \end{pmatrix}. \qquad (4)$$

The moments of inertia of the entire satellite are A, B, C and $\boldsymbol{\omega} = (\omega_1, \omega_2, \omega_3)$ is the angular velocity. The symmetric flywheel with the moments of inertia A_F, $B_F = C_F$ is rotating about the body-fixed $\bar{e}_1$-axis with angular velocity w.

The frame e_1, e_2, e_3 in which Eq. (3) holds and the frame $\bar{e}_1$, $\bar{e}_2$, $\bar{e}_3$ in which Eq. (4) is valid are connected by the linear transformation

$$\bar{e}_j = a_{ij}e_i. \qquad (5)$$

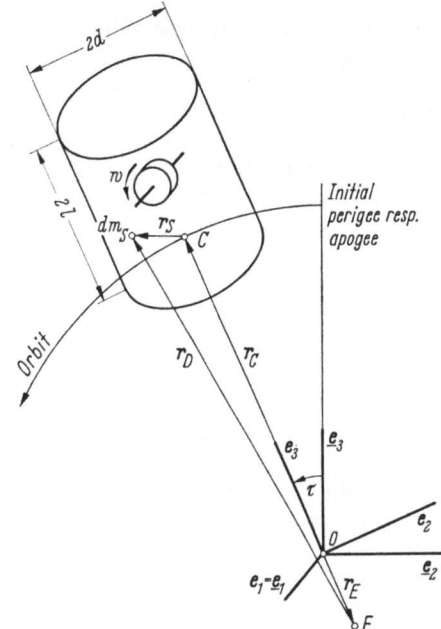

Fig. 1. Schematic of physical and geometrical situation.

It is shown in the paper of HOFER [4] that the integrals in Eqs. (1) and (2) only can be evaluated in closed form if the entire satellite is a homogeneous, cylindrical body and

$$a_{ij} = \delta_{ij} \qquad (6)$$

where δ_{ij} is the KRONECKER symbol.

The mentioned result is in the case of the e_3-axis coinciding with the symmetry axis:

$$-\Gamma m_E \int\limits_{m_S} \frac{\boldsymbol{r}_D}{r_D^3}\, dm_S = \begin{pmatrix} 0 \\ 0 \\ I_3 \end{pmatrix} \qquad (7)$$

with

$$I_3 = -\Gamma m_E m_S \frac{2l + \sqrt{(r - l)^2 + d^2} - \sqrt{(r + l)^2 + d^2}}{d^2 l} \qquad (8)$$

and

$$-\Gamma m_E \int\limits_{m_S} \frac{\boldsymbol{r}_s \times \boldsymbol{r}_D}{r_D^3}\, dm_S = 0. \qquad (9)$$

The cylindrical body is of diameter $2d$ and length $2l$. The distance r is given by the vector $\boldsymbol{r} = \boldsymbol{r}_C - \boldsymbol{r}_E$ and characterizes the satellite orbit relative to the attraction centre as shown in Fig. 1.

Equations (1), (3), (7) and (8) describe the differential equations of the satellite orbit motion:

$$2\dot{r}\dot{\tau} + r\ddot{\tau} = 0, \qquad (10)$$

$$\ddot{r} - r\dot{\tau}^2 = -\Gamma m \frac{2l + \sqrt{(r - l)^2 + d^2} - \sqrt{(r + l)^2 + d^2}}{d^2 l} \qquad (11)$$

where use is made of the identity $mr_C = m_E r$ with the total mass $m = m_E + m_S$. Further, no objection turns out to the assumption that the orbit plane remains in the inertial frame.

Without loss of generality the initial conditions can be assumed as

$$r(0) = r_0, \quad \dot{r}(0) = 0, \quad \tau(0) = 0, \quad \dot{\tau}(0) = v_0/r_0 \tag{12}$$

i.e. a horizontal injection into orbit with distance r_0 and velocity v_0. The initial conditions $r(0) = r_0$ and $\dot{\tau}(0) = v_0/r_0$ can be combined to the generalized eccentricity

$$E = \frac{r_0 v_0^2}{\Gamma m} - 1. \tag{13}$$

It will be shown that $-1 < E < \infty$ is admissible for some geometrical body parameters whereas the Kepler eccentricity is limited to $-1 < e < 1$.

Introducing the dimensionless time

$$t^* = \sqrt{\frac{\Gamma_m}{r_0^3}}\, t \tag{14}$$

and the normalized variables

$$R = r/r_0, \quad D = d/r_0, \quad L = l/r_0 \tag{15}$$

the first integral of Eq. (10) is

$$\tau' = \frac{\sqrt{1+E}}{R^2}. \tag{16}$$

The prime (') means differentiation with respect to time t^*. Regarding Eq. (16) it follows from Eq. (11):

$$R'' = \frac{1+E}{R^3} - \frac{2}{D^2} + \frac{1}{D^2 L}\left\{\sqrt{(R+L)^2 + D^2} - \sqrt{(R-L)^2 + D^2}\right\}. \tag{17}$$

In the case $D \to 0$, $L \to 0$ Eq. (16) reduces to the second Kepler law

$$\tau' = \frac{\sqrt{1+e}}{R^2}, \tag{18}$$

and Eq. (17) to

$$R'' = \frac{1+e}{R^3} - \frac{1}{R^2} \tag{19}$$

with the solution

$$R = \frac{1+e}{1 + e\cos\tau}, \tag{20}$$

the first Kepler law. Thereby, the generalized eccentricity E is identical with the Kepler eccentricity e.

The satellite attitude is determined by Eqs. (5) and (6). The corresponding angular velocity is obtained from the motion of frame e_1, e_2, e_3:

$$\omega = \begin{pmatrix} \dot{\tau} \\ 0 \\ 0 \end{pmatrix}. \tag{21}$$

Equation (2) with Eqs. (4), (9) and (21) is valid if

$$\dot{w} = -\frac{A}{A_F}\ddot{\tau}. \tag{22}$$

The flyhweel motion (22) will be excited by the attitude control device. Only on circular orbits the right side of Eq. (22) is identically zero and possibly no flywheel is necessary.

3. Classical treatment of the problem

For small ratios $D \ll 1$, $L \ll 1$ the solution for the orbit will still be similar to a Kepler orbit. Therefore a description by an ellipse of slowly varying elements seems to be reasonable. In celestial mechanics, six elements are used to locate the orbit in space and the satellite on the orbit. Two elements describe the plane of motion: Ω the right ascension of ascending node and γ the inclination. These elements can be disregarded since the orbit plane is fixed in an inertial frame as shown in section 2. Two other elements characterize the shape of the orbit in the plane: $2a$ the major axis and e the eccentricity. One element gives the orientation of the orbit within the plane of motion: ω the argument of perigee. One more element, the epoch t_0, is necessary to specify the time instant at which the satellite passes a certain orbit location. From Eq. (12) follows for the epoch $t_0 = 0$.

So, there are three essential elements left: a, e, ω. The slow variation of these elements can be calculated from wellknown formulas of the celestial perturbation method. Following MOULTON [6], it can be written

$$\dot{a} = \frac{2e \sin \tau}{n \sqrt{1 - e^2}} R^* + \frac{2a \sqrt{1 - e^2}}{nr} S^*, \tag{23}$$

$$\dot{e} = \frac{\sqrt{1 - e^2}}{na} \sin \tau \ R^* + \frac{\sqrt{1 - e^2}}{na^2 e} \left(\frac{ap}{r} - r \right) S^*, \tag{24}$$

$$\dot{\omega} = -\frac{\sqrt{1 - e^2}}{nae} \cos \tau \ R^* + \frac{\sqrt{1 - e^2}}{nae} \left(1 + \frac{r}{p} \right) \sin \tau \ S^* - \frac{r \sin (\omega + \tau) \cot \gamma}{na^2 \sqrt{1 - e^2}} W^*, \tag{25}$$

with the third Kepler law

$$n^2 = \frac{\Gamma m}{a^3} \tag{26}$$

and the orbit parameter $p = (1 - e^2) a = (1 + e) r_0$.

The perturbation acceleration has the three coordinates W^*, S^*, R^* in the frame e_1, e_2, e_3. From Eq. (7) follows $W^* = S^* = 0$ and

$$R^* = I_3 - \frac{\Gamma m}{r^2}. \tag{27}$$

In Eq. (27) Γm is written instead of Γm_E since in Eqs. (23) to (25) r appears instead of r_C and the identity $m r_C = m_E r$ holds. The expansion of Eq. (27) in a power series gives, regarding $D \ll 1$, $L \ll 1$:

$$R^* = -\frac{\Gamma m}{r^4} \left(l^2 - \frac{3}{4} d^2 \right). \tag{28}$$

Because of $S^* = 0$, the small variations Δa and Δe are joint by

$$\Delta a = \frac{2ae}{1 - e^2} \Delta e. \tag{29}$$

Hence, it is sufficient to evaluate Eq. (24). Using Eqs. (15), (24), (28), and the dimensionless time t^* one obtains

$$e' = -\frac{\sqrt{1 + e}}{R^4} \left(L^2 - \frac{3}{4} D^2 \right) \sin \tau. \tag{30}$$

Now, the second Kepler law, Eq. (18), is used to substitude the independent variable t^* in Eq. (30) by the true anomaly τ. The result can be written by means of the first Kepler law, Eq. (20), as

$$\frac{de}{d\tau} = -\frac{L^2 - \frac{3}{4} D^2}{(1 + e)^2} (1 + e \cos \tau)^2 \sin \tau. \tag{31}$$

According to the perturbation method the variations of the elements are so slow that all elements on the right hand side of Eqs. (23) to (25) can be regarded as constant. Doing this in Eq. (31), one gets the variation Δe for a limited time given by the true anomaly T

$$\Delta e = -\frac{L^2 - \frac{3}{4} D^2}{(1 + e)^2} \left\{ (1 - \cos T) + e(1 - \cos^2 T) + \frac{e^2}{3} (1 - \cos^3 T) \right\}. \tag{32}$$

It turns out that there is only a short-periodic perturbation with orbit period regarding the elements of the orbit shape, a and e. This effect is plausible because of a constant system energy.

The behaviour of the element ω is given by Eq. (25). The same approach as used just before leads to

$$\omega' = \frac{L^2 - \frac{3}{4} D^2}{e(1 + e)^2} (1 + e \cos \tau)^2 \cos \tau. \tag{33}$$

The variation $\Delta\omega$ for a limited true anomaly T is

$$\Delta\omega = \frac{L^2 - \frac{3}{4}D^2}{e(1+e)^2}\left\{eT + \left(1 + \frac{3}{4}e^2\right)\sin T + \frac{e}{2}\sin 2T + \frac{e^2}{12}\sin 3T\right\}. \tag{34}$$

Now, a secular and a short-periodic perturbation are obvious. The secular perturbation of ω leads to a progression of the apsides: the orientation of the orbit within the plane of motion is varying slowly. The progression of apsides during half an orbit amounts

$$\Delta\omega_{\text{half orbit}} = \pi\frac{L^2 - \frac{3}{4}D^2}{(1+e)^2}. \tag{35}$$

The periodic variations of a, e, and the progression of apsides will vanish if the satellite satisfies the condition of a spherical inertia ellipsoid

$$D:L = 2/\sqrt{3}. \tag{36}$$

So, the perturbance method gives the result that sufficiently small satellites will show the progression of apsides as essential effect.

4. First integral of motion

With increasing ratios D, L greater deviations from the Kepler orbit are to be expected. Then, the elements a, e, ω are no more suitable to characterize the orbit shape. Further, the third Kepler law will be invalid since the orbit is no ellipse: the orbit period T is no more correlated to the major axis $2a$. So, it is necessary to introduce the radius R_A of apogee or perigee respectively, the corresponding true anomaly τ_A, and the orbit period T^* as characteristics. These characteristics are defined by

$$R_A = R(t_1^*), \quad \tau_A = \tau(t_1^*), \quad T^* = 2t_1^* \tag{37}$$

where the time t_1^* follows from the condition

$$R'(t_n^*) = 0, \quad n = 0, 1, 2, \ldots \tag{38}$$

Equation (38) is trivially satisfied by $t_0^* = 0$ due to the initial conditions (12). Therefore the apogee/perigee time instant t_1^* is the first nontrivial zero of radius velocity R'. Since the orbit motion is a conservative process, the generalized Kepler orbit is represented completely by the trajectory between the initial state (12) and the first succeeding apogee/perigee which is described by the orbit characteristics (37). Further, it is obvious that $t_n^* = nt_1^*$ and $R(t_{2n}^*) = 1$, $R(t_{2n+1}^*) = R_A$, $\tau(t_n^*) = n\tau_A$.

Note that the special case of circular orbits is excluded here. It can be treated separately according to Hofer [4]. However, circular orbits also are obtained easily below by imbedding considerations.

In contrary to Kepler orbits, the orbit period T^* of generalized Kepler orbits, defined by Eq. (37), is not corresponding to the time of one revolution $T_{2\pi}$ because of $\tau_A \neq \pi$. I.e., the characteristic T^* is rather uninteresting and therefore will not be treated in detail here. However, in the circular orbit case Eq. (16) leads to

$$\frac{2\tau_A}{T_{\text{circ}}^*} = \frac{2\pi}{T_{2\pi}^*} = \sqrt{1 + E_{\text{circ}}}. \tag{39}$$

For that reason Duboshin [2] and Hofer [4] don't need the characteristic τ_A and can use the orbit period $T_{2\pi}$.

For ratios $D \ll 1$, $L \ll 1$ approximations of the characteristics R_A and τ_A can be given using the results of section 3.

From $r_A = a(1 + e)$, $r_0 = a(1 - e) = const$ and Eq. (29) follows

$$R_A = \frac{1+e}{1-e} + \frac{1+e}{(1-e^2)^2}\Delta e. \tag{40}$$

Further, the statement yields

$$\tau_A = \pi + \Delta\omega. \tag{41}$$

Replacing T in Eqs. (32) and (34) by Eq. (41), one gets

$$R_A = \frac{1+e}{1-e} - \frac{2}{3}\frac{(1+e)(3+e^2)}{(1-e^2)^2}\left(L^2 - \frac{3}{4}D^2\right), \tag{42}$$

$$\tau_A = \pi + \pi\frac{L^2 - \dfrac{3}{4}D^2}{(1+e)^2}. \tag{43}$$

For beam-shaped satellites ($D = 0$) the characteristic R_A is decreasing with increasing beam length L while τ_A is increasing. Disk-shaped satellites ($L = 0$) show a completely reverse behaviour: The characteristic R_A is increasing and τ_A is decreasing with increasing disk diameter D. Satellites with a spherical inertia ellipsoid ($D:L = 2/\sqrt{3}$) represent the indifferent case. Of course, these prognoses will only be trustful for sufficiently small satellites.

For arbitrary ratios D, L the characteristics R_A and τ_A have to be computed numerically from Eqs. (16) and (17) under consideration of Eq. (38). Since this is a rather lengthy approach — the integration has to be done for each set of parameters E, D, L — it is of great interest to look for a first integral of Eq. (17). One gets

$$\int_0^{R'} R''R'\, dt^* = \int_1^{R}\left\{\frac{1+E}{R^3} - \frac{2}{D^2} + \frac{1}{D^2 L}\left(\sqrt{(R+L)^2 + D^2} - \sqrt{(R-L)^2 + D^2}\right)\right\}dR. \tag{44}$$

The integration of Eq. (44) results in

$$R'^2 = (1+E)\frac{R^2-1}{R^2} - \frac{4}{D^2}(R-1) - \frac{1}{L}\ln\frac{R-L+\sqrt{(R-L)^2+D^2}}{R+L+\sqrt{(R+L)^2+D^2}}\frac{1+L+\sqrt{(1+L)^2+D^2}}{1-L+\sqrt{(1-L)^2+D^2}}$$

$$+\frac{1}{D^2 L}\left\{(R+L)\sqrt{(R+L)^2+D^2} - (R-L)\sqrt{(R-L)^2+D^2}\right.$$

$$\left. - (1+L)\sqrt{(1+L)^2+D^2} + (1-L)\sqrt{(1-L)^2+D^2}\right\}$$

$$\triangleq (R-1)\, f(R, E, D, L). \tag{45}$$

Zeroing Eq. (45) and considering Eq. (38)

$$f(R_A, E, D, L) = 0 \tag{46}$$

follows as an implicit formula for the characteristic R_A. The function f as defined by Eq. (45) can be simplified greatly in the cases of beam-shaped and disk-shaped satellites. For $D = 0$

$$f(R_A, E, 0, L) = (1+E)\frac{R_A+1}{R_A^2} - \frac{2}{L(R_A-1)}\,\text{arcoth}\,\frac{R_A - L^2}{L(R_A-1)} \tag{47}$$

holds, and in the case $L = 0$:

$$f(R_A, E, D, 0) = (1+E)\frac{R_A+1}{R_A^2} - \frac{4}{D^2} + \frac{4}{D^2(R_A-1)}\left(\sqrt{R_A^2+D^2} - \sqrt{1+D^2}\right). \tag{48}$$

The behaviour of the generalized Kepler orbit characteristic R_A, given by Eq. (46), has been investigated in detail in the mentioned cases $D = 0$ and $L = 0$ and for cylindrical satellites with shape-ratios $D:L = 1/\sqrt{10}$; 0.8061; 1; $2/\sqrt{3}$; $\sqrt{10}$. The results are shown in Figs. 2 to 8. With the exception of the case $L = 0$, the ratio L is chosen as the free parameter. Therefore in Figs. 2 to 7, the region in the lower right has to be excluded because of the kinematic constraint

$$R_A > L. \tag{49}$$

For ratios $D \ll 1$, $L \ll 1$ one recognizes the results of the perturbation method. Further, the said prognoses are valid for arbitrary ratios D, L in the case of beam-shaped and disk-shaped satellites. But additional nonlinear effects arise: diverse solutions R_A can exist for one set E, D, L and satellites with a spherical inertia ellipsoid ($D:L = 2/\sqrt{3}$) are no more indifferent.

Figures 2 and 3 are examples for satellites with two, one or no solutions R_A while Figs. 4 to 8 demonstrate satellites with a unique solution R_A. In the case of diverse solutions R_A there are diverse possible orbits in which the satellite with unique parameters and initial conditons (12) can move. But since the satellite motion is well determined at most one orbit is stable. This fact will be sharpened in Section 5. If no solution R_A exists, the satellite falls down to the gravitational attraction centre or escapes before reaching an apogee/perigee point. Some trajectory but no orbit will appear then.

For the further discussion it is useful to classify the satellite behaviour in disk-like/beam-like if $R_A \gtrless R_{A\,\text{Kepler}}$. Figure 6 shows that satellites with $D:L = 2/\sqrt{3}$ have an absolutely disk-like behaviour with increasing length. At the same time the shape ratio $D:L = 2/\sqrt{3}$ is the upper boundary for satellites with changing behaviour, for satellites with $D:L < 2/\sqrt{3}$ behave beam-like if $L \ll 1$. Figure 5 presents satellites with the geometrically simple shape $D = L$. In this case a typical changing behaviour occurs from beam-like to disk-like with increasing length. The lower boundary for the changing is $D:L = 0.8061$. Figure 4 shows satellites with $D:L = 0.8061$, the behaviour is absolutely beam-like. However, Fig. 4 is decorated by the fact that the $E = 0$ curve is passing through the point $R_A = 1$, $L = 1$. To calculate the mentioned shape ratio Eq. (46) has to be treated for the critical value $R_A = 1$. A limit consideration gives

$$f(1, E, D, L) = 2(1 + E) - \frac{4}{D^2} + \frac{2}{D^2 L}\left(\sqrt{(1 + L)^2 + D^2} - \sqrt{(1 - L)^2 + D^2}\right) = 0. \qquad (50)$$

In the special case $E = 0$, $L = 1$ one gets from Eq. (50), neglecting the trivial solution $D = 0$, the cubic equation

$$D^3 - 2D^2 - 4D + 4 = 0 \qquad (51)$$

with the positive root $D = 0.8061$.

Finally, Fig. 8 gives a presentation of disk-shaped satellites. In contrary to satellites with a finite length there exists no kinematic boundary for disk-shaped satellites. The parameter region is chosen as $0 < D < 10$ and generalized eccentricity values are treated until $E = -0.999$. It has to be mentioned that disk-shaped satellites with nonsmall values D are rather academic in contrary to beam-shaped satellites. But this limit case is essential for understanding the various effects.

Kepler orbits exist only for a limited injection speed v_0 which is determined by $-1 < e < 1$, see Eq. (13). For $e = e_{\text{escape}} = 1$ one gets $R_A \to \infty$ and a parabolic orbit. The generalized Kepler orbits will be limited in the same way. The boundary can be characterized by the generalized eccentricity E_{escape} which belongs to an infinite characteristic R_A. From Eq. (46) follows the defining equation $f(R_A \to \infty, E_{\text{escape}}, D, L) = 0$ which after passing to the limit can be rewritten as

$$E_{\text{escape}} = \frac{1}{L} \ln \frac{1 + L + \sqrt{(1 + L)^2 + D^2}}{1 - L + \sqrt{(1 - L)^2 + D^2}} - \frac{1}{D^2 L}\left\{4L - (1 + L)\sqrt{(1 + L)^2 + D^2}\right.$$
$$\left. + (1 - L)\sqrt{(1 - L)^2 + D^2}\right\} - 1. \qquad (52)$$

In Fig. 9 the trend of E_{escape} is shown. The extreme values are $D = 0$, $L = 1$: $E_{\text{escape}} \to \infty$ and $D \to \infty$, $L = 0$: $E_{\text{escape}} \to -1$. Therefore, the generalized eccentricity can assume values with $-1 < E < \infty$.

5. Stability of motion

The generalized Kepler orbits given by Figs. 2 to 8 have to be investigated on stability. It is essential to say that only orbital stability and not stability in the sense of Lyapunov is of interest. The classical Kepler orbits show orbital stability and are instable in the sense of Lyapunov. Orbital stability considerations can be done by treating Eq. (17). This equation represents an autonomous nonlinear vibration system. Therefore phase plane methods can be applied which make use of the potential energy of the system. In Fig. 10 a schematic of

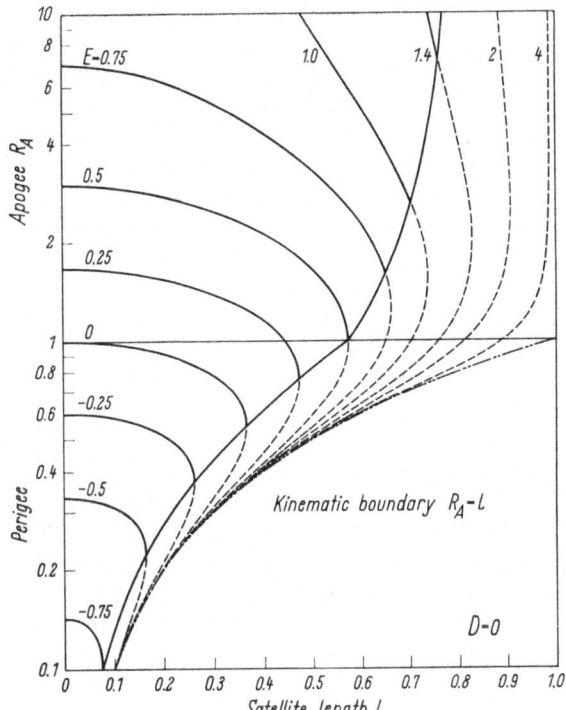

Fig. 2. Characteristic R_A of beam-shaped satellites. The broken lines are located in the instable region.

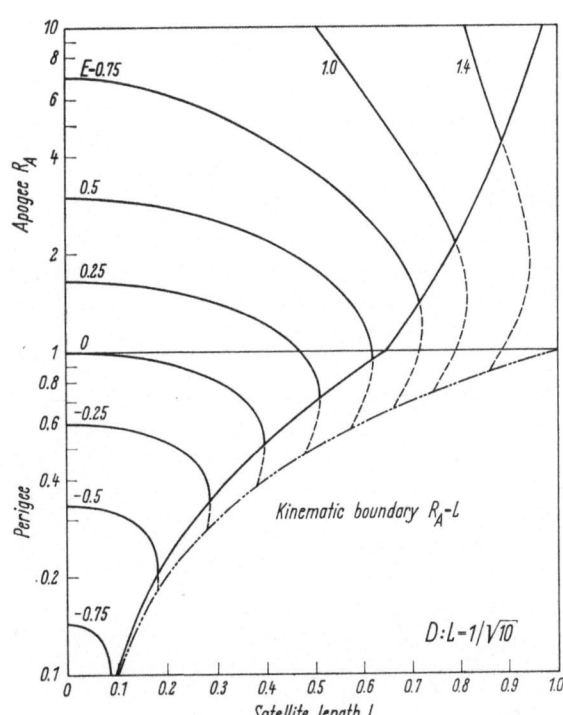

Fig. 3. Characteristic R_A of cylindrical satellites $(D:L = 1/\sqrt{10})$. The broken lines are located in the instable region.

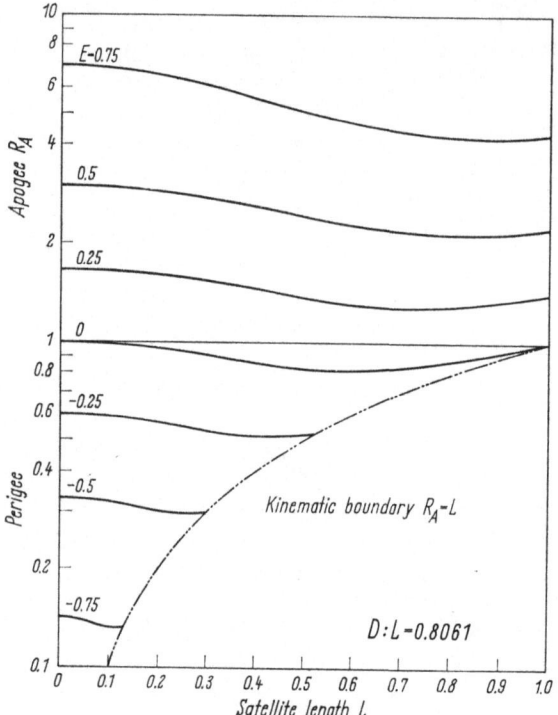

Fig. 4. Characteristic R_A of cylindrical satellites
($D:L = 0.8061$).

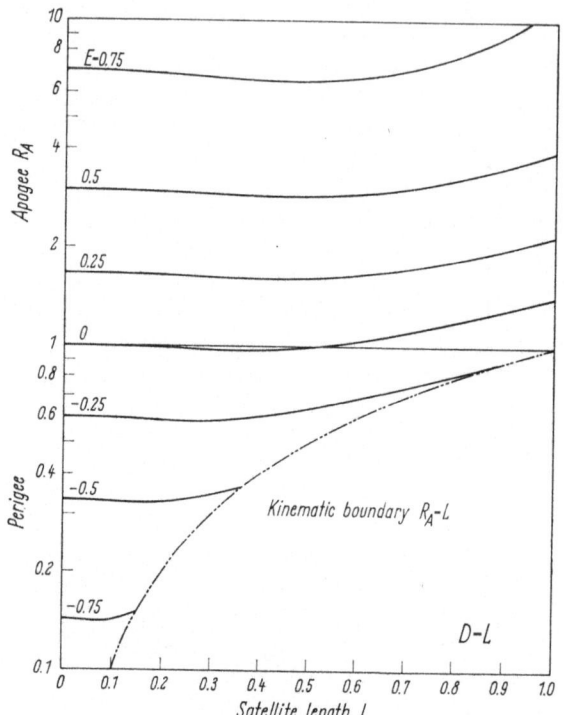

Fig. 5. Characteristic R_A of cylindrical satellites
($D = L$).

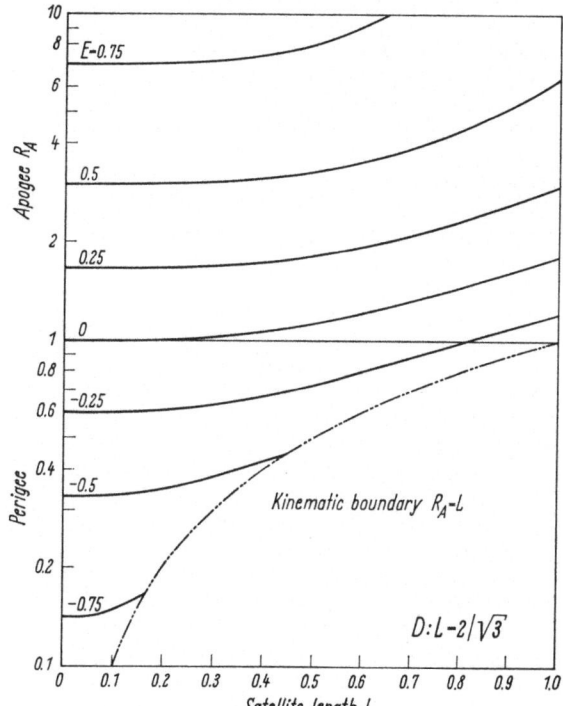

Fig. 6. Characteristic R_A of cylindrical satellites with spherical inertia ellipsoid.

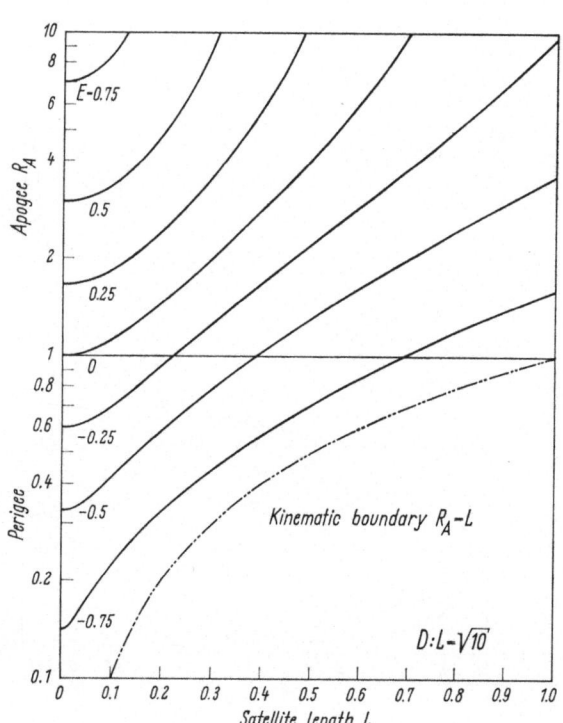

Fig. 7. Characteristic R_A of cylindrical satellites $(D:L = \sqrt{10})$.

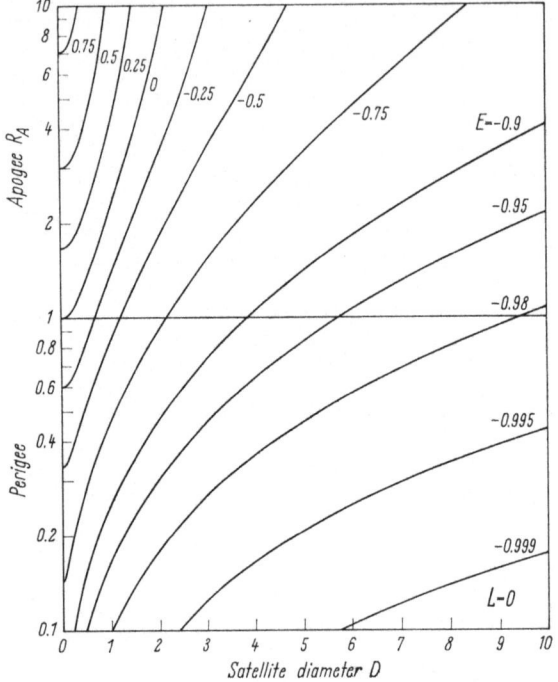

Fig.8. Characteristic R_A of disk-shaped satellites.

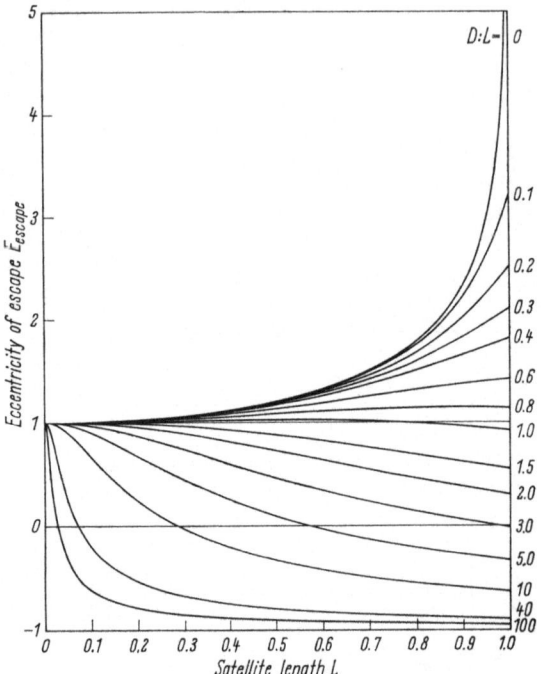

Fig.9. Initial condition parameter of escaping cylindrical satellites.

$V_{\text{pot}} - V_0 = -V_{\text{kin}}$ is drawn. V_{kin} and V_{pot} respectively denote the kinetic and potential normalized energy of the radial motion component which is conservative; hence $V_{\text{kin}} + V_{\text{pot}} = V_0 = const.$ Due to Eq. (45) the following identity holds:

$$V_{\text{pot}} - V_0 = -(R - 1)\, f(R, E, D, L). \qquad (53)$$

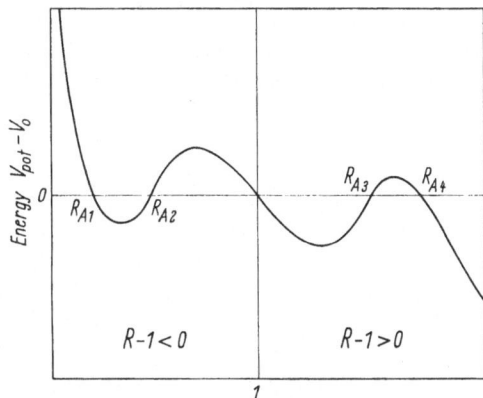

Fig. 10. Schematic of energy function.

Note that the expression (53) vanishes for all values R_A which are candidates for the characteristic R_A^* of stable orbits (if existing at all). From Fig. 10 one reads that an orbit with characteristic R_A will be stable ($R_A = R_A^*$) if the energy $V_{\text{pot}} - V_0$ takes only nonpositive values between the trivial solution $R = 1$ (initial condition) and $R = R_A$.

Assuming for the beginning an energy function (53) with not vanishing slope at point $R = 1$, at most one stable orbit exists for each set E, D, L. Depending on the sign of the slope, there will be possible values R_A^*: either the smallest apogee radius R_{A+} or the greatest perigee radius R_{A-}. So, only two candidates are left for the characteristic R_A^*:

$$R_{A+,-}^* = \min_m \pm (R_{Am} - 1), \qquad m = 1, 2, \ldots \qquad (54)$$

Here, R_{Am} denotes all solutions R_A numbered as shown in Fig. 10. The stability condition for the remaining R_{A+}, R_{A-} simply reads as

$$\text{sgn}\,(R_A^* - 1)\, R''(1, E, D, L) = 1, \qquad (55)$$

where the identity $\partial(V_{\text{pot}} - V_0)/\partial R = 2R''$ is noted. From Eq. (17) it is clear that $R'' = R''(R, E, D, L)$.

If only two solutions R_A exist ($m = 2$) the stability condition

$$\text{sgn}\,(1 - R_A)\, R''(R_A, E, D, L) + \text{sgn}\,(R_A - 1)\, R''(1, E, D, L) = 2 \qquad (56)$$

holds and Eq. (54) can be disregarded. The proof of Eq. (56) follows from geometrical considerations concerning the energy function in Fig. 10 and some analytical properties of function (45).

A detailed numerical investigation resulted in the following classification for the number of diverse charactristics R_A:

$$m = 2, \qquad 0 < D\!:\!L < 0.3603,$$

$$m > 2, \qquad 0.3603 < D\!:\!L < 0.4097,$$

$$m = 1, \qquad 0.4097 < D\!:\!L < \infty.$$

In the region $0.4097 < D\!:\!L < \infty$ there is $R_A = R_A^*$ because of $m = 1$. Stability condition (55) is satisfied. The region $0 < D\!:\!L < 0.3603$ is governed by stability condition (56). However, the intermediate region requires the operation indicated by Eq. (54) and the application of condition (55).

For the special shape ratios $D:L = 0$ and $D:L = 1/\sqrt{10}$ the stability boundaries are shown in Figs. 2 and 3. The instable characteristics R_A are given by the broken lines. A stability diagram is presented in Fig. 11. The wide instable region for $D = 0$ is shrinking with increasing value $D:L$ and vanishes with $D:L = 0.4097$. Only one stable orbit exists if $D:L > 0.4097$.

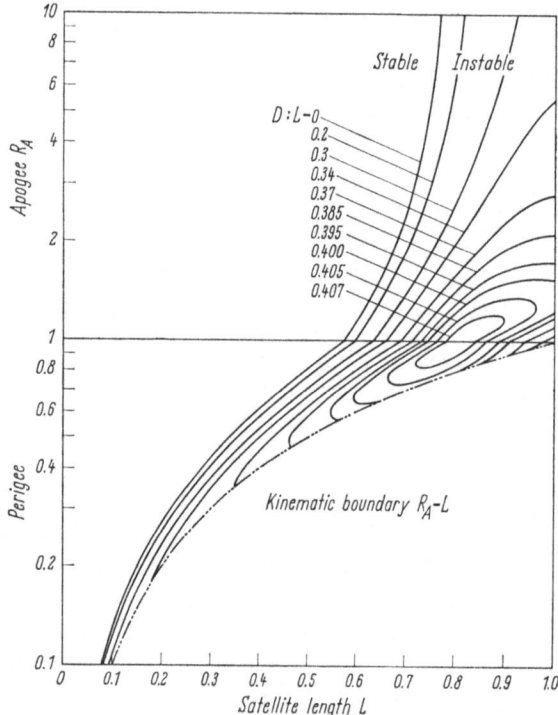

Fig. 11. Stability diagram of cylindrical satellites.

If more than two solutions R_A exist ($m > 2$) the typical nonlinear jump phenomenon may occur. Indeed, in the intermediate region $0.3603 < D:L < 0.4097$ jumps can be observed. For $D:L = 0.395$, Fig. 12 shows various energy functions, calculated by Eq. (53). A jump is caused by the gradual raising of the local maximum above the zero level. So very small variations of the parameters E, D, L may lead to a remarkable change of the stable orbit characteristic, e.g. from $R_A^* = 1.7715$ to $R_A^* = 1.1010$ for changing L from $L = 0.991$ to $L = 0.992$. On the other side, the corresponding R_A diagram, Fig. 13, illustrates that a jump appears if curves $E = const$ have points of inflection with vertical slope.

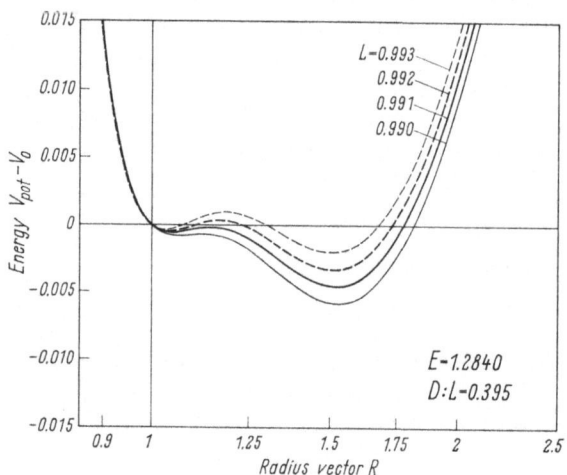

Fig. 12. Energy function illustrating jump phenomenon.

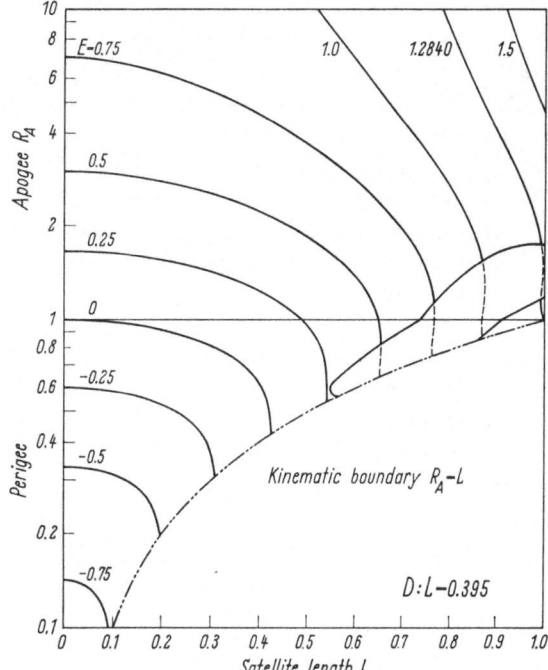

Fig. 13. Characteristic R_A of cylindrical satellites. The broken lines are located in the instable region. Illustration of jump phenomenon.

The assumption of not vanishing slope of the energy function at point $R = 1$ can be dropped now. If Eq. (53) yields the multiple solution $R = 1$ then $R_A = 1$ appears as nontrivial candidate for the orbit characteristic R_A^*. It is easy to see that circular orbits are necessarily stable if the curvature of the energy function is negative. Therefore, the stability condition can be formulated as

$$\operatorname{sgn} \frac{\partial R''}{\partial R}\bigg|_{R=1} = 1. \tag{57}$$

Anyhow, the application of condition (57) is not necessary since the case $R_A = 1$ is imbedded in the results of Eq. (55).

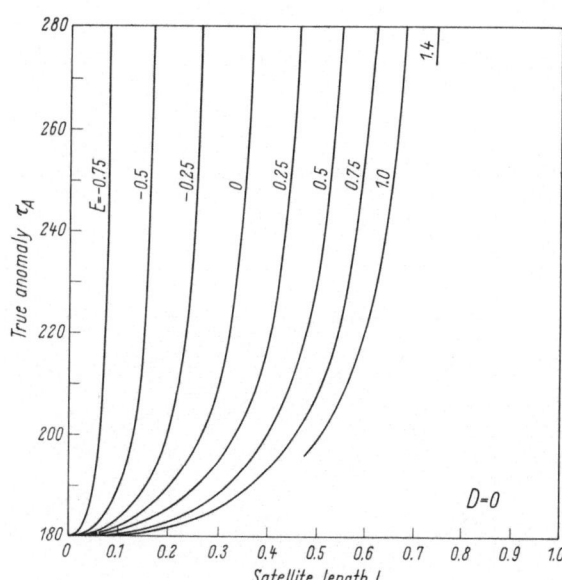

Fig. 14. Characteristic τ_A of beam-shaped satellites.

6. The complete motion

In Section 4, it has been shown that two characteristics R_A, τ_A are necessary to describe the orbit. The behaviour of R_A has been investigated in detail. Therefore, the approximation formula (43) is the only information on τ_A. To get further information, Eqs. (16) and (17) have to be integrated numerically. That was done on the TR4 digital computers, at the University of Stuttgart and at Leibniz computer centre, Munich.

The results of the numerical integration are presented in Figs. 14—17. For ratios $D \ll 1$, $L \ll 1$ the approximations by the perturbation method can be verified. Obviously, only the stable orbit can be found by a stable numerical integration method. No instable branches of τ_A curves appear. In Figs. 2, 4, 6, 8 and Figs. 14—17 the parameters E and $D : L$ are just the same. Therefore, R_A curves and τ_A curves correspond completely. R_A is limited to $0.1 \leq R_A \leq 10$ in Figs. 2—8. From this it results that some τ_A curves don't start at zero.

Figure 14 presents the beam-shaped satellite $D = 0$. It is obvious that all τ_A curves have vertical asymptotes if the length L is reaching the stability boundary. So, regarding Eq. (16) and the corresponding limited characteristic R_A, one can deduce an infinite period T^* for orbits

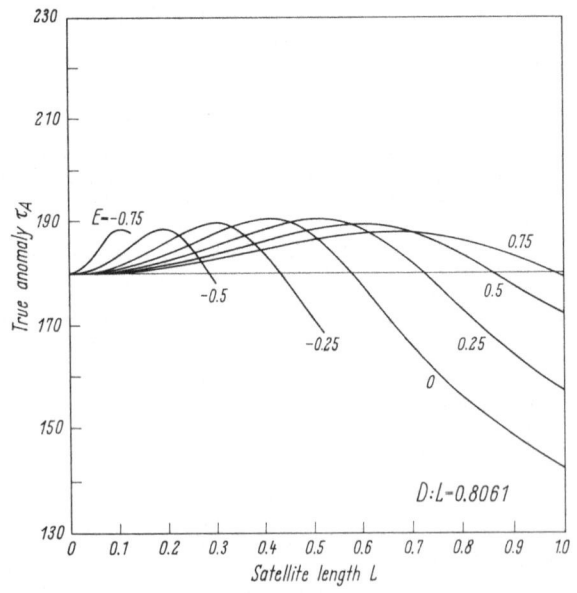

Fig. 15.   Characteristic τ_A of cylindrical satellites $(D : L = 0.8061)$.

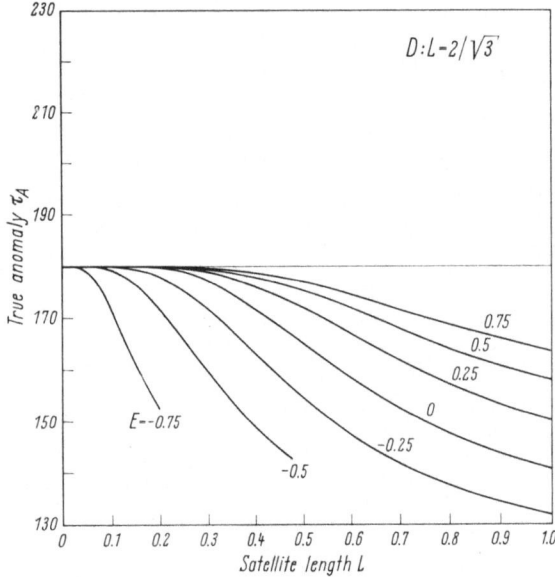

Fig. 16.   Characteristic τ_A of cylindrical satellites with spherical inertia ellipsoid.

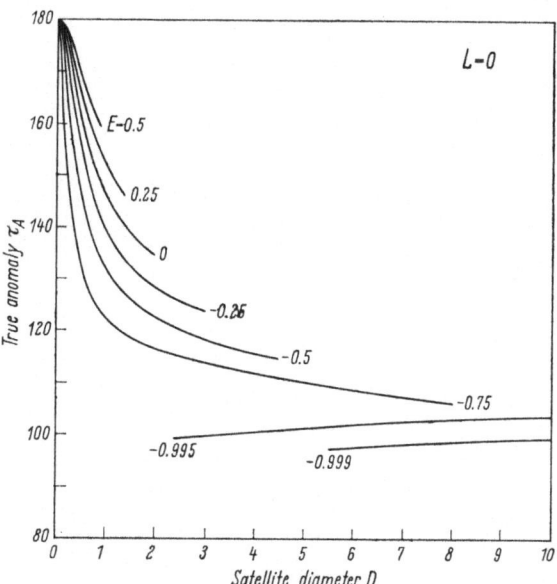

Fig. 17. Characteristic τ_A of disk-shaped satellites.

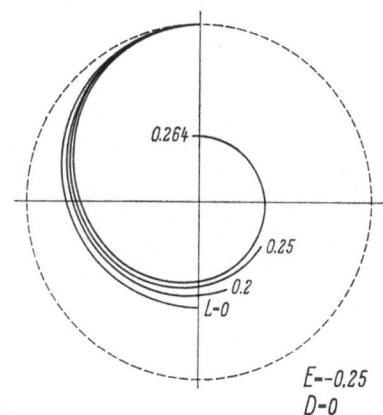

Fig. 18. Initial state ($E = -0.25$) to perigee trajectories of beam-shaped satellites.

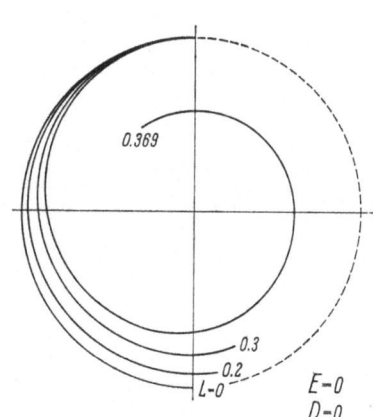

Fig. 19. Initial state ($E = 0$) to perigee trajectories of beam-shaped satellites.

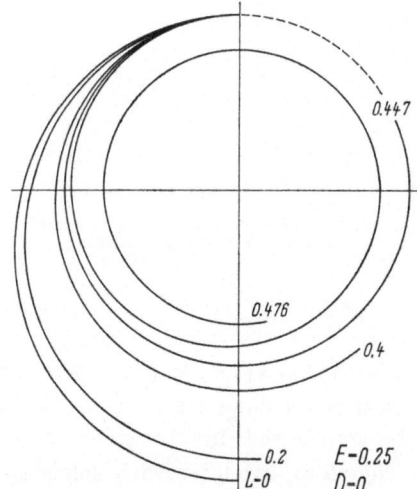

Fig. 20. Initial state ($E = 0.25$) to apogee/perigee trajectories of beam-shaped satellites.

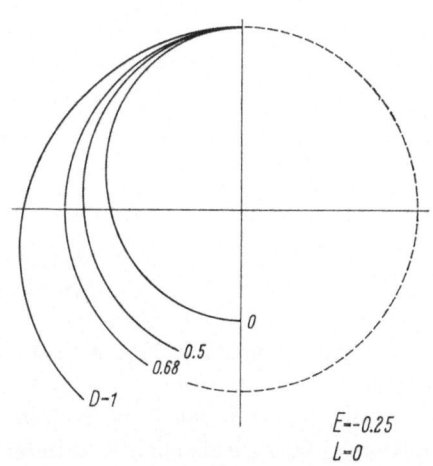

Fig. 21. Initial state ($E = -0.25$) to apogee/perigee trajectories of disk-shaped satellites.

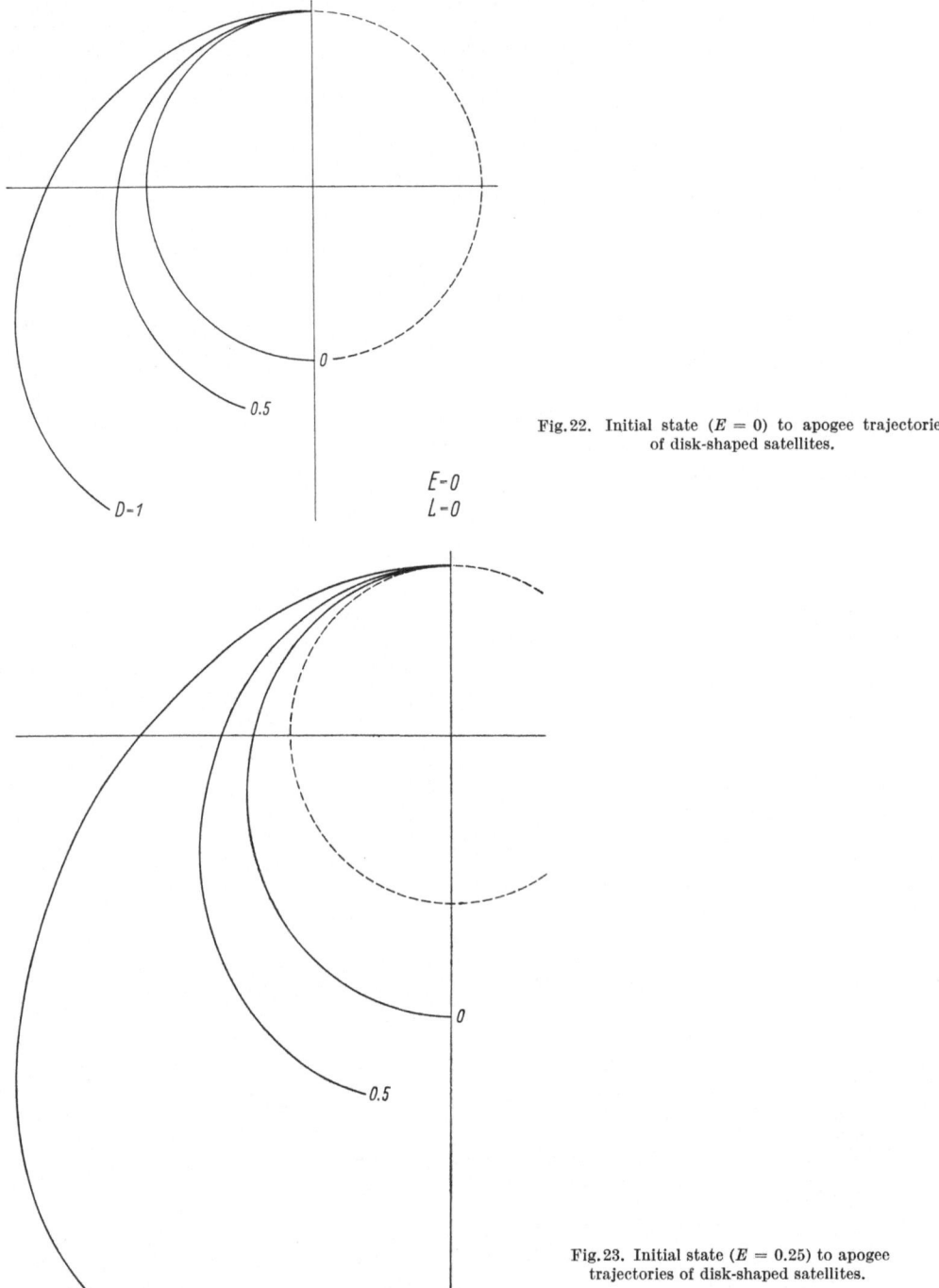

Fig. 22.　Initial state $(E = 0)$ to apogee trajectories of disk-shaped satellites.

Fig. 23.　Initial state $(E = 0.25)$ to apogee trajectories of disk-shaped satellites.

on the stability boundary. Figure 15 shows satellites with shape ratio $D:L = 0.8061$. The characteristic τ_A changes from values $\tau_A > \pi$ to $\tau_A < \pi$ after reaching a maximum. The classification disk-like/beam-like from Section 4 cannot be transferred directly since the analogue definition $\tau_A \lessgtr \pi$ yields slightly different results. Therefore, satellites with spherical inertia ellipsoid $(D:L = 2/\sqrt{3})$ behave disk-like with respect to R_A and τ_A, see Fig. 16. Finally, Fig. 17 gives a presentation for disk-shaped satellites. For $D \to \infty$ a horizontal asymptote $\tau_A \simeq 90°$

seems to exist which is independent of the injection speed. But a more detailed investigation of this academic case has not been accomplished.

The characteristics R_A, τ_A allow to obtain a general view of all possible orbits. Nevertheless, it is of interest to get an idea what real orbits are alike. For that purpose it is sufficient to calculate the trajectories between the initial state and the first succeeding apogee/perigee. In Figs. 18—20 trajectories for beam-shaped satellites are shown for various beam lengths. The greatest length is chosen near the stability boundary to demonstrate the enormous increase of characteristic τ_A. Figure 20 contains a nontrivial circular orbit. Remembering Eq. (39), the corresponding value τ_A is no direct measure for the shortening of the circular orbit period $T_{2\pi}$ which has been found by DUBOSHIN [2]. Figures 21—23 demonstrate trajectories of disk-shaped satellites. The greatest diameter is arbitrarily chosen as $D = 1$. A circular orbit appears in Fig. 21. Again, the corresponding characteristic τ_A cannot be reduced immediately to the prolongation of the orbit period $T_{2\pi}$ as calculated by HOFER [4].

An extreme example for the jump phenomenon is given in Fig. 24 by use of the data of Fig. 12. The satellite length is varied in equidistant steps. For $L = 0.990$ and $L = 0.991$ similar orbits are obtained. However, the variation from $L = 0.991$ to $L = 0.992$ results in quite another orbit. For $L = 0.992$ and $L = 0.993$ the orbits are similar again. If R_A jumps appear, τ_A jumps will happen too. This is a consequence of Eq. (16). In fact τ_A jumps from $\tau_A = 1342°$ to $618°$.

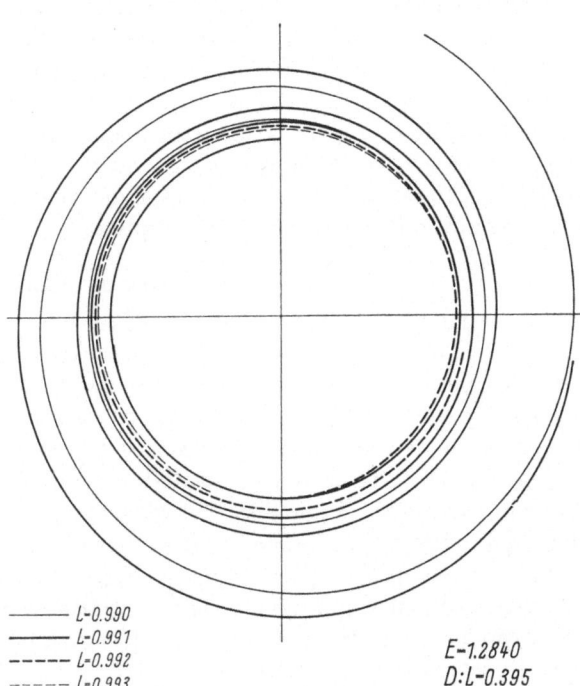

Fig. 24. Initial state ($E = 1.2840$) to apogee trajectories of cylindrical satellites ($D:L = 0.395$). Illustration of jump phenomenon.

——— $L-0.990$
——— $L-0.991$
- - - - $L-0.992$
- - - - $L-0.993$

$E-1.2840$
$D:L-0.395$

References

1. SUTTON, G. W., and F. W. DIEDERICH: Synchronous rotation of a satellite at less than synchronous altitude. AIAA J. 5, 813—815 (1967).

2. DUBOSHIN, G. N.: One particular case of the general problem of the translational-rotational motion of two bodies. Astronom. J. 36, 153—163 (1959).

3. MAGNUS, K.: Der Stabsatellit in einem radialsymmetrischen Schwerefeld. Z. Flugwiss. 11, 233—241 (1963).

4. HOFER, E.: Partikuläre Lösungen der allgemeinen Grundgleichungen für Bahn- und Drehbewegungen von Satelliten. Ing. Arch. 34, 264—274 (1965).

5. MAGNUS, K.: Drehbewegungen starrer Körper im zentralen Schwerefeld. In: Applied Mechanics, Proc. of the XIth International Congress of Applied Mechanics, Munich 1964, Berlin/Heidelberg/New York: Springer 1966, pp. 88—98.

6. MOULTON, F. R.: Einführung in die Himmelsmechanik, Teubner 1927.

Instability of jets, threads, and sheets of viscous fluid

By

G. I. Taylor

University of Cambridge, U.K.

The instability which I propose to discuss is well-known in the form it assumes when a very viscous fluid like thick oil or honey falls in a stream onto a plate. A short distance above the plate the stream begins to wave or rotate in spiral form. It is possible to calculate the diameter at points in a *steady* falling stream of viscous fluid, and it is found that it thickens up very rapidly a short distance above the plate which stops the motion. The reason for the instability is clear. If the stream is very thin the longitudinal compression acts in the same way as end compression in a thin elastic rod. The rod becomes unstable at a certain load, and less force is needed to move the ends towards one another when the rod is bent than when it is straight. This is called the Euler instability. If however the stream is thick it may be that the work required to produce a given amount of longitudinal compressive strain is less if it remains straight than if it bends. Figures 1 and 2 show thin and thick streams of glycerine falling onto a plate. In Fig. 2 the stream spreads out so steadily that a reflection of the upper part of the stream can be seen in the fluid spreading on the plate.

Electrical versus mechanical instability

My attention was drawn to this kind of instability by observing the behaviour of a thin jet of viscous fluid drawn from the end of a metal capillary tube by electric forces. That thin fluid jets produced in this way are very unstable is well known. Figure 3 shows two photographs of an electrically driven jet (LANE and GREEN 1956) emerging from the tip of a tube 0.22 mm in diameter; on the left is a time exposure and on the right a spark exposure of the same jet. It has been assumed that the violent instability is due to the mutual repulsion of charges carried in different parts of the jet, and indeed when the jet breaks into drops this seems a reasonable explanation. ZELENY (1917) found that similar effects occur with viscous jets, but in that case the length of the straight part, which in Fig. 3 is only 0.3 mm long, could extend to nearly 1 cm before instability appeared. It is true that a charged thread at the axis of a conducting cylinder can be proved to be unstable but the spark photograph of Fig. 3 does not suggest that the jet is unstable when first formed.

In Zeleny's and Lane's apparatus the end of the tube from which the jet emerged was placed near a charged plate, but no precautions were taken to control the distant field, on the implicit assumption that when the distance between tube and plate is small the potential field far from this gap has little effect on the jet. In apparatus which I described some years ago (1965) the distant field was much more uniform, and no instability or scatter was observed even after the jet broke into drops. Figure 4 is an example. The drops into which the jet broke owing to the Rayleigh varicose instability moved along the axis of the 1 mm tube from which the jet was pulled by the electric field.

While experimenting with glycerine streams produced electrically, I found that the stream could sometimes be completely steady for its whole length (up to 5.5 cm in my apparatus), but by altering some of the physical conditions I could photograph with an exposure of a millisecond a stream which seemed to vanish at a definite point (Fig. 5). Of course it did not in fact disappear; it must have started waving at the point indicated by the arrow. I therefore

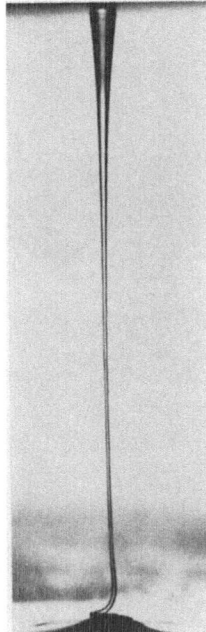

Fig. 1. Thin stream of glycerine falling on a plate.

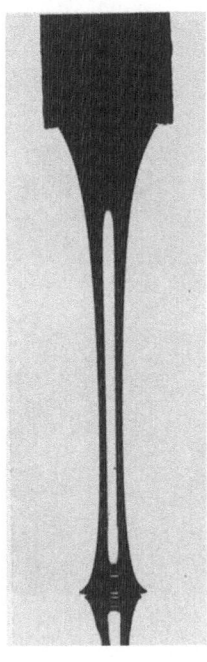

Fig. 2. Thick stream of glycerine falling on a plate.

Fig. 4. Tellus oil pulled from tube in apparatus designed for constant electric field.

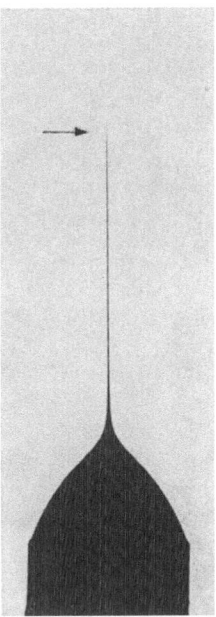

Fig. 5. Electrically driven glycerine jet, 1-ms exposure.

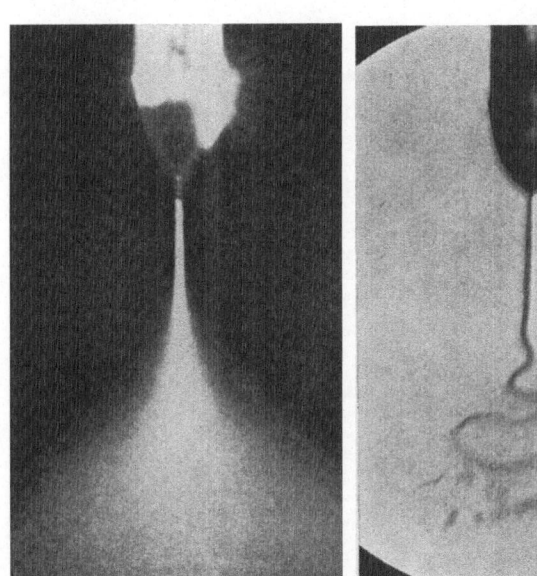

Fig. 3. Water pulled from a tube by electric force (LANE and GREEN 1956).

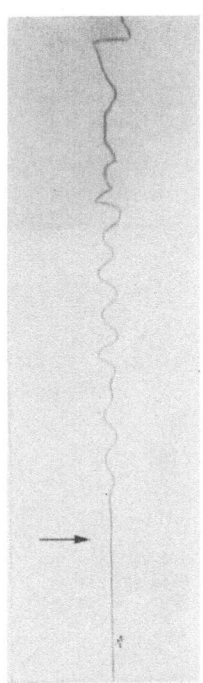

Fig. 6. Electrically driven glycerine jet, 2-μs exposure.

48 A*

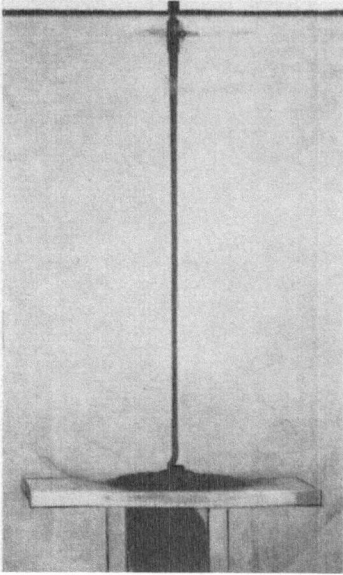

Fig. 7. Coloured glycerine falling through water.

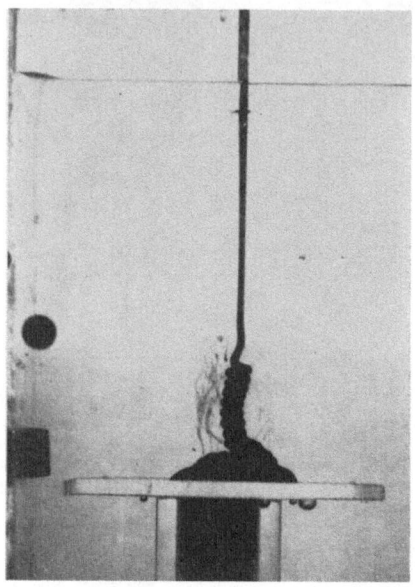

Fig. 8. Glycerine falling into water above round mark, salt solution of specific gravity 1.237 below.

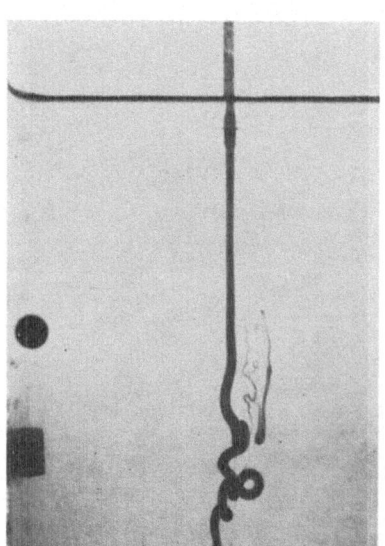

Fig. 9. Glycerine falling into fluid of specific gravity 1.225 above round mark, 1.237 below.

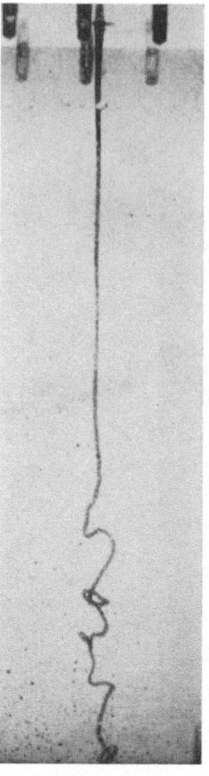

Fig. 10. 1/2-mm jet in specific gravity 1.225 above, 1.237 below.

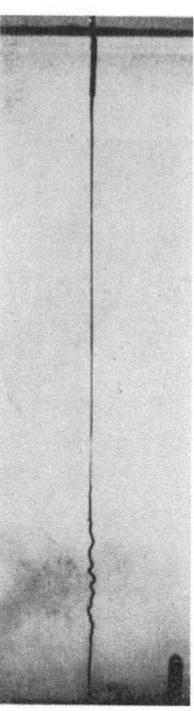

Fig. 11. 1/2-mm jet in specific gravity 1.225 above, 1.230 below.

took a photograph of the same jet under the same conditions using a flash of 2-μs duration (Fig. 6). It will be seen that oscillations begin at the point marked by the arrow in Fig. 6. They do not increase rapidly as would be expected if there were an instability due to reaction between the electric field and the jet. The jet becomes crooked to a limited extent. It was this that made me think the instability was purely mechanical and due to longitudinal compression of a viscous stream. This would explain the extreme straightness of the jet before it began to oscillate, as well as the limited bending afterwards, provided the instability began in a steady jet at or near the point where the diameter reached its minimum, because that is the place where the longitudinal stress due to viscosity changes sign. I therefore measured the diameters along a downward pointing jet similar to that of Figs. 5 and 6 and found that the instability did not occur till just after the jet had reached a minimum diameter. The reason for using jets directed downwards is that the only force which could contribute to a change in sign of the viscous stress is then the aerodynamic drag on the moving column.

Gravity driven jets of viscous fluid

The difference between the instability of glycerine falling onto a plate (Fig. 1) and the instability of an electrically driven jet seems to be that the deceleration which produces the instability is much more sudden in the former case and the fluid piles up immediately after the compression sets in. To obtain a small deceleration in a falling stream many methods are available, but the simplest is to let the fluid fall into a fluid of less density than itself and then, at the point where the instability is to be established, to alter the density of the surrounding fluid so that the gravitational driving force decreases. Figure 7 shows coloured glycerine (specific gravity 1.255) falling in water onto a plate. The jet is quite straight till it reaches a point a few jet diameters above the plate where it begins to oscillate like the jet of Fig. 1. On filling the tank up to the level of the round mark on the left (Fig. 8) with fluid of specific gravity 1.237 and the remainder carefully with water, it will be seen that the glycerine begins to oscillate in a spiral manner near the level of the change in density. The downward velocity has in fact changed in the ratio $d_1/\pi d_2$, where d_1 is the diameter of the jet and d_2 that of the spiral. In Fig. 8 this reduction is about 12:1 and the change in the driving force is from 1.255 g — 1.000 g = 0.255 g to 1.255 g — 1.237 g = 0.018 g, i.e., the driving force has been reduced in the ratio 14:1. On repeating the experiment with a surrounding specific gravity of 1.225 above and 1.237 below, the driving force is changed from 1.255 g — 1.225 g = 0.030 g to 0.018 g, i.e., the driving force is reduced in the ratio of only 1.67:1. Figure 9 shows this case. The change in jet velocity at the level of density change is much less than in Fig. 8; in fact the length of the coiled portion is about twice that contained in the same vertical distance in the uncoiled portion. A finer jet, 1/2-mm diameter, is shown in Figs. 10 and 11. In Fig. 10 the glycerine is falling into the same density distribution as in Fig. 9. Figure 11 shows glycerine falling through a fluid for which the specific gravity is 1.225 at the top and 1.230 at the bottom. Thus the driving force at the level of density change is only reduced from 1.255 g — 1.225 g = 0.030 g to 1.255 g — 1.230 g = 0.025 g. Thus the force changes in the ratio 1.2:1. The crumpling in Fig. 11 is much less than in Fig. 10.

Crumpling of threads of viscous fluids in the absence of gravity

The analogy between the equations of motion of a viscous fluid and that of an incompressible elastic solid is well-known. It was pointed out to me by BROOKE BENJAMIN that when a thin rod is subjected to end compression it forms an elastica and that Love's *Theory of Elasticity* contains a picture of the set of elasticas. I therefore found a very viscous fluid known as SAIB (Sucrose Acetate Isobutyrate) and stretched a thread of it between two sticks. I then laid this thread on the surface of a dish of mercury and either pushed the sticks towards one another or laid two matches on the thread and pushed them towards one another, and photographed the thread. Figure 12 shows some of these photographs together with some of Love's calculated elasticas. The comparison would not be expected to be at all exact because this method of producing the threads does not make them with uniform diameter.

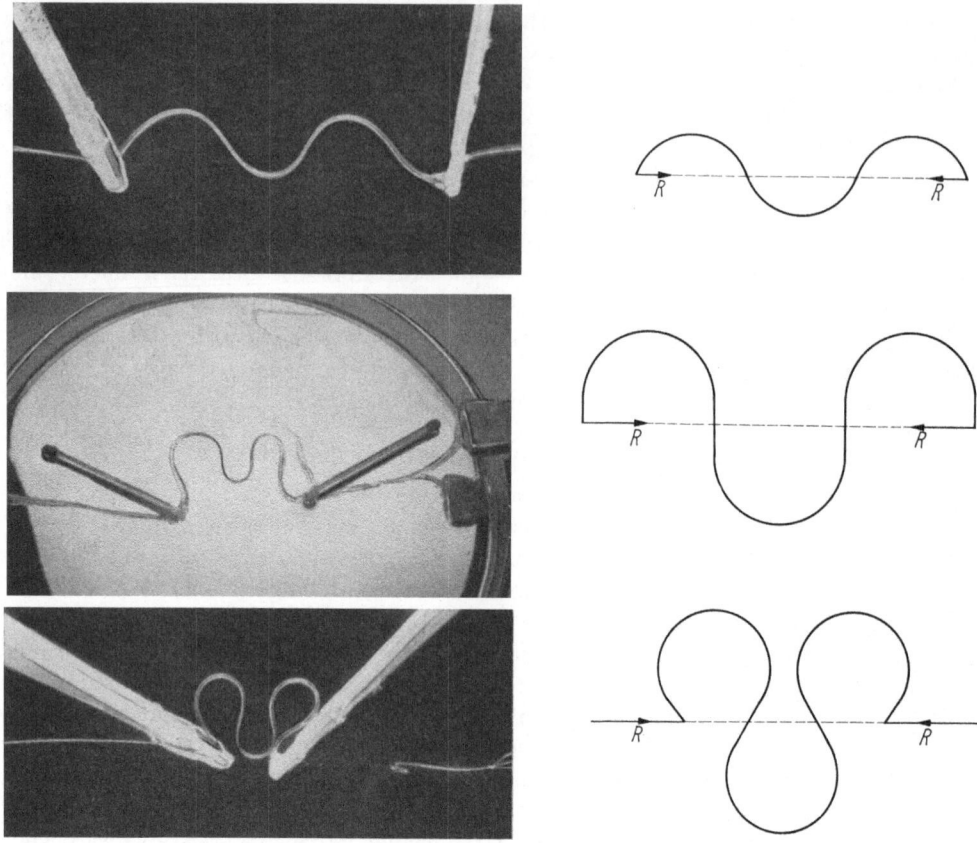

Fig. 12. Threads of SAIB floating on mercury compared with Love's calculated elasticas.

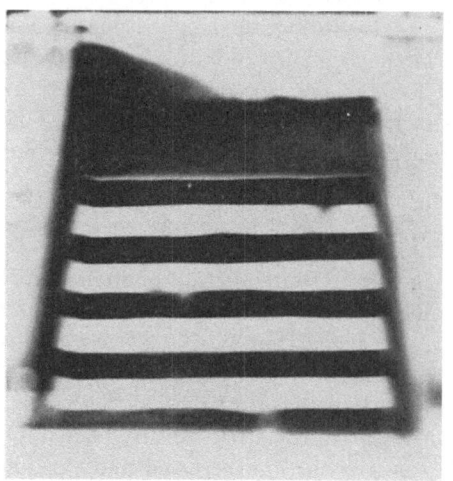

Fig. 13. Reflections of striped card in layer of fluid (40 poise, 1-cm thick) when $-\varepsilon \cong 0.1\ \mathrm{sec}^{-1}$.

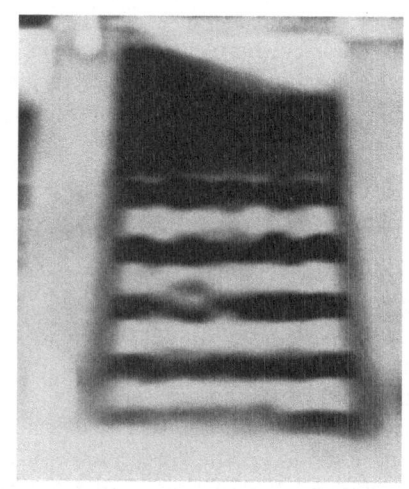

Fig. 14. Reflection when $-\varepsilon \cong 0.4\ \mathrm{sec}^{-1}$.

Fig.15. 15-cm disc rotating in syrup
layer at 1 revolution in 2.7 sec.

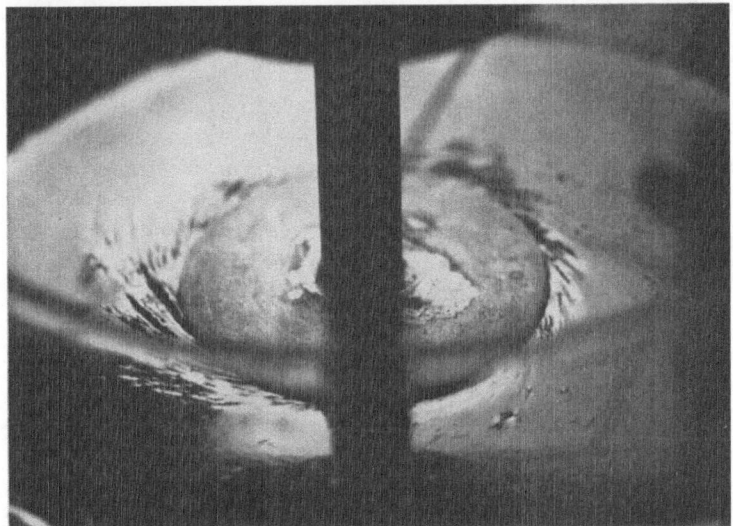

Fig.16. 15-cm disc rotating in syrup
layer at 1 revolution in 2.4 sec.

Fig.17. 15-cm disc rotating in syrup
layer at 1 revolution in 1.4 sec.

49 a*

Instability of sheets of viscous fluid floating on an immiscible non-viscous fluid

In the absence of gravity a sheet of viscous fluid of thickness h under end load due to rate of extension ε is unstable for all wave lengths when the total viscous compressive stress per unit width $-4\mu\varepsilon h - T$ is positive. Here T is the sum of the surface tensions of both sides of the sheet. As $-\varepsilon$ increases, instability of wave length λ is possible as soon as $-4\mu\varepsilon h - T$ becomes positive; but the rate of growth is a maximum when the wavelength is

$$\lambda = 2\pi \left(\frac{\varrho g}{-4\mu\varepsilon h - T} \right)^{-1/2}, \tag{1}$$

ϱ being the density of the lower fluid (Benjamin, private communication).

Though it has been difficult to realize the physical conditions under which this formula is legitimately applicable because of the difficulty of obtaining a thin viscous sheet floating on an immiscible fluid, some experiments were made in which SAIB was floated on concentrated brine and syrup on carbon tetrachloride. The thickness of these sheets could not be less than about 0.6 to 1.0 cm because the equilibrium depth of the lower surface of a large bubble of the upper liquid in the lower was of order 1 cm, and if a hole was pierced through it, it would not close up again. Two methods were used in these qualitative experiments, and in each case no instability was observed till the rate of compressive strain reached a certain value and then waves suddenly appeared. In the first method the layer was spread between two vertical walls which penetrated a short distance into the lower liquid. The sides of the viscous sheet were contained by means of vertical rubber sheets joining the ends of the vertical walls so that when the walls approached one another at speed U the rate of strain was undirectional and equal to $-U/L = \varepsilon$, where L was the distance apart at the instant considered. To observe the waves a horizontally striped card was set up and photographed obliquely by light reflected from the surface on the viscous layer. Figures 13 and 14 are photographs of the stripes taken when $-\varepsilon$ was roughly 0.1 sec^{-1} and 0.4 sec^{-1} respectively. The viscosity was about 40 poise and the surface tensions at the upper and lower surfaces about 30 and 15 dyne cm^{-1} respectively. The thickness of the layer was about 1 cm, so that in Fig. 13, $-4\mu\varepsilon h$ was about 16 dyne cm^{-1} while in Fig. 14 it was roughly 64 dyne cm^{-1}. Thus it seems that waves appeared between the conditions in which $-4\mu\varepsilon h$ was 16 and 64, and since T was roughly $30 + 15 = 45$ dyne cm^{-1} it seems that they appeared in the range of ε where they would be expected.

The other qualitative experiment was to immerse a horizontal disc of 5.5-cm diameter into some syrup floating on carbon tetrachloride contained in a glass vessel of 15-cm diameter. The depth of the syrup was 1 cm. On rotating the disc the surface remained flat till the angular speed was 1 revolution in 2.7 secs. Figure 15 shows it in this condition. Figure 16 shows the situation at 1 revolution in 2.4 secs. Short waves can be seen at about 45° to the periphery, but most of the surface appears smooth. Figure 17 shows the surface at 1 revolution in 1.4 secs. When this experiment was made it was thought that a skin with some rigidity might have formed, but the surface flattened leaving no visible trace of the wave pattern in a few seconds after stopping the rotation. It may well be that some surface viscosity is present, but the complete disappearance of the pattern suggests that there was no element of rigidity. Wave crests of 45° would be perpendicular to the direction of maximum compressive strain so that the expected phenomenon seems to be like that shown in Fig. 17. The suddenness with which the waves appear gives the impression that the rotating disc presents a true instability problem.

References

Lane, W. R., and H. L. Green 1956: Surveys in Mechanics, Ed. by G. K. Batchelor and R. M. Davies, Cambridge University Press, p. 192.

Taylor, G. I.: Proc. Roy. Soc. A 291, 145—158 (1966).

Zeleny, J.: Phys. Rev. X, 1—6 (1917).

The interaction of local buckling and column failure of thin-walled compression members

By

A. van der Neut

Technological University, Delft

List of symbols

a	wave amplitude of flange (imperfection)	$\alpha = a/h$	waviness parameter
b	flange width	ε	compressive strain, or overall specific shortening $(-\Delta L/L)$
c	half column depth, radius of gyration		
e_0	wave amplitude of column axis (imperfection)	$\eta = d(P/P_l)/d(\varepsilon/\varepsilon_l)$, stiffness reduction factor	
h	thickness of flange	$\xi = \pi x/L$	
x	longitudinal coordinate	$(\ )' = d(\ )/dP/P_l$	
K	compressive load of column	suffix b	refers to (b)ending buckling load
L	column length	suffix f	refers to (f)ailure load
P	compressive force of flange	suffix l	refers to (l)ocal buckling load
W	column deflection	suffix E	refers to (E)uler column load

1. Introduction

Thinwalled compression members, as used in aircraft structures (Fig. 1), are subject to column failure (bending), torsional buckling and local buckling.

Torsional buckling is not being envisaged in this paper.

A rather widespread opinion is that optimum design requires equality of the local buckling load K_l and the Euler buckling load K_E. During a symposium on post-buckling behaviour [1] KOITER and SKALOUD contested this opinion by arguing that local buckling phenomena should promote column failure.

Local buckling does not exhaust the load carrying capacity. Then a better compromise seems to be $K_E > K_l$, where the failure load K_f might exceed K_l. However, the experience of experimentalists is that with $K_E > K_l$ failure can occur below K_l. A further observation is that test specimen collapse explosively, as evidenced by the sharp bang and the severity of the destruction. Indeed, as will be shown in this paper, explosive collapse, known to occur with shells due to non-linearity, may be a characteristic of this buckling process as well. Then it is to be expected that the strength is sensitive to imperfections and that, consequently, failure loads obtained from identical test specimen would show quite some scatter. Two types of imperfection may have importance: departures from flatness of the composing elements of the structure and non-straightness of the column axis.

Deviations from flatness remove the discontinuity between prebuckling and postbuckling behaviour. Therefore, even when the applied compressive stress is below the "local buckling"-stress, the analysis of the buckling problem needs knowledge of the bending stiffness of the compression member in its post-buckling state. This knowledge can be acquired numerically for a member of given cross section, but even for the simplest sections it puts quite a problem.

The present paper is a first reconnaissance of the shell-like behaviour of these structures and of its quantitative aspects. Then complexities presented by practical configurations need not be encompassed. A model has been taken, which is representative for actual structures as to the phenomena under consideration and which is easy to handle analytically.

This model is shown in Fig. 1. It consists of 2 load carrying flanges, width b, thickness h, connected at distance $2c$ by webs, which are rigid in shear and laterally but which have no longitudinal stiffness. The webs offer simple support to the flanges. In this way the normal force carrying flanges have boundary conditions which are easy to take into account.

The investigation has been subdivided into 4 stages: each next one including one aspect in addition:

1. The perfect structure: the flanges are flat and the column axis is straight. The column failure load K_b (better K_b/K_l) is established as a function of K_E/K_l. It appears that the equilibrium at $K_b/K_l = 1$ is instable in the range $1 < K_E/K_l < 1.725$.

2. The column with imperfect flanges, the column axis is straight. Again K_b/K_l is established as a function of K_E/K_l. It appears that K_b is quite sensitive to the amount of imperfection when K_E/K_l is close to unity, and that the unfavourable effect of imperfection on K_b exists over the whole range of K_E/K_l up to about 2.

3. The same configuration as considered in part 2. The character of the equilibrium at the bifurcation point K_b is investigated by establishing the load gradient for small deflection of the column axis. It appears that the equilibrium is instable (explosive collapse) over a range of K_E/K_l, the extent of which depends upon the amount of imperfection. Sensivity to non-straightness of column axis can be expected.

4. The column with imperfect flanges and imperfect column axis. The failure load K_f is established as dependent on the slenderness parameter K_E/K_l, the waviness of the flanges and the deviation from straightness of the column axis. It appears that the reduction of load carrying capacity because of non-straightness is slight.

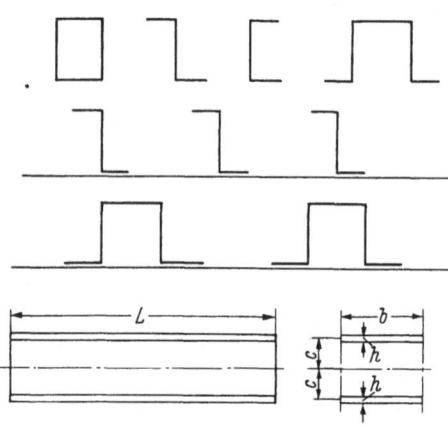

Fig. 1. Structural cross sections and model.

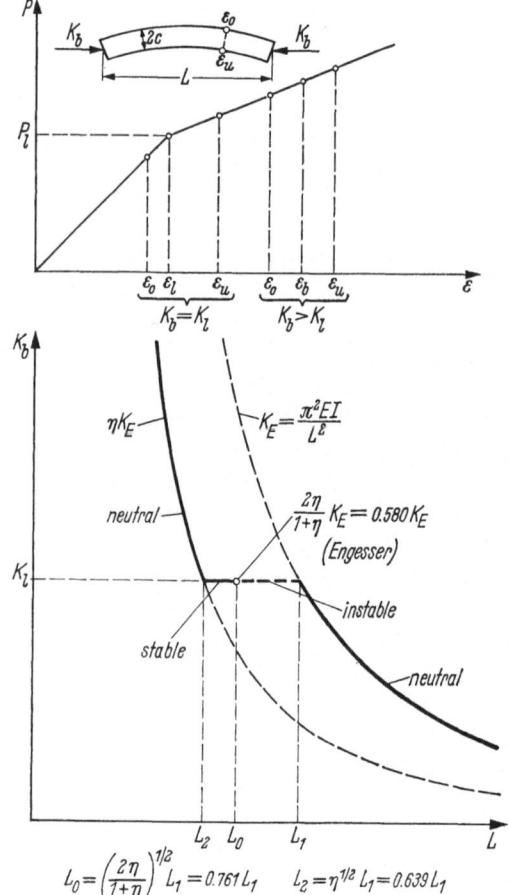

Fig. 2. Column buckling load $K_b(L)$.

$$L_0 = \left(\frac{2\eta}{1+\eta}\right)^{1/2} L_1 = 0.761 L_1 \qquad L_2 = \eta^{1/2} L_1 = 0.639 L_1$$

2. The perfect column

For large slenderness of the column $K_E < K_l$. Then the stiffness of the flanges is $E\,bh$ and the bending stiffness of the column is EI.

With small slenderness K_E might exceed K_l. The flanges are in the postbuckling state and their stiffness against incremental compressive strain $\Delta\varepsilon$ is $\eta E\,bh$. Therefore the bending stiffness under axial load is ηEI and the column strength is

$$K_b = \eta K_E.$$

Fig. 2 gives the two curves of K_E and K_b versus L. The upper part shows the flange load P versus strain ε. The slope at $P > P_l$ decreases slowly with increasing P/P_l, but for $\varepsilon/\varepsilon_l < 3$ its variation is very slight, almost equal to 0.4083, which is the slope at $P = P_l$ [2].

Superimposing deflection at the load $K = K_l$ one flange gets increased strain and opposes to the strain increment with the stiffness $\eta E\,bh$, whereas the other flange returns to the unbuckled condition, offering the stiffness $E\,bh$ to its strain increment. Then the bending stiffness of the column is $\dfrac{2\eta}{1+\eta}EI$ (the Engesser "double-modulus"-formula). The column is in neutral equilibrium at K_l when the column length is $L_0 = \left(\dfrac{2\eta}{1+\eta}\pi^2EI/K_l\right)^{1/2}$. For $L_2 < L < L_0$ the equilibrium at K_l is stable. However for $L_1 > L > L_0$ the equilibrium is instable, collapse will proceed explosively.

The curve of Fig. 2 transforms into the one of Fig. 3 by replacing the abscissa L by L^{-2} or K_E/K_l. This graph is composed of 3 straight lines. It shows that in the range $1 < K_E/K_l < 1.725$, which is being used very much in aeronautical structure, the perfect column collapses explosively at $K = K_l$. Within this K_E/K_l-range imperfection will have the effect to reduce the buckling strength K_b to values below K_l.

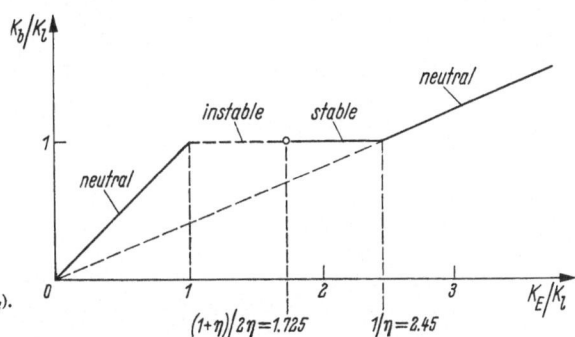

Fig. 3. Column buckling load $K_b(K_E)$.

3. The buckling load with imperfect flanges

The initial waviness z of the middle surface of a flange can be developed into a series of functions corresponding to the various buckling modes. The behaviour of the flange is mainly governed by the term which corresponds to the mode pertaining to the smallest buckling stress. Therefore the waviness has been assumed to be given by

$$z = a \cos \pi y/b \sin \pi x/b. \tag{3.1}$$

Instead of the wave amplitude a we use the waviness parameter

$$\alpha = a/h. \tag{3.2}$$

The relation between P and ε for this imperfect plate strip, simply supported at its edges has been established by using the Ritz-Galerkin approximate solution of the non-linear plate equation, thereby taking the deflection in the shape of the buckling mode. Since we are interested only in the behaviour at ε in the vicinity of ε_l this approximation is sufficiently accurate. Some results have been given in Fig. 4.

49 b*

The stiffness of the flange to strain increment at its edges caused by bending of the column is

$$S = dP/d\varepsilon. \tag{3.3}$$

The reduction of stiffness with respect to the stiffness of the flat flange is given by the reduction factor

$$\eta = d(P/P_l)/d(\varepsilon/\varepsilon_l). \tag{3.4}$$

Figure 5 shows η versus P/P_l (or K/K_l) for various values of α. A continuous curve replaces the broken line which relates to the perfect flange.

The investigation has been confined so far to the case in which the two flanges have equal α. Then the column axis will remain straight under the load K until the buckling load K_b is

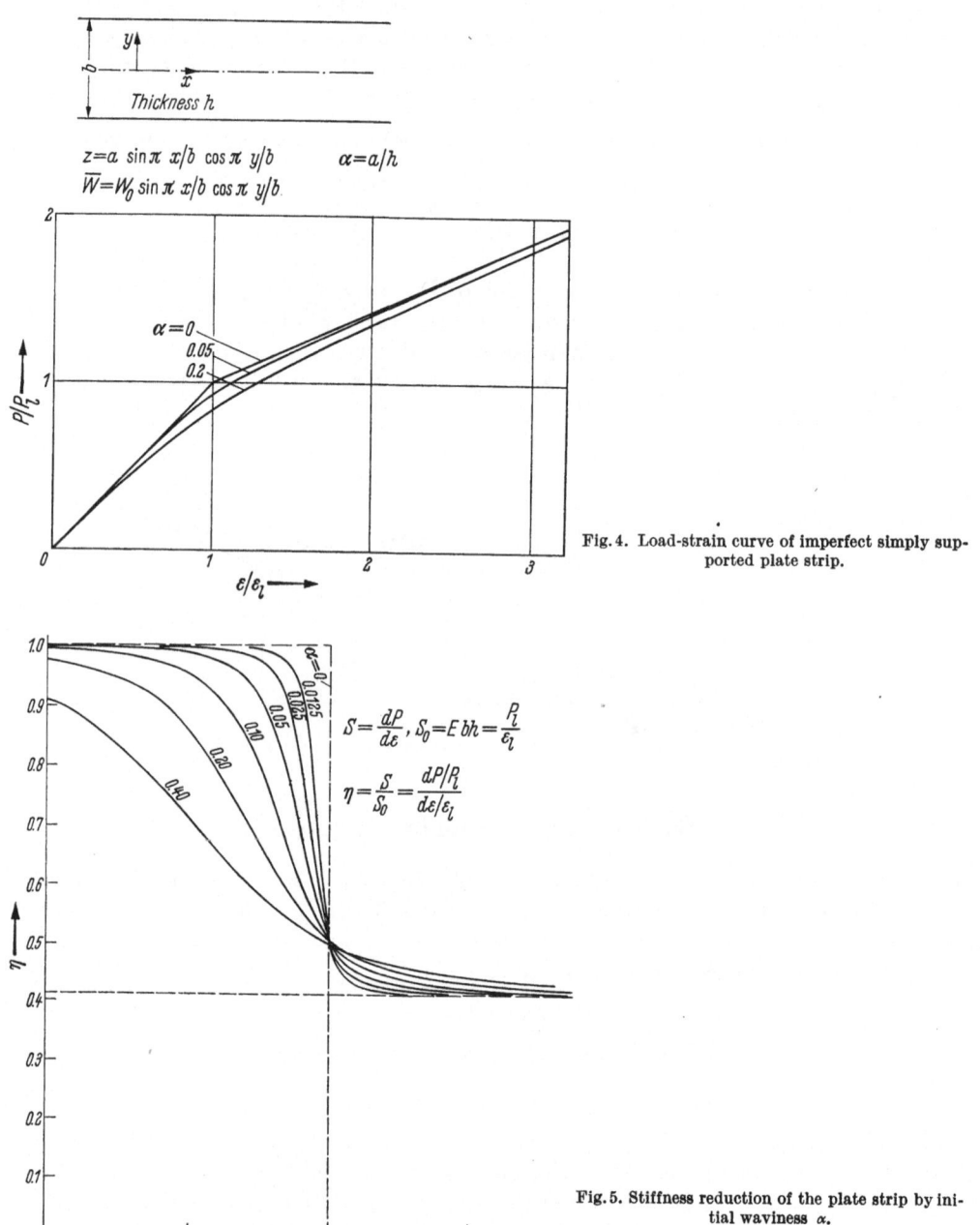

$z = a \sin \pi\, x/b\, \cos \pi\, y/b$ $\alpha = a/h$

$\overline{W} = W_0 \sin \pi\, x/b\, \cos \pi\, y/b$

Fig. 4. Load-strain curve of imperfect simply supported plate strip.

$S = \dfrac{dP}{d\varepsilon},\ S_0 = E\,bh = \dfrac{P_l}{\varepsilon_l}$

$\eta = \dfrac{S}{S_0} = \dfrac{dP/P_l}{d\varepsilon/\varepsilon_l}$

Fig. 5. Stiffness reduction of the plate strip by initial waviness α.

reached. The bending stiffness against infinitesimal deflection at the load K is $\eta_K EI$. With this bending stiffness the deflected column is in equilibrium if $K = \eta_K K_E$. Therefore a load K is the buckling load K_b when

$$\frac{K/K_l}{\eta(K/K_l)} = \frac{P/P_l}{\eta(P/P_l)} = \frac{K_E}{K_l}. \tag{3.5}$$

K_E is a measure of the column length. Then (3.5) gives the relation between column length L and buckling load K_b in dependence of α, η being a function of α. Evaluation of (3.5) by means of the data contained in Fig. 5 yields the relation between K_b/K_l and K_E/K_l for the various values of α (Fig. 6).

The broken line for $\alpha = 0$ is the degeneration of the smooth curves for $\alpha \neq 0$. It appears that $K_b < K_l$ when $K_E/K_l < 2$. The reduction in strength because of α is rather important in the vicinity of $K_E/K_l = 1$: for $\alpha = 0.0125$ 10%, for $\alpha = 0.05$ 17%, for $\alpha = 0.2$ 30%. This confirms the conjecture of imperfection sensitivity in the instable part of the perfect-column-curve (Figs. 3 and 4).

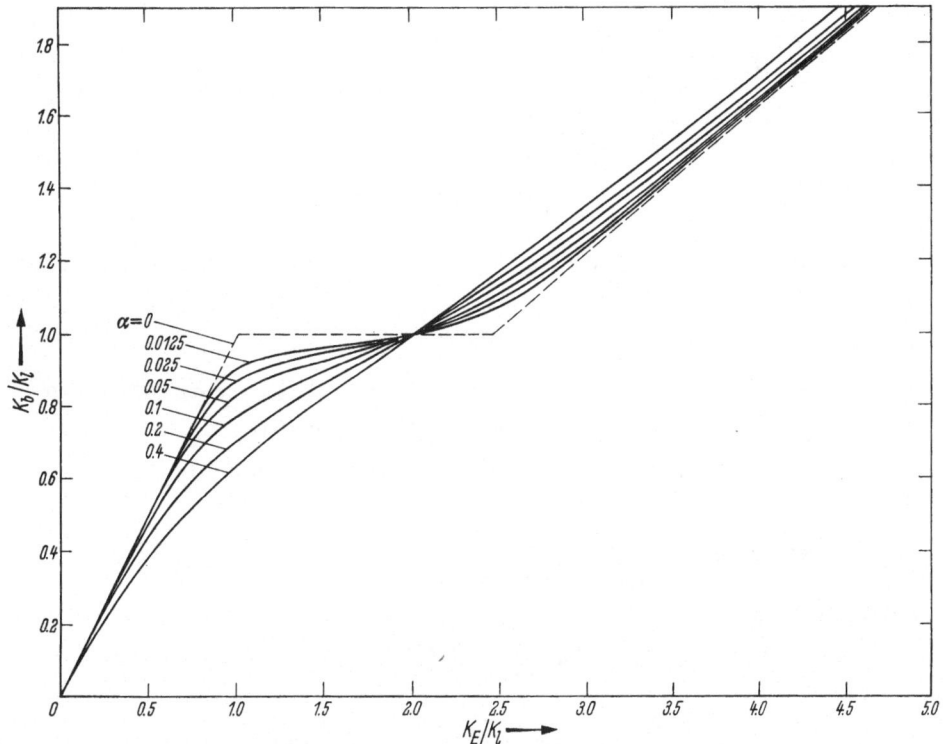

Fig. 6. Buckling curves of columns with imperfect flanges.

4. The character of the equilibrium at the buckling load

The question to be considered is: how must the load change in order to maintain equilibrium when the column deflects slightly ?

The importance of the problem area of the character of the equilibrium at the bifurcation point has been stressed first by W. T. KOITER [3].

The equilibrium load at the deflection W is

$$K = K_b + k. \tag{4.1}$$

If k would appear to be negative the equilibrium at the buckling load K_b would be instable and imperfection of the column axis would cause a failure load $K_f < K_b$.

Since the column is symmetrical with respect to the line of action of K — the waviness of the flanges being equal and the column axis being straight — the equilibrium load K must be equal for $W = W_1$ and $W = -W_1$. So, if K is affected by W, k must be proportional to an even power of W, possibly W^2. It will appear that dK/dW^2 is finite. Then the order of magnitude of quantities to be taken into account is $(W/c)^2$.

In addition to dK/dW^2 the derivative $dK/d(-\Delta L)$ will be established (where $-\Delta L$ is the shortening of the column).

In the deflected position the normal forces in upper and lower flange (Fig. 7) are respectively:

$$P_o = \frac{1}{2} K \left(1 - \frac{W}{c}\right), \qquad P_u = \frac{1}{2} K \left(1 + \frac{W}{c}\right). \tag{4.2}$$

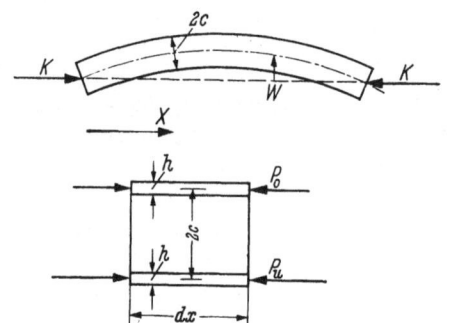

Fig. 7. Flange forces of the deflected column.

The bending moment is

$$M = (P_o - P_u) c = -KW. \tag{4.3}$$

So as to obtain the differential equation of W P_o and P_u must be expressed in the edge strains ε_o and ε_u, which are related to W by

$$\frac{d^2 W}{dx^2} = (\varepsilon_o - \varepsilon_u)/2c. \tag{4.4}$$

We need the relation between P and ε in the direct vicinity of $P_b = \frac{1}{2} K_b$, which can be given by the Taylor-expansion, breaking it off after the third derivative:

$$P = P_b + \left(\frac{dP}{d\varepsilon}\right)_b (\varepsilon - \varepsilon_b) + \frac{1}{2} \left(\frac{d^2 P}{d\varepsilon^2}\right)_b (\varepsilon - \varepsilon_b)^2 + \frac{1}{6} \left(\frac{d^3 P}{d\varepsilon^3}\right)_b (\varepsilon - \varepsilon_b)^3. \tag{4.5}$$

Hence

$$\frac{dP}{d\varepsilon} = \eta Ehb = \left(\frac{dP}{d\varepsilon}\right)_b + \left(\frac{d^2 P}{d\varepsilon^2}\right)_b (\varepsilon - \varepsilon_b) + \frac{1}{2} \left(\frac{d^3 P}{d\varepsilon^3}\right)_b (\varepsilon - \varepsilon_b)^2. \tag{4.6}$$

From (4.5) and (4.6) follows

$$P_o - P_u = (\varepsilon_o - \varepsilon_u) \left[\frac{1}{2} \left(\frac{dP}{d\varepsilon}\right)_o + \frac{1}{2} \left(\frac{dP}{d\varepsilon}\right)_u - \frac{1}{12} \left(\frac{d^3 P}{d\varepsilon^3}\right)_b (\varepsilon_o - \varepsilon_u)^2\right]. \tag{4.7}$$

Using

$$\frac{dP/P_l}{d\varepsilon/\varepsilon_l} = \eta \quad \text{and} \quad \frac{d^3 P/P_l}{d(\varepsilon/\varepsilon_l)^3} = \eta^2 \eta'' + \eta \eta'^2,$$

where

$$(\ )' = d(\ )/d(P/P_l)$$

we obtain from (4.3), (4.4), (4.7)

$$M = \left[\frac{1}{2} (\eta_o + \eta_u) - \frac{1}{3} (\eta^2 \eta'' + \eta \eta'^2)_b \frac{c^2}{\varepsilon_l^2} \left(\frac{d^2 W}{dx^2}\right)^2\right] EI \frac{d^2 W}{dx^2}. \tag{4.8}$$

Since η_o and η_u as well as W are functions of x the bending stiffness $[\cdots] EI$ is a function of x. $\eta_o + \eta_u$ has still to be expressed in W. We do this by using again a Taylor-expansion:

$$\eta = \eta_b + \left(\frac{d\eta}{dP}\right)_b p + \frac{1}{2} \left(\frac{d^2 \eta}{dP^2}\right)_b p^2 \tag{4.9}$$

(breaking off after the second derivative, which is in harmony with (4.5) because $d^3 P/d\varepsilon^3$ correlates with $d^2 \eta/dP^2$), where p is the difference between P and P_b.

$$p_o = \frac{1}{2} \left[k - (K_b + k) \frac{W}{c}\right], \qquad p_u = \frac{1}{2} \left[k + (K_b + k) \frac{W}{c}\right]. \tag{4.10}$$

Retaining terms of the order W^2/c^2, (4.9) yields

$$\frac{1}{2}\,(\eta_o + \eta_u) = \eta_b + \eta_b'\,\frac{k}{k_l} + \frac{1}{2}\,\eta_b''\left(\frac{K_b}{K_l}\right)^2\frac{W^2}{c^2}\,.$$

Substituting this result and (4.8) into (4.3) the differential equation becomes:

$$(\eta_b + \theta_b)\,EI\,\frac{d^2W}{dx^2} + (K_b + k)\,W = 0, \tag{4.11}$$

where

$$\theta_b = \eta_b'\,\frac{k}{K_l} + \frac{1}{2}\,\eta_b''\left(\frac{K_b}{K_l}\right)^2\frac{W^2}{c^2} - \frac{1}{3}\,(\eta^2\eta'' + \eta\eta'^2)_b\,\frac{c^2}{\varepsilon_l^2}\left(\frac{d^2W}{dx^2}\right)^2. \tag{4.12}$$

θ_b is of the order W^2/c^2, there fore small compared to η_b and so is k compared to K_b. Therefore the bending stiffness deviates only slightly from $\eta_b EI$ and W will be very close to $\overline{W} = W_0\sin\xi$, the solution of (4.11) for $\theta_b = 0$. Therefore we put

$$W = W_0(\sin\xi + \omega), \qquad \xi = \pi x/L. \tag{4.13}$$

Substitution of (4.13) into (4.11) yields

$$\begin{aligned}
&\eta_b\,EI\,\frac{d^2\overline{W}}{dx^2} + K_b\overline{W} + \theta_b\,EI\,\frac{d^2\overline{W}}{dx^2} + k\overline{W}\\
&\quad + \left[\eta_b\,EI\,\frac{d^2\omega}{dx^2} + K_b\omega + \theta_b\,EI\,\frac{d^2\omega}{dx^2} + k\omega\right]W_0 = 0.
\end{aligned} \tag{4.14}$$

The sum of the first two terms vanishes. The order of the 3rd and 4th terms is $(W_0/c)^2K_E W_0$, of the 5th and the 6th terms $K_E\omega W_0$ of the 7th and 8th terms $(W_0/c)^2 K_E\omega W_0$.

Hence ω is of the order $(W_0/c)^2$, the 7th and 8th terms are negligible, hence (4.14) reduces to

$$\frac{d^2\omega}{d\xi^2} + \omega = \left(\theta_b - \frac{k}{K_E}\right)\frac{1}{\eta_b}\sin\xi. \tag{4.15}$$

Neglecting in (4.12) again terms of the order $(W_0/c)^4$ and using

$$\frac{\pi^4 c^4}{\varepsilon_l^2 L^4} = \left(\frac{K_E}{K_l}\right)^2$$

the differential equation for ω (4.15) becomes

$$\frac{d^2\omega}{d\xi^2} + \omega = -\left[\left(\frac{K_l}{K_E} - \eta_b'\right)\frac{k}{K_l} + \left(-\frac{1}{2}\,\eta'' + \frac{\eta'^2}{\eta}\right)_b\frac{1}{3}\frac{K_b^2}{K_l^2}\frac{W_0^2}{c^2}\sin^2\xi\right]\frac{1}{\eta_b}\sin\xi. \tag{4.16}$$

Its solution is

$$\omega = \frac{1}{24}\left(-\frac{1}{2}\,\eta\eta'' + \eta'^2\right)_b\frac{K_E^2}{K_l^2}\frac{W_0^2}{c^2}\sin^3\xi \tag{4.17}$$

and

$$\frac{dK/K_l}{d(W_0^2/c^2)} = -\frac{1}{4}\left(\frac{-\frac{1}{2}\,\eta\eta'' + \eta'^2}{\eta}\frac{1}{K_l/K_E - \eta'}\right)_b\frac{K_E^2}{K_l^2}\,. \tag{4.18}$$

Since $\eta' < 0$ (see Fig. 5) the equilibrium at K_b is instable if $\eta'' < 2\dfrac{\eta'^2}{\eta}$.

Then (4.17) yields $\omega > 0$, which means that the curvature of W near the middle of the span is greater than the curvature of $\overline{W}$.

The column shortening $-\Delta L$ consists of two parts: 1, the average shortening of the two flanges; 2, the shortening because of the slope dW/dx. Using a Taylor-expansion of ε into p (compare with 4.9) we find

$$\frac{dK/K_l}{d(-\Delta L/\varepsilon_l L)} = \eta_b\left[1 - \frac{(K_l/K_E - \eta')^2}{-\frac{1}{2}\,\eta\eta'' + \eta'^2}\right]_b^{-1}. \tag{4.19}$$

The various possibilities for the type of bifurcation are listed in Fig. 8: the first two cases represent instable equilibrium at K_b, the third one stable.

50*

Figure 9 shows for a number of α-values the tangent to the K-shortening-curve at the bifurcation, when K is the buckling load K_b (The origins of the ordinate of the various curves are shifted. The K_E/K_l-value pertaining to K_b/K_l can be read from Fig. 6.)

It appears that instability exists when $\varepsilon/\varepsilon_l < 1$, e.g. (from Fig. 6) with $\alpha = 0.0125$ for $K_E/K_l < 1.6$, with $\alpha = 0.2$ for $K_E/K_l < 1.4$. The sharpness of the peak decreases with increasing imperfection α. At $\alpha = 0.2$ the instability has almost vanished.

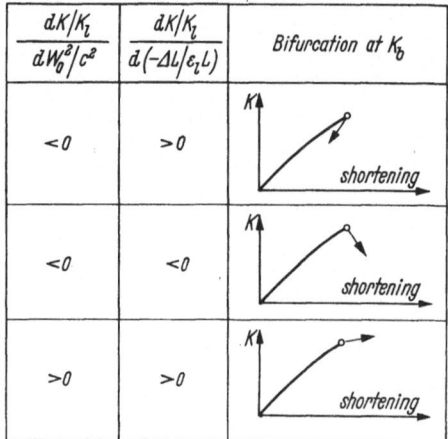

$\dfrac{dK/K_l}{dW_0^2/c^2}$	$\dfrac{dK/K_l}{d(-\Delta L/\varepsilon_l L)}$	Bifurcation at K_b
< 0	> 0	
< 0	< 0	
> 0	> 0	

Fig. 8. Types of bifurcation.

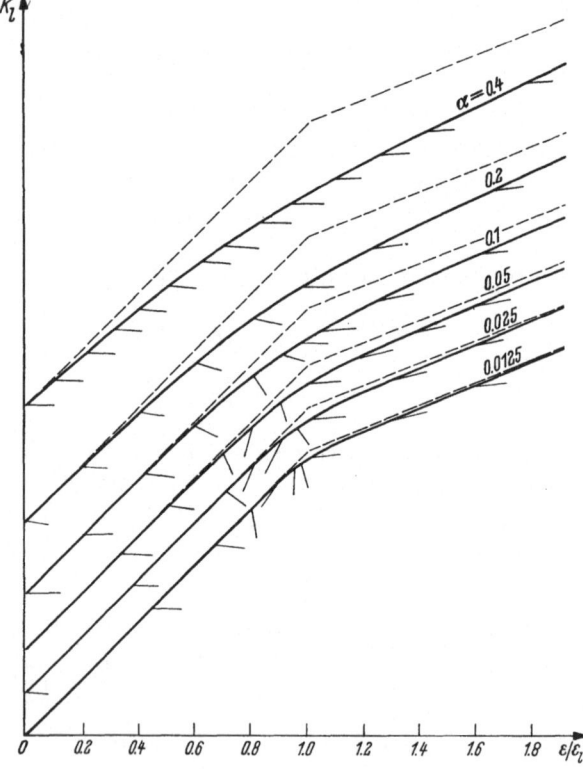

Fig. 9. P-ε curves and tangents to the load-shortening curve at the bifurcation.

5. Imperfection of the column axis

The possibility of instable equilibrium at K_b suggests that the failure load will be sensitive to imperfection of the column axis. For obvious reason we assume the deviation of the centre of the cross section (excentricity) to be

$$e = e_0 \sin \xi. \tag{5.1}$$

Any load K causes a deflection W. The K-W-curve will have a maximum $K_f < K_b$, which is the failure load. We want to establish the strength reduction $(1 - K_f/K_b)$ as a function of e_0/c.

The analysis is analogous to that of Chapter 4. One difference is that the Taylor-expansion of P in ε uses ε_K as its reference point instead of ε_b, where ε_K pertains to $P = \frac{1}{2}K$. The differential equation, analogous to (4.11), (4.12), is

$$(\eta_K + \theta_K)\, EI\, \frac{d^2 W}{dx^2} + K(W + e) = 0, \tag{5.2}$$

where

$$\theta_K = \frac{1}{2}\, \eta_K'' \frac{K^2}{K_l^2} \left(\frac{W+e}{c}\right)^2 - \frac{1}{3} (\eta^2 \eta'' + \eta \eta'^2)_K \frac{c^2}{\varepsilon_l^2} \left(\frac{d^2 W}{dx^2}\right)^2. \tag{5.3}$$

In this case we do not succeed in finding the exact solution because W is finite. An approximate solution $\overline{W}$ is obtained by using the Ritz-Galerkin method, thereby taking

$$\overline{W} = W_0 \sin \xi. \tag{5.4}$$

The validity of the Taylor-expansion, as used, is confined to small deviations of P_o and P_u from $\frac{1}{2}K$, therefore to finite values $\left(\dfrac{W+e}{c}\right)^2 \ll 1$. Then θ_K is small in comparison to η_K and (5.2) contains terms of quite different order. The large terms can be eliminated by means of the exact solution of (5.2) for $\theta_K = 0$, which is $W_{0K}\sin\xi$, where

$$W_{0K} = \frac{e_0}{K_K/K - 1}, \qquad K_K = \eta_K K_E \quad \text{(Vianello)}, \tag{5.5}$$

and by putting

$$W_0 = W_{0K}\,(1+\gamma), \tag{5.6}$$

where γ should be small compared to unity.

Evaluating the Ritz-Galerkin condition we obtain the cubic equation in γ:

$$\gamma^3 + a\gamma^2 + b\gamma + c = 0, \tag{5.7}$$

where

$$a = \frac{y^2 + 2y - 3z}{y^2 - z}, \qquad b = \frac{2y - 3z + 1 + x}{y^2 - z}, \qquad c = \frac{1 - z}{y^2 - z}, \tag{5.7a}$$

$$x = \frac{8}{3}\left(1 - \frac{K}{K_K}\right)^3 \frac{K_l^2}{K^2}\left(\frac{\eta}{\eta''}\right)_K \frac{c^2}{e_0^2}, \qquad y = \frac{K}{K_K}, \qquad z = \frac{2}{3}\left(1 + \frac{\eta'^2}{\eta\eta''}\right)_K. \tag{5.7b}$$

The parameters are $\dfrac{K}{K_l}$ and $\dfrac{e_0}{c}$; the latter one affecting only the coefficient b. For given values of these parameters γ can be solved from (5.7).

For a given value of e_0/c K_f can be determined from the condition that (5.7) has 2 equal roots. The condition is

$$X^3 + \frac{3}{4}X^2 - 3CX - C(C+2) = 0, \tag{5.8}$$

where

$$X = -3b/a^2 \quad \text{and} \quad C = \frac{27}{2}c/a^3. \tag{5.8a}$$

Then

$$\gamma_1 = \gamma_2 = \frac{2C + X}{1 + X}\cdot\frac{a}{6}. \tag{5.9}$$

The Eq. (5.8) permits to answer the question how much excentricity e_0/c is needed so that a given value of K is the failure load K_f. Then a and c are known and solving X from (5.8) we find b, x and finally e_0/c.

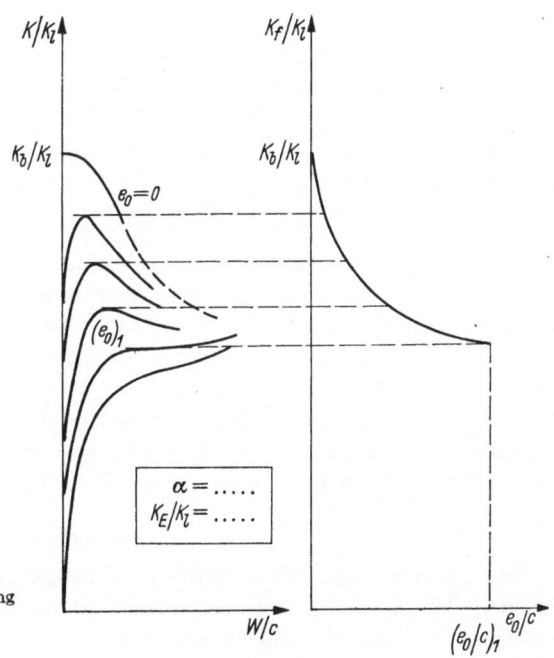

Fig. 10. Development of strength reduction with increasing excentricity.

Figure 10 gives a qualitative picture of the K-W-curves for various excentricities e_0 and the $K_f - e_0$ curve. With the excentricity $(e_0)_1$ the maximum of the K-W-curve becomes a point of inflection: 3 roots of (5.7) coincide. At $e_0 > (e_0)_1$ the maximum has vanished. The conditions for 3 equal roots are

$$C = \frac{1}{2}, \qquad X = -1. \tag{5.10}$$

Then

$$\gamma_1 = \gamma_2 = \gamma_3 = -\frac{1}{3}a. \tag{5.11}$$

From $C = \frac{1}{2}$ follows K which is the minimal value of K_f; thereafter follows $(e_0/c)_1$ from $X = -1$.

Results of the numerical evaluation are shown in Fig. 11. It gives the strength reduction $1 - \frac{K_f}{K_b}$ as a function of $\frac{e_0}{c}$ for two values of K_E/K_l. Curves for equal K_E/K_l and different α almost coincide, so that one single curve sufficiently represents the range $0 < \alpha < 0.1$. It

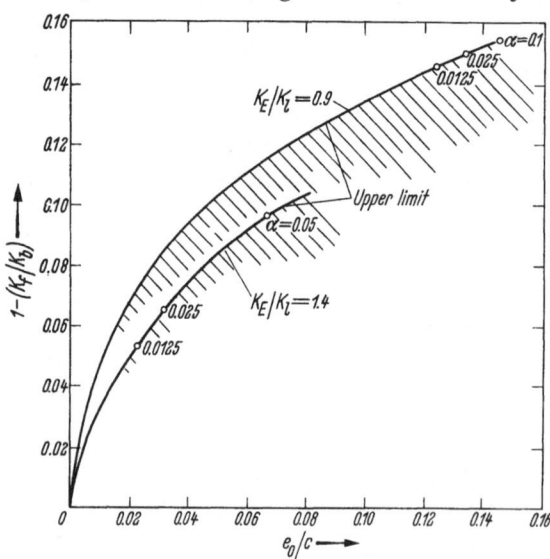

Fig. 11. Reduction of column strength by excentricity.

should however be recalled that the effect of α on K_b/K_l is significant (see Fig. 6). $(e_0/c)_1$ depends very much on α and increases with increasing α. The approximate position of $(e_0/c)_1$ has been indicated.

The validity of the results depends on the condition that the Taylor-expansion of $P(\varepsilon)$, as used, represents sufficiently accurate the actual $P - \varepsilon$-curve. This condition appears not to be fulfilled when P_o and P_u differ from $\frac{1}{2}K_f$ more than 15 to 20%. Therefore the validity of the curves is restricted to small values of e_0/c not exceeding 2%. However the stiffness $dP/d\varepsilon$, as following from the Taylor-expansion, turns out to be smaller than the actual stiffness. Then the analysis overestimates the effect of excentricity. The actual strength reduction caused by e_0/c is smaller than the computation predicts. Below the curves a hatched region has been indicated, representing the incertainty.

Provided the stresses remain within the elastic region, imperfection of the column axis appears to have a minor effect upon the load carrying capacity; it will maximally be in the order of 10%. The main reduction stems from initial waviness of the flanges.

6. Conclusions

The reduced longitudinal stiffness of plates in the post buckling state affects the column strength. Columns without imperfections fail explosively at the local buckling load K_l over a certain range of K_E/K_l-values above unity. (K_E is Euler column load.)

Initial waviness α of the composing thinwalled elements of columns reduces the column strength K_b to values below K_l even when K_E is much greater than K_l. For K_E/K_l close to unity this reduction is significant. Over a range of K_E/K_l values, the magnitude of which depends on α, failure is exlosive.

Initial curvature of the column axis gives a further reduction of the failure load K_f, which is not very significant.

These conclusions have been obtained by investigating columns, the section of which is composed of 2 simply supported flanges having equal waviness and which behave elastically. The investigation should be continued on asymmetric sections, f.i. the same model where the imperfections of the two flanges are different.

Acknowledgment

The author is indebted to J.J.MEYER, student of Aeronautical Engineering at the Technological University, Delft, for his careful programming and evaluation of the computations.

References

1. CAMPUS, F., and MASSONET, CH.: Comportement postcritique des plaques utilisées en construction métallique, Colloque international tenu à l'Université de Liège les 12 et 13/11/'62, Mémoires de la Société Royale des Sciences de Liège, 5me série, tome VIII, fascicule 5. Interventions de W. T. Koiter et M. Skaloud pages 64—68, 103, 104.

2. HEMP, W. S.: The theory of flat panels buckled in compression. Aeronautical Research Council, Reports and Memoranda, No. 2178, June 1945.

3. KOITER, W. T.: On the stability of the elastic equilibrium (in Dutch) Doctor thesis, Technological University, Delft, 1945.

On the curves of synthesis in plane, instantaneous kinematics

L. S. Woo   and   **F. Freudenstein**

IBM New York Scientific Center,   Columbia University,

New York   New York

1. Introduction

The kinematic geometry of the infinitesimally separated positions of a moving plane has been the subject of several relatively recent studies [1, 2, 3, 6, 9, 10, 11, 12, 13]. Amongst other matters, these are concerned with the loci of points, which may be identified with the paths of kinematic elements on the links of mechanisms selected to synthesize the motion. Classical kinematics (Burmester Theory) is concerned with circular paths, because of their association with the paths of the axes of the turning elements on links functioning as cranks. In recent times, investigations have been concerned with general paths of technical significance, such as conic sections [1, 3, 10] and cycloidal motions [9]. For the present study, we shall describe a single, uniform approach, which yields the loci for these and many other curves of technical interest.

The approach is based on an application of E. Cesàro's developments in intrinsic geometry [4, 8] combined with the theory of instantaneous invariants of O. Bottema and G. R. Veldkamp [2, 12, 13]. Some of the theoretical considerations for plane kinematics in terms of differential geometry may be found in H. G. Guggenheimer's recent text on differential geometry [7], while a rigorous mathematical treatment of one-parameter transformations is given in reference [5]. The feasibility of the approach appears to be due, basically, to the fact that the intrinsic equation of a plane curve (the equation relating radius of curvature to arc length) is naturally suited to curvature analysis formulated in terms of the centrodes.

The method, which will be described in detail later can be summarized in terms of the following steps:

a) **The curvature ratios,** $\lambda i = \varrho_i/\varrho$ $(i = 1, 2, 3, \ldots, m)$ where ϱ denotes the radius of curvature of the curve and ϱ_i that of the ith evolute, characterize a plane curve at a given point up to $(m + 3)$-point contact. Following earlier studies by G. R. Veldkamp [12, 13] and one of the authors [6], λ_1 and λ_2 are expressed as functions of the instantaneous invariants of the motion (a_i, b_i) and the coordinates (x, y) of the generating point in the moving plane. An expression of this nature has been developed also for λ_3.

b) Following E. Cesàro [4] we determine the contact invariant, $C(\lambda_1, \lambda_2, \ldots) = 0$, of the family of curves of the trajectory of interest. In general, for a family of curves, whose intrinsic equation has n independent parameters, the contact invariant is an identity in the first $(n + 1)$ curvature ratios. For many transcendental curves, whose intrinsic equation is simple, the invariant is obtainable by direct differentiation and elimination. We shall show also how to find the contact invariant for any algebraic curve with independent or homogeneously related coefficients.

c) Substitution of the curvature ratios $\lambda_i(x, y, a_i, b_i)$ into the contact invariant, $C(\lambda_1, \lambda_2, \ldots, \lambda_{n+1}) = 0$, yields the superosculation curve, the locus of points in the moving plane having $(n + 4)$-point contact with the reference curve.

d) If the further requirement is added that the generating point is to have $(n + 5)$-point contact with the reference curve, the contact may be called one of stationary superosculation. To find such points, if they exist, we take the derivative of the contact invariant with respect to the angle of contingency of the curve. This derivative, which must also vanish. will be a function of the first $(n + 2)$ curvature ratios. Substituting the curvature ratios expressed in terms of x y, a_i and b_i into this function, we obtain a locus of points having stationary super-osculation. The real intersections, if any, of this locus and the supercosulation curve, will in general yield a finite number of discrete points having $(n + 5)$-point contact with the reference curve. These will be called the Burmester points of the motion. For the case of straight-line and circular-arc generation, these reduce to the Ball point and the classical Burmester points.

We now consider these steps in greater detail.

2. Determination of curvature ratios

The following method is due to G. R. VELDKAMP [12, 13]. Consider two planes V and v in continuous relative motion defined in terms of fixed (X, Y) and moving (x, y) cartesian coordinates, as shown in Fig. 1.

The motion of the moving plane may be defined by regarding a, b as functions of φ:

$$\left. \begin{array}{l} X(\varphi) = x \cos \varphi - y \sin \varphi + a(\varphi) \\ Y(\varphi) = x \sin \varphi + y \cos \varphi + b(\varphi) \end{array} \right\} . \tag{1}$$

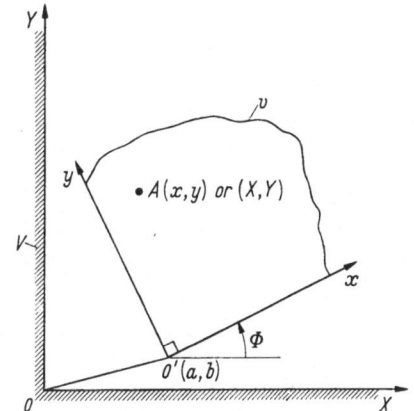

Assuming that a and b are analytic functions of φ at an arbitrary reference position, φ_0, the derivatives of $X(\varphi)$ and $Y(\varphi)$ with respect to φ exist at φ_0.

Without loss of generality, we may choose canonical coordinates, such that $\varphi_0 = 0$ and the two frames V and v coincide in the reference position; furthermore the origin lies at the instantaneous center, $[a(0) = 0, b(0) = 0]$, the x-axis along the common tangent of the centrodes, and a unit inflection circle lies in the half plane $y < 0$, [12]. Under these conditions it is known [12] that the instantaneous invariants a_0, b_0, a_1, b_1 and a_2 vanish [where $a_0 = a(0)$, $b_0 = b(0)$, $a_i = d^i a/d\varphi^i$ and $b_i = d^i b/d\varphi^i$] and that $b_2 = -1$.

The curvature, $\varkappa$, of the path of any point $A(x, y)$ in v is given by

Fig. 1. Relative motion of two planes.

$$\varkappa = \frac{X'Y'' - X''Y'}{(X'^2 + Y'^2)^{3/2}} \tag{2}$$

where the primes denote differentiation with respect to φ. G. R. VELDKAMP [13] has shown that the curvature and its first two derivatives with respect to φ at $\varphi = 0$, are given by:

$$\left. \begin{array}{l} (x^2 + y^2)^{3/2} \varkappa = f \\[2mm] (x^2 + y^2)^{5/2} \dfrac{d\varkappa}{d\varphi} = -g \\[2mm] (x^2 + y^2)^{7/2} \dfrac{d^2\varkappa}{d\varphi^2} = -h \end{array} \right.$$

where

$$\begin{array}{l} f = x^2 + y^2 + y \\[1mm] g = [(a_3 - 3) x + b_3 y] f - y(a_3 x + b_3 y) \\[1mm] h = 5xg + (x^2 + y^2) \tau \end{array} \right\} \tag{3}$$

where

$$\begin{array}{l} \tau = [(a_4 + 4b_3) x - (4a_3 - b_4 - 5) y + 3] f \\[1mm] \qquad - y[a_4 x + (b_4 - 1) y]. \end{array}$$

Assuming that $A(x, y)$ is not a singular point and that the curvature is finite, it is not difficult to show that the curvature ratios, λ_i, are algebraic functions of x, y and the instantaneous invariants (a_i, b_i).

Let $k_i = 1/\varrho_i$ denote the curvature of the ith evolute to the curve at A. Then it is well known [13] that

$$\left.\begin{aligned} & \varkappa_{i+1} \frac{d\varkappa_i}{ds_i} + \varkappa_i^3 = 0, \text{ where} \\ & s_i = \text{arc length on the } i\text{th evolute} \\ & (ds_i = d\varrho_{i-1}) \end{aligned}\right\} . \tag{4}$$

We shall illustrate the computation of λ by means of this recursive equation. Now

$$\lambda_1 = \varrho_1/\varrho = \varkappa/\varkappa_1 .$$

From (4),

$$\lambda_1 = \frac{-1}{\varkappa^2} \frac{d\varkappa}{ds} = - \frac{1}{\varkappa^2} \frac{d\varkappa}{d\varphi} \bigg/ \frac{ds}{d\varphi} . \tag{5}$$

Substituting $\dfrac{ds}{d\varphi} = (x^2 + y^2)^{1/2}$ and $\varkappa_1$ from (3) into [5], we have

$$\lambda_1 = \frac{g}{f^2} . \tag{6}$$

The computation of λ_2 follows along similar lines and according to G. R. Veldkamp [13] is given by

$$\lambda_2 = \frac{3g^2 + xfg + fh}{f^4} . \tag{7}$$

The computation of λ_3, which is needed later on, is considerably more lengthy. The result is:

$$\lambda_3 = \frac{1}{f^6} \left[15g^3 + 10xfg^2 + 10fgh + 3xf^2h + f^2l \right.$$
$$\left. - f^2g(x^2 - y^2)(b_3x - a_3y + 2y - 3) - 4gf^2y^2 \right]$$

where

$$(x^2 + y^2)^{9/2} \frac{d^3\varkappa}{d\varphi^3} = -l$$

and

$$l = 10xh - 5g(x^2 + y^2)(b_3x - a_3y + 2y + 1) - 15x^2g - \varepsilon(x^2 + y^2),$$

where

$$\varepsilon = (x^2 + y^2)(-a_5x - 2b_4x + 2x) - x(a_3y - b_4y + y + a_3)$$
$$+ 4a_3x(-a_3y + 2y + 1) - 3f(-3a_3x + b_4x + 3). \tag{8}$$

Equations (6), (7), (8) express the curvature ratio as algebraic functions of the coordinates (x, y) of the generating point and the instanteneous invariants (a_i, b_i) of the motion.

Next we need to find the contact invariant.

3. Determination of contact invariant

Let the origin of coordinates be chosen at the generating point, A, on a plane curve, with the x-axis along the tangent, the arc length along the curve being denoted by s. This is the familiar moving frame of Frenet. Then in the neighborhood of the origin, $x(s)$ and $y(s)$ may be expanded in a power series in s. This series will be convergent provided the curve is analytic at A. Assuming the curvature to be finite, the first four terms of the series may be readily evaluated and are given by:

$$
\left.\begin{aligned}
x(s) &= s - \frac{s^3}{6\varrho^2} + \frac{\varrho_1 s^4}{8\varrho^4} + \frac{1}{120}\frac{(\varrho^2 + 4\varrho_1\varrho_2 - 15\varrho_1^2)}{\varrho^6} s^5 + \cdots \\
y(s) &= \frac{s^2}{2\varrho} - \frac{\varrho_1}{6\varrho^3} s^3 - \frac{(\varrho^2 + \varrho\varrho_2 - 3\varrho_1^2)}{24\varrho^5} s^4 + \\
&\quad + \frac{1}{120}\frac{(\varrho^2\varrho_3 + 10\varrho_1\varrho_2\varrho - 15\varrho_1^3 + 6\varrho^2\varrho_1)}{\varrho^7} s^5 + \cdots
\end{aligned}\right\}. \tag{9}
$$

Where $\varrho = \dfrac{ds}{d\psi}$ is the radius of curvature at A and ψ denotes the angle of contingency.

It is natural to inquire now whether the curvature derivatives $\varrho_1, \varrho_2, \ldots$ of all orders are independent, or if there exists an equation, which defines the relation between a finite number of derivatives for a particular curve.

In the case of curves which have an explicit natural equation, we simply eliminate the constants of the equation by successive differentiation. For example, for the logarithmic spiral $\varrho = cs + d$, where c, d are parametric constants, and we have $\lambda_1 = \dfrac{d\varrho}{ds} = c$, $\lambda_2 = \dfrac{d\varrho_1}{ds} = c^2$; hence $C(\lambda_1, \lambda_2) = \lambda_2 - \lambda_1^2 = 0$.

In the case of algebraic equations, although no explicit natural equation may exist except in differential form, the contact invariant can be found as follows:

The degree of an algebraic curve is invariant under translation and rotation of the coordinates. If we transform the coordinates to the Frenet form, the equation of the curve will remain of the form

$$
\sum_{i,j} a_{ij}x^i y^j = 0, \text{ and there is no constant term}. \tag{10}
$$

Substituting (9) into (10), we obtain a power series in s, which we designate as $w(s)$ and the coefficients of which are linear in the parameters, a_{ij}. The power series $w(s)$ is also convergent, provided $x(s)$ and $y(s)$ are convergent. Hence, $w(s)$ vanishes identically within the radius of convergence. If the total number of parametric coefficients, a_{ij}, is equal to n, we equate the first n non-zero coefficients appearing in the power series in $w(s)$ to zero. This will give a system of n homogeneous, linear, simultaneous equations in the unknowns a_{ij}. The necessary condition for the existence of a solution is that the determinant of the coefficients vanishes. Since the coefficients are rational functions of $(\varrho, \varrho_1, \varrho_2, \ldots, \varrho_n)$, a defining relation can be established. The sufficiency of the condition can also be justified from the theory of one-parameter Lie groups [5].

For conic sections, for example, the Frenet form is

$$
y = a_{20}x^2 + a_{02}y^2 + a_{11}xy. \tag{11}
$$

Substituting (9) into (11), we obtain a power series in s. Equating the first four coefficients of $w(s)$ (s^2, s^3, s^4, s^5) to zero, we obtain four simultaneous equations in (a_{02}, a_{20}, a_{11} and $a_{01} = -1$). The determinant of the coefficient matrix must vanish:

$$
\begin{vmatrix}
\dfrac{1}{2}, & 0, & 0, & \dfrac{-1}{2\varrho} \\[2ex]
0, & 0, & \dfrac{1}{2\varrho}, & \varrho_1/6\varrho^3 \\[2ex]
\dfrac{-1}{6\varrho^2}, & \dfrac{1}{8\varrho^2}, & \dfrac{-\varrho_1}{3\varrho^3}, & \dfrac{-1}{24}(\varrho^2 + \varrho\varrho_2 - 3\varrho_1^2) \\[2ex]
\dfrac{2\varrho_1}{16\varrho^4}, & \dfrac{-\varrho_1}{12\varrho^4}, & \dfrac{-(\varrho^2 + \varrho\varrho_2 - \varrho_1^2)}{24\varrho^5}, & \dfrac{(-\varrho^2\varrho_3 + 10\varrho\varrho_1\varrho_2 - 15\varrho_1^3 + 6\varrho\varrho_1^2)}{120\varrho^7}
\end{vmatrix} = 0. \tag{12}
$$

Expanding and collecting terms, we obtain

$$
-9\varrho^2\varrho_3 + 45\varrho\varrho_1\varrho_2 - 36\varrho^2\varrho_1 - 40\varrho_1^3 = 0.
$$

51*

Dividing by ϱ^3 and substituting $\lambda_i = \dfrac{\varrho_i}{\varrho}$, we obtain the contact invariant:

$$C(\lambda_1, \lambda_2, \lambda_3) = 9\lambda_3 - 45\lambda_1\lambda_2 + 36\lambda_1 + 40\lambda_1^3 = 0. \qquad (13)$$

It is clear that the same method is applicable to algebraic curves with coefficients related by a homogeneous algebraic equation. The contact invariant can be defined [4] as the lowest-order identity in the curvature and its derivatives, which is characteristic of a family of curves. For a family of curves whose intrinsic equation has n independent parameters, the contact invariant may involve no more than the first $(n + 1)$ curvature ratios. For we could differentiate the contact invariant an arbitrary number of times with respect to the angle of contingency and obtain further relations involving curvature ratios of higher order. For example, Eq. (13) is the only one (for conic sections) which involves only the first three curvature ratios. This is the meaning of the term "lowest order" in the above definition.

Table 1 lists nineteen of the technically more important plane curves, their intrinsic equation, their contact invariant and their parameters for osculation to a prescribed curve. The intrinsic equations of these and many other curves can be found in [4, 14].

In Case 1 (straight line), the superosculation curve (inflection circle) has 3-point contact with the reference curve.

In Cases 2, 11, 12, 14, 15, and 17, the superosculation curve has 4-point contact with the reference curves.

In Cases 3, 4, 5, 6, 8, 9, 10, 13, and 16, the curve has 5-point contact and in cases 18 and 19 it has 6-point contact.

For the cases involving 5-point contact (excepting the tractrix), the contact invariant is of the form

$$\alpha\lambda_2 + \beta\lambda_1^2 + \gamma = 0 \qquad (\alpha \neq 0). \qquad (14)$$

The fact that this is the form of the invariant for many different families of curves, has already been noted by Cesàro [4].

Following the derivations in Ref. [4] and [14], we illustrate the computations of the entries in Table 1 with respect to the latter two Cases (18 and 19).

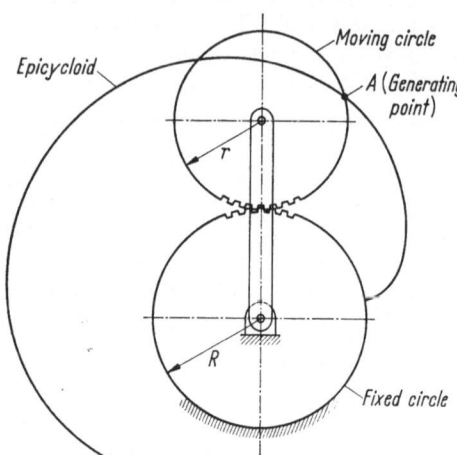

Fig. 2. Epicycloid with generating point A.

Cycloidal curves (Case 18) are generated by points on the circumference of a circle rolling inside (hypocycloids) or outside (epicycloids) a fixed circle (Fig. 2). This case, then, may be regarded as a special case of general cycloidal motion [9] (in which the generating point is not restricted to the circumference of the moving circle).

As shown in (14), the intrinsic equation of a cycloidal curve is

$$\frac{s^2}{a^2} + \frac{\varrho^2}{b^2} = 1 \qquad (15)$$

where

$$a = \frac{4r(R+r)}{R} \quad \text{and} \quad b = \frac{4r(R+r)}{R+2r} \qquad (16)$$

and r, R are regarded as positive for external gearing.

Differentiating (15) with respect to the angle of contingency, ψ, we have

$$\frac{s}{a^2} + \frac{\varrho_1}{b^2} = 0. \qquad (17)$$

Further differentiation yields

$$\frac{\varrho_2}{\varrho} = -\frac{b^2}{a^2}$$

and

$$\frac{\varrho\varrho_3 - \varrho_1\varrho_2}{\varrho^2} = 0.$$

or

$$C(\lambda_1, \lambda_2, \lambda_3) = \lambda_3 - \lambda_1\lambda_2 = 0. \tag{19}$$

Equation (19) is the contact invariant. For an osculating cycloidal curve, which has 5-point contact with the reference curve, we solve for s, a^2 and b^2 in terms of ϱ, ϱ_1 and ϱ_2 using Eqs. (17), (18) and (19):

$$\left.\begin{aligned} s &= \varrho\varrho_1/\varrho_2 \\ a^2 &= \frac{\varrho^3}{\varrho_2}\left(\frac{\varrho_1^2}{\varrho\varrho_2} - 1\right) \\ b^2 &= -\varrho^2\left(\frac{\varrho_1^2}{\varrho\varrho_2} - 1\right) \end{aligned}\right\}. \tag{20}$$

For the general conic section (Case 19), the Frenet form of the curve, as given previously, is:

$$y = a_{20}x^2 + a_{02}y^2 + a_{11}xy. \tag{11}$$

Substituting $x(s)$ and $y(s)$ from Eq. (9) into Eq. (11), and equating the coefficients of like powers of s (s^2, s^3 and s^4), we can solve for the parameters a_{20}, a_{11}, and a_{02}:

$$\left.\begin{aligned} a_{20} &= 1/\varrho \\ a_{11} &= -2\varrho_1/3\varrho^2 \\ a_{02} &= (9\varrho^2 - 3\varrho\varrho_2 + 5\varrho_1^2)/9\varrho^3 \end{aligned}\right\}. \tag{21}$$

The contact invariant, obtained earlier as Eq. (13), can be found by adding the condition obtained by equating the coefficient of s^5.

To obtain the equation of the conic section osculating to a given curve (with five-point contact), we substitute Eq. (21) into Eq. (11):

$$(3\varrho x - \varrho_1 y)^2 + (9\varrho^2 + 4\varrho_1^2 - 3\varrho\varrho_2)\, y^2 = 18\varrho^3 y. \tag{22}$$

In the case of a parabola, $4a_{20}a_{02} = a_{11}^2$ and Eq. (21) yields

$$C(\lambda_1, \lambda_2) = 3\lambda_2 - 4\lambda_1^2 - 9 = 0,$$

while in the case of an equilateral hyperbola, $a_{02} + a_{20} = 0$ and $C(\lambda_1, \lambda_2) = 3\lambda_2 - 5\lambda_1^2 - 18 = 0$ (see Table 1).

Certain feasibility considerations follow directly from the contact invariant. For example, for the vertex of a parabola, $\lambda_1 = 0$, $\lambda_2 = 3$ and $\lambda_3 = 0$. It is not feasible, therefore, to match both λ_1 and λ_2 values at the vertex with any point on a cycloid (Case 9), epicycloid or hypocycloid (Case 10), (for which λ_2 is negative).

It is sometimes possible to obtain the contact invariant for curves of transcendental form by a double expansion in power series. For examples if the curve is given in the form $y = f(x)$, y may be expanded in a power series in x; thereafter, x and y may be expanded in a power series in the arc length s and coefficients of s equated. The functions $y = \sin x$ and $y = e^x$ etc. may be treated in this manner. However, the contact invariant may be so involved that the corresponding superosculation curve is of little, except theoretical significance.

4. Superosculation curves

Substitution of Eqs. (6), (7), (8) for λ_1, λ_2 and λ_3 into the equation of the contact invariant yields the superosculation curve, $S(x, y) = 0$ in the moving plane. A table of superosculation curves for many curves of technical interest are shown in Table 1. The algebraic geometry of a number of these is summarized in Table 2.

Consider, for example, the class of curves, having five-point-contact superosculation, the equation of the contact invariant being of the form (14) $(\alpha\lambda_2 + \beta\lambda_1^2 + \gamma = 0)$.

Cases (3), (4), (5), (6), (8), (9), (10), (13) and (16) in Table 1 belong to this category.

Table 1. *Superosculation curves,* $S(x, y) = 0$

(I) Case	Curve	(II) Equation in $f(\varrho, s)$ or $f(x, y)$ form	(III) Contact Invariant $C(\lambda_1, \lambda_2, \ldots) = 0$	(IV) Osculating Curve $(\varrho, \varrho_1, \varrho_2 \ldots \text{given})$	(V) Superosculation Curve Equation $S(x, y) = 0$	(VI) Order of Contact	(VII) Name
1	Straight Line	$\varrho = \infty$	$\varrho = \infty$	$\varrho = \infty$	$f = 0$	3	Inflection Circle*
2	Circle	$\varrho = a$	$\lambda_1 = 0$	$\varrho = a$	$g = 0$	4	Cubic of Stationary Curvature*
3	Involute	$\varrho^2 = 2aS$	$\lambda_2 = 0$	$a = \varrho_1, \ S = \varrho^2/2\varrho_1$	$3g^2 + xfg + fh = 0$	5	Involute Superosculation Curve
4	Logarithmic Spiral	$\varrho = aS$	$\lambda_2 - \lambda_1^2 = 0$	$a = \varrho_1/\varrho, \ S = \varrho^2/\varrho_1$	$2g^2 + xfg + fh = 0$	5	Log-spiral Superosculation Curve
5	Catenary	$\varrho = a + \dfrac{S^2}{a}$	$3\lambda_2 - 3\lambda_1^2 - 4 = 0$	$a = \dfrac{4\varrho^3}{4\varrho^2 + \varrho_1^2}, \ S = \dfrac{a\varrho_1}{2\varrho}$	$6g^2 + 3xfg + 3fh - 4f^4 = 0$	5	Catenary-1 Superosculation Curve
6	Catenary of Constant Stress	$\varrho = \dfrac{1}{2}a\left(e^{S/a} + e^{-S/a}\right)$	$\lambda_2 - 2\lambda_1^2 - 1 = 0$	$a^2 = \dfrac{4\varrho^4}{\varrho^2 + 4\varrho_1^2}, \ \cosh\dfrac{S}{a} = \dfrac{2\varrho}{a}$	$g^2 + xfg + fh - f^4 = 0$	5	Catenary-2 Superosculation Curve
7	Tractrix	$\varrho^2 = a^2\left(e^{2S/a} - 1\right)$	$\lambda_2^2 + 4\lambda_1^2 - 2\lambda_1^2\lambda_2 = 0$	$a = \dfrac{1}{2}\left(\varrho_1 + \sqrt{\varrho_1^2 - 4\varrho^2}\right)$ $S = \dfrac{a}{2}\ln\left(1 + \dfrac{\varrho^2}{a^2}\right)$	$(3g^2 + xfg + fh)(g^2 + xfg + fh) + 4g^2f^4 = 0$	5	Tractrix Superosculation Curve
8	Cornu's Spiral	$\varrho S = a^2$	$\lambda_2 - 3\lambda_1^2 = 0$	$a^2 = -\varrho^3/\varrho_1, \ S = -\varrho^2/\varrho_1$	$xg + h = 0$	5	Cornu-spiral Superosculation Curve
9	Ordinary Cycloid	$\varrho^2 + S^2 = a^2$	$\lambda_2 + 1 = 0$	$a = (\varrho^2 + \varrho_1^2)^{1/2}, \ S = -\varrho_1$	$3g^2 + xfg + fh + f^4 = 0$	5	Ordinary Cycloid Superosculation Curve
10	Cycloidal curve with prescribed gear ratio (Epi-or Hypocycloid)	$\varrho^2(1 - 2n)^2 + S^2 = a^2$, where n = gear ratio, planet to sun gear	$(1 - 2m)^2\lambda_2 + 1 = 0$	See Case 18	$(1 - 2n)^2(3g^2 + xfg + fh) + f^4 = 0$	5	Superosculation Curves for Cycloidal Curves with Given gear Ratio
11	Parabola-1	$y + a_{00} + a_{10}x + a_{20}x^2 = 0$			$y(a_3x + b_3y) + 3x(x^2 + y^2 + y) = 0$	4	Cubic of P1 Curvature*

Table 1 (continued)

52*

(I) Case, Curve	(II) Equation in $f(\rho, s)$ or $f(x,y)$ form	(III) Contact Invariant $C(\lambda_1, \lambda_2, \ldots) = 0$	(IV) Osculating Curve ($\rho, \rho_1, \rho_2 \ldots$ given)	(V) Superosculation Curve Equation $S(x,y) = 0$	(VI) Order of Contact	(VII) Name
12 Parabola-2	$y + a_{00} + a_{10}x + a_{02}y^2 = 0$			$x(a_3x + b_3y) - 3(1+y)(x^2 + y^3 + y) = 0$	4	Cubic of $P2$ Curvature
13 General Parabola	$\sum_{i,j}^{0,1,2} a_{ij}x^i y^j = 0$ and $4a_{02}a_{20} = a_{11}^2$	$3\lambda_2 - 4\lambda_1^2 - 9 = 0$	$y = \frac{1}{2} \times \left(\frac{1}{\rho}x^2 + \frac{\rho_1 y^2}{9\rho^3} - \frac{2\rho_1 xy}{3\rho^2}\right)$	$5g^2 + 3xfg + 3fh - 9f^4 = 0$	5	Parabola Superosculation Curve
14 Equilateral Hyperbola-1	$y + a_{00} + a_{10}x + a_{11}xy = 0$			$3(x^2 + y^2 + y)(x^2 - y^2 - y) + 2xy(a_3x + b_3y) = 0$	4	Quartic of H-1 Curvature*
15 Equilateral Hyperbola-2	$y + a_{00} + a_{10}x + a_{20}(x^2 - y^2) = 0$			$3x(x^2 + y^2 + y)(2y + 1) + (a_3x + b_3y)(x^2 - y^2) = 0$	4	Quartic of H-2 Curvature*
16 General Equilateral Hyperbola	$\sum_{i,j}^{0,1,2} a_{ij}x^i y^j = 0$ and $a_{20} + a_{02} = 0$	$3\lambda_2 - 5\lambda_1^2 - 18 = 0$	$y = \frac{1}{2} \times \left(\frac{1}{\rho}x^3 - \frac{1}{\rho}y^2 - \frac{2\rho_1 xy}{3\rho^2}\right)$	$4g^2 + 3xfg + 3fh - 18f^4 = 0$	5	Equilateral Hyperbola Superosculation Curve
17 Ellipse (Type $E\lambda$) of given proportions	$y + a_{00} + a_{10}x + a_{20}(x^2 + \lambda y^2) = 0$, $\lambda > 0$			$3(x^2 + y^2 + y)[(1 - \lambda)xy - \lambda x] + (a_3x + b_3y)(y^2 + \lambda x^2) = 0$	4	Quartic of Type-$E\lambda$ Elliptic Curvature*
18 General Cycloidal Curve (Epi- or Hypocycloid)	$\frac{S^2}{a^2} + \frac{\rho^2}{b^2} = 1$; a, b as in Eq. (16)	$\lambda_3 - \lambda_1\lambda_2 = 0$	$S = \rho\rho_1/\rho_2$, $a = \frac{\rho^3}{\rho^2}\left(\frac{\rho_1^2}{\rho\rho_2} - 1\right)$, $b^2 = -\rho^2\left(\frac{\rho_1^2}{\rho\rho_2} - 1\right)$	$12g^3 + 9xfg^2 + 9fgh + 3xf^2h + f^2l - (x^3 + y^2)(b_3x - a_3y + 2y - 3)f^2g - 4y^2f^2g = 0$	6	General Epi- or Hypocycloid Superosculation Curve
19 General Conic Section	$\sum_{i,j}^{0,1,2} a_{ij}x^i y^j = 0$	$9\lambda_3 - 45\lambda_1\lambda_2 + 36\lambda_1 + 40\lambda_1^3 = 0$	$(3\rho x - \rho_1 y)^2 + (9\rho^2 + 4\rho_1^2 - 3\rho\rho_2)y^2 = 18\rho^3 y$	$40g^3 + 9[5xfg^2 + 5fgh + 4gf^4 + 3xfg^2h + f^2l - 4y^2f^2g - (x^2 + y^2)(b_3x - a_3y + 2y - 3)f^2g] = 0$	6	Quintic of Conic-Section Curvature*

* Indicates previously known results.

Table 2. *General algebraic and geometric properties of superosculation curves, curves of stationary superosculation and the curve $\lambda_3 = 0$*

Case	(I) Curve	(II) Contact Invariant Form of Equation	(III) Order	(IV) Circularity	(V) Terms of Lowest Order	(VI) Multiplicity at Pole	(VII) Multiplicity of Tangency at Pole — Pole Tangent	(VII) Multiplicity of Tangency at Pole — Pole Normal	(VIII) Multiplicity at Ball Point	(IX) Multiplicity of Tangency to Inflection Circle at Ball Point
1	5-point Contact Superosculation Curve with Contact Invariant as in (II)	$\alpha\lambda_2 + \beta\lambda_1^2 + \gamma = 0$, $\alpha \neq 0$, $\gamma \neq 0$	8	4	$(3\alpha + \gamma)\, y^4 + (12\alpha + 9\beta)\, x^2 y^2$	4	2	0	1	1
2	5-point Contact Superosculation Curve with Contact Invariant as in (II)	$\alpha\lambda_2 + \beta\lambda_1^2 = 0$, $\alpha \neq 0$	7	3	$3\alpha y^4 + (12\alpha + 9\beta)\, x^2 y^2$	4	2	0	1	1
3	Stationary Superosculation Curves for Cases 1 and 2	$\alpha\lambda_3 + (\alpha + 2\beta)\,\lambda_1\lambda_2 + 2\gamma\lambda_1 = 0$, $\alpha \neq 0$	11	5	$-(42\alpha + 18\beta + 6\gamma)\, xy^5 - (96\alpha + 72\beta)\, x^3 y^3$	6	3	1	2	1
4	$\lambda_3 = 0$ Curve (also Contact Invariant of the Involute of an Involute)	$\lambda_3 = 0$	11	5	$-33xy^5 - 60x^3 y^3$	6	3	1	2	1
5	Superosculation Curve for Epicycloidal or Hypocycloidal Curves	$\lambda_3 = \lambda_1\lambda_2$	11	5	$-24(xy^5 + x^3 y^3)$	6	3	1	2	1
6	Special Form of Case 1 for Parabola	$3\lambda_2 - 4\lambda_1^2 - 9 = 0$	8	4	Product of $x^2 y$ and terms of degree 2	5	1	2	1	1
7	Special Form of Case 3 for Parabola	$3\lambda_3 - 5\lambda_1\lambda_2 - 18\lambda_1 = 0$	11	5	Product of xy^2 and terms of degree 4	7	2	1	2	1

Using Eqs. (6), (7), Eq. (14) becomes

$$S(x, y) = (\beta + 3\alpha) g^2 + \alpha x f g + \alpha f h + f^4 = 0, \tag{23}$$

where f, g, and h are defined in Eq. (3). Equation (23) is the superosculation curve.

In the general case, when $\gamma \neq 0$, this is an algebraic equation of order 8 and circularity 4. It has a quadruple point at the pole, with the pole tangent counting twice as one of its tangents. When γ vanishes, the equation is tricircular and of order 7, the tangency and multiplicity at the pole remaining unchanged. The curve passes through the Ball point, at which it is tangent to the inflection circle.

The lowest-order terms in (23) are of degree 4. These are

$$(3\alpha + \gamma) y^4 + (12\alpha + 9\beta) x^2 y^2. \tag{24}$$

For the special case of the parabola (Case 13), $3\alpha + \gamma = 12\alpha + 9\gamma = 0$. The superosculation curve has multiplicity 5 at the pole, at which the lines $x = 0$ and $y = 0$ are respectively doubly and singly tangent to the curve (see Table 2, Case 6).

In the particular case when the motion is generated via general cycloidal motion, for instance, the instantaneous invariants a_{2n} and b_{2n-1} vanish (12). A rational, parametric representation for y and x^2 can then be obtained by the substitution $t = (x^2 + y^2)/y$. The superosculation curve then has the following parametric form:

where
$$\left.\begin{aligned}
y &= \frac{At + B}{A + C} \\
x^2 &= y(t - y) \\
A &= (3\alpha + \beta) [(a_3 - 3) t - 3]^2 + 6\alpha(t + 1) [(a_3 - 3) t - 3] \\
B &= 3\alpha t(t + 1)^2 \\
C &= \alpha t(t + 1)^2 (4a_3 - b_4 - 5) + \alpha t(t + 1) (b_4 - 1) - \gamma(t + 1)^4
\end{aligned}\right\}. \tag{25}$$

With the appropriate choice of the numerical values of α, β and γ, Eq. (25) applies to any of the cases in Table 1 for which the contact invariant is of the form $\alpha\lambda_2 + \beta\lambda_1^2 + \gamma = 0$.

These results are illustrated in Section 9.

5. Determination of curves of stationary superosculation

Let us denote this curve by $S'(x, y) = 0$. It is obtained by taking the derivative of the contact invariant with respect to the angle of contingency. This is readily obtained with the aid of the relation

$$\frac{d\lambda_i}{d\psi} = \frac{d}{d\psi} (\varrho_i/\varrho) = \lambda_{i+1} - \lambda_i\lambda_1.$$

Again consider, for example, the class of curves whose contact invariant is $\alpha\lambda_2 + \beta\lambda_1^2 + \gamma = 0$.

The derivative of the contact invariant can be reduced to

$$\alpha\lambda_3 + (\alpha + 2\beta) \lambda_1\lambda_2 + 2\gamma\lambda_1 = 0. \tag{26}$$

Substituting Eqs. (6), (7), (8) for the curvature ratios into (26), we obtain the corresponding stationary superosculation curve in the moving plane:

$$\begin{aligned}
S'(x, y) = {}&15\alpha g^3 + 10\alpha x f g^2 + 10\alpha f g h + 3\alpha x f^2 h + \alpha f^2 l \\
&- \alpha(x^2 + y^2) (b_3 x - a_3 y + 2y) f^2 g + \alpha(3x^2 - y^2) f^2 g \\
&+ (\alpha + 2\beta) (3g^3 + x f g^2 + f g h) + 2\gamma g f^4 = 0.
\end{aligned} \tag{27}$$

This is an algebraic curve of order 11 and circularity 5. It has a point of multiplicity 6 at the pole, with the pole tangent counting as a three-fold tangent. The lowest-order terms are of degree 6 and they are:

$$-(42\alpha + 18\beta + 6\gamma) xy^5 - (96\alpha + 72\beta) x^3 y^3. \tag{28}$$

The curve passes twice through the Ball point, at which it is tangent to the inflection circle.

For the special case of the parabola (Case 13, Table 1), $S'(x, y) = 0$ has multiplicity 7 at the pole. At this point the line $x = 0$ is a single tangent and $y = 0$ is a double tangent (Table 2, Case 7).

6. The Burmester points

The number of Burmester points are given by the real intersections other than the pole and the Ball point, of $S(x, y) = 0$ and $S'(x, y) = 0$. For example, for the class of curves whose contact invariant equation is $\alpha\lambda_2 + \beta\lambda_1^2 + \gamma = 0$, $S(x, y)$ is given by Eq. (23) and $S'(x, y)$ is given by Eq. (27).

When $\gamma \neq 0$ the number of Burmester points is then generally equal to $(88 - 2 \times 4 \times 5 - 4 \times 6 - 2 \times 3 - 3)$ or 15. When $\gamma = 0$, the number of Burmester points is generally equal to $(77 - 2 \times 3 \times 5 - 4 \times 6 - 2 \times 3 - 3)$ or 14. For the special case of the parabola (Table 1, Case 13), the number of Burmester points is equal to $(88 - 2 \times 4 \times 5 - 5 \times 7 - 2 \times 2 - 3)$ or 6, in agreement with O. Bottema [3].

Table 2 includes a summary of the algebraic geometry of some superosculation curves and related loci.

7. Synthesis of mechanisms for higher-order path generation

The synthesis can be identified according to three types:

Type 1: This refers to the previous discussion. The superosculation curve is used as the locus of points, which satisfy the contact invariant. In addition, the stationary superosculative curve, together with the superosculation curve, determine the Burmester points of the motion.

Type 2: This refers to the specification described in Ref. [6]. Here the curvature ratios are matched at a particular point of a curve without consideration of the superosculation curve. The same problem has been considered in detail in Ref. [13].

Type 3: This is intermediate between the foregoing. Here we match one or more of the curvature ratios locally and in addition impose the restriction that the generating point must lie on the superosculation curve. For example, for 5-point contact superosculation, we look for points on the superosculation curve with a prescribed value of λ_1. If, however, the contact invariant is of the form $\lambda_2 + f(\lambda_1) = 0$, this particular case reduces to Type-2.

In all of these cases, the synthesis involves the determination of the points of intersection of appropriate curves of synthesis (the superosculation curve and/or the curves $\lambda_i = $ constant). Knowing the order of the curves and their geometrical characteristics (Table 2), the potential number of these intersections can be determined.

Once the synthesis is completed, it is possible to compute the precision of the motion. In Type-1 synthesis, for example, we proceed as follows:

Using the Frenet form of the equation of the curve $[f(x, y) = 0]$, the error may be regarded as the difference in y of the two motions in the vicinity of the precision point. If the difference in the lowest curvature ratios, which is not matched between reference curve and mechanism is $\Delta\lambda_k$, Eq. (9) shows that the error, L, is approximately given by

$$L = \frac{\Delta\lambda_k}{(k + 2)!} \varrho \left(\frac{s}{\varrho}\right)^{k+2}. \tag{29}$$

Since the error determination requires knowledge of a curvature ratio whose order is one higher than the curvature ratio of highest order appearing in the contact invariant, the ratio λ_3, as derived earlier, is frequently needed in this connection.

8. Illustrations

We illustrate the results by four examples.

Example 1: Involute superosculation curve in the case of cardioid motion.

Figure 3 shows the involute superosculation curve [Eq. (25)] for cardioid motion $(a_3 = \dfrac{1+R}{R}, b_4 = \dfrac{R^3-1}{R^2(R-1)}$ as shown in G. R. VELDKAMP [12]). For the involute (Case 3, Table 1, $\alpha = 1, \beta = \gamma = 0$), the superosculation curve may be regarded as attached to the planet gear. Any point on the superosculation curve (and attached to the planet gear) will have at least 5-point contact with the involute. Different points on the superosculation curve correspond to precision points having different pressure angles.

Example 2: Logarithmic spiral approximated by cardanic motion with 5-point contact.

In this case Eq. (25) is used with $\alpha = 1, \beta = -1, \gamma = 0$ (Table 1, Case 4), while for cardanic motion, $a_3 = 0$ and $b_4 = -1$ when the sun gear radius is -1. The superosculation curve breaks up and its finite portion is the line $y = -0.5$. Suppose we choose to approximate the logarithmic

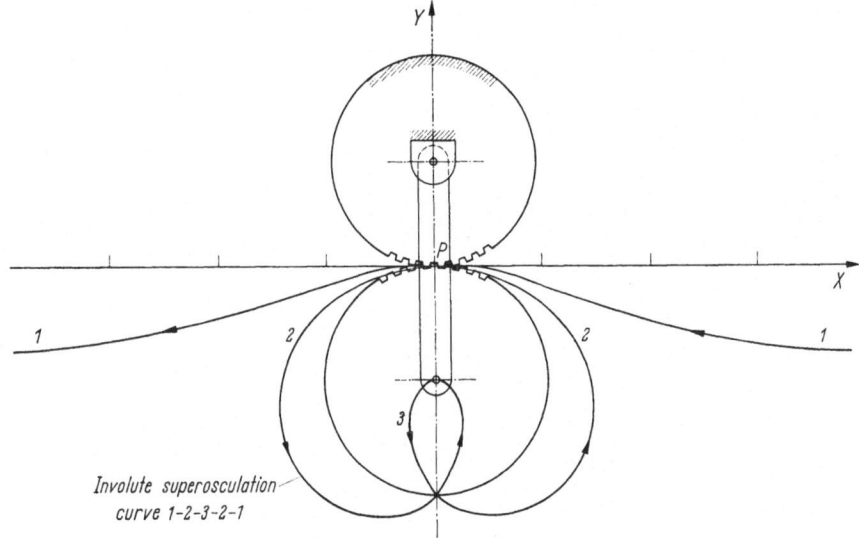

Fig. 3. Involute superosculation curve for cardioid motion.

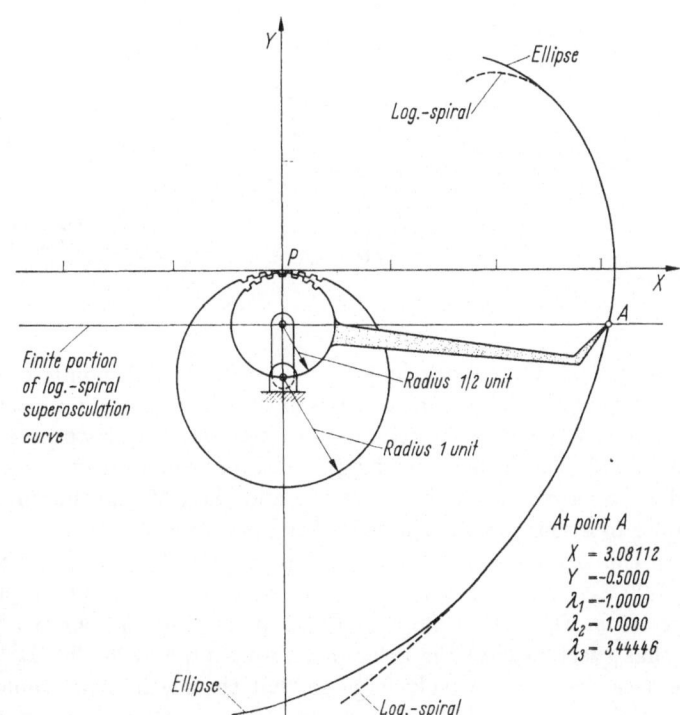

Fig. 4. Five-point contact approximation of logarithmic spiral via cardanic motion.

spiral $\varrho = (1/2)(1 + 2s)$ at the point $s = 1$ (this curve simulates the upper-leg centrode for the relative motion at the human knee joint). By computerized scanning of points on the line $y = -0.5$, we locate the point A (3.08112, -0.5) for which $\lambda_1 = -1$, $\lambda_2 = 1$. Using Eq. (8), we find $\lambda_3 = -3.44446$. The generated curve (ellipse) closely matches the log spiral as shown in Fig. 4. According to Eq. (29) the deviation between the curves is given by $L = \frac{\Delta\lambda_3}{120}\left(\frac{s}{\varrho}\right)^5$. In this case $\Delta\lambda_3 = 2.44$, and $L \approx \left(\frac{s}{6.5}\right)^5$ inches. For example, at a point one inch away from the precision point, the error is very nearly 10^{-4} inches.

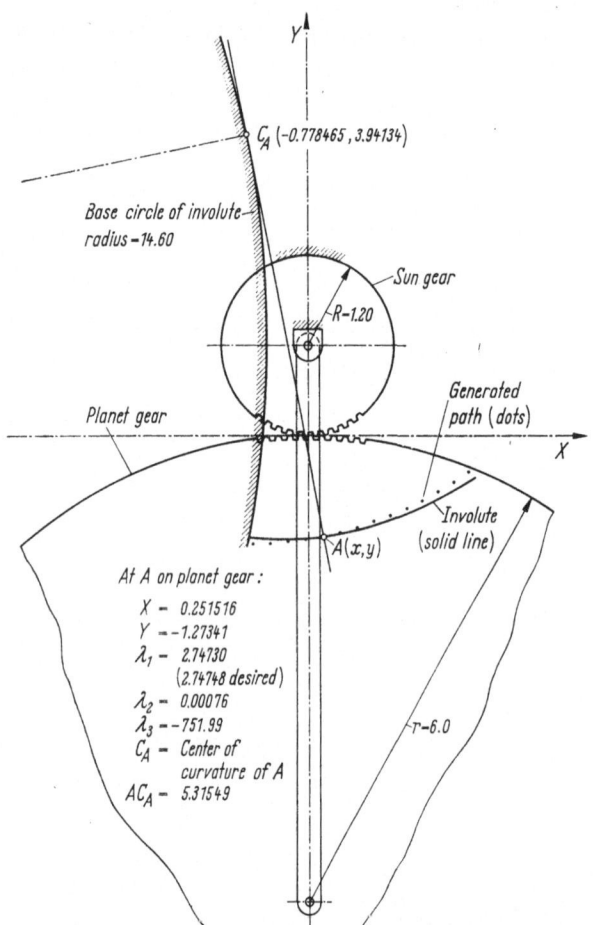

Fig. 5. Involute gear-tooth profile: five-point contact approximation via cardioid motion.

Example 3: Gear tooth (involute) profile at pitch point: order approximation[1].

In recent times the approximation of the involute has been suggested for wheel dressing (A. H. Candee) and as a reference in optical-comparator measurements (R. Unterberger, R. Koller). The former has suggested approximation via a cardioid and described a 4-point contact approximation. K. H. Sieker and the latter authors use the motion of a point on the coupler of a four-bar linkage with 4-point contact.

In this example, we consider a 5-point matching at a 20° pressure angle ($\lambda_1 = \cot 20° = 2.74748$, $\lambda_2 = 0$) via 5:1 external differential gearing. The superosculation curve is again obtained from Eq. (25). As shown in Fig. 5, point A on this curve, obtained by computerized scanning, has (very nearly) the desired λ-values ($\lambda_1 = 2.74730$, $\lambda_2 = 0.00076$) and $\lambda_3 = -751.99$. The base circle is relatively large so that the mechanism would not be very economical in the utilization of space. However, it gives a very close (5-point contact) approximation to the involute. If only 4-point matching is needed ($\lambda_1 = 2.74730$), an infinite choice of mechanisms is obtainable by using any point on the λ_1-curve as the generating point.

Example 4: Hypocycloidel dwell mechanisms[2].

This is an example of Type-3 synthesis. For an in-line swinging-block mechanism in the symmetry position, the instantaneous invariants are obtainable from Ref. [12] [$b_3 = 0$, $a_3 = 3$, $a_4 = 0$, $b_4 = \dfrac{1}{r_1 \varrho_1}(r_1 - 3 + r_1 \varrho_1)$ where $r_1 \varrho_1 = r_1 - \varrho_1$; ϱ_1 = distance between fixed pivots; r_1 = distance between moving-crank point and swinging-block pivot].

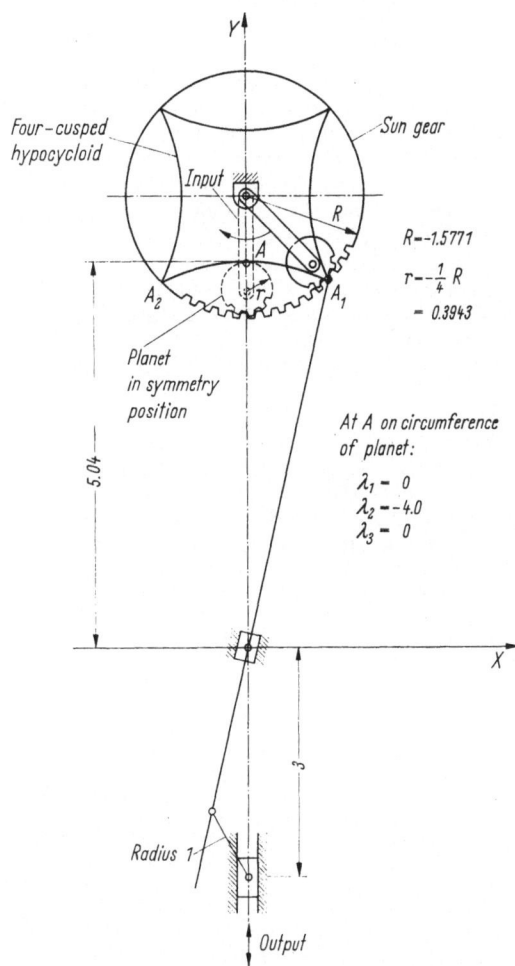

Fig. 6. Hypocycloidal dwell mechanism.

For a hypocycloid of 4 cusps with point A (Fig. 6) on the circumference and in the symmetry position. (Table 1, Case 10), we have $\lambda_1 = 0$, $\lambda_2 = -4$, $\lambda_3 = 0$.

We now wish to match these λ-values using a point on the floating link of a swinging-block mechanism in the symmetry position. In this position, the λ_1-curve ($\lambda_1 = 0$) breaks up into the pole normal (the axis of symmetry) and the pole tangent. Scanning points on the pole normal, when $\varrho_1 = -3$, $r_1 = -2$ (say) we find $OA = 5.04$ (Fig. 6).

A hypocycloidal dwell mechanisms can then be obtained by using the arm of the differential-gear system as driver and replacing the fixed-crank pivot of the swinging-block mechanism by a slider, as shown in Fig. 6. During the motion of A, the slider M will be at a dwell for 1/4 of the arm rotation (between A_1 and A_2). The remaining motion will be variable.

The error in the coincidence of tne hypocycloidal path and the swinging-block trajectory is too small for visual detection in Fig. 6.

9. Conclusion

Together with a computer-aided evaluation of the curvature ratios and the curves of synthesis, the theory of higher-contact approximations using the instantaneous invariantes and intrinsic geometrical invariants can be applied to a variety of motions of technical interest.

---

[1] This subject is discussed in the following references: Candee, A. H.: Kinematic geometry of gear teeth, Chilton 1961; Sieker, K. H.: Einfache Getriebe, 2nd Ed., Fuessen: Winter 1956; Unterberger, R.: VDI-Berichte 77, 107—111 (1964); Koller, R.: Antriebstechnik 3, 9, 338—342 (1964).

[2] See, for example, Tuttle, S. B.: Mechanisms for Engineering Design, Wiley 1967.

References

1. Bereis, R.: Aufbau einer Theorie der ebenen Bewegung mit Verwendung komplexer Zahlen, Oest. Ing. Archiv 5, 246—266 (1951).
2. Bottema, O.: On instantaneous invariants Proc. Int'l. Conference on Mechanisms, Yale University, New Haven, Conn.: Shoestring Press 1961, pp. 159—164.
3. — On some loci in plane instantaneous kinematics, J. Mechan. 2, 2, 141—146 (1967).
4. Cesàro, E.: Vorlesungen über natürliche Geometrie, Leipzig: Teubner 1901.
5. Cohen, A.: An introduction to the Lie theory of one-parameter groups, New York: Stechert 1931.
6. Freudenstein, F.: Higher path-curvature analysis in plane kinematics, ASME Trans. 87B, 1965, pp. 184—190.
7. Guggenheimer, H.: Differential geometry, New York: McGraw-Hill 1963.
8. Kowalewski, G.: Allgemeine natürliche Geometrie, Leipzig: Teubner 1901.
9. Sandor, G. N.: On infinitesimal cycloidal kinematic theory of planar motion, ASME Trans. 88E, J. Appl. Mech., Dec. 1966, 927—933.
10. Sandor, G. N., and F. Freudenstein: Higher-order plane motion theories in kinematic synthesis, Trans. ASME 89B, J. Eng'g. Ind., 1967, 223—230.
11. Steward, G. C.: On the cardinal points in p'ane kinematics, Phil. Trans. Royal Soc. London, Series A, 244, 1951, 19—46.
12. Veldkamp, G. R.: Curvature theory in plane kinematics, Groningen: Wolters 1963.
13. — Some remarks on higher curvature theory, ASME Trans. 89B, Feb. 1967, 84—86.
14. Wieleitner, H.: Spezielle ebene Kurven, Leipzig: Goeschen 1908.

List of Other Contributed Papers

(15-minute presentation)

The following is a listing of the shorter contributed papers, arranged alphabetically according to first author. The location of at least one of the authors (the one who presented the paper) will be found in the List of Participants, except in a few instances in which the presentation was made by a colleague.

ACHENBACH, J. D., and G. HERRMANN, Composite materials as continua with complex microstructure.

ALBLAS, J. B., A general theory of magneto-elastic-interactions.

ALEXOPOULOS, C. C., and J. F. KEFFER, Influence of a turbulent wake upon a stratified flow.

ALLISON, I. M., The testing and analysis of photoelastic model pressure vessels.

AMAZIGO, J. C., Buckling under axial compression of long cylindrical shells with random axisymmetric imperfections.

AMES, W. F., S. Y. LEE and J. N. ZAISER, Nonlinear vibration of a traveling threadline.

ANTON, IOAN, and OCTAVIAN POPA, Sensitivity to cavitation of the hydrofoil cascades.

APPLEBY, E. J., Constitutive equations of function type and their status in modelling a class of real materials.

ARBOCZ, J., and C. D. BABCOCK, Jr., The effect of general imperfections on the buckling of cylindrical shells.

ARIARATNAM, S. T., and T. K. SRIKANTAIAH, Stability of linear systems with random parametric coupling.

ARUTIUNIAN, N. KH., and B. L. ABRAMINA, Some contact problems for a semispace and an elastic layer with a vertical cylindrical cavity.

AUGUSTI, G., Instability of struts subject to radiant heat.

AZMEH, M., Theoretical and experimental investigations of highly cambered flex-wings.

BĂLAN, S., and S. RĂUTU, The field of application of chromoplasticity.

BARKER, L. M., and R. E. HOLLENBACH, Compressive and rarefaction wave propagation in fused quartz.

BARTA, T. A., A contribution to an engineering theory of thin elastic shells (Part I — Linear theory).

BATAILLE, J., G. DUPEYRAT et B. LEFUR, Etude des instabilités des écoulements polyphasiques dans un milieu poreux ou entre deux plaques parallèles.

BEER, J. M., N. A. CHIGIER, R. MELLOR and M. J. WHEATLEY, Droplet trajectories in liquid sprays.

BENTALL, R. H., and K. L. JOHNSON, Stresses in an elastic strip passing between rollers.

BEVIR, M. K., and J. A. SHERCLIFF, Theory of electromagnetic flowmeters with non-uniform fields.

BJØRNØ, L., A comparison between measured pressure waves in water arising from electrical discharges and detonation of small amounts of chemical explosives.

BLACKADAR, A. K., and H. TENNEKES, Asymptotic similarity in neutral, barotropic, planetary boundary layers.

BLUMAN, G. W., and J. D. COLE, The general similarity solution of the heat equation.

BODNER, S. R., and A. BERKOVITS, Incremental material behavior in creep buckling problems.

BOLEY, B. A., Melting of a slab under space and time dependent heating.

BOOCOCK, D., and L. MAUNDER, Vibration of a symmetric tuning fork.

BRAKHAGE, H., The solution of the subsonic lifting surface problem by a method of discrete singularities.

BROADBENT, E. G., A class of two-dimensional flows with heat addition.

BROWN, E. H., Virtual work theorems for frame stability problems.

BRUN, L., Sur la séparation des énergies dans les systèmes linéaires évolutifs.

BRUN, R., J. P. GUIBERGIA et R. MARMEY, Couche limite laminaire et relaxation de vibration à l'aval d'une onde de choc mobile.

BUDIANSKY, B., and D. G. M. ANDERSON, Numerical shell analysis-nodes without elements.

BÜRGER, W., Weak discontinuities of hyperbolic differential systems, with applications to the mechanics of fluids and gases.

BURNAT, M., The method of Riemann invariants for multidimensional nonelliptic systems.

BUSSE, F. H., Bounds on the transport of mass and momentum by turbulent flows.

CARRIER, G. F., Stochastically controlled phenomena in dynamical systems.

CASAL, P., Potentiels de mouvements en fluides compressibles et en magnétodynamique.

CAULK, R. H., and W. H. SCHWARZ, Ultrasonics in fluids with memory.

CHAKRABORTY, B. B., On the stability of a perfectly conducting liquid jet in the presence of a magnetic field.

CHEN, P. J., The growth of acceleration waves of arbitrary form in homogeneously deformed elastic materials.

CHEN, P., and B. ROTH, A unified method for the finitely and infinitesimally separated position problems of spatial kinematic-synthesis.

CHENG, D. H., and H. J. THAILER, On bending of curved circular tubes.

CHEREPANOV, G. P., On crack propagation in solids.

CHILVER, A. H., and K. C. JOHNS, Coupled modes of elastic buckling of imperfect systems.

CHITALEY, A. D., and F. A. McCLINTOCK, Elastic-plastic mechanics of steady crack growth under anti-plane shear.

CHRISTIANSEN, S., and E. HANSEN, Stress concentration around holes in cylindrical shells.

CLARK, R. A., An expansion bellows problem.

COHEN, H., and C. N. DeSILVA, A direct nonlinear theory of elastic membranes.

COMTE-BELLOT, G., et J. P. SCHON, Excitation paramétrique dans les anémomètres à fils chauds à intensité constante.

CONTENSOU, P., Optimisation du vol plané dans un vent horizontal variable.

COUPRY, G., Influence d'un écoulement supersonique sur le comportement vibratoire d'un cylindre circulaire mince.

CRANDALL, S. H., On the distribution of maxima in the response of an oscillator to random excitation.

CRISP, J. D. C., The static equilibria of cylindrical blood vessels and their stability.

CRISTESCU, N., On viscoplastic constitutive equations.

CROCHET, M. J., On constitutive relations for a mixture of two simple materials.

CUMBERBATCH, E., Self-focusing in non-linear optics.

DARROZES, J. S., Etude de la diffusion dans un mélange gazeux.

DAVIES, R. M., A statistical mechanical approach to the behavior of muscles.

DAVIS, R. T., Laminar incompressible flow past the parabola and the paraboloid.

DEHOUSSE, M. E., Concerning the influence of the spring mass of a linear oscillator in a gravity field.

DIAZ, E., The membrane theory for shells formed by second order surfaces utilizing hypercomplex analytic functions.

DILLON, O. W., Jr., and J. KRATOCHVIL, Thermodynamics of perfectly plastic solids.

DOKMECI, M. C., On a nonlinear theory of multilayer sandwich shells and plates.

DRUCKER, D. C., The application of mechanics to the problem of brittle fracture.

DUBEY, R. N., Instabilities in thin tubes.

DUGAN, J. P., On the viscous drag of bodies moving near a free surface.

DURAND, M., Sur l'extension d'un critère de rupture fragile aux matériaux orthotropes; application aux ruptures par fatigue.

EBBENI, J., Meaning of diffraction effects in Moiré analysis of strains.

ELDER, J. W., Weak coupling models of thermal turbulence.

EUVRARD, D., J. HUBERT and G. TOURNEMAINE, Comportement asymptotique de l'écoulement à grande distance d'un obstacle se déplacant à la vitesse du son.

FARELL, C., On the linearized theory of the wave resistance of a submerged spheroid.

FAURE, R., Le phénomène Bethenod.

FAYED, M. H., and G. D. GALLETLY, Stress analysis of irregularly shaped bodies of revolution.

FERRANDON, J., Flambement et vibrations des structures.

FINLAYSON, B. A., and L. E. SCRIVEN, Convective instability by active stress.

FÖRSCHING, H., Free vibrations of axisymmetrically loaded orthotropic conical shells.

FRANCIS, P. H., Elasto-plastic wave propagation in a rate sensitive finite rod having a thermal gradient.

FRÝBA, L., Vibration of solids under random load nonstationary in space and time.

FUNG, Y. C., The mechanical properties of living soft tissues and some analytic problems associated with them.

GARDNER, C. S., J. M. GREEN, M. D. KRUSKAL, and R. M. MIURA, Nonlinear wave propagation in dispersive systems.

GARG, S. K., Numerical solutions for spherical elastic-plastic wave propagation.

GERMAIN, J. P., and C. MARCOU, Etude théorique et experimentale des houles complexes et des phénomènes pesants dans les canaux a houle.

GERSTEN, I., Theory of laminar and turbulent wall jets along highly curved walls including heat transfer.

GHEORGHITZA, ST. I., On the marginal stability in porous media.

GIANGRECO, E., and M. COMO, New developments on the theory of stability of structures.

GILMAN, P. A., and E. R. BENTON, A combined Ekman-Hartmann boundary layer.

GLADWELL, G. M. L., Forced tangential and rotatory vibration of a rigid circular indenter on a semi-infinite solid.

GOLDENVEIZER, A. L., On foundation of the shell theory and estimation of its errors.

GRADOWCZYK, M., Analytical models for free-surface flows over erodible beds.

GRAEBEL, W. P., A study of non-steady disturbances in Couette flow.

GREEN, S. J., and C. J. MAIDEN, The behavior of aluminium and aluminium alloy under biaxial stress loading at strain rates to 10^2/sec.

HAFTKA, R., and J. SINGER, The buckling of discretely stiffened conical shells.

HALL, I. M.. Swirling flow in pipes and nozzles.

HALL, M. G., The boundary layer over an impulsively-started flat plate.

HANAGUD, S., Dynamic locking and thermodynamic effects on finite amplitude shock wave propagation in locking solids.

HANSON, R. K., and D. BAGANOFF, Determination of shock-heated nitrogen reaction rates from pressure measurements.

HAYASHI, K., Rigidity and natural frequencies of thermally loaded swept wings.

HAYASHI, T., ,Study of intracranial pressure caused by head impact.

HINDELANG, F. J., Coupled relaxation of polyatomic molecules in gasdynamic flow.

HODGE, P. G., Jr., C. T. HERAKOVICH and R. B. STOUT, Elastic-plastic torsion of solid and hollow bars.

HUANG. N. C., Axisymmetric dynamic snap-through of elastic clamped shallow spherical shells.

HUNG, J. Y., Stability of a miniature switch.

HUNT, J.C. R., and G. S. S. LUDFORD, Three-dimensional MHD duct flows with strong transverse magnetic fields: I-Obstacles in a constant area channel.

HUTCHINSON, J. W., Stress and strain field at the tip of a crack in an elastic-plastic material.

IMAI, I., A new approximation for the viscous flow near the trailing edge of a flat plate.

INOUE, J., On the self-synchronization of mechanical vibrators.

ISHERWOOD, D. P., and J. G. WILLIAMS, The effect of stress-strain properties on notched tensile failure.

ISHLINSKY, A. YU., On the equations of inertial navigation.

JAHSMAN, W. E., A reexamination of the Kolsky technique as a means of measuring dynamic material behavior in uniaxial stress.

JOHNSON, J. N., Dislocation dynamics and steady plastic wave profiles in metals.

KALISZKY, S., Approximate solutions for impulsively loaded inelastic structures and continua.

KALKER, J. J., Transient rolling contact phenomena.

KALNINS, A., Thermal instability of layered spherical disks.

KAMPÉ DE FÉRIET, J., Mécanique statistique des milieux continues dont les equations du mouvement sont linéaires.

VAN KEMPEN, H. P. M., and M. G. F. PEUTZ, A contribution to the theory of elasticity for multilayer-systems.

KERIMOV, K. A., and K. A. RAKHMATULIN, Studies of elastic threads with longitudinal and transverse impact.

KHARLAMOV, P. V., On some problems of the dynamics of a rigid body.

KLEINSTEIN, G., and L. TING, Optimum asymptotic solutions for heat conduction and fluid dynamics problems.

KLEPACZKO, J., The strain rate behavior of iron in pure shear.

KO, D. R., and T. KUBOTA, Finite disturbance effect on a laminar incompressib lewake behind a flat plate.

KONDO, J., Interaction of blunt bodies in supersonic flow.

KOVASZNAY, L. S. G., and V. KIBENS, Detailed measurements in the intermittent zone of a turbulent boundary layer.

KRASILSHIKOVA, E. A., Three-dimensional problems with moving boundaries in a transient supersonic aerodynamics of a thin airfoil.

KRUPKA, V., Local problems of thin-walled shells.

KUIPER, R. A., nad D. BERSHADER, Shock tube study of wave interactions in ionizing argon.

KUNIN, I. A., Non-local theories of solids.

LANDIS, F., and R. G. BACKSHALL, Boundary layer velocity distribution in turbulent swirling pipe flow.

LECKIE, F. A., and R. K. PENNY, Some observations on the creep deformations of structures.

LEE, M. S., and C. S. HSU, A τ-decomposition method of stability analysis for retarded dynamical systems.

LEE, R. S., and H. K. CHENG, On the outer-edge problem of a hypersonic boundary layer.

LEGENDRE, R., Ecoulement plan réversible d'un fluide compressible autour d'un profil.

LEHMANN, TH., Anisotropy and second-order-effects in plasticity.

LEUTHEUSSER, H. J., and V. H. CHU, Characteristics of turbulent plane Couette flow.

LEW, H. S., Formulation of a statistical equation for blood motion.

LIEBOWITZ, H., Failure of notched columns.

LIN, T. H., and M. ITO, A theory of fatigue crack nucleation based on slip stresses and strains.

LIPPMANN, H., A Cosserat-type theory of plastic flow.

LOITSIANSKI, L. G., Approximate methods in boundary layer theory.

LUKASIEWICZ, S., Equations of the technical theory of shells with the effect of transfer shear deformations.

LUNKIN, YU. P., and G. B. KOLESHKO, Laminar boundary layer with simultaneous vibrational and dissociation relaxations.

LUXTON, R. E., Generation of a uniform shear flow in a wind tunnel.

MAHRENHOLTZ, O., and H.-J. VON BREDOW, A phenomenon of flow in systems of elastic, valveless pipes.

MAJERUS, J. N., Restrictive conditions on the finite elastic strain energy function and a compatible function for compressible polymeric materials.

MANDEL, J., Sur diverses conséquences d'un principe restreignant les formes possibles pour les équations de comportement.

MANDL, G., and R. F. LUQUE, Fully developed plastic shear flow of granular materials.

MANN-NACHBAR, P., and W. NACHBAR, A stochastic approach to buckling of circular cylindrical shells under uniform axial compression.

MARÉCHAL, J., Etude de la deformation plane d'un écoulement turbulent.

MAXWORTHY, T., Aligned fields magneto-fluid dynamic flow over axisymmetric bodies.

MAZET, R., Analyse de trois courbes d'écrouissage d'un dural a 200°C.

McDEVITT, J. B., and J. A. MELLENTHIN, Some effects of mass addition on the hypersonic aerodynamic characteristics of cones.

McIVOR, I. K., Dynamic snap-through of a plastically deforming cylindrical panel.

McQUILLEN, E. J., and M. A. BRULL, Dynamic fully-coupled thermoelastic response of thin cylinders to thermal inputs.

MEDICK, M. A., Torsional waves in elastic bars of non-circular section.

MICHEL, R., Resultats sur les couches limites et les flux de chaleur turbulents en écoulement supersonique accéléré.

MIKLOWITZ, J., A method for solving elastic waveguide problems involving non-mixed edge conditions, and an application.

MINDLIN, R. D., Crystal lattice theory of torsion of a rectangular bar of simple cubic structure.

MÖHRING, W. F., E.-A. MÜLLER and F. F. OBERMEIER, Sound generation by unsteady flow as a singular perturbation problem.

MOK, C. H., The effective dynamic properties of a fiber-reinforced material and the propagation of sinusoidal waves.

MOON, F. C., and Y.-H. PAO, Vibration and dynamic instability of a beam-plate in a magnetic field.

MOREAU, J. J., La notion de sur-potentiel et les Liaisons unilatérales en élastostatique.

MÜSCHNER, W. H., On a nonlinear unsteady theory of the submerged hydrofoil.

NAHAVANDI, A. N., G. BOHM and R. BUCHHEIM, Numerical stability analysis for the flexural vibration of beams with shear effects.

NAKAMURA, I., and Y. FURUYA, The velocity profile in the skewed boundary layer on the rotating bodies in axial flow.

NAPOLITANO, L. G., and D. R. CRAWFORD, Diffusion-thermo-chemical effects in sound waves.

NEREM, R. M., L. A. CARLSON and R. A. GOLOBIC, Radiation gas dynamic coupling behind reflected shock waves in air.

NEUBER, H., On combined micro and macro support effects of stress concentration.

NIGUL, U., Regions of effective application of methods of three-dimensional and two-dimensional analyses of transient stress waves in shells and plates.

NIILER, P. P., The dynamics of the gulf stream.

NIKOLAEVSKII, V. N., and E. F. AFANASIEV, On some behaviour of media with microstructure of continuous particles.

NIKOLSKII, A. A., Uniform inertial motions of continuum and rarefied gases.

OLHOFF, N., Optimal design of vibrating circular plates.

OLSZAK, W., and Z. BYCHAWSKI, Energetic foundation of transition phenomena in elasto-visco-plasticity.

OSHIMA, K., Laboratory simulation of ionospheric aerodynamics.

OSWATITSCH, K., and A. FROHN, The velocity around a lifting delta wing in supersonic flow with sonic leading edges.

OVSJANNIKOV, L. V., Nonstationary motions of incompressible fluid with free surface.

PARKER, D. F., Rapid deflections of a stretched string, as an example of cumulative interactions between non-linear waves.

PERRONE, N., and R. KAO, A general relaxation technique for solving non-linear problems—Example: Large deflection solution of arbitrary membrane.

PERZYNA, P., On physical foundations of viscoplasticity.

PHILLIPS, J. W., Stress waves produced by brittle fracture of rods.

PRAGER, W., Variational principles for elastic plates with relaxed continuity reguirements.

PROSCURIAKOV, M. N., On the dynamic stability of some non-linear systems under random external loading.

RAO, A. K., Stress concentrations at interface corners.

REHFIELD, L. W., Single-mode buckling and postbuckling of continuous structures.

REISSNER, E., and F. Y. M. WAN, Rotationally symmetric stress and strain in shells of revolution.

RIGAUT, F., Etude analogique d'écoulements transsoniques par la méthode de l'hodographe.

RIM, K., and D. L. BARTEL, Optimal synthesis of spatial structures with inequality constraints on stress and geometry.

ROBERT, A. J., Principe et description d'un photoélasticimétre tridimensionnel.

ROE, P. L., A momentum analysis of lifting surfaces in inviscid supersonic flow.

ROHLIN, L., Creep caused by intermittent load.

ROMITI, A., Shock wave precursor measurements by electrostatic probes.

ROTT, N., Higher-order boundary-layer theory for thermally-induced acoustic oscillations.

RUDINGER, G., Shock waves in mathematical models of the aorta.

RUMYANTSEV, V. V., On the stability of steady motions.

SADOWSKY, M. A., S. L. PU and M. A. HUSSAIN, Elastic bodies with internal polarization.

SAWCZUK, A., Upper bounds to shakedown loads for shells.

SCHÄFER, M., und D. BEHR, Anwendung der Iterationsmethode in der Theorie nichtlinearer Schwingungs-systeme auf funktionalanalytischer Grundlage.

SCHAPERY, R. A., and J. S. NOEL, Cavity growth in nonlinear viscoelastic material under triaxial tension.

SCHMIDT, B., Electron beam density measurements in shock waves in argon.

SCHMIDT, R., A nonlinear theory of multisandwich shells.

SCHMIEDER, L., Theory of shells with shear deformations in covariant notation.

SCHNEIDER, W., Necessary and sufficient conditions for the properties of a fluid to permit mechanical simi-larity.

SCIAMMARELLA, C. A., and G. DI CHIRICO, Moiré—holographic technique for three-dimensional stress ana-lysis.

SEDOV, L. I., On the concept of stress tensor for models of continuous media with internal degrees of freedom.

ŠEJVL, M., Théorie du fraisage interrompu des roues hypoïdes à dents courbées épicycliques et à contact linéaire précis.

SELLS, C. C. L., Subcritical flow past a lifting aerofoil section.

SERENSEN, S. V., and V. P. KOGAEV, Mechanical fracture as a random time process.

SHAPIRO, A. H., M. Y. JAFFRIN and S. L. WEINBERG, On some fluid mechanical aspects of ureteral function.

SHEMYAKIN, E. I., On stress-waves in solids.

SHEN, M. C., Theory of three-dimensional free surface flow in a channel of arbitrary cross section.

SHIEH, R. C., On local stability of the equilibrium of Voigt-Kelvin solids subjected to nonconservative forces.

SINITSYN, A. P., A. F. SMIRNOFF and S. T. STELMAKH, Vibrations and stability of the shells due to the loads and temperature.

SMITH, S. H., The creation of surface waves by viscous forces.

SOBOYEJO, A. B. O., Application of stochastic processes to the problem of creep failure in engineering struc-tures at elevated temperatures.

SONNTAG, G., Laws of models concerning embedded corrugated metal pipes (CMP).

SPENCE, D. A., and H. OCKENDON, Non-linear wave propagation in a relaxing gas.

STALLYBRASS, M. P., An equivalence between certain mixed boundary value problems in continuum me-chanics.

STEARMAN, R., An experimental study on the aeroelastic stability of thin cylindrical shells.

STEVENS, K. K., and R. M. EVAN-IWANOWSKI, Parametric resonance of viscoelastic columns.

STRUMINSKI, V. V., Investigation of stability of laminar flows and transition flow processes.

SUNAKAWA, M., K. ICHIDA and Y. SUNAGA, An experimental investigation on the buckling pressure of spherical shells.

SVARDAL, A., and S. TJØTTA, Acoustic streaming.

TABARROK, B., On duality in the oscillations of framed structures.

TANG, S.-C., and D. H. Y. YEN, Interaction of a plane acoustic wave with an elastic spherical shell.

TATSUMI, T., and K. GOTOH, The transition process and the equilibrium state in a free shear layer.

TAYLOR, J. E., and E. F. MASUR, A global condition for optimal structural design.

THOMAS, P. D., R. J. CONTI and C. H. KRUGER, Flow over slender bodies with massive blowing.

TILLMAN, S. C., The buckling of shallow spherical caps.

TING, T. C. T., Interaction of shock waves due to combined two shear loadings.

TONTI, E., Variational principles in elastostatics.

TSIRK, A., Elastic response of rectangular plates to moving load-masses.

VAGLIO-LAURIN, R., and M. HOFFERT, Supersonic laminar flow over a convex separation edge.

VALENSI, J., D. GUFFROY, J. MARCILLAT et B. ROUX, Effet de l'incidence et du nombre de Reynolds sur l'écoulement hypersonique autour d'un cône circulaire.

VALID, R., Une méthode de calcul des structures en plasticité et en fluage isothermes.

VAUGHAN, H., and A. FLORENCE, Plastic flow buckling of cylindrical shells due to impulsive loading.

VEKUA, I. N., On construction of solutions of equations of the spherical shell.

WAECHTER, R. T., Wing-body interaction in supersonic airfoil theory.

WARBURTON, G. B., and A. M. J. AL-NAJAFI, Free vibration of thin cylindrical shells with a discontinuity in the thickness.

WARREN, W. E., Geometrical stress singularities in the linear theory of electrostriction.

WEDEMEYER, E. H., Stability of helical flows in the annulus between two coaxial cylinders.

WEHRLI, CH., On the stability of nonuniformly rotating viscoelastic shafts.

WEMPNER, G., and G. PATRICK, Numerical methods for nonlinear analyses of thin shells.

WENGLARZ, R. A., and T. R. KANE, Gyroscopic drifts associated with rotational case vibrations.

WIGINTON, C. L., C. DALTON and W. T. SNYDER, Laminar flow in the entrance region of a rectangular duct.

VAN WIJNGAARDEN, L., On the propagation of gravity waves on a flow with nonuniform velocity distribution.

WILLIS, D. R., B. B. HAMEL and J. T. LIN, Development of the molecular distribution function on centerline of a free jet solids expansion.

WNUK, M. P., Delayed fracture in viscoelastic-plastic solids.

WU, E. M., Fracture criteria for orthotropic plates under compression and shear.

YEUNG, K. W., and C. P. YU, Convective stability of a polytropic fluid layer in a magnetic field.

ZDRAVKOVICH, M. M., Smoke observations of laminar and transitional wakes of the group of cylinders.

ZEYTOUNIAN, R. KH., Ecoulements faisant intervenir à la fois convection naturelle et forcée.

ZIEGLER, F., Wave propagation in laminated random media.

ZIEREP, J., Minimum drag bodies and the reverse—flow theorem at sonic speed.

ZYCZKOWSKI, M., Optimum structural design in rheology.